Handbook of Research on Smart Technology Applications in the Tourism Industry

Evrim Çeltek
Gaziosmanpasa University, Turkey

IGI Global
DISSEMINATOR OF KNOWLEDGE

A volume in the Advances in Hospitality, Tourism, and the Services Industry (AHTSI) Book Series

Published in the United States of America by
IGI Global
Business Science Reference (an imprint of IGI Global)
701 E. Chocolate Avenue
Hershey PA, USA 17033
Tel: 717-533-8845
Fax: 717-533-8661
E-mail: cust@igi-global.com
Web site: http://www.igi-global.com

Library of Congress Cataloging-in-Publication Data

Names: Çeltek, Evrim, 1977- editor.
Title: Handbook of research on smart technology applications in the tourism industry / Evrim Çeltek, editor.
Description: Hershey PA : Business Science Reference, [2020] | Includes bibliographical references and index. | Summary: "This book examines the strategic, tactical, and operational perspectives of smart technologies in the tourism industry"-- Provided by publisher.
Identifiers: LCCN 2019035775 (print) | LCCN 2019035776 (ebook) | ISBN 9781799819899 (hardcover) | ISBN 9781799819905 (ebook)
Subjects: LCSH: Tourism--Information technology. | Hospitality industry--Information technology. | Artificial intelligence--Business applications | Internet of things.
Classification: LCC G156.5.I5 H36 2020 (print) | LCC G156.5.I5 (ebook) | DDC 910.285--dc23
LC record available at https://lccn.loc.gov/2019035775
LC ebook record available at https://lccn.loc.gov/2019035776

This book is published in the IGI Global book series Advances in Hospitality, Tourism, and the Services Industry (AHTSI) (ISSN: 2475-6547; eISSN: 2475-6555)

British Cataloguing in Publication Data
A Cataloguing in Publication record for this book is available from the British Library.

For electronic access to this publication, please contact: eresources@igi-global.com.

Advances in Hospitality, Tourism, and the Services Industry (AHTSI) Book Series

Maximiliano Korstanje
University of Palermo, Argentina

ISSN:2475-6547
EISSN:2475-6555

Mission

Globally, the hospitality, travel, tourism, and services industries generate a significant percentage of revenue and represent a large portion of the business world. Even in tough economic times, these industries thrive as individuals continue to spend on leisure and recreation activities as well as services.

The Advances in Hospitality, Tourism, and the Services Industry (AHTSI) book series offers diverse publications relating to the management, promotion, and profitability of the leisure, recreation, and services industries. Highlighting current research pertaining to various topics within the realm of hospitality, travel, tourism, and services management, the titles found within the AHTSI book series are pertinent to the research and professional needs of managers, business practitioners, researchers, and upper-level students studying in the field.

Coverage

- Destination Marketing and Management
- Hotel Management
- Health and Wellness Tourism
- International Tourism
- Sustainable Tourism
- Casino Management
- Service Design
- Service Training
- Food and Beverage Management
- Travel Agency Management

IGI Global is currently accepting manuscripts for publication within this series. To submit a proposal for a volume in this series, please contact our Acquisition Editors at Acquisitions@igi-global.com or visit: http://www.igi-global.com/publish/.

The Advances in Hospitality, Tourism, and the Services Industry (AHTSI) Book Series (ISSN 2475-6547) is published by IGI Global, 701 E. Chocolate Avenue, Hershey, PA 17033-1240, USA, www.igi-global.com. This series is composed of titles available for purchase individually; each title is edited to be contextually exclusive from any other title within the series. For pricing and ordering information please visit http://www.igi-global.com/book-series/advances-hospitality-tourism-services-industry/121014. Postmaster: Send all address changes to above address.

Titles in this Series

For a list of additional titles in this series, please visit: www.igi-global.com/book-series

Global Opportunities and Challenges for Rural and Mountain Tourism
Devkant Kala (University of Petroleum and Energy Studies, India) and Satish Chandra Bagri (Hemwati Nandan Bahuguna Garhwal University, India)
Business Science Reference • © 2020 • 375pp • H/C (ISBN: 9781799813026) • US $225.00

Organizational Behavior Challenges in the Tourism Industry
Şule Aydin (Nevsehir Haci Bektas Veli University, Turkey) Bekir Bora Dedeoglu (Nevsehir Haci Bektas Veli University, Turkey) and Ömer Çoban (Nevsehir Haci Bektas Veli University, Turkey)
Business Science Reference • © 2020 • 524pp • H/C (ISBN: 9781799814740) • US $225.00

Handbook of Research on Social Media Applications for the Tourism and Hospitality Sector
Célia M.Q. Ramos (University of Algarve, Portugal) Cláudia Ribeiro de Almeida (University of Algarve, Portugal) and Paula Odete Fernandes (Polytechnic Institute of Bragança, Portugal)
Business Science Reference • © 2020 • 400pp • H/C (ISBN: 9781799819479) • US $295.00

New Trends and Opportunities for Central and Eastern European Tourism
Puiu Nistoreanu (The Bucharest University of Economic Studies, Romania)
Business Science Reference • © 2020 • 338pp • H/C (ISBN: 9781799814238) • US $225.00

Managing, Marketing, and Maintaining Maritime and Coastal Tourism
Mónica Morais de Brito (CEGOT, Portugal & Instituto Politécnico de Beja, Portugal) Alvaro Dias (Universidade Lusófona de Humanidades e Tecnologias, Portugal & ISCTE, Instituto Universitário de Lisboa, Portugal) and Mafalda Patuleia (Universidade Lusófona de Humanidades e Tecnologias, Portugal)
Business Science Reference • © 2020 • 325pp • H/C (ISBN: 9781799815228) • US $225.00

Global Developments in Healthcare and Medical Tourism
Sudip Paul (North-Eastern Hill University, India) and Sharad Kumar Kulshreshtha (North-Eastern Hill University, India)
Medical Information Science Reference • © 2020 • 341pp • H/C (ISBN: 9781522597872) • US $245.00

Multilevel Approach to Competitiveness in the Global Tourism Industry
Sérgio Jesus Teixeira (Higher Institute of Administration and Languages, Portugal) and João Matos Ferreira (University of Beira Interior, Portugal)
Business Science Reference • © 2020 • 393pp • H/C (ISBN: 9781799803652) • US $225.00

IGI Global
DISSEMINATOR OF KNOWLEDGE

701 East Chocolate Avenue, Hershey, PA 17033, USA
Tel: 717-533-8845 x100 • Fax: 717-533-8661
E-Mail: cust@igi-global.com • www.igi-global.com

Dedicated to my mother Ayşe.

Editorial Advisory Board

List of Contributors

Table of Contents

Section 3
Smart Tourism

Detailed Table of Contents

With more and more people traveling worldwide (the number doubling in the last 20 years), tourist destinations are now more than ever trying to maintain and enhance their competitiveness in the global market. In this regard, novel business models combined with state-of-the-art technology can play a crucial role in not only satisfying the increasing tourism demand but also ensuring a sustainable growth to avoid the deteriorating effects on both the social and natural habitat. For these reasons, the Slovenian government included tourism as one of the priority areas for investment. Hence, the Tourism 4.0 initiative was launched to unite tourism stakeholders with high-tech companies and unify their scattered ideas, experiences, knowledge, and expertise. The ambition is to position Slovenia as a top destination for sustainable tourism with high economic value. This chapter introduces the basic concepts behind Tourism 4.0 and how it relates to technologies for an enhanced tourism experience.

One of the most important points of consideration that will ensure the sustainability and profitability of businesses in the digitized world market is the experience offered to customers. Experience in business preferences, especially of the digitized customers, is quite specific. This chapter discussed the contribution of smart technologies to customer experience for restaurants, and emphasized its significance. Moreover, the subjects of artificial intelligence, smart technology, and QR code were addressed based on customer experience. The contribution offered to customers was emphasized by giving examples of smart technology applications used in restaurant businesses to improve customer experience. The chapter will contribute theoretically to the subject which has not been adequately studied in the literature.

Tourism companies, like other companies producing goods and/or services, practice different strategies to achieve competitive advantage and satisfy the needs and wants of customers/guests. The most important recent strategy is electronic customer relationship management (e-CRM). In today's competitive environment, a successfully implemented e-CRM enhances a customer centric culture that is key to tourism growth and competitive advantage. For tourism companies, e-CRM succeeds only if it is totally integrated strategically and operationally in the business and in the information system of the company. This chapter focuses on e-CRM concept, its importance in tourism, dimensions of e-CRM, major challenges of e-CRM and its opportunities for companies, its implementation in tourism, e-CRM Technologies used in tourism, e-CRM success factors, and the reasons of e-CRM failures.

The development of service robotics comes along with the development of industrial robotics. Service robots are designed to perform professional job tasks as well as for service users in areas of everyday

The rapid increase in the population has caused problems in the correct use and management of resources in cities. Solutions to these problems have been sought based on knowledge management. In today's digital age, the concept of smartness of cities has been put forward together with the web-based applications. It has changed the expectations and needs of tourists and residents, especially in the tourism sector in the service sector. Smart tourism destinations, which are seen as a solution to the expectations and needs of the changing tourists and local people, bring holistic innovations covering all the stakeholders in the tourism ecosystem. In this direction, it is thought that knowledge management makes the right use of resources obligatory and contributes to sustainable tourism understanding. For this purpose, in the chapter, suggestions were made to provide maximum benefit from web-based projects that are carried out in qualitative direction of knowledge management and smart tourism destinations.

Chapter 18
Smart Municipalities in Tourism 391
Nurettin Ayaz, Karabük University, Turkey
Bayram Akay, Burdur Mehmet Akif Ersoy University, Turkey

Tourism is a fragmented and information-oriented sector covering tour operators, travel agencies, hotel sales representatives, associations, meeting offices, transportation, car rental, airlines, cruise, souvenirs, restaurants, hotels/motels, and entertainment. In the tourism sector, technology is an important requirement for the interconnection of all these fragmented organizations, activities, and elements, and for the satisfaction of tourists. However, the effective use of technology in a destination requires the leadership of a leader. This chapter approaches the smart municipality applications oriented to tourism sector with the example of Turkey, a well-known destination around the world, from the perspective of municipalities, which is regarded as a democratic leader at local level and provides a general framework regarding the studies on an innovative, environment-friendly, solution-oriented, and more livable urban life. Suggestions are provided for smart technologies to improve the quality of life of local citizens and to the satisfaction of tourists.

Chapter 19
A New Concept in Tourism: Smart Tourism Destinations 414
Gamze Coban, Faculty of Tourism, Nevsehir Haci Bektas Veli University, Turkey
Şule Aydın, Faculty of Tourism, Nevsehir Haci Bektas Veli University, Turkey

This chapter provides an insight into the topic of smart destinations. By adopting smart technologies, tourism destinations, as well as cities, gain more opportunities to offer better quality of life for residents and visitors. Smart cities aim to improve resource management, sustainability, and living conditions in urban environments by utilizing ICTs. The concept of smart tourism destination, deriving from smart city, refers to the use of technology in tourism destinations to increase the service quality and tourist satisfaction by focusing on tourists' expectations. In other words, smart tourism destinations aim to integrate technology into the destination for these purposes. This chapter presents the concepts of smart city, smart tourism, and smart destination. The emergence of smart city and smart destination concepts and the issues and challenges they might face are discussed. In addition to some future research directions, a brief discussion on potential controversies is presented.

Smart city concept is emerging as a new venture in the tourism sector especially in a country like India. This chapter analyzes the tourist experience in Jaipur city of Rajasthan which is emerging as a popular tourist destination taken over for development under smart city mission. The chapter adopted empirical approach, but to some extent it also used secondary data sources. The authors analyzed tourist experience by using smart attraction, smart mobility, smart amenities, smart ancillary, smart safety, and smart available packages as important dimensions of smart tourism destination. Regarding tourist experience, the chapter noticed smart available packages, smart safety, and smart attractions are providing a pleasant experience to tourists in contrast with remaining aspects of smart tourism destination. The result of the study provides implication for tourism marketer and policy maker to take appropriate steps to enhance the tourist experience by improving smart facilities & services.

The chapter is divided into two differentiated parts. The first includes a brief technical review of the concepts of Territorial Intelligence and Intelligent Tourist Destination, situating them in the context of the new tourist destination planning paradigms. This part ends with a first approach to the Spanish case and the progress of its public tourism policies towards these concepts. The second part goes deeper into the Intelligent Tourist Destination model applied in Spain, driven by the public administration, paying special attention to its official standardisation process, and the indicators adopted to that end. Finally, the chapter ends with some brief conclusions.

Foreword

Technology has brought exciting changes to the life of all of us. Whatever role we have as managers, teachers, consumers, students, citizens; technology changed our way of decision making, communicating, learning, entertaining… Tourism is also one of the industries that has been effected a lot from the technology. All around the world tourists may find the best hotels for themselves, reserve a room, search the easiest routes to reach a touristic destination, may learn about the other visitors' comments, may have a lingual support or interactive experience in a museum, may get a personalized excursion etc. This list is endless and what makes it possible is the smart technologies. Technology today is not only so inclusive; it is very smart also. Smart technologies like mobile devices, digital platforms, big data, open data, internet of things (IoT), three-dimensional printing, social interactive robots, artificial intelligence, virtual reality, augmented reality, blockchain, NFC, QR codes are transforming tourism industry and creating new opportunities, new challenges.

This book which is about the smart technology use in tourism provides very valuable contribution to anyone who has been interested in tourism as a researcher, student, and a decision maker. It includes three sections as; smart technology applications; Industry 4.0 and customer experience; smart tourism (smart destination, smart city, smart municipality and smart tourism planning). The importance of smart technologies in tourism industry in enhancing the customer experience is emphasized heavily in the book. I believe "Handbook of Research on Smart Technology Applications in the Tourism Industry" will be an essential reference source for the tourism researchers.

I appreciate the work of Associate Professor Evrim Çeltek in leading and editing this book and also appreciate the authors who have put their knowledge and effort to this book. I hope the book will increase the awareness, interest and research on the use of technology in the tourism industry.

Sevgi Ayşe Öztürk
Anadolu University, Turkey

Preface

Smart technologies enhance customer experience, increases the amount of data gained through connected devices, and widens the scope of analytics. This provides a range of exciting marketing possibilities such as selling existing products and services more effectively, delivering truly personalized customer experiences, and potentially creating new products and services. Handbook of Research on Smart Technology Applications in the Tourism Industry are an essential reference source that discusses the use of smart technologies in tourism, as well as its importance in enhancing the customer experience. Featuring research on topics such as augmented reality, virtual reality, QR code, big data, artificial intelligence, gamification, blokchain, industry 4.0, smart tourism, smart destinations, smart cities, smart municipalities, electronic customer relationship management, robots, near field communication (NFC), opinion mining and internet of things (IoT), this book is ideally designed for tourism business professionals, marketing managers, academicians, researchers, and graduate-level students seeking coverage on the use of smart technologies in tourism.

This book aims to show how smart technologies can influence the consumer relationship in the field of tourism industry. And present featuring tools and techniques applicable to the sectors under analysis.

Objective of the book is to bring together related academicians who dedicate themselves to find solutions to the problems scientifically and objectively and provide space for them to claim their opinions, put forward ideas or models and contribute to the field with their enlightening viewpoints.

The book is designed as three sections: smart technology applications; Industry 4.0 and customer experience; smart tourism (smart destination, smart city, smart municipality and smart tourism planning).

Now let's have an outlook over the contents of chapters in sequence.

Progress and Development of Virtual Reality and Augmented Reality Technologies in Tourism: A Review of Publications from 2000 To 2018

Progress and development of tourism technologies, applications available through smart devices provide an increasing diversity in tourism marketing. Augmented reality, virtual reality applications with the smart technologies, which are considered as new marketing communication instruments, increase awareness and help in terms of getting information about tourist and giving information to the tourist. Thus, this study aims at providing an assessment regarding the progress of AR and VR researches published in hospitality and tourism journals. In particular, the chapter aims to determining the current fields, topics and research methods of AR and VR articles published in 32 hospitality and tourism research journals between 2000-2018, comparing the subjects and research methods of 32 journals and discussing the changes in topic areas throughout the years, to offer suggestions for further research on AR and VR for hospitality and tourism journals.

Blockchain Technology and Its Uses in the Tourism Industry

The public and private sectors will undergo a significant transformation with the use of blockchain technology, and this potential of blockchain technology will be influential in all areas of life. In addition, blockchain technology can help ease the integration of the Internet of Things, augmented reality, and artificial intelligence applications. The question of how such a technology that is in the process of development will be applied in areas such as taxation, notary operations, and banking arises simultaneously. Studies related to blockchain technology have mainly been carried out in the fields of finance, logistics, banking, and education. However, there is a paucity of studies on blockchain technology in the tourism industry, which has an important role in the global economy. Therefore, the potential of using blockchain technology in the tourism industry is evaluated in this chapter. For this purpose, the features, advantages, and disadvantages of blockchain technology are explained. The potential uses of blockchain technology in the tourism industry are then discussed.

Opinion Mining In Tourism: A Study On "Cappadocia Home Cooking" Restaurant

In the tourism sector, online tourist reviews analysis is one of the methods to evaluate the products and services offered by businesses and understand the needs of tourists. These reviews take place in social networks and e-commerce sites in parallel with the developments in information and communication technologies. Tourists generate these reviews during or after their use of the products or services. In the literature, these reviews are referred to as UGC (User Generated Content) or eWOM (electronic word-of-mouth). The scientific evaluation of the textual contents in tourist reviews is done by text mining, which is a sub-area of data mining. This study consists of two main parts. The first section discusses the methods and techniques of opinion mining or sentiment analysis. In addition, aspect-based sentiment analysis and techniques to be used in the application are discussed. In the second part, a case study was carried out using aspect-based sentiment analysis method. In the application "Cappadocia home cooking" restaurant used tourist reviews.

Near Field Communication (NFC) Applications in the Tourism Industry

The use of near field communication (NFC), a type of wireless technology that allows data transfer by bringing two devices closer together, has been increasing very rapidly. NFC offers great ease to its users through its contactless operation and provides the optimum environment for wireless communication with other technologies. Users with NFC-enabled mobile phones can share websites, contact information, phone numbers, music files, videos, and photos and can also unlock doors and can use their phones as mobile wallets to make purchases safely and easily. In recent years, with the increasing use of technology, it is clear that NFC technology is promising for the tourism industry. In this chapter, background on NFC technology is provided. The features of NFC technology and its advantages and disadvantages for the tourism industry are evaluated. NFC technology applications (data sharing, payments, tickets, keys, etc.) for the tourism industry are explained. Finally, examples of NFC applications in the tourism industry are discussed for both tourists and sector stakeholders.

QR Code Applications in Tourism

With the development of technology new innovative approaches emerge day by day. With the internet and mobile devices being used more actively, many applications provide great convenience to our lives. This situation has brought about many changes and developments when it is evaluated in terms of both businesses and consumers. When it is considered in terms of enterprises, it can be said that applications that will increase customer satisfaction and loyalty have been started to be used for the purpose of being different from the others in competition conditions, being adopted in the market, interiorise a sustainable policy and making more profit. When it is considered in terms of customers it can be said that there are some technological applications that will save time, increase customer satisfaction and loyalty in tourism enterprises as well as in all other businesses. One of these applications is the use of QR (Quick Response code).

Big Data, Artificial Intelligence and Their Implications in Tourism Industry

Tourism businesses use AI and big data to connect guests creatively and meet their expectations with personalized service. Big data enables tourism professionals to learn more about their customers. And the more they know the better experience they can offer to customers. As it provides real value, AI has already become an integral part of operations, and this trend will continue in the future. Tourism businesses use AI tools to reduce operating costs and maintenance bills as in many other sectors. AI-oriented marketing has already been widely used in the hospitality industry. Moreover, as long as technology evolves and becomes more complex, tourism professionals will find more ways and methods to implement big data and AI to satisfy customers, and AI will continue to transform the tourism industry. The use of artificial intelligence and big data applications are discussed in this chapter. Properties, advantages and problems of artificial intelligence and big data are discussed, and some examples are given from the perspective of tourism industry.

Internet of Things in Tourism: A Proposal of the Information System for Cappadocia Hot-air Ballooning

With the revolution of Industry 4.0, the technologies that enter our daily lives are based on smart devices, applications and platforms with internet connection. A wide range of these technologies, collecting under one umbrella is known as IoT (Internet of Things). In this context, the study will consist of three parts. In the first part of the study, the stages of a touristic travel in smart tourism destinations are evaluated by considering IoT architecture. The technologies used in these phases and their contributions to the tourism sector and tourists are examined. In the implementation section, an IoT-based information system is proposed for Cappadocia hot air balloon tours. The main purpose of the system is to determine whether the appropriate weather conditions are formed before the flight in hot air balloon flights. The proposed system will allow for the automation and evaluation of data already collected using traditional methods. With the implementation of the system; work and time savings can be achieved, and more accurate measurements will make safe flights.

Virtual Reality Applications in Tourism

Through modern ages, tourism becomes a vital part of human life. With the advance of technology, tourism movements gain new tools. Those tools gave tourism an edge on management, marketing, education and economic areas. Also, on demand side, new types of tourists with their brand new demands and motivations became an important factor for who is responsible for tourism supply. One of the most noticeable tools that new technology era has brought is virtual reality applications. Despite dating back to 1950s, VR (virtual reality) has gained an interesting reputation in last few years with the advances of related technologies. VR applications are used in many sectors like construction, military, communication, marketing, education etc. Tourism industry is one of them of course. Especially, VR is used as a both tool and product in hotel & accommodation sector, recreation & entertainment, transportation, destination marketing and a touristic product itself.

Digital Gamification in Tourism Industry

The use of digital platforms allows people to play various types of games to not only meet their personal needs, but also to keep people's minds away from the strenuous work tempo of daily life. The game is an action that allows people to socialize and gain different experiences and to have a pleasant time and is limited by certain fixed rules for a purpose, including the outcome of winning and losing as a result of these rules. This action led to the emergence of a concept called gamification along with digitalizing technology. Gamification is the use of game elements, game design and mechanics in the game environment that enable users to socialize and have fun in non-game environments. In this study; literature on game, gamification, gamification theories, elements and models, player's types, tourism and game, gamification in tourism marketing have been reviewed. As a result of the literature review, the relationship between tourism and gamification has been explained and sample gamification models in tourism industry have been included. In the last part of the study, based on literature review and sample gamification applications, a new model has been designed and suggestions for digital gamification studies in tourism have been presented.

Technological Developments, Industry 4.0 and its Effects on Tourism Sector

Technological developments in recent years have been affecting the lives of both people and societies more rapidly than in the past. Developments in the field of communication, robotics, transportation etc. are called the 4th Industrial Revolution or Industry 4.0 in the industrial sector. Technological developments have created great changes in the services sector as well as in the industrial sector. Industry 4.0, it has also led to changes in the transformation of the tourism sector and is likely to occur in future processes. The purpose of this study is to examine the impact of Industry 4.0 on the tourism sector.

Expansion of Technology Utilization through Tourism 4.0 in Slovenia

With more and more people traveling worldwide (the number doubling in the last 20 years), tourist destinations are now more than ever trying to maintain and enhance their competitiveness in the global market. In this regard, novel business models combined with state-of-the-art technology can play a crucial role in not only satisfying the increasing tourists demand but also ensuring a sustainable growth in order to avoid the deteriorating effects on both the social and natural habitat. For these reasons, the Slovenian government included tourism into one of the priority areas for investment. Hence, the Tourism 4.0 initiative was lunched with the aim of uniting the tourism stakeholders with high-tech companies and unify their scattered ideas, experiences, knowledge and expertise. The ambition is to position Slovenia as a top destination for sustainable tourism with high economic value. In the following chapter, the authors will introduce the basic concepts behind Tourism 4.0 and how it relates to technologies for an enhanced tourists experience.

Customer Experience in the Restaurant Industry: Use of Smart Technologies

One of the most important points of consideration that will ensure the sustainability and profitability of businesses in the digitized world market is the experience offered to customers. Experience in business preferences, especially of the digitized customers, is quite specific. In this study, the contribution of smart technologies to customer experience was discussed for restaurants, and its significance was emphasized. Moreover, the subjects of artificial intelligence, smart technology and QR code were addressed based on customer experience. Finally, the contribution offered to customers was emphasized by giving examples of smart technology applications used in restaurant businesses in order to improve customer experience. It is thought that this study will contribute theoretically to the subject which has not been adequately studied in the literature.

Electronic Customer Relationship Management in Tourism

Tourism companies like other companies producing goods and/or services, practicing a number of different strategies to achieve competitive advantage and satisfy the needs and wants of customers/guests. The most important of these strategies in the recent period is electronic customer relationship management (e-CRM). In todays competitive environment, a successfully implemented e-CRM that enhances a customer centric culture is a key to the tourism growth and create competitive advantage. For tourism companies, e-CRM is successful only if it is totally integrated strategically and operationally in the business and in the information system of the company. This chapter will focus on e-CRM concept, its importance in tourism, dimensions of e-CRM, major challenges of e-CRM and its opportunities for companies, its implementation in tourism, e-CRM Technologies used in tourism, e-CRM success factors and the reasons of e-CRM failures.

In Which Department(s) the Robots might be Employed-Evidences from Turkish University Students

The development of service robotics comes along with the development of industrial robotics. Service robots are designed to perform professional job tasks as well as for service users in areas of everyday life. One of the fields for service robots application is tourism. Aim of this study to determine the opinions of the university students who have tourism education, towards tasks may be performed by the robots. The research question was "in which hotel department the robots might be employed". In total, 396 valid surveys were collected from the students, Mann Whitney U and Kruskal Wallis-H Tests were conducted to test the hypotheses. It was found that gender, department, and grade are significantly effective on the students' opinions towards tasks might be performed by service robots.

Smart Hotels and Technological Applications

Technological developments and demographic changes have started to encourage the hotel businesses to use automation systems and adapt the concept of smart hotel. Smart hotels use state-of-the-art technology to provide customers with different and more technology oriented experiences. Robot works as receptionist, concierge, bellboy and housekeeper in such hotels. These applications are observed to minimize service failures and save on costs. For instance, chatbots created in smart hotels by means of artificial intelligence respond to customers' problems and demands quickly and accurately. Virtual reality applications used in the marketing of smart hotels make it possible for guests to take a virtual tour of the cultural heritage sites of the destinations. In addition, Internet of Things (IoT) is another technology that is often used in smart hotels to offer customers a more comfortable holiday and get instant data. Likewise, in smart hotels, self-service software technology is often used and enables customers to perform many tasks such as check-in/out, getting hotel room keys, etc.

Smart Applications in Tourism

Technology process that started with the reservation provided many services in accommodation establishments such as online room selection, room colour selection, air conditioning adjustment, temperature and light control, virtual scenery creation, online service. Another area that is affected by technology is food and beverage services. Food and beverage businesses also benefit from technology while serving their guests. Many services are provided through technological revolutions such as online reservation, online service, smart tables, robots and so on. Today technology also affects museums and museum services positively. With virtual reality museums offer visitors the opportunity to visit their works online. Online ticket sales also save visitors time. Also, credit cards and ticket sales provide economic convenience to the guests. With the help of technology, heat and light adjustments of museums can be made and it helps to protect the works exhibited in these museums.

Smart Tourism, Smart City and Smart Destinations as Knowledge Management Tools

The rapid increase in the population has caused problems in the correct use and management of resources in cities. Solutions to these problems have been sought based on knowledge management. In the digital age we are in, the concept of smartness of cities has been put forward together with the web based applications. It has changed the expectations and needs of tourists and residents especially in the tourism sector in the service sector. Smart tourism destinations, which are seen as a solution to the expectations and needs of the changing tourists and local people, bring holistic innovations covering all the stakeholders in the tourism ecosystem. In this direction, it is thought that knowledge management makes the right use of resources obligatory and contributes to sustainable tourism understanding. For this purpose, in the study, suggestions were made to provide maximum benefit from web-based projects that are carried out in qualitative direction of knowledge management and smart tourism destinations.

Smart Municipalities in Tourism

Tourism is a fragmented and information-oriented sector covering tour operators, travel agencies, hotel sales representatives, associations, meeting offices, transportation, car rental, airlines, cruise, souvenirs, restaurants, hotels/motels and entertainment. In the tourism sector, technology is an important requirement for the interconnection of all these fragmented organizations, activities and elements and for the satisfaction of tourists. However, the effective use of technology in a destination requires the leadership of a leader. This study approaches the smart municipality applications oriented to tourism sector with the example of Turkey, a well-known destination around the world, from the perspective of municipalities, which is regarded as a democratic leader at local level and provides a general framework regarding the studies on an innovative, environment-friendly, solution-oriented and more livable urban life. Suggestions are provided for smart technologies to improve the quality of life of local citizens and to the satisfaction of tourists.

A New Concept in Tourism: Smart Tourism Destinations

The purpose of this chapter is to provide an insight into the topic of smart destinations. By adopting smart technologies, tourism destinations, as well as cities, gain more opportunities to offer better quality of life for residents and visitors. Smart cities aim to improve resource management, sustainability, and living conditions in urban environments by utilizing ICTs. The concept of smart tourism destination, deriving from smart city, refers to the use of technology in tourism destinations in order to increase the service quality and tourist satisfaction by focusing on tourists' expectations. In other words, smart tourism destinations aim to integrate technology into the destination for these purposes. In this chapter, the concepts of smart city, smart tourism, and smart destination are presented. The emergence of smart city and smart destination concepts and the issues and challenges they might face are discussed. In addition to some future research directions, a brief discussion on potential controversies is presented.

Tourist Experience in Smart City Destination of Jaipur (Rajasthan)

Smart city concept is emerging as new venture in tourism sector especially in a country like India. The present chapter focus to analyze the tourist experience in Jaipur city of Rajasthan which is emerging as popular tourist destination and taken over for development under smart city mission. Although, chapter adopted empirical approach, but to some extent it also used secondary data sources. The authors used to analyze tourist experience by using smart attraction, smart mobility, smart amenities, smart ancillary, smart safety and smart available packages as important dimensions of smart tourism destination. Regarding tourist experience, the chapter noticed smart available packages, smart safety and smart attractions are providing pleasant experience to tourists in contrast with remaining aspects of smart tourism destination. The result of the study provides implication for tourism marketer and policy maker to take appropriate steps to enhance the tourist experience by improving smart facilities and services.

Intelligent Tourist Destinations and Their Application to Public Policies: The Spanish Case

The chapter is divided into two differentiated parts. The first includes a brief technical review of the concepts of Territorial Intelligence and Intelligent Tourist Destination, situating them in the context of the new tourist destination planning paradigms. This part ends with a first approach to the Spanish case and the progress of its public tourism policies towards these concepts. The second part goes deeper into the Intelligent Tourist Destination model applied in Spain, driven by the public administration, paying special attention to its official standardization process, and the indicators adapted to that end. Finally, the chapter ends with some brief conclusions.

Smart Tourism Planning: Geographical Evidence from Poland

The main focus of the chapter is to discuss geographical context of smart tourism planning through the analysis of relations between spatial planning, tourism development, and technology. The main question addressed in this enquiry is whether development of technology and tourism respects sustainable spatial planning. Thus, geographical influence of tourism and ICT on spatial planning was investigated. Research is carried out in Poland as tourist destinations in the country are substantially different regarding their tourist attractions, infrastructure, and development. Geographically weighted regression (GWR) was applied to identify regions where spatial planning is significantly influenced by tourism and technology development as well. Both mentioned phenomena were recognized as competing in terms of spatial planning. Tourism and technology development did not support spatial planning parallely. Hence, establishing and developing smart tourism planning was confirmed as challenging from the geographical perspective.

Evrim Çeltek
Gaziosmanpaşa University, Turkey

Section 1
Smart Technology Applications

Chapter 1
Progress and Development of Virtual Reality and Augmented Reality Technologies in Tourism:
A Review of Publications From 2000 to 2018

Evrim Çeltek
https://orcid.org/0000-0002-2722-5256
Gaziosmanpaşa University, Turkey

ABSTRACT

Progress and development of tourism technologies and applications available through smart devices provide an increasing diversity in tourism marketing. Augmented reality, virtual reality applications with the smart technologies, which are considered as new marketing communication instruments, increase awareness and help in terms of getting information about tourists and giving information to the tourist. Thus, this chapter provides an assessment regarding the progress of AR and VR researches published in hospitality and tourism journals. In particular, the chapter determines the current fields, topics, and research methods of AR and VR articles published in 32 hospitality and tourism research journals between 2000-2018, comparing the subjects and research methods of 32 journals and discussing the changes in topic areas throughout the years, to offer suggestions for further research on AR and VR for hospitality and tourism journals.

INTRODUCTION

Progress and development of tourism technologies, applications obtainable through smart devices provide an improving diversity in promotion of tourism. Marketers have developed additional inventive practices

DOI: 10.4018/978-1-7998-1989-9.ch001

to draw in customers in the last decade. These new promoting tools are virtual reality, augmented reality, applications with the smart technologies, which are thought as new promoting communication instruments, increase awareness and help in terms of getting info regarding the tourist and giving info to the tourist.

This chapter aims at examining the development and progress of augmented and virtual reality technologies in articles printed in hospitality and tourism journals between 2000 and 2018. This chapter reviews articles published in last eighteen years. Using a wide range of sources, primarily in tourism literature, it comprehensively reviews and analyses previous studies within the context of AR and VR applications in tourism. The paper additionally projects future developments in AR and VR, and demonstrates essential changes which will probably influence the structure of tourism industry. A significant contribution of the paper is its summary of the research and development efforts that have been endeavored within the field, and therefore the challenges tourism researchers are, and can be, facing. The article reports on a study that analyzed the published AR and VR papers in thirty two research journals in tourism, and examined the trend of AR and VR research based on the publications excluding book reviews, research notes, reports, commentaries, and case studies. The research topics and methods and industry sectors of the chosen articles were analyzed and discussed through content analysis. The majority AR and VR related publications in tourism and hospitality journals were searched and compiled from three online databases. The research topics and methods and industry sectors of the selected articles were analyzed and discussed through content analysis. Almost all AR and VR related publications in tourism and hospitality journals were searched and compiled from three online databases. They were 1) ScienceDirect (www.sciencedirect.com); 2) Emerald Management eJournals (www.emeraldinsight.com); and 3) Sage Journals (http://online.sagepub.com).

Thus, this study aims at providing an assessment regarding the progress of AR and VR researches published in hospitality and tourism journals. In particular, the study aime at:

- Determining the current fields, topics and research methods of AR and VR articles published in 32 hospitality and tourism research journals between 2000-2018,
- Comparing the subjects and research methods of 32 journals and discussing the changes in topic areas throughout the years,
- To offer suggestions for further research on AR and VR for hospitality and tourism journals.

BACKGROUND

Augmented Reality and Virtual Reality

The process of taking and integration digital info either with a live streaming video or with the user's actual, real time atmosphere is called augmented reality (AR). AR is used by smart phones equipped with GPS technology, and it allows users to pinpoint a location to regulate device orientation (Taylor, 2013). This is such a technology that enables blending new information into a current picture. (Berryman, 2012; Craig, 2013; Höllerer and Feiner, 2004;). The fundamental aim of an AR system is to supplement the real world with 3D virtual objects that exist in the same space as the real world to enrich user's perception of and interaction with the real world (Azuma et al., 2001).

AR enables smart phone and tablet users to point their phones' or tablet's cameras at certain objects (Linaza et al., 2012; Azuma, 1997) – whether it is a print advertising or maybe a cup of coffe or brochures

/magazines - to trigger a 3D video. AR lets tourism businesses unpredictably integrate the computer based digital world into the real world appealing to technology fans typically skeptical of customary advertising strategies (Craig, 2013).AR appeals to each single sense or all senses as it is not restricted to a specific display technology, like Head Mounted Displays (HMDs) or the sense of sight. As a technique, AR mixes a live view in real-time with virtual computer-generated pictures, making real-time augmented experiences of reality. While applying this method AR enhances the users' perception of reality and their surrounding atmosphere (Kounavis et al., 2012). Thus, Hassan and Rahimi (2016) defined AR as an advanced level of virtual reality that combines reality with computer-simulated imageries within the real medium. This definition was supported by Jung et al. (2015) and Dadwal and Hassan (2015) who considered AR as combination of computer simulations of digital imageries in real environments.

Tourists can benefit from mobile augmented reality applications in several aspects including searching for, sharing or exchanging info and helpful tips also as comments on a place or destination within a large network. Hence, contact among different users, e.g. tourists during this case, can be increased and experience sharing among tourists is developed (Russell, 2012). In addition, mobile augmented reality applications can enable tourists protect themselves from irrelevant or overdose of information by tailoring their wants and adjusting the settings of the applications they're using upon request. It's helpful for tourists as it provides them with a huge amount of knowledge regarding historical sites and museum exhibitions, and it might be overwhelming for them to get info they significantly search for. For this reason, museums, heritage sites, cities and tourist professionals generally might organize and convey info in layers or upon request considering tailored needs of tourists with regards to their interests, age, occupation, education level and so on. Monitoring their visit regarding their personal want, they may maximize from the enjoyment of the trip and turn it into a remarkable experience (Berryman, 2012; Carmigniani et al., 2011). AR has become an accepted and valuable tool for tourism industry, mostly due to its ability to enable tourists with restricted data of an area to naturally and realistically experience it (Martínez-Graña et al., 2013; Chung et al., 2015). Additionally, AR will produce personalized and tailored info (Kounavis et al., 2012; Kourouthanassis et al., 2015), serving to minimise frustration caused by information-overload (Chung et al., 2015), and so facilitate high-quality tourism (Rey-López et al., 2011).

Augmented reality is a kind of real-time and interactive experience created by combining the physical elements we perceive in the world around us with data from computer-based graphics, video, audio, GPS, etc. Through the hardware and software used in augmented reality applications, it is enriched by adding virtual data on the real world we perceive around us. So "Increased''. Augmented reality, unlike virtual reality, does not involve the user in a computer-based virtual simulation in a way that detaches the user from reality, but enriches the existing reality. Augmented reality is an improved version of reality created by superimposing computer-generated images on the user's real-world perception. Augmented reality can be applied more easily to everyday life, as it does not completely break the user away from his / her physical reality as in virtual reality. AR platforms used in tourism are personal computers with webcams, kiosks, digital signage, window displays, smart phones and tablets, AR glasses and head-mounted displays. (Craig, 2013; Çeltek, 2015; Santosa and Gook, 2012; Krevelen, Poelman, 2010; Yung and Lattimore, 2017; Chung et al., 2015; Schmalsteig et al., 2011).

Augmented reality is divided into four groups, depending on which method the virtual and real views are superimposed: marker based, markerless, projection based, and superimposition based (Chung et al., 2015; Schmalsteig et al., 2011; Çeltek, 2015):

Figure 1. AR platforms and types of AR
Source: Adapted from Craig, 2013; Çeltek, 2015; Santosa and Gook, 2012; Krevelen, Poelman, 2010; Yung and Lattimore, 2017; Chung et al., 2015; Schmalsteig et al., 2011.

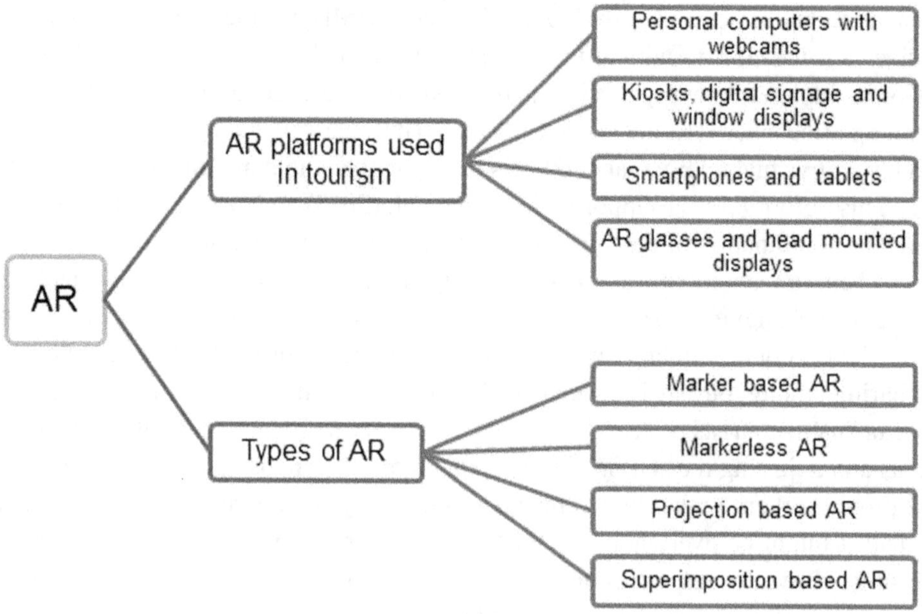

- In a **marker based augmented reality,** a camera and a QR code or similar pointer are used. Detecting these markers by image processing methods requires faster and less processing power, while the position and orientation of the marker is calculated and superimposed on the virtual content.
- In the **markerless augmented reality**, technologies such as GPS, speed meter, accelerometer is used instead of pointer. This technique, which uses virtualization according to the user's location, comes to the fore with the most applications by the developers with the widespread use of smart phones.
- In **projection based augmented reality**, virtualization can be realized in two or three dimensions. In the two-dimensional version, an artificial light is projected onto a physical plane to determine how the user interacts with it, while in the other version with laser plasma technology a three-dimensional interactive hologram can be formed.
- In **superimposition based augmented reality**, object recognition is used. The related object is recognized and superimposed on its virtual reality image.

Simply defined as 'a computer-generated world' virtual reality (VR) can both be quite simplistic and embracing. The basic component of any VR system is expected to contain a computer-generated world but ideally it is expected to surround the participant perceptually, (Pan and Hamilton 2018). Represents three-dimensional (3D), interactive, computer-generated environments are represented by virtual reality application. These environments will either be models of real or imagined worlds, and they aim at representing data through a synthetic experience. The VR technology merged from various disciplines, as well as multimedia, data-base design, psychology, computer graphics, electronics, real-time and distributed systems, robotics, cybernetics, and telepresence (Gurau, 2007). VR is described as the use of a computer-generated three dimensional atmosphere, that is named a 'virtual environment' (VE), one can navigate and probably interact with others, within the form of real-time simulation of one or additional five senses of the user. 'Navigate' stands for the ability to move around and explore the virtual environment, and 'interact' is that the ability to select and move objects inside the VE (Guttentag, 2010).VR has the potential to revolutionize the marketing and sales of tourism once it's considered from a marketing perspective (Williams and Hobson, 1995). VR can't only be used to project and manage a destination; however it may be used to market a destination (Cheong, 1995). The accessibility of low-priced VR viewers like Google Cardboard and plenty of tourism-related VR content create it possible for anyone to experience virtual tours of cities and tourism attractions all around the world. VR carries limitless potential for mass virtual visitation to actual tourism destinations (Tussyadiah et al. 2018).

Virtual reality is a three-dimensional image or artificial environment created by combining interactive software and hardware that makes the user feel physically interacted in a real environment. Virtual reality is divided into three groups according to the participation rate (not to be understood as the number of participants): non-immersive, semi-immersive and fully immersive (Halarnkar et.al., 2012; Galeano, 2017;):

- In non-immersive virtual reality, several sensory organs are stimulated so as not to interfere with the user's environmental reality. The user enters this three-dimensional virtual environment via a portal or window.
- In semi-immersive virtual reality, the user is partially integrated into the three-dimensional virtual environment. As with flight simulations, the user is heavily involved in the virtual environment, but is not completely separate from his reality.
- In fully-immersive virtual reality, all the senses of the user are stimulated by a piece of hardware that attaches to the head of the user and combines with the virtual environment to completely cover the user's field of view. Fully-immersive virtual reality delivers realistic user experiences.

When virtual reality is considered as a whole system, the importance of the three main components emerges. • A computer with high processing capacity, • Head Set, • Input Devices (such as joysticks, gloves, pedals, motion platforms such as Virtuix Omni). The head set also consists of three main components. These include sensors (magnetometers, accelerometers, gyroscopes), lenses and display screen (RealityTechnologies.com, 2018; Novák-Marcinčin and Kuzmiaková, 2009; Bamodu, Ye, 2013). Figure 2 summarize the key elements in a virtual reality system used in tourism and types of virtual reality used in tourism.

Recently, some researchers studied on the advantages of VR regarding the tourism context. As for the tourists' point of view, they stated that main advantages of VR comprise augmentation of tourism experiences (Moorhouse, TomDieck, and Jung, 2018; Bonetti, Warnaby, and Quinn, 2018;); facilitation of immersive, dealing with, social, and amusing experiences (e.g. Castro et al., 2017; Tromp, 2017; Guttentag, 2010; Jung, Lee, Chung, and TomDieck, 2018;), furthermore because of the potential to supply accessible tourism for all (Guttentag, 2010; Williams and Hobson, 1995). As to the point of views of companies and destinations adopting VR, factors including promoting, sales and distribution (Williams and Hobson, 1995; Huang et al., 2016; Moorhouse et al., 2018; Gibson and O'Rawe, 2018), further revenue generation (Tromp, 2017; Radde, 2017), as sustainability and also the protection of heritage (Williams and Hobson, 1995; Guttentag, 2010;) were taken because the advantages of VR. Studies on

Figure 2. Key elements in a virtual reality, types of virtual reality
Source: Adapted from Halarnkar et.al., 2012; Galeano, 2017; RealityTechnologies.com, 2018; Novák-Marcinčin and Kuzmiaková, 2009; Bamodu, Ye, 2013.

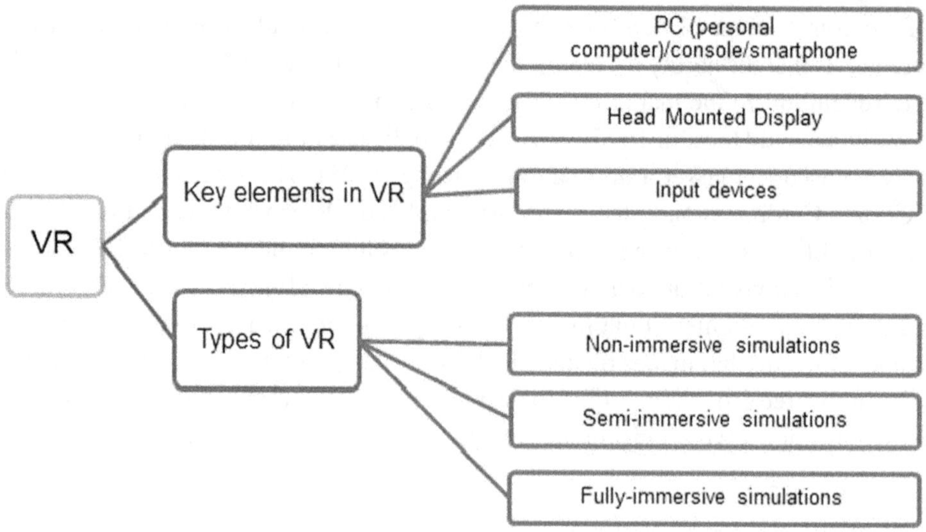

VR have additionally stated its potential for promoting cultural heritage and arts from the perspective of tourism (Jung et al., 2016; Jung et al., 2017; Chen, Pan, and Zhang, 2012; Marasco et al., 2018). Table 1 summarizes the advantages and challenges of AR and VR in tourism.

MAIN FOCUS OF THE CHAPTER

Data Collection

The review process is similar to the one of Yung and Lattimore (2017), who used the method for their systematic quantitative literature review on virtual reality and augmented reality analysis in tourism. The five-step review protocol, adapted from Yung and Lattimore (2017)'s systematic quantitative literature process, consists of (1) assigning review aims and formulating analysis questions; (2) distinguishing search terms, databases, and literature selection criteria; (3) searching the databases for the literature and screening search outcomes against the standards before refining exclusion and inclusion criteria; (4) appraising literature quality and relevance, making correspondence analysis, structuring outline tables through extracting relevant information; and (5) synthesizing and reporting findings. The data were collected in January and April 2018. Yang et al. (2017) suggested seven academic databases as the most significant electronic databases for tourism, and identified from previous tourism review publications including Science Direct (Elsevier), Web of Science, EBSCO Host (Hospitality and tourism Complete), ProQuest, Emerald, Sage and Scopus. On the other hand, all AR and VR connected publications in tourism and hospitality journals were searched in this study, and data were gathered from three online databases. They were 1) Emerald Management eJournals (www.emeraldinsight.com); 2) ScienceDirect (www.sciencedirect.com); 3) Sage Journals (http://online.sagepub.com). These databases were selected

Table 1. Advantages and challenges of AR and VR in Tourism

	ADVANTAGES	CHALLENGES	AUTHOR/S
AUGMENTED REALITY	• Brands become memorable as customers react positively to fun, • As user is in charge of the navigation, it personalizes communication per user • In contrast to conventional media, there is no limitation of space or time. • It provides opportunity for customers to share their tailored content with others. • It requires much lower cost compared to conventional media. • It supplies tourists with access to location-based information about the neighbourhood. • Makes possible to access the various contents, which are well-timed and updated. • Has flexibility in delivering texts, videos, or images. • Provides interactive additional explanations which are integrated with map-based services and additional information. • Rises brand awareness. • Brand reputation. • Geo targeting. • Interactive marketing • Personalization, Viral marketing, Socialization, Emotional Connection, Repeat Engagement • Provides filtered information tailored to the needs of traveller and aims at creating a more personal experience,	• AR is a technologically complex service. • AR doesn't have any standards. • Tall buildings may limit the performance of existing GPS systems in smart phones (it may happen even indoors). • It includes privacy related problems. • Ethical issues and user issues. • It depends on approval or participation of the user. • There is lack of awareness and understanding of people on it.	Chung et al. (2015); Fritz, et al. (2005); Kennedy-Eden and Gretzel (2012); Han and Jung (2014); Lashkari et al. (2010); Tom Dieck and Jung (2017); Chou and ChanLin (2012); Höllerer and Feiner (2004); Yovcheva et al. (2013); Han et al. (2014); Weber (2014); Xu et al. (2017); Celtek (2015); Fino et al. (2013); Marimon et al. (2014); Wang et al. (2013); Wasko (2013); Kounavis et al. (2012); Kourouthanassis et al. (2015); Lee (2012); Selvam et al. (2016); Cranmer, (2017); He, Wu, Li (2018).
VIRTUAL REALITY	• Practical advertising. • Promotes people visit to real museum, and functions as indirect evidence of marketing instrument for destinations. • Facilitates tourists with the opportunity to see previews on destinations and their respective attractions and facilities. • Sustainability and the protection of heritage. • VR systems can be used for direct marketing as well. • Try and buy. • Customer engagement. • Increasing brand awareness. • Brand reputation. • Gamification • Full immersion	• It depends on approval or participation of the user. • Its effects on health are still unidentified. • People don't know how to run it.	Guttentang (2010); Tavakoli, Mura (2015); Xu, Buhalis, Weber (2017); Tussyadiah, et al., (2018); Xu, et al., (2016); Castro et al., (2017); Tromp, (2017); Bonetti, Warnaby and Quinn, (2018); Moorhouse, Tom Dieck,and Jung, (2018); Gibson and O'Rawe, (2018); Huang et al., (2016); Moorhouse et al., (2018); Williams and Hobson, (1995);

due to the limitations of university database access. The articles reviewed were limited to tourism journals. The journals which were not related to tourism (engineering, marketing, technology journals) included VR and AR topics in tourism were also excluded, as this research aims at determining the progress of VR and AR in tourism related journals. Relevant keywords and also the terms "tourism", "vacation", "destination", "hospitality" and "museum" were used in the search to enable the current state of AR and VR related findings and discussion within the tourism and hospitality literature to be given. To safe increase and guarantee the quality and effectiveness of the review, solely original research articles printed in English-language peer reviewed journals were considered. After filtering the databases with tourism journals, 32 tourism related journals were found. Table 2 gives the details of reviewed journal names.

In accordance with the aim of the study, "augmented reality" or "AR" or "virtual reality" or "VR" or "virtual world" or "virtual environment" was searched in keywords, abstracts and titles, to receive relevant literature. Having completed the literature search, the articles were screened considering the inclusion and exclusion criteria. As the next step, 2 selection criteria were applied. Firstly, literature review studies, research notes, book reviews and editor prefaces were all excluded, and solely full-length papers printed in tourism and hospitality journals were included within the analysis. Then, only papers published between 2000 and 2018 were analyzed to make sure that the information presented was up-to-date. Articles were searched and relevant papers for this study were gathered. Every identified paper was reviewed carefully by the researcher and determined whether to include it in the study or not. Data collection was conducted three times, and every one sets of selected papers were intercompared for reliability. 193 articles and abstracts including "augmented reality" or "AR" or "virtual reality" or "VR" or "virtual world" in titles as keywords, were found from the tourism journals. The decision to include a paper was primarily based on its direct relevance to the subject of AR and VR applications in tourism and hospitality. In addition, excluded papers, the overall topic of that weren't related to AR and VR. At the end of the data collection process, a final sample of 62 papers was generated for analysis but 4 articles weren't open to access, and finally 58 articles were analyzed. As for date distribution, the numbers of studies published in 2008, 2010, 2011, 2012, 2013, 2014, 2015, 2016, 2017 and 2018 were 1, 3, 2, 3, 3, 5, 6, 4, 18 and 13, respectively. The papers were drawn from 16 journals, and they fell into three major categories:

(1) AR related papers (14 article)
(2) VR related papers (37 article)
(3) Both AR and VR related papers (7 article).

Hard copies and on-line versions of the articles were accessed and they were read in detail by the researcher. The paper was downloaded for further analysis if the objective of it had been related to any AR and VR elements. A complete of 58 printed articles were found to be relevant to AR and VR perspectives at the end of the database search. Then, a content analysis process was applied to those articles. The abstract of each article was read by the researcher to find out whether it was related to the study during the selection process. Then, so as to further confirm their relevance, full-length articles of the identified abstracts were read completely again twice. Solely full-length papers and research notes were analyzed during this study, as they were refereed articles. Thus, book reviews, abstracts, editor prefaces, internet columns and conference reports were excluded. Table 3 shows the distribution of articles.

Content analysis was used to analyze information among the chosen articles, publication year, publication journal, research design, data collection methodology, statistical technique, target industry sector, subject category and application platform for each article were all recorded. The research was designed in three types of research: quantitative, qualitative, and mixed. Subject category, statistical technique, information collection technique and target industry application were classified into many categories. The classifications were done through a review of previous studies. All articles were analyzed considering the following items. Table 2 shows the variables accompanied by the appropriate definitions.

Table 2. Description of analysis variables

VARIABLES	DESCRIPTION
Publication year	The reviewed articles included only papers published between 2000 and 2018 to ensure the information presented was up-to-date.
Journal name	32 Journals Reviewed for VR and AR Related Articles. The journals are listed as; 1. Annals of Tourism Research, 2. Cornell Hospitality Quarterly, 3. International Journal of Contemporary Hospitality Management, 4. International Journal of Culture, 5. Tourism and Hospitality Research, 6. International Journal of Event and Festival Management, 7. International Journal of Hospitality Management, 8. International Journal of Museum Management and Curatorship, 9. International Journal of Tourism Cities, 10. International Journal of Tourism Management, 11. Journal of Destination Marketing & Management, 12. Journal of Hospitality & Tourism Research, 13. Journal of Hospitality and Tourism Insights, 14. Journal of Hospitality and Tourism Management, 15. Journal of Hospitality and Tourism Technology, 16. Journal of Hospitality, 17. Journal of Hospitality, Leisure, Sport & Tourism Education, 18. Journal of Outdoor Recreation and Tourism, 19. Journal of Tourism Analysis, 20. Journal of Tourism Futures, 21. Journal of Vacation Marketing, 22. Museum Management and Curatorship, 23. Sport, Business and Management: An International Journal, 24. The Cornell Hotel and Restaurant Administration Quarterly, 25. The Tourist Review, 26. Tourism and Hospitality Research, 27. Tourism Economics, 28. Tourism Management Perspectives, 29. Tourism Review, 30. Tourist Studies, 31. Worldwide Hospitality and Tourism Themes, 32. Tourism Management
Time periods	As the present study took a long period, the whole study period was divided into three sub-periods: 2000-2005, 2006-2011 and 2012-2018 to better understand the evolution of research focus over time
Article subject	Articles related with AR or VR or both AR and VR
Application platform	AR and VR application platforms in tourism; marketing, guiding, gamification and education.
Research design	Research design types; qualitative, quantitative and mixed.
Statistical techniques	Statistical techniques are descriptive, univariate, multivariate, SEM/CFA/path analysis and other statistics.
Target industry	Target industries are *Hotel/lodging:* includes specific AR and VR applications for hotels, *Destination/ DMO:* includes specific AR and VR applications for destinations, *Restaurant/food service:* includes specific AR and VR applications for restaurants, *Travel Agent/Online Travel Agent:* includes specific AR and VR applications for travel agents, *Airlines*: includes specific AR and VR applications for airlines, *General hospitality/tourism industry:* it includes articles which target all of the tourism sector applications and solutions, do not target a specific tourism business like hotel, restaurant, travel agent etc., *Museums:* includes specific AR and VR applications for museums, *Tourism education*: includes specific AR and VR applications for tourism schools whichare used by students and lecturers.
Subject category	Subject categories are; Decision making/purchase intention, Motivation, Quality, Characteristics/ segmentation, WOM/satisfaction/complaint, Technology acceptance, Information search, Experience, Technology adoption, Marketing effectiveness, AR and VR design/evaluation, Consumer relationship/ evaluation, Web site design/evaluation, Branding/image, Advertising/promotion

Data Analysis

The data were analyzed in three steps. First, descriptive statistics were applied to demonstrate the overall status quo of AR and VR related articles printed in the 32 tourism and hospitality journals. Second, considering the long research period, the whole study was broken into three sub-periods: 2000-2005, 2006-2011 and 2012-2018 to better understand the evolution of research focus in time (Morosan et al., 2014; Harrington et al., 2014; Leung, Xue and Bai, 2015). The sub-periods were determined by focusing on both the total number of articles printed in every year and subjective judgment of the authors. CA was used to find out the significant changes of AR and VR related researches across these three sub-periods. Finally, another CA was conducted to examine the publication preferences among the 32 different journals. CA could be a statistical technique used for comparisons among objects and attributes simultaneously based on categorical/nominal data. It's based on perceptual maps including a group of objects and attributes displayed graphically in a joint space based directly on the association of objects and attributes. Objects fall in close proximity when they have high association in a perceptual map (Hair et al., 2010; Leung, Xue and Baic, 2015).

Results

32 journals published in the field of tourism have been examined. Articles about AR and VR were found in 16 journals. The number of articles examined was 58 (see Table 3). According to the results of the content analysis, AR and VR related articles were mostly written in 2017 and 2018. The articles on AR and VR have started to be written since 2008. No articles written before 2008 were found. This suggests that the articles on AR and VR in the academic field are beginning to gain importance since 2008. Most AR and VR articles were published in JHLSTE (12 articles), TM (8 articles) and JDMM (7 articles), respectively. The articles were written on VR with a maximum number of 37 articles. The number of articles written about AR was 14. This shows that the articles on VR in the field of tourism are more intense. The number of articles containing both AR and VR topics was 7. The majority of the articles focused on the field of marketing with 35 articles and education area with 12 articles. Looking at the data collection and research characteristics of the articles, the research was designed both qualitative (28 articles) and quantitative (27 articles). Surveys were mostly used as data collection tools. Descriptive statistics were often used as statistical techniques. The sector targeted by the articles is mostly destination and general tourism. The subjects of articles focused on technology adoption, WOM/satisfaction/complaint and experience respectively.

Topical Review

Correspondence analysis is based on the assumption that similarities or differences between the categories of variables are expressed in terms of distances and which subcategories are more similar than the other categories and that the similarities between the subcategories of the variables in the cross tabulation (the table formed by the variables, the intervening table or the compliance table) to be displayed graphically in size. If there are three or more categorical variables, multiple correspondence analysis is applied (Özdamar, 2004). Since the number of categorical variables in the present study was more than three, multiple correspondence analysis was applied. In this frame, the "publication period" (3), "subject category" (16), "targeted industry" (9) questions among the questions in the examination form,

Table 3. Article frequencies

		Frequency	Percent
Publication year	2008	1	1,7
	2010	3	5,2
	2011	2	3,4
	2012	3	5,2
	2013	3	5,2
	2014	5	8,6
	2015	6	10,3
	2016	4	6,9
	2017	18	31,0
	2018	13	22,4
Total		58	100,0
Journal name	ATR	3	5,2
	IJCHM	4	6,9
	IJHM	3	5,2
	JDMM	7	12,1
	JHLSTE	12	20,7
	JHTM	2	3,4
	JHTR	2	3,4
	JHTT	5	8,6
	JTF	2	3,4
	JTR	4	6,9
	JVM	1	1,7
	TE	1	1,7
	THR	1	1,7
	TM	8	13,8
	TMP	2	3,4
	TR	1	1,7
Total		58	100,0
Article subject	VR	37	63,8
	AR	14	24,1
	AR and VR	7	12,1
Total		58	100,0
Application platform	Marketing	35	60,3
	Guiding	9	15,5
	Gamification	2	3,4
	Education	12	20,7
Total		58	100,0

continued on following page

Table 3 Continued

		Frequency	Percent
Research design	Qualitative	28	48,3
	Quantitative	27	46,6
	Mixed	3	5,2
Total		58	100,0
Data collection	Interview	10	17,2
	Survey	22	37,9
	Experiment	9	15,5
	Document/online document	16	27,6
	Interview and survey mixed	1	1,7
Total		58	100,0
Statistical techniques	Descriptive	21	36,2
	Univariate	5	8,6
	Multivariate	8	13,8
	SEM/CFA/path analysis	10	17,2
	Other statistics	14	24,1
Total		58	100,0
Target industry	Hotel/lodging	2	3,4
	Destination/DMO	21	36,2
	Restaurant/food service	2	3,4
	General hospitality/tourism industry	16	27,6
	Museums	4	6,9
	Tourism education	13	22,4
Total		58	100,0
Subject category	Decision making/purchase intention	2	3,4
	Motivation	1	1,7
	Quality	3	5,2
	Characteristics/segmentation	3	5,2
	WOM/satisfaction/complaint	12	20,7
	Technology acceptance	4	6,9
	Information search	2	3,4
	Experience	7	12,1
	Technology adoption	13	22,4
	Marketing effectiveness	4	6,9
	AR and VR design/evaluation	6	10,3
	Web site design/evaluation	1	1,7
Total		58	100,0

Table 4. Table of correspondence analysis of the general variables (periods, subject category and target industry)

Model Summary					
Dimension	Cronbach's Alpha	Variance Accounted For			
		Total (Eigenvalue)	Inertia	% of Variance
1	,664	1,793	,598	59,780
2	,630	1,724	,575	57,474
Total		3,518	1,173	
Mean	,647[a]	1,759	,586	58,627
a. Mean Cronbach's Alpha is based on the mean Eigenvalue.				

including category numbers in parentheses, and multiple correspondence analysis were applied to see how the combination would be, and in the direction of the answers given, the results of the analysis of the 3X16X9 versatile table are presented in Table 2 for each variable and for each dimension. However, the inertia obtained as a result of the analysis shows that the data reveals 117% of the total variance. 59% of them describe dimension 1 and 57% describe dimension 2. The eigenvalues give a complete measure of the correspondence between the two-dimensional graphs obtained by the real graph. In this respect, it can be said that the correspondence between the two-dimensional graphs obtained by the real graph (1,173%) is very good.

The graph obtained from the analysis is shown below (Figure 3). The publication years have been examined by dividing into the three turnovers (2000-2005, 2006-2011 and 2012-2018). According to this graph, the articles were written during the publication period 2012-2018 focused on WOM/satisfaction/complaint/perception; AR and VR design/evaluation; experience; information search and motivation subjects. The sectors targeted by the articles are museums and tourism education. The articles aiming at tourism education included AR and VR design/evaluation. The articles aiming at museums examined tourists' AR and VR experience in the museum. The articles written during the 2006-2011 period included quality, marketing effectiveness, decision making / purchasing and using intention, technology adaptation in tourism industry, technology acceptance by customers/tourists and customer characteristics/segmentation subjects. The target sectors of the articles were destination/DMO, restaurant/food service, and general tourism. The articles targeting the general tourism sector examined the adaptation of AR and VR applications to tourism industry and the characteristics of tourists using AR and VR. The articles targeting the destination sector examined the impact of the use of AR and VR on destination choice and purchase intent. When these results are evaluated, article publications are concentrated in 2006-2011 and 2012-2018 publication period. This shows that AR and VR researches in tourism areas experience entrance and growth stages. During the entrance period (2006-2011) the researches focused on technology adaptation, marketing benefit and customer and sector perspectives in the general tourism, destination and restaurant sectors. During the growth (2012-2018) period of the researches, the focus of the museum and tourism education was on the customer's perspective with experience, satisfaction, design development and motivation.

Figure 3. Graph of correspondence analysis of the general variables of the articles (periods, subject category and target industry)

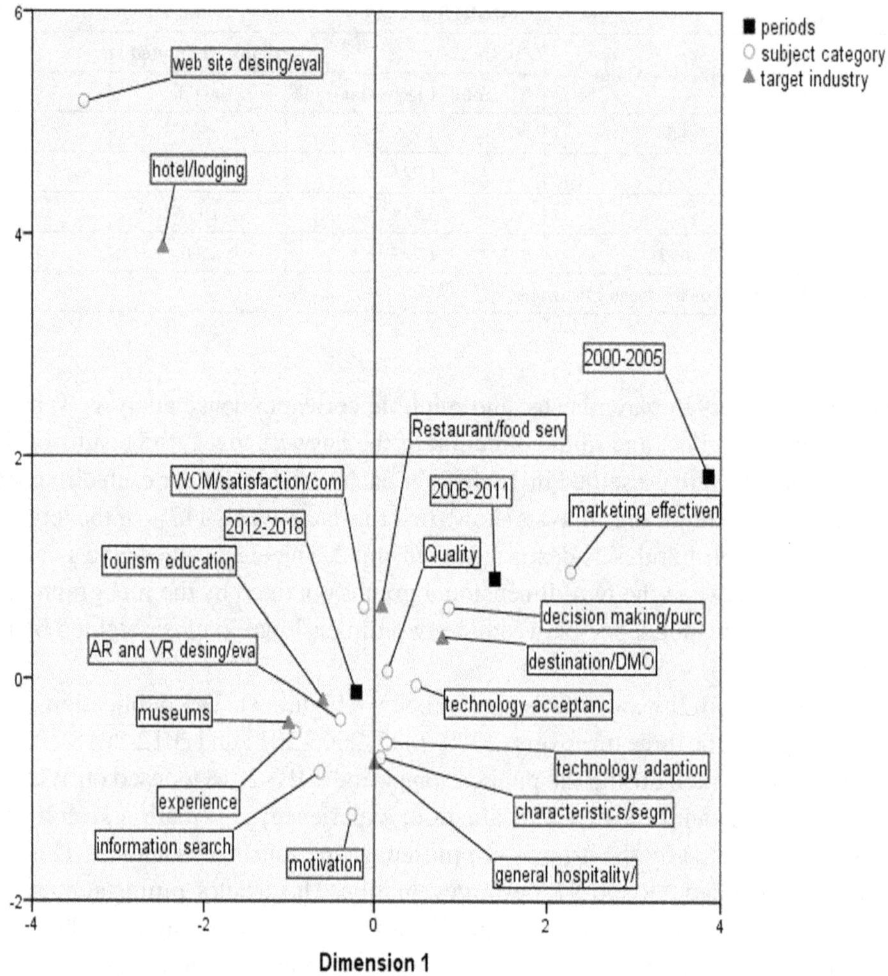

Methodological Review

Since the number of categorical variables in the study was more than three, multiple correspondence analysis was applied. In this frame, the "application platform" (4), "data collection" (5), "journal name" (16), research design (3), statistical techniques (5) questions among the questions in the examination form, including category numbers in parentheses, Multiple correspondence analysis was applied to see how the combination would be, and in the direction of the answers given, the results of the analysis of the 4X5X16X3X5 versatile table are presented in Table 3 for each variable and for each dimension. However, the inertia obtained as a result of the analysis shows that the data reveals 107% of the total variance. 58% of them describe dimension 1 and 49% describe dimension 2. The eigenvalues give a complete measure of the correspondence between the two-dimensional graphs obtained by the real graph. In this respect, it can be said that the correspondence between the two-dimensional graphs obtained by the real graph (1,07%) is very good.

Table 5. Table of Correspondence Analysis of the General Variables (application platform, data collection, journal name, research design, statistical techniques)

Model Summary				
Dimension	Cronbach's Alpha	Variance Accounted For		
		Total (Eigenvalue)	Inertia	% of Variance
1	,821	2,912	,582	58,247
2	,743	2,468	,494	49,357
Total		5,380	1,076	
Mean	,785[a]	2,690	,538	53,802
a. Mean Cronbach's Alpha is based on the mean Eigenvalue.				

The graph obtained from the analysis is shown below (Figure 4). In this graph, data collection tool, journal name, research design, statistical techniques and application areas of AR and VR were examined. According to this graph, univariate was used as a statistical technique in TM and JHTR journals' articles. The research design used in these articles is mixed. TE, TR, JTR, JHTT, JHLSTE, JHTM, TMP and IJHM were used as quantitative method. Surveys and experiments were also used as data collection tools in these journals. Multivariate (multivariate statistical techniques such as factor analysis, cluster analysis, regression, discriminant analysis, logistic regression, multidimensional scaling and CAE) and SEM/CFA/path analysis. Structural equation modeling (SEM)/confirmatory factor analysis (CFA)/(path analysis should belong to multivariate statistical techniques in nature) were the statistical techniques used in the analysis of journals as they appeared to gain increasing popularity in the tourism research. Thus, structural equation modeling /confirmatory factor analysis/path analysis were separated from multivariate statistical techniques to making an individual category. Hence, comparing structural equation modeling with other multivariate statistical techniques, multivariate statistical techniques were actually the most popular statistical techniques in these journals' AR and VR related articles. The articles in the journals focus on AR and VR practices in the area of guidance.

Qualitative research has been used in the articles of IJCHM, JDMM, JTF, JVM, ATR and THR. The data collection tools used in the articles were interviews and document/online document. The most common statistical techniques used in the articles were descriptive (Descriptive statistical techniques such as mean, frequency and correlation) and other statistics (Other statistical techniques including mathematical modeling, algorithm, analytical hierarchical process and neural network analysis). As for the area of marketing, gamification and education, articles in the journals focused on AR and VR practices.

Figure 4. Graph of correspondence analysis of the general variables of the articles (application platform, data collection, journal name, research design, statistical techniques)

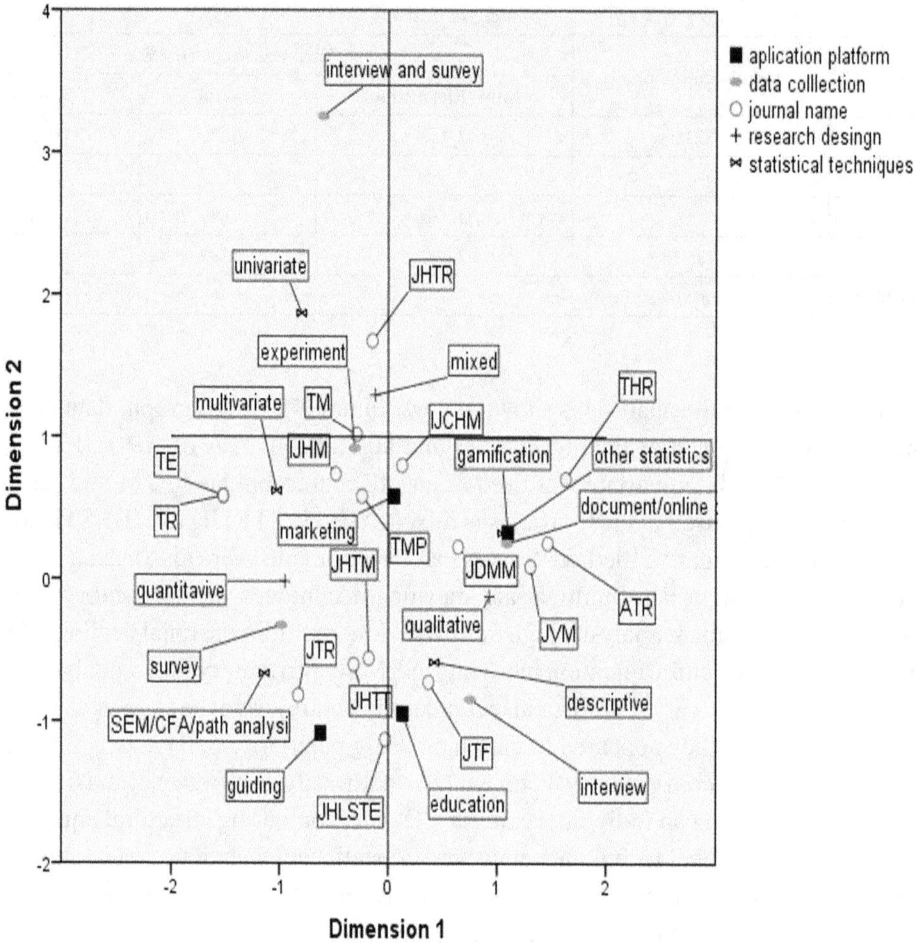

SOLUTIONS AND RECOMMENDATIONS

Regarding the AR and VR, a number of applications have been developed. Today, both in the form of pilot or research projects and commercially used applications, AR and VR applications have been released in tourism industry. This study aims at providing contribution to the progress of augmented reality (AR) and virtual reality (VR) research through the review of journals published in32 hospitality and tourism journals and providing suggestions for further research.

The two-dimensional perceptual map indicates that articles in 2012-2018 range mostly examined AR and VR applications from the perspective of customers. The focus of the work in this period was customer satisfaction, customer complaints and customer experience, use and user motivation in applications (AR and VR). Further studies should examine AR and VR applications from the perspective of tourism enterprises. These studies should examine the use of AR and VR applications in different and specific tourism establishments, difficulties, costs and application methods.

The two-dimensional perceptual map indicates that articles in the 2012-2018 range included the design and development of AR and VR applications in tourism education; in the museums, the focus is on customer experience. Further studies should focus on the design and development of AR and VR applications for destinations, restaurants, travel agencies, airlines and museums. In addition, studies should examine the customer experience of AR and VR applications in other tourism areas.

Looking at the data collection and research characteristics of the articles, the research was designed both qualitatively (28 articles) and quantitatively (27 articles). The survey method was used in the majority of the articles reviewed in the study. Interview or focus group interview may be applied in future studies. These studies have the potential to gather in-depth information on AR and VR applications. Descriptive statistics are often used as statistical techniques. Further studies should use multivariate statistical techniques.

The two-dimensional perceptual map indicates that the articles reviewed examined AR and VR applications related to destination and general tourism (it includes articles which target whole of the tourism sector applications and solutions, did not target a specific tourism business like hotel, restaurant, travel agent etc.,). The subject of articles focused on technology adoption, WOM/satisfaction/complaint and experience respectively. AR and VR have the potential to create an information system for museums, accommodation businesses, travel agencies. It also has the potential to provide information about tourism products. Further studies should focus on AR and VR applications in museums, accommodation establishments, travel agencies and food and beverage companies. Thus, it will be possible to determine customer satisfaction and usage in these areas.

When these results are evaluated, article publications concentrated in 2006-2011 and 2012-2018 publication period. This shows that AR and VR researches in tourism areas are experiencing entrance, growth stages. During the entrance period (2006-2011) the researches focused on technology adaptation, marketing benefit and customer and sector perspectives in the general tourism, destination and restaurant sectors. During the growth (2012-2018) period of the researches, the focus of the museum and tourism education was on the customer's perspective with experience, satisfaction, design development and motivation. Further studies should focus both customer and sector perspectives on technology adaption by tourist and businesses in different tourism fields.

FUTURE RESEARCH DIRECTIONS

The present study had some limitations as well. It collected research articles from 32 tourism and hospitality journals only. The classifications and trends generated within the study might not be generalized to all or any AR and VR research in hospitality and tourism disciplines. The results provide a single glimpse into a rapidly changing dynamic environment due to the continuous publication of new articles. Therefore, future research might take these limitations into account. For instance, additional business and marketing journals may be included in review.

CONCLUSION

Several inadequacies were identified considering the specific nature of AR and VR related articles.. As it was indicated by the two-dimensional perceptual map, AR and VR research in the hospitality and tourism fields experienced introduction and growth stages. The research focus shifted from business perspective and customer perspective to customer perspectives. Articles on AR and VR were mostly published in 2017 and 2018. AR and VR related articles in the tourism journals as less. The lack of articles on AR and VR in tourism journals is a major shortcoming. This is why AR and VR issues have gained importance in recent years. The articles written onthis subject will increase in the coming years. Within the 32 journals, 16 journals published AR and VR related articles in the field of tourism. At the following years, AR and VR related publications may spread in other journals.

The articles were written about VR with a maximum of 37 articles. The number of articles written about AR is 14. This shows that the articles on VR in the field of tourism are more intense. This is because AR is a technology that develops after VR. In the future, AR-themed studies might increase. The majority of the articles focused on the field of marketing with 35 articles and education area with 12 articles. The majority of the articles reviewed in this study focused on marketing characteristics of AR and VR applications and their use in marketing. AR and VR have the potential to be applied in different areas besides tourism marketing. AR and VR studies should focus on guidance, gamification and education in the field of tourism in the future.

REFERENCES

Azuma, R., Baillot, Y., Behringer, R., Feiner, S., Julier, S., & MacIntyre, B. (2001). Recent advances in augmented reality. *IEEE Computer Graphics and Applications*, *21*(6), 34–47. doi:10.1109/38.963459

Azuma, R. T. (1997). A survey of augmented reality. *Presence (Cambridge, Mass.)*, *6*(4), 355–385. doi:10.1162/pres.1997.6.4.355

Bamodu, O., & Ye, X. (2013). Virtual Reality and Virtual Reality System Components. *Proceedings of the 2nd International Conference On Systems Engineering and Modeling* (ICSEM-13). 10.2991/icsem.2013.192

Berryman, D. R. (2012). Augmented reality: A review. *Medical Reference Services Quarterly*, *31*(2), 212–218. doi:10.1080/02763869.2012.670604 PMID:22559183

Bonetti, F., Pantano, E., Warnaby, G., Quinn, L., & Perry, P. (2018, June). Augmented Reality in Real Stores: Empirical Evidence from Consumers' Interaction with AR in a Retail Format. In 4th International AR &VR Conference 2018: The Power of AR & VR for Business.

Carmigniani, J., Furht, B., Anisetti, M., Ceravolo, P., Damiani, E., & Ivkovic, M. (2011). Augmented reality technologies, systems and applications. *Multimedia Tools and Applications*, *51*(1), 341–477. doi:10.100711042-010-0660-6

Castro, J. C., Quisimalin, M., Córdova, V. H., Quevedo, W. X., Gallardo, C., Santana, J., & Andaluz, V. H. (2017). Virtual reality on e-Tourism. In *International conference on information theoretic security*, Singapore: Springer, pp. 86-97.

Celtek, E. (2015). Augmented Reality Advertisements in Tourism Marketing. In N. Takıran, & R. Yılmaz (Eds.), Handbook of Research on Effective Advertising Strategies in the Social Media Age. pp. 125-146. Hershey, PA: IGI Global.

Chen, S., Pan, Z., & Zhang, M. (2012). A virtual informal learning system for cultural heritage. In Z. Pan, A. D. Cheok, W. Mueller, M. Chang, & M. Zhang (Eds.), *Transactions on edutainment VII* (pp. 180–187). Berlin, Germany: Springer International Publishing. doi:10.1007/978-3-642-29050-3_16

Cheong, R. (1995). The virtual threat to travel and tourism. *Tourism Management, 16*(6), 417–422. doi:10.1016/0261-5177(95)00049-T

Chou, T.-L., & ChanLin, L.-J. (2012). Augmented Reality Smartphone Environment Orientation Application: A Case Study of the Fu-Jen University Mobile Campus Touring System. *Procedia: Social and Behavioral Sciences, 46*, 410–416. doi:10.1016/j.sbspro.2012.05.132

Chung, N., Han, H., & Joun, Y. (2015). Tourists' intention to visit destination: Role of augmented reality applications for heritage site. *Computers in Human Behavior, 50*, 588–599. doi:10.1016/j.chb.2015.02.068

Craig, A. B. (2013). Understanding Augmented Reality, Concepts and Applications. (1st Ed.). Imprint: Morgan Kaufmann.

Cranmer, E. E. (2017). *Developing an augmented reality business model for cultural heritage tourism: the case of Geevor Museum* (Unpublished Doctoral thesis). Manchester Metropolitan University.

Dadwal, S. S., & Hassan, A. (2015). The Augmented Reality Marketing: A Merger of Marketing and technology in Tourism. In N. Ray (Ed.), *Emerging Innovative Marketing Strategies in the Tourism Industry* (pp. 63–80). Hershey, PA: IGI Global. doi:10.4018/978-1-4666-8699-1.ch005

Fino, E., Martín-Gutiérrez, J., Fernández, M., & Davara, E. (2013). Interactive Tourist Guide: Connecting Web 2.0, Augmented Reality and QR Codes. *Procedia Computer Science, 25*, 338–344. doi:10.1016/j.procs.2013.11.040

Fritz, F., Superregui, A., & Linaza, M. (2005). Enhancing Cultural tourism experiences with Augmented Reality technologies. In *Proceedings 6th International Symposium on Virtual Reality, Archaeology and Cultural Heritage VAST*, pp. 1-5, Pisa, Italy, November, San Sebastian, Spain, The Eurographics Association.

Galeano, G. (2017). *Design and development of a Mixed Reality application in the automotive field.* (Unpublished Master Thesis). Department of Control and Computer Engineering, Politecnico di Torino.

Gibson, A., & O'Rawe, M. (2018). Virtual reality as a travel promotional tool: Insights from a consumer travel fair. In T. Jung, & M. tom Dieck (Eds.), *Augmented reality and virtual reality, Progress in IS* (pp. 93–107). Cham, Switzerland: Springer. doi:10.1007/978-3-319-64027-3_7

Gurau, C. (2007). Virtual Reality Applications in Tourism, Information and Communication Technologies in Support of the Tourism Industry (pp. 180–197). Hershey, PA: IGI Global. doi:10.4018/978-1-59904-159-9.ch007

Guttentag, D. A. (2010). Virtual reality: Applications and implications for tourism. *Tourism Management, 31*(5), 637–651. doi:10.1016/j.tourman.2009.07.003

Hair, J. F. Jr, Black, W. C., Babin, B. J., & Anderson, R. E. (2010). *Multivariate Data Analysis* (7th ed.). Upper Saddle River, NJ: Pearson.

Halarnkar, P., Shah, S., Shah, H., Shah, H., & Shah, A. (2012). A Review on Virtual Reality. *IJCSI International Journal of Computer Science Issues*, *9*, 6–1.

Han, D.-I., Jung, T., & Gibson, A. (2014). Dublin AR: implementing augmented reality in tourism. In Z. Xiang, & I. Tussyadiah (Eds.), *Information and Communication Technologies in Tourism* (pp. 511–523). New York: Springer International Publishing.

Harrington, R. J., Chathoth, P., Ottenbacher, M., & Altinay, L. (2014). Strategic management research in hospitality and tourism: Past, present and future. *International Journal of Contemporary Hospitality Management*, *26*(5), 778–808. doi:10.1108/IJCHM-12-2013-0576

Hassan, A., & Rahimi, R. (2016). Consuming 'innovation' in tourism: Augmented Reality as an innovation tool in Digital Tourism Marketing. In N. Pappas, & I. Bregoli (Eds.), Global Dynamics in Travel, Tourism, and Hospitality (pp. 130–147). Hershey, PA: IGI Global. doi:10.4018/978-1-5225-0201-2.ch008

He, Z., Wu, L., & Li, X. (2018). (R) (2018). When art meets tech: The role of augmented reality in enhancing museum experiences and purchase intentions. *Tourism Management*, *68*, 127–139. doi:10.1016/j.tourman.2018.03.003

Höllerer, T. H., & Feiner, S. K. (2004). Mobile augmented reality. In H. Karimi, & A. Hammad (Eds.), *Telegeoinformatics: Location-Based Computing and Services*. Taylor & Francis Books Ltd.

Huang, Y. C., Backman, K. F., Backman, S. J., & Chang, L. L. (2016). Exploring the implications of virtual reality technology in tourism marketing: An integrated research framework. *International Journal of Tourism Research*, *18*(2), 116–128. doi:10.1002/jtr.2038

Jung, T., TomDieck, M. C., Lee, H., & Chung, N. (2016). Effects of virtual reality and augmented reality on visitor experiences in museum. In A. Inversini, & R. Schegg (Eds.), *Information and communication technologies in tourism 2016: Proceedings of the international conference in Bilbao, Spain*, pp. 621–635, February 2-5, 2016. Cham, Switzerland: Springer International Publishing. 10.1007/978-3-319-28231-2_45

Jung, T., TomDieck, M. C., Moorhouse, N., & TomDieck, D. (2017). Tourists' experience of virtual reality applications. Paper presented at 2017 *IEEE International Conference on Consumer Electronics (ICCE)*. doi:. 2017.788928710.1109/ICCE

Jung, T., Chung, N., & Leue, M. (2015). The determinants of recommendations to use augmented reality technologies: The case of a Korean theme park. *Tourism Management*, *49*, 75–86. doi:10.1016/j.tourman.2015.02.013

Kennedy-Eden, H., & Gretzel, U. (2012). A taxonomy of mobile applications in tourism. *Ereview of Tourism Research*, *10*(20), 47–50.

Kounavis, C., Kasimati, A., & Zamani, E. (2012). Enhancing the tourist experience through mobile augmented reality: Challenges and prospects. *International Journal of Engineering Business Management*, *4*(10), 1–6.

Kourouthanassis, P., Boletsis, C., Bardaki, C., & Chasanidou, D. (2015). Tourists responses to mobile augmented reality travel guides: The role of emotions on adoption behavior. *Pervasive and Mobile Computing*, *18*, 71–87. doi:10.1016/j.pmcj.2014.08.009

Lashkari, A., Parhizkar, B., & Mohamedali, M. (2010). Augmented Reality Tourist Catalogue Using Mobile Technology. In *Proceedings IEEE, Second International conference on Computer Research and Development*, pp. 121-125, *Kuala Lumpur, Malaysia*. Washington, DC: IEEE.

Lee, K. (2012). Augmented reality in education and training. *TechTrends, 56*(2), 13–21. doi:10.100711528-012-0559-3

Linaza, M., Marimon, D., Carrasco, P., Alvarez, R., Montesa, J., Aguilar, S. R., & Diez, G. (2012). Evaluation of mobile augmented reality applications for tourism destinations. In M. Fuchs, F. Ricci, & L. Cantoni (Eds.), *Information and Communication Technologies in Tourism*. Wien, Austria: Springer-Verlag.

Marasco, A., Buonincontri, P., Niekerk, M., Orlowski, M., & Okumus, F. (2018). Exploring the role of next-generation virtual technologies in destination marketing. *Journal of Destination Marketing & Management, 9*, 138–148. doi:10.1016/j.jdmm.2017.12.002

Marimon, D., Sarasua, C., Carrasco, P., Alvarez, R., Montesa, J., Adamek, T., . . . Gasco, P. (2018). *Mobi AR: Tourist experiences through mobile augmented reality*. Retrieved from http://www.researchgate.net/publication/228979424_MobiAR_Tourist_Experiences_through_Mobile _Augmented_Reality

Martínez-Graña, A., Goy, J., & Cimarra, C. (2013). A virtual tour of geological heritage: Valourising geodiversity using Google Earth and QR code. *Computers & Geosciences, 61*(12), 83–93. doi:10.1016/j.cageo.2013.07.020

Moorhouse, N., TomDieck, M. C., & Jung, T. (2018). Technological Innovations Transforming the Consumer Retail Experience: A Review of Literature. In T. Jung, & M. TomDieck (Eds.), Augmented Reality and Virtual Reality. Progress in IS. Cham, Switzerland: Springer.

Morosan, C., Bowen, J. T., & Atwood, M. (2014). The evolution of marketing research. *International Journal of Contemporary Hospitality Management, 26*(5), 706–726. doi:10.1108/IJCHM-11-2013-0515

Novák-Marcinčin, J., & Kuzmiaková, M. (2009). Basic Components Of Virtual Reality. *Annals of the University of Petroşani Mechanical Engineering (New York, N.Y.), 11*, 175–182.

Özdamar, K. (2004). *Paket Programlar İle İstatistiksel Veri Analiz*. Eskişehir, Turkey: Kaan Kitabevi.

Pan, X., & Hamilton, A. F. D. C. (2018). Why and how to use virtual reality to study human socialinteraction: The challenges of exploring a new research landscape. *British Journal of Psychology, 109*(3), 1–23. doi:10.1111/bjop.12290 PMID:29313958

RealityTechnologies.com. (2018). *The Ultimate Guide to Understanding Virtual Reality (VR) Technology*, Retrieved from https://www.realitytechnologies.com/virtual-reality/Reality

Rey-López, M., Barragáns-Martínez, A., Peleteiro, A., Mikic-Fonte, F., & Burguillo, J. (2011). MoreTourism: Mobile recommendations for tourism. In *IEEE International Conference on Consumer Electronics (ICCE)*. pp. 347-348. IEEE.

Santosa, M., & Gook, L. B. (2012). ARkanoid: Development of 3D game and handheld augmented reality. *International Journal Of Computational Engineering Research*, 2(4), 1053–1059.

Schmalstieg, D., Langlotz, T., & Billinghurst, M. (2011). Augmented reality 2.0. In *Virtual Realities* (pp. 13–37). Vienna, Austria: Springer. doi:10.1007/978-3-211-99178-7_2

Selvam, A., Yap, T. T.-V., Ng, H., Tong, H.-L., & Ho, C.-C. (2016). Augmented Reality for Information Retrieval Aimed at Museum Exhibitions using Smartphones. *Journal of Engineering and Applied Sciences (Asian Research Publishing Network), 100*(3), 635–639.

Tavakoli, R., & Mura, P. (2015). Journeys in Second Life' – Iranian Muslim women's behaviour in virtual tourist destinations. *Tourism Management, 46*, 398–407. doi:10.1016/j.tourman.2014.07.015

Taylor, B. (2013). Augmented reality applications – from the consumer to the business. Retrieved from http://www.creativeguerrillamarketing. com/augmented-reality/augmented-reality-applications-from the -consumer-to-the-business/

tom Dieck, M. C., & Jung, T. H. (2017). Value of Augmented Reality at Cultural Heritage sites: A Stakeholder Approach. *Journal of Destination Marketing & Management, 6*(2), 110–117. doi:10.1016/j.jdmm.2017.03.002

Tromp, P. (2017). *How virtual reality will revolutionize the hospitality industry*. Retrieved from https://www.hospitalitynet.org/opinion/4080737.html

Tussyadiah, I. P., Wang, D., Jung, T. H., & tom Dieck, M. C. (2018). Virtual reality, presence, and attitude change: Empirical evidence from tourism. *Tourism Management, 66*, 140–154. doi:10.1016/j.tourman.2017.12.003

Wang, X., Kim, M., Love, P., & Kang, S.-C. (2013). Augmented Reality in built environment: Classification and implications for future research. *Automation in Construction, 32*, 1–13. doi:10.1016/j.autcon.2012.11.021

Wasko, C. (2013). What teachers need to know about augmented reality enhanced learning. *TechTrends, 57*(1), 17–21. doi:10.100711528-013-0672-y

Weber, J. (2014). Information and communication technologies in tourism 2014. In R. Baggio, M. Sigala, A. Inversini, & J. Pseonen (Eds.), *ENTER. Dublin, Ireland*. London, UK: Springer.

Williams, A. P., & Hobson, J. S. P. (1995). Virtual reality and tourism: Fact or fantasy? *Tourism Management, 16*(6), 423–427. doi:10.1016/0261-5177(95)00050-X

Xu, F., Buhalis, D., & Weber, J. (2017). Serious games and the gamification of tourism. *Tourism Management, 60*, 244–256. doi:10.1016/j.tourman.2016.11.020

Xu, F., Tian, F., Buhalis, D., Weber, J., & Zhang, H. (2016). Tourists as mobile gamers: Gamification for tourism marketing. *Journal of Travel & Tourism Marketing, 33*(8), 1124–1142. doi:10.1080/10548408.2015.1093999

Yang, E. C. L., Khoo-Lattimore, C., & Arcodia, C. (2017). A systematic literature review of risk and gender research in tourism. *Tourism Management, 58*, 89–100. doi:10.1016/j.tourman.2016.10.011

Yovcheva, Z., Buhalis, D., & Gatzidis, C. (2013). Engineering Augmented Tourism Experiences. In L. Cantoni, & Z. Xiang (Eds.), Information and Communication Technologies in Tourism (pp. 24–35). Berlin, Germany: Springer.

Yung, R., & Khoo-Lattimore, C. K. (2017). New realities: A systematic literature review on virtual reality and augmented reality in tourism research. *Current Issues in Tourism*. doi:10.1080/13683500.2017.1417359

KEY TERMS AND DEFINITIONS

Augmented Reality: The process of taking and integration digital info either with a live streaming video or with the user's actual, real time atmosphere is called augmented reality (AR).

Fully-Immersive VR Simulations: Provide the most immersive implementation of virtual reality technology. Hardware such as head-mounted displays and motion detecting devices are used to appeal to all of the senses of a user in a fully-immersive simulation.

Marker Based AR: It can be processed by using the camera on the smart device to scan a visual marker such as a QR code attached to an advert.

Markerless AR: This type of AR doesn't need a marker and it usually requires a location or position. The AR content generally works from the GPS system or compass of smart devices. This technology is more device dependent in terms of capacity.

Non-Immersive VR Simulations: The least immersive applications of virtual reality technology. In a non-immersive simulation, only a limited number of user's senses are stimulated, allowing peripheral awareness of the reality outside the virtual reality simulation.

Semi-Immersive VR Simulations: Provide a more immersive experience, although the user is partly immersed in the virtual setting. Semi-immersive simulations are similar to the technologies found in flight simulation in many aspects.

Virtual Reality: Represents three-dimensional (3D), interactive, computer-generated environments are represented by virtual reality application. These environments will either be models of real or imagined worlds, and that they aim at representing data through a synthetic experience.

Chapter 2
Blockchain Technology and Its Uses in the Tourism Industry

Yunus Topsakal
Adana Alparslan Türkes Science and Technology University, Turkey

ABSTRACT

The public and private sectors will undergo a significant transformation with the use of blockchain technology, and this potential of blockchain technology will be influential in all areas of life. In addition, blockchain technology can help ease the integration of the Internet of Things, augmented reality, and artificial intelligence applications. The question of how such a technology that is in the process of development will be applied in areas such as taxation, notary operations, and banking, arises simultaneously. Studies related to blockchain technology have mainly been carried out in the fields of finance, logistics, banking, and education. However, there is a paucity of studies on blockchain technology in the tourism industry which has an important role in the global economy. Therefore, the potential of using blockchain technology in the tourism industry is evaluated in this chapter. For this purpose, the features, advantages, and disadvantages of blockchain technology are explained. The potential uses of blockchain technology in the tourism industry are then discussed.

INTRODUCTION

"Blockchain" was mentioned in Satoshi Nakamoto's original bitcoin article published in 2008. Although "blockchain" was not used in this article as a word, a technology component underlying crypto coins was defined there as a series of data blocks chained together cryptographically (Nakamoto, 2008). For that reason, bitcoins were the first application of blockchain technology (Iansiti and Lakhani, 2017).

Nowadays, with the production of various crypto coins like bitcoins and the exchange of currency through stock markets, the rate of investing in crypto coins is increasing rapidly all over the world. The basis of crypto coin exchange is a decentralized validation mechanism. Bitcoin (Nakamoto, 2008), crypto coin, and blockchain concepts are often confused with each other. A bitcoin is a crypto coin that is independent of a central government and provides payment anywhere in the world with low transmission costs without the need for the approval of a central authority, providing certain privacy to its owner

DOI: 10.4018/978-1-7998-1989-9.ch002

(Mukhopadhyay et al., 2016). Blockchain technology, on the other hand, can be defined as a distributed ledger, a management system that is widely used in bitcoin transfers today. Blockchain infrastructure is not only created for crypto coin transfers.

A blockchain is a simple, clear, transparent, distributed, and time-stamped global account book, data file, or simple database that can be understood by everyone. Since 2009, with the release of bitcoins, all transactions are digitally preserved (Çarkacıoğlu, 2016). With the popularity of crypto coins such as bitcoins, blockchain technology, which includes crypto coins, has become remarkable. Blockchain applications are among the developments that are predicted to have a profound impact on business life in the near future (Soylu, 2018).

Structurally, a blockchain is a series of information blocks that are securely chained. New blocks are defined when participants create new blocks of information or change existing information about an asset (for example, transaction ledgers, status changes, new market prices, or new owners being entered). After the first block, the newly created valid blocks are securely chained to the previous trusted block. Thus, reliable audit evidence is created by guaranteeing the reliability of the blocks (Hong Kong Monetary Authority, 2016).

The main goal of this chapter is to examine the concept of blockchains, which provide a secure exchange of numerical values and serve as the basis for crypto coins, and to determine possible application areas in the tourism industry in the near future. In this context, information is given about blockchains and the features, advantages, and disadvantages of blockchain technology. The areas in which blockchain technology is used are examined and the existing and possible future usage areas for the tourism industry are determined.

BACKGROUND

An Overview of Blockchain Technology

As a general definition, Tian (2016) states that the essence of blockchains is a technical plan of a reliable database that is maintained collectively by decentralized and reliable methods. In order to better understand the operation of blockchain technology, a simplified description of the bitcoin blockchain, the first implementation of the analogy-derived blockchain technology by Nikolai Hampton, is given below (Hampton, 2016);

A bitcoin blockchain can be thought of as a physical book containing a ledger of about ten minutes of transactions with a bitcoin crypto coin on each page. When a page is full of new transactions, it is signed with a unique serial number, stamped, and affixed to the book. In this analogy, blocks of pages represent the connection between blocks of serial numbers. The serial number is a product of transactions on that page, and the serial numbers of adjacent pages are locked together with a mathematical function to form a solid chain of pages. This makes it impossible to change one of the transactions without changing the serial number of the pages, and therefore the connection between that page and the next page. To change a transaction in a book, it is necessary to remove all pages after the transaction and fill them with new transactions, create new serial numbers, and paste all pages into the book. Users of this book always consider the book with the most pages as the actual book. Therefore, the book continues to grow with the addition of a new page every ten minutes, and so a person who wishes to successfully rewrite a transaction history in the book will have to work faster than the rest of the community. In other words, the amount of effort required for a person to do this operation far surpasses what a single person can do, so this structure is very safe.

A blockchain is segmented into smaller data structures called "blocks". The information contained in a block can be listed as follows (Lin and Liao, 2017):

1. Header: It contains metadata about the block, such as:
   a. Hash: The digital fingerprint or unique identifier of the current block.
   b. Hash of Previous Block: Used as a point of reference for linking/chaining blocks and to create the current block's hash. It makes it possible to track all of the previous blocks in the blockchain.
   c. Time Stamp: Digital time when the block is generated.
2. Data: Details of the transaction performed (payment transaction ledger, voting transaction ledger, etc.).
3. Other Information: Nonce value, block signature, etc.

A blockchain is stored on many different computers instead of being stored on a single server. Each computer in a blockchain is called a "miner", and as the number of miners increases, the consensus mechanism of the blockchain becomes stronger (Beck, 2018). The crypto coin called bitcoin is the best known application of blockchain technology. Bitcoins allow users to transfer money or shop over the Internet without intermediaries such as banks or governments. In 2015, another crypto coin system, Ethereum, was released. Ethereum, which also uses a blockchain database, has introduced a smart contract system. The smart contract system has enabled complex money transfers or transactions based on certain conditions. Smart contracts are also stored in a distributed manner on all computers within the blockchain and must be verified by each computer (miner). Thanks to Ethereum smart contracts, it is possible to produce tokens in the system and enable users to perform various transactions (Sehra et al., 2018).

The most striking feature of blockchain technology is that none of the users involved in a particular blockchain can play a central role or make a single change to the generated data. Any member who has access to a blockchain can create new blocks, which cannot be modified or deleted by others. Because the information of each new block created is stored in all the computers in the blockchain, and since each block contains the information of the previous one, a change being made requires the information in the subsequent blocks to be changed (Kim & Justl, 2018).

Although blockchain technology is a new technology, it continues to develop rapidly. It can be said that this technology has made great progress since the idea was put forward by Satoshi in 2008. The developmental stages of blockchains have taken place in the literature within three categories as blockchains 1.0, 2.0, and 3.0. Swan (2015) defined these phases as follows: Blockchain 1.0 refers to crypto currencies with cash-related applications such as money transfers, wire transfers, and digital payments. Blockchain 2.0 is an application that provides solutions for stocks, bonds, futures, loans, and smart contracts and has features that can eliminate intermediaries rather than simple payment transactions. Blockchain 3.0 comprises applications beyond the money and financial markets, especially in the fields of government, health, science, culture, and arts (Swan, 2015).

Blockchains allow the global distribution and decentralization of all types of transactions between all parties (Swan, 2015). In addition to the infrastructure that a blockchain offers as a distributed ecosystem currency, it can also create a decentralized system that can be used in place of existing corporate platforms for electronic payments, with many different application areas. Blockchain systematics can replace the currently used network protocols for security, identity management, decentralized structures, and asset management. Features that meet the requirements arising from these applications can be added to the blockchain. Here it would be appropriate to give brief information about central, multicenter, and distributed system structures (Zheng et al, 2018):

- **Central Systems:** A network structure in which all users connect to a central server.
- **Multicenter Systems:** A peer-to-peer communication network structure between nodes in which many servers connect. Devices on the network communicate with each other and manage connections between various nodes. This is a system that has a structure that works on multicentric information in order to realize the objectives, not as a result of a central effect of the components.
- **Distributed systems:** Systems in which the processes are distributed among the components and the actions are coordinated by transmitting the information. The components interact with each other to achieve a common goal. There is no central point of failure because each node can act as a server.

Blockchain is divided into three different types according to access status. These types are;

Open Blockchains

Open blockchains are networks with wide distribution such as Bitcoin (Peters & Panayi, 2015). Authorization from any authority is not required for participation in such blockchain networks. Anyone on earth can download software of the relevant blockchain protocol to establish a node on their own devices and become a miner (Usta & Doğantekin, 2017). In these networks where no authorization is required for participation, miners can create transactions, monitor existing transactions, and verify transactions. In such networks, transactions can be monitored transparently, while anonymity is maintained. Examples of such blockchain systems are Bitcoin, Etherum, LiteCoin and Monero. While the amount of transactions from which account to which account is visible to everyone. It is known by the owners of these accounts to whom it belongs (Buterin, 2015).

Allowed Blockchains

These are the systems that can control the permissions of the activities that can be done by the people who will participate in the chain. In such blockchain systems, participation in the blockchain network is not public. It is necessary to get permission from the approval authority to join the network and verify the transaction (Natarajan et al., 2017). They are generally blockchain networks established between corporations and companies that have come together for certain purposes. The network is managed by a group of representatives of these companies. It must be a member of this union established to be included in the network and contracts must be done. In such systems, issues such as which nodes will verify the operations and which ones will be controlled are also taken by common decisions. Since the systems are closed to the outside and officially recognizing each other, high-resource reconciliation methods such as proof of work are not used (Buterin, 2015).

Private Blockchains

It is smaller than other blockchains. They are used by organizations that trust users in their network and contain confidential information. It is a privatelly established blockchain system within institutions or companies. These systems are managed by the company. The ability to read new transactions in the blockchain network may be open to anyone while subjecting permission to create new transactions, which is entirely under the authority of the authorizing authority (Buterin, 2015).

Basic Components of Blockchain

As the Internet democratizes information, the block chain is expected to democratize some values (Tapscott & Tapscott, 2017). The block chain may relate to the management of all forms of money, wealth, intellectual property and different values in which the government is expected to protect the public interest. Some institutions act as intermediaries in payments, exchanges and settlement, and play one or more critical tasks in improving the proper functioning of the financial system. These roles generally include financial, operational and legal risk management as well as a management structure for customers and markets in which they provide services. In addition, cryptography can be used to encrypt transaction information in the account so that only certain participants can decode the details of each transaction. Encryption used for this purpose can be an important tool in situations requiring confidentiality (Mills et al., 2016). Although it is difficult to determine the functions of the block chain exactly like the functions of money, different functions come to the fore according to different usage areas. In general, block chain services consist of three main components (Hyperledger Whitepaper, 2016); Peer-to-Peer Protocol (P2P Protocol), Distributed Ledger Technology (DLT), Consensus Protocol.

Peer-to-Peer Protocol (P2P Protocol)

A blockchain is a digital ledger in which users can view the information inside and it is possible to be public. User parties are not responsible for the ledgers in open blockchains, and anyone can use them. It can be held by thousands of miners in different parts of the world in a distributed network, not by a person or organization. As a decentralized model of trust-based peer-to-peer transactions, blockchain technology refers to transactions that do not have the most basic level of intermediary (Gupta, 2017). However, it provides the possibility of transition (from person-to-person, person-to-machine, machine-to-machine) in all decentralized and unreliable transactions on a global basis for all kinds of interactions and transactions. It is possible to make use of a different functioning and structure that is not yet foreseen but which can affect the use of existing established power relations and hierarchies and which can easily benefit the society (Swan, 2015).

Distributed Ledger Technology

The blockchain, in its most recognized form, is used in distributed ledger technology for electronic payments (Xu et al., 2016). However, not all distributed ledger technology applications need to use blockchain technology (Mills et al., 2016). The blockchain for distributed ledger technology is only one of many data structures. Distributed ledger technology manages blockchain by applying three key features (Hyperledger Whitepaper, 2016);

- It effectively computes the encrypted transaction summary of each data set after each block.
- It effectively transmits a minimal delta change to a data set where a partner is out of sync.
- Each participant minimizes the amount of stored data required for the operation of distributed ledger technology.

Consensus Protocol

Consensus mechanisms allow for the secure updating of a distributed shared state and have become an active research topic for the last three decade. The common method used to ensure fault tolerance in a distributed system is to distribute the shared state to the participants in the network (Çarkacıoğlu, 2016). The replicated shared state is updated by the state machine executed on all copies according to the state transition rules specified previously. This technique is called state machine replication. State replication ensures that the state does not disappear in case one or more nodes collapse. The state machine allows all nodes to process the same data. Nodes that process the same data will probably produce the same output in the end. As a result, it is possible to reach a final agreement on the change of state through the consensus protocol (Müller and Hasic, 2016). Replicas also communicate with each other to form a consensus and to accept the state as a whole after a state change has taken place. In a blockchain-based system, the shared state forms the blockchain, and the state transition rules are the rules envisaged by the blockchain protocol (Narayanan et al., 2016). Consensus in a distributed system is difficult. The reconciliation algorithms should be able to withstand failures of nodes, network segmentation, message delays, unordered messages, corrupted messages, and other risks. It also has to deal with nodes that intentionally attempt to perform malicious operation. Each algorithm generates the necessary set of assumptions for synchronization, while synchronization provides the necessary set of assumptions for message security, malfunction, malicious nodes, security and performance of exchanged messages. Consensus for a blockchain network means that all nodes in the network allow the blockchain to accept a consistent overall state (Baliga, 2017: 9). In other words, the state in which miners approve the majority provide consensus.

A consensus protocol has three key features that can be determined for its applicability and effectiveness (Baliga, 2017: 5):

- Security: If all nodes produce the same output and the outputs generated by the nodes are valid according to the protocol rules, a consensus protocol is determined to be secure. This is also called shared state consistency.
- Viability: The consensus protocol ensures that all non-failing nodes involved in the compromise will likely contribute and viability.
- Fault Tolerance: The consensus protocol provides fault tolerance as long as it is possible to recover from the failure of a participating node.

The Characteristics of Blockchain

In the digital world, the way of regulate and maintain administrative control has to change. At this point, the blockchain promises to solve this problem. A blockchain is an open, distributed registry that can effectively, verifiable and permanently ledger transactions between the two parties. The ledger itself can be programmed to trigger transactions automatically (Ianisti & Lakhani, 2017).

Distributed Database

Blockchain technology is distributed database structure (Iansiti & Lakhani, 2017; Yli-Huumo et al, 2016; Xu et al, 2016). The database is scattered between nodes. While some researchers state that these are computers, Lemieux (2016) consider them as part of the system as a general definition (Iansiti & Lakhani, 2017).

Blockchain technology can access the entire database of nodes in the system, but cannot control the data stored in the blockchain of a single node. Each new entry in the blockchain is verified by all nodes in the system without the need for any intermediary (Iansiti & Lakhani, 2017).

End-to-End Communication

Instead of using any central structure for communication between the parties, individual nodes transmit and store information directly to each other in a peer-to-peer network (Nakamoto, 2008; Yli-Huumo et al., 2016). Because of the consensus among the nodes in the blockchain system, no center is needed (Pilkington, 2016). Information in the blockchain is stored by all nodes in BitShares (Iansiti & Lakhani, 2017). Some authors claim that the blockchain is not stored by all nodes but can be used by all nodes (Nakamoto, 2008).

Transparency

The concept of transparency in blockchain technology is achieved by the fact that all blocks can see all operations (Iansiti & Lakhani, 2017). This indicates that it is more transparent than a central system in which it is managed by a third party (Yli-Huumo et al., 2016).

Many sources say that blockchain technology is unattended (Tian, 2016), as Nakamoto stated in his report. Iansiti and Lakhani (2017) states that the blockchain is a nickname. Unlike cash transactions where transactions are made by anonymous individuals, Bitcoin transactions are always associated with an account. Each user has a unique address, which is a nickname for identification, and transactions take place through these nicknames (Iansiti & Lakhani, 2017).

Irreversibility of Records

Blockchain technology uses computational algorithms and approaches to ensure that stored state cannot be change (Iansiti & Lakhani, 2017). The term chain stems from all new states linked to the previous states in a chronological order on the chain. The blockchain stores all transactions within the blockchain so far. It is impossible to change the information inside an old block without breaking the chain (Iansiti & Lakhani, 2017).

Computational Logic

The digital nature of record keeping depends on the computational logic of the blockchain operations and is essentially programmed. Therefore, users can create algorithms and rules that automatically initiate transactions between nodes (Ianisti & Lakhani, 2017). Each node in the blockchain has the same transaction history. Therefore, the point of failure does not exist and eliminates downtime. The transaction records are stored sequentially in blocks. The successive blocks of stored transaction records are also added sequentially to the existing blockchain. Therefore, fraud is extremely difficult. Blockchain-based systems can be developed at lower costs compared to existing systems of similar security level by combining technology elements such as distributed structure and reconciliation protocol (METI, 2017).

The Benefits and Challenges of Blockchain Technology

Blockchain consists of a data block produced based on cryptography theory (Nakamoto, 2008). In addition, the system offers a structure that allows transactions between individuals without the need for a reliable third party. Anyone can see the entire transaction history. The complete transaction history ensures the validity of each virtual coin and all virtual coins can be tracked from the moment they are created. In addition, it provides backward transparency by providing resolution thanks to its technology. Prevents changes to the current ledger. This eliminates the need for management in the system. This process provides low-cost transactions (Beck et al, 2016). For instance, Gibbs & Yordchim (2014) suggested that the adoption of Bitcoin by small businesses would play an important role in the development of tourism in their study in Taylan.

Security is one of the most important advantages of blockchain technology. Ledger cannot be modified after a transaction has been added to the open ledger book and published across the entire network. This is because each block is linked to its predecessor. Protected against any change or damage to any block in the chain. Since the code of the changed block changes, the integrity of the chain will be lost, and the next block will not address the changed block. Since the integrity of the chain is lost, the chain will not be valid and will not be approved by other users on the network (Mooney, 2011).

Blockchain is a manipulation-safe system where ledgers are irreversible and cannot be emulated by one-way cryptographic hash functions. Although security is a relative concept, we can say that blocks are relatively secure because users can only transfer data if they have a private key. Private keys are used to generate a signature for each transaction a user sends. This signature is used to verify that the transaction comes from the user and also prevents it from being changed by other users after the transaction is executed (Tasca & Tessone, 2017). In a cyber attack, attackers need to take over the majority of nodes in the network, in order to take over the system, which has a distributed structure, which makes this possibility quite low (Mugla et al., 2017).

Yli-Huumo et al. (2016) and Swan (2016) state that blockchain technology faces some technical challenges; business volume, latency, size and bandwidth, security, resource consumption, availability, release, difficult forks and multiple chains and privacy. We see that standards have not yet been established at the point of implementing blockchain technology. Open source system; the fact that it is constructed by many different software groups in different ways in line with different ideals prevents the creation of a standard. All companies use their own infrastructure and the system they have established, there is no standard infrastructure. In order to solve this problem, in the coordination of LINUX Open Source community; group of 54 major technology companies such as IBM, Cisco, Fujitsu and J.P. Morgan, Accenture companies established an open source community called kod Hyperledger (Hyperledger, 2018).

The basic features and benefits of blockchain can help predict the sectors that can be use blockhain technology. Key features and benefits include;

- No need for intermediary third parties
- Low transaction costs

- Improved cash flow
- Cannot change transactions of the past
- Transparency and tracking of transactions

The analysis in the Gartner report on Blockchain technology is shown below (Furlonger & Valdes, 2017);

Strengths

- Distributed flexibility and control
- Decentralized network
- Open source
- Security and cryptography
- Proof of works
- Local asset creation
- Dynamic and fluid exchange

Weaknesses

- Lack of registry interoperability
- Poor user experience
- Lack of tested technology
- Deficiencies in developer tools
- Wallet and key management
- Lack of talent and costs
- Lack of trust in new technology providers

Opportunities

- Low transaction costs
- Acceleration of business processes and effectiveness
- Reduction in fraud transactions
- Reduction in systemic risks
- Monetary democratization
- New business models
- Rationalization of applications

Threats

- Legal limitations
- Political actors
- Technology errors
- Enterprise integration limits
- Different blockchain systems

- Registry conflicts
- Lack of management

MAIN FOCUS OF THE CHAPTER

Potential Usage of Blockchain Technology

Blockchain implementations have a wide area of applications, ranging from supply chains and logistics (Iansiti & Lakhani, 2017; Tian, 2016) to the energy sector (Lavrijssen & Carrilo, 2017), the healthcare sector (Hoy, 2017), and the music industry (Rethink Music Initiative, 2015).

There are open source studies on the development of blockchains and related technologies. One of them is Hyperledger, an umbrella project for developers supported by the Linux Foundation. Hyperledger was created in December 2015 with the participation of 17 different companies and currently has over 130 members. The aim of this project is to establish robust and efficient standards for the development of open source blockchain projects. In addition, a modular blockchain structure will be established and connections with different global account ledgers will be established through interfaces (Gupta, 2017).

Secure messaging can be provided for Internet applications via blockchain applications, especially for devices with limited resources that do not have firewall software against intrusion attacks for security purposes. Software companies can place smart contract software in embedded systems to enable software updates to be performed remotely and safely (Christidis & Devetsikiotis, 2016).

In supply-chain management, transactions can be performed on a common blockchain without the approval of a reliable center and payments can be automated after the delivery phase. The monitoring of these transactions can be monitored transparently by the parties at every stage with blockchain technology. It is not possible for one of the parties in the system to delete records or change them retroactively (Watanabe et al., 2015).

In the field of health, blockchain technology can be utilized for medical records of patients, safe transfer between parties, and the detection and prevention of drug fraud. The Estonian government launched a project called Guardtime in 2011 and made the health platform work on blockchain technology (Mettler, 2016). Distributed database-based security approaches are used with RFID systems in monitoring and approving cultivation, spraying, and packaging processes in agriculture and agricultural applications (Tian, 2016).

In the software development sector, which requires the cooperation of more than one party, it is also possible to monitor and manage the transactions to be carried out between parties by automating them with sequential and multiple validation systems with smart contract transfers. Xu et al. (2017) conducted the classification and comparison of blockchain and blockchain-based systems in the design and evaluation of software architectures. A list of applications of blockchain technology is given below (Usta & Doğantekin, 2017);

- Digital identity
- Know your customer
- Meeting capital needs for enterprises
- Global payment systems
- Property and accident insurance compensation process

- Fundraising and management
- Syndication loan
- Internet of Things blockchain
- Voting by proxy
- Automated adaptation mechanism
- Copyright registration systems
- Supply chain management
- Public and health records and tenders
- Land registry systems
- Copy product protection
- Military command chains
- Notary

Many banks and financial institutions are exploring possible uses of blockchain technology other than crypto coins and investing in innovative solutions. As a result of these researches, some of the possible financial uses published in various sources are (Deloitte, 2015; Cognizant, 2016; Evans, 2015):

- Payment transactions
- Money transfers
- Buying / selling platforms
- Clearing management
- Authorization and verification
- Digital identity management
- Document management
- Islamic banking applications

Blockchain technology has innovative uses in public as well, and early applications have begun to emerge. Some areas where blockchain technology can be used in the public sector are as follows (Cognizant, 2016):

- Voting
- Document management
- Energy distribution
- Smart contracts
- Digital identity
- Digital passport
- Social security system
- Tax system

The first project that implemented blockchain technology was a financial application that enabled money transfers. For that reason, the banking sector is an important potential sector for blockchain technology. Today, banks meet their technology and software needs with their central information systems. Banks perform all monetary transactions through their own network devices, servers, software, and databases. Communication with other banks or regulatory agencies such as the central bank is conducted through integration services. The need for reliable institutions and integration with intermediary banks increases both time and costs in banking transactions. Blockchain technology is seen as a new technology to address these problems and processes of the banking sector. For that reason, banking sector representatives are rapidly investing in this field and trying to move their possible scenarios from centralized systems to distributed systems. As it is a developing technology, the applications are being carried out meticulously and cautiously. It is predicted that the costs of the banking sector will be reduced by using blockchain technology. According to research in the *Financial Times*, integration with this technology will save about $20 billion in third-party costs (Kaminska, 2015). Pilkington & Crudu (2016), in their study, mentioned that Moldova's Bitcoin converting power can accelerate the flow of investment to the country and be effective in combating system failures.

Current Usage of Blockchain Technology

Blockchains can be used in the voting process. In the near future, voting will be possible in a more secure and transparent way thanks to blockchain technology. When voting through the blockchain system, the voter will vote anonymously and will be able to follow the outcome of the vote throughout the election. Moreover, the possibility of changing and destroying votes in a blockchain is eliminated. Elections with blockchain technology are also expected to reduce costs and change the minds of voters who do not go to the polls due to concerns about voting security. Today, various entities have begun to emerge from the concept of voting through blockchains. One of them is FollowMyVote. The FollowMyVote system promises a transparent election experience, where the entire electoral process will take place on a blockchain. FollowMyVote's system further improves security by using elliptic curve cryptography. Simply put, cryptography provides security and privacy while the public ledger adds transparency and accountability. A decentralized system provides FollowMyVote with the appropriate security protocols necessary for large-scale elections (https://followmyvote.com/#).

Notary operations have started to be done via blockchain technology. The notary public authority appointed by the state is an institution that registers transactions that should be registered by law, such as securing the rights of individuals and certifying transactions of purchase and sale. All of these registration and approval procedures can be performed via blockchain technology without the need for a notary who carries out notarial activities. This has become possible because the records kept in the blockchain cannot be changed, corrupted, stolen, or destroyed. The number of blockchain structures that provide notary public services is increasing day by day. BlockSign, Stampd, and Stampery are the leading blockchain applications in this field (https://decentralize.today/).

BlockSign allows any document. To sign something, someone uploads the document and adds his or her signature, which is called a signature stamp. This signature stamp includes the signer's name and email address. If other people need to sign the document, they are asked to do the same with their signature stamps. BlockSign takes the document and turns it into a single 32-digit string of letters and numbers. That string of 32 digits is then written into a record book called the blockchain (https://blocksign.com).

For Stampd, the user chooses a particular document or other electronic file for timestamping on a selected public blockchain. The file is processed in the user's browser to derive the SHA256 hash, which is essentially a one-way conversion of the arbitrary large file to a fixed 32-byte data digest. The hash is transmitted to Stampd's servers. A cryptocurrency transaction is then carried out in order to post this hash on the chosen blockchain. In this way, the hash is permanently recorded on the blockchain and linked to a particular point in time. This hash can only be linked to the original content of the user's electronic file, thus also linking that file with the particular timestamp. It is noted that this link can be verified by anyone without reference to any centralized service or authority (https://stampd.io).

Medicalchain uses blockchain technology to securely store health records and maintain a single version of the truth. Different parties, such as doctors, hospitals, laboratories, pharmacists, and health insurers, can request permission to access a patient's record to serve their purposes and record transactions on the distributed ledger. Medicalchain provides solutions to today's health record problems. The platform is built to securely store and share electronic health records (https://medicalchain.com).

Current Usage of Blockchain in Tourism Industry

The future of tourism industry with developing blockchain technology focuses on four main points. These are simplified and safer passenger identification, improved baggage tracking, user-friendly customer loyalty programs and ease of payments between travel agencies and airlines industry (Amadeus, 2017). Some of blockchain applications in tourism sector are given below.

Beenest: One of blockchain's most innovative uses is the Beenest home sharing platform. The technology here is used to get contact between customers and homeowners. Beenest mission is to reinvent business travel, starting with accommodations. Beenest helps travellers find high quality, conveniently located hotels and homes in major cities around the world (https://www.beenest.com).

CheapAir: CheapAir is team of 125 travel enthusiasts use blockchain technology, a boatload of airfare data, and superior customer service to make it easier to buy affordable airline tickets (https://www.cheapair.com).

ShoCard: Although still in its development stage, ShoCard platform is expected to pave the way for a decentralized ID database using a standard format to enable travel companies to quickly and easily verify customer identity. ShoCard is a digital identity card you use through a mobile app. It's the one identity verification system that works the way consumers and businesses need it to for security, privacy, and always-on fraud protection (https://shocard.com).

Trippki: An example of a customer loyalty reward system. Qualified customers are assigned 'TRIP' tokens to stay in a hotel and these tokens are permanently registered in the block chain; this also does not expire and can be used at any time (https://trippki.com/).

TUI start to use its own blockchain project called *"Bed-Swap"* to carry its stocks and flexible sales margins in real time according to the current level of demand.

Winding Tree: The winding tree platform is an example of baggage tracking through perhaps the foremost baggage chain today and also includes a booking element. The absence of third parties means that all transactions can benefit from greater transparency, making reservation and tracking easy and secure. Winding Tree is a marketplace for travel companies. Winding Tree directly connects OTAs and suppliers of travel. It is an open-source ecosystem of smart contracts, interoperable API standards, and harmonized data structures that enable decentralized distribution and discovery of travel products and services (https://windingtree.com/#).

Pegasus Airlines uses the blockchain the first in the travel industry in Turkey. Pegasus launched 'BlockChain' in Sabiha Gökçen Airport. Flight descriptions and updates via Pegasus 'Flight Control System' will instantly be displayed on the screens of Sabiha Gökçen Airport Operations Control Center. Operational information such as door, luggage band, bellows usages, which are created in Istanbul Sabiha Gökçen Airport systems over the same BlockChain infrastructure, will be transferred to Pegasus systems instantly (https://pegasys.tech/).

SOLUTIONS AND RECOMMENDATIONS

Blockchain technology has been rapidly entering daily life since the introduction of bitcoins in 2008. Today, it can be found in every field, from finance to the tourism sector. One of the main reasons for this is that there is no need for third-party agents with blockchain technology, and its reliability is high. The tourism industry is one of the industries that adapt rapidly to technological developments. New-generation technologies such as robotics, the Internet of Things, artificial intelligence, sensors, cognitive technologies, nanotechnology, quantum computing, wearable technologies, augmented reality, intelligent signaling, intelligent robots, big data, and 3D have led to the fourth industrial revolution, often referred to as Industry 4.0. The use of Industry 4.0 technologies in business life, communication, and education has changed business environments and lifestyles. Therefore, the concept of Tourism 4.0 emerged because the tourism industry was rapidly affected by technology and began to adapt to these technologies. In particular, with the increase in the number of Generation Z tourists, new technologies are being used in the tourism sector. In this context, the concept of super-smart tourists, or Tourist 5.0, started to be talked about. It is inevitable that blockchain technology will affect the tourism industry. Although some blockchain applications have started to be used in the tourism industry, blockchain technology is not fully used in tourism sub-sectors. In this context, considering the literature, suggestions for the use of blockchains in the tourism industry are presented. These recommendations are as follows:

- The biggest benefit that blockchain technology can provide to tourism may be smart contracts, because hotels have a wide range of contracts with travel agencies, travel agencies with airline companies, travel agencies with overseas travel agencies, and hotels with suppliers. Most of these contracts are renewed every year. These contracts can be concluded quickly without the need for intermediaries with smart contracts. Thus, both time and money can be saved.
- With the digital identification numbers generated by the blockchain, tourists can quickly pass through airports without having to deal with passports issues. Taking into consideration the length of stay of a tourist entering a hotel with this digital identification number and his or her credit card expenditures (by using big data analysis and algorithms), his or her smart phone can be used to suggest points of interest.
- The luggage of tourists can be taken from their homes before they go on holiday and left in the rooms of the facilities where the tourists will stay before they arrive with blockchain technology. Nowadays, Generation Z travelers particularly see the process of delivering baggage, picking up baggage, and carrying baggage to be a chore.
- Hotels and travel agencies in particular are mostly working with foreign currency. Receiving payments with virtual money such as bitcoins could be important for tourism businesses in terms of fast money. One of the problems that may occur at this point may be virtual currency exchange rates. However, Generation Z tourists in particular may turn to businesses that offer the possibility of booking with virtual money.
- Today, most hotels offer loyalty programs for their guests. These guests are given bonuses and gifts such as a free overnight stay with these programs. Guests of international chain hotels can use the free nights in all hotels in the chain. Small and medium-sized accommodation facilities in certain destinations may develop a joint loyalty program with blockchain technology.
- Tourism enterprises constitute an industry that works intensively with government departments. Lists of hotel guests must be shared with the police or local authorities every day. Hotels also shop

intensively with many suppliers. This shows that hotels are heavily involved in invoicing issues. In this context, developing a joint blockchain application with the government or suppliers could reduce the workload of both the hotels themselves and the government or the suppliers.

FUTURE RESEARCH DIRECTIONS

In this study, block chain technology and some blockchain applications in use have been examined and some suggestions have been made about the potential of using blockchains in the tourism industry. In this context, this study is a conceptual study. Although blockchain technology has the potential to be used in the tourism industry, it is not widely used there yet. A model proposal study could be realized for the tourism industry with blockchain technology. Blockchain technology can be examined separately for tourism sub-sectors such as hotels, restaurants, and travel agencies. In addition, the perceptions and expectations of the tourism sector regarding blockchains can be determined by field studies or surveys with managers and tourists.

CONCLUSION

Blockchain technology has not yet been fully utilized in the tourism industry. However, it is currently being used in the aviation industry in general, and the aviation sector is a sector that triggers innovation in the tourism industry. Particularly, the participation of Generation Z tourists in tourism mobility makes it necessary to adapt to such innovations in the tourism industry. People who participate in the pace of today's busy work and life rhythms do not want to waste time on holiday on details such as luggage, passport control, and reservation issues. New-generation technologies can now simplify unnecessary time-consuming tasks. Some technological applications, such as tourists' check-in at hotels in kiosks and the opening of room doors with smart phones, have started to take place in hotels. Similarly, hotel managers have started to perform hotel operations faster and cheaper with robots and smart applications. Blockchain technology, which entered business life with the introduction of bitcoins, has great potential to make things faster and more reliable in tourism enterprises. In order to provide more convenient and reliable services to tourists and to reduce the workload within the sector itself, sector managers, local governments, and national governments should implement blockchain technology in the tourism industry.

REFERENCES

Amadeus. (2017). *Blockchain: harnessing its potential in travel.* Retrieved from https://amadeus.com/en/insights/research-report/blockchain-harnessing-its-potential-in-travel

Baliga, A. (2017). *Understanding blockchain consensus models*. Santa Clara, CA: Persistent Systems Ltd.

Beck, R. (2019). Beyond bitcoin: The rise of blockchain World. *Computer*, *2018*(February), 54–58.

Beck, R., Czepluch, J. S., Lollike, N., & Malone, S. (2016, June). *Blockchain-the gateway to trust-free cryptographic transactions*. Paper presented at the meeting of 24th European Conference on Information Systems, Istanbul, Turkey.

Buterin, V. (2015). *On public and private blockchains*. Retrieved from https://blog.ethereum.org/2015/08/07/on-public-and-private-blockchains/

Çarkacıoğlu, A. (2016). *Crypto-money bitcoin*. Ankara, Turkey: Capital Market Board Research Department.

Christidis, K., & Devetsikiotis, M. (2016). Blockchains and smart contracts for the internet of things. *IEEE Access Journal*, *4*, 2292–2303. doi:10.1109/ACCESS.2016.2566339

Cognizant. (2016). *Blockchain in banking: a measured approach*. Retrieved from https://www.cognizant.com/whitepapers/Blockchain-in-Banking-A-Measured-Approach-codex1809.pdf

Deloitte. (2015). *Blockchain distrupting the financial services industry*. Retrieved from https://www2.deloitte.com/content/dam/Deloitte/ie/Documents/FinancialServices/IE_Cons_Blockchain_1015.pdf

Furlonger, D., & Valdes, R. (2017). *Practical blockchain: a gartner trend insight report*. Retrieved from https://haas.campusgroups.com/htc/get_file?eid=13961189757744lf06512fc062b0a63e

Gibbs, T., & Yordchim, S. (2014). Thai perception on Bitcoin value. *International Journal of Social, Behavioral, Educational, Economic Business, and Industrial Engineering*, *8*(7), 2334–2336.

Gupta, M. (2017). *Blockchain for dummies*. Hoboken, NJ: John Wiley & Sons.

Hampton, N. (2016). *Understanding the blockchain hype: Why much of it is nothing more than snake oil and spin*. Retrieved from https://www.computerworld.com.au/article/606253/understanding-blockchain-hype-why-much-it-nothing-more-than-snake-oil-spin/

Hong Kong Monetary Authority. (2016). *Whitepaper on distributed ledger technology*. Hong Kong: Hong Kong Monetary Authority.

Hoy, M. B. (2017). An introduction to the blockchain and its implications for libraries and medicine. *Medical Reference Services Quarterly*, *36*(3), 273–279. doi:10.1080/02763869.2017.1332261 PMID:28714815

Hyperledger. (2018). *Hyperledger community*. Retrieved from https://www.hyperledger. org/community

Hyperledger Whitepaper. (2016). *Hyperledger Whitepaper v2.0.0*. Retrieved from https://github.com/hyperledger/hyperledger/wiki/Whitepaper-WG

Iansiti, M., & Lakhani, K. (2017). The truth about blockchain. *Harvard Business Review*, *95*(1), 118–127.

Kaminska, I. (2015). *Blockchain promises back-office ledger revolution*. Retrieved from https://www.ft.com/content/7aad0826-638c-11e5-9846-de406ccb37f2

Kim, K., & Justl, J. M. (2018). Potential antitrust risks in the developement and use of blockchain. *Journal of Taxation and Regulation of Financial Institutions*, *31*(3), 5–16.

Lavrijssen, S., & Carrilo, A. P. (2017). Radical prosumer innovations in the electricity sector and the impact on prosumer regulation. *Sustainability*, *9*(7), 1207. doi:10.3390u9071207

Lemieux, V. L. (2016). Trusting records: Is blockchain technology the answer? *Records Management Journal, 26*(2), 110–139. doi:10.1108/RMJ-12-2015-0042

Lin, I. C., & Liao, T. (2017). A survey of blockchain security ıssues and challenges. *International Journal of Network Security, 19*(5), 653–659.

METI. (2017). *Evaluation forms for blockchain-based systems*. Tokyo, Japan: Ministry of Economy, Trade, and Industry, Information Economy Division Commerce and Information Policy Bureau.

Mettler, M. (2016). Blockchain technology in healthcare: The revolution starts here. In *Proceedings of IEEE 18th International Conference on E-Health Networking, Applications and Services* (pp. 1-3). Munich, Germany. IEEE. 10.1109/HealthCom.2016.7749510

Mills, D., Wang, K., Malone, B., Ravi, A., Marquardt, J., Chen, C., . . . Baird, M. (2016). Distributed ledger technology in payments, clearing, and settlement. *Finance and Economics Discussion Series, 2016*(95), 1-36.

Mooney, C. (2011). The truth about fracking. *Scientific American, 305*(5), 80–85. doi:10.1038cientific american1111-80 PMID:22125868

Mugla, E. K., Akba, M. F., Katip, I., & Karaarslan, E. (2017, October). *Blockchain based cyber security systems*. Paper presented at the International Conference on Information Security & Cryptology, Ankara, Turkey.

Mukhopadhyay, U., Skjellum, A., Hambolu, O., Oakley, J., Yu, L., & Brooks, R. (2016). A brief survey of cryptocurrency systems. In *Proceedings of 14th Annual Conference on Privacy, Security and Trust* (pp. 745-752), Auckland, New Zealand

Müller, C., & Hasic, D. (2016). *Blockchain: Technology and Applications*. Salzburg, Austria: University of Salzburg Department of Computer Sciences.

Nakamoto, S. (2008). *Bitcoin: A peer-to-peer electronic cash system (technical report)*. Retrieved from www.bitcoin.org

Narayanan, A., Bonneau, J., Felten, E., Miller, A., & Goldfeder, S. (2015). *Bitcoin and Cryptocurrency Technologies*. Princeton, NJ: Princeton University.

Natarajan, H., Krause, S., Karla, G., & Helen, L. (2017). *Distributed Ledger Technology (DLT) and blockchain*. Washington, D.C.: World Bank Group. doi:10.1596/29053

Peters, G., & Panayi, E. (2015). *Understanding modern banking ledgers through blockchain technologies: future of transaction processions and smart contracts on the internet of money*. Retrieved from https://arxiv.org/pdf/1511.05740.pdf

Pilkington, M. (2016). Blockchain technology: principles and applications. In F. Xavier Olleros & M. Zhegu (Eds.), *Research handbook on digital transformations* (pp. 1–39). Northampton, MA: Edward Elgar.

Pilkington, M., & Crudu, R. (2016). *Tourism 2.0 and e-Governance in the Republic of Moldova - Can the Blockchain Help?* Retrieved from http://papers.ssrn.com/sol3/papers.cfm?abstract_id=2732350

Rethink Music Initiative. (2015). *Rethink music report on transparency*. Retrieved from http://www.rethink-music.com/news/rethink-music-report-on-tranparency

Sehra, A., Cohen, R., & Vic, A. (2018). On cyrptocurrencies, digital assets and private money. *Journal of Payments Strategy & Systems*, *12*(1), 13–32.

Soylu, A. (2018). Industry 4.0 and new approaches in entrepreneurship. *Pamukkale University Journal of Institute of Social Sciences*, *32*, 43–57.

Swan, M. (2015). *Blockchain: blueprint for a new economy*. Sebastopol, CA: O'Reilly Media.

Swan, M. (2016, October). *Decentralized finance: blockchains, prediction, and valuation*. Paper presented at the Economist and Finance Disrupted, New York, NY.

Tapscott, D., & Tapscott, A. (2017). *Realizing the potential of blockchain a multistakeholder approach to the stewardship of blockchain and cryptocurrencies*. Geneva, Switzerland: World Economic Forum.

Tasca, P., & Tessone, C. J. (2017). Taxonomy of blockchain technologies. principles of ıdentification and classification. *Ledger*, *4*, 1–39.

Tian, F. (2016). An agri-food supply chain traceability system for China based on RFID blockchain technology. In *Proceedings of 13th International Conference on Service Systems and Service Management* (pp. 1-6), Kunming, China. Academic Press.

Usta, A., & Doğantekin, S. (2017). *Blockchain 101*. İstanbul, Turkey: Kapital Medya Hizmetleri.

Watanabe, H., Fujimura, S., Nakadaira, A., Miyazaki, Y., Akutsu, A., & Kishigami, J. J. (2015). Blockchain contract: A complete consensus using blockchain. In *Proceedings of IEEE 4th Global Conference on Consumer Electronics* (pp. 577-578). Las Vegas, NV. 10.1109/GCCE.2015.7398721

Xu, X., Pautasso, C., Zhu, L., Gramoli, V., Ponomarev, A., Tran, A. B., & Chen, S. (2016). The blockchain as a software connector. In *Proceedings of 13th Working IEEE/IFIP Conference on Software Architecture* (pp. 182-191). Venice, Italy. IEEE.

Xu, X., Weber, I., Staples, M., Zhu, L., Bosch, J., Bass, L., ... Rimba, P. (2017). A taxonomy of blockchain-based systems for architecture design. In *Proceedings of IEEE International Conference on Software Architecture* (pp. 243-252). Gothenburg, Sweden. 10.1109/ICSA.2017.33

Yli-Huumo, J., Ko, D., Choi, S., Park, S., & Smolander, K. (2016). Where is current research on blockchain technology? - a systematic review. *PLoS One*, *11*(10), 1–27. doi:10.1371/journal.pone.0163477 PMID:27695049

Zheng, Z., Xie, S., Dai, H., Chen, X., & Wang, H. (2018). Blockchain challenges and opportunities: A survey. *International Journal of Web and Grid Services*, *14*(4), 352–375. doi:10.1504/IJWGS.2018.095647

ADDITIONAL READING

Guerra, A. G. (Ed.). (2019). *Organizational transformation and managing innovation in the Fourth Industrial Revolution*. Hershey, PA: IGI Global. doi:10.4018/978-1-5225-7074-5

Horst, T., & Roman, B. (Eds.). (2019). *Business transformation through blockchain*. Newyork, NY: Palgrave Macmillan.

Mezghani, K., & Aloulou, W. (Eds.). (2019). *Business transformations in the era of digitalization*. Hershey, PA: IGI Global. doi:10.4018/978-1-5225-7262-6

Önder, I., & Horst, T. (2018). Blockchain and tourism: Three research propositions. *Annals of Tourism Research*, *72*(C), 180–182. doi:10.1016/j.annals.2018.03.005

Sabri, E. (Ed.). (2019). *Technology optimization and change management for successful digital supply chains*. Hershey, PA: IGI Global. doi:10.4018/978-1-5225-7700-3

Shi, N. (Ed.). (2019). *Architectures and frameworks for developing and applying blockchain technology*. Hershey, PA: IGI Global. doi:10.4018/978-1-5225-9257-0

Thomason, J., Bernhardt, S., Kansara, T., & Cooper, N. (Eds.). (2019). *Blockchain technology for global social change*. Hershey, PA: IGI Global. doi:10.4018/978-1-5225-9578-6

Wuehler, M., Solomon, L., Rene, M., Kedar, I., Paul, R. A., & Joseph, J. B. (2018). *Blockchain: a practical guide to developing business, law, and technology solutions*. Newyork, NY: McGraw-Hill.

KEY TERMS AND DEFINITIONS

Bitcoin: A crypto coin that is independent of a central government and provides payment anywhere in the world with low transmission costs without the need for the approval of a central authority, providing certain privacy to its owner.

Blockchain: A simple, clear, transparent, distributed, and time-stamped global account book, data file, or simple database that can be understood by everyone.

Consensus Protocol: A process in computer science used to achieve agreement on a single data value among distributed processes or systems.

Crypto Coins: A series of data blocks chained together cryptographically.

End-to-End Communication: A design framework in computer networking.

Industry 4.0: A new phase in the Industrial Revolution that focuses heavily on interconnectivity, automation, machine learning, and real-time data.

Peer-to-Peer (P2P) Protocol: A decentralized communications model in which each party has the same capabilities and either party can initiate a communication session.

Chapter 3
Opinion Mining in Tourism: A Study on "Cappadocia Home Cooking" Restaurant

Ibrahim Akın Özen
https://orcid.org/0000-0003-1172-5448
Faculty of Tourism, Nevşehir Hacı Bektas Veli University, Turkey

Ibrahim Ilhan
https://orcid.org/0000-0002-6614-9356
Faculty of Tourism, Nevşehir Hacı Bektas Veli University, Turkey

ABSTRACT

In the tourism sector, online tourist reviews analysis is one of the methods to evaluate the products and services offered by businesses and understand the needs of tourists. These reviews take place in social networks and e-commerce sites in parallel with the developments in information and communication technologies. Tourists generate these reviews during or after their use of the products or services. In the literature, these reviews are referred to as UGC (User Generated Content) or eWOM (electronic word-of-mouth). The scientific evaluation of the textual contents in tourist reviews is done by text mining, which is a sub-area of data mining. This chapter discusses the methods and techniques of opinion mining or sentiment analysis. In addition, aspect-based sentiment analysis and techniques to be used in the application are discussed. A case study was carried out using aspect-based sentiment analysis method. In the application "Cappadocia home cooking" restaurant used tourist reviews.

INTRODUCTION

In tourism sector, online tourist reviews analysis is one of the methods of evaluating the products and services offered by businesses and of understanding the needs of tourists. These reviews take place in social networks and e-commerce sites in parallel with the developments in information and communication technologies. Tourists generate these reviews during or after their use of the products or services. In the literature, these reviews are referred to as UGC (User Generated Content) or eWOM (electronic

DOI: 10.4018/978-1-7998-1989-9.ch003

word-of-mouth). For businesses, these reviews are of understanding the emotions of tourists and at the same time being able to solve future problems (He, Zha, & Li, 2013). In addition, these reviews are becoming more and more important for other tourists to develop their travel plans (Lei & Law, 2015). The most important reason for this situation is that tourists are more confident in the tourist's reviews that have been experienced before, rather than the information obtained from the other information sources available on the internet. Tourists consider such reviews impartial and honest (Nowacki, 2019). Before tourists decide to buy, they give importance to the ideas of other tourists (Misner & Devine, 1999). The scientific evaluation of the textual contents in tourist reviews is done by text mining, which is a sub-area of data mining. Text mining is defined as the process of extracting the implicit knowledge from textual data (Feldman & Sanger, 2006). The text classification, clustering, and association are the typical tasks of text mining (T. Jo, 2019).

Reviews on the web are collected under two categories. These categories are information that expresses facts or opinions. Facts express objective sentences that contain no emotions. Opinions are about people's natural and general subjective definitions of events and assets (Agarwal & Mittal, 2016).

Many studies in the literature have been conducted using NLP (natural language processing) method of text mining of the texts reflecting the facts. Text analyzes that express opinions are very limited. Opinion mining (OM) or sentiment analysis (SA) is a field of study that analyzes people's opinions and thoughts about assets (products, services, etc.) through texts. In other words, sentiment analysis performs the classification of an opinion as positive, neutral or negative (Mostafa, 2013). In previous studies, opinion mining is also referred to as sentiment analysis (Can & Alatas, 2017; Liu, 2012; Medhat, Hassan, & Korashy, 2014).

This study consists of two main parts. The first section discusses the methods and techniques of opinion mining or sentiment analysis. In addition, aspect-based sentiment analysis (ABSA) and techniques to be used in the application are discussed. In the second part, a case study was carried out using aspect-based sentiment analysis method. In the application "Cappadocia home cooking" restaurant tourist reviews are used.

BACKGROUND

In this part, opinion mining or sentiment analysis, sentiment analysis techniques, opinion mining in tourism, challenges of opinion mining in tourism are defined and explained.

Opinion Mining or Sentiment Analysis

The concepts Opinion Mining (OM), Sentiment Analysis (SA) and Subjectivity Analysis are broadly used as synonyms. According to some researchers, these concepts are explained as follows:

An opinion might possibly be merely thought as a positive or negative sentiment, view, attitude, emotion, or assessment about an entity (product, person, event, business or subject) or an aspect of this entity from a consumer or group of users (Serrano-Guerrero, et al., 2015)..

Opinions are ordinarily subjective expressions that exposit fill's sentiments, appraisals or feelings toward entities, events and their properties. A little amount of opinionated texts was available before the recent World Wide Web expansion. (Wiebe & Riloff, 2011).

Opinion mining is used to identify subjective information from text (Kaur & Chopra, 2016).

Opinion mining or sentiment analysis is a study that analyzes people's ideas and thoughts about assets such as products and services in the text. (Agarwal & Mittal, 2016).

Sentiment analysis instead known as opinion mining is definite as a technique of identifying the positive/negative course of a text (Nasim & Haider, 2017).

As can be seen from the above definitions, SA or OM is often used interchangeably. However, OM makes polarity detection, whereas SA focuses on emotion recognition. Because the identification of sentiment is often exploited for detecting polarity, the two fields are usually combined under the same umbrella or even used as synonyms.(Cambria, et al., 2013).

The online textual information is of two sorts: Facts and opinions claim. Facts are goal sentences regarding the entities and don't display any sentiments. Opinions are subjective in character and generally describe the people's sentiments towards events and entities. The majority of the existing research with all the available on-line text continues to be emphasized about the truthful data in a variety of natural language processing (NLP) jobs, e.g., information retrieval, text classification (Forman, 2003) and so forth. Research on digesting the opinionated sentences continues to be very limited because of a large numbers of challenges mixed up in field (Cambria, et al., 2013).

OM or SA research is categorized as document level (Wilson, Wiebe, & Hoffmann, 2005), sentence level (Meena & Prabhakar, 2007) and aspect / feature-level sentiment analyzes (Y. Jo & Oh, 2011).

Document-Level Sentiment Analysis: Document-level sentiment analysis classifies an assessment document as containing positive or negative polarity. It views a record as an individual unit (Agarwal & Mittal, 2016). This level is linked to the job called document-level sentiment classification. Nevertheless, if a document presents several sentences coping with different aspects or entities, then your sentence level is more desirable (Serrano-Guerrero, et al., 2015).

Sentence-Level Sentiment Analysis: Sentence-level sentiment analysis is closely related to document-level sentiment analysis. Because the document has a structure consisting of sentences (Wilson, et al., 2005). Sentiment analysis at the sentence level expresses the opinion or sentiment expressed in the sentence. Both the document-level and the sentence-level sentiment analysis do not detect what exactly people liked and did not like (Agarwal & Mittal, 2016).

Aspect-Based Sentiment Analysis (ABSA): In opinion mining, it is often not enough to determine the opinion or polarity of the opinion that is mentioned in a text alone. It is also necessary to find out on which entities the opinion is expressed. Aspect-based sentiment analysis is the process of finding out for which entities the opinions in a document or sentence are specified (Liu, 2012). Aspect based sentiment analysis is also called feature-level opinion mining (Hu & Liu, 2004; Thet, Na, & Khoo, 2010). A text example applied aspect-based sentiment analysis can be given as follows;

For instance, "although the service isn't that good, I still like the food"; with this example, "service" and "food" are two entities regarding which thoughts and opinions are indicated. Aspect-based sentiment analysis model, first of all, recognizes these entities, and furthermore, opinions regarding these entities will be identified. (Hu & Liu, 2004; Liu, 2012).

Sentiment Analysis Techniques

There are two main approaches in current scientific studies in the field of sentiment analysis. These are machine learning and lexicon sentiment analysis approaches.

Table 1. Online dictionary databases

Name	Number of words and phrases	Analysis Group	Opinion words	Sentiment Score	Web
Liu and Hu lexicon	Around 6800	Words and phrases	Positive, negative and subjectivity	-1 and +1	https://www.cs.uic.edu/~liub/FBS/sentiment-analysis.html#lexicon
SentiWord Net v3	117000 synonymous	Nouns, verbs, adjectives and adverbs	Positive, negative and objectivity	-1 and +1	https://sentiwordnet.isti.cnr.it/
Natural Language Processing (SentiWords)	Around 155,000 English	Words and phrases	Positive, negative	-5 and +5	http://www2.imm.dtu.dk/pubdb/views/publication_details.php?id=6010
WordStat	9164 negative -4847 positive English	Word patterns	Positive, negative	Rule Based	https://provalisresearch.com/products/content-analysis-software/wordstat-dictionary/sentiment-dictionaries/
SenticNet	50000	Natural language concepts	Pozitif-negatif	-1 and +1	https://sentic.net/
The Whissell Dictionary of Affect in Language	348000	Spoken English	Word	Standard deviation	https://www.god-helmet.com/wp/whissel-dictionary-of-affect/index.htm

Source: (Davydova, 2017)

Machine Learning Approaches

In machine learning methods, algorithms and linguistic features are used to make text classification. The machine learning approach is examined in two categories as supervised and unsupervised learning methods.

The supervised learning method aims to model the function that includes these data and results, using data that has already been observed and whose results are known (labeled). Unsupervised learning includes methods based on the discovery of patterns hidden from unlabeled data, unlike supervised learning (Agarwal & Mittal, 2016). Regarding supervised methods, Maximum Entropy (ME), Support Vector Machines (SVM), Naive Bayes (NB) are a few of the most typical approaches used (Chenlo & Losada, 2014; Rushdi Saleh, et al., 2011). While semi-supervised and unsupervised methods are suggested when it's impossible with a preliminary group of tagged documents/opinions to categorize the others of things (Xianghua, et al., 2013). Besides, hybrid methods, combining supervised and unsupervised techniques, or semi-supervised techniques even, can be utilized to categorize sentiments.

Lexicon-Based Approaches

Two subclassifications can be found here: Dictionary-based and Corpus-based approaches.

Dictionary-Based Approach: The dictionary-based approach depends on uncovering opinion seed words, and then searches the dictionary of their synonyms and antonyms. The dictionary-based approach focuses on compiling word lists based on synonyms and antonyms for every word. This process begins by obtaining a small set of words with known positive or negative emotion and an algorithm is utilized to increase this list using on-line dictionaries (Buzova, Sanz-Blas, & Cervera-Taulet, 2018). For this purpose, many dictionaries that give positive, negative or neutral opinions to words or phrases on the internet have been developed. In this study, some of the known online dictionary databases are presented in the Table 1. In this study, we will look at most known dictionary databases.

Corpus-Based Approaches: Corpus-based approaches mainly depend on the method to see the polarity of the words (Agarwal & Mittal, 2016). These dictionaries are produced from a mates of seed thoughts and opinions cost that grows throughout the operation of incidental status by implementation of the employment of either listing or semantic techniques (Serrano-Guerrero, et al., 2015). Making use of the corpus-based strategy alone isn't as effectual as the dictionary- based approach since it is hard to get ready an enormous ensemble to cover almost all English terms, but this method includes a main advantage that will help to discover domain and context particular opinion phrases and their orientations utilizing a domain name corpus (Medhat, et al., 2014).

Opinion Mining in Tourism

In the literature, the first attempt on opinion mining (sentiment analysis) is document-level by Turney (2001). The first study on aspect/feature-based sentiment analysis was conducted by Hu and Liu (2004).

In recent years, important studies have been carried out on the opinion mining in the domain of tourism (Dolnicar & Otter, 2003).

Ttripadvisor.com reviews of 2510 hotel customers in Sarasota, Florida were evaluated through the text mining method. Online reviews of satisfied and dissatisfied customers were compared. According to the findings of the research, it revealed some common categories used in both positive and negative reviews, including workplace (eg hotels, restaurants and clubs), rooms, furniture, members and sports. The desire to recommend a hotel to others refers to the intangible aspects of hotel stays, such as staff members, more often than dissatisfied customers. On the other hand, dissatisfied customers often mention tangible aspects such as hotels accommodation and financing (Berezina, et al., 2016).

Xiang, et al., (2017) compared online reviews on three major online review platforms (TripAdvisor, Expedia and Yelp) in terms of information quality for all hotels in Manhattan, New York. They used Latent Dirichlet Allocation (LDA), one of the text analysis methods. They discovered five main topics: Basic Service, Value, Landmarks / Attractions, Food- Experience and Core Item.

Marrese-Taylor, Velásquez, & Bravo-Marquez (2014), argue that the aspect-based opinion mining approach applied in the literature will not be applied in the field of tourism. The basis of the claim is that the products in the field of tourism do not consist entirely of physical products. Researchers have analyzed TripAdvisor reviews for the Los Lagos region of Chile with their new model. The new model confirmed that it was particularly effective in determining the emotional aspect of interpretations.

Xiang, et al., (2015) were looking for semantic relationships between words to deconstruct hotel consumer experiences by combining aspect extraction approaches and also statistical analysis, factor and regression analysis.

Tsytsarau & Palpanas (2012) have examined methods related to sentiment analysis and opinion mining in their studies in recent years and they proposed machine-learning, dictionary-based, semantic and statistical approaches. Schuckert, Liu, & Law (2015) suggested using supervised machine learning algorithms (Naive Bayes, Support Vector Machine, and N-gram) in their online review analysis.

Table 2. Restaurant reviews and ABSA related studies

References	Method	Data source
(Brody & Elhadad, 2010)	Aspect Detection and Sentiment Analysis.	Online restaurant reviews from Citysearch New York
(Zhang, et al., 2011)	Sentiment Analysis	Restaurant reviews
(Afzaal & Usman, 2016)	Aspect-Based Opinion Classification	Twitter (tourist place)
(Afzaal, Usman, & Fong, 2019)	Aspect-Based Sentiment Classification	TripAdvisor restaurants reviews
(Lei & Law, 2015)	Content Analysis and Taxonomy	TripAdvisor Reviews on Restaurants in Macau
(Nasim & Haider, 2017)	Aspect-Based Sentiment Analysis	Yelp restaurant review
(Amalia, Putri, & Alamsyah, 2017)	Opinion Mining	TripAdvisor (hotel-restaurant)
(Weismayer, Pezenka, & Gan, 2018)	Aspect-Based Sentiment Detection	TripAdvisor (hotel-restaurant)
(Xiang, et al., 2015)	Text Analytics	Expedia (hotel-restaurant)
(Alkalbani, et al., 2017)	Opinion Mining	Cloud service
(Saeidi, et al., 2016)	Aspect Based Sentiment Analysis	SentiHood
(Blair-Goldensohn, et al., 2008)	Aspect-Based Summarization	Local service and restaurant

Schmunk, et al., (2013) compared dictionary-based and machine learning methods in opinion mining. They found that SVM method in machine learning methods provides very important results in determining sentiment polarity and sentiment classification.

In the literature, the studies using aspect-based sentiment analysis technique in restaurant reviews are presented in Table 2.

In this case study, aspect-based sentiment analysis will be applied to restaurant reviews.

Challenges of Opinion Mining in Tourism

Opinion mining conducted in the field of tourism is provided from different data sources. These data are collected from three primary data sources. **(1) UGC data**: The first and most important data source consists of the reviews of tourists about products and services. Such reviews are called user-generated context (UGC) such as online review data and online image data. **(2) Device and sensor data:** Data collected from devices and sensors used in tourist areas such as GPS, mobile data and Bluetooth data, Rfid data, Wifi data. **(3) Operational data:** Operational data provided during tourism activities such as web search data, on-line booking data, webpage visiting data (Li, et al., 2018).

The data obtained from these data sources are used in different tourism studies. UGC data is generally used in tourist sentiment analysis, tourist behavior analysis, tourism marketing and tourism recommendation research. Device data has gained a significant advantage in investigating the spatial-temporal behavior of tourists. Operational data is rarely used in tourism research since it is under state and private sector protection (Li, et al., 2018).

Although it provides significant advantages with the data set used in tourism research, it has some disadvantages. These disadvantages are related to data quality, privacy and data cost. Table 3 presents the advantages and disadvantages of the data sets used in tourism research.

Table 3. Several types of data sets used in tourism research

Dataset Source	Dataset type	Topics focused on tourism research	Advantages	Disadvantages
UGC data	Tourists online textual data	Opinion mining for tourists Sentiment analysis Tourist behavior analysis Tourist recommendations	Low cost; Multiple information inclusion	Data quality problem
	Online picture data			
Device and Sensor data	Devices GPS data	Data possibility in tourism Tourist spatial-temporal behavior Tourist tracking systems Tourism reference Effect estimation of weather on tourism	Global High precision Crowded indoor availability	High cost Privacy concerns Small-range coverage
	Mobile roaming data			
	Bluetooth data			
	RFID data			
	WIFI data			
	Weather data			
Operational data	Operations WEB search data	Tourism demand prediction Search engine optimization	Low cost; Reflecting community attention	Data privacy problem
	Transaction data	Tourist behavior analysis; Tourism promotion	Operations in tourism marketplaces	

Source: Adapted from (Li, et al., 2018)

Data quality: Although there is sufficient data set in tourism research, quality is seen as a problem. The factors affecting data quality in tourism research are explained below.

1. **Fake Reviews:** Tourists can create fake reviews in online textual data. This reduces the reliability of the data set (Filieri, 2016; Kapucugil & Özdağoğlu, 2015).
2. **Irony Reviews:** Tourist reviews can include sarcastic remarks and irony. A very challenging job in extracting opinion is irony detection (Medhat, et al., 2014; Mostafa, 2013; Serrano-Guerrero, et al., 2015).
3. **Multiple Domains or Languages:** Extracting opinions from textual data from multiple languages and domains may require special algorithms and techniques (Boiy & Moens, 2009; Rushdi Saleh, et al., 2011; Serrano-Guerrero, et al., 2015).
4. **Information Loss:** Reduction methods applied in high volume data sets may cause loss of information (Marjani, et al., 2017).
5. **High Performance Requirements:** Dynamically generated data on devices require high performance. In addition, methods supported by advanced analysis provide interactive graphics on laptops, desktops, or mobile devices such as smartphones and tablets (Marjani, et al., 2017).

Data Privacy: Due to confidentiality, there are barriers to access to data sets created during the operations of tourism stakeholders (tourists, online travel agencies, hotels, government sectors) (Li, et al., 2018).

Data Cost: Data collection devices (e.g., GPS loggers and Bluetooth sensors) in tourism research constitute high cost (Li, et al., 2018).

MAIN FOCUS OF THE CHAPTER

In this part methodology, software and components used in research, data collection and generation, research model are defined and explained.

Methodology

Aspect-based sentiment level and unsupervised machine learning technique were used in the study. When non-structural data are used in sentiment analysis, unsupervised learning technique yields successful results (Brody & Elhadad, 2010). There are many restaurants serve local dishes accompanied by training in Cappadocia Tourism Region. The subject of this research is one of them, the Cappadocia Home Cooking Restaurant serving regional cuisine for tourists in region. The reviews of tourists on TripAdvisor about products and services offered will be evaluated. The main purpose of the study is to analyze tourists' reviews of "Cappadocia Home Cooking Restaurant" on "Tripadvisor.com" using aspect-based sentiment analysis method and the sample is limited to a single restaurant.

Software and Components Used in Research

RapidMiner: In 2001, YALE (Yet Another Learning Environment) is the first software that was announced. It can be used in a very wide area with its hundreds of add-on options as well as keeping up with the current developments. It has the capacity to perform 99% of the processes such as machine learning, data mining and text mining. Rapid Miner is written in Java and it can work with other languages / environments such as Python, Weka or R. In addition to the paid and commercial version, there is also a community version that is promised to be free forever and is distributed under the AGPL license. It is also possible to obtain an academic license for use for academic purposes (Hofmann & Klinkenberg, 2013).

Aylien API: It is an API application that can perform semantic analysis of texts obtained from social media and blogs. It can be preferred in field dependent text analysis. It is an API specialized in aspect-based sentiment analysis such as cars, hotels, airlines and restaurants. In this research, Aspect-based sentiment analysis component presented by Aylien was used. The component is compatible with RapidMiner software. The component is built into Rapidminer software. The component called Analyze Aspect-based sentiment evaluates tourist comments at sentence level. The component identifies a positive, negative, and neutral view for each entity in the sentence.

This api is offered by Aylien company and 1000 comments are made daily free of charge. Since 174 datasets were used for this research, the analysis was made free of charge (Alkalbani, et al., 2017).

Data Collection and Generation

The data set used in the analysis was obtained from the TripAdvisor website by web scraping (web harvesting or web data extraction). Web scraping technique can be defined as the process of collecting data from web sites with the help of computer programs or APIs (Thelwall, 2001).

Figure 1. Model for analysis of tourists' reviews

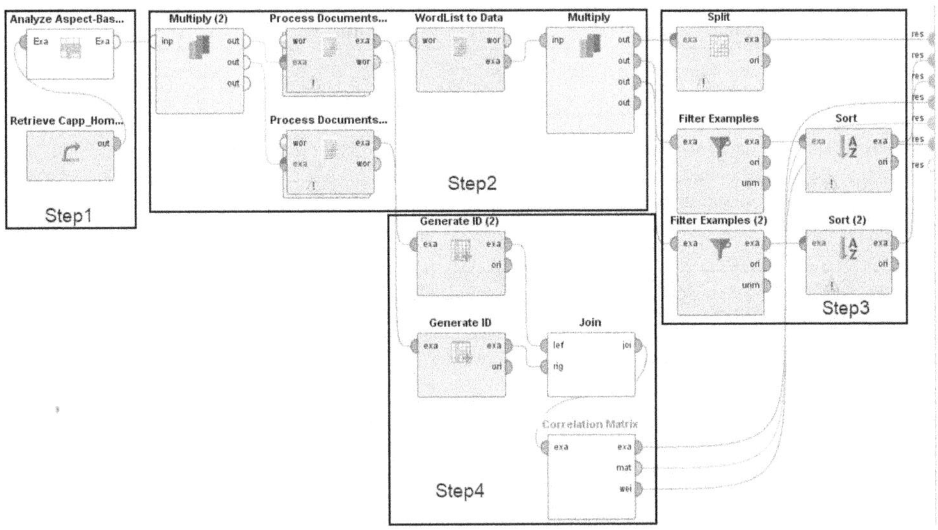

As a result, the data set consisting of tourist reviews containing 174 lines was recorded in the excel file. RapidMiner text analysis program has the ability to import excel files.

Research Model

A model was created for the processing of the data set that was transferred to RapidMiner software. The model to be applied and its contents are given in Figure 1.

The analysis model created in the RapidMiner software in Figure 1 consists of 5 steps. These steps are as follows:

Step 1: Aspect-based sentiment analysis.
Step 2: Prepping the results. Realize the top aspects mentioned and their sentiment (positive, negative or neutral).
Step 3: Splitting and filtering results.
Step 4: Correlation analysis on the words and aspects.
Step 5: Visualization of findings.

Step 1: Aspect-based sentiment analysis: In this step, the excel file obtained during data collection and generation was transferred to RapidMiner software. There are 174 lines of reviews for each tourist in the transferred excel file. Aspect based sentiment analysis (ABSA) was applied to the data set via Aylien component. ABSA is the most important component of this model. By using this component, the sentiment polarity (positive, negative, neutral) of tourist reviews was determined.

The output obtained as a result of ABSA applied to tourist reviews is given in table 4. According to Table 4, the assets included in the reviews were identified and evaluated as positive, negative or neutral according to the tourist perspective.

Table 4. Example of the result obtained from ABSA

Review	Rating	Aspects
My mom and I last minute decided we wanted to do a cooking class in Cappadocia. …	5	food: positive, desserts: positive, value: negative, drinks: positive
A spectacular restaurant find in an out of the way little sleepy village….	5	food: positive, value: positive, drinks: positive
Tolga and his wife Tuba and his mother welcomed us into their home where we were able to experience a fun lesson in simple Turkish home cooking….	5	food: positive, menu: positive, staff: positive
We had wonderful time during the cooking class. Owners were amazing, took great care of us and made us feel at home!...	5	food: positive, menu: positive, staff: positive
We had such a good night with Tolga and his family, learning to cook new dishes and making new friends..	5	food: positive, location: positive, value: positive
The food is very homey, and quite different from the food you would get in the restaurants. Every dish was delicious and we were fairly happy with the food! The portion is huge and definitely more than enough.	4	food: positive, value: negative, staff: positive

Step 2: Prepping the results: Data set tokenization was performed using the Process Documents from Data component. This component is used to create word vectors by means of text processing operators. Before running tokenization, the data was duplicated using the Multiplication Operator to perform two parallel types of analysis using the same data (Figure 2).

In Figure 2, the document processing stage took place in 2 stages. These were; (a) Process documents from data and (b) Process documents from data.

a) **Process Documents From Data:** In the new column obtained as a result of Aspect-based sentiment analysis in Table 3, if the "aspect: polarity" pair has a feature, then 1 weight will be assigned. "Binary Term Occurrences" parameter is used in the "process documents from data (a)" component (Figure 3).

b) **Process Documents From Data:** Under the process documents from data (2) component, the text processing process consists of 5 steps. "Term Occurrences" parameter is used in the "process documents from data (b)" component (Figure 4).

The functions of the subcomponents of Figure 4 are described below.

- Step 1: Tokenize: This component breaks down the review text into words and phrases. The component's parameters are specified as non-letters, character specification, and regular expression. Non-letter mode was used in this study.
- Step 2: Transform Cases: In this component, all words are converted to small characters in order to ensure the integrity of meaning in the review text. For instance, if the review text contains terms like "wonderful" or "Wonderful" or "WONDERFUL", then each one of these words are changed into the same case and so are all treated the same.
- Step 3: Filter tokens (by context): Meaningless signs in the text are removed.

Figure 2. Multiply operator

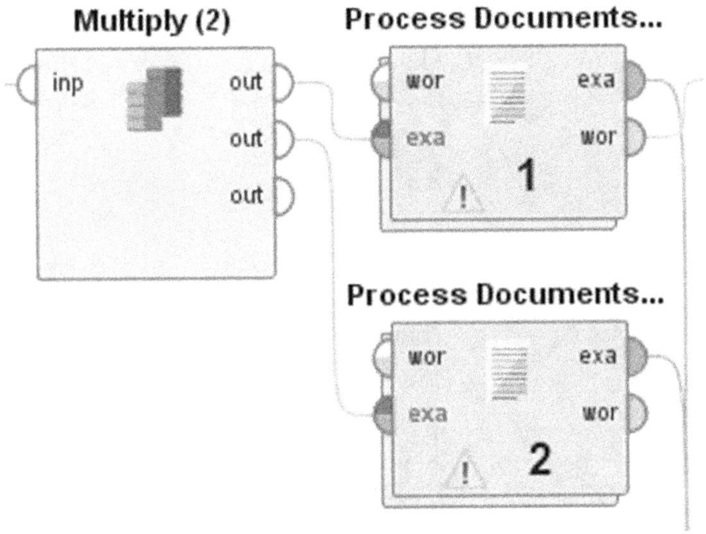

Figure 3. Process documents from data (1)

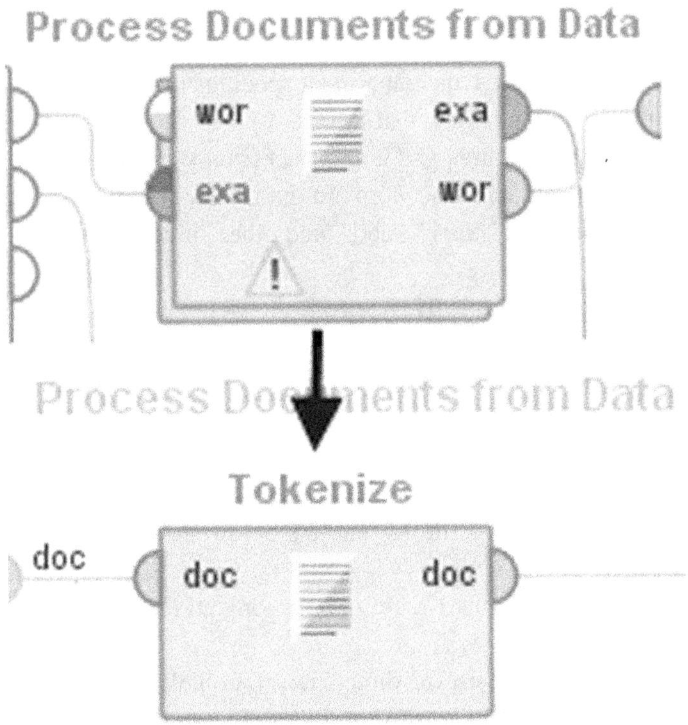

Figure 4. Process documents from data (2)

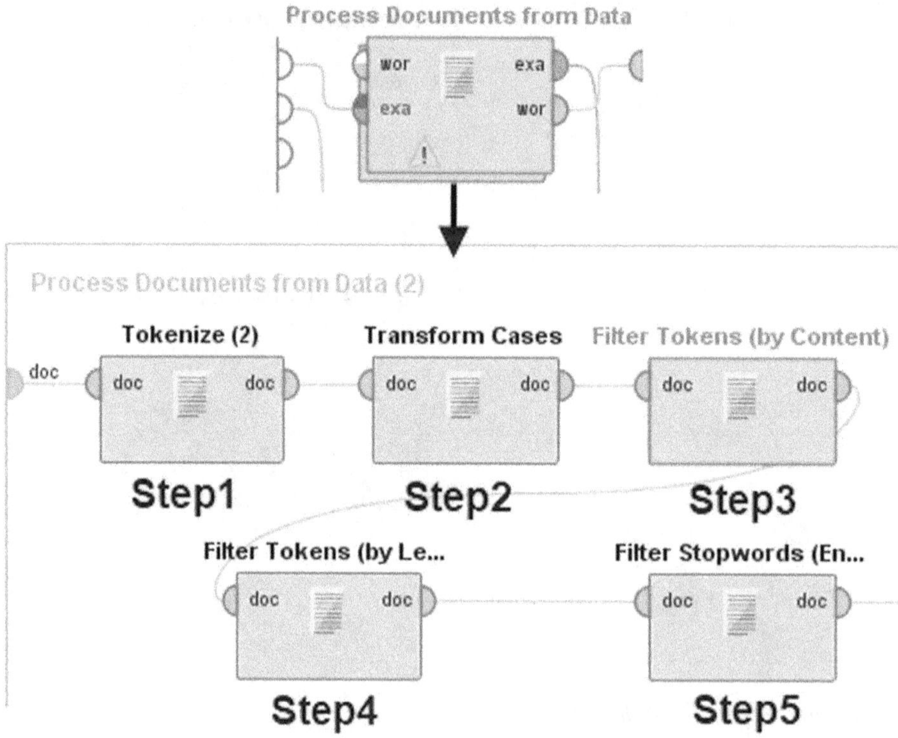

- Step 4: Filter Tokens (by length): This component specifies the minimum and maximum number of meaningless characters in the review text.
- In this research, words less than three and larger than twenty-five characters were removed.
- Step 5: Filter stop words (English): The words in the reviews but not adding meaning to the sentence were removed. Such as "a," "this," "and," and other similar terms.

Step 3: Splitting, filtering and sort results: Figure 5 shows the components related to the splitting, filtering and sort of the results obtained.

ABSA analysis applied to the review texts was separated with split component (Table 4). The ":" sign is selected in the split component parameter. In Table 5, word_1 (aspect) and word_2 (polarity) are separated into columns.

Negative and positive polarity properties were classified using the filter component. As a result, it is ranked according to the number of repetitions. Table 6 below shows the positive aspect and Table 7 shows the negative aspect.

In Table 6, the most repetitive and positive outlooks among all the reviews of tourists are presented in 12 categories.

According to the results of the analysis, the dimensions that have caused the tourists' positive point of view have emerged (Table 6). From the positive point of view of the tourists, the fact that the "food" dimension is considerably high shows that it creates value suitable for the real purpose of the enterprise. Tourists seem to be pleased with the local dishes on offer. In the second place, it was seen that the dimen-

Figure 5. Splitting, filtering and sorting

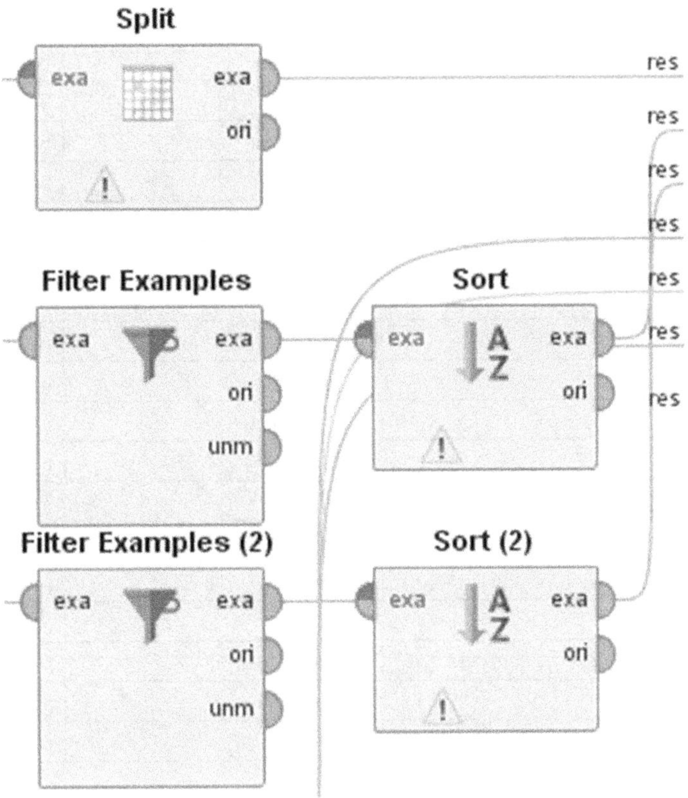

sion which caused the positive point of view of the tourists was the "staff". The high "staff" dimension indicates that the employees in the enterprise perform their duties well. Tourists were pleased with the staff. The third perceived positive dimension is the "value". It is understood that according to the value dimension, the expectations of tourists are met according to the cost they endure. The fourth and fifth positive dimensions show the satisfaction of tourists from the "desserts" and "drinks" offered. A positive menu dimension indicates that the content of the menu is appropriate. Other positive dimensions are "reservation", "location" "ambience", "busyness", "cleanness", and "quietness". Although these dimensions are low, they lead to a significant positive perspective for the enterprise.

In Table 7, the most repetitive and negative outlooks among all the reviews of tourists are presented in 10 categories.

According to the results of the analysis, the dimensions that caused the negative point of view of the tourists for the business have emerged (Table 7). It is seen that tourists have a very low level of negative view of the business. While the "food" dimension of the establishment leads to a low negative point of view, it is seen that the tourists do not have a negative point of view regarding the menu dimension. Although the "value" dimension leads to the negative perspective in the first place, it is not significant when compared to the "value" dimension in the positive perspective.

Table 5. Example set (splitting, filtering)

In Documents	Total	Word1	Word2
12	12	ambience	positive
10	10	busyness	positive
5	5	cleanliness	positive
25	25	desserts	positive
24	24	drinks	positive
163	163	food	positive
15	15	location	positive
23	23	menu	positive
2	2	quietness	positive
20	20	reservations	positive
38	38	staff	positive
29	29	value	positive
1	1	desserts	positive
2	2	drinks	positive
2	2	food	positive
1	1	menu	positive
1	1	reservations	positive
1	1	value	positive

Table 6. Sorted dataset (positive aspect)

Row No.	Word	In Documents	Total
1	food: positive	163	163
2	staff: positive	38	38
3	value: positive	29	29
4	desserts: positive	25	25
5	drinks: positive	24	24
6	menu: positive	23	23
7	reservations: positive	20	20
8	location: positive	15	15
9	ambience: positive	12	12
10	busyness: positive	10	10
11	cleanliness: positive	5	5
12	quietness: positive	2	2

Table 7. Sorted dataset (negative aspect)

Row No.	Word	In Documents	Total
1	Value: negative	8	8
2	Busyness: negative	5	5
3	Location: negative	5	5
4	Food: negative	4	4
5	Reservations: negative	3	3
6	Staff: negative	1	1
7	Ambience: negative	1	1
8	Cleanliness: negative	1	1
9	Drinks: negative	1	1
10	Facilities: negative	1	1

Step 4: Correlation Analysis: Correlation analysis results. The correlation coefficients between the offered product, service categories and related words are shown in the Table 8. The values with high correlation coefficient are considered in the Table 8.

Correlation coefficients (r) in Table 8 will be evaluated. Evaluation of correlation coefficients determines the weak (0 <r <= 0.3), moderate (0.3 <r <= 0.7) and strong (0.7 <r <= + 1) positive relationships between the variables (Gürbüz & Şahin, 2014).

According to the results obtained from the correlation analysis, the main categories (positive and negative) obtained from tourist comments and the words associated with these categories are shown in the correlation table.

- **Positive Aspect-Based Sentiment:** The reason for the positive aspect of tourists towards desserts seems to be a strong correlation with the word "dessert" (r = 0.862). In Table 6, where positive dimensions are presented, "desserts" have a positive effect in the fourth place. But it is seen that the word "dessert", which affects the size of the desserts, has the highest correlation in the correlation table. In the positive dimension of sweets, the most repeated word was "dessert". This is thought to be due to the fact that tourists do not know the local dessert names offered in the menu. Therefore, the correlation coefficient was found to be high. It is understood that the positive aspect of tourists towards drinks is significantly correlated with the word "tea" (r = 0.824). The reason why the word "tea" constitutes a significant positive perspective in terms of drinks is that the region's teas are offered to tourists. Especially teas made from dried fruits are offered to tourists in the region. The positive aspect of the menu appears to be significantly correlated with the word's "local cuisine", "stuffed", "soup" (r = 0.793, r=0,517, r= 0,406). It is seen that the tourists' point of view of the menu is due to the local cuisine. It seems that soup and stuffing are the most popular local cuisine dishes. It is understood that the soup and stuffing that are unique to the region have gained the appreciation of tourists. The positive aspect towards the staff appears to be significantly correlated with the word "friendly" (r = 0.777). Another point of view that tourists consider the business as positive is that the business is "pro-friend". Business staffs have been successful in communicating with tourists. In this way, tourists see themselves as part of that business. The

Table 8. Negative and positive aspect of the words related to the correlation table

	Desserts Positive (r)	**Drinks pos.**	**Menu pos.**	**Staff pos.**	**Ambience: pos.**	**Reser-vation pos.**	**Facilities Neg.**	**Cleanliness: neg.**
dessert	0,862							
tea		0,824						
local cuisine			0,793					
stuffed			0,517					
soup			0,406					
friendly				0,777				
atmosphere					0,772			
book						0,648		
things							0,484	
actually								0,482

positive aspect for reservation appears to be significantly correlation with the word "book" (r = 0.648). Tourists consider the booking facilities sufficient.

- **Negative Aspect-Based Sentiment:** The negative aspect of the facilities seems to be moderate correlation with the word "things" (r = 0.484). The negative aspect for "cleanliness" appears to be moderately correlated with the word "actually" (r = 0.482). As a result of the analysis, although the tourists have a positive view of the business in many ways, they think that the business is generally not clean and the facilities are insufficient.

Step 5: Visualization of Findings: In Rapidminer software, visualization of the findings can be done using graphics or advanced graphics in the results tab.

SOLUTIONS AND RECOMMENDATIONS

In this research, tourist reviews of "Cappadocia home cooking" a local food and beverage business in Cappadocia, were analyzed. Tourist reviews were collected automatically from the TripAdvisor website. Text mining method and ABSA technique were used in the analysis. As a result of the analysis, positive aspects consisting of 15 categories were also determined. In addition, the negative aspect consists of 10 categories. The main factors that cause positive opinions in tourists have been local "foods", "drinks" and "friendly" behaviors. Moreover, the reason for negative opinions were seen as "value". Correlation levels between categories and words were also determined. The most correlated words that cause positive opinion among tourists were determined as "dessert" and "tea". According to the results of the analysis, businesses will provide the opportunity to evaluate themselves according to the tourism perspective. Thus, companies will be able to develop a strategy that considers their positive and negative aspects.

Tourism is an information-intensive sector (Doolin, Burgess, & Cooper, 2002). The tourism sector should consider online tourist reviews to improve their products and services. Extracting meaningful relationships from these reviews provides an important source of information for all stakeholders. Tourist reviews from other online platforms can be analyzed by considering other stakeholders in the tourism sector.

Online tourist reviews and opportunities for stakeholders can be listed as follows;

Opportunities for destinations;

- Opportunity to understand and interpret the destination
- Opportunity to understand and interpret tourist needs

Opportunities for tourism businesses;

- Opportunity to evaluate products and services offered to tourists: Discovering negative aspects, discovering positive aspects, opportunity to manage customer relationships.
- Opportunity to understand and evaluate competitors: Providing product and service advantage, product and service development or improvement, providing a price advantage (price determination).

Opportunities for tourists;

- Opportunity to make travel plans.

FUTURE RESEARCH DIRECTIONS

Aspect-based sentiment analysis is a method that can express opinion for each entity or phenomenon that exists in the document. Previous studies have mostly taken into account the opinion of sentence or document. This generalization makes it difficult to discover information in unstructured texts.

In text mining, opinion extraction is divided into two areas: Domain dependent and domain independent. Domain dependent opinion extraction can be performed by creating a dictionary about the area to be analyzed. In this direction, it is aimed to develop the research. In the future, a special dictionary will be used for local restaurants. Developing the domain dependent dictionary for local restaurants is presented to the attention of researchers.

CONCLUSION

In this study, the methods of determining opinions from online tourist reviews were examined. Determination of opinions from textual content is done using sentiment analysis (alternatively known as opinion mining) techniques. Sentiment analysis is applied at document-level, sentence-level and aspect-based sentiment levels.

Aspect-based sentiment level is thought to be a more effective technique than other levels (Medhat, et al., 2014; Nasim & Haider, 2017). The most important reason for this is that the text analysis using ABSA technique has the ability to determine opinions for each entity/event in the document. Therefore, aspect-based sentiment level was used in the case study.

REFERENCES

Afzaal, M., & Usman, M. (2016). A novel framework for aspect-based opinion classification for tourist places. *The 10th International Conference on Digital Information Management, ICDIM 2015*, (Icdim), 1–9. 10.1109/ICDIM.2015.7381850

Afzaal, M., Usman, M., & Fong, A. (2019). Predictive aspect-based sentiment classification of online tourist reviews. *Journal of Information Science*, *45*(3), 341–363. doi:10.1177/0165551518789872

Agarwal, B., & Mittal, N. (2016). *Prominent feature extraction for sentiment analysis* (N. Mittal, Ed.)., doi:10.1007/978-3-319-25343-5

Alkalbani, A. M., Gadhvi, L., Patel, B., Hussain, F. K., Ghamry, A. M., & Hussain, O. K. (2017). Analysing cloud services reviews using opining mining. *Proceedings - International Conference on Advanced Information Networking and Applications, AINA*, 1124–1129. 10.1109/AINA.2017.173

Amalia, N., Putri, S., & Alamsyah, A. (2017). *Opinion Mining of Tripadvisor Review Towards Five-Star Hotels in Bandung City.*, *4*(1), 4.

Berezina, K., Bilgihan, A., Cobanoglu, C., & Okumus, F. (2016). Understanding Satisfied and Dissatisfied Hotel Customers: Text Mining of Online Hotel Reviews. *Journal of Hospitality Marketing & Management*, *25*(1), 1–24. doi:10.1080/19368623.2015.983631

Blair-Goldensohn, S., Neylon, T., Hannan, K., Reis, G. A., McDonald, R., & Reynar, J. (2008). Building a sentiment summarizer for local service reviews. *Workshop on NLP in the Information Explosion Era.*

Boiy, E., & Moens, M. F. (2009). A machine learning approach to sentiment analysis in multilingual web texts. *Information Retrieval*, *12*(5), 526–558. doi:10.100710791-008-9070-z

Brody, S., & Elhadad, N. (2010). An Unsupervised Aspect-Sentiment Model for Online Reviews. *HLT '10 Human Language Technologies: The 2010 Annual Conference of the North American Chapter of the Association for Computational Linguistics.*

Buzova, D., Sanz-Blas, S., & Cervera-Taulet, A. (2018). Does culture affect sentiments expressed in cruise tours' eWOM? *Service Industries Journal*, 1–20. doi:10.1080/02642069.2018.1476497

Cambria, E., Schuller, B., Xia, Y., & Havasi, C. (2013). New avenues in opinion mining and sentiment analysis. *IEEE Intelligent Systems*, *28*(2), 15–21. doi:10.1109/MIS.2013.30

Can, U., & Alatas, B. (2017). Duygu Analizi ve Fikir Madenciliği Algoritmalarının İncelenmesi. *Int. J. Pure Appl. Sci*, *3*(1), 75–111.

Chenlo, J. M., & Losada, D. E. (2014). An empirical study of sentence features for subjectivity and polarity classification. *Information Sciences*, *280*, 275–288. doi:10.1016/j.ins.2014.05.009

Davydova, O. (2017). Sentiment Analysis Tools Overview, Part 1. Positive and Negative Words Databases. Retrieved June 15, 2019, from https://medium.com/@datamonsters/sentiment-analysis-tools-overview-part-1-positive-and-negative-words-databases-ae35431a470c

Dolnicar, S., & Otter, T. (2003). Which Hotel attributes Matter? A review of previous and a framework for future research. *Proceedings of the 9th Annual Conference of the Asia Pacific Tourism Association (APTA)*. Academic Press.

Doolin, B., Burgess, L., & Cooper, J. (2002). Evaluating the use of the Web for tourism marketing: A case study from New Zealand. *Tourism Management, 23*(5), 557–561. doi:10.1016/S0261-5177(02)00014-6

Feldman, R., & Sanger, J. (2006). The Text Mining Handbook. In The Text Mining Handbook. doi:10.1017/CBO9780511546914

Filieri, R. (2016). What makes an online consumer review trustworthy? *Annals of Tourism Research, 58*, 46–64. doi:10.1016/j.annals.2015.12.019

Forman, G. (2003). An Extensive Empirical Study of Feature Selection Metrics for Text Classification George. *Journal of Machine Learning Research*. doi:10.1162/153244303322753670

Gürbüz, S., & Şahin, F. (2014). Sosyal bilimlerde araştırma yöntemleri. In Seçkin Yayıncılık. Ankara, Turkey.

He, W., Zha, S., & Li, L. (2013). Social media competitive analysis and text mining: A case study in the pizza industry. *International Journal of Information Management, 33*(3), 464–472. doi:10.1016/j.ijinfomgt.2013.01.001

Hofmann, M., & Klinkenberg, R. (2013). RapidMiner: Data mining use cases and business analytics applications. Boca Raton, FL: CRC Press.

Hu, M., & Liu, B. (2004). Mining and summarizing customer reviews. *Proceedings of the 2004 ACM SIGKDD International Conference on Knowledge Discovery and Data Mining - KDD '04*, 168. 10.1145/1014052.1014073

Jo, T. (2019). *Text Mining*. Cham, Switzerland: Springer; doi:10.1007/978-3-319-91815-0

Jo, Y., & Oh, A. H. (2011). Aspect and sentiment unification model for online review analysis. *Proceedings of the Fourth ACM International Conference on Web Search and Data Mining - WSDM '11*, 815. 10.1145/1935826.1935932

Kapucugil, A., & Özdağoğlu, G. (2015). Text Mining as a Supporting Process for VoC Clarification. *Alphanumeric Journal, 3*(1). doi:10.17093/aj.2015.3.1.5000105108

Kaur, A., & Chopra, D. (2016). Comparison of text mining tools. *2016 5th International Conference on Reliability, Infocom Technologies, and Optimization, ICRITO 2016: Trends and Future Directions*, 186–192. 10.1109/ICRITO.2016.7784950

Lei, S., & Law, R. (2015). Content analysis of TripAdvisor reviews on restaurants: A case study of Macau. *Journal of Tourism, 16*(1), 17–28. Retrieved from http://search.ebscohost.com/login.aspx?direct=true&db=hjh&AN=111435361&site=ehost-live

Li, J., Xu, L., Tang, L., Wang, S., & Li, L. (2018). Big data in tourism research: A literature review. *Tourism Management, 68*, 301–323. doi:10.1016/j.tourman.2018.03.009

Liu, B. (2012). Sentiment Analysis and Opinion Mining. *Synthesis Lectures on Human Language Technologies*, *5*(1), 1–167. doi:10.2200/S00416ED1V01Y201204HLT016

Marjani, M., Nasaruddin, F., Gani, A., Karim, A., Hashem, I. A. T., Siddiqa, A., & Yaqoob, I. (2017). Big IoT Data Analytics: Architecture, Opportunities, and Open Research Challenges. *IEEE Access: Practical Innovations, Open Solutions*, *5*, 5247–5261. doi:10.1109/ACCESS.2017.2689040

Marrese-Taylor, E., Velásquez, J. D., & Bravo-Marquez, F. (2014). A novel deterministic approach for aspect-based opinion mining in tourism products reviews. *Expert Systems with Applications*, *41*(17), 7764–7775. doi:10.1016/j.eswa.2014.05.045

Medhat, W., Hassan, A., & Korashy, H. (2014). Sentiment analysis algorithms and applications: A survey. *Ain Shams Engineering Journal*, *5*(4), 1093–1113. doi:10.1016/j.asej.2014.04.011

Meena, A., & Prabhakar, T. V. (2007). Sentence Level Sentiment Analysis in the Presence of Conjuncts Using Linguistic Analysis. In Advances in Information Retrieval (pp. 573–580). doi:10.1007/978-3-540-71496-5_53

Misner, I., & Devine, V. (1999). *The world's best-known marketing secret: building your business with word-of-mouth marketing*. Retrieved from https://scholar.google.com.tr/scholar?hl=tr&as_sdt=0%2C5&q=The+world's+best+known+marketing+secret%3A+Building+your+business+with+word-of-mouth+marketing.&btnG=

Mostafa, M. M. (2013). More than words: Social networks' text mining for consumer brand sentiments. *Expert Systems with Applications*, *40*(10), 4241–4251. doi:10.1016/j.eswa.2013.01.019

Nasim, Z., & Haider, S. (2017). ABSA Toolkit: An Open Source Tool for Aspect Based Sentiment Analysis. *International Journal of Artificial Intelligence Tools*, *26*(06). doi:10.1142/S0218213017500233

Nowacki, M. (2019). World Cities' Image in TripAdvisor Users' Reviews. *Ereview of Tourism Research*, *16*(2–3). Retrieved from https://journals.tdl.org/ertr/index.php/ertr/article/view/327

Rushdi Saleh, M., Martín-Valdivia, M. T., Montejo-Ráez, A., & Ureña-López, L. A. (2011). Experiments with SVM to classify opinions in different domains. *Expert Systems with Applications*, *38*(12), 14799–14804. doi:10.1016/j.eswa.2011.05.070

Saeidi, M., Bouchard, G., Liakata, M., & Riedel, S. (2016). *SentiHood: Targeted Aspect Based Sentiment Analysis Dataset for Urban Neighbourhoods*. 1546–1556. Retrieved from http://arxiv.org/abs/1610.03771

Schmunk, S., Höpken, W., Fuchs, M., & Lexhagen, M. (2013). Sentiment Analysis: Extracting Decision-Relevant Knowledge from UGC. In Information and Communication Technologies in Tourism 2014 (pp. 253–265). doi:10.1007/978-3-319-03973-2_19

Schuckert, M., Liu, X., & Law, R. (2015). Hospitality and Tourism Online Reviews: Recent Trends and Future Directions. *Journal of Travel & Tourism Marketing*, *32*(5), 608–621. doi:10.1080/10548408.2014.933154

Serrano-Guerrero, J., Olivas, J. A., Romero, F. P., & Herrera-Viedma, E. (2015). Sentiment analysis: A review and comparative analysis of web services. *Information Sciences*, *311*, 18–38. doi:10.1016/j.ins.2015.03.040

Thelwall, M. (2001). A web crawler design for data mining. *Journal of Information Science*, *27*(5), 319–325. doi:10.1177/016555150102700503

Thet, T. T., Na, J. C., & Khoo, C. S. G. (2010). Aspect-based sentiment analysis of movie reviews on discussion boards. *Journal of Information Science*. doi:10.1177/0165551510388123

Tsytsarau, M., & Palpanas, T. (2012). Survey on mining subjective data on the web. *Data Mining and Knowledge Discovery*, *24*(3), 478–514. doi:10.100710618-011-0238-6

Weismayer, C., Pezenka, I., & Gan, C. H.-K. (2018). Aspect-Based Sentiment Detection: Comparing Human Versus Automated Classifications of TripAdvisor Reviews. In Information and Communication Technologies in Tourism 2018 (pp. 365–380). doi:10.1007/978-3-319-72923-7_28

Wiebe, J., & Riloff, E. (2011). *Finding mutual benefit between subjectivity analysis and information extraction*. IEEE Transactions on Affective Computing; doi:10.1109/T-AFFC.2011.19

Wilson, T., Wiebe, J., & Hoffmann, P. (2005). Recognizing contextual polarity in phrase-level sentiment analysis. *Proceedings of the Conference on Human Language Technology and Empirical Methods in Natural Language Processing - HLT '05*, 347–354. 10.3115/1220575.1220619

Xiang, Z., Du, Q., Ma, Y., & Fan, W. (2017). A comparative analysis of major online review platforms: Implications for social media analytics in hospitality and tourism. *Tourism Management*, *58*, 51–65. doi:10.1016/j.tourman.2016.10.001

Xiang, Z., Schwartz, Z., Gerdes, J. H. Jr, & Uysal, M. (2015). What can big data and text analytics tell us about hotel guest experience and satisfaction? *International Journal of Hospitality Management*, *44*, 120–130. doi:10.1016/j.ijhm.2014.10.013

Xianghua, F., Guo, L., Yanyan, G., & Zhiqiang, W. (2013). Multi-aspect sentiment analysis for Chinese online social reviews based on topic modeling and HowNet lexicon. *Knowledge-Based Systems*, *37*, 186–195. doi:10.1016/j.knosys.2012.08.003

Zhang, Z., Ye, Q., Zhang, Z., & Li, Y. (2011). Sentiment classification of Internet restaurant reviews written in Cantonese. *Expert Systems with Applications*. doi:10.1016/j.eswa.2010.12.147

ADDITIONAL READING

Alaei, A. R., Becken, S., & Stantic, B. (2019). Sentiment analysis in tourism: Capitalizing on big data. *Journal of Travel Research*, *58*(2), 175–191. doi:10.1177/0047287517747753

Cambria, E., Das, D., Bandyopadhyay, S., & Feraco, A. (Eds.). (2017). *A practical guide to sentiment analysis*. Cham, Switzerland: Springer International Publishing. doi:10.1007/978-3-319-55394-8

Dwivedi, R. K., Aggarwal, M., Keshari, S. K., & Kumar, A. (2019). Sentiment Analysis and Feature Extraction Using Rule-Based Model (RBM). In *International Conference on Innovative Computing and Communications* (pp. 57-63). Springer, Singapore. 10.1007/978-981-13-2354-6_7

Liu, B. (2012). Sentiment analysis and opinion mining. *Synthesis lectures on human language technologies, 5*(1), 1-167.

Salloum, S. A., AlHamad, A. Q., Al-Emran, M., & Shaalan, K. (2018). A survey of Arabic text mining. In *Intelligent Natural Language Processing: Trends and Applications* (pp. 417–431). Cham: Springer. doi:10.1007/978-3-319-67056-0_20

Shirsat, V. S., Jagdale, R. S., & Deshmukh, S. N. (2019). Sentence Level Sentiment Identification and Calculation from News Articles Using Machine Learning Techniques. In *Computing, Communication and Signal Processing* (pp. 371–376). Singapore: Springer. doi:10.1007/978-981-13-1513-8_39

Vyas, V., & Uma, V. (2019). Approaches to sentiment analysis on product reviews. In *Sentiment Analysis and Knowledge Discovery in Contemporary Business* (pp. 15–30). IGI Global. doi:10.4018/978-1-5225-4999-4.ch002

KEY TERMS AND DEFINITIONS

Aspect-Based Sentiment Analysis: is the level of determining opinions in the text analyzed. At this level of analysis, the sentiment polarity is determined separately for each entity or event in the document. Used for detailed document analysis.

Machine Learning: Machine Learning is the general name of computer algorithms (Decision Trees, Naïve Bayes, Logistic Regression, Random Forest) that model a given problem according to the data obtained from the problem environment. Since it is an intensively studied subject, many approaches and algorithms have been proposed.

NLP (Natural Language Processing): The natural language processing tools can be used to facilitate the SA process. It gives better natural language understanding and thus can help produce more accurate results of SA.

Sentiment Polarity: It is the expression that determines the sentimental aspect of an opinion. In textual data, the result of sentiment analysis can be determined for each entity in the sentence, document or sentence. The sentiment polarity can be determined as positive, negative and neutral.

Supervised Learning: Machine learning is one of the methods. The data is taken from systems that operate on the principle of response to the effect and organized in the input-output order.

SVM (Support Vector Machine): It is one of the most effective and simple machine learning methods used in classification. For classification, it is possible to separate the two groups by drawing a boundary between the two groups in one plane. Where this boundary is drawn should be the farthest from the members of both groups. SVM determines how to draw this limit.

Unsupervised Learning: Machine learning is one of the methods. It aims to explore groups within the data that are either non-class or not.

Chapter 4
Near Field Communication (NFC) Applications in the Tourism Industry

Pınar Çelik Çaylak
https://orcid.org/0000-0003-0599-4086
Akdeniz University, Turkey

ABSTRACT

The use of near field communication (NFC), a type of wireless technology that allows data transfer by bringing two devices closer together, has been increasing very rapidly. NFC offers great ease to its users through its contactless operation and provides the optimum environment for wireless communication with other technologies. Users with NFC-enabled mobile phones can share websites, contact information, phone numbers, music files, videos, and photos and can also unlock doors and can use their phones as mobile wallets to make purchases safely and easily. In recent years, with the increasing use of technology, it is clear that NFC technology is promising for the tourism industry. This chapter provides background on NFC technology. The features of NFC technology and its advantages and disadvantages for the tourism industry are evaluated. NFC technology applications (data sharing, payments, tickets, keys, etc.) for the tourism industry are explained. Finally, examples of NFC applications in the tourism industry are discussed for both tourists and sector stakeholders.

INTRODUCTION

Technological developments show their effects in a wide range of areas from manufacturing to the service sector. Along with the technological innovations that are of interest to all sectors and the widespread use of information communication technologies, in the tourism sector, which follows technological developments closely, the use of new technologies has become almost mandatory. Technological developments are changing our thought processes while simultaneously introducing new concepts in all areas of life.

Near field communication (NFC) has recently become very popular in the field of technology. In today's increasingly interconnected world, this fast and innovative technology allows you to safe interaction with

DOI: 10.4018/978-1-7998-1989-9.ch004

the world around you with a simple touch. NFC is a contactless processing technology that is found in billions of smartphones, tablets, and other electronic devices and is being added to new devices almost every day. NFC is a technology standard that allows two NFC-compatible devices that are close to each other to communicate over wireless technology at close range but without contact. Founded by Nokia, Sony, and NXP, the NFC Forum is working on the development and dissemination of this technology.

NFC is seen as an ideal technology to strengthen short-distance connections between two devices due to the rapid expansion of devices in line with the technology in question (Hardy et al., 2010). Considering that the tourism sector is based on cooperation between a wide range of services and products, the benefits of technological developments are quite high in this sector (Neuhofer et al., 2015). For this reason, mobile technologies are considered as one of the most interesting developments affecting the tourism industry. In particular, NFC is perceived as a major technological development that has significant impact on both tourism businesses and tourism research (Buhalis & Law, 2008). NFC technology provides great opportunities for destinations, hotels, and tourism businesses to improve the efficiency and quality of existing services (Vitezić et al., 2015). The concept of smart tourism together with the emergence of smart tourists demands more technology. NFC, as one of the smart tourism tools, is especially able to meet the expectations of tourists because of its easy usage and safety.

The use of NFC in tourism is relatively new compared to other technologies, but it offers new applications for tourists, such as E-ticketing, data exchange, E-wallets, boarding passes, and ratings. Once a smartphone is placed close to an NFC tag, information such as text, imagery, or soundtracks will automatically appear. No app downloads are necessary for NFC tags to be activated by smartphones. For example, tourists can use their mobile phones to get information about transportation schedules, waiting times for arrivals, ticket purchases, and so on from panels placed at bus stops. When a tourist brings his/her phone closer to a historical artifact exhibited in a museum, he/she can learn everything about it, or he/she can pay by phone in a restaurant where he/she eats.

The aim of this study is to examine NFC technology and to give information about its current usage in the tourism sector. In this context, information is given about NFC technology compared with other wireless technologies, as well as NFC operating modes, NFC communication modes, and the advantages and disadvantages of NFC. The importance of the use of NFC technology in tourism and its current usage areas are illustrated with examples.

BACKGROUND

Near Field Communication (NFC)

With the expansion of mobile devices, innovative developments of technologies are emerging rapidly (Canadi et al., 2010). One of them, Near Field Communication (NFC), is a growing technology among automatic identification technologies. NFC, a new generation of wireless communication technology, developed in late 2002 in partnership with Sony and Philips and was adopted by ISO / IEC as standard on 8 December 2003. NFC technology provides close-range communication between electronic devices that comply with NFC standards (Coskun et al., 2011).

Due to the NFC technology is composed of an interface and protocol developed on RFID, it is compatible with RFID technology (Bilginer and Ljunggren, 2011). NFC is short-range, high frequency, low-bandwidth and wireless technology between two NFC-enabled devices (Coskun et al., 2011). NFC technology provides 13.56 MHz (high frequency) and up to 424 Kbit/s (low bandwidth) secure data communication (Grassie, 2007). A high frequency of 13.56 MHz, actually used by Radio Frequency Identification (RFID). However, even if RFID is capable of receiving and transmitting more than a few meters, the NFC is limited in the very close range (Coskun et al., 2011). In NFC technology, transmission is triggered by bringing two NFC-compatible devices closer to each other's. NFC-based transactions with a short transmission range are more secure (Dubey et al, 2011).

While RFID allows unidirectional communication, NFC also allows bidirectional communication (Olivieri, 2015). NFC offers great convenience to its users thanks to its bidirectional and unidirectional data communication capability and contactless usage and offers the optimum environment for wireless communication with other technologies. NFC provides a shorter transmission interval and a lower bandwidth of data communication compared to other RFD-based technologies such as Bluetooth, WiFi and ZigBee. (Özdenizci et al., 2011, Lange and Steck, 2014).

It can be said that NFC technology has many advantages. First, smartphones have a dual role, read, and store this information. NFC technologies also support the entire RFID system because they are generated by RFID technology. In addition to the easy-to-use application that is open to everyone, NFC technology is well protected against hacking in communication between devices, as the devices cannot operate if the devices are more than four centimeters away (Ok et al., 2010). The other advantage of NFC technology, referred to as "dematerialization", is that all information included in physical devices (credit cards, access keys, badges, etc.) is stored in the smartphone. Other advantages of NFC include the speed of information exchange (useful, for example, for cash spending or transport) and the convergence of the offline and online world (by making available more information about the products in a physical store) (Palumbo and Dominici., 2015). NFC technology was originally designed for applications such as payment and ticketing, however, it is used in many fields such as entertainment and social applications, access to control, content distribution, peer-to-peer data/money model transfer, smart advertising, tourism applications etc. (Garrido et. al. 2010).

Comparison With Other Wireless Technologies

NFC and Radio Frequency Identification (RFID)

NFC and RFID technologies are based on the same standards. Although NFC is based on RFID technology, it does not use the long-range feature of RFID technology. RFID is a process in which substances are uniquely identified using radio waves, and NFC is a subset specialized in the RFID technology family (Thrasher, 2013).

RFID technology carries one-way data from label to reader. In NFC technology, both bidirectional and unidirectional data transmission is performed (Karabulut, 2018). Furthermore, it is another prominent element that the arrangement for data exchange is no longer a card reader or a typical RFID, but a smartphone (Arcese et. al., 2014). The most important feature that separates NFC from RFID is that the bidirectional device transmits information (Anokwa, 2007).

In this context, two important differences between NFC and RFID can be mention.

- While RFID can only send data to the reader, NFC allows two-way communication between reader and tag.
- RFID can operate at distances up to 100 m, while NFC operates at a maximum of 10 cm.

Table 1. Comparison NFC and RFID

	NFC	RFID
Set-up time	<0,1ms	<0,1 ms
Range	Up to 10 cm	Up to 1 m
Usability	Human centric, Easy, Intuitive fast	Item centric, Easy
Selectivity	High, given security	Partly given
Use cases	Pay, get access, share, initiate service, easy set up	Item tracking
Consumer experience	Touch, wave, simply connect	Get information

Source: Dubey et al, 2011: 320.

The following table provides a quick comparison of NFC and RFID.

NFC and Quick Response Code (QR) Code

QR codes are a variation of matrix codes (Canadi et al., 2010). The Quick Response Code (QR Code), a 2D barcode type, is used to access information through a mobile phone. QR codes are used to quickly encode and decode data (Rouillard, 2008) QR codes are pixelated squares. A matrix of dots or two-dimensional barcodes connected to a specific web address, loaded with strategic information to connect to reader applications for mobile devices (Ávila et al., 2015).

When NFC and QR code technology are compared, it is required to have an NFC enabled device to use NFC technology but QR code can be used directly by all smartphones with the right application. In NFC technology, the communication mode can be passive or active, while in the QR code the communication mode is only passive (Erkiz, 2018). Although NFC offers faster, easier, more secure transactions and options, QR codes market awareness is higher than NFC. This is because QR codes are more cost-effective than NFC (Sabella, 2016). Differences between NFC and QR code show in Table 2.

NFC and Bluetooth

NFC also transmits radio waves like as Bluetooth (Lange and Steck, 2014). NFC and Bluetooth are short-range communication technologies that are incorporated into mobile phones. The main purpose of Bluetooth is to connect devices up to 10 meters in part, but some classes can communicate up to 100 meters (Berged, 2008).

NFC is set up faster than the standard Bluetooth connection, but operates at a lower speed. Besides the most distinctive feature that distinguishes NFC technology from Bluetooth is that data transfer can be initiated by triggering two devices each other (Karabulut, 2018).

NFC is easier to use than Bluetooth because Bluetooth must activate the device to set up Bluetooth communication (Ghosh, 2015). This is another advantage of NFC technology that comes from ease of use. Bluetooth brings users the need for manual installation and pairing with other smartphones. However, NFC connects two smartphones instantly and provides fast signaling. Differences between NFC and Bluetooth show in Table 3.

The difference between NFC and Bluetooth (Joan, 2012);

Table 2. Differences between NFC and QR code

NFC	QR Code
The NFC interface is integrated into almost all mobile phones.	To read the QR code, a special application is needed on the mobile phone.
NFC technology has been standardized worldwide and offers a wide range of applications in areas such as industry, logistics, marketing, automotive	The QR-code can optically downgrade the customer layout
No special application required	The QR code is susceptible to dirt which reduces the first-pass read rate significantly.
NFC is based on ISO14443 standard or ISO15693 and according to the International Transmission standard, the frequency is 13.56 MHz.	QR code data cannot be changed so that they are static and cannot be changed.
Maximum distance up to 5 cm and fast transmission establishment	Long website URLs affect a very large QR code, which leads to poor read rates.
Each NFC chip has a unique ID number worldwide. Thus, each product is unique, retractable and becomes an original.	The QR Code can be processed in two steps, first opening the scanning program and second scanning the code.
A virtual 100% first pass reading rate	The cost of QR codes is low.
NFC provides the possibility to add, read and modify data on the chip at any time.	Once created and cannot be changed again

Source: "NFC-Technology". n.d.

- NFC has a much shorter range than Bluetooth.
- Bluetooth is much faster than NFC.
- Bluetooth consumes more power than NFC.
- Bluetooth requires devices to be paired while NFC does not.
- NFC works with passive devices while Bluetooth does not.
- NFC has a different set of uses than Bluetooth.

In addition, the two technologies can work together to create a better connection between wireless devices. Combination that Bluetooth and NFC are shown in Figure 1.

Table 3. Comparison NFC and Bluetooth

	NFC	Bluetooth
Set-up time	<0,1ms	~ 6 sec
Range	Up to 10 cm	Up to 30 m
Usability	Human centric, Easy, Intuitive fast	Data centric, Medium
Selectivity	High, given security	Who are you?
Use cases	Pay, get access, share, initiate service, easy set up	Network for data exchange, headset
Consumer experience	Touch, wave, simply connect	Configuration needed

Source: Dubey, 2011: 320.

Figure 1. Combination that Bluetooth and NFC
Source: NFC Forum, 2019

NFC-enabled smartphones are also becoming increasingly important for customers' shopping journey. "Retail Everywhere" has emerged and NFC and Bluetooth complement each other to facilitate pull- and push-based customer interactions (NFC and Bluetooth, n.d.).

NFC and WIFI

Wi-Fi (Wireless Fidelity) is a communication technology that supports bidirectional data transfer, such as NFC. With Wi-Fi, a network can be created wherever needed, and many people can join the network at the same time. NFC is a technology designed for private data transmission, while Wi-Fi is a technology designed for group communication (Sabella, 2016). NFC and Wi-Fi work differently and have different functions however both belong to information transmission technology.

NFC Operating Modes

NFC technology enables data communication in four different operating modes. These are (a) reader / writer mode, (b) card emulation mode (c) peer-to-peer mode (p2p mode) and (d) Wireless charging mode ("What are the operating", n.d.).

- **Card Emulation Mode:** In this mode, the NFC compatible device assumes the function of a contactless card (López-de-Ipiña et al., 2010). So an NFC enabled mobile device acts as a contactless smart card such as debit card, credit card, access cards, loyalty card etc. and can be used to manage payment systems based on different methods like Mifare, Visa payWave, MasterCard Pay Pass or American Express Express Pay (Ruiz et al., 2016). When using the card emulation mode, the

Figure 2. NFC operating modes
Source: Erkiz,2018

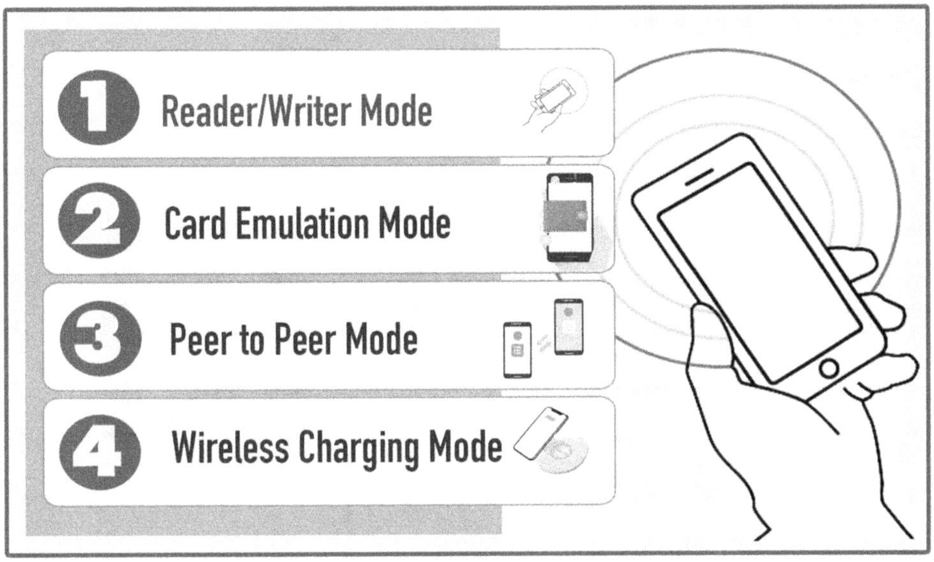

NFC enabled mobile device operates as a passive device that receives the required power from the reader's RF field (Coskun et al., 2011).

Thanks to the magnetic field created by the NFC reader (13.56 MHz.), the information on the device is transferred to the reader by touching the NFC compatible mobile device (Figure 2,3). Card emulation mode, which provides secure data communication, provides many contactless payment systems and electronic ticketing systems (Özdenizci et al., 2011). Multiple smart card applications can be stored on a mobile device (Coskun et al., 2011). Some of the applications supported by this mode are; payment applications, customer loyalty advantage applications, mobile ticketing applications, access control applications, ID / passport applications.

- **Reader/Writer Mode:** This mode enables an NFC mobile device to communicate with an NFC tag. The purpose of communication is to read or write data on a label from a mobile phone (Coskun et al., 2011). That is, this mode allows NFC devices to access data from an embedded RFID tagged object (López-de-Ipiña et al., 2010).

Reader / writer mode, which operates in accordance with the ISO14443 standard and RF (Radio Frequency) interface, starts by touching the NFC-enabled mobile device to the NFC-compliant passive tag (NFC tag). Thus, the NFC-enabled mobile device can read the data in the NFC tag, modify the data stored in the tag, or write new data to the tag (Figure 4). Smart Poster applications are one of the common examples of NFC applications that use reader / writer mode (Özdenizci et al., 2011). A typical example is to read smart posters, for example, to open a specific website. When the smartphone holds the chip placed inside the smart poster, a mobile web page opens and tickets to the event can be purchased or donated directly.

Figure 3. NFC Card Emulation Mode
Source: NFC Forum, 2019

- **Peer to Peer Mode:** In this mode, two NFC devices can exchange data from each other (Coskun et al., 2011). Peer-to-Peer mode is standardized on the ISO/IEC 18092 standard (López-de-Ipiña et al., 2010). Both mobile phones have integrated power and each uses their own energy while remaining active in this mode (Coskun et al., 2011). In this mode, there is a communication between two NFC-enabled mobile devices. This means that data, contacts and photos can be transferred from one mobile device to another ("NFC-Technology", n.d.). For example, you can share Bluetooth or WiFi link set-up parameters or you can exchange data such as virtual business cards or digital photos.

In this mode bidirectional half-duplex communication is performed in a similar manner to other modes. Therefore, device has to listen to the other while it is transmitting and can begin transmitting data after the first one is finished (Coskun et al., 2011).

- **In Wireless Charging Mode:** Wireless Charge technology is also known as charging through induction. Small IoT devices such as a Bluetooth headset, fitness tracker or smartwatch can be charged with the contactless transfer of up to 1 W of power ("What are the operating", n.d.). The NFC wireless charging technology, like other wireless charging technologies, operates based on an inductive connection between the two spiral antennas and the tuning networks that keep the

Figure 4. Reader/Writer Mod
Source: NFC Forum, 2019

Figure 5. Peer to Peer Mode
Source: NFC Forum, 2019

Figure 6. Wireless Charging mode
Source: NFC Forum, 2019

antenna circuits close to the their resonance frequencies (Strommer et al., 2012). It has become more widespread, especially with the support of many smartphone manufacturers.

NFC Communication Modes

NFC technology has two communication modes, active communication mode and passive communication mode, and these modes determine how two NFC-enabled devices communicate with each other (Chang,

2013). In NFC technology, communication begins with a signal from an initiator and is completed by the response of a target device that receives and responds to the initial signal from the initiator (Erkiz, 2018). An NFC device must have an internal power supply, such as a mobile phone, to be considered active, while a passive device does not have an internal power supply. The ability of NFC enabled devices to act as both passive and active modes makes NFC technology unique among contactless communication technologies. (Lange and Steck, 2014).

- **Active Communication Mode:** Both devices generate a radio frequency (RF) signal for transmitting data without pairing, just as in other technologies (Cerruela, 2016). The advantages of using the active mode are that it has a high transmission rate (above 1 MBit/s) and theoretically possible to work over longer distances (up to 20 cm). To avoid collisions, only the sending device emits an electromagnetic field, while the receiving device closes its field while listening (Erkiz, 2018).
- **Passive Communication Mode:** Only one of the devices generates the radio frequency field, while the passive device acts as the receptor. The second device, which is passive one (NFC Tag), uses a technique called "charge modulation" to transmit information from the active device or initiator (Cerruela, 2016). Namely, when operating in passive mode, only the initiating device (initiator) sends RF energy to the target to energize. The target responds by modulating this energy to send data to the initiator but cannot generate its own RF field (Erkiz, 2018).

NFC Standards and Standardization Organizations

Various elements such as smart cards, mobile phones, readers, labels work together in NFC technology. Standards are defined by various governing bodies to ensure accessibility, security issues and interoperability of these elements (Erkiz, 2018). The ISO / IEC, ECMA and NFC Forum offers a wide range of standards defining NFC protocols. As with many other communication technologies, these protocols are placed on top of each other (Olivieri, 2015). Standards are explained in the Figure 7 below.

Some stakeholders in the NFC ecosystem set standards or work in standard groups to contribute to the development of NFC technology. The NFC forum is the most important organization in the NFC ecosystem that contributes to NFC technology. The Near Field Communication (NFC) Forum is a non-profit industry association that supports the use of NFC short-range wireless interaction in consumer electronics, mobile devices and PCs. NFC Forum was established by Nokia, Sony and NXP on March 18, 2004 to develop and disseminate NFC technology (NFC Forum, 2019). The task of the NFC Forum is to develop specifications, to ensure interoperability between devices and services, and to increase the use of NFC technology by training the market on NFC technology (Egger, 2013). The NFC Forum has developed a very stable framework for developing applications, interoperable trouble-free solutions and secure transactions. The NFC Forum also coordinated the work of dozens of organizations by setting up committees and working groups (Arcese et. al., 2014).

Figure 7. NFC Standard
Source: NFC Everywhere, 2017

Standard	Subject	Relations with NFC
EMVCo	Payment	It provides rules for NFC systems that accept payments or act as payment cards and discusses the compatibility of level 1 interface modules
FeliCa	Contactless Smartcard	Developed by Sony, FeliCa is a contactless RFID smart card system that complies with JIS: X6319-4 and is also included as a condition for compliance with the NFC Forum specification.
GlobalPlatform	Secure Element	It specifies multiple application architecture to secure the elements used to protect operations on NFC systems.
ISO/IEC 7816	Contact smartcard	Defines a communication format that complies with NFC and ISO/ IEC 14443. Most ISO/IEC 14443 contactless cards use the ISO/IEC 7816-4 instruction set.
ISO/IEC 10373-6	Proximity Card	Defines test methods specific to proximity cards and objects.
ISO/IEC 14443	Proximity Card	It defines the most commonly used standard for objects and readers in applications such as payment, transport, identification, etc.
ISO/IEC 15693	Vicinity Card	It defines a contactless card that can be read at a distance of up to one meter compared to proximity cards. The NFC Forum Type 5 Label is based on the ISO / IEC 15693 standard and provides an expected reading range with mobile phones slightly longer than Type 2 Tags.
ISO/IEC 18000-3M3	Item-level RFID	Identifies an EPC Global Gen2 HF reader with an air interface at 13.56 MHz, the same operating frequency as NFC. Used for high stackable labels with fast bulk reading.
ISO/IEC 18092	NFC Interface and Protocol	It defines Near Field Communication and includes parts of ISO / IEC 14443 and FeliCa.
MIFARE	Contacless Smartcard	It refers to a contactless smart card format compatible with NFC and incorporates proprietary technologies based on various levels of ISO / IEC 14443 A.
MISRA-C	Automotive	It provides guides to NFC developers for C programming in automotive and was developed by the Engine Industry Software Reliability Association.
NFC Forum Specification	NFC Devices	It defines interoperability requirements in NFC applications.

MAIN FOCUS OF THE CHAPTER

NFC in Tourism

Mobile services support various situations in daily life with the spread of mobile Internet and better equipped mobile (Canadi et al., 2010). NFC technology can be considered as one of the ICT developments that has a great potential for travel and tourism (Pesonen and Horster, 2012). Although most ICT developments are not designed directly for tourism, they nonetheless have a major impact on tourism as a whole, NFC being one of them (Smith, 2015).

NFC technology also enables paperless travel, so that mobile phones can do everything tourists need at their destination, and for those traveling to specific technologically advanced places. NFC technology can reduce the number of operations a user needs to perform, thus enhancing the user's experience and facilitating adoption as well as providing a safe and easy way to share information (Dragović et al., 2018).

NFC can have a significant and positive impact on the travel and tourism industry (Egger, 2013; Pesonen and Horster, 2012). This is because NFC can provide a range of possible functions that can be applied in the tourism industry, such as mobile use, information provision, access authorization, object identification, system management, and location-based services (Egger, 2013; Smith, 2015).

There are many ways to use NFC technology in the tourism industry. One of the most widely used applications of NFC technology in tourism is mobile payment (Dragović et al., 2018). NFC tags provide information to tourists and thus tourists do not have to carry travel guides with them when visiting tourist attractions. NFC can be used at touristic attractions; for example, tourists can scan information points in museums, allowing their phones to communicate by voice and maybe even provide translations in their own language. NFC tags in cities can create tourist routes for tourists, providing information about their location and information about historical sites, connecting with mobile apps, and showing the next important point. Smart posters and smart maps can be embedded with NFC tags. Tourists scan the POI on a smart map and get information about the tourist location on their devices (Put and Smith, 2016). NFC smart posters can be viewed at train stations, and if a user moves his phone over a station name, he will instantly connect to the train website, preloaded with the current location (Smith, 2015).

Uses of NFC in the tourism sector may be listed as follows (Dragović et al., 2018):

- Making mobile payments with debit cards and credit cards or devices using smart wallet software (Android Pay, Apple Pay, Google Wallet).
- Providing information through smart posters or labels on cultural objects in museums.
- Providing access to hotel rooms using smartphones as keys.
- Free and passwordless access to Wi-Fi networks by scanning tags with the necessary information.
- Discount cards for loyal guests that can be operated by smartphone.
- Payments for city transport cards using smartphones.
- NFC-based virtual destination cards on mobile phones while maintaining the functionality of the smart destination cards.
- Providing location-based information.

Benefits of NFC in Tourism

As a smart tourism tool, NFC bridges the gap between the physical and digital worlds so that when a tourist visits a tourist attraction he or she can admire it in the physical world but also use a smartphone to learn more about the attraction digitally. As NFC offers tangible added value to the tourism sector, customers will demand it and NFC-based service delivery will be a competitive advantage (Egger, 2013). Many authors have reported on the advantages of NFC technology.

Ok et al. (2010) stated that the advantages of NFC are as follows: (a) mobile devices can be used both to store information and as NFC readers; (b) NFC technology works harmoniously with existing RFID structures, tags, and contactless smart cards; (c) it is easy to use because users do not need to know anything about the technology; (d) the short transmission range ensures safety.

NFC supports a "passive communication model," thus providing great energy-saving advantages. This means that devices do not create any RF fields and they can only complete communications when one end device is powered (Csapodi and Nagy, 2007).

According to Put and Smith (2016), NFC technology has the ability to enhance the tourism experience and increase value for both the destination and the tourist. Some of the benefits for destinations are

Figure 8. Benefits of NFC
Source: Egger, 2012

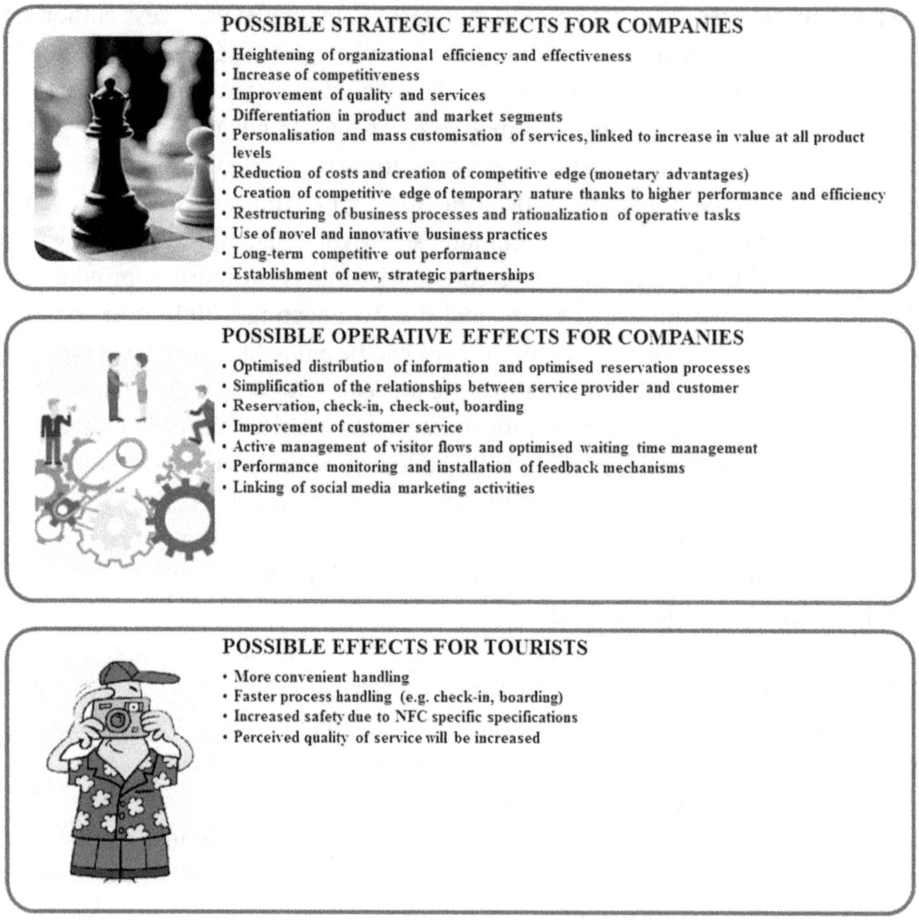

economic rewards, big data, and differentiation keys. On the other hand, they listed the benefits for tourists as including translation services, reduced expenses, navigation, and support for sustainable visitors.

The performance of NFC technology is better than traditional payment cards and mobile methods (Smith, 2015). The advantage of paying with NFC is that the transaction is done quickly and no errors are made at the terminals. At the same time, it is also very secure because each transaction generates a unique identification number (Dragović et al., 2018).

The most important strategic and operational functionalities of NFC for companies and the most important benefits for tourists are presented in the figure below (Egger, 2012).

According to some authors, there are many disadvantages and threats of NFC alongside these benefits. Kostakos and O'Neill (2007) stated that one barrier in using NFC is that users cannot easily type or read other information on their mobile devices while NFC is occurring.

Arcese et al. (2014) conducted a SWOT analysis for NFC technology and listed its strengths, weaknesses, opportunities, and threats. According to them, there are two weaknesses of NFC: first, the high investment on the POS, SIM, and device, and second, the need to reach critical mass before gaining a foothold in the market. They also stated some threats, like many competitor technologies, contactless

Figure 9. SWOT Analysis for NFC
Source: Arcese et al.,2014

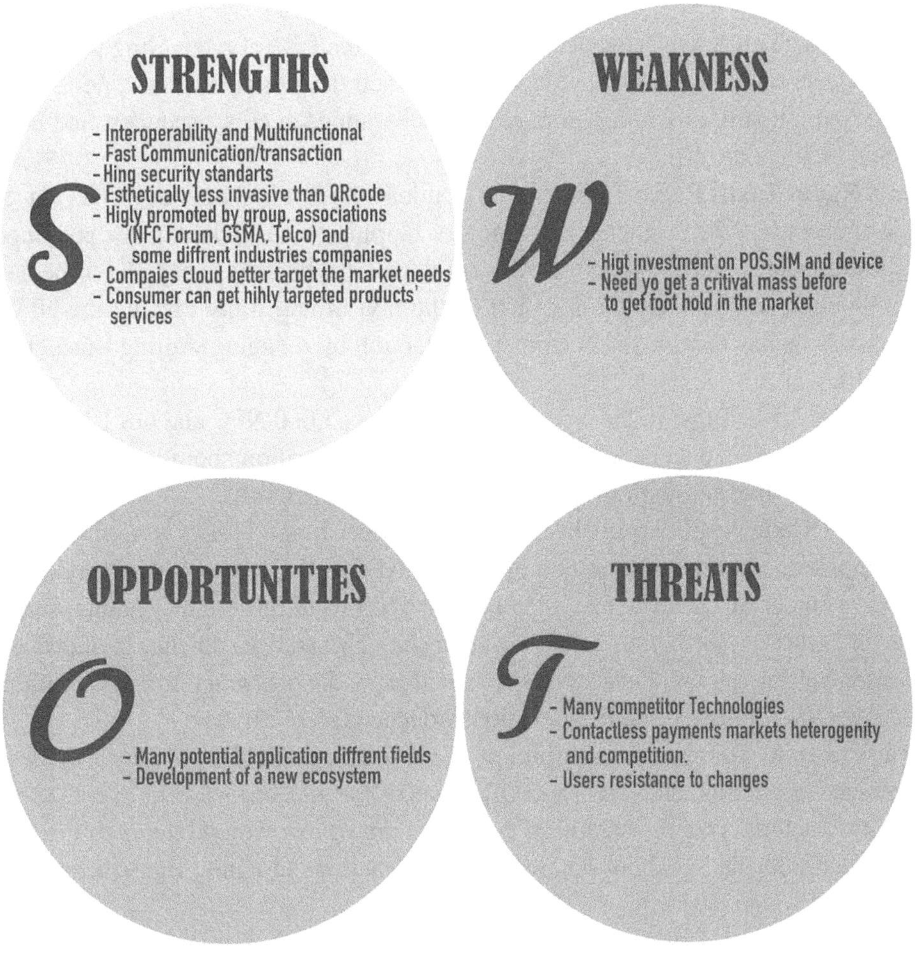

payments, market heterogeneity, and competition. According to them, finally, the customer's resistance to change can be cited as one of the main threats to the spread of NFC technology.

NFC Case Studies in Tourism

In this part of the study, NFC trials and sample applications in the tourism sector are presented. The focus of this section will be on some NFC use cases selected from the tourism sector.

- ***Scandinavian Airlines System-SAS:*** Scandinavian Airlines System-SAS has been using NFC technology since 2011. They have produced smart tags that can be used by all loyal passengers at the airport, from check-in and check-out to simple baggage checks. These tags are placed in shops, bars, doors, and many other places at the airport. During the trial period, it was revealed that customers enjoyed the product and that SAS met customers' expectations. Foreseeing the future of NFC, the company was one of the first airlines to offer frequent travelers a smart solution in the form of the SAS Smart Pass, making their travels even more trouble-free (NFC Reference Guide, 2013).

- **Japan Airlines "Tap & Go" Service:** "Tap & Go" is the ease of a boarding pass based on NFC. Using NFC technology, it allows passengers to use their devices at various "one-touch" points until they arrive on board, providing more efficient operations at entry, security, and boarding gates (NFC Reference Guide, 2013).
- **London's Radio Taxis:** Radio Taxis, one of London's largest taxi operators, has placed NFC and QR code stickers in 2,500 vehicles. Scanning or tapping these stickers gives passengers a direct link to the company's app and then provides a link that can be used to search for the nearest Radio Taxi in one click and also see the location of the taxi in real time. Placing the NFC tags in its fleet, Radio Taxis has maximized customer satisfaction by reducing waiting times for passengers (Clark, 2014a).
- **Rio de Janeiro NFC Tags:** In the city of Rio de Janeiro, 5,000 NFC and QR labels were installed at bus stops to provide local people and tourists with information about bus routes and times and information about local attractions and destinations (Clark, 2014b).
- **NFC and QR Code Applications in Ardèche, France:** In the French region of Ardèche, 5,000 hotels, campsites, and resorts were equipped with NFC and QR tags to enable tourists to access multimedia content of local interest. One hundred NFC and QR contact points were also established at the tourist information offices in Ardèche. The added contents included guided tours and treasure hunting games. This service is provided by the regional tourism development board, L'Agence de Développement Touristique de l'Ardèche (Dyer, 2013a).
- **Madhya Pradesh Tourism Development Corporation:** India's Madhya Pradesh Tourism Development Corporation uses NFC and QR codes to provide tourists with information on historical sites. By touching a tag or scanning a code, visitors who download the mobile application can obtain detailed information about popular tourist attractions in Hindi, English, Japanese, French, and Mandarin (Boden, 2013a).
- **Barcelona Contactless Project:** There are NFC and QR contact points at more than 8,000 locations in Barcelona, Spain, which provide information on local services directly to the smartphones of tourists and citizens. The Barcelona Contactless project is one of 32 Barcelona smart city projects. City hall, the bus network, bus stops, and subway entrances and exits are equipped with NFC touch points. In addition, hotels, hostels, tourist offices, and tourist sites are equipped with NFC contact points. For example, when you are in a pharmacy, you can see information about the pharmacy by touching the NFC contact point, or you can get information about surrounding pharmacies when that pharmacy is closed (Boden, 2014).
- **Sydney Hotels with NFC Touch Points:** NFC contact points have been established in Sydney, Australia, allowing tourists to get directions to local attractions from over 200 hotels, rental car companies, and information centers in the city (Dyer, 2013b).
- **Aimia Hotel, Mallorca:** Aimia Hotel in the Port de Sóller resort of Mallorca uses NFC to provide guests with information on local services at the touch of a mobile phone. A panel with an NFC tag was placed in the reception area of the hotel. With this panel, customers can learn about public transport, places to visit, and weather forecasts. In addition, maps can be downloaded with the application and information such as the hotel's Wi-Fi password and restaurant menu can be obtained (Boden, 2013b).

- **Sydney NFC and QR Code Applications:** The Government of South Wales has set up NFC and QR applications for an interactive walking tour at 37 historic sites in The Rocks, Sydney's oldest district. Tourists touch NFC tags or scan QR codes to learn about these 37 historical sites in The Rocks (Clark, 2012a).
- **Grenoble (French City) NFC System:** The French city of Grenoble uses a local NFC payment system for citizens and tourists to pay for tram tickets and ski passes (Clark, 2012a).
- **Spanish City of Caceres:** Tourist information centers in Caceres, Spain, provide Google Nexus S NFC phones to tourists to provide access to museums, promotional vouchers, and information about the city's attractions (Brown, 2011).
- **New York's LaGuardia Airport:** Passengers traveling through New York's LaGuardia Airport now benefit from a technology-efficient F&B experience that leverages the latest consumer technologies and payment options. New York's LaGuardia Airport features a 10,000-square-meter dining room at Delta Air Lines' Terminal C, offering self-service check-out from over 100 iPads and NCR, which passengers can use to browse and order from menus. Customers can also pay for contactless readers using Apple's NFC-enabled "touch and go" system (iPads, Apple Pay, 2015).
- **Clarion Hotel in Stockholm:** The Clarion Hotel was the world's first hotel to use NFC technology to change room keys, and in 2011, NFC technology started to be used in mobile key solutions. Using the RFID-based basic locking systems infrastructure, the hotel allowed customers to receive room keys on NFC-compatible smartphones, skipping check-in (or reducing the time) and going directly to their rooms. Before arriving at the hotel, a welcome message is sent, along with a reminder message to the customer to check into the room with his or her mobile phone. After check-in, the key to the hotel room is wirelessly sent to the phone and the customer is allowed to go directly to the room without checking in at the front desk (Clark, 2010a).
- **Museum of London:** Visitors to the museum can learn more about specific exhibitions, access coupons, book tickets, and connect via social media. Visitors can learn more about the various objects on display through NFC-enabled devices, buy tickets for future exhibitions, and "like" or "follow" the museum on various social media platforms (Clark, 2011).
- **China Eastern Airlines and China Southern Airlines**: China Eastern Airlines and China Southern Airlines offer NFC-enabled electronic luggage tags that allow passengers to check their luggage and then track their location on their mobile phones. The service will be available in September 2019 at the Beijing Daxing International Airport. The size of the e-tag is similar to the size of the iPhone 8 and it will be like an ID card for the suitcase. When passengers arrive at the airport, they will be able to check their suitcases by e-tags and self-service machines and scan a barcode to check their suitcase status (Clark, 2019a).
- **Böen Wines, California:** Böen Wines uses NFC labels included on all bottle caps to enable customers to easily find out where and how their wine was made, get advice on food pairings, and share their experiences on the platform and through social media (Clark, 2009b).
- **Sydney City Card Alipay:** With Alipay payments, which Chinese tourists are used to and prefer, tourists will have access to suggestions on what to see, where to shop, and how to visit (Clark, 2019c).
- **Ocean Medallion:** Tourists on Carnival Corporation cruise ships receive a wearable device with NFC and Bluetooth Low Energy (BLE). This device allows tourists to access a wide range of services, including checking in, unlocking rooms, accessing personalized entertainment, ordering food and drinks, and paying (Boden, 2017).

SOLUTIONS AND RECOMMENDATIONS

Today's tourist, who wants to travel more individually, demands more technology. The use of technology simultaneously makes tourists feel more independent and increases their experience. They do not want to lose time at the front desk during check-in. For this reason, opening doors with messages sent to NFC-compatible phones is perfectly matched to their expectations. The use of NFC tags when exploring a city step by step, visiting a museum, or passing through an airport is also among their preferred ways to travel. Although the use of NFC applications in the tourism sector is increasing day by day, many tourism enterprises do not have sufficient resources to develop their own NFC applications. In particular, airlines and chain hotels use NFC technology, while small and medium-sized enterprises are limited. These small businesses could also use more cost-effective NFC tags.

Today's tourists, who are generally willing to use technology, sometimes resist technology. In particular, making payments with NFC technology is not so common in some countries yet. NFC will break this resistance over time because it is the most reliable wireless technology.

FUTURE RESEARCH DIRECTIONS

In this chapter, NFC technology and its usage in the tourism sector have been examined. Sample applications of NFC technology in the tourism sector have been given. The use of NFC technology is increasing with each passing day in all sectors. As it provides convenience to tourists and meets the demands of the new generation of tourists, its use is increasing steadily in the tourism sector. Using a mobile phone for opening room doors, paying in restaurants, getting information about historical artifacts in a museum, and obtaining boarding passes all respond to the expectations of today's tourists and allow them to make fast transactions. In this study, cases of NFC applications in the tourism industry have been given. Cases of similar usage of NFC technology in the tourism industry are expected to increase in the future. The NFC technologies generally examined in this study can be examined separately in more detail for hotels, travel agencies, restaurants, airlines, airports, and attractions. In addition, studies could be carried out on tourist satisfaction and the expectations of tourists from NFC applications.

CONCLUSION

In this study, NFC technology and its structure, operating modes, communication modes, and usage and applications in the tourism industry have been examined from a general perspective. NFC technology promises added value, security, simple operation, and easy transmission to the user. Tourists who demand more technology are using NFC technology extensively. NFC technology also provides paperless travel, allowing mobile phones to be used for many things a tourist needs. NFC technology, which provides a safe and easy way to share information, can increase the user's experience by reducing the number of tourist transactions.

As a result, it is seen that the interest in the application of NFC technology in the tourism industry has increased with the application of NFC technologies and the increase in the success of these experiences. The increasing importance of NFC technology in the tourism industry also reveals innovative solutions. The diversity of NFC implementations in the tourism industry promises that new NFC implementations will continue to emerge in the future. In addition, the increase in the number of devices that support NFC businesses have started to allow NFC technology to be used more profitably. This heralds the introduction of NFC applications even in small and medium-sized enterprises.

REFERENCES

Anokwa, Y., Borriello, G., Pering, T., & Want, R. (2007, March). A user interaction model for NFC enabled applications. In *Proceedings of the Fifth Annual IEEE International Conference on Pervasive Computing and Communications Workshops (PerComW'07),* White Plains, NY. IEEE.10.1109/PERCOMW.2007.18

Arcese, G., Campagna, G., Flammini, S., & Martucci, O. (2014). Near field communication: Technology and market trends. *Technologies*, *2*(3), 143–163. doi:10.3390/technologies2030143

Ávila, A. L., Lancis, E., García, S., Alcantud, A., García, B., & Muñoz, N. (2015). Smart destinations report: Building the future. SEGITTUR. Retrieved from https://www.segittur.es/opencms/export/sites/segitur/.content/galerias/descargas/documentos/Report-on-smart-destinations-en.pdf

Berget, Ø. (2008). *Investigation of security features in Near-field communication (NFC).* (Unpublished Master's thesis). University of Oslo, Norway.

Bilginer, B., & Ljunggren, P. L. (2011) Near Field Communication. (Unpublished Master's thesis). Lund University, Sweden.

Boden, R. (2013a). Madhya Pradesh to offer NFC info. Retrieved from https://www.nfcworld.com/2013/04/29/323771/madhya-pradesh-to-offer-nfc-info/

Boden, R. (2013b). Spanish hotel delivers guest information via NFC. Retrieved from https://www.nfcworld.com/2013/07/24/325127/spanish-hotel-delivers-guest-information-via-nfc/

Boden, R. (2014). Barcelona equips 8,000 city locations with NFC and QR touchpoints. Retrieved from https://www.nfcworld.com/2014/12/05/333113/barcelona-equips-8000-city-locations-nfc-qr-touchpoints/

Boden, R. (2017). Carnival unveils NFC and BLE wearable for cruise ship passengers. Retrieved from https://www.nfcworld.com/2017/01/05/349260/carnival-unveils-nfc-ble-wearable-passengers-access-personalised-cruise-ship-services/

Brown, C. (2011). Spanish city to lend NFC phones to tourists. Retrieved from https://www.nfcworld.com/2011/04/17/37061/spanish-city-to-lend-nfc-phones-to-tourists/

Buhalis, D., & Law, R. (2008). Progress in information technology and tourism management: 20 years on and 10 years after the Internet—The state of eTourism research. *Tourism Management*, *29*(4), 609–623. doi:10.1016/j.tourman.2008.01.005

Canadi, M., Höpken, W., & Fuchs, M. (2010). Application of QR Codes in Online Travel Distribution. In U. Gretzel, R. Law, & M. Fuchs (Eds.), *Information and Communication Technologies in Tourism 2010* (pp. 137–148). Vienna, Austria: Springer. doi:10.1007/978-3-211-99407-8_12

Cerruela García, G., Luque Ruiz, I., & Gómez-Nieto, M. (2016). State of the art, trends and future of bluetooth low energy, near field communication and visible light communication in the development of smart cities. *Sensors (Basel), 16*(11), 1968. doi:10.339016111968 PMID:27886087

Chang, H. H. (2013). *Everyday NFC: near field communication explained.* Bellevue, WA: Coach Seattle Incorporated.

Clark, S. (2010). NFC phones replace room keys and eliminate check-in at Swedish hotel. Retrieved from https://www.nfcworld.com/2010/11/03/34886/nfc-keys-hotel-sweden/

Clark, S. (2011). Museum of London adds NFC. Retrieved from https://www.nfcworld.com/2011/08/16/39129/museum-of-london-adds-nfc/

Clark, S. (2012a). Sydney picks NFC and QR codes to guide visitors around The Rocks. Retrieved from https://www.nfcworld.com/2012/07/02/316609/sydney-picks-nfc-and-qr-codes-to-guide-visitors-around-the-rocks/

Clark, S. (2012b). Grenoble builds NFC system. Retrieved from https://www.nfcworld.com/2012/04/04/314972/grenoble-builds-nfc-system/

Clark, S. (2014a). London taxi firm promotes app downloads with NFC and QR stickers. Retrieved from https://www.nfcworld.com/2014/05/20/329236/london-taxi-firm-promotes-app-downloads-nfc-qr-stickers/

Clark, S. (2014b). Rio gets 5,000 NFC tags. Retrieved from https://www.nfcworld.com/2014/06/20/329851/rio-gets-5000-nfc-tags/

Clark, S. (2019a). Chinese airlines to offer NFC baggage tags at Beijing airport. Retrieved from https://www.nfcworld.com/2019/08/19/363939/chinese-airlines-to-offer-nfc-baggage-tags-at-beijing-airport/

Clark, S. (2019b). California winemaker to use NFC to provide buyers with a 'farm-to-table' experience. Retrieved from https://www.nfcworld.com/2019/08/07/363802/california-winemaker-to-use-nfc-to-provide-buyers-with-a-farm-to-table-experience/

Clark, S. (2019c). Tourism Australia adds visitor card to Alipay. Retrieved from https://www.nfcworld.com/2019/02/04/360292/tourism-australia-adds-visitor-card-to-alipay/

Coskun, V., Ok, K., & Ozdenizci, B. (2011). *Near field communication (NFC): From theory to practice.* UK: John Wiley & Sons.

Csapodi, M., & Nagy, A. (2007, July). New Applications for NFC Devices. Paper presented at the meeting 16tj IST Mobile &Wireless Communications Summit, IEEE, and Budapest, Hungary. 10.1109/ISTMWC.2007.4299077

Dragović, N., Stankov, U., & Vasiljević, Đ. (2018). Contactless Technology as a Factor of Tourism Industry Development-A Review of Current Practices and Future Directions. *Economic Themes, 56*(2), 179–202. doi:10.2478/ethemes-2018-0011

Dubey, A. K., Giri, M., Sahare, M., & Dubey, A. K. (2011, June). Step-up analysis and generalization approach for Trusted NFC application Development for enhancing real time use Location. In *Proceedings of the 2011 International Conference on Communication Systems and Network Technologies*, IEEE. Katra, Jammu. 10.1109/CSNT.2011.73

Dyer, K. (2013a) French region provides multimedia content to tourists via NFC. Retrieved from https://www.nfcworld.com/2013/10/15/326406/french-region-provides-multimedia-content-tourists-via-nfc/

Dyer, K. (2013b). The Informed Tourist equips Sydney hotels with NFC touchpoints. Retrieved from https://www.nfcworld.com/2013/10/07/326228/informed-tourist-equips-sydney-hotels-nfc-touchpoints/

Egger, R. (2013). The impact of near field communication on tourism. *Journal of Hospitality and Tourism Technology*, *4*(2), 119–133. doi:10.1108/JHTT-04-2012-0014

Erkiz, K. S. (2018) Yakın Alan İletişim Teknolojisinin Sağlık Alanında Kullanımı Üzerine Araştırma ve Bir Uygulama Örneği, *(NFC)* (Unpublished Master's thesis). Gazi University, Turkey.

EverywhereN. F. C. (2017). Retrieved from https://www.nxp.com/docs/en/supporting-information/BL-STI-NFC-is-everywhere-Hannibal-Su.pdf

ForumN. F. C. (2019). Retrieved from https://nfc-forum.org/

Garrido, P. C., Miraz, G. M., Ruiz, I. L., & Gómez-Nieto, M. Á. (2010). A near field communication tool for building intelligent environment using smart posters. *International Journal of Computers and Communications*, *4*(1), 9–16.

Ghosh, S., Goswami, J., Kumar, A., & Majumder, A. (2015, May). Issues in NFC as a form of contactless communication: A comprehensive survey. In *Proceedings International Conference on Smart Technologies and Management for Computing, Communication, Controls, Energy and Materials (ICSTM)*, IEEE. Chennai, India. 10.1109/ICSTM.2015.7225422

Grassie, K. (2007). Easy handling and security make NFC a success. *Card Technology Today*, *19*(10), 12–13. doi:10.1016/S0965-2590(08)70134-8

Hardy, R., Rukzio, E., Holleis, P., Broll, G., & Wagner, M. (2010, September). *MyState: Using NFC to share social and contextual information in a quick and personalized way.* Paper presented at the *12th ACM international conference adjunct papers on ubiquitous computing*, Copenhagen, Denmark iPads, Apple Pay and self-service at heart of LaGuardia Airport upgrades (2015). Retrieved from https://www.futuretravelexperience.com/2015/04/ipads-apple-pay-self-service-heart-laguardia-airport-upgrades/

Joan, B. (2012) Difference Between NFC and Bluetooth, Retrieved from http://www.differencebetween.net/technology/hardware-technology/difference-between-nfc-and-bluetooth/

Karabulut, Z. E. (2018). *Mobil sistemler üzerinde, biyometrik, NFC ve konum bilgilerini kullanarak kişi tanıma* (Unpublished Master's thesis). İstanbul Ticaret University, Turkey.

Kostakos, V., & O'Neill, E. (2007, March). NFC on mobile phones: issues, lessons and future research. Paper presented at Fifth Annual IEEE International Conference on Pervasive Computing and Communications Workshops (PerComW'07). IEEE. New York. 10.1109/PERCOMW.2007.84

Lange, P., & Steck, T. (2014). Near Field Communication-Its adoption process and technology acceptance. (Unpublished Master's thesis). Lund University, Sweden.

López-de-Ipiña, D., Díaz-de-Sarralde, I., & Zubía, J. G. (2010). An ambient assisted living platform integrating RFID data-on-tag care annotations and twitter. *Journal of Universal Computer Science, 16*(12), 1521–1538.

Neuhofer, B., Buhalis, D., & Ladkin, A. (2014). A typology of technology-enhanced tourism experiences. *International Journal of Tourism Research, 16*(4), 340–350. doi:10.1002/jtr.1958

NFC and Bluetooth: The Perfect Pair. (n.d.). Retrieved from https://nfc-forum.org/nfc-and-bluetooth/

NFC Reference Guide for Air Travel. (2013). Retrieved from http://nfc-forum.org/wp-content/uploads/2013/12/2013_10_28_NFC_Reference_Guide_for_Air_Travel_FINAL.pdf

NFC-Technology. (n.d.) Retrieved from https://www.smart-tec.com/en/auto-id-world/nfc-technology

Ok, K., Coskun, V., Aydin, M. N., & Ozdenizci, B. (2010, November). Current benefits and future directions of NFC services. In *Proceedings International Conference on Education and Management Technology*, IEEE. Cairo, Egypt. 10.1109/ICEMT.2010.5657642

Olivieri, S. J. (2015). *An Investigation of Security In Near Field Communication Systems* (Unpublished doctoral dissertation). Northeastern University, Boston, MA.

Özdenizci, B., Ok, K., Aydın, M. N., & Coşkun, V. (2011) Yakın Alan İletişimi Teknolojisi. *Türkiye Bilişim Vakfı Bilgisayar Bilimleri ve Mühendisliği Dergisi, 4*(1).

Palumbo, F., & Dominici, G. (2015) Unraveling the complexity of tourist experience with nfc technology and mobile wallets. In S. Erçetin, & S. Banerjee (Eds.), Chaos, complexity, and leadership 2013 (pp. 189-196). Springer Proceedings in Complexity. Cham, Switzerland: Springer. doi:10.1007/978-3-319-09710-7_16

Pesonen, J., & Horster, E. (2012). Near field communication technology in tourism. *Tourism Management Perspectives, 4*, 11–18. doi:10.1016/j.tmp.2012.04.001

Put, V. W., & Smith, R. (2016). Smart Tourism Tools: Linking Technology to the Touristic Resources of a City, Retrieved from www.semanticscholar.org

Rouillard, J. (2008, July). *Contextual QR codes*. Paper presented at *2008 The Third International Multi-Conference on Computing in the Global Information Technology*, IEEE. Athens, Greece.

Ruiz, I. L., García, G. C., & Gómez-Nieto, M. Á. (2016). Red Thread. An NFC Solution for Attracting Students and Engaging Customers. In *Ubiquitous Computing and Ambient Intelligence* (pp. 263–274). Cham, Switzerland: Springer. doi:10.1007/978-3-319-48799-1_31

Sabella, R. R. (2016). *NFC for Dummies*. Hoboken, NJ: John Wiley & Sons.

Smith, R. (2015). *SMART tourism tools: Linking technology with the touristic resources of city destinations* (Unpublished doctoral dissertation). NHTV Breda University of Applied Sciences, The Netherlands.

Strommer, E., Jurvansuu, M., Tuikka, T., Ylisaukko-Oja, A., Rapakko, H., & Vesterinen, J. (2012, March). NFC-enabled wireless charging. In *Proceedings 4th International Workshop on Near Field Communication, IEEE.* Helsinki, Finland. 10.1109/NFC.2012.17

Thrasher, J. (2013). RFID vs. NFC: What's the Difference? Retrieved from https://blog.atlasrfidstore.com/rfid-vs-nfc

Vitezić, V., Car, T., & Šimunić, M. (2015, April). *Managing Innovative Technology in the Hotel Industry--Response to Growing Consumer Preferences.* Paper presented at 3rd International Scientific Conference Tourism in Southern and Eastern Europe. Opatija, Croatia.

What are the operating modes of NFC devices? (n.d.). Retrieved from https://nfc-forum.org/resources/what-are-the-operating-modes-of-nfc-devices/

ADDITIONAL READING

Anokwa, Y., Borriello, G., Pering, T., & Want, R. (2007). "A user interaction model for NFC enabled applications," in *Fifth Annual IEEE Int. Conf. on Pervasive Computing and Communications Workshops*, pp. 357-361. 10.1109/PERCOMW.2007.18

Broll, G., Siorpaes, S., Rukzio, E., Paolucci, M., Hamard, J., Wagner, M., & Schmidt, A. (2007). Supporting Mobile Service Usage through Physical Mobile Interaction, *5th Annual IEEE International Conference on Pervasive Computing and Communications*, White Plains,NY, USA. 10.1109/PERCOM.2007.35

Franco, P. (2015). *Understanding bitcoin, Cryptography engineering and economics, Wiley Finance series*. Cornwall, UK: Wiley.

Kindberg, T., & Barton, J.; Morgan et al. (2002). People, Places, Things: Web Presence for the Real World, *Proc. WMCSA2000, in MONET* vol. 7, no. 5.

Madlmayr, G., Kleebauer, P., Langer, J., & Scharinger, J. (2008). Lecture Notes in Computer Science: Vol. 5183. *"Secure Communication between Web Browsers and NFC Targets by the Example of an e-Ticketing System", E-Commerce and Web Technologies* (pp. 1–10).

Pesonen, J., & Horster, E. (2012, October). Near field communication technology in tourism. *Tourism Management Perspectives*, *4*, 11–18. doi:10.1016/j.tmp.2012.04.001

KEY TERMS AND DEFINITIONS

Bluetooth: An open wireless technology standard for transmitting data from mobile phones, computers and other electronic devices over short distances.

Near-Field Communication (NFC): Standards-based short-range wireless connectivity technology that makes life easier and more convenient for consumers around the world by making it simpler to make transactions, exchange digital content, and connect electronic devices with a touch.

NFC Forum: The most important organization in the NFC ecosystem that contributes to NFC technology. NFC Forum is a non-profit industry association that supports the use of NFC short-range wireless interaction in consumer electronics, mobile devices and PCs.

Radio Frequency Identification (RFID): Used to detect and monitor all kinds of living and inanimate objects from a distance without touching them. RFID a technology used electromagnetic fields to automatically identify and track tags attached to objects so as to facilitate a contactless payment.

Quick Response Code (QR Code): A variation of matrix codes, are a 2D barcode type and are used to access information via a mobile phone.

Chapter 5
QR Code Applications in Tourism

Eda Özgül Katlav
https://orcid.org/0000-0002-4168-909X
Nevşehir Hacı Bektaş Veli University, Turkey

ABSTRACT

With the development of technology, new innovative approaches emerge day by day. With the internet and mobile devices being used more actively, many applications provide great convenience to our lives. This situation has brought about many changes and developments when it is evaluated in terms of both businesses and consumers. Enterprises are using applications that will increase customer satisfaction and loyalty to be different from competitors, adopt in the market, interiorize a sustainable policy, and make more profit. Some technological applications will save customers time and increase satisfaction and loyalty in tourism enterprises and all other businesses. One of these applications is the use of QR codes.

INTRODUCTION

The impact of technology on our lives has led to improvement of innovative tools and applications designed to facilitate daily routines. Quick Response (QR) codes are a tool widely utilized to improve customer experience in service sector, especially in tourism. Innovative technologies have considerably changed human lives including travel experiences. In tourism sector, technological developments are being utilized in order to save time for customers and to increase satisfaction and loyalty. QR code, one of these applications, is utilized by several institutions in tourism sector.

Having been common in the 2000s, QR code has begun to be utilized in many areas across the globe. Although there had been some concerns that consumers' interests in QR code wouldn't be long, a reverse situation emerged: These codes have been utilized in several parts of lives such as our beverage, menu of restaurants, notifications of customer satisfaction/complaint etc. Moreover, the biggest QR code of the world has been made by utilizing trees in China and a hotel in Dubai has been designed as QR code. Consumers' interests in these codes continue to increase. For this reason, these codes are being improved each passing day and widely used by customers.

DOI: 10.4018/978-1-7998-1989-9.ch005

These codes are used in companies such as accommodation, transportation, food and beverage etc. and also museums so as to give information or make promotion. For instance, information about local history, culture, flora and fauna belonging to destination can be given through QR code and this is its usage with informative aim. Besides, QR codes can be utilized to display more information about national parks, zoos, museums, and also local tourist attractions and exhibitions in museums. Customers in destinations can scan QR code to buy tickets at couch or train stations and in this way these codes contribute to paperless ticket system avoiding long queues. Furthermore, the codes are used to give information about calories and recipes of foods in menu of restaurants, message boards in lobby, concierge table in hotels (Emek, 2012) or to make extra discount during happy hour in activities carried out inside the hotel (Landman, 2011). QR codes have been also integrated into boarding passes of airways. In this section, there are QR code and its practices in tourism sector and also its practice areas in tourism with examples. Moreover, the focus is on making QR codes one of parts of general experience belonging to consumers who visit a destination or a tourism company and how tourism sector use them in various ways.

BACKGROUND

The most significant point of QR code which has got involved in our lives due to latest developments in mobile technology is to accelerate information flow. Instead of memorizing or noting an advertisement, an information letter etc. which are seen in a newspaper, at a mall, in a shop or while walking on the street, this information can be immediately reached, recorded, and utilized at will thanks to an easy application in a smart phone and its camera. In these regards, QR code is regarded as one of the most effective visual factors to accelerate information flow in digital world.

QR code is defined as "pictural bridges which can be basically embedded in physical environment" (Coleman, 2011) and is an innovative extension of standard barcode (Bi, Feng, Liu, & Wang, 2008). In the year 1994, QR code was developed by a Japanese company named "Denso Wave" in order to follow vehicles in production process and perform an inventory in automobile industry (Cata, Patel, & Sakaguchi, 2013; Narayanan, 2012). Without demanding any charges or imposing restriction to generate codes, the company has permitted unlimited access across the globe. Therefore, QR code is widely utilized not only in its original country, Japan, but also in several countries (Kan, Teng, & Chou, 2009). Since codes can include so much information in them and attract consumers' interest, they have begun to be utilized over time in many different areas from product tracking to entertainment or from marketing to interior product labelling (Shin et al., 2012).

As seen in Figure 1, the code is made of square black modules on a white background to code a text, URLs (Uniform Resource Locator) to reach web pages or other data.

QR code is similar to Universal Product Code (UPC) (Crompton, Lafrance, & Van 't Hooft, 2012). Because of its lines and blanks, UPC is called 1D and QR codes are called 2D. 2D barcodes have been inclusively developed to code large amount of information in several applications. Universal Product Code (UPC) has a capacity for maximum 20-digit word. However, QR Code has a capacity for a few hundred times more information with a few dozens. QR code can handle numerical and alphabetic characters, Kanji, Kana, Hiragana, symbols, binary, and control codes. Up to 7,089 characters can be coded in one symbol. As seen in Figure 2, QR Codes can store so much data in the picture in them (Crompton et al., 2012; Finžgar & Trebar, 2011).

Figure 1. Quick Respond (QR) code
Source: www.br.qr-code-generator.com/

Since QR code has information both horizontally and vertically, the same amount of data can be downsized to approximately one-tenth of a traditional barcode (See Figure 3).

It is easy (Brabazon et al., 2014) and free to generate QR code by utilizing internet browsers and smart phones. It is possible to generate codes in smart phones or on websites by downloading QR code application under the leadership of the company Denso Wave. It is quite easy to generate QR code by utilizing smart phones. In order to see the information in the generated code, it is necessary to download the application which can read the code. There are many applications developed for this purpose. The application uncovers data on QR code with help of camera and displays it by transforming it into a text which users can read. QR code can include any communication information, short text, URL belonging to any web site etc. (Niklas & Böhm, 2011).

QR codes are utilized for several purposes. They give information about product, label and price like traditional barcode system; in addition, it is possible to generate the code for video, audio, name card, e-mail, SMS etc. Companies utilize QR codes containing web site of a brand, promotion info or information about any product; in this way, they enable their brands to interact with consumers through different ways. For example, companies make window shoppers take purchasing decision with help of a small discount through QR codes and in this way they turn them into paying customers (Bayrak Meydanoglu, 2016).

Figure 2. Capacity of QR code

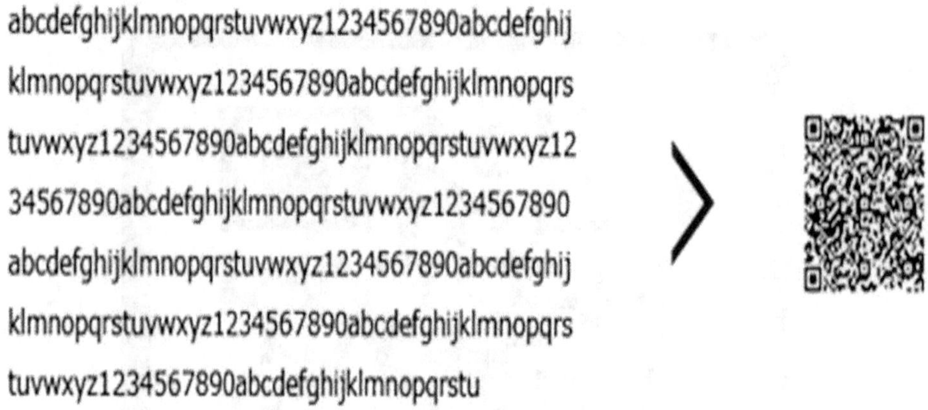

MAIN FOCUS OF THE CHAPTER

Generating QR Code

Having been popular in the last years, QR codes can be found in every part of life. They are frequently utilized in advertising boards, show windows, restaurant, hotels, transportation companies. However, there are some questions about this situation. Are these codes beginning of a new trend consumer relations? How can be utilized for the sake of customer satisfaction? QR (Quick Response) Codes are used to divert an offline consumer to an online mobile platform. Consumers can scan the 2D barcode with a mobile device, upon which they are diverted to a website landing page or access all other sorts of information, including web addresses, personal or professional contact information, or Facebook action / promotion pages (Landman, 2011).

QR code application is a simple and practical application preferred by customers to save time (Zasadna, 2014, p. 110) This preference has caused companies to utilize the application and provide convenience to their customers. QR codes can contain communication information of company, information about its web page, its promotion and its products and services, and they make it possible to reach consumers through mobile devices (Bayrak Meydanoglu, 2016, p. 291). Besides, QR codes can be utilized for every kind of aims, from invoice processing to loyalty programs, thanks to increase in their usage areas and development of opportunities for quick and simple communication and interaction (Çelik & Topsakal, 2017). It is significant to utilize and practice QR code applications in tourism companies in terms of meeting needs of modern world and adapting innovations. QR code applications increase their prefer-

Figure 3. Size of traditional barcode and QR code

ability with their simplicity and universality, their familiarity and area of usage which are enhancing day by day (Cinnioğlu & Demirdelen, 2018, p. 102)

QR codes draw customers' attention and provide them with a different experience. Nevertheless, companies need to offer consumers added value in order to utilize these codes since they ought to demonstrate a remarkable approach so that consumers show interest towards QR codes and scan them with help of application. By utilizing the application, companies should motivate consumers with help of a special offer to consumers through the approach "Get prize". However, consumers have begun to be suffocated by messages intensively sent to them in a form of either SMS or mail in recent times. For example, a hotel company could also use it as a promotional tool to offer a discount on drinks in the bar or dinner in the restaurant and offer 10%-25% off during certain hours. It is possible to say the QR Code Happy Hour has arrived (Landman, 2011).

Having been first used by Denso Wave in Japan in 1994, QR code applications are one of the most innovative methods in mobile marketing (Bamoriya, 2014). Quick Response Code (QR Code) is made of square black modules on a white background. In 2010, the company opened patent rights of QR codes to public free of charge for the sake of humanity and thus QR codes began to be utilized in various areas (Zhang et al. 2012, p. 817). The current usage of QR codes demonstrates great variety from commercial tracking systems to entertainment, from product marketing to interior product labeling (Aktaş, 2012). Being a two-dimensional system, QR code application is activated by scanning the labels on mobile devices and the scanning code is rapidly deciphered (Watson, McCarthy, & Rowley, 2013).

In accordance with wishes and expectances of consumers, QR codes with new features are being generated day by day because most people consider capacity of QR codes is inadequate and they don't seem well and attracting. Therefore, several companies are trying to generate more attracting codes (Kiryakova, Angelova, & Yordanova, 2013). The company Denso Wave has also generated different kinds of QR codes in order to meet the expectations. These codes are QR Code Model 1 and Model 2, Mictro QR Code, iQRCode, SQRC Code and Frame QR Code. These vary in information capacity, security and formal characteristics ("What is a QR Code? | QRcode.com | DENSO WAVE," n.d.)

There is a new phase in design developments: designer QR Codes (an embedded logo or QR Codes with images). This is possible due to their ability of error correction. Certain modules in QR Code can be changeable but still they are readable. Designer codes are more attracting and this is an additional motivation for users to scan them. Examples of designer QR Codes are BBC, Walt Disney, Coca Cola, Nestea, Pepsi and et al (Kiryakova et al., 2013).

QR coding is utilized by several services: production (for marking tool), advertisement, marketing, branding (logos), service and product coding, discounts, e-cards, Google Maps in tourism, e-banking. Sonic Notify, RFID labels and NFC can be named among the alternative technologies of QR coding. A license is not required for usage of QR coding. It has been defined and approved by ISO standards (Zasadna, 2014). QR codes are an innovative extension of standard barcode on almost all products in the world and besides on foods, books and tickets (Brabazon, Winter, & Gandy, 2014). QR Codes can include different kinds of data such as calendar event, communication information, e-mail address, geographical location, phone number, SMS, text, URL etc. QR Codes are scanned and read by special software – QR Reader/ Scanner. Smart phones can work with various readers/ scanners, so users can use their preferred application. There are many free applications for different mobile operating systems – QR Droid, QR Barcode Scanner, iNigma, ScanLife, AT&T Code Scanner, Scan For iPhone, QRafter, Quickmark, Nokia Barcode Reader, Google Zxing, Blackberry Messenger and others (Kiryakova et al., 2013).

Today's companies widely utilize QR code practices arousing consumers' interests in accordance with their marketing strategies. For this reason, QR codes are common in journals, on advertisement boards, cans of cereal, cans of beverage, weekly advertisement papers and in other marketing environments. A qualitative focus group research in Japan has concluded that loyal customers would like to use QR codes to reach introductory information and discounted products (Okazaki, Navarro & Campo, 2013). These codes are utilized by many different institutions and companies so as to reach more information about either an object or event or to know wishes, complaints or satisfaction of consumers (For example; museums, restaurants, hotels, transportation companies, food companies etc.) (Cata, Patel, & Sakaguchi, 2013).

There are disadvantages of QR codes. Consumers may not know how to use QR code application, there may be more than one application to scan QR code or they may scan an inadequate QR code page. Such bad experiences may discourage the consumers to use other QR code applications (Cata, Patel, & Sakaguchi, 2013; Probst, 2012). QR code technology has completely redefined potential benefits of signboards and functions. Instead of being a single product with a constant message conveyed in a language, signboard serves an interactive world as a portal, so people can reach a much more detailed content with a single sign. Tourism sector is generally regarded as an innovative action (Buhalis & Law, 2008). The fundamental characteristic of QR code system is to enable deep embedded material inside a web site to be reached through smart phone. In order to support this, it is required to generate necessary web pages and to integrate them into a web site. The most important point to take into account is that tools through which this content can be translated into one of more than 120 languages are now available, and for this reason the ability of the media to communicate with people speaking no English is magnificent. While technologies of movable electronic device keep being improved, current systems are generally adequate to integrate them into communication product platform of general visitor just like total purchase and usages of local and foreign visitors (Emek, 2012).

QR Code Usage in Companies

With the rapid development of technology, mobile technological devices (mobile phone, tablet etc.) have begun to be utilized for marketing activities. High development potential of mobile technologies provides companies with an adequate environment for their marketing activities. Therefore, companies utilize mobile marketing methods through SMS, MMS, voice, mobile banner ads, location-based marketing, mobile games and QR codes. (Kiryakova et al., 2013). QR codes have drawn attention of both consumers and companies; hence, its usage purpose has come to a different point from its initial one. Have been originally generated just for product tracking, these codes are widely utilized now in marketing activities of several different sectors. They are used in daily life not only for fliers and name cards but also as a payment system. Besides they are utilized for various business purposes at factories and in logistics operations ("What is a QR Code? | QRcode.com | DENSO WAVE," n.d.)

Companies adapt these code systems into their advertising activities in order to both draw attention of customers and demonstrate them that they are familiar with new technologies. However, there are some points to which companies pay attention while utilizing these codes. Otherwise, it is inevitable for them to be unsuccessful. Unsuccessful QR code applications can discourage customers to use them. For this reason, companies need to pay attention to the six steps demonstrated in Figure 4 in order to be successful while using QR codes.

The six steps demonstrated in Figure 1 are explained as follows (Kiryakova, Angelova, & Yordanova, 2013):

Figure 4. QR code practice steps of companies
Source: Kiryakova, Angelova, & Yordanova (2013)

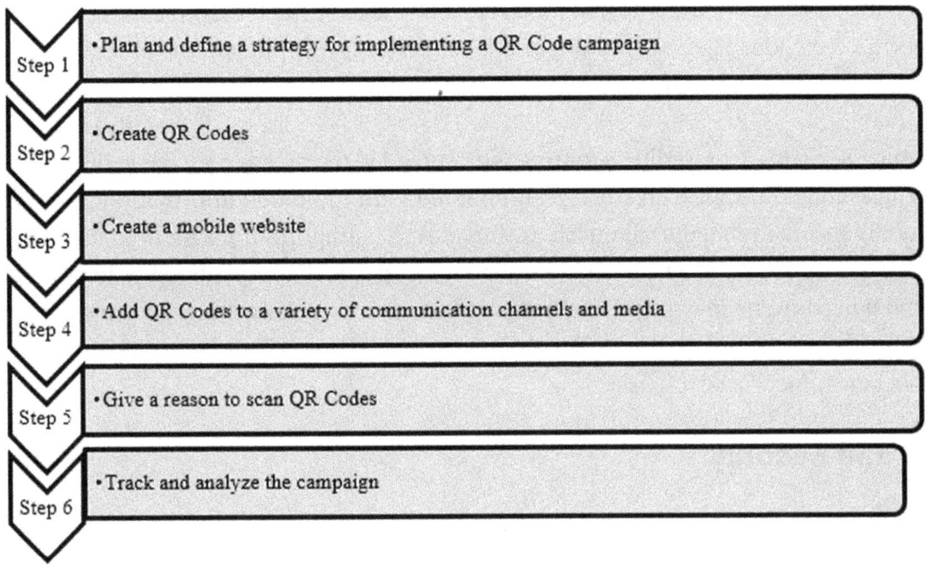

- In the first step, it is important for companies to utilize the technology and to describe the target group to be successful in the QR code activity. Companies which would like to prefer this method need to know whether the target group has mobile devices such as mobile phone/tablet or not. When companies decide to use these codes, it is necessary to plan their marketing strategies and practice it in a correct way.
- In the second one, QR codes are generated by companies. They don't need to have any information or training in order to generate the codes. They can be generated free of charge thanks to many internet pages or mobile applications. After deciding which page/mobile application to generate codes, companies need to attach importance to details such as information in the codes or its size in order to scan it rapidly and easily. Because several QR code applications generated by many companies are unsuccessful due to the fact that they have too much information.
- The third step is about internet pages to which the codes direct users. QR codes direct users to a web page after scanning. Companies ought to generate a mobile compatible version of their own web pages instead of directing directly to them since internet pages sometimes do not seem well in mobile devices.
- The fourth step is the phase in which the locations of QR codes are determined. Although the phase is quite significant, several companies skip it and those which can't practice it in a correct way become unsuccessful. For example, QR codes which are located on advertisement boards on highways are hard to be scanned and quite dangerous.
- The fifth step focuses on the fact that it is necessary to give a motivation to customers in order to encourage them to scan the codes. First customers pay attention to these codes and scan every code they see but later customers seeing them everywhere lose their attention to the codes. For this reason, it is necessary to inform them about what they gain when they scan the codes. For

example, There should be a motivation such as "Scan us on Facebook and get 30% discount for SPA" or "Scan it to get a beverage during Happy Hour"

- In the sixth and last step, companies analyze how successful the application is and how many people scan the code. They also evaluate how effective the application is at reaching the target of companies and how successful the QR code application is.

The fact that QR codes are rapidly becoming widespread in the recent period can decrease customers' attention to these codes. Because customers bombarded with too much information everywhere would like to reach only the one related to themselves. Besides, it is important for them to have devices which enable to reach this information and ability to utilize it; in addition, companies should build an appropriate system and it needs to be in working order. If companies can combine their own usage purposes with QR code application in accordance with the above-mentioned steps, they can both reach their business target and meet customers' expectations.

QR Codes and Security

Nowadays, QR codes are used for a wide range of transactions from withdrawals to check-in, making life easier. Although QR codes are widely used, there are concerns about the use of these codes. Because using QR code provides many conveniences for businesses and consumers, there are some security problems and risks associated with these codes (Bani-hani, Wahsheh, & Al-Sarhan, 2014). QR codes can be manipulated into different styles (Kieseberg et al., 2011). According to Krombholz et al., (2014) there are attacks on the malicious use of QR codes. The most common attack on social media is social engineering. Social engineers are interested in stealing people's information and data. Attackers use QR codes to redirect fraudulent websites that mask legitimate websites to steal sensitive personal information such as usernames, passwords, credit card information. Social engineers create a new QR code by completely modifying the QR codes or by modifying the QR code modules (black- white / white - black), and those who scan the code are directed to the malicious link.

Attackers can access the login passwords, contact information, photos, videos, banking accounts, email and social networking accounts of users or take full control of mobile devices and use microphones, cameras, GPS and even smartphone devices as part of their subsequent attacks. Other possible attack scenarios can be performed using malicious QR codes. Security and protection techniques are increasingly needed to overcome these security threats.

The main problem with QR codes is that they cannot be read directly by people and can only be read using certain tools (scanning devices). Kaspersky Lap first detected a malicious QR code in September 2011. In this attack, as mentioned above, when the user scanned the code, he was directed to a website and then a malicious file was downloaded to the user's device without user information. In addition, there are malicious websites in the Trojan image that contains QR codes for mobile applications and can send text messages (Bani-hani et al., 2014; Narayanan, 2012). As a result of these problems, research shows that users have concerns about interacting with QR codes.

Narayana (2012) has developed some suggestions for the safe use of QR codes:

- QR codes are difficult to understand at first glance. Therefore, it is possible to see if the code is secure by downloading an application that can provide a preview before scanning the codes.

- Legitimate QR codes never ask for personal information. Therefore, users should not fill in the forms that request personal information on the opened page after scanning the code.
- Businesses that develop QR codes should provide information about how users will use them and whether the code will redirect to a URL, phone number, or SMS at the location of the codes.
- After scanning the QR code, users should add https to the beginning of the URL without clicking the URL.
- Users should be warned not to provide credit card information.
- Businesses should prefer a color designed code rather than ordinary code. Because it will be more difficult for the attackers to imitate these codes.

QR codes are likely to be misused by attackers in the future for different purposes. QR codes are a very convenient tool for intruders to infringe users' privacy and initiate many attacks targeting smartphone devices. Therefore, there is an urgent need for common security rules and algorithms that can prevent attackers using QR codes, maximize the security of smartphones and protect users' privacy. However, awareness of both businesses and consumers using this type of code should be increased. Because, in a research, the awareness about security of the users using QR code was investigated with the survey technique and it was determined that the users had a low level of security awareness (Bani-hani et al., 2014). There are also applications developed to direct users to a secure site after QR codes have been scanned. For example, the Norton Snap QR code reader developed by Symantec Norton protects users from online threats. This application checks the security of URLs encoded in QR codes and advises users whether the website is secure (Bani-hani et al., 2014, p. 2).

QR Code Usage in Tourism Sector

In parallel with rapidly improving technology, consumers' expectations, their purchasing habits, and their wishes and needs have begun to become different (Kim, Park, & Morrison, 2008). These developments and changes have considerable affected tourism sector in which there is an intensive competition and they play a significant role in development of modern tourism. Companies in tourism sector try to carry out different practices from classical service delivery in order to meet the changing wishes or needs of consumers, offer them different experiences, increase the perceived customer value and strengthen physical evidence factors. Tourism companies benefit from innovations of technological developments while carrying out these practices. In addition, tourism companies have begun to attach importance to innovation activities in both production and marketing processes. One of these innovation activities is QR code applications which have begun to be utilized in several sectors and different areas (Cinnioğlu & Demirdelen, 2018).

QR code applications are used to increase customer experience in tourism companies. QR codes are used in hotels (more information about activities for tourists inside and outside the hotel and about main tourist attractions of the destinations), in food and beverage companies (so that customers can observe how food is cooked or how food seems before it is ordered) or travel companies, catalogues, banners (QR codes give more information about a certain offer, but in the same time they encourage customers to purchase a certain tour package with pictures and videos).

QR codes provide the opportunity to increase visitor participation in museums, hotels, destinations and archaeological sites. Benefits of using QR code in tourism sector are as follows (Çeltek, 2017; Marakos, 2015):

- To link on a website, where guests will receive additional information about hotels, museums, airplanes, trains, restaurants, destinations, etc.
- Informing tourists about upcoming events.
- Get information about guest expectations and feedback from social media pages (Instagram, Facebook, Twitter).
- Direct connection to obtain tickets or information about airplanes, trains, buses or events,
- To use it as a part of an interactive game for tourist.
- Provides access to videos on cooking stages and recipes of dishes cooked by celebrity chefs in restaurants.
- To reach the videos about the cooking stages and recipes of the dishes cooked in the restaurants,
- Enroll guests in loyalty programs and enable them to connect faster.
- Enables them to quickly communicate customer satisfaction and complaints to the company,
- Linking with a tour guide in museums where tourists can watch videos, audio files, maps, etc.

At the same time the usage aim of QR codes in companies is to offer new experiences to tourists with smart phone (Firoiu & Croitoru, 2013). This code, whose usage has become widespread, provides a rapid and simple interaction between tourists and destination. Information boards and promotional materials utilize QR codes to ease invoicing processes and to enrich loyalty campaigns. The codes provide an opportunity for multilingual introduction of historical and natural attractions belonging to the destinations (Çelik &Topsakal, 2017). Customers' awareness about QR code applications should be raised before they can use them. Their attentions should be drawn by making such announcements: "Scan this Code for our Latest Restaurant, Shopping and Tourist Tips. QR code practices in tourism sector are carried out in many various areas. The codes are widely utilized in accommodation companies, food and beverage companies, transportation companies and museums, especially in Destination Management Organizations.

QR Code Usage in Destinations

Technological developments can be observed in all adequate areas in tourism sector. Before stating explanation of tourism companies about QR code practices, it is necessary to touch on usage of technology and technological devices in destinations because destinations which are willing to gain competitive advantage in tourism, to enhance customer experiences demonstrate progress in parallel with technology. In addition to destinations, companies in these destinations support these practices. The fact that destinations have become in harmony with technology has led to the concept of The Smart Tourism Destinations (STD). With technology being embedded on all organizations and entities in these destinations, STD enrich tourist experiences. By applying smartness concept to address tourists' needs before, during and after their trip, destinations can increase their competitiveness level. In STD, Many applications about augmented reality, electronical travel guide, customer complaint/recommendation are utilized (Çelik & Topsakal, 2017).

QR code practices in STD are mentioned in this part. QR codes in destinations are utilized for introduction and giving information about activities in destinations to tourists. They prefer them to reach information about historical and natural attractions in that area through mobile devices. Besides, it is possible with help of these codes to offer information in many languages. Tourists are informed about the destinations by putting these codes in bus stops, buses, libraries, and at the city square in destinations. Several countries attach importance to QR code practice, and projects and practices are launched

Figure 5. The "Thailand Scan Me" project
Source: (www.mots.go.th, 2019)

within the scope of tourism policies. In this regard, The Ministry of Tourism and Sports in Thailand launched a project called "Thailand Scan me" in 2016. In the project, QR codes were placed in more than 300 points throughout the country to provide local and foreign tourists with information about the tourist attractions. Being in charge of the project, the Minister of Tourism and Sports has stated that they aim at giving information about touristic attractions, restaurants, hotels, shopping malls in his country. Tourists coming to Thailand reach the necessary information about the country by scanning codes with help of devices such as mobile phone or tablet (www.mots.go.th).

QR codes are used in order for introduction, giving information etc. in many famous destinations in the world such as Macau, New York, Macau and they offer an excellent tourist experience. Such practices can make positive contribution to destination image in terms of intent of recommendation, tourist satisfaction etc. There will likely be many various practices in coming years with development of QR codes.

QR Code Application in Accommodation Companies

Since today's advancement and development in technology is much faster compared to past, Companies follow these changes in order to keep up with the era, gain competitive advantage and provide customer satisfaction. Buhalis and Law (2008, p.609) have stated that tourism companies make progress in parallel with technology. Accommodation companies, a building stone of tourism sectors, have been rapidly adapting to shifts and developments in technology. The impact of technology on accommodation companies can be demonstrated as follows: Key sistem was once utilized to open hotel rooms but as a result of technology advancements, companies have begun to use card system while opening the rooms. However, the fact that cards in this system can't sometimes open rooms can be a problem for guests. On the other hand, accommodation companies put QR codes on these cards in these days so they can eliminate the risk of being unread.

Accommodation companies benefit from QR codes for various purposes. The codes are utilized to give information to customers and increase intra-company sales. Therefore, it is possible to observe the codes in elevators, on concierge tables and information boards. Companies use QR codes to give customers information about the hotel, in which restaurant they have dinner or when they can utilize pool. After customer scan QR codes in the place where they would like to take information, they are directed to hotel web pages or the related information and so they can get information about activities in hotel,

Figure 6. QR code welcome message
Source: Rizwan, 2017

restaurant and menu, activity recommendation etc. It is possible to give information outside the hotel just like inside it by using these codes. Customers can get information places to see in destination/province where the hotel is placed, local restaurants and regional activities (Emek, 2012, p. 5).

On the other hand, QR codes are also utilized to convey customer complaints, requests and wishes. Instead of giving an URL to customers on whom customers evaluate the hotel and the service by writing their complaints and wishes, hotel companies offer QR codes which customers can reach much quicker. Scanning this code through their mobile phones or tablets, customers are directed to the related site and give answers to questions through which they can state their satisfaction/dissatisfaction. For example, San Sebastian Hotel de Londres, The Signature Hotel in Saudi Arabia and HTEL Serviced Apartments in Amsterdam, which all utilize QR code for customer satisfaction, have practiced it and received many positive feedbacks from their customers (Landman, 2011).

Figure 7. Capsule hotel in China
Source: http://www.chinadaily.com.cn/china/2017-07/11/content_30068314.htm

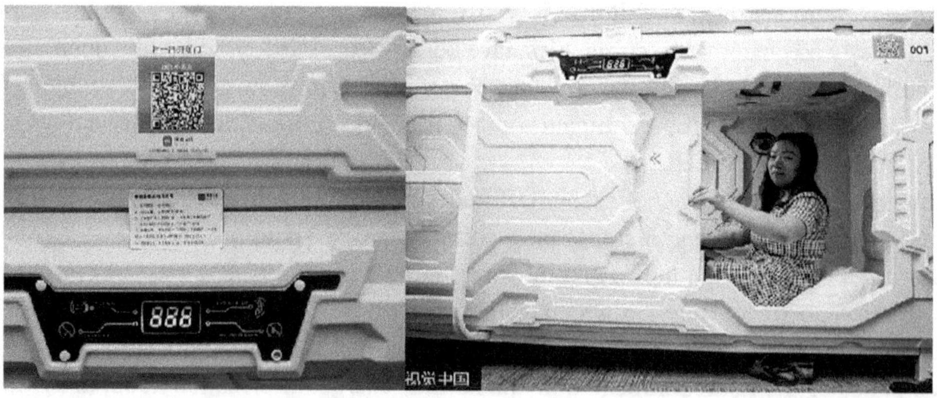

Another example The Westin Bayshore Hotel, Canada, affiliated with the Marriott Hotel, puts a fancy plate of cookies in the rooms to welcome guests. The cookies on this plate contain information about the QR Code, Facebook and Twitter, and guests receive a personalized welcome message (Rizwan, 2017). When the hotel guest scans the QR code, the URL of the hotel page opens and the guests are directed to the hotel's main web page. Thus, guests will be informed about the hotel activities and the hotel.

QR codes are used to pay accommodation businesses. For example QR codes are used to rent capsule hotels, which are used in China for 30 minutes mid-day nap. Customers who scan the codes on the capsule can stay in the capsule by paying (www.chinadaily.com.cn, 2017)

QR codes have different usage areas in accommodation businesses. It is possible to come across a building or a hotel room which is designed as a QR code. A hotel room designed by Dutch Antonie Peters at Modez Hotel in the Netherlands is completely covered with QR codes from wall to pillow to curtain to nightstand (Figure 8). In addition, each QR code in the room contains different information (www.scandit.com, 2019) .

QR Code Practice in Food and Beverage Companies

Qr code application was first utilized by certain innovative pioneer restaurants in developed countries and then other companies began to include the code in their promotion activities and business processes. These companies benefit from the technology to accelerate their business processes, to meet customers' expectations in a rapid way and to give information about their business.

Food and beverage companies are the top users of QR code practice among tourism companies. The companies considerably decrease durations of ordering and delivering it to restaurant thanks to the practice and they reduce number of employees by half. In order to use this system, companies first transform their menus into e-menus and then put a link or application in these e-menus, so they enable customers to reach menus on their table in a short time. Further to that, advantages like detailed information about foods-beverages in the menu, nutritional value, calories and recipes of them increase usage of QR codes in food and beverage companies (Emek, 2012, p. 5). For example, an internet site performing service with the name E-menu (intelligent and cloud restaurant) adapt menus of companies to this system. Companies transfer menus of food and beverage companies into electronic media and they use QR codes in processes from the phase of ordering to the that of payment (http://emenu.cc, 2018).

It has been determined in researches about QR code usage in food and beverage companies that these codes are utilized in order to provide customer satisfaction, to offer customers more different and quality service, and create the difference compared to their rival companies. Besides, according to research results, practice areas of these codes are generally menus and web sites of restaurants (Cinnioğlu & Demirdelen, 2018, p. 107).

Having 14 luxurious hotel chains in places like England, London, Manchester and Heatrow Airport, Radisson Edwardian uses QR code to increase interest in their restaurants. The hotel put a QR code on a part of their menu cards in their hotel restaurants. Customers of the hotel go to a web page by scanning the code and reach a certain content about restaurant or product on Radisson web site (May, 2011; www.radissonbluedwardian.com, n.d.)

With help of QR codes in menu, customers can be directed to a page of the site including instructions of the chef who made the design of food, food recipe and video show, food of the month or chef's choice. Radisson hotel manager has stated that brand, restaurant and food items have "promotion priority"; for this reason, integration of QR codes have become "a powerful tool" to demonstrate products to guests (May, 2011).

Figure 8. Design hotel room with QR codes
Source: https://www.scandit.com/the-qr-code-hotel-room

Figure 9. QR Code Practice in e-menu of food and beverage companies
Source: http://emenu.cc/desktop/en/weixindiancan

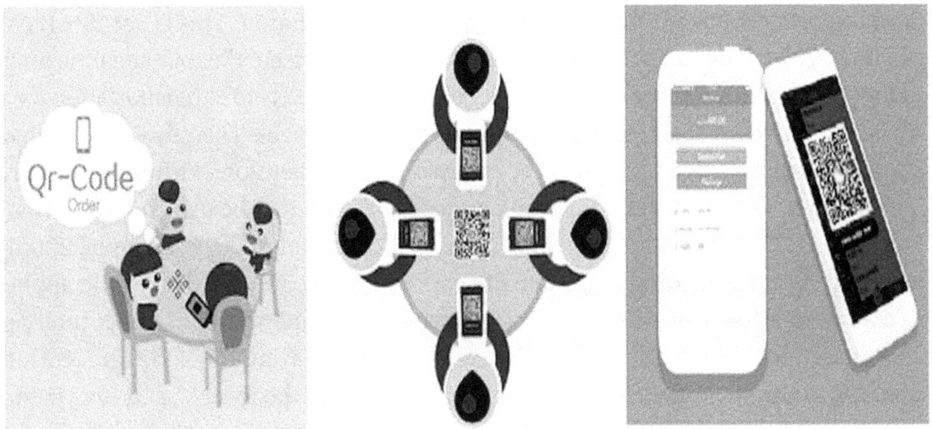

There are advantages and disadvantages of QR code usages in food and beverage companies. Advantages of QR code practice can be stated as follows (Mohammad & Azam, 2015, p. 8): companies don't have to employ too many people and they enable employees to save time. They also provide customers with a different electronic menu which is suited for their expectations and wishes. Customers don't feel under time pressure to order and it is possible to decrease order mistakes. They enable companies to make menu and price changes in an easier way. The possibility to pay electronically can prevent mistakes and carry out payments. As a result, it provides guests with a different experience. Disadvantages of the practice is as follows: building this system can be costly. It requires internet usage inside the company. Customers may have security problems and the most important one is that it may lead to troubles in communicating with customer in person (Mohammad & Azam, 2015, p. 8).

Figure 10. QR Code Practice in Radisson Hotel
Source https://www.radissonbluedwardian.com/feature.do?feature=press_qrcodes2011

QR Code Application in Transportation Companies

One of the most important elements of travel for tourists is transportation. This is because the transportation activity in tourism includes starting from the arrival of the tourist to the tour and returning (Sorupia, 2005). Immediately after a tourist decides to travel, he / she buys tickets to a transportation means (plane, ship, train, bus, etc.) that can reach the destination and plans his travel. Therefore, planning transportation activities takes tourists time and makes them more difficult. Therefore, airline, bus, train and so on. m-tiketting / e-tiketting, QR code and NFC applications have started to be used in public transportation vehicles (Eken & Sayar, 2014; Guirao, Eugenia López, & Comendador, 2015; Sorupia, 2005). By using the QR codes in transportation companies, tourists can scan the code with the mobile application during the travel and have all kinds of information (departure and landing times, travel route, price, etc.) and can make transactions (purchase, cancellation, etc.) (Eken & Sayar, 2014; Finžgar & Trebar, 2011). For this reason, QR codes facilitate the transportation activities of the tourists and save them time. It is possible to see more examples of the use of QR codes in airlines that are pioneers in using innovations in transportation companies. The way QR codes are used by other transport companies is similar to that of airline companies. For this reason, in this section, examples of airline businesses are discussed.

Today, several airline companies utilize QR code for promotion and facilitating boarding processes. Airline companies include the codes in their processes in order to save time for customers (Eken & Sayar, 2014). Customers prefer to use these systems which facilitate their pre-travel processes and save time. Today, these codes are utilized by several airline companies for processes such as check-in and updating reservation information. In order to do that, customers first make check-in online by utilizing devices such as mobile devices/computers/tablets and generate e-ticket. QR code involving passenger and flight information on e-ticket is created and sent to e-mail address of customers. When customers come to boarding area or in security areas, they pass them without any problems by scanning the QR code to scanners (Emek, 2012). Besides, airline companies utilize the codes for various promotions and campaigns. For example, Turkish Airlines launched a campaign by using QR code application in Olympic Games in 2012. The advertisement designed by Turkish Airlines can be scan in Figure 11.

Turkish Airlines has transformed national flags into QR codes and then placed them on digital bus shelters all over London to engage Olympic fans. Besides, Turkish Airlines has also added a game o these codes. Those who find and scan the QR Code with their mobile devices automatically check-in to that flag and get the chance to win a ticket to one of Turkish Airlines' 200 destinations. Different countries represented in Olympic Games have been associated with destinations of Turkish Airlines. Even if every person participating in this doesn't win a ticket, at least he or she gets information about destinations of Turkish Airlines flights (www.airlinetrends.com).

Another airline company using QR code application is Jet Airways Airlines. Jet Airways, one of Indian airline companies, benefits from QR code application for its in-plane journal Jetwings, name card, press advertisements and various ones of its communication channels including its own Facebook page. The company would like to share information about its services, the latest updates of products and content with customers through QR codes. Another airline company named CityJet compete with its rivals in social media by utilizing Quick Response (QR) codes. CityJet Airlines encourages passengers to find and scan 18 QR code hidden in airports, Facebook and Twitter channels. CityJet directs them to Facebook application of Cityjet in which they can register themselves for a chance to gain a range of prizes and to get clues about where the next is. The winner of this competition is the one who has gained virtual kilometers. Airline customers have found the application very successful (Fox, 2012).

Figure 11. QR code practice of Turkish Airlines

QR Code Application in Museums

There is many information in museum or art gallery. It is very important that these are readable and attractable. QR code, one of the tools which can realize it, is an application which attracts visitors' attention. With help of QR codes, it is possible to make a detailed explanation about art pieces and objects in museums and art galleries (Emek, 2012). Therefore, QR codes are utilized in museums, art galleries and outdoor museums for giving information, education and attracting attention. Museum managements prefer to use QR Codes as a tool to attract new target groups of visitors, mainly young people and feedbacks of these applications are positive. According to result of a survey carried out around the globe to evaluate web pages for Museums and Web Conference in 2016, physical museum visits, number of web site visitors and visitors of social media have increased 48%, 50% and 82,3% respectively, together with developments in information and communication technologies (Solima & Izzo, 2018, p. 3).

QR codes are used to attract visitors in museums and archaeological sites (Marakos, 2015, p. 62). Same time in order to increase the interaction with visitors, museums utilize QR code or augmented reality applications (Emek, 2012). The applications build a link among pieces through games and small quizzes inside museums and make museum visit more attracting (Kiryakova et al., 2013). For example, QR code application is utilized in Serbian National Museum "MUZZEUM". In the museum, a project was carried out in order to put an application about augmented reality into practice and augmented reality was combined with QR code. Within this context, a digital library was generated and art pieces in the museum were transferred to the digital library. In the digital store, pictures, videos and audios of art pieces can be found (Jevremovic & Petrovski, 2012).

Figure 12. MUZZEUM application – ARQR code with 3D object overlay
Source: Jevremovic & Petrovski, 2012

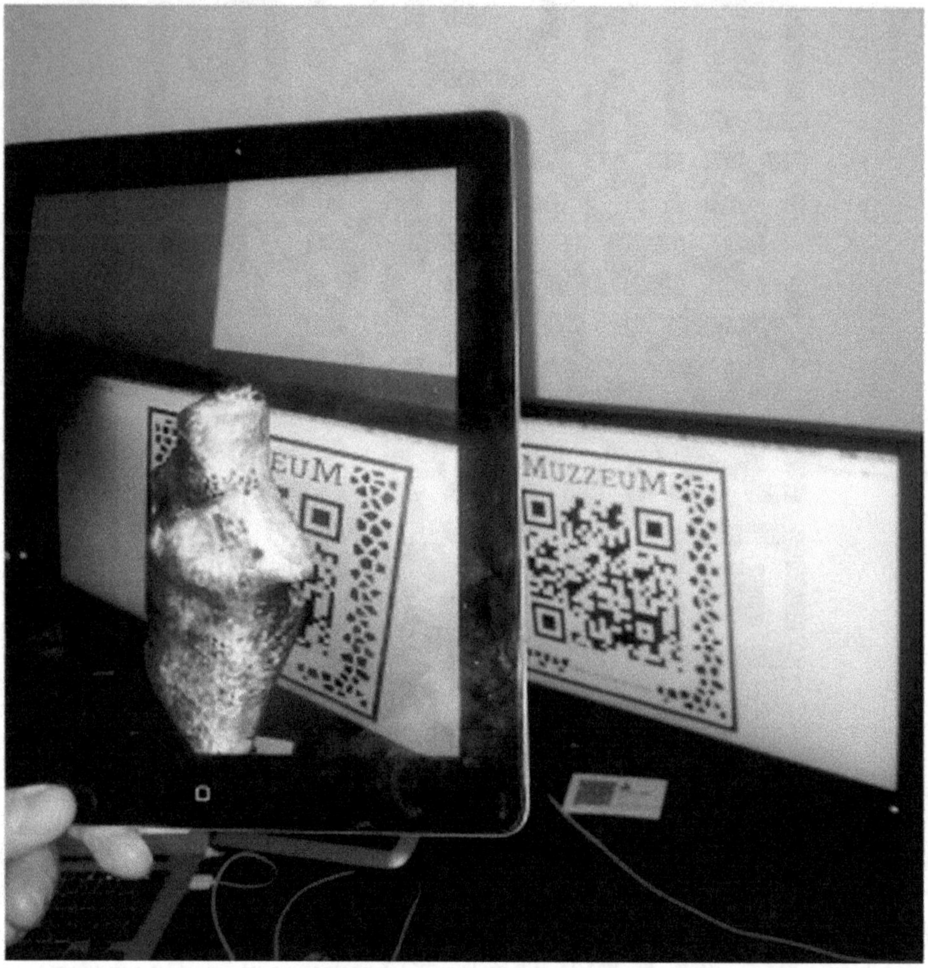

ARQR, a combination of QR code and augmented reality (AR) was generated by creating a data base. The visuality of augmented reality in museum (ARQR) was scanned by utilizing mobile vehicles. In order to prevent robbery of significant pieces in museum, art pieces are stored in a physically remote storehouse. As seen Picture X, when scanning QR codes places in the museum, the 3D visuality of augmented reality is observed on mobile devices of visitors (Ishak et al., 2014, p. 36; Jevremovic & Petrovski, 2012).

In Cleveland Art Museum and Archeology Museum in Bologna, QR codes are utilized to try audio tours of their collections. QR codes are put in their art advertisements which are poster size and these posters are found in eleven completely different locations in Cleveland. The smaller ones have been utilized in different and various public areas. QR code directs visitors to web page of museum and art galleries, increases internet traffic and enables smart phone to perceive tours (İlhan & Çeltek, 2016).

Mercedez-Benz Museum also uses the QR code to inform visitors. A brochure published by the museum contains the QR code and visitors are directed to the museum's home page by scanning it (www.mercedes-benz.com, n.d.).

Figure 13. QR Code in Mercedenz Benz Museum
Source:https://www.mercedes-benz.com/content/dam/brandhub/mercedes-benz/classic/generalimages/MBMUS_Information_2018_ENG_Update_ICv2_NETTO_x3.pdf

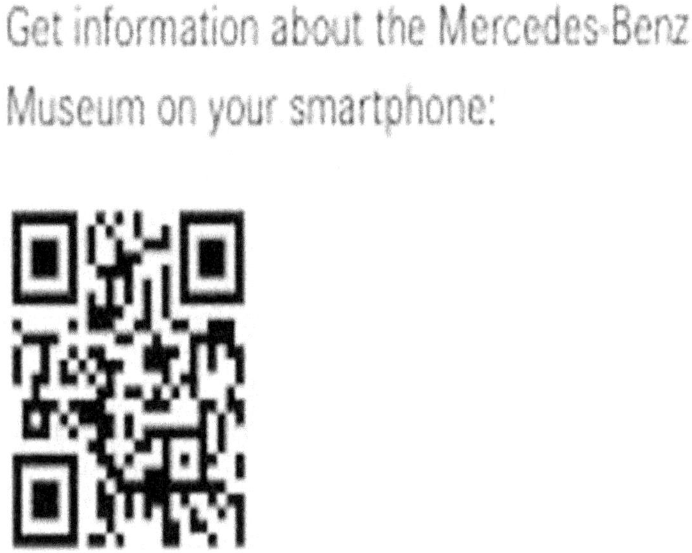

The Wolfsoniana Modern Art Museum in Genoa has implemented a pilot application for intelligent systems. The purpose of this application is to provide visitors with a new and immersive experience when visiting the exhibition. In application, through the automated identification technologies such as Near Field Communication (NFC) and QR codes, visitors can interact with their surroundings through their own devices, such as smartphones or tablets. With this application, it is aimed to provide information to visitors according to their language, age, device preferences for the works of art in the museum.

As a result, QR codes are utilized for the purposes such as giving information inside museum and increasing interests in museums outside museum. Thanks to usage of technological devices, visitors get

Figure 14. Some screenshots of the Wolfsoniana App illustrating the NFC mode
Source: Caridi, Coccoli, & Volpi, 2013

informed about museums, art galleries etc., get detailed information about art pieces by scanning the related QR code while visiting museum, have additional information about period of pieces. Besides, applications such as video and audio increase memorability of pieces.

SOLUTIONS AND RECOMMENDATIONS

It is possible to come across many applications related to the use of QR codes in the tourism sector. Codes are used especially in marketing activities within the tourism sector. The most important reason for the use of QR code in the tourism sector is to meet changing customer expectations, to provide convenience in service, to eliminate service failures (especially in restaurants) and to improve service quality. Although using QR code has provided some convenience for businesses, it is thought that it will cause some problems in the future. For example, especially in restaurants, the QR code shortens the work processes to be carried out by the employees, consumers can order and leave without any personnel assistance. However, as a result of such an application, tourism businesses may reduce the number of personnel working in the service or do not employ any service staff at all. In such a case, the feedback received from the customer through the business personnel may be reduced and this practice, which is thought to provide customer satisfaction, may create customer dissatisfaction. In order to avoid this situation, enterprises can employ a few staff members who will warmly welcome customers and explain how the system is used and frequently receive feedback from customers about this application.

Some criticisms of the use of QR codes are that consumers will be less interested in these codes if they are exposed to these codes. Tourism businesses should refer to these codes for situations and business processes that really save customers time. Because consumers are willing to scan QR codes in the beginning, they may become insensitive to scanning the codes in time. In order to remove this problem, tourism businesses should carefully select the information they will place in the QR codes and give the customers clues as to what information they will access when they scan these codes.

Although there has been no significant security gap regarding QR codes, some attackers have abused them. Attackers place a different URL in the QR code and can obtain customers' personal information using the business name. As a result, there is a suspicion about scanning QR codes in customers. Tourism businesses should state that they do not require personal information and credit card information from their customers and should warn them never to pay through these codes. In addition, businesses can provide customers with protection from malicious applications by recommending a reliable QR code application for android and ios operating systems.

As a result, businesses should inform their customers about how to use them before using QR code applications. They should give little clues about the advantages that these codes will give them and encourage them to scan the codes. Although there are some views that the intense interest in these codes will diminish over time, codes are constantly differentiated, new designs are produced and measures for security of codes are increased.

FUTURE RESEARCH DIRECTIONS

The use of QR code in the tourism sector has become widespread. As mentioned above, businesses use these codes for information and publicity in accommodation, food and beverage, transportation businesses, museums and galleries and tourism destinations. In the future, many studies can be conducted on the use of QR codes in the tourism sector for both businesses and consumers. Information about the advantages and disadvantages of using QR code is very limited. Therefore, in the future researches, the advantages / disadvantages of using these codes can be investigated first. However, it may be another research topic whether businesses achieve their intended use of these codes. In addition, consumers' approaches to using codes, their willingness, the benefits or damages provided by the codes can be investigated. Studies on the effect of demographic characteristics such as gender, cultural difference, age may also be conducted in the use of QR codes.

CONCLUSION

As a result of developments in mobile technology, QR code which is an improved version of traditional barcodes has found usage area in different sectors and companies because QR codes offer several opportunities in terms of companies and consumers. While companies utilize the application in order to provide customer satisfaction and loyalty, to gain acceptance in the market, customers have adopted QR code applications in order to save time. Tourism sector follows the developments in the worlds and is quite successful to adopt them. The facts that preferences and expectations of tourists are changing and competition factors are becoming tough cause companies to use new applications. Both destinations and tourism companies offer customers different experiences to receive a more share from tourism. Instead of directly giving information to consumers, companies use promotion tools, through which they include their customers in processes and offer them different experiences, as a marketing strategy. Besides, saving time while having experiences attracts also consumers' interest. QR codes can be utilized in every area, from in the beginning of tourist's travel to in purchasing flight ticket and commenting about destination or company to which tourists go until the end of their travel. Just as these codes can give information about historical and cultural touristic attractions, restaurants, hotels, they are utilized to save time, provide customer satisfaction and loyalty in tourism companies such as hotel. If destinations and companies in tourism sector popularize these applications and include different technological devices in their processes, this enables companies to have advantage to meet customers' expectations and to come to the fore in competition.

REFERENCES

Aktaş, C. (2012). Medya Yakınsaması: Hızlı Yanıt Veren Kod Aracılığıyla Geleneksel Gazetenin, Çevrimiçi Gazete ile Artan Rekabet Potansiyeli Üzerine Bir Tartışma. *International New Media Conference*, 118–128.

Bamoriya, H. (2014). QR Code Based Marketing in Indıa and Japan. *Singidunum Journal of Applied Sciences, 11*(2), 20–29. doi:10.5937jas11-5504

Bani-hani, R. M., Wahsheh, Y. A., & Al-Sarhan, M. B. (2014). Secure QR Code System. In *Proceedings 2014 10th International Conference on Innovations in Information Technology (IIT)*. Doi:, 1–6.10.1109/Innovations.2014.6985772

Bayrak Meydanoglu, E. (2016). *Türk Üniversite Öğrencilerinin Kare Kod Reklamciliğa Karşi Tutumlari Attitudes Of Turkish University Students Towards QR Code Advertising. Journal of Yasar University* (Vol. 11). Retrieved from https://dergipark.org.tr/download/article-file/343663

Bi, L., Feng, Z., Liu, M., & Wang, W. (2008). Design and Implementation of the Airline Luggage Inspection System Base on Link Structure of QR Code. In *Proceedings 2008 International Symposium on Electronic Commerce and Security* (pp. 527–530). IEEE. 10.1109/ISECS.2008.200

Brabazon, T., Winter, M., & Gandy, B. (2014). *Digital Wine: How QR Codes Facilitate New Markets for Small Wine Industries*. Springer. doi:10.1007/978-981-287-059-9

Buhalis, D., & Law, R. (2008). Progress in information technology and tourism management: 20 years on and 10 years after the Internet—The state of eTourism research. *Tourism Management, 29*(4), 609–623. doi:10.1016/j.tourman.2008.01.005

Cata, T., Patel, P. S., & Sakaguchi, T. (2013). Article ID 813339, 7 pages. *Cancer, 2013*. doi:10.5171/2013

Çelik, P., & Topsakal, Y. (2017). Akıllı Turizm Destinasyonları: Antalya Destinasyonunun Akıllı Turizm Uygulamalarının İncelenmesi Smart Tourism Destinations: Review of Smart Tourism Applications of Antalya Destination. *Journal of Travel and Hospitality Management, 14*(3), 149–166. Retrieved from http://dergipark.gov.tr/download/article-file/389164

Çelik, P., & Topsakal, Y. (2017). *Akıllı Turizm Destinasyonları: Antalya Destinasyonunun Akıllı Turizm Uygulamalarının İncelenmesi Smart Tourism Destinations: Review of Smart Tourism Applications of Antalya Destination. Journal of Travel and Hospitality Management* (Vol. 14). Retrieved from https://dergipark.org.tr/download/article-file/389164

Çeltek, E. (2017). QR Code Advertisements in Tourism Marketing. In In Narrative Advertising Models and Conceptualization in the Digital Age (pp. 269–289). Hershey, PA: IGI Global. doi:10.4018/978-1-5225-2373-4.ch015

Cinnioğlu, D., & Demirdelen, H. (2018). Kare Kodların (QR Code) Restoran İşletmelerinde Kullanımının Belirlenmesi: İstanbul Örneği (pp. 100–108).

Coleman, J. (2011). QR Codes: What Are They and Why Should You Care? *Kansas Library Association College and University Libraries Section Proceedings, 1*(1), 16–23. doi:10.4148/culs.v1i0.1355

Crompton, H., Lafrance, J., & Van 't Hooft, M. (2012). QR Codes 101. *Learning and Leading with Technology*, 22–25. Retrieved from www.sparqcode.com/static/maestro

Eken, S., & Sayar, A. (2014). A smart bus tracking system based on location-aware services and QR codes. *INISTA 2014 - IEEE International Symposium on Innovations in Intelligent Systems and Applications, Proceedings*, 299–303. IEEE. 10.1109/INISTA.2014.6873634

Emek, M. (2012). *Usage of QR Code in Tourism Industry. In 6th World Conference for Graduate Research in Tourism*. Fethiye, Turkey: Hospitality and Leisure; Retrieved from http://hdl.handle.net/11376/108

Finžgar, L., & Trebar, M. (2011). Use of NFC and QR code identification in an electronic ticket system for public transport. In *Proceedings SoftCOM 2011, 19th International Conference on Software, Telecommunications, and Computer Networks*, (iii), 1–6.

Firoiu, D., & Croitoru, A.-G. (2013). Tourism and Tourism Infrastructure From the Perspective of Technological Changes. *Romanian Economic Business Review*, *8*(2), 93–103. Retrieved from http://search.proquest.com/docview/1448007631?accountid=35465%5Cnhttp://sfx.dbc.dk/dbc-45DBC_UCN?url_ver=Z39.88-2004&rft_val_fmt=info:ofi/fmt:kev:mtx:journal&genre=article&sid=ProQ:ProQ%3Apqrl&atitle=TOURISM+AND+TOURISM+INFRASTRUCTURE+FROM+THE+PERSPECTIVE

Guirao, B., Eugenia López, M., & Comendador, J. (2015). *New QR Survey Methodologies to Analyze User Perception of Service Quality in Public Transport: The Experience of Madrid. Journal of Public Transportation* (Vol. 18). Retrieved from https://scholarcommons.usf.edu/cgi/viewcontent.cgi?article=1460&context=jpt

http://emenu.cc. (2018). No Title. Retrieved from http://emenu.cc/desktop/en/weixindiancan

İlhan, I., & Çeltek, E. (2016). Mobile Marketing : Usage of Augmented Reality in Tourism Mobile Marketing : Usage of Augmented Reality in Tourism Mobil Pazarlama: Turizmde Artırılmış Gerçeklik Kullanımı. *Gaziantep University Journal of Social Sciences*, (December). doi:10.21547/jss.256721

Ishak, I., Sidi, F., Affendey, L. S., Sani, N. F. M., Hamzah, A. S., & Bawon, P. (2014). Mobile plant tagging system for urban forest eco-tourism using QR code. *Proceedings - 2013 International Conference on Advanced Computer Science Applications and Technologies, ACSAT 2013*, 37–41. 10.1109/ACSAT.2013.15

Jevremovic, V., & Petrovski, S. (2012). MUZZEUM - Augmented Reality and QR codes enabled mobile platform with digital library, used to Guerrilla open the National Museum of Serbia. *Proceedings of the 2012 18th International Conference on Virtual Systems and Multimedia, VSMM 2012: Virtual Systems in the Information Society*, 561–564. 10.1109/VSMM.2012.6365977

Kan, T.-W., Teng, C.-H., & Chou, W.-S. (2009). Applying QR code in augmented reality applications, *1*(212), 253. doi:10.1145/1670252.1670305

Kieseberg, P., Leithner, M., Mulazzani, M., Munroe, L., Schrittwieser, S., Sinha, M., & Weippl, E. (2011). QR code security, 430. doi:10.1145/1971519.1971593

Kim, D., Park, J., & Morrison, A. M. (2008). A Model of Traveller Acceptance of Mobile Technology 1, *407*, 393–407.

Kiryakova, G., Angelova, N., & Yordanova, L. (2013). QR Codes in The Business World. *Trakia Journal of Sciences*, *11*, 370–376.

Kiryakova, G., Angelova, N., & Yordanova, L. (2013). QR Codes in The Business World. *Trakia Journal of Sciences*, *11*, 370–376.

Krombholz, K., Frühwirt, P., Kieseberg, P., Kapsalis, I., Huber, M., & Weippl, E. (2014). QR code security: A survey of attacks and challenges for usable security. Lecture Notes in Computer Science (Including Subseries Lecture Notes in Artificial Intelligence and Lecture Notes in Bioinformatics), 8533 LNCS, 79–90. doi:10.1007/978-3-319-07620-1_8

Landman, P. (2011). How hotels can use QR Codes for marketing and customer service. Retrieved from https://www.phocuswire.com/How-hotels-can-use-QR-Codes-for-marketing-and-customer-service

Marakos, P. (2015). Implementing QR code in museums and archaeological sites. In *International Workshop on Virtual Archaeology* (pp. 23–26). Museums & Cultural Tourism; Retrieved from http://vamct.syros.aegean.gr/2015/abstracts2015.pdf#page=62

May, K. (2011). Hotel chain enhances dining experience with QR codes. Retrieved from https://www.phocuswire.com/Hotel-chain-enhances-dining-experience-with-QR-codes

Mohammad, S., & Azam, A. (2015). A model for ordering in restaurant based on QR Code without presence of a waiter at the table Introduction, *2*(12), 1–9.

Narayanan, A. S. (2012). QR Codes and Security Solutions. *International Journal of Computer Science and Telecommunications*, *3*(7), 1–4. Retrieved from http://www.ijcst.org/Volume3/Issue7/p13_3_7.pdf

Niklas, S., & Böhm, S. (2011). Applying Mobile Technologies for Personnel Recruiting –an Analysis of User-Sided Acceptance Factors. *International Journal of EBusiness and EGovernment Studies*, *3*(1), 169–178.

Okazaki, S., Navarro, A., & Campo, S. (2013). Cross-Media Integration Of Qr Code: A Preliminary Exploration. Journal of Electronic Commerce Research, 14(2), 137–148.

Probst, A. (2012). The Expectations of Quick Response (QR) Codes in Print Media: An Empirical Data Research Anthology Number of QR Codes Printed in the Top 100 Magazines (Jan-Dec 2011). *UW-L Journal of Undergraduate Research XV*, 1–13. Retrieved from https://www.uwlax.edu/urc/jur-online/pdf/2012/probst.ali.pdf

Rizwan, Z. (2017). No Title. Retrieved from https://scanova.io/blog/blog/2017/01/18/qr-codes-hotels/

Solima, L., & Izzo, F. (2018). QR Codes in cultural heritage tourism: New communications technologies and future prospects in Naples and Warsaw. *Journal of Heritage Tourism*, *13*(2), 115–127. doi:10.1080/1743873X.2017.1337776

Sorupia, E. (2005). Rethinking the Role of Transportation in Tourism. *Proceedings of the Eastern Asia Society for Transportation Studies*, 1767–1777. Retrieved from http://citeseerx.ist.psu.edu/viewdoc/download?doi=10.1.1.578.2905&rep=rep1&type=pdf

Watson, C., McCarthy, J., & Rowley, J. (2013). Consumer attitudes towards mobile marketing in the smart phone era. *International Journal of Information Management*, *33*(5), 840–849. doi:10.1016/j.ijinfomgt.2013.06.004

What is a QR Code? QRcode.com DENSO WAVE. (n.d.). Retrieved from https://www.qrcode.com/en/about/

www.chinadaily.com.cn. (2017). No Title. Retrieved from http://www.chinadaily.com.cn/china/2017-07/11/content_30068314.htm

www.mercedes-benz.com. (n.d.). No Title. Retrieved from https://www.mercedes-benz.com/content/dam/brandhub/mercedes-benz/classic/generalimages/MBMUS_Information_2018_ENG_Update_ICv2_NETTO_x3.pdf

www.mots.go.th. (2019). No Title. Retrieved from https://www.mots.go.th/mots_en57/ewt_news.php?nid=3362&filename=index

www.radissonbluedwardian.com. (n.d.). No Title. Retrieved from https://www.radissonbluedwardian.com/feature.do?feature=press_qrcodes2011

www.scandit.com. (2019). No Title. Retrieved from https://www.scandit.com/the-qr-code-hotel-room/

Zasadna, K. O. (2014). *QR Coding and Alternative Technologies. Finansovyy prostir [Financial space]* (Vol. 3). Retrieved from http://privatbank.

ADDITIONAL READING

Boob, A., Shinde, A., Rathod, D., & Gaikwad, A. (2014). Qr Code Based Mobile App and Business Process Integration. *International Journal of Multidisciplinary and Current Research*, *2*, 1014–1017.

Buhalis, D., & Amaranggana, A. (2012). Smart tourism destinations. In *Information and communication technologies in tourism 2014* (pp. 553–564). Cham: Springer.

Canadi, M., Höpken, W., & Fuchs, M. (2010). Application of QR codes in online travel distribution. *Information and Communication Technologies in Tourism*, *2010*, 137–148.

Çeltek, E. (2017). QR Code Advertisements in Tourism Marketing. In *Narrative Advertising Models and Conceptualization in the Digital Age* (pp. 269–289). IGI Global. doi:10.4018/978-1-5225-2373-4.ch015

Dickinson, J. E., Ghali, K., Cherrett, T., Speed, C., Davies, N., & Norgate, S. (2014). Tourism and the smartphone app: Capabilities, emerging practice and scope in the travel domain. *Current Issues in Tourism*, *17*(1), 84–101. doi:10.1080/13683500.2012.718323

Kiryakova, G., Angelova, N., & Yordanova, L. (2013). QR codes in the business world, *Trakia Journal of Sciences, 11*, (pp.370–376). Retrieved from http://www.uni-sz.bg

Miralles, P. (2010). Technological innovation, a challenge for the hotel sector. *UPGRADE: The European Journal for the Informatics professional, 11*(2), (pp.33-38).

Pillai, A. E., Prakash, D., Al-Marhoobi, N. A., & Shrivastava, M. (2017). *Application of QR Codes in Tourism Industry: A Review of literature*. Retrieved from www.ijcta.com

Shin, D. H., Jung, J., & Chang, B. H. (2012). The psychology behind QR codes: User experience perspective. *Computers in Human Behavior*, *28*(4), 1417–1426. doi:10.1016/j.chb.2012.03.004

Tardivo, G., Scilla, A., & Viassone, M. (2015). A new frontier in the satisfaction of the cultural tourist: the QR code. In Hospitality, Travel, and Tourism: Concepts, Methodologies, Tools, and Applications (pp. 1356-1374). IGI Global. doi:10.4018/978-1-4666-6543-9.ch078

KEY TERMS AND DEFINITIONS

Augmented Reality (AR): An enhanced version of reality created by the use of technology to overlay digital information on an image of something being viewed through a device (such as a smartphone camera).

Barcode: A code consisting of a group of printed and variously patterned bars and spaces and sometimes numerals that is designed to be scanned and read into computer memory and that contains information (such as identification) about the object it labels.

Mobil Application: A mobile application, most commonly referred to as an app, is a type of application software designed to run on a mobile device, such as a smartphone or tablet computer.

Near Field Communication (NFC): A technology for digitally transmitting information over short distances (usually between a smartphone and another device) using radio waves.

Smart Tourism Destination: One that responds to the requirements of the present global and mobile elites by facilitating access to tourism and hospitality products, services, spaces and experiences through ICT-based tools and where a healthy social and cultural environment can be found through a focus on the city's social and human capital.

Tourism Business: Business tourism is the provision of facilities and services to the millions of delegates who annually attend meetings, congresses, exhibitions, business events, incentive travel and corporate hospitality.

Uniform Resource Locator (URL): Encoded in QR code to provide more information about products.

Universal Product Code: A combination of a bar code and numbers by which a scanner can identify a product and usually assign a price.

Chapter 6
Big Data, Artificial Intelligence, and Their Implications in the Tourism Industry

Evrim Çeltek
https://orcid.org/0000-0002-2722-5256
Tourism Business and Hotel Management College, Gaziosmanpasa University, Turkey

Ibrahim Ilhan
https://orcid.org/0000-0002-6614-9356
Faculty of Tourism, Nevşehir Hacı Bektas Veli University, Turkey

ABSTRACT

Tourism businesses use AI and big data to connect guests creatively and meet their expectations with personalized service. Big data enables tourism professionals to learn more about their customers, and the more they know, the better experience they can offer to customers. As it provides real value, AI has already become an integral part of operations, and this trend will continue. Tourism businesses use AI tools to reduce operating costs and maintenance bills as in many other sectors. AI-oriented marketing has already been widely used in the hospitality industry. Moreover, as long as technology evolves and becomes more complex, tourism professionals will find more ways and methods to implement big data and AI to satisfy customers, and AI will continue to transform the tourism industry. Properties, advantages, and problems of artificial intelligence and big data are discussed in this chapter, and some examples are given from the perspective of the tourism industry.

INTRODUCTION

Big data is defined as large amount of data sets that cannot be analyzed and managed with traditional data processing tools (Xu et al., 2019). It is briefly described as 5V: Volume, Velocity, Variety, Verification and Value (Atalay and Çelik, 2017). Big data is based on the society's ability to utilize knowledge in new ways to produce useful insights or value-creating goods and services (Schönberger and Cukier,

DOI: 10.4018/978-1-7998-1989-9.ch006

2013: 11). Artificial intelligence (AI) is an area of computer science focusing on the creation of intelligent machines that work and react as humans do. Artificial intelligence is the simulation of human intelligence processing through machines, especially computer systems (Şener, 2019). These processes include learning (the acquisition of information and rules, and use of information), reasoning (use of rules to achieve approximate or conclusive results), and self-correction. Specific AI applications include expert systems, speech recognition and machine vision.

Tourism businesses use AI and big data to connect guests creatively and meet their expectations with personalized service. Big data enables tourism professionals to learn more about their customers. And the more they know the better experience they can offer to customers. As it provides real value, AI has already become an integral part of operations, and this trend will continue in the future. Tourism businesses use AI tools to reduce operating costs and maintenance bills as in many other sectors. AI-oriented marketing has already been widely used in the hospitality industry. Moreover, as long as technology evolves and becomes more complex, tourism experts will find more ways and methods to implement big data and AI to satisfy customers, and AI will continue to transform the tourism industry.

The use of artificial intelligence and big data applications are discussed in this chapter. Properties, advantages and problems of artificial intelligence and big data are discussed, and some examples are given from the perspective of tourism industry.

BACKGROUND

Big Data

Big data is defined as large amount of data sets that cannot be analyzed and managed with traditional data processing tools (Xu et al., 2019). It is briefly described as 5V: Volume, Velocity, Variety, Verification and Value (Atalay and Çelik, 2017). Gartner's 3V definition which is used to describe big data in many sources is as follows: It has large volume, velocity and / or large variety. They are information assets that require new forms of information processing that will enhance our decision-making capabilities and improve insight and process optimization (Çakırel, 2016). Big data is based on society's ability to utilize knowledge in new ways to produce useful insights or value-creating goods and services (Schönberger and Cukier, 2013: 11).

Big data collected from; social media shares, photos, blogs, video, text and recorded files, such as high volume and variety of data that can be processed and defined meaningful (Kudyba, 2014).

Monino and Sedkaoui (2016) defined big data as the term used when the volume of data used for the organization reaches a critical level, and new technological storage, processing and usage methods and approaches are required for it.

Big data is defined as large amount of information that cannot be processed by traditional methods, and comes from various sources such as website server records, internet logs, and cell phone communication records in another definition (Snijders et al., 2012: 1-5).

Big data can be defined as data that cannot be easily processed by traditional tools (Ward and Barker, 2013). Dülger states that big data (2015); is the situation in which all kinds of corporate data are digitally coded by human beings and machines, and personal data that emerge through internet and social media shares are converted into meaningful and processible format.

In 2011, McKinsey Global Institute used the term big data to refer to tools that retrieve, store, manage, make available, and analyze data sets that typical and traditional database software cannot (Manyika et al., 2016).

Big data is a term used to describe data sets beyond the storage, management, and processing capacity of commonly used programs. The combination of enormous dimensions of big data and complexity of the analysis needed to benefit from it has led to the development of new class technologies and tools to manage them. In fact, big data describes both the type of managed data and the technology used to store and processes it. Most of these technologies come from technologies developed by companies such as Google, Amazon, Facebook and LinkedIn, while dealing with incredibly large social media data (Cackett, 2016).

Big data is not only volumetric magnitude, but also it is the data collected from different sources such as social media shares, blogs, photos, videos, log files need to be made meaningful and processible. In addition, these data are generated and stored at an ever-increasing rate in terms of volume and type (Atalay and Çelik, 2017).

Tools That Provide Big Data

A variety of big data technologies are used in tourism industry. These technologies and what they are used for are given in the following:

- **Internet of Things:** The Internet of Things is a concept used for creating online networks in part by placing chips, sensors, and communication modules in everyday objects, and data on anything that surrounds people. Dataization represents a fundamental enrichment in human understanding. With the big data, the world can now be seen as a universe of knowledge (Schönberger and Cukier, 2013). The increase in these devices connected to and communicating with each other causes data explosion. Large data appears as a result of data explosion and it causes important problems regarding data analysis. New approaches and methods are developed in the analysis as well as in storing and making big data available. Data mining and text mining are the most important of these methods.
- **Data Mining:** Data mining can also be defined as the discovery of information from data. In data mining, the aim is to extract information from large amounts of data by using automatic and partial automatic methods. Data mining uses algorithms from various disciplines such as statistics, artificial intelligence, and computer science to develop models from data, to find interesting structures or recurring themes. It combines all the technologies that can analyze the database in order to find useful information and possible, meaningful and useful relationships in the data (Monino and Sedkaoui, 2016). Considering data mining in the context of big data, it can be defined as the discovery of confidential information in the big data obtained by using reliable statistical techniques. Data mining is utmost important as it is one of the most basic methods in the analysis of big data.
- **Text Mining:** Text mining is a technique allowing to automate processes to derive major trends in large volumes of text content and to evaluate statistical engagement on different topics (Monino and Sedkaoui, 2016). Text mining is a very important statistical technique especially for analyzing big data in web form. Sometimes, data obtained in the form of text from social media, is defined

as social media mining, and is the one obtained on the web, is defined as web mining and both has their own specific methods, mainly based on data mining and text mining.

- **Natural Language Processing (NLP):** Natural language processing is one of the most important steps in analyzing the texts obtained for institutions and organizations investing in big data technologies. The fact that these institutions and organizations have very large text data, and the need to analyze these texts has made them the best companies in the natural language processing (Oğuzlar, 2011).
- **Google MapReduce**: MapReduce is both a programming model, and an application associated with the processing and creation of large data sets (Dean and Ghemawat, 2016).
- **Hadoop:** Hadoop can be defined as a large data infrastructure that includes a storage system which distributes processing tools. Hadoop is designed to process large amounts of data, from structural and non-structural terabytes to petabytes. Hadoop works with the cluster structure that enables ordinary servers to come together. Servers can be dynamically added and removed from the cluster structure. The main purpose of Hadoop software is not only to process data quickly but also to divide large data into pieces and make them workable. It is an open source library developed in Java programming language (Monino and Sedkaoui, 2016).
- **NoSQL (not only SQL):** NoSQL receives data of different types and sizes, and allows searching through these data. Such database designs require more processing and storage space. Since traditional database solutions are not sufficient together with big data, NoSQL systems are used in almost all big data applications (Schönberger and Cukier, 2013).

Characteristics of Big Data

There are five components in the formation of big data concept. These concepts, called Variety, Velocity, Volume, Verification and Value, are expressed as 5V deriving from the initial letters of the English expression.

Variety: Variety refers to the forms of the data. Big data comes in different forms such as structured, unstructured or semi-structured, including text data, image, audio, video and sensor data. Big data systems need to be flexible enough to handle such a variety of data (Bahga and Madisetti, 2019).

Velocity: Velocity of data refers to how fast the data is generated. Data generated by certain sources such as social media data or sensor data can arrive at very high velocities, (Aktan, 2018). Velocity is another important characteristic of big data and the primary reason for the exponential growth of data. High velocity of data results in the volume of data accumulated to become very large, in short span of time. Some applications can have strict deadlines for data analysis (such as trading or online fraud detection) and the data needs to be analyzed in real-time. Specialized tools are required to ingest such high velocity data into the big data infrastructure and analyze the data in real-time (Bahga and Madisetti, 2019). The rate big data is produced is very fast. This velocity is increasing day by day. The more the velocity increases, the more the diversity and number of transactions increase.

Volume: Big data may be a data type whose volume is so massive that it might not fit on one machine, and therefore, specialized tools and frameworks are needed to store, process and analyze such information. For instance, social media applications process billions of messages every day, and industrial and energy systems generate terabytes of sensor data every day, etc. Volumes of data generated by modern IT, industry, Internet of things, and alternative systems are growing exponentially driven by lowering prices of data storage, and process architectures and also the need to extract valuable insights from the data to enhance business processes, efficiency and service to customers (Bahga and Madisetti, 2019). Organizations collect information from a variety of sources, including business transactions, social media, sensors or direct machines. Storing them was a big problem in the past, now this problem has been reduced with new technologies (such as Hadoop).

Verification: Within the information density, verification is another component that makes the data secure during the flow. Monitoring information at a security level may be visible to the right people or it may remain confidential (Göksu, 2014).

Value: Value of data refers to the quality of data for the intended purpose. It is the most important component as the data creates a value. Big data should be creating added value for the enterprise after the data production and processing layers. It should have immediate impact on decision-making processes and be ready to make right decision at right time (Göksu, 2014).

Artificial Intelligence

Artificial intelligence (AI) is an area of computer science that emphasizes the creation of intelligent machines working and reacting like humans. Artificial intelligence is the simulation of human intelligence processes with machines, especially computer systems (Şener, 2019). These processes include learning (the acquisition of information and rules, the use of information), reasoning (use of rules to achieve approximate or conclusive results), and self-correction. Specific AI applications include expert systems, speech recognition and machine vision.

Artificial intelligence or AI refers to intelligent performance of computers or machines. Artificial intelligence, as a concept, emerged in 1950s. However, with the development of technology, artificial intelligence applications have become more reliable for businesses. Basically, AI refers to computers or machines that perform traditional tasks through a cognitive function. It is related to concepts such as automation and big data (Revfine.com, 2019). Nowadays, from the framework of customer data, artificial intelligence can be used in a variety of applications from basic customer service to personalization tasks, to further problem solving, direct messaging and even in sales processes with a wide range of functions.

AI is the name given to all man-made systems including learning, and discovering. In short, AI is automatic systems that can generate algorithms (Köroğlu, 2017).

Artificial intelligence is a science that tries to gain intelligence to machines by transforming people's decision-making and problem-solving skills into algorithms in order to enable the computer to display intelligent behaviors. According to M. Minksy (1995) AI is the science of creating machines that are thought and designed by humans and exhibit intelligent behaviors (Keleş, Keleş and Akçetin, 2017).

The term artificial intelligence was first described by John McCarthy as the science and engineering of making intelligent machines, especially intelligent computer programs (Demirhan, Kılıç, İnan, 2010).

Artificial intelligence, when done by human beings called intelligence (intelligent behavior) is processed by the machine (Tektaş, Akbaş, Topuz, 2002).

Computers using artificial intelligence are equipped with human intelligence capacities such as information acquisition, perception, vision, thinking and decision making. The first study on artificial intelligence was made by McCulloch and Pitts (Uygunoglu, Yurtcu, 2006).

The term AI was first used by Minsky and McCarthy in 1956 at a Machine Intelligence conference in the United States. Since then, it has been used in various applications such as computer hardware and software; vision, recognition, pattern and speech perception in robots; learning, cognitive modeling, accessing to information, information search and game theory (Bilge, 2007).

Artificial intelligence; is a set of software and hardware systems that have many abilities such as human behaviors, digital logic, motion, speech and sound perception. In other words, artificial intelligence makes computers think like humans. It would be wrong to consider artificial intelligence under a single heading. Machine learning and deep learning are concepts that constitute artificial intelligence. Machine learning is an algorithm allowing the machine to derive logical and rational results from the data provided. For example, in a store, an algorithm is written about the customer's shopping receipt data. This algorithm provides information that customers who buy chips also receive coke. Chips shelves and coke cabinets are brought closer to each other and an increase is seen in such a situation. Machine learning appeared in the 1980s and has become more popular with the use of data mining (Arıkan, 2019). Deep learning was introduced in 2010s. Using large data, it doesn't compute on a single layer, but on many layers at one time. It even discovers the parameters you need to define in machine learning. Perhaps it is a system that can make evaluations with better parameters. For example; we need to separate a picture of a banana with an orange. In machine learning, we were trying to introduce the experiences of mankind to the machine through parameters. If its color is orange it is probably orange, yellow is banana. However, deep learning can learn this difference on its own. Only by showing orange and banana paintings to the deep learning system, it will create its own rules, and will realize that color and shape are the main distinguishing features to reveal the differences. Thus, it can perform its operations by creating its own discriminatory abilities without the need of basic human abilities (Şener, 2019).

Artificial Intelligence Technologies

In the tourism industry, a variety of artificial intelligence technologies are used in artificial intelligence applications. These technologies and what they are used for are given in the following (Udentify Blog, 2017, Arıkan, 2019):

- **Natural Language Generation:** Generating text from computer data. Currently, it is used at customer service, report generation and summarizing business intelligence analysis.
- **Speech Recognition:** It formats and transforms human speech into a convenient way for computer applications. It is currently used in interactive voice response systems and mobile applications.
- **Virtual Agents:** These are advanced systems that can simply connect with people from chatbots. Today it is used in customer service and support, and is used as a smart hotel room or house manager.
- **Machine Learning Platforms:** It is used to provide models, compute algorithms, develop APIs and develop training toolsets, provide data to implement models, and design, train and deploy to other machines. Mostly it includes forecasting or classification of tourism data.
- **Deep Learning Platforms:** A special type of machine learning that consists of artificial neural networks with multiple layers of abstraction. Today, it is primarily used in model recognition and classification applications supported by very large data sets.
- **Decision Management:** Engines that add rules and logic to AI systems and are used for initial setup / training and ongoing maintenance and adjustment. This technology is used in a wide variety of corporate applications and for untouched and automatic decision-making.
- **Biometrics:** Enables more natural interactions among people and machines, including images, touch recognition, speech, and body language. Currently, it is primarily used in market research and security.

- **Robotic Process Automation:** Used to automate human action to support effective business processes. It is currently used at areas where people are too expensive or inefficient to carry out a task or operation.
- **Text Analytics and NLP (Natural language processing):** Natural language processing (NLP) is used in text analysis to understand sentence structure and meaning, emotion and intention through statistical and machine learning methods. Among the areas include it are fraud detection and security, a wide variety of automated assistance and applications used to scan unstructured data.

MAIN FOCUS OF THE CHAPTER

Big Data Applications in Tourism

Tourism businesses can use big data technologies to obtain valuable information such as better understanding of behavior of tourists, identifying developing preferences and needs, and monitoring the geographical location of tourists. For example, it is possible to recommend hotels, restaurants and events to tourists regarding their preferences, online behavior and geographical location (Elisabeth et al., 2013). Some big data applications also allow monitoring the effectiveness of policies and regulations in tourism. In fact, it is recommended in reducing traffic problems in cities or in facilitating access to crowded places such as airports. They also support drivers to make decisions depending on their current location, destination, preferences, and other changing parameters (such as weather or accidents) (Yazıcı et al., 2013). Big data applications and their implementations in tourism are given under the following headings.

Providing New and Competitive Products and Services: Creators of database-based products and services are online travel agencies, travel search firms and leading technology providers in travel industry. For example, Amadeus has developed Featured Results and Extreme Search capabilities for its customers to enhance the travel search experience. Travel search website Hipmunk has developed new features such as Agony Index to rate airline flights and Ecstasy Index for hotel searches. KAYAK, a travel search site, has developed a predicted price offering (Davenport, 2013).

Better Decision Support - Internal Operations: By analyzing macroeconomic and weather data, airlines can forecast consumer demands (Davenport, 2013). Hilton, a US multinational tourism company uses modern data architecture to gain insights from data and to serve customers better. Hilton focuses on using data analytics to get a 360-degree view of each customer including booking data, customer profile data, and even details of how customers use the hotel's facilities. Gaining these insights allows Hilton to recognize its customers better and in turn considers them more valuable. This leads to more revisits and happier guests (Brar, 2019).

Personalization of Services: MGM Resorts International is a tourism company that has successfully implemented big data analysis for enhancing its personalization strategy. The company used Facebook's big data tools successfully, resulting in a 300% increase in revenue over a three-year period. The springboard of MGM's personalized services is enrollment of customers to company's "M Life program": once the customer's data are registered into program's database, MGM Resorts is able to track customer's movements in all of its 19 properties. Nick Mattera, MGM Resorts International's social media manager, states, "if tourists start looking at room rates on MGM.com and then leave to compare prices to a competing site, MGM can take them to their site MGM.com with a relevant Facebook offer" (Karampatsou, 2018). British Airways (BA) uses an intelligent 'Know Me' feature to provide personalized search results to customers. In this impressive big data case study, BA identified that their customer base largely consists of busy, time-pressed professionals who require fast, concise results. Therefore, 'Know Me' uses in-depth data analysis to provide relevant and targeted proposals to consider. BA received a huge amount of positive feedback from customers who were satisfied with the company's understanding their travel needs (Feliu, 2019). Big data tools allow a deeper knowledge and insight in traveler's needs and preferences, offering a more holistic approach designing a customer centric strategy. Tourism companies are now able to process and analyze structured and unstructured data thanks to big data. In addition, big data contains a great amount of multifarious and useful information about tourists who were inaccessible for companies so far (Karampatsou, 2018).

Revenue Management – Improving Pricing Strategy: Marriot International, a hotel chain, is the best example of using big data to estimate optimal price for its rooms. The strategy focuses on developing relevant algorithms for faster and more accurate data analysis, and it enables revenue management to be accessible on the Internet, as well as expanding at all levels of the hotel chain's operations such as restaurants, catering and meeting areas. Furthermore, Marriot's data scientists are not only limited to analyzing internal data to predict demand, they also combine customers' data with semi-structured and unstructured data collected from external sources, such as forecasts and event schedules. Marriot International's this strategy allows it to optimize its services and products and adjust the hotel's price accordingly (Karampatsou, 2018). The travel search site Kayak.com effectively uses big data tools to ensure consistent price optimization. Algorithms, analytical models and machine learning enable the company to form a competitive pricing strategy by monitoring and analyzing rivals' prices. Kayak.com launched its new forecasting model, Flight Price Forecasting in 2013. This tool is used to predict whether the price of a particular flight will increase or decrease within seven days. Thus, it allows the travel search site to deliver better targeted products and services and to drive new revenue streams in the long run and improve revenue management (Karampatsou, 2018, Davenport, 2013). Air France-KLM uses big data to develop a new revenue management system on all airlines. The system contains much more data than previous systems, including all relevant passenger data for last two years. The system calculates and optimizes revenue for origin/destination routes and pricing based on passenger profiles. It also anticipates the possibility of cancellation and no show on flights and thus how much overbooking to be allowed. The system runs on the big data operating system called Hadoop. Tests have shown a significant improvement in revenue management performance over existing systems (Davenport, 2013).

Advantages and Problems of Big Data

Big data focuses on what to do with it rather than how much information it has. It retrieves data from any source and analyzes them to find answers that help save money and time, new project development and optimized proposals, as well as making smart decisions (Eğer, 2019). Combined with strong analysis, the followings can be achieved through big data in tourism industry (Sheoran, 2017, Eğer, 2019, Davenport, 2013):

- Identify real-time errors and identify the core causes of problems.
- Generating sale coupons based on customers' buying habits.
- Revenue out flow and cancellations.
- Multi-channel complaining.
- Demand forecasting.

- Help to recalculate risks in the processes before decision making.
- Increase productivity and reduce costs.
- Service improvements
- Much more effective new product development
- Effective decision-making after handling the results (example, understanding and measuring the customer)
- Inventory management.
- Customer trustworthiness.
- Helping in the travel industry for better customer relationships.
- The personalization of company's services.

In addition to providing us with unprecedented opportunities, big data also raises issues that need to be addressed (https://proente.com, 2019):

- Data Privacy: Big data generated data contains a lot of information about private lives. A balance must be struck between the amount of confidential data released and the convenience of applications and services supported by big data.
- Data Security: Even if it is reassuring to know that data is being held for a particular purpose, how confident can their security be?
- Data Discrimination or Customer Discrimination: When everything is known about customers, discrimination can be made against people based on the data about the customers' lives.
- Evaluating the big data components together is a very comprehensive process.
- When the generated data is obtained from many different environments, it is difficult to integrate it in different formats.
- It is very difficult to meet the software and hardware requirements of big data transactions.
- Planning how big data stacks will be managed in the future requires extensive work and planning.
- Preventing the breaches of security and reliability in the flow of data and creating related authorization strategies is another important issue.
- Tourism businesses face lack of skilled personnel in terms of big data analytics.

Artificial Intelligence Applications in Tourism

Voice recognition, synthetic speech, image processing, autonomous learning and reasoning capabilities are what artificial intelligence offers to businesses. Businesses use artificial intelligence in various applications such as interacting with new customers, analyzing the growing data volume, attracting the target audience, accelerating the processing of information, synthesis, influencing the decision-making process, ensuring the permanency of customer relations (Balkan, 2019). Artificial intelligence plays an increasingly important role in tourism sector because of its ability to perform human functions. This enables tourism businesses to save significant amount of money, eliminate human-based error, and provide superior service. Artificial intelligence applications and examples in tourism are given under the following headings.

Big Data Analysis: Another way of using AI in the tourism industry is data analysis. Artificial intelligence can be used to quickly sort large amounts of data and draw important conclusions about customers or prospects. One of the businesses benefitting from the advantages of AI is the Dorchester

Collection hotel chain, which uses the Metis AI platform. Using artificial intelligence, the hotel collected information about customers through surveys and online reviews. AI analyzed this collected information to draw conclusions about overall performance (Revfine.com, 2019). EasyJet invested in an artificially intelligent algorithm that determines seat pricing automatically, depending on demand. Furthermore, the system can also analyze historical data to predict demand patterns up to a year in advance. These analytics can also impact decision-making about new routes, schedules, and codeshare alliances in advance (Feliu, 2019). For example, Southwest Airlines have teamed up with NASA to improve airline safety constantly. By using intelligent algorithms, Southwest and NASA have created an automated system that can crunch an enormous amount of data to flag anomalies and prevent accidents (Feliu, 2019). In order to use artificial intelligence, big data is needed first. Without it, artificial intelligence cannot work and predictive statistics and analyzes cannot be done. Therefore, in tourism sector, it is necessary to collect big data via software. Then, with the collected data, it can make predictions for the future.

Marketing: AI and big data allow hotels to micro-target their customers. Each customer is unique, so it is important to create personalized offers to each customer. Not everyone responds to discounts; some guests are more motivated by offers that deliver a unique experience such as local food or microbrew tastings or an upgrade to VIP status. With the big data tools, marketers can match the customer with the right offer, and AI can ensure delivery in right time exactly to the right person (hospitalitytech.com, 2018).

Customer Relationship Management: With machine learning solutions, artificial intelligence can provide valuable data about customers through messages, voice or video calls, media broadcasts, various pages which have people pass sometime being captured there and comments on web pages. Tourism businesses benefitting from this data can communicate in terms of the characteristics and expectations of existing and target customers. In order to get to know customers better, it is important to know what they are interested in. However, it is impossible to work with a person to access and evaluate these data. Artificial intelligence is indispensable for these processes. In addition, by using this data in the future, brands that will know their customers much better will be able to offer unique offers to their customers (Pehlivan, 2019).

Chatbots and Messaging: The use of virtual assistants, or chatbots, as bases for system interactions with humans is attracting people's interests gradually. This is due to the fact that human languages offer a natural way of communication, and methods and tools allowing computers to analyze and process such languages are available now (Bozic et al, 2019). While online travel agencies like Expedia are actively using artificial intelligence for customer service, Booking.com has launched a chat tool that automates online travel searches. Expedia reminds customers on how many people call to change reservations using artificial intelligence to simplify the process. In the future, users will be able to make transactions via chatbot to change or cancel a reservation easily without having to talk to any of the users (ie without a call center) or logging in the website. By processing big data with artificial intelligence, it is possible to make intelligent predictions preferred by each passenger (such as seat in flight or accommodation preferences) and turn this information into reservations that meet expectations quickly and comfortably. Business travelers can easily benefit from this virtual assistant service as they travel frequently and consistently. This time saving means saving money as well, which is an important priority for business travelers (Savaş, 2018, Saulat, 2018).

Robotic Concierge and Virtual Concierge: Hipmunk, Expedia, Skyscanner and Cheapflights also use artificial intelligence for customer service. Hilton hotels have gone a step further by serving a real robot called "Connie" who helps guests at check-in and advises on local attractions and things to do. Connie, a small robot standing at the hotel reception desk, learns to interact with visitors and provide them with a personalized experience. It has access to WayBlazer's travel information as well to help customers get the answers they need (Savaş, 2018). Hotels in New York and Tokyo use robots to perform check-in and other tasks. Today, virtual concierge applications make the hotel room more comfortable for tourists by adjusting the room temperature and delivering pillows according to the preferences of the guests (hospitalitytech.com, 2018).

Facial Recognition With Additional Heft From Blockchain: Travel requires repeated review of travel documents by different groups of people. There are complex boarding and liquidation procedures (especially for cruise ships). Face recognition technology puts an end to these strenuous paper processes. With facial recognition, tourists can move smoothly during airports, immigration, customs and board aircrafts without any need for travel documents examined at every step. Integrated with face recognition technology blockchain, customers can pay with a simple face scan in restaurants and duty-free shops. Blockchain technology enables reliable data to be used to complete transactions (Saulat, 2018).

Advantages and Problems of Artificial Intelligence

Using artificial intelligence, tourism businesses collected information about customers through surveys and online reviews. AI analyzed this collected information to draw conclusions about overall performance. With artificial intelligence technologies, marketers can segment consumers and create different marketing content for each segment. With personalized campaigns, both customers become satisfied and this increases marketing efficiency (Campaign Türkiye, 2017).

Analysis and reporting can be carried out in a short time using artificial intelligence technologies. Managers can see the big picture and determine their strategies accordingly with the help of artificial intelligence.

Search Engine Optimization: Adgorithms Company developed artificial intelligence called Albert. It is capable of bidding, integration and management (Campaign Türkiye, 2017).

Artificial intelligence reduces the risk of making mistakes to almost zero. The data is processed and analyzed with greater accuracy. It is used as a virtual assistant. This saves customer's time.

Nowadays, AI technologies have many business advantages. However, there are reasons that make it difficult to adopt and implement in tourism businesses; these are (Udentify Blog, 2017):

- Lack of a defined business model.
- Not having the necessary skills.
- Necessity to invest in data to create predictive applications.
- No budget.
- What is needed to implement an AI system stays unclear..
- Failure to prove AI systems.
- Not knowing the right process and not having the ability to manage.

SOLUTIONS AND RECOMMENDATIONS

Today, businesses use artificial intelligence to interact with new customers, analyze growth volume and attract potential customers. In addition, AI applications are used to process, synthesize information, make decision-making process fast and efficient, improve customer relations and ensure continuity.

Tourism businesses can use big data technologies to obtain valuable information such as better understanding of the behavior of tourists, identifying developing preferences and needs, and monitoring the geographical location of tourists. For example, it is possible to recommend tourists to hotels, restaurants and events according to their preferences, online behavior and geographical location (Elisabeth et al., 2013).

Voice recognition, synthetic speech, image processing, autonomous learning and reasoning capabilities are what artificial intelligence offers to businesses. Businesses use artificial intelligence in various applications including interacting with new customers, analyzing the growing data volume, attracting the target audience, accelerating the processing of information, synthesis, influencing the decision-making process, ensuring the continuity of customer relations (Balkan, 2019). Artificial intelligence plays an increasingly important role in tourism sector, thanks to its ability to perform human functions. This enables tourism businesses to save significant money, eliminate human error, and provide superior service.

Artificial intelligence is used in tourism industry for big data analysis, marketing, customer relationship management, chats and messaging, robotic and virtual concierge and face recognition applications.

Big data applications are used in tourism industry for providing new and competitive products and services, better decision support-internal operations, personalization of services, and revenue management -improving pricing strategy.

FUTURE RESEARCH DIRECTIONS

In this chapter, use of big data and AI and some big data and AI applications in tourism industry have been examined. In this context, this study is a conceptual study. Big data and AI technologies have the potential to be used in tourism industry. A model proposal study on big data and AI technology is recommended for tourism industry. Big data and AI technologies can be studied on independently for tourism sub-sectors such as hotels, restaurants, and travel agencies. In addition, the perceptions and expectations of the tourism sector regarding big data and AI can be determined through field studies or surveys with managers and tourists.

CONCLUSION

Big data is defined as large amount of data sets that cannot be analyzed and managed with traditional data processing tools (Xu et al., 2019). Artificial intelligence (AI) is an area of computer science focusing on the creation of intelligent machines that work and react as humans do. Artificial intelligence is the simulation of human intelligence processing through machines, especially computer systems (Şener, 2019). Tourism businesses use AI and big data to connect guests creatively and exceed their expectations with personalized service. Big data enables tourism professionals to learn more about their customers. And the more they know the better experience they can give to customers. AI is already becoming an integral part of operations as it provides real value, and this trend will continue in the future. Tourism businesses use AI tools to reduce operating costs and maintenance bills as in many other sectors. AI-oriented marketing is already used in the hospitality industry widely. However, as technology evolves and becomes more complex, hotel experts will even find a variety of ways and methods to implement big data and AI to satisfy customers, and AI will continue to transform the tourism industry.

Big data and data analytics are changing the theory and practice of tourism industry. Smart tourism and allied companies are using these technologies to anticipate customer needs, rewrite how they meet customer expectations, redefine customer engagement, and achieve new levels of customer satisfaction. In doing so, these firms are creating a new basis for the award of customer loyalty.

REFERENCES

Aktan, E. (2018). Big Data: Application Areas, Analytics and Security Dimension. *Bilgi Yönetimi Dergisi, 1*(1), 1–22. doi:10.33721/by.403010

Arıkan, M. (2019). *Yapay zeka nedir?* Retrieved from https://www.mediaclick. com.tr/blog/yapay-zeka-nedir

Atalay, M., & Çelik, E. (2017). Artificial Intelligence And Machine Learning Applications In Big Data Analysis. *Mehmet Akif Ersoy Üniversitesi Sosyal Bilimler Enstitüsü Dergisi, 9*(22), 155–172. doi:10.20875/makusobed.309727

Bahga, A., & Madisetti, V. (2019). *Big Data Science and Analytics: A Hands-On Approach*. Published by VPT.

Balkan, E. (2019). *Yapay Zeka, Büyük Veri ve Pazarlama Devrimi.* Retrieved from https://kobitek.com/yapay-zeka-buyuk-veri-ve-pazarlama-devrimi

Bilge, U. (2007). *Tıpta Yapay Zeka ve Uzman Sistemler*. Türkiye Bilsim Dernegi Kongresi.

Blog, U. (2017). *Top 10 Yapay Zeka (Artificial Intelligence) Uygulamaları.* Retrieved from http://blog.udentify.co/04/2017/en-cok-ilgi-ceken-yapay-zeka-uygulamalari/

Bozic, J., Tazl, O. A., & Wotawa, F. (2019). Chatbot Testing Using AI Planning. In *Proceedings 2019 IEEE International Conference on Artificial Intelligence Testing (AITest)*, Newark, CA. 10.1109/AITest.2019.00-10

Brar, K. (2019). *Hotels using big data to check out guests.* Retrieved from https://www.tnp.sg/news/views/hotels-using-big-data-check-out-guests

Cackett, D. (2016). *Information Management and Big Data, A Reference Architecture.* White paper. Redwood Shores: Oracle Corporation.

Çakırel, Y. (2016). İşletmelerde Büyük Veri, *Kırklareli Üniversitesi İktisadi ve İdari Bilimler Fakültesi Dergisi, 5*(1).

Davenport, T. H. (2013). *At the Big Data Crossroads: turning towards a smarter travel experience.* Amadeus.

Dean, J., & Sanjay, G. (2016). *MapReduce: Simplified Data Processing on Large Clusters*. Google, Inc. Web

Demirhan, A., Kılıç, Y. A., & İnan, G. (2010). Tıpta yapay zeka uygulamaları. *Yoğun Bakım Dergisi, 9*(1), 31–41.

Dülger, Ü. (2015). *Stratejik Büyük Veri Yönetiminin Yatırımlar Üzerindeki Etkileri* (Unpublished master dissertation). İstanbul University, Turkey.

Eğer, Ö. (2019). *Big Data'nın (Büyük Veri) Endüstriyel Kullanımı.* Retrieved from https://www.endustri40.com/big-datanin-buyuk-veri-endustriyel-kullanimi/

Elisabeth, E., Nock, R., & Célimène, F. (2013). Demonstrator of a Tourist Recommendation System. In V. Bhatnagar, & S. Srinivasa (Eds.), Big Data Analytics. (pp. 171-175) BDA. Lecture Notes in Computer Science, vol. 8302. Cham, Switzerland: Springer. doi:10.1007/978-3-319-03689-2_11

Feliu, C. (2019). *Big data case study: 5 relevant examples from the airline industry.* Retrieved from https://blog.datumize.com/5-relevant-examples-of-a-big-data-case-study-from-the-airline-industry

Göksu, C. (2014). *Datawarehouse Türkiye.* Retrieved from http://datawarehouse.gen.tr/big-datanedir-geleneksel-veri-yonetimine-etkisi-ne-olur/ https://proente.com

Karampatsou, M. (2018) Big Data in Tourism, (Unpublished master's thesis). School Of Economics, Business Administration, & Legal Studies, Greece.

Keleş, A., Keleş, A., & Akçetin, E. (2017). Pazarlama Alaninda Yapay Zekâ Kullanim Potansiyeli Ve Akilli Karar Destek Sistemleri. *Electronic Turkish Studies, 12*(11).

Köroglu, Y. (2017). *Yapay Zeka'nın Teorik ve Pratik Sınırları.*

Kudyba, S. (2014). *Big Data, Mining and Analytics: Components of Strategic Decision Making. Publisher.* Taylor Francis. doi:10.1201/b16666

Manyika, J., Chui, M., Brown, B., Bughin, J., Dobbs, R., Roxburgh, C., & Byers, A. H. (2011). *Big data: The next frontier for innovation, competition, and productivity. McKinsey Global Institute.* MGI.

Monino, J.-L., & Sedkaoui, S. (2016). Big Data, Open Data, and Data Development: Vol. 3. *London, UK: ISTE, Ltd.* Hoboken, NJ: John Wiley.

Oğuzlar, A. (2011). *Temel Metin Madenciliği.* Bursa, Turkey: Dora.

Pehlivan, B. (2019). *Dijital Pazarlamada yapay zeka.* Retrieved from http://www.yapayzekatr.com/2019/03/03/dijital_pazarlamada_yapay_zeka/

Proente. (2019). *Big Data (Büyük Veri) Nedir?* Retrieved from https://proente.com/big-data-buyuk-veri-nedir/

Revfine.com. (2019). *How to use Artificial Intelligence in the Hospitality Industry.* Retrieved from https://www.revfine.com/artificial-intelligence-hospitality-industry/

Saulat, A. (2018) *Four Ways AI is Re-imagining the Future of Travel.* Retrieved from https://www.mindtree.com/blog/four-ways-ai-re-imagining-future-travel

Savaş, G. (2018). *Yapay Zeka İle Turizmde Dönüşümün 5 Örneği / Uygulamasi.* Retrieved from http://www.turizmdosyasi.com/yapay-zeka-ile-turizmde-donusumun-5-ornegi-uygulamasi-h15034.html

Schönberger, V. M., & Cukier, K. (2013). *Büyük Veri - Yaşama, Çalışma ve Düşünme Şeklimizi Dönüştürecek Bir Devrim. Çev. Banu Erol.* İstanbul, Turkey: Paloma.

Şener, S. (2019). *Yapay Zeka, Makine Öğrenimi ve Derin Öğrenme Arasındaki Farklar.* Retrieved from https://www.endustri40.com/yapay-zeka-makine-ogrenimi-ve-derin-ogrenme-arasindaki-farklar/

Sheoran, S. K. (2017). Big data: A big boon for tourism sector. *International Journal of Research in Advanced Engineering and Technology, 3*(1), 10–13.

Tektaş, M., Akbaş, A., & Topuz, V. (2002). *Yapay zeka tekniklerinin trafik kontrolünde kullanilmasi üzerine bir inceleme.* Retrieved from http://www.trafik.gov.tr/icerik/bildiriler/ pdf/C4-7.pdf

Türkiye, C. (2017). *Yapay Zeka ve Pazarlama – II.* Retrieved from https://www. campaigntr.com/yapay-zeka-ve-pazarlama-ii/

Uygunoğlu, T., & Yurtcu, Ş. (2006). Yapay Zeka Tekniklerinin İnşaat Mühendisliği Problemlerinde Kullanımı. *Yapı Teknolojileri Elektronik Dergisi, 2*(1), 61–70.

Xu, F., Nash, N., & Whitmarsh, L. Big data or small data? A methodological review of sustainable tourism. *Journal of Sustainable Tourism.* doi:10.1080/09669582.2019.1631318

Yazici, M., Kamga, C., & Singhal, A. (2013). A big data driven model for taxi drivers' airport pick-up decisions in New York City. In *IEEE International Conference on Big Data*, pp. 37–44. IEEE. 10.1109/ BigData.2013.6691775

ADDITIONAL READING

Albuquerque, H., Costa, C., & Martins, F. (2018). The use of Geographical Information Systems for Tourism Marketing purposes in Aveiro region (Portugal). *Tourism Management Perspectives, 26*, 172–178. doi:10.1016/j.tmp.2017.10.009

Connelly, R., Playford, C. J., Gayle, V., & Dibben, C. (2016). The role of administrative data in the big data revolution in social science research. *Social Science Research, 59*, 1–12. doi:10.1016/j.ssresearch.2016.04.015 PMID:27480367

Edwards, D., & Griffin, T. (2013). Understanding tourists' spatial behaviour: GPS tracking as an aid to sustainable destination management. *Journal of Sustainable Tourism, 21*(4), 580–595. doi:10.1080/09 669582.2013.776063

Graham, M., & Shelton, T. (2013). Geography and the future of big data, big data and the future of geography. *Dialogues in Human Geography, 3*(3), 255–261. doi:10.1177/2043820613513121

Hawelka, B., Sitko, I., Beinat, E., Sobolevsky, S., Kazakopoulos, P., & Ratti, C. (2014). Geo-located Twitter as proxy for global mobility patterns. *Cartography and Geographic Information Science, 41*(3), 260–271. doi:10.1080/15230406.2014.890072 PMID:27019645

Kitchin, R. (2013). Big data and human geography opportunities, challenges and risks. *Dialogues in Human Geography, 3*(3), 262–267. doi:10.1177/2043820613513388

Luo, X., Dong, L., Dou, Y., Zhang, N., Ren, J., Li, Y., ... Yao, S. (2017). Analysis on spatial-temporal features of taxis' emissions from big data informed travel patterns: A case of Shanghai, China. *Journal of Cleaner Production*, *142*, 926–935. doi:10.1016/j.jclepro.2016.05.161

Shoval, N., & Isaacson, M. (2007). Tracking tourists in the digital age. *Annals of Tourism Research*, *34*(1), 141–159. doi:10.1016/j.annals.2006.07.007

Snijders, C., Matzat, U., & Reips, U. D. (2012). Big data: Big gaps of knowledge in the field of internet science. *International Journal of Internet Science*, *7*(1), 1–5.

KEY TERMS AND DEFINITIONS

Artificial intelligence: Artificial intelligence (AI) is an area of computer science that emphasizes the creation of intelligent machines that work and reacts like humans.

Big Data: Big data is defined as a large amount of data sets that cannot be analyzed and managed with traditional data processing tools.

Data Mining: Data mining can also be defined as the discovery of information from data.

Deep Learning Platforms: A special type of machine learning consisting of artificial neural networks with multiple layers of abstraction. Today, it is primarily used in model recognition and classification applications supported by very large data sets.

Internet of Things: The Internet of Things is a concept of creating online networks in part by placing chips, sensors, and communication modules in the everyday objects, and data in everything that surrounds people.

Machine Learning Platforms: It is used to provide models, compute algorithms, develop APIs and develop training toolsets, provide data to implement models, and design, train and deploy to other machines. Mostly it includes forecasting or classification of tourism data.

Natural Language Processing (NLP): Natural language processing is one of the most important steps in analyzing the texts obtained for institutions and organizations investing in big data technologies.

Text Mining: Text mining is a technique that makes it possible to automate processes to derive major trends in large volumes of text content and to evaluate statistical engagement on different topics.

Chapter 7
Internet of Things in Tourism: A Proposal of the Information System for Cappadocia Hot-Air Ballooning

Ibrahim Akın Özen
https://orcid.org/0000-0003-1172-5448
Nevşehir Hacı Bektas Veli University, Turkey

ABSTRACT

With the revolution of Industry 4.0, the technologies that enter our daily lives are based on smart devices, applications, and platforms with internet connection. A wide range of these technologies collected under one umbrella is known as IoT (internet of things). This chapter evaluates the stages of a touristic travel in smart tourism destinations by considering IoT architecture. The technologies used in these phases and their contributions to the tourism sector and tourists are examined. In the implementation section, an IoT-based information system is proposed for Cappadocia hot air balloon tours. The main purpose of the system is to determine whether the appropriate weather conditions are formed before the hot air balloon flights. The proposed system allows for the automation and evaluation of data already collected using traditional methods. With the implementation of the system; work and time savings can be achieved, and more accurate measurements will make safe flights.

INTRODUCTION

Three major industrial revolutions have emerged in the development of modern industry. Today, the fourth industrial revolution called Industry 4.0 has entered. Industry 4.0 is an ecosystem that links data, people, processes, services, systems (I-scoop, 2018). Industry 4.0 is also referred to as smart industry, smart factory or smart production industry. Smart production systems can be defined as fully integrated and co-operative production systems that respond in real time to meet changing customer needs and changing demands and conditions in factories and supply networks (Zheng, et al., 2018).

In the Industry 4.0 revolution, although production technologies constitute the main focus, it is possible to talk about a series of transformations affecting other interconnected sectors at the digital production

DOI: 10.4018/978-1-7998-1989-9.ch007

stage. It is thought that digital transformation will be realized by using third party technologies. Third party technologies that constitute the source of digital transformation are Information Technology, Operational Technology, Internet of Things, IoT Devices, Sensors, Robotics, Data, Artificial Intelligence, Intelligent Decentralized Production, Self-optimizing systems (I-scoop, 2018; Sap, 2018).

Considering that the technologies used in the Industry 4.0 revolution are based on an IP address (Internet Protocol), applications consist of a wide range. IoT is used as the term Internet of Things.

This digital transformation, which was mentioned with the Industry 4.0 revolution, had a significant impact on the tourism sector. In this context, by using IoT technologies, the concepts of smart tourism, smart city, smart destinations were introduced (Boes, Buhalis, & Inversini, 2016; Buhalis & Amaranggana, 2013; Khan, et al., 2017). Smart tourism includes tourist activities supported by smart technology. Smart tourism is therefore defined as a tourism system that takes advantage of intelligent technology to create, manage and deliver intelligent tourist experiences (Buhalis & Amaranggana, 2013).

The concept of smart city is based on ICT (Information and Communication Technology) infrastructure, which includes the internet of objects, big data analysis and cloud computing. In a smart city, all urban activities must be connected. This structure can be established with the help of the internet of objects. Smart tourism, unlike the smart city, focuses not only on the local people but also on the tourist experiences, and the quality of life of the locals as well as the mobility, ease of access and sustainability of resources are also important (Gretzel, et al., 2015).

The smart destination is defined as an innovative tourist destination built on the latest technological infrastructure. In the literature, it is seen that the concepts of smart tourism and smart destinations are generally used in the same sense. (Blaser, 2019; Gretzel, et al., 2015; D. Wang, Li, & Li, 2013).

In this context, the study will consist of two parts. In the first part of the study, the stages of a touristic travel in smart tourism destinations are evaluated by considering IoT architecture. In the implementation section, an IoT-based information system is proposed for Cappadocia hot-air balloon tours. The main purpose of the system is to determine whether the appropriate weather conditions are formed before the flight in hot-air balloon flights. In the conclusion section, suggestions were made to the private sector, government and destination management office to practice the implementation.

BACKGROUND

In this part, IoT, smart tourism destinations, IoT architecture and design in smart tourism destinations, challenges and IoT, benefits of IoT in tourism are defined and explained.

Internet of Things (IoT)

The traditional internet connects machines to machines, server to machine or web pages to web pages. The IoT refers to the networked interconnection of everyday objects, tools, devices or computers (Chaouchi, 2013).

The IoT is described as a network that can connect any object with the internet based on a protocol intended for exchanging info and communication among numerous smart devices to be able to achieve monitoring, tracking, administration and location recognition objectives (C. Wang, et al., 2014).

The IoT concentrates on the recognition of 3 main ideas, namely things-oriented, internet-oriented and semantic-oriented (Saleem, et al., 2017). The things-oriented idea involves smart devices, such as RFID tags, sensors, actuators, video cameras, laser scanning devices, the Global Positioning System (GPS) and Near Field Communication (NFC). The internet-oriented concept allows communication amongst smart devices through various communication technologies, including WiFi, Z.Wave, ZigBee Bluetooth and cellular marketing communications. and attaches them to the internet. The semantic-oriented concept understands a variety of applications with the help of smart devices. Over the past couple of years, the IoT technology offers gained significant attention in a variety of applications, and has allowed to get the interconnection of the internet to various network embedded devices used in everyday life (Chen, et al., 2012).

It provides automated the operation of numerous systems, just like medical, travelling, military services, appliances for the home, security, monitoring, agriculture and power plants. In a few areas, IoT devices include transceivers, tiny controllers and protocols, allowing their connection with other devices as well as with external organizations to allow the realization of completely computerized systems (Zanella, et al., 2014).

IoT In the Future

According to many reports, the use of IoT is expected to increase significantly in the future. According to the leading companies and research companies of the sector, the increase in the future smart devices and the investments to be made in these devices are explained below.

Internet of Things (IoT) sensors and devices are reported to grow at an annual growth rate of 23% from 2015 to 2021. It is also reported that by 2021, 21 billion mobile phones will be connected to the Internet (Ericsson, 2016). According to the 2017 report, predicted that IoT devices and sensors will reach 20.4 billion by 2020 and that service providers and end users will spend $ 2 trillion to acquire IoT devices (Gartner, 2017). According to the 2018 report, IHS Markit estimates that 15.4 billion IoT devices in 2015 will reach 30.7 billion in 2020 and 75.4 billion in 2025 (IHS Markit, 2018). Furthermore, McKinsey estimates that the total size of the IoT market in 2015 has reached $ 900 million and that it will increase by 32.6% in 2020 to $ 3.7 billion (McKinsey, 2016).

Smart Tourism Destinations

Many studies in the tourism literature have covered the conceptual framework for smart tourism destinations. Some of these studies are as follows.

Smart tourism destination is defined as touristic areas in which all stakeholders can be exchanged in real time through technological platforms by linking information about tourism activities. These integrated platforms should have multiple touch points accessible from a variety of end-user devices that will support the creation and facilitation of real-time tourism experiences and increase the efficiency of managing tourism resources (Buhalis & Amaranggana, 2013).

The main purpose of creating a smart tourism destination is to interact between the visitor and the destination through a sensitive interface for the solution of specific needs. Smart cities are thought to be the basis of smart tourism destinations. (Khan, et al., 2017). From this point of view, smart tourism destinations can be formed by converting the components defined for smart cities into tourism applications (Buhalis & Amaranggana, 2013) and the dimensions of smart tourism destinations are defined as smart people, smart mobility, smart living and smart environment (Buhalis & Amaranggana 2013; Cohen, 2012).

Smart tourism destinations are composed of components such as cloud services, Internet of Things (IoT) and End User Internet Service System. An integral characteristic of cloud services is usually that it offers web-browser based usage of a selection of technological equipment including applications, data and software. The IoT provides smart destinations with support regarding information sophistication and administration as it pertains to automation and control. The end-user services systems support users with equipment and applications in order to gain access to the services linked to tourism. Included in these are applications that are oriented towards serving vacationers which enable usage of items and support services like the usage of payment program and interfaces; telecommunication interfaces and devices; cellular connections that consist of hotspots and various other such services; and so on (Zhang, Li, & Liu, 2012).

Koo et al. (2016) stated that the main purpose of smart tourism destinations is to create unique experiences for tourists and visitors. Such experiences derive from a highly effective transformation of user data into solutions that relate with specific needs. Koo et al. further explain that collective intelligence reaches the crux of system interfaces and applications that underlie technological platforms, which are instrumental in the creation of user experiences. Likewise, Sigala & Marinidis, (2012) identified system advancements (for example web 2.0) founded on an understanding management perspective as integral in managing the efficacy of processes that derive from transformations of stakeholders' tacit knowledge to explicit knowledge. This permits such smart systems to effectively use knowledge also to configure solutions, and address situational needs of the users. Therefore, it might be argued that collaborative systems work as such interfaces give a basis for better communication and coordination (Micera, et al., 2013).

As mentioned earlier, smart tourism destinations can use the technological infrastructure of smart cities. In this respect, there is an extensive research on smart cities in tourism literature.

Naphade et al. (2011) proposed a model of seven components for smart tourism destinations. These components include government services, transport, water and energy, health services, education, public safety and other core ICT systems. Furthermore, Glebova, (2014) conceptualized the smart city with 5 dimensions: Intellectual transport systems, public security, energy consumption control and management, environmental ICT and protection. Alternately, Bellini, et al. (2014) suggested the model predicated on a data-oriented IT perspective: Data sources, administration, street-guide, point-of-interest, local public transport, sensors, temporal, and metadata. Anthopoulos (2015) categorized smart cities into five levels: Natural environment, hard infrastructure–non-ICT-based, hard infrastructure–ICT-based, services and soft infrastructure and people.

The flagship of smart transformation in destinations is access to real-time information by all stakeholders throughout the destination. Providing real-time access to information can be achieved by means of IoT architecture designs by using smart devices and services by virtue of IoT technologies.

IoT Architecture and Design in Smart Tourism Destinations

The basic requirement in IoT system architecture is that objects must be linked with each other and with other devices and services. This provides a real-time flow of information, allowing you to dynamically manage events. When creating an IoT architecture, it should be taken into consideration that heterogeneous devices can work with each other.

In this study, IoT architecture is discussed from a touristic travel point of view (Figure 1). Figure 1 is composed of two main parts. The first section describes the multiple layers of IoT architecture and the functions of these layers.

Figure 1. IoT architecture in smart tourism destinations

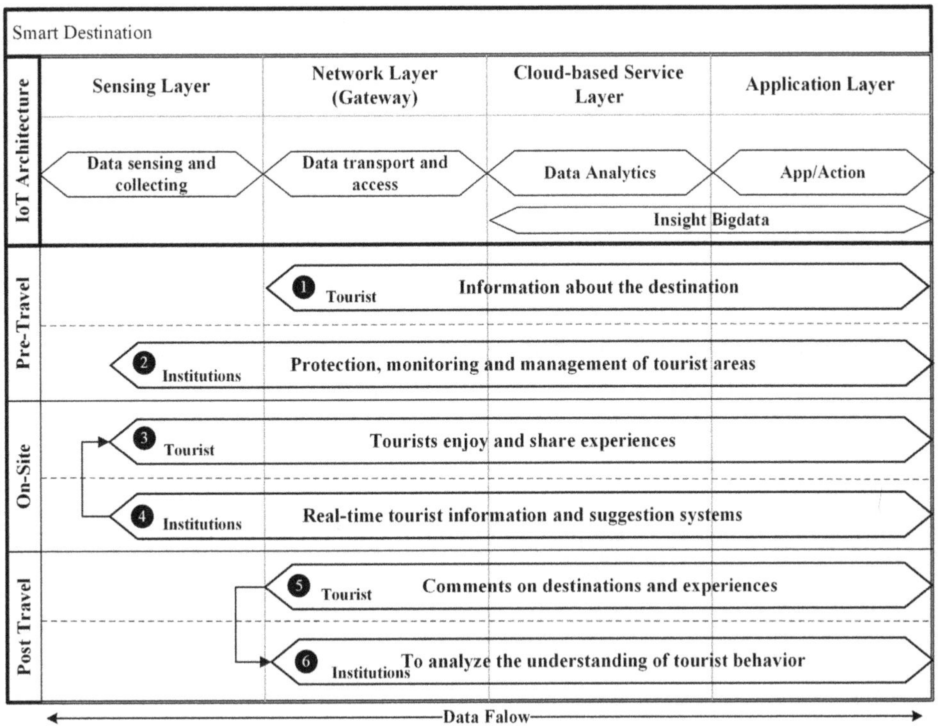

These layers are; sensing layer, network layer, cloud-based service layer and application layer. (Serpanos & Wolf, 2018; Sinha & Park, 2017; S. Zheng, et al., 2013). Sensing layer is integrated with available hardware objects to sense the statuses of things. Network layer is the infrastructure to support over wireless or wired connections among things. Cloud-Based Service layer is to create and manage services required by users or applications. Interfaces layer consists of the interaction methods with users or applications (Li, Xu, & Zhao, 2015). The technologies used in these layers will be described in detail in the following sections. The second section describes the IoT technologies used in all stages of a touristic journey and the functions it provides.

According to Figure 1, a touristic travel experience in smart tourism destinations is shown in 3 stages in terms of technology use (Çelik & Topsakal, 2017; Barbara Neuhofer, Buhalis, & Ladkin, 2014; Solima, 2016; T. Zhang, et al., 2019). These stages are expressed as pre-travel, on-site and post travel. There are two main actors in these stages.

In Figure 1, bars 2, 4 and 6 show the institutions that offer technology in smart tourism destinations such as private sector, government agencies, destination management offices (DMO). On the other hand, tourists using the IoT-based technological services offered at the destination are shown with bars 1, 3 and 5. The descriptions of the bars according to the figure 1 are described below.

- **Pre-Travel (Tourists/Fig1.bar1):** Tourists are exploring the necessary services for destinations and tours and seeking information. Tourists use social media, augmented reality, virtual reality and websites to get preliminary information about destinations (İlhan & Çeltek, 2016). At this stage, travel plans and reservations can be made. A set of information (text, images, audio, video, information, maps, etc.) about destinations, products and services can be accessed. Thus, the selection experience for tourists is significantly enhanced (Barbara Neuhofer, et al., 2014, p. 345; Solima, 2016). This section covers the network, cloud-based service and application layer of IoT architecture.

- **Pre-Travel (Institutions-Fig1.bar2):** Institutions (private sector, government, destination management offices) can develop environmental warning systems, destination management and monitoring systems to ensure destination security by using IoT sensors (Perkins & Thorns, 2001; Zhao, et al., 2013; R. Zheng, et al., 2016; S. Zheng, et al., 2013). In addition, destination promotion services can be provided through web services. This section can be performed using all layers of IoT architecture.
- **On-Site (Tourists-Fig1.bar3/ institutions-Fig1.bar4):** It is the stage where technology is most used for tourists and institutions. With real-time information and suggestion systems provided by institutions, tourist experiences can be made easier, accessible and rich (Buhalis & Amaranggana, 2013; Gavalas, et al., 2014; B. Neuhofer, 2014).

At this stage, technologies such as mobile applications, Global Positioning System (GPS) and location-based services (LBS) come to the forefront. Functions such as getting directions to the places to be traveled, determining the places of interest, locating the points of interest, researching these places via social media, reading the comments of other tourists and communicating are used (Angelakis, et al., 2016; Gavalas, et al., 2014; Noguera, et al., 2012; Schmidt-Belz, et al., 2002). These stages can be performed using all layers of IoT architecture.

- **Post Travel (Tourists-Fig1.bar5/ institutions-Fig1.bar6):** In this stage, tourists can make comments and evaluations about their experiences in destinations through social media and blogs. These comments, especially through social networks, constitute an important source of information for institutions. In this way, they will have the opportunity to provide return on the products and services they offer. The big data obtained at this stage can be analyzed using data mining and text mining techniques (He, Zha, & Li, 2013; Khan, et al., 2017; Liu, 2012; Prameswari, et al., 2017).This stage covers the cloud-based service and application layers of IoT architecture.

All the stages described above will be explained in the following section in terms of IoT architecture and the technologies used in this architecture.

Sensing Layer

IoT is likely to be considered an all over the world physical inner-connected network, where things are connected seamlessly and may end up being controlled remotely. In the sensing layer, the smart systems on tags or sensors can automatically sense the surroundings and exchange data among devices. In recent years, advanced sensor technologies have made it more flexible and accessible. This has led to significant improvements in environmental monitoring through IoT applications. Everything in IoT keeps an electronic identity and will be very easily tracked inside the digital domain name. The technique of designated unique identification to one thing is named a universal unique identifier (UUID). In particular, UUID is crucial to effective solutions application in an enormous network just like IoT. The identifiers might make reference to names and addresses. (Li, et al., 2015).

Table 1. IoT Sensors

IoT-Sensor Layer Smart Tag / Sensor	Maximum Coverage Range	Access/ Readers	Information	Connectivity
QR Code	1.5 - 2.5 feet	Very few smartphones have inbuilt QR code apps	Information is constant and cannot be personalized.	One-to-one
RFID	3 - 4 feet	Needs special readers	Information is constant and embedded in the tag.	One-to-one
BLE Beacons	Up to 330 feet	All smartphones with Bluetooth 4.0+ (Bluetooth Smart)	Information is dynamic and can be personalized based on location, frequency, time of day.	One-to-one or One-to-many
Intelligent Sensor	Manufacturer Defined	Wired or wireless	Environment data (position, humidity, Wind Direction, Wind Speed, Temperature, Rain Gauge)	Sensor to Machine or Embedded Edge Device

Source: Adapted from (Li, et al., 2015)

Table 1 includes IoT sensors, coverage areas, access, information set and connection options.

Network Layer (Gateway)

The network layer in IoT, connects all things and allows them to be familiar with their surroundings. With the network layer, things may talk about info with the linked things, which is vital to intelligent occasions management and digesting in IoT. Furthermore, the networking layer is with the capacity of aggregating info from existing IT infrastructures; data may then become transmitted to decision-making models for the high-level complex solutions (Li, et al., 2015; Sinha & Park, 2017; R. Zheng, et al., 2016).

There are three distinct types of networks which are involved in IoT solution deployments from a device perspective. They are (1) device to device communication (sensor and actuator network or an edge device network), (2) device to gateway network, and (3) device to cloud service (Sinha & Park, 2017). Device to Device Communication: Message exchanges between devices on a local area network. Device to Gateway Communication: Message exchanges between a device and an internet gateway. Device to Cloud Service Communication: Message exchanges between a device and a cloud service or a gateway device and a cloud service.

A great IoT gateway (see Table 2) must support multiple communication protocols to link sensor/actuator network to cloud services. In general, a sensor network protocol and cloud service communication are certainly not homogeneous. Therefore, a gateway device uses various protocols (Sigfox, ZigBee, WiFi) to transfer the data collected from the sensors to the application layer.

There are various criteria that should be considered in choosing communication alternatives for IoT devices and gateways. Whenever we deal with end devices including sensors and actuators, conversation distance, ordering and construction complexity, battery-life, and application topology really are a minimal group of criteria to become evaluated. When a device needs to connect to edge devices and cloud solutions via a wide area network or mobile networks (3G / 4G / 5G), the connection and maintenance costs will be part of the evaluation elements (Sinha & Park, 2017).

Table 2. IoT Communication Protocols

	Networ Layer/ Protocol	Maximum Coverage Range	Frequency	Standard	Connectivity
Short range wireless	Bluetooth	50–150 m	2.4 GHz	Bluetooth 4.2 core specification	Device to Device Device to Edge Device
	NFC Near Field Communication	Very short ranges (20 cm)	13.56 MHz	ISO/IEC 18000-3	Device to Device Device
	ZigBee	10 m	2.4 GHz	IEEE802.15.4	Device to Device Device to Edge Device
	WiFi	30 m	2.4 GHz 5.8 GHz 5.0 GHz	802.11/a/b/g/n/ac	Device to Device Device to Edge Device
	Z Wave	30 m	900 MHz	Z-Wave Alliance	Device to Device Device to Edge Device
	6LoWPAN (IPv6 over Low-Power Wireless)	100 m	2.4 GHz	In development	Device to Device Device to Edge
Long range wireless	3G/4G/5G	5 km max for GSM 200 km max for HSPA	900/1800/1900/2100 MHz	GSM/GPRS/EDGE (2G), UMTS/HSPA 3G), LTE (4G)	Device to Cloud Edge device to Cloud
	Sigfox	30–50 km (rural) 3–10 km (City)	900 MHz	Sigfox	Device to Cloud Edge device to Cloud
	LoRaWAN	2–5 km (urban) 15 km (suburban area)	867–869 MHz (Europe) 902–928 MHz (North America)	LoRaWAN	Device to Cloud Edge device to Cloud

Source: Adapted from (Sinha & Park, 2017, p. 31)

Cloud-Based Service Layer

Data from IoT appliances have to be collected and stored in a secure manner. IoT devices have the power to cross the Internet indirectly or directly via applications and share resources with the Cloud (Alabdulsalam, et al., 2018). Offered storage services in the cloud stem from cloud computing. Cloud computing is an on-demand network that provides access to a joint pool of configurable computing resources (servers, services, applications, storage, and networks) which, can be provided and released quickly with minimal service provider interaction or management effort (Quick, Martini, & Choo, 2013). Cloud computing is a measured service and a client will pay for service which is contingent on usage. A pay as you go model is integrated typically into cloud computing which allows the end user to receive an efficient and cost-effective service. Rates of data traffic of the usage of cloud hardware are monitored by an automated meeting service. Also incorporated is the maintenance fee, cloud clients are obligated to pay a maintenance fee for the cloud service whether the service is used or not. The costs are associated with managing the virtual cloud environment. There is also the option for limited but free cloud provider services. There is a multitude of cloud storage hosting providers, many offer free cloud storage services as seen with Dropbox, and Google Drive (Quick, et al., 2013).

Cloud computing transfers storage and processing power away from the end user to a point on the cloud or internet that is more managed and centralized. Cloud computing offers not only storage services but also a music and video player, document and image editing, and e-mail sending capacity. The development of a broader distribution of free software, quicker networks, and advanced virtualization technology in an information technology infrastructure has popularized cloud computing. Multiple if not most cloud storage services permit users to access cloud storage with a smartphone or with a PC. With the budding technological advancements and technological age, access via a device such as a smartphone has only helped to spread cloud storage services further (Chung, et al., 2012; L. Wang, et al., 2010).

Cloud computing has multiple deployment models available like the private cloud infrastructure, the general public cloud infrastructure, the hybrid cloud, and the city cloud. The private cloud infrastructure is habitually run by single organizations for the precise usage of that organization. The general public cloud infrastructure will be utilized by everyone and is run by companies like Microsoft, Thinkspeak and Amazon. Community cloud and hybrid cloud are believed less common, taking characteristics from both public and private cloud infrastructures. The Cloud's deployment models contain multiple characteristics that distinguish it from traditional computing paradigms. Cloud computing characteristics permit scalability, and flexibility, plus a large number of resources which can be on demand. Resources provided by the cloud such as storage, editing, etc., have evolved and also have been built-into large-scale applications to increase efficiency. Predicated on the kind of resources provided, cloud-computing services could be split into (SaaS) software as a service, (IaaS) infrastructure as a service and (PaaS) platform as a service (Bhardwaj, Jain, & Jain, 2010).

Software as a Service (SaaS)

SaaS is an amenity widely known in cloud computing. *SaaS* offers applications that were explicitly created by the provider. The applications are predominantly run as web-based and installed by the provider. SaaS allows the least amount of flexibility for the user. SaaS does not permit users to deploy user software applications or to specify hardware, which means users simply access and use the presented product.

SaaS users do not have to install and run applications but can instead access what is needed via the internet. SaaS users have no control over which operating system to use for applications (Lampropoulos, Siakas, & Anastasiadis, 2018).

Infrastructure as a Service (IaaS)

IaaS can deliver on-demand general computing resources consisting of storage or virtualized servers. This is typically intended for personal purposes or business, to store data like images, documents and music files (Chung et al., 2012). End users have the ability to configure personal virtual hardware, storage capabilities, and processor type. IaaS also allows the user to specify the type of operating system to be used in the personal cloud environment. A good example of the Infrastructure as a Service will be the Amazon Elastic Cloud Computer (EC2). The Amazon Elastic Cloud Computer is a web service that offers many features such as configuring storage capacity to owner preference, providing security, network functionality and more (Amazon, 2019).

Platform as a Service (PaaS)

PaaS offers related features to Infrastructure as a Service, providing virtualized servers where the user can form personal applications or work on existing applications. The company maintains the operating-system that works on the virtualized server. The provider can be held in charge of the maintenance of the server hardware. A number of software programming interfaces or development systems are typically supplied by the company so the user can form and operate personal applications generally via the web.

IoT and cloud services are tightly integrated for such reasons that cloud services present accessibility on a worldwide scale which permits users to discover and locate a connected device instantly. IoT devices such as smartphones, tablets, smartwatches and more could be immediately located (Zunnurhain, 2016).

Application Layer

It may be the last stage of data representation. Global administration to any software is offered in this coating based on the related info prepared in the last layer. This level represents the many applications that are created according to prepared data outcomes, like smart towns, smart homes, wise healthcare, smart vehicular tracking and transport and many more.

That's where user applications are deployed. Each software comprises two interfaces: A front-end user interface which represents the access stage for users or items to connect to the system, and a back-end user interface which links this coating to all of those other platform and allows the application form to become fulfilled by the low layer (Nitti, et al., 2017).

Challenges and Internet of Things

IoT components are applied using divergent protocols and technologies. As a total result, these parts have complex configurations and poor style (Jindal, Jamar, & Churi, 2018). Technological challenges can be a reflection of four parameters:

Vulnerabilities

The impact of the IoT data being taken and sold can range in the amount of seriousness for the buyer nevertheless, lack of privacy all together represents an over-all weakness in security. Systems that hinge on the cloud service provider for security reasons face huge dangers. Availability and protection of general cloud services are reliant upon the protection of the application form programming user interface (API). API's certainly is a software program which allows communication to stream between multiple applications (Chou, 2013).

Cloud providers strive to make sure that security is usually well incorporated to their support model. Nevertheless, if individuals are relying on poor API's, it exposes businesses and people to a variety of security problems. People that utilize uncovered API's to interact and manage cloud services can possess personal/private communication become manipulated, personal/private data could be stolen, and even more if malicious celebrations gain entry through the API publicity (Chou, 2013). Security problems likely to happen regarding this vulnerability will end up being connected with confidentiality, integrity, and availability.

Confidentiality

Confidentiality is a core principle of security, and as such includes IoT protection. Confidentiality requires that data not be disclosed to unauthorized entities. It is a standard way of identifying if you are who you say you are and that you have the right to view the information that is being accessed. Data regularly held in confidentiality are usually military data, security credentials, patient data and more. Confidentiality can be achieved by implementing security passwords, security clearance, encryption and even more (Abomhara & Koien, 2015).

Integrity

Integrity in the IoT steps the standing of data; it's the assurance that data are not tampered with. Predicated on the IoT program that's being used there exists a specific integrity necessity. A good example of integrity within the IoT is usually a remote individual monitoring program using integrity checks to fight random mistakes that may occur because of info sensitivities. Integrity looking at is an activity that involves evaluating data in before and after condition determining if adjustments in data happened. Changes that happen in data can stem from the increased loss of details or manipulation. Integrity checks is seen at particular points in something such as turning something on / off (Abomhara & Koien, 2015).

Availability

An IoT device or the device's user should be in a position to access services at any time whenever required (Abomhara & Koien, 2015). Availability necessity is a combined mix of a sectors/consumers capability to gain access to a technologies support along with the technologies capability to provide services (Abomhara & Koien, 2015; Lele, 2019). For example, the types of systems and devices within the IoT possess different availability requirements.

Benefits of IoT In Tourism

The main objectives of using IoT-based information systems in destinations are:

1. Facilitation and enrichment of tourist experiences
2. Ensure the safety of destinations and tourists
3. Protecting natural and cultural assets in destinations
4. Making use resources in destination efficiently and effectively

For these purposes, IoT based information systems used in destinations and their explanations are given in the table.

Table 3. IoT based systems used in tourism

IoT-based system	Explanations
Geographical Information Systems (GIS)	Information systems that can capture, store, manage, manipulate, analyses, integrate and display large amounts of geographical data.
Global Positioning System (GPS)	Satellite-based navigation system that provides positioning navigation and timing services to users in any weather conditions around the world 24 hours a day.
Intelligent Transport System (ITS)	Telematic systems which provide detailed information on traffic information from independent locations, traffic guidance and dynamic routing.
Location-based Services (LBS)	This can collect and deliver information to and from a mobile device depending on the automatic location of the user. The aim of LBS is to provide targeted information to the user based on his/her geographic location. Such information includes but is not limited to places to visit, eat and stay as well as emergency and health services.
Virtual Reality and Augmented Reality (VR/AR)	Anyone can experience the culture, history and other points of tourist interests in a visual and interactive manner without actually visiting the destination. An example of this include on-line guided tours of museums and heritage sites where the visitor can experience the destination without actually visiting the destination.
Weather, Climate and Ocean Change Forecasting System (WCOCFS)	This information can be useful for bidding for events, making decisions about proposed development, putting measures in place for hazards and risks associated with bad weather, provide tourists with updated information, energy management and other issues.
Destination Management System (DMS)	A system that consolidates and distributes a comprehensive range of tourism products through a variety of channels and platforms.
Environment Management Information Systems (EMIS)	A combination of computer hardware, software and professional services that integrates disparate information about environmental issues to manage the environmental function within an organization. It systematically gathers, analyses and reports business information related to environmental management such as waste tracking and emissions. This allows a company to track, refine and improve its environmental practice.
Gamification	A process of taking approaches, techniques and technologies developed in the sphere of game design and applying them to non-game functions typically to engage the gamers in real-world applications.

Source: Adapted from (Ali & Frew, 2013, p. 65)

MAIN FOCUS OF THE CHAPTER

Case Study: IoT Based Weather Forecast System for Cappadocia Hot-air Balloon Flights

Methodology and Study Area

The study area is the Göreme National Park in Cappadocia, Turkey. Cappadocia is an important tourism region. Göreme National Park, which is located in the region, was included in the UNESCO World Heritage List in 1985 due to its natural and cultural characteristics. As of 2019, there are 1092 natural and cultural regions worldwide registered in the UNESCO World Heritage List. Of these, 845 are cultural, 209 are natural, and 38 are mixed (cultural / natural) regions (UNESCO, 2019). Only 38 places around the world are among both cultural and natural regions. Göreme National Park is located within the mentioned 38 places. It is very important tourists visiting the region. One of the most important tourist activities is hot-air balloon tours.

Table 4. Cappadocia region tourist hot-air balloon flight and tourist numbers (2014-2018)

Years	2014	2015	2016	2017	2018
Number of Flights	27.006	28.817	15.674	19.500	26.000
Number of Tourists	474.595	498.886	228.805	330.000	535.500

Data Source: (SHGM, 2018).

Hot-air balloon tours are different from other touristic products due to the necessity of special weather conditions, the need for organization and the desire of the region where the flight takes place.

Considering the issues listed, few destinations in the world are suitable for hot-air balloon tours. There are valleys formed by wind and floods in the Cappadocia Region in millions of years. Weather rivers are formed in these valleys due to the temperature difference between day and night. These weather rivers are suitable weather conditions for hot-air balloons. Thus, the hot-air balloon can move horizontally and vertically between the valleys. So Cappadocia is known as an important hot-air balloon flight center in the world (Özen, 2019).

In the last 20 years, important investments have been made in hot-air balloon tourism in Cappadocia. These investments in Turkey have managed to stand out in the world. There are 258 balloons, 205 licensed balloon pilots and 210 ground crew chiefs from 25 registered balloon companies in the Cappadocia region.

The number of flights of 25 balloon companies in Cappadocia is given in Table 4.

Balloon companies operating in Cappadocia provided a balloon experience to 2,115,186 tourists with 116,997 flights between 2014-2018 (SHGM, 2018).

Cappadocia Hot-air Balloon Flight Area

Cappadocia balloon flight area is divided into three sectors. Figure 2 shows the balloon flight area sectors of the Cappadocia Region.

According to Figure 2; Sector A refers to the region between the northern boundary of the balloon flight area and the 38 ° 40 '25" K parallel. Sector B refers to the region between 38° 40° '25" K and 38° 35 '44" K parallel. Sector C refers to the area between the 38 ° 35 '44" K parallel and the southern boundary of the balloon flight area. Visually from the 38° 40 '25" K parallel Çavuşin Primary School Intersection in front of Zelve Open Air Museum 'and' Devrent Viewing Area; 38° 35 '44" K parallel to the south passes through 'Nevşehir Closed Prison Ort and Ortahisar - İbrahimpaşa stabilized road junction (SHGM, 2015).

Hot-air Balloon Flights and Coordination in Cappadocia

Demand and supply for the balloon flights increased after the 2000s in the Cappadocia Region. As a result, the Cappadocia Slot Service Center (CSSC) was established to arrange balloon flights. In order to coordinate the flights of the balloon companies, CSSC introduced a slot (flight permit allocated to the balloon companies at a specified time). When the meteorological conditions in the region are appropriate, each company can perform two flights per day, morning and noon. It is planned by the CSSC to determine how many balloons each balloon company will fly and how many passengers will fly.

Figure 2. Cappadocia hot-air balloon flight area
Source: (SHGM, 2019)

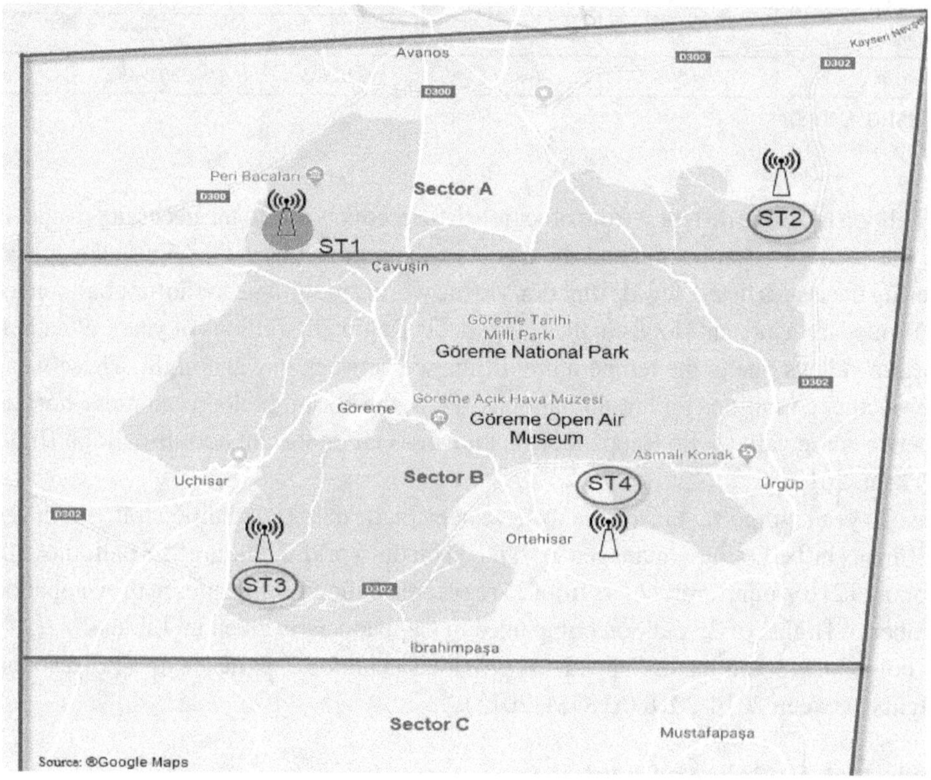

Balloon Flights and Meteorological Assessment (Current Status)

Balloon flights in Cappadocia Region are subject to weather conditions. In hot-air balloon flights, wind direction and intensity are considered as important factors. For the balloons to take off in the region, it is recommended that the wind force on the flight site does not exceed 27 km / h. (Ultramagic, 2008).

For this reason, the Meteorological Evaluation Group (MEG), which consists of 12 balloon pilots, was formed by CSSC. Every day of the week, Five MEG pilots carry out physical assessments in the flight area using traditional methods of wind direction and intensity and make preliminary assessments. Each pilot informs the center of his / her assessment (flyable or non-flyable) in the area of responsibility. CSSC evaluates the pilot evaluations by majority vote and decides on flight permits or cancellations for balloon flights. Meteorological evaluations are published on the web address "http://shm.kapadokya.edu.tr". Signs for flights on the web page are marked as red flag (not flyable), yellow flag (preparation is possible but not flyable) and green flag (flyable). Figure 1 shows the flight permits of the flight regions of the web site, indicated by the flag icon.

In the current situation, meteorological measurements made by manpower by traditional methods will provide more reliable and instant measurements through the IoT based information and communication system proposed. The following sections will provide details of the IoT based information and communication system proposed.

IoT Based Weather Forecast System Proposal

The main purpose of the application is to propose an IoT based information and communication system where the wind direction, intensity, temperature and humidity values are measured and evaluated with the help of IoT based weather stations to be installed in the hot-air balloon flight area of Cappadocia (Figure 2). In line with this main objective, the following sub-objectives are planned to be realized:

1. To provide security in balloon flights by using IoT based systems and services in hot-air balloon flight area.
2. To determine the flight times of hot-air balloons for balloon companies and to publish them with the help of web services.
3. To speed up the decision-making process for balloon flights by making instant weather forecasts based on online data.
4. To make daily, monthly and annual statistical evaluations by means of recorded weather forecast data.
5. To use automated real-time data instead of data collected using labor using traditional methods to forecast weather on balloon flights, reducing accident risks for balloon flights.

IoT-Based System Application Area

The system consists of four meteorological evaluation stations to be established in the A, B and C sectors located in the Cappadocia hot-air balloon flight area (Figure 1). The place where the stations will be installed is determined by the CSSC. Since the area of application is rural, the energy needs of the system will be provided with solar cells.

Structure of the system

The system consists of 5 main sections. Power Supply -Weather Station, Sensor Layer, Network Layer, Could-Based Service Layer, Application Layer. Figure 3. shows a block diagram of the structure system.

The descriptions of the IoT hardware and services in the block diagram of the system shown in Figure 3 are presented below.

In weather forecasting stations, the power supply of the IoT devices used will be obtained from solar panels. Accordingly, devices 1,2,3,4 in figure 3 form the power supply section. The functions of these devices in the system are as follows:

1. **Mexxsun 100-Watt 12v Polycrystal Solar Panel:** It converts the solar energy into electrical energy. It will be provided to use the DC / DC converter to be used to charge the battery in the system or to feed the system directly.
2. **Mexxsun CM20D DC / DC Converter:** It is used to reduce or increase the voltage level from the solar battery to the charging voltage of the battery.
3. **Vigor NP 12-17 Battery:** 12 Volt 17 Amps battery. It will be used to provide the necessary power to operate the system. In addition, the energy to be produced from the solar panels will be stored in the battery.

Figure 3. IoT based weather forecasting station block diagram

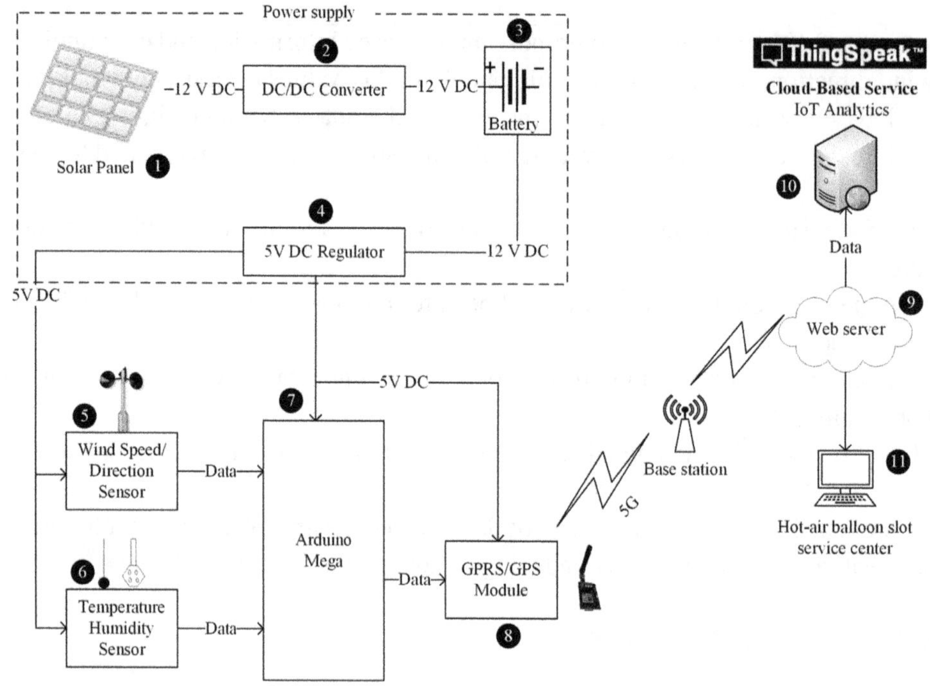

4. **LM2596 5V Regulator:** It is the element that will provide the 5-volt DC voltage needed for the operation of the Arduino and sensors. The 12-volt DC voltage from the battery is reduced to 5 volts DC voltage.

In weather forecasting stations, the sensor layer is comprised of sensors and microcontrollers in which the data from the environment is collected. Accordingly, the devices 5, 6, 7 of FIG. 3 are included in this section. The functions of these devices in the system are presented below;

5. **Davis Anemometer:** It consists of high stability and accuracy sensors used for measuring wind direction and intensity. The wind direction is transmitted by the analog output and the wind speed is transmitted by the digital output.
6. **SHT21 Temperature and Humidity Sensor:** It is the industry standard temperature and humidity sensor. The calibrated and linear sensor outputs the I2C protocol. The sensor consists of a capacitive type humidity sensor in a single CMOSens chip and a temperature sensor in an analog-digital structure. This structure provides low power consumption, high precision and long-term use.
7. **Arduino Mega:** Arduino Mega is a microprocessor card based on ATMEGA2560. It has 54 digital input / output pins, 16 analog inputs, 16 MHz crystal oscillator, one USB connection, one power input. They can be operated with low power consumption thanks to sleep modes. The sensor data obtained from the environment is collected, controlled and distributed via this card.

In weather forecasting stations, the network layer will allow data from sensors to be transferred to cloud-based services. The devices and services 8, 8 in Fig. 3 form the network layer. The functions of these devices and services in the system are presented below;

8. **Telit GE910-GNSS GPRS / GPS Module:** It is one of the Arduino modules. It connects the cellular networks via the sim card on it and transfers the Arduino data to the cloud environment via base stations.
9. **Web Server:** Private server where data from weather stations are stored and published. The number 10 cloud-based web service layer in Figure 3 enables the analysis of data from IoT-based sensors.
10. **ThingSpeak:** ThingSpeak is an IoT analytics platform service that allows you to aggregate, visualize, and analyze live data streams in the cloud. With MATLAB analytics inside ThingSpeak, you can write and execute MATLAB code to perform preprocessing, visualizations, and analyses (ThingSpeak, 2019). With this service, the data obtained from 4 weather stations (Fig.2, ST1, ST2, ST3, ST4) are evaluated with MATLAB application and the required weather conditions for balloon flights will be calculated.

Figure 3 is the app where the hot-air balloon flight permits are managed and published via the web server of the number 11.

11. **Hot-Air Balloon Slot Service Center:** Refers to the web service center where flight permits are issued for the balloon companies in line with the data calculated in the Thing Speak service (wind direction, intensity, temperature, humidity). As a result of the assessment of the weather conditions, hot-air balloon flights are allowed, canceled or are on hold.

SOLUTIONS AND RECOMMENDATIONS

In this study, IoT based system is proposed for hot-air balloon flights. Installing the proposed system will help prevent hot-air balloon accidents caused by weather conditions. The most important factor determining hot-air balloon flights is wind speed. By virtue of the proposed system, the wind speed in the flight area will be measured by sensitive sensors. The measured values will be saved to the cloud-based services. In this way, weather forecasts can be made using scientific methods.

The most important problem in IoT-based environmental management systems is the collection and transmission of environmental data to the cloud. The most important reasons for this are the inadequacy of infrastructure (IoT devices need energy and coverage in wireless networks) especially in rural tourism areas. Therefore, the energy needs of IoT devices in rural areas can be solved by using green energy sources (wind and solar energy). In addition, cellular networks (4g / 5g) can be used to transmit the collected data to the network medium.

Traditional environmental data collection methods have some problems. For example, inadequate workarounds, heavy labor and poor sustainability. It is difficult to meet real-time monitoring and management requirements in large tourist areas. Therefore, IoT is a good choice for performing long-term, multi-factor environmental monitoring.

The IoT-based information system proposal described in detail in the implementation section of this study will serve as an example for other tourism activities. It will guide the researchers who will work on destination security in academic studies. In addition, this study was brought to the attention of smart destination executives (private sector, public institutions and destination management offices) in order to implement the proposed system.

FUTURE RESEARCH DIRECTIONS

IoT technologies and smart tourism destinations are a multi-disciplinary field of study. For this reason, it is necessary to collaborate with researchers working in the field of system analysis, software developers and electronics while conducting research on this subject.

In the literature of tourism, studies for smart destinations are theoretical rather than practical. Therefore, it does not inspire practitioners enough. The most important limitation of this research is the insufficient examination of system analysis at the user level. In the future, the proposed system will be revised using the system modeling language UML (Unified Modeling Language).

CONCLUSION

Smart tourism destinations are tourism regions that focus on improving the tourist experience supported by the Internet of Things and cloud computing. The main objective of smart tourism destinations is to focus on the needs of tourists in the products and services offered to tourists by using modern technologies considering service quality and activity management. (Huang, Yuan, & Shi, 2012).

The priorities of the construction of smart tourism destinations should be to improve the travel experience of tourists, to facilitate the efficient allocation of tourism resources, to integrate tourism suppliers at both micro and macro levels, and to ensure that local people benefit from these developments.

In the first part of this study, the stages of a touristic travel in smart tourism destinations are evaluated considering IoT architecture. The technologies used in these phases and their contributions to the tourism sector and tourists are examined. In the application section, an IoT based information system is proposed for Cappadocia hot-air balloon tours. The main purpose of the system is to determine whether the appropriate weather conditions are formed before the flight in hot-air balloon flights. The proposed system will allow for the automation and evaluation of data already collected using traditional methods. With the implementation of the system; work and time savings can be achieved, and more accurate measurements will make safe flights. The establishment of tourism activities in touristic areas can be achieved by using IoT based smart systems. Thus, the tourist experience can be enriched by making the tourist experiences in suitable conditions and making the tourists feel in a safer environment. Technology-supported safe tourism areas should be seen as an important step in creating smart tourism destinations.

REFERENCES

Abomhara, M., & Koien, G. M. (2015). Cyber Security and the Internet of Things: Vulnerabilities, Threats, Intruders, and Attacks. *Journal of Cyber Security and Mobility*. doi:10.13052/jcsm2245-1439.414

Alabdulsalam, S., Schaefer, K., Kechadi, T., & Le-Khac, N. A. (2018). *Internet of things forensics – Challenges and a case study*. IFIP Advances in Information and Communication Technology; doi:10.1007/978-3-319-99277-8_3

Ali, A., & Frew, A. J. (2013). Information and communication technologies for sustainable tourism. In Information and Communication Technologies for Sustainable Tourism. doi:10.4324/9780203072592

Amazon. (2019). Amazon EC2. Retrieved from https://aws.amazon.com/tr/ec2/

Angelakis, V., Tragos, E., Pöhls, H. C., & Kapovits, A. (2016). *Designing, Developing, and Facilitating Smart Cities*. Designing, Developing, and Facilitating Smart Cities; doi:10.1007/978-3-319-44924-1

Anthopoulos, L. G. (2015). Transforming City Governments for Successful Smart Cities. In Transforming City Governments for Successful Smart Cities. doi:10.1007/978-3-319-03167-5

Bellini, P., Benigni, M., Billero, R., Nesi, P., & Rauch, N. (2014). Km4City ontology building vs data harvesting and cleaning for smart-city services. *Journal of Visual Languages and Computing*, *25*(6), 827–839. doi:10.1016/j.jvlc.2014.10.023

Bhardwaj, S., Jain, L., & Jain, S. (2010). Cloud Computıng: A Study Of Infrastructure As A Servıce (IaaS). *International Journal of Engineering*.

Blaser, M. (2019). *Development of a general maturity model for Smart Tourism Destinations*.

Boes, K., Buhalis, D., & Inversini, A. (2016). Smart tourism destinations: ecosystems for tourism destination competitiveness. *International Journal of Tourism Cities*. doi:10.1108/IJTC-12-2015-0032

Buhalis, D., & Amaranggana, A. (2013). Smart Tourism Destinations. In Information and Communication Technologies in Tourism 2014. doi:10.1007/978-3-319-03973-2_40

Çelik, P., & Topsakal, Y. (2017). Akıllı Turizm Destinasyonları: Antalya Destinasyonunun Akıllı Turizm Uygulamalarının İncelenmesi. *Seyahat ve Otel İşletmeciliği Dergisi*, *14*(3), 149–166. doi:10.24010oid.369951

Chaouchi, H. (2013). The Internet of Things: Connecting Objects to the Web. In *The Internet of Things*. Connecting Objects to the Web; doi:10.1002/9781118600146.ch1

Chen, X., Sun, L., Zhu, H., Zhen, Y., & Chen, H. (2012). Application of internet of things in power-line monitoring. *Proceedings of the 2012 International Conference on Cyber-Enabled Distributed Computing and Knowledge Discovery, CyberC 2012*. 10.1109/CyberC.2012.77

Chou, T.-S. (2013). Security Threats on Cloud Computing Vulnerabilities. *International Journal of Computer Science and Information Technology*. doi:10.5121/ijcsit.2013.5306

Chung, H., Park, J., Lee, S., & Kang, C. (2012). Digital forensic investigation of cloud storage services. *Digital Investigation*, *9*(2), 81–95. doi:10.1016/j.diin.2012.05.015

Ericsson. (2016). *Ericsson Mobility Report: On the pulse of the networked society*. Retrieved from https://www.ericsson.com/assets/local/mobility-report/documents/2016/Ericsson-mobility-report-june-2016.pdf

Gartner. (2017). Gartner Says 8.4 Billion Connected “Things” Will Be in Use in 2017, Up 31 Percent From 2016. doi:10.1017/CBO9781107415324.004

Gavalas, D., Konstantopoulos, C., Mastakas, K., & Pantziou, G. (2014). Mobile recommender systems in tourism. *Journal of Network and Computer Applications*, *39*(1), 319–333. doi:10.1016/j.jnca.2013.04.006

Glebova, I. S. (2014). Assessment of Cities in Russia According to the Concept of “Smart City” in the Context of the Application of Information and Communication Technologies. *Mediterranean Journal of Social Sciences*. doi:10.5901/mjss.2014.v5n18p55

Gretzel, U., Werthner, H., Koo, C., & Lamsfus, C. (2015). Conceptual foundations for understanding smart tourism ecosystems. *Computers in Human Behavior*, *50*, 558–563. doi:10.1016/j.chb.2015.03.043

He, W., Zha, S., & Li, L. (2013). Social media competitive analysis and text mining: A case study in the pizza industry. *International Journal of Information Management*, *33*(3), 464–472. doi:10.1016/j.ijinfomgt.2013.01.001

Huang, X. Kai, Yuan, J. Zheng, & Shi, M. Yu. (2012). Condition and Key Issues Analysis on the Smarter Tourism Construction in China. *Communications in Computer and Information Science*. doi:10.1007/978-3-642-35286-7_56

I-scoop. (2018). The Internet of Things (IoT) - essential IoT business guide. Retrieved from https://www.i-scoop.eu/internet-of-things-guide/

İlhan, İ., & Çeltek, E. (2016). Mobile Marketing: Usage of Augmented Reality in Tourism. *Gaziantep University Journal of Social Sciences*, *15*(24217), 581–599. doi:10.21547/jss.256721

Jindal, F., Jamar, R., & Churi, P. (2018). Future and Challenges of Internet of Things. *International Journal of Computer Science and Information Technology*, *10*(2), 13–25. doi:10.5121/ijcsit.2018.10202

Khan, M. S., Woo, M., Nam, K., & Chathoth, P. K. (2017). Smart city and smart tourism: A case of Dubai. *Sustainability (Switzerland)*, *9*(12). doi:10.3390u9122279

Koo, C., Shin, S., Gretzel, U., Hunter, W. C., & Chung, N. (2016). *Conceptualization of Smart Tourism Destination Competitiveness*. Asia Pacific Journal of Information Systems; doi:10.14329/apjis.2016.26.4.561

Lampropoulos, G., Siakas, K., & Anastasiadis, T. (2018). Internet of Things (IoT) in Industry: Contemporary Application Domains, Innovative Technologies and Intelligent Manufacturing. *International Journal of Advances in Scientific Research and Engineering*, *4*(10), 109–118. doi:10.31695/IJASRE.2018.32910

Lele, A. (2019). Cloud computing. In *Smart Innovation*. Systems and Technologies; doi:10.1007/978-981-13-3384-2_10

Li, S., Da Xu, L., & Zhao, S. (2015). The internet of things: A survey. *Information Systems Frontiers*, *17*(2), 243–259. doi:10.100710796-014-9492-7

Liu, B. (2012). Sentiment Analysis and Opinion Mining. *Synthesis Lectures on Human Language Technologies*, *5*(1), 1–167. doi:10.2200/S00416ED1V01Y201204HLT016

Markit, I. H. S. (2018). Telecommunications Industry Solutions - Market Data, Analysis; Forecast Reports; Services IHS Markit. Retrieved from https://ihsmarkit.com/industry/telecommunications.html

McKinsey. (2016). Internet of Things: The IoT opportunity - Are you ready to capture a once-in-a-lifetime value pool? *Hong Kong IoT Conference*, (June). 10.1007/978-1-4419-8237-7

Micera, R., Presenza, A., Splendiani, S., & Del Chiappa, G. (2013). *SMART Destinations: new strategies to manage tourism industry*. Shiuma et al.

Naphade, M., Banavar, G., Harrison, C., Paraszczak, J., & Morris, R. (2011). Smarter cities and their innovation challenges. *Computer*, *44*(6), 32–39. doi:10.1109/MC.2011.187

Neuhofer, B. (2014). The Technology Enhanced Tourist Experience. In Information and Communication Technologies in Tourism (pp. 90–96).

Neuhofer, B., Buhalis, D., & Ladkin, A. (2014). A Typology of Technology-Enhanced Tourism Experiences. *International Journal of Tourism Research*, *16*(4), 340–350. doi:10.1002/jtr.1958

Nitti, M., Pilloni, V., Giusto, D., & Popescu, V. (2017). *IoT Architecture for a sustainable tourism application in a smart city environment*. Mobile Information Systems; doi:10.1155/2017/9201640

Noguera, J. M., Barranco, M. J., Segura, R. J., & Martínez, L. (2012). A mobile 3D-GIS hybrid recommender system for tourism. *Information Sciences*, *215*, 37–52. doi:10.1016/j.ins.2012.05.010

Özen, A. (2019). Kapadokya Bölgesinde Yaşanan Sıcak Hava Balon Deneyiminin Davranışsal Niyetler Üzerindeki Etkilerini Belirlemeye Yönelik Bir Araştırma. *Verimlilik Dergisi*, *1*, 165–198.

Perkins, H. C., & Thorns, D. C. (2001). Gazing or performing? Introduction: The tourist performance. *International Sociology*, *16*(2), 185–204. doi:10.1177/0268580901016002004

Prameswari, P., Surjandari, I., & Laoh, E. (2017, November). Mining online reviews in Indonesia's priority tourist destinations using sentiment analysis and text summarization approach. In *Proceedings 2017 IEEE 8th International Conference on Awareness Science and Technology (ICAST)*, 121–126. 10.1109/ICAwST.2017.8256429

Quick, D., Martini, B., & Choo, K. K. R. (2013). Cloud Storage Forensics. In Cloud Storage Forensics. doi:10.1016/C2013-0-09718-6

Saleem, Y., Crespi, N., Member, S., Husain Rehmani, M., & Copeland, R. (2017). Internet of Things-aided Smart Grid: Technologies, Architectures, Applications, Prototypes, and Future Research Directions. 1–30. IEEE.

Sap. (2018). What is the Internet of Things? IoT Technology. SAP. Retrieved from https://www.sap.com/trends/internet-of-things.html#pdf-asset=e825c3a3-c27c-0010-82c7-eda71af511fa&page=1

Schmidt-Belz, B., Nick, A., Poslad, S., & Zipf, A. (2002). Personalized and location-based mobile tourism services. *Workshop on "Mobile Tourism Support Systems" in Conjunction with Mobile HCI*, 14. 10.1016/j.canlet.2008.11.010

Serpanos, D., & Wolf, M. (2018). IoT System Architectures. In Internet-of-Things (IoT) Systems (pp. 7–15). doi:10.1007/978-3-319-69715-4_2

SHGM. (2015). Nevşehir Kapadokya Balon Uçuş Sahası Slot Uygulama Talimatı. Retrieved from Nevşehir Kapadokya Balon Uçuş Sahası Slot Uygulama Talimatı website: http://web.shgm.gov.tr/documents/sivilhavacilik/files/mevzuat/sektorel/talimatlar/SHT-BALON-SLOT.pdf

SHGM. (2018). Sivil Havacılık Genel Müdürlüğü, Kapadokya Bölgesi Sıcak Hava Balon İstatistikleri, Resmi Yazısı (No. 29156034-401.04.99/E.136). Ankara, Turkey.

SHGM. (2019). Cappadocia Slot Service Center. Retrieved from http://shm.kapadokya.edu.tr/

Sigala, M., & Marinidis, D. (2012). E-Democracy and Web 2.0: A Framework Enabling DMOS to Engage Stakeholders in Collaborative Destination Management. *Tourism Analysis*, *17*(2), 105–120. doi:10.3727/108354212X13330406124052

Sinha, S. R., & Park, Y. (2017). Building an effective IoT ecosystem for your business. In Building an Effective IoT Ecosystem for Your Business. doi:10.1007/978-3-319-57391-5

Solima, L. (2016). Smart Tourism Destination from IoT Perspective: Adaptive Orientation System. *International New Challenges and Boundaries In Tourism: Policies, Innovations, and Strategies*, (June), 654–666.

ThingSpeak. (2019). ThingSpeak Documentation. Retrieved from https://www.mathworks.com/help/thingspeak/index.html?s_tid=CRUX_lftnav

Ultramagic. (2008). Technical support - Ultramagic. Retrieved from https://ultramagic.com/technical-support/

UNESCO. (2019). UNESCO World Heritage Centre - World Heritage List. Retrieved from http://whc.unesco.org/en/list/

Wang, C., Li, X., Liu, Y., & Wang, H. (2014). The research on development direction and points in IoT in China power grid. *Proceedings - 2014 International Conference on Information Science, Electronics, and Electrical Engineering, ISEEE 2014*. 10.1109/InfoSEEE.2014.6948106

Wang, D., Li, X., & Li, Y. (2013). China's "smart tourism destination" initiative: A taste of the service-dominant logic. *Journal of Destination Marketing & Management*, *2*(2), 59–61. doi:10.1016/j.jdmm.2013.05.004

Wang, L., Von Laszewski, G., Younge, A., He, X., Kunze, M., Tao, J., & Fu, C. (2010). Cloud computing: A perspective study. *New Generation Computing*, *28*(2), 137–146. doi:10.100700354-008-0081-5

Zanella, A., Bui, N., Castellani, A., Vangelista, L., & Zorzi, M. (2014). *Internet of things for smart cities*. IEEE Internet of Things Journal; doi:10.1109/JIOT.2014.2306328

Zhang, L., Li, N., & Liu, M. (2012). *On the basic concept of smarter tourism and its theoretical system*. Tribune Tourism.

Zhang, T., Wei, W., Fu, X., Hua, N., & Wang, Y. (2019). Exploring the roles of technology, people, and organization in building a tourism destination experience: Insights from the 2nd USA-China tourism research summit and Industry Dialogue. *Journal of Destination Marketing & Management*, *0–1*(February). doi:10.1016/j.jdmm.2019.03.001

Zhao, J., Zheng, X., Dong, R., & Shao, G. (2013). The planning, construction, and management toward sustainable cities in China needs the Environmental Internet of Things. *International Journal of Sustainable Development and World Ecology*, *20*(3), 195–198. doi:10.1080/13504509.2013.784882

Zheng, P., Wang, H., Sang, Z., Zhong, R. Y., Liu, Y., Liu, C., & Xu, X. (2018). Smart manufacturing systems for Industry 4.0: Conceptual framework, scenarios, and future perspectives. *Frontiers of Mechanical Engineering*, 1–14. doi:10.100711465-018-0499-5

Zheng, R., Zhang, T., Liu, Z., & Wang, H. (2016). An EIoT system designed for ecological and environmental management of the Xianghe Segment of China's Grand Canal. *International Journal of Sustainable Development and World Ecology*, *23*(4), 372–380. doi:10.1080/13504509.2015.1124470

Zheng, S., Xiong, X., Vause, J., & Liu, J. (2013). Real-time measurement of wind environment comfort in urban areas by Environmental Internet of Things. *International Journal of Sustainable Development and World Ecology*, *20*(3), 254–260. doi:10.1080/13504509.2013.779616

Zunnurhain, K. (2016). Vulnerabilities with internet of things. *Proceedings of the International Conference on Security and Management (SAM)*, 83. The Steering Committee of The World Congress in Computer Science, Computer….

ADDITIONAL READING

Balandina, E., Balandin, S., Koucheryavy, Y., & Mouromtsev, D. (2015, July). IoT use cases in healthcare and tourism. In *2015 IEEE 17th Conference on Business Informatics* (Vol. 2, pp. 37-44). IEEE. 10.1109/CBI.2015.16

Balandina, E., Balandin, S., Koucheryavy, Y., & Mouromtsev, D. (2015, November). Innovative e-tourism services on top of Geo2Tag LBS platform. In *2015 11th International Conference on Signal-Image Technology & Internet-Based Systems (SITIS)* (pp. 752-759). IEEE. 10.1109/SITIS.2015.11

Guo, Y., Liu, H., & Chai, Y. (2014). The embedding convergence of smart cities and tourism internet of things in China: An advance perspective. [AHTR]. *Advances in Hospitality and Tourism Research*, *2*(1), 54–69.

Jovicic, D. Z. (2019). From the traditional understanding of tourism destination to the smart tourism destination. *Current Issues in Tourism*, *22*(3), 276–282. doi:10.1080/13683500.2017.1313203

Kim, H. C., & Kim, Y. S. (2016). Smart tourism information system using location-based technology. *International Journal of Software Engineering and Its Applications*, *10*(11), 11–24. doi:10.14257/ijseia.2016.10.11.02

Li, C., Chen, D., Wu, D., & Su, X. (2015). Design of an EIoT system for nature reserves: A case study in Shangri-La County, Yunnan Province, China. *International Journal of Sustainable Development and World Ecology*, *22*(2), 184–188. doi:10.1080/13504509.2014.943329

Tripathy, A. K., Tripathy, P. K., Ray, N. K., & Mohanty, S. P. (2018). iTour: The future of smart tourism: An IoT framework for the independent mobility of tourists in smart cities. *IEEE Consumer Electronics Magazine*, 7(3), 32–37. doi:10.1109/MCE.2018.2797758

KEY TERMS AND DEFINITIONS

Analytics: In the IoT, analytics can be defined as the results achieved by analyzing data obtained through various intelligent devices.

Cappadocia Hot-air Balloon Tours: Cappadocia is known around the world as one of the best places to fly with hot-air balloons. The spectacular surrealistic landscapes combined with excellent flying conditions allow the balloons to gently drift over and between fairy chimneys, pigeon houses hewn into the unique rock formations, orchards and vineyards through impressive valleys, each with distinctive rock formations, colors and features and then float up over rippled ravines for breathtaking views over the region.

Cappadocia Slot Service Center (CSSC): The official organization of Cappadocia hot-air balloon tours.

Edge Device: Nearly all IoT businesses have 'things' or edge devices. Edge devices could be smart or made smart.

Gateway: Gateways could be independent boxes or small processing units that get embedded into existing edge devices.

IoT: The internet is not limited to the use of desktops, laptops or smartphones and it implies also the use of interlinked objects on an ecosystem such as smart automobiles, watches, jewelry, glasses among others.

ThingSpeak: ThingSpeak is an IoT analytics platform service that allows you to aggregate, visualize, and analyze live data streams in the cloud.

UML (Unified Modeling Language): System modeling language. It consists of 3 parts. Behaviour Diagram, Interaction Diagram ve Structure Diagram.

Chapter 8
Virtual Reality Applications in Tourism

Vahit Oguz Kiper
Sakarya University of Applied Sciences, Turkey

Orhan Batman
Sakarya University of Applied Sciences, Turkey

ABSTRACT

Through modern ages, tourism becomes a vital part of human life. With the advance of technology, tourism movements gain new tools. Those tools gave tourism an edge on management, marketing, education, and economic areas. Also, on demand side, new types of tourists with their brand-new demands and motivations became an important factor for those responsible for tourism supply. One of the most noticeable tools that new technology era has brought is virtual reality applications. Despite dating back to the 1950s, VR (virtual reality) has gained an interesting reputation in the last few years with the advances of related technologies. VR applications are used in many sectors like construction, military, communication, marketing, education, etc. Tourism industry is one of them of course. Especially, VR is used as both a tool and product in hotel & accommodation sector, recreation & entertainment, transportation, destination marketing, and a touristic product itself.

INTRODUCTION

Virtual reality (VR) plays a vital role in many sectors nowadays due to its technological and practical advantages. VR is a risk-free application for many jobs and lowering costs for some others. And sometimes, VR makes it possible what looks impossible for the other way (Craig, Sherman, & Will, 2009). Tourism industry with its hospitality, travel, transportation, marketing, management, recreation segments and sectors now use the benefits of VR like other sectors and industries (Sharda, 2010)

Since tourism is a service-oriented industry, it's argued if VR would really be useful for tourism. However, recent developments on VR technology, changing demands of Z generation and challenge of competition in the market created a space of use for VR with IT (information technology) in this industry

DOI: 10.4018/978-1-7998-1989-9.ch008

(Xiang, Magnini, & Fesenmaier, 2015). It is now important to take full advantage of VR and understand how to successfully apply it in tourism.

The main aim of this chapter is to investigate how to use VR application in tourism industry. Also, it is beneficial to understand how VR works itself and its components. Like many other things, VR comes with its side effects and weaknesses. This chapter also investigates that weaknesses to maximize the productivity of VR use in tourism.

First of all, virtual reality is defined with its historical development process in next subtitles. While making research about this chapter, its naturally seen that almost all literature were prepared in the name of natural or applied sciences mentality by the researchers of those fields. For tourism researchers, it might be a little tricky to fully understand and assimilate knowledge about this literature. With that, this chapter is prepared in a social sciences point of view, focusing and analyzing knowledge by social sciences researchers' eye. Instead of applied sciences' quantitative definitions and equations, this chapter aims to look the subject by an angle of qualitative and social view. It is believed that this will make it easier for social sciences researchers to contribute to the literature by further researches.

BACKGROUND

Virtual reality basically provides a medium that is in fact unreal for its user which they may experience unlimited scenarios with a computer system (Hobson & Williams, 1995). In another definition, virtual reality presents ultimate interface between computer applications and humans in a three-dimensional graphical world (Stone, 1995). Contrary to these two definitions which use "computerized" word, The Oxford Dictionary (OED) first made the definition of VR in 1980 as a synthesized reality with computerized clothing (Whyte, 2002, p. 2). There is a reason why VR systems called synthesized and that reason is of images that participants see in displays are usually generated by computers. They might as well be almost-real version of external world or they might also be fully creative and fictional.

Virtual reality dates back more than it's expected. In 1962 a vision director Morton Heilig developed first VR simulator named Sensorama Simulator (Bostan, 2007). In 1963, Sutherland developed a sketchpad which is also considered as the foundation of computer aided drafting system (CAD) and this development gave Sutherland the title: father of virtual reality systems. In fact, more and more computerized systems were developed since then based on Sutherland's model and today we reach the realistic and multifunctional VR systems (Craig, Sherman, & Will, 2009).

VR systems can be basically classified into 3 categories such as immersive (using Head Mounted Displays), semi-immersive and non-immersive VR systems (Gutierrez, Vexo, & Thalmann, 2008, p. 2). Full immersive VR systems are most realistic ones and comes with the highest price as expected. In fully immersive mediums, users body stay in the real world but almost all her/his senses are concentrated into virtual world. User freely roam around, experience and travel in immersive mediums. There might be a large zone for user to really move parallel to her/his movement in virtual world or as a space saving solution, there might be a walking mill similar to treadmill. All those motions are transferred via the avatar of the user (Ventrella, 2000).

In semi-immersive systems, there might be an HMD but these systems are usually fit for basic uses with lower freedom for user. In public malls, there are examples of semi-immersive VR systems. User wear an HMD helmet in front of a flat screen, seeing a recorded tape of a VR image, feeling like she/he is really inside of that medium. User might look around and may interact with limited objects by the help of input devices called gloves (like a mouse works for a computer or a joystick work for a video game). Non-immersive VR systems are the cost friendly ones with the most limited freedom. They are barely different from watching a movie with 3-D glasses. User wears special glasses for VR application and follows a screen that displays a recorded image. Non immersive VR systems could be named as visually rich videos that helps watching person to better understand what is displayed (Craig, Sherman, & Will, 2009).

VR has 3 characteristics as a medium: being interactive, being spatial and being a real-time application. Hence, if it would be otherwise, VR would be completely useless or non-different from older technology devices. There are already 3-D glasses to watch movies, play games, watch video without interacting or participating with a spatial zone or non-real-time (recorded) (Gutierrez, Vexo, & Thalmann, 2008).

Technological Developments Through Virtual Reality

4.8 billion of people over earth have a mobile smart phone while only 4.2 billion of them have a toothbrush (Loureiro, 2018). This only shows the commitment and dedication of human being into technology itself, perhaps this commitment might be turning into a crazy habit, even though that does not change the fact. The young generation which is also known as Z generation is now a growing in number group and also rising in spending statistics among tourist flows (Buffa, 2015). Z generation is also commonly known as millennials who were born into technology age and technological developments. They are mostly die-hard followers of new smart technologies especially tablets and telephones. Early of this age, there was a conflict between Z generation young members and their parents about overusing of smart devices (Serra, Ribes, & Baidal, 2019).

Since we know Z generation is addicted to new technologies and they are increasingly becoming new tourist type and also they are comfortable to pay a little more than their ancestors for experiences, it clearly tells that it must be considered what they demand. There is also another fact that should be taken into consideration. Millennial generation was grown up by computer games. They closely missed the times of video games and they had realistic and mostly simulating computer games and that already makes them really familiar with virtual reality concepts. Especially, games like Hitman, Grand Theft Auto series and Call of Duty series are fine examples of virtual games that you control an avatar, make him move freely, act independently in an unreal and virtual medium.

Technology is really becoming an emerging driver and fundamental force for tourist destinations (Kuflik, Wecker, Lamir, & Stock, 2015). Developing 5G improvements will surely increase the speed and possibilities of communication and technology. Companies like Facebook, Google and Microsoft are believed to make large investment in VR applications to develop even further technologies on that field. On the other hand, companies and institutions like NASA and IBM, spend great budgets to improve VR application that they already use (Patel & Cardinali, 1994; Retz, 2019; Thompson, 2019; Microsoft, 2019).

Developments of VR can be traced back to 1950s, but the key development of evolution was occurred in USA in 1980s (Stone, 1995). Many scientists and institution aid the development of VR. The need of such a technology in military, surgery and construction increased the speed of developing of VR systems (Whyte, 2002). Besides that, there were other factors that affects the improvement of VR. Movies like Surrogates, Inception and The Matrix are thought by authors to contribute to popularity and possibility of VR applications. Surrogates movie told the story of human kind were attached to machines at their homes and they are all living in a virtual world by their avatars with interacting to each other in real time. Well known Matrix series had a similar story which whole world is connected to machines and they are all transferred into a virtual world even without realizing it. Inception movie were a little different by its story which told other realities might be possible by dreaming.

Authors discuss the birth of VR and they go back to 1950's as it's mentioned before. However, if we go further back, one can date back the birth of VR through development of optic physics, and here is why: following the developments on optic physics field, scientists developed microscope and telescope. These two helped people to see far away and see really close. After that, merging optic knowledge with electrics, scientist developed video players and televisions. They were surely revolutionary inventions in their time. With screens showing us displays, computers were no longer calculators for us that give only numeric outputs. With more development on computer systems, here we have VR systems in use today.

Virtual reality is not only an entertainment tool. Flight simulators are also considered of natural developers of VR. Boeing Company is known as an early user of flight simulators to train its pilots. With that simulator, a realistic environment was created for pilot candidates looking just like a cockpit of a plane with almost all hardware and physical conditions (such as plane weight, wind, fly physics etc.) so that pilots can face possibilities of a flight in a risk-free situation (Valentino, Christian, & Joelianto, 2017).

Advantages, Disadvantages, and Main Component of Virtual Reality

VR gives participants almost complete freedom to create, sketch and imagine in a 3-D world (Craig, Sherman, & Will, 2009, p. 301) which is a fantastic opportunity for creative designers who enjoys creating. It is obviously tremendous to achieve and experience things that would be impossible in real life such as going to the Moon or South Pole (Kim, 2005). Indeed, the story of Superman is a well-known phenomenon that inspires lots of people to fly freely around the world or even travel between planets.

According to research observations and findings it is logical to address the challenge of unlimited walking or moving inside virtual world due to restricted area in real world during using VR application (Steinicke, Bruder, Jerald, Frenz, & Lappe, 2010). Even though VR systems try to solve this issue with treadmills, this does not bring a permanent solution. As another weakness, a badly designed interface would cause some trouble to user such as discomfort and sickness as a negative effect (Kim, 2005).

A typical VR system should contain a computer to process the data, a graphical card to show graphical contents, an input device/devices to allow participant to control and a viewing output (Riva, and others, 2007). One of the most important input device is position/motion sensor. They track the position of the user. There are several ways to do so. Electromagnetic, mechanical, optical, ultrasonic and muscular devices can track the movement of a user (Craig, Sherman, & Will, 2009, p. 18). Head mounted displays (HMD) usually employs two different display for each eye to provide isolated images (Kim, 2005). A reason that makes HDMs are nice is that they isolate eyes from external world, participant only see what she/he is seen on VR display, and those devices are usually equipped with tracking input so that VR processor can track the user's head motions, where the user look, turn his/her head, even lean or roll his/her head.

For sound display, there is mainly 2 types of use; headphones and loudspeakers. Headphones are more useful for semi-immersive and non-immersive VR systems and some HMD devices also include headphones for user. On the other hand, loudspeakers are more useful for fully immersive applications because they are better for depth of sound with surrounding effect for a more realistic experience. There is also haptic displays that is about the sense of touch. Haptic output display is connected to participant's skin or directly her/his muscular system. For example, in VR application, if a user is pushed towards right, haptics output device also pushes the participant through right to make it more realistic (Whyte, 2002).

Unfortunately, no practical output devices exist for simulating taste, smell, wind or other exotic human senses. For smelling simulation there are devices called olfactory which contains an odorant that controlled by the VR data processor and diffuses odor in different concentration and ratios. Food simulator is also problematic. There are taste-food simulators depend on chewing motion and includes some chemicals assigned to give some flavor into mouth. The chewing is made by a rubber positioned close to user's mouth and activated during need of chewing and tasting. Yet it might not be possible to tell that taste-food simulator devices are a hit success right now (Kim, 2005).

MAIN FOCUS OF THE CHAPTER

Virtual reality applications find itself a growing space of use in recent years and it seems like to keep this rapid growth (Gutierrez, Vexo, & Thalmann, 2008). As a part of services industry, tourism requires high knowledge and capability of management, planning, marketing and developing products for staying competitive among instantly changing consumer demands and behaviors (Fyall & Garrod, 2005).

This book chapter aims to understand the basic use and importance of VR applications in tourism industry. It simply both asks and answers the questions of how VR is related to tourism, how VR could be useful for tourism purposes, how VR can not really be a tool for tourism and how to merge VR applications with tourism activities.

Since VR applications are actually topic of natural and applied sciences, most of the researches are made according to natural sciences mentality. This brings the possibility of having difficulties for who research or work in tourism industry to adjust VR applications into tourism activities. Because of that, this chapter aims to give social sciences researchers a wider edge to understand and discuss the topic for further researches and implementations.

THE USE OF VIRTUAL REALITY IN TOURISM

Virtual reality finds itself a wide range of use due to its practical use and advanced possibilities that it serves to its users. Tourism industry is one of them and day by day tourism benefits more and more by VR applications. They are mainly spotted at education, marketing and sustainability subject of tourism. One of the reasons that makes VR useful is its capability to go beyond barriers of time and space with allowing collaboration (Craig, Sherman, & Will, 2009). VR applications are also used in demonstrating technical competence, design review and marketing (Whyte, 2002, p. 74).

Virtual Reality in Tourism Education

Managers who are responsible for sale teams and the one who struggle to find time and place, benefit from VR by creating training sessions (Upadhay & Khandelwal, 2016). Training a qualified and a well adopted employee is a real struggle in tourism business because tourism is a really service-oriented business and requires not mechanization but man power. Almost all tourism businesses, at least the ones which aims to survive in competitive market give effort to train their employees (Koçel, 2018). Well trained employees are somewhere out there but there are certainly competitors looking for them like hunters. So one way or another it comes to training own employees to maximize their productivity for the business.

Willingness to train employees do not solve all problems because training comes with two costs: time and price. A manager must have technical tools to train an employee and also spend most of her/his time on trainings (Ertuğral, Aslan, & Balık, 2015). By using VR in tourism education, one manager can train much more employees at a time while interacting with them without need of actual tools or equipment. In fact, the manager or the instructor do not have to be there while training. Even better, there is no obligation for a manager or instructor to be. There might be a program developed for this purpose only and it can handle training itself. Vogel and others (2004) also reports that data transfer via VR in the name of education and training is way more successful than conventional methods.

Extreme tourism types sound exotic for almost everyone but when it comes to actually do it, most would surely hesitate. Basic routines for tourism industry (such as housekeeping, check-in, services, flight attendance etc.) is being taught in tourism schools and colleges for many years without serious problems but how can one teach about kayaking, golf, sailing or agricultural tourism without actual equipment? VR applications can be useful for teaching extreme and alternative tourism types to tourism students with risk-free situations and at least in practicable ways. This also works in two ways. Not to repeat similar ideas again in next parts, it's a must to say that extreme tourism activities are much more achievable and reachable for tourists via VR applications. One who hesitates to experience an extreme tourism type might well try to experience it in VR application and may become a regular fan after overcoming hesitations.

A research among master students on e-Tourism program in Universitia della Svizzerra Italiana showed VR is useful for tourism education. According to research results, students who were selected for pilot application used VR sessions in addition to their classes and then they filled a questionnaire for their VR experiences (Marchiori & Cantoni, 2015).

Virtual Reality for Tourist Destinations

Today cities become really competitive among each other and have a complex structure and that requires to aid services and activities with information and communication technologies (ICT) (Liberato, Alen, & Liberato, 2017). Results tell that when tourist use VR experience, it enhances their perception about destination image and this helps destination to setup an emotional link with tourists (Pantona & Servidio, 2001). VR also might be quite handful for regional and urban planning (Haklay, 2003).

In literature it's seen that it's discussed if VR would be a replacement for actual tourist guides (Ersu, 2018). It is true that many electronic and new-tech products are used as tour guides in many destinations. That itself is quite controversial in many aspects. But a whole replacement of tourist guides with VR is looking highly irrelevant. Augmented reality applications might indeed take place of tour guides but VR does not look like an opposition right now.

Theme parks are small and unique destinations that attract tourists and there are many of them around the world. One well known example is Disneyland that welcomes millions of visitors each year. However, building and running a theme park or a theme city is not an easy task. Enormous investment brings huge risks and this is why there is not a theme park in every corner. With creating theme parks in VR systems, investment cost would be much lower and risk will be lower, too. Participants would spend some time in virtual theme park and would probably pay the investment back much quicker (Lukas, 2013).

In almost all tourism-introduction books, it's said that destination is a main part of tourism activity and one must travel to a destination from where she/he lives. Unfortunately, this reality excludes some of the society: handicapped and elderly people. Since handicapped and elderly people have some struggle to move to destinations, VR application can bring the destination to them. Museums, artificial sites, cities, extreme tourism activities can be reachable by VR systems for handicapped and elderly people.

Virtual Reality in Cultural Heritage and Environment Protection

Using VR applications in museums come with some advantages. There is no limitation of visiting hours, unlimited access and no limit of maximum visitors. Using VR applications basically turns museums into a timeless and placeless open access place of attractiveness (Aydoğan, 2017). VR systems can be useful to protect and enrich cultural heritage (Sürücü & Başar, 2016) and also might be handful to decrease the harm of tourism activities (Dilek, Kızılırmak, & Dilek, 2018).

VR applications can be used to visit places that do not exist anymore. There were great wonders on earth that is known through old books or paintings. People kept being curious about these places and most of the generations missed the chance of seeing them. By re-creating them in VR, people may have a chance to visit Colossus of Rhodes, Temple of Artemis or Lighthouse of Alexandria. It would be fascinating to roam in Gardens of Babylon.

Not only non-existing cultural heritage sites, but ancient destinations and civilizations would attract quite interest to visit in VR applications. Visiting Egypt at times of Giza Pyramid is being built, Ottoman Empire's Istanbul at its best age, ancient Greek with enlightening philosophers, forgotten civilization of Aztecs and Mayans with glorious city of Tenochtitlan. Going a little further, it would be a tremendous virtual experience to visit the age of dinosaurs.

Virtual Reality for Marketing and Developing Touristic Products

After increasing competition in the market, most business feel the pressure of developing unique products and services to sell in new market called "experience economy" (Aluri, 2017). Competition might sometimes be cruel for businesses. Especially in tourism industry which has a highly fluid and elastic type of demand. This makes tourism enterprises really vulnerable to demand changes. Because there will always be better quality service somewhere and there are always products just like yours (Faulkner & Valerio, 2000; Kozak, Kozak, & Kozak, 2017, p. 94).

There are plenty of tourism types nowadays and it is getting even impossible to follow them or categorize them properly. Researchers try to follow or anticipate tourist demands according to tourism types such as agricultural tourism, space tourism, wine tourism etc. However, before even researchers discuss a new tourism trend, it might get old and old-fashioned in the market. There is a possible new tourism trend named as "simulation tourism" by the authors of this chapter that might quite fit well with the content of this topic. It is thought to be useful to mention about it in case it might turn on new horizons for other researchers.

In simulation tourism, tourists are motivated to be a participant of a simulation, so to speak. There are quite familiarities between the motivation of using a VR application. If that would come true, and simulation tourism became a reality, then VR would be the vital part of it. Survival themed TV shows are shown in many countries. Only a few lucky participants join these shows. There might be others as well to experience a survival struggle and they would be the possible guests of simulation tourism who are dependent on VR applications.

It is complicated to discuss the possibility of using VR application in transportation. There are research papers mentioning about using of VR before flight and cruise sales basically showing customers about their seats, cabin and other amenities. Even though it is practical and possible to apply, it is not considered by authors that it would make a significant impact on sales. It is not expected from a customer to say: "Wow! That seat is really amazing, i must fly with that company instead of taking a bus or walking the whole road" after seeing a plane seat before a flight.

Still, this does not mean that VR application is no use for transportation. A motorcycle manufacturing company Kawasaki have already used VR motorcycle application for a better and safer ride. This was aimed to teach how to ride a bike and that might inspire some new travel agencies specialized on virtual tours and travels (Innoactive, 2016). Though, as long as VR does not transport people from somewhere to another place, not virtually but really, the use of VR in transportation would be limited.

VR already found a place for itself in hospitality sector. This is actually familiar with VR use in real estate sector. In real estate sector, contractors try to attract customers by VR applications that they see inside and outside of a real estate before purchasing it and even before it is constructed. Some hotels and travel agencies began to work same way by showing their hotels, restaurants, pools, entertainment amenities, spa centers and etc. to tourists. Before VR, tourists used to see hotels only by brochures or by visual images from internet or they had to take the shot by trusting the advice of a relative. With VR involved, tourists now gather more information about a hotel before vacation (Cooper & Macneil, 2015; Yung & Lattimore, 2019).

In production and marketing of services, there is certain line of things to do. A product is produced before production and a service is served during consumption (Usta, 2016, p. 114). In a shorter version, a business handles the cost of a service or product before it sells it. If it is not planned or anticipated correctly, then the business will begin to lose profits. With VR application, a product or a service can be tested in a virtual medium by producers and pilot consumers. This would give a solid idea about how consumers react to product. When enterprises develop a new product or producing a regular product or a service, they feel the stress of selling enough product/service at once to balance cost/profit equation (Doswell, 1997, p. 56-58). With VR involved, it will reduce the stress of unknown selling amount and cost/profit balance. This will also help the business to follow a straighter sale price.

Undertaking the cost of a product or service before selling is not only a risk for businesses. Tourists also take a risk by purchasing a product/service before consumption. If only tourists had a chance to see what product/service looks like and if they could experience it virtually, that would give them an idea about product/service would look like.

Another important task for businesses is to monitor consumer behavior closely. It is a real challenge that many businesses fail despite giving effort and also many researchers try to contribute to literature about how to watch and analyze tourist behavior before, during and after purchasing decisions. Since putting someone on every one of tourists during their travel would be quite disturbing and costly, businesses try more indirect approaches for this purpose. As a recommendation it could be said that VR application would give an edge to business owners and managers to monitor consumer behaviors closely and with more details because every action and decision of a consumer would be easily recorded without any additional effort or disturbing the consumer.

Issues, Controversies, Problems About Virtual Reality in Tourism

Using an HMD device for a long time might cause headache, nausea and the weight of HMD is also another issue for a user (Craig, Sherman, & Will, 2009). Companies that work on production of VR systems try to solve that issue but no radical solution was found yet. For augmented reality, mobile phones and some smart glasses could be well enough but when it comes to VR applications, it is still complex and dependent of input output devices.

The biggest obstacle against use of VR is still the natural willing of human kind about actually feeling, seeing and touching while experiencing a touristic product and just visiting a tourist destination (Özgüneş & Bozok, 2017). A great deal has been taken in the name of realistic graphical contents, though, the sense of taste and smell is still a tough challenge for VR systems productions to overcome.

Gastronomy is a shining tourism trend for many destinations with new tourist types demand to enjoy and experience what they eat and this brings the question of what is the place of VR systems in gastronomy tourism. Some can mention about using VR applications in training and marketing of gastronomy tourism and even experiencing it. Nevertheless, it seems like it's an effort of forcing VR application too much to fit in that area. Since gastronomy training needs to be specialized with some kitchen tools, it would require extremely advanced and realistic technological development to exchange real training with virtual one. Likewise, promoting or experiencing a traditional food seems insignificant right now (Pablo & others, 2019)

One possible problem for using VR applications in tourism marketing is the difference between expectations coming from virtual medium and reality. A VR application might exceed the reality while promoting a tourism product or a destination or a tourism facility. Even a user might misunderstand what she/he experienced during VR application and might get disappointed when she/he really consumes or buys a product or a service. These two situations would decrease the trust of a tourist to the business, destination or towards a product. Likewise, Jones (2006) in his paper mentions about differences of virtual reality and actual reality although significant development in VR technology about catching up reality of life experiences. On the other hand, Dilworth (2010) supporting that idea by pointing out although the great improvement, there is still quite distance between virtual reality and actual reality which could mislead people about real life who use VR applications.

Awareness and costs are another struggle for VR systems to overcome. Most of managers, trainers, specialists, entrepreneurs, planners, marketers or destination stakeholders are unaware of how effective a VR application can be despite VR is highly useful for direct or indirect employee training (Rodriguez, Gutierrez, & Sanchez, 2012). On the other hand, it must be acknowledged that VR systems are not cheap or affordable for many of mentioned above. However, it is possible and logical to say that both situations will change positively in time while decision makers will be more aware of VR usefulness and costs decrease.

SOLUTIONS AND RECOMMENDATIONS

Anyone who remembers early computers would also recall that they were giant machines looking like a big electric substation but now they are perfectly fit in tablets and even in our telephones. Just like computers, VR systems should be much smaller and ergonomic not to make people feel like wearing a war helmet from medieval times and not to make them carry a heavy thing on their head. That looks like one of the first improvements that HMDs should have.

A big question mark is clearly seen as human being seeks for real feelings and senses by its nature. Since changing the genetic codes of human being is nor ethical neither possible right now, it would be wise to simply make VR applications more realistic and avoid situation that conflicts human beings' natural instincts of seeking for reality. To protect the reliability of both VR systems and products/services that is meant to be promoted by VR applications, nothing more than a product/service itself can present to a customer must be promoted within VR applications. Trust is a hard to build and easy to demolish attitude among people.

It's understood that VR systems must promote itself before it promotes other products because of still lacking of awareness. Investors and enterprises about VR systems might be focusing too much on developing newer version of technology. Although technological development is must, marketing and promoting VR systems whilst developing them is necessary to make funding for new development attempts.

FUTURE RESEARCH DIRECTIONS

This chapter is focused basically how to use VR applications in tourism industry. Actual practices and possible way of use are tried to be explained whilst arguing possible obstacles and controversies are tried to be spotted. All innovations and investments are made based on needs so it might contribute to subject to further investigate how to use VR applications in tourism industry.

Also, this chapter ignores the economic aspects of use of VR applications and rather only focuses giving advices on possible utilization. For this reason, a research to measure economical/beneficial aspects of VR application which were mentioned in this chapter would also contribute to the current literature. Since VR applications come with a price by developing efforts, marketing them is also important to fund these development researches. For this reason, empirical researches on field of possible markets are needed to give developers a perspective of possible income.

CONCLUSION

Human being has an unlimited structure of needs and technological improvements develop in time to fulfill that needs. The more the technology develops, the more it will take place in people's life. Like all other industries and sectors, tourism businesses must pursue the race between humankind and technology. Because eventually a newer technology will take old one's place, and competitors in the market will give effort to get one step ahead of their rivals. It is a requirement to follow both technological development and demands of consumers.

Virtual reality is not a brand-new product neither a shocking discovery. Human race have never been satisfied of what she/he is or what she/he has. Since they have the ability to imagine, to dream and to create with that, there is a one-way journey so far and it will keep that way. As mentioned earlier, scientific developments on optic physics and it's meeting with electricity returned people as telescopes, televisions, augmented and virtual reality devices. Now it's seen that it evolves through extended reality concept. This development will keep its pace and tourism destinations and enterprises should catch this pace up.

REFERENCES

Aluri, A. (2017). Mobile Augmented Reality (MAR) Game As A Travel Guide: Insights From Pokemon Go. *Journal of Hospitality and Tourism Technology*, *8*(1), 55–72. doi:10.1108/JHTT-12-2016-0087

Aydoğan, D. (2017). Virtual Museums in the Context of Virtual Reality and Simulation. *E-Journal of New Media, 1*(2), 137–148. doi:10.17932/IAU.EJNM.25480200.2017.1/2.137-148

Bostan, B. (2007). *Sanal Gerçeklikte Etkileşim*. Marmara Universitesi Sosyal Bilimler Enstitüsü İletişim Bilimleri Anabilim Dalı Yayınlanmamış Doktora Tezi.

Buffa, F. (2015). Young Tourists and Sustainability: Profiles, Attitudes and Implications for Destination Strategies. *Sustainability, 7*(10), 14042–14062. doi:10.3390u71014042

Cooper, M., & Macneil, N. J. (2015). Virtual Reality Mapping: IT Tools for the Divide Between Knowledge and Action in Tourism. *Tourism Recreation Research, 30*(3), 61–68. doi:10.1080/02508281.2005.11081487

Craig, A. B., Sherman, W. R., & Will, J. D. (2009). *Developing Virtual Reality Applications*. Morgan Kaufmann Publishers-Elsevier.

Dilek, N. K., Kızılırmak, İ., & Dilek, S. E. (2018). Virtual Reality or Just Reality? A Swot Analysis of the Tourism Industry. *Journal of Tourismo, 4*(1), 67–74.

Dilworth, J. (2010). Realistic Virtual Reality and Perception. *Philosophical Psychology, 23*(1), 23–42. doi:10.1080/09515080903533942

Doswell, R. (1997). *Tourism: How Effective Management Make the Difference*. Butterworth-Heinemann.

Ersu, Ö. (2018). Profesyonel Turist Rehberliğinde Dijital Dönüşüm: Mesleğin Yakın Geleceği Arttırılmış Gerçeklik ve Sanal Gerçeklik Uygulamaları. *Uluslararası Turizm, İşletme. Ekonomi Dergisi, 2*(2), 578–586.

Ertuğral, S. M., Aslan, S., & Balık, M. (2015). Turizm İşletmelerinde İnsan Kaynakları Yönetimi. In O. Akova, İ. Kızılırmak, & H. Tanrıverdi (Eds.), *Turizm İşletmeciliği Temel Kavramlar ve Uygulamalar* (pp. 177–204). Ankara, Turkey: Detay Yayıncılık.

Faulkner, B., & Valerio, P. (2000). An Integrative Approach to Tourism Demand Forecasting. In C. Ryan, & S. Page (Eds.), Tourism Management Towards the New Millennium (pp. 45-57). Pergamon.

Fyall, A., & Garrod, B. (2005). From Competition to Collaboration in the Tourism Industry. In W. F. Theobald (Ed.), Global Tourism 3rd Edition (pp. 52-74). Elsevier Butterworth-Heinemann. doi:10.1016/B978-0-7506-7789-9.50009-1

Gutierrez, M., Vexo, F., & Thalmann, D. (2008). *Stepping into Virtual Reality*. Springer. doi:10.1007/978-1-84800-117-6

Haklay, M. E. (2003). Virtual Reality and GIS. In P. Fischer, & D. Unwin (Eds.), *Virtual Reality in Geography* (pp. 47–56). London, UK: Taylor and Francis.

Hobson, J. S., & Williams, A. P. (1995). Virtual Reality: A New Horizon for the Tourism Industry. *Journal of Vacation Marketing, 2*(1), 125–135.

Innoactive. (2016). *Mobile VR in Marketing – Kawasaki presents the Ninja ZX-10R 360° App*. Retrieved from https://innoactive.de/mobile-vr-in-marketing-kawasaki-presents-the-ninja-zx-10r-360-app/

Jones, S. (2006). Reality© And Virtual Reality ©. *Cultural Studies*, *20*(2-3), 211–226. doi:10.1080/09502380500495692 PMID:9548022

Kim, G. J. (2005). *Designing Virtual Reality Systems The Structered Approach*. London, UK: Springer.

Koçel, T. (2018). *İşletme Yöneticiliği*. İstanbul, Turkey: Beta.

Kozak, N., Kozak, M. A., & Kozak, M. (2017). *Genel Turizm İlkeler Kavramlar*. Ankara, Turkey: Detay Yayıncılık.

Kuflik, T., Wecker, A. J., Lamir, J., & Stock, O. (2015). An Integrative Framework for Extending the Boundaries of the Museum Visit Experience: Linking the Pre, During and Post Visit Phases. *Information Technology & Tourism*, *15*(1), 14–47. doi:10.100740558-014-0018-4

Liberato, P., Alen, E., & Liberato, D. (2017). Smart Tourism Destination Triggers Consumer Experience: The Case of Porto. *European Journal of Management and Business Economics*, *27*(1), 6–25. doi:10.1108/EJMBE-11-2017-0051

Loureiro, A. (2018). There Is A Forth Industrial Revolution: The digital Revolution. *Worldwide Hospitality and Tourism Themes*, *10*(6), 740–744. doi:10.1108/WHATT-07-2018-0044

Lukas, A. S. (2013). *The Immersive Worlds Handbook Designing Theme Parks and Consumer Spaces*. New York: Focal Press.

Marchiori, E., & Cantoni, L. (2015). Including Augmented Reality In Tourism Education Programs. In P. Sheldon, & C. H. Hsu (Eds.), Tourism Education: Global Issues and Trends (pp. 115-134). Emerald Insight (Published Online). doi:10.1108/S1571-504320150000021006

Microsoft. (2019). *Microsoft Store*. Retrieved from https://www.microsoft.com/en-us/store/b/virtualreality

Özgüneş, R. E., & Bozok, D. (2017). Turizm Sektörünün Sanal Rakibi (Mi?): Arttırılmış Gerçeklik. *Uluslararası Türk Dünyası Turizm Araştırmaları Dergisi*, *2*(2), 146–160.

Pablo, P., Ester Gonzalez, S., Redouane, K., Jaime, R., Ignacio, B., Francisco, P., & Alvaro, V. (2019). Immersive Gastronomic Experience with Distributed Reality. *Workshop on Everyday Virtual Reality*. Osaka, Japan: WEVR 2019.

Pantona, E., & Servidio, R. (2001). An Exploratory Study of the Role of Pervasive Environments for Promotion of Tourism Destinations. *Journal of Hospitality and Tourism Technology*, *2*(1), 50–65. doi:10.1108/17579881111112412

Patel, H., & Cardinali, R. (1994). Virtual Reality Technology in Business. *Management Decision*, *32*(7), 5–12. doi:10.1108/00251749410068111

Retz, M. (2019, May 1). *Facebook Wants to Bring Your Entire Body Into Virtual Reality*. Retrieved from https://edition.cnn.com/2019/05/01/tech/facebook-vr-avatar/index.html

Riva, G., Gagglioli, A., Villiani, D., Preziosa, A., Morganti, F., Strambi, L., ... Vezzadini, L. (2007). An Open Source Virtual Reality Platform for Clinical and Research Applications. *Virtual Reality Second International Conference ICVR* (pp. 699-707). Beijing, China. 10.1007/978-3-540-73335-5_76

Rodriguez, J., Gutierrez, T., & Sanchez, J. E. (2012). Training of Procedural Tasks Through the Use of Virtual Reality and Direct Aids. In C. L. Lanyi (Ed.), Virtual Reality and Environments (pp. 43-68). Croatia: InTech. doi:10.5772/36650

Serra, F. F., Ribes, J. F., & Baidal, J. A. (2019). Smart Destinations and Tech-Savvy Millennial Tourists: Hype versus Reality. *Tourism Review, 74*(1), 63–81. doi:10.1108/TR-02-2018-0018

Sharda, N. (2010). Tourism Informatics Visual Travel Recommender Systems, Social Communities, and User Interface Design. Hershey, PA: IGI Global. doi:10.4018/978-1-60566-818-5

Steinicke, F., Bruder, G., Jerald, J., Frenz, H., & Lappe, M. (2010). Visiting Tourist Landmarks in Virtual Reality Systems by Real-Walking. In N. Sharda (Ed.), Tourism Informatics: Visual Travel Recommender Systems, Social Communities, and User Interface Design (pp. 180-193). Hershey, PA: IGI Global. doi:10.4018/978-1-60566-818-5.ch011

Stone, R. (1995). The Reality of Virtual Reality. *World Class Design to Manufacture, 2*(4), 11–17. doi:10.1108/09642369310091106

Sürücü, O., & Başar, M. E. (2016). Kültürel Mirası Korumada Bir Farkındalık Aracı Olarak Sanal Gerçeklik. *Artium, 4*(1), 13–26.

Thompson, S. (2019, March 12). *VR Applications: 21 Industries Already Using Virtual Reality*. Retrieved from https://virtualspeech.com/blog/vr-applications

Upadhay, A. K., & Khandelwal, K. (2016). Virtual Reality: Adding Immersive Dimension to Sales Training. *Human Resource Management International Digest, 26*(4), 42–45. doi:10.1108/HRMID-01-2018-0014

Usta, Ö. (2016). *Turizm Genel ve Yapısal Yaklaşım*. Ankara, Turkey: Detay Yayıncılık.

Valentino, K., Christian, K., & Joelianto, E. (2017). Virtual Reality Flight Simulator. *Internetwoking Indonesia Journal, 9*(1), 21–25.

Ventrella, J. (2000). Avatar Physics and Genetics. In J. C. Heudin (Ed.), *VW virtual Worlds Second International Conference* (pp. 107-118). Paris, France: Springer. 10.1007/3-540-45016-5_11

Vogel, J., Bowers, C., Meehan, C., Hoeft, R., & Bradley, K. (2004). Virtual Reality for Life Skills Education: Program Evaluation. *Deafness & Education International, 6*(1), 39–50. doi:10.1179/146431504790560636

Whyte, J. (2002). *Virtual Reality and the Built Environment*. Architectural Press.

Xiang, Z., Magnini, V., & Fesenmaier, D. (2015). Information Technology and Consumer Behaviour in Travel and Tourism: Insights From Travel Planning Using the Internet. *Journal of Retailing and Consumer Services, 22*, 244–249. doi:10.1016/j.jretconser.2014.08.005

Yung, R., & Lattimore, C. K. (2019). New Realities: A Systematic Literature Review on Virtual Reality and Augmented Reality in Tourism Research. *Current Issues in Tourism, 22*(17), 2056–2081. doi:10.1080/13683500.2017.1417359

ADDITIONAL READING

Aydoğan, D. (2017). Virtual Museums in the Context of Virtual Reality and Simulation. *E-Journal of New Media*, *1*(2), 137–148. doi:10.17932/IAU.EJNM.25480200.2017.1/2.137-148

Craig, A. B., Sherman, W. R., & Will, J. D. (2009). *Developing Virtual Reality Applications*. Morgan Kaufmann Publishers-Elsevier.

Gutierrez, M., Vexo, F., & Thalmann, D. (2008). *Stepping Into Virtual Reality*. Springer. doi:10.1007/978-1-84800-117-6

Kim, G. J. (2005). *Designing Virtual Reality Systems The Structered Approach*. London: Springer.

Lyrtas, M., Pablos, P. O., Damiani, E., & Dial, L. (2011). *Digital Culture and E-Tourism*. IGI Global.

Pablos, P. O., Tennyson, R., & Zhao, J. (2012). *Global Hospitality and Tourism Management Technologies*. Business Science Reference-IGI Global Publishing. doi:10.4018/978-1-61350-041-5

KEY TERMS AND DEFINITIONS

Destination: The place where tourism activities occur or the place where tourist go for tourism motivations.

Head Mounted Display: A kind of helmet that is worn for using virtual reality application to see images or hear sounds or gave head movement inputs to VR processor.

Immersive: Explains a situation that completely or almost completely involves a person by the person's senses.

Medium: A specific environment of a specific thing or event.

Space Tourism: A tourism type that involves people who are motivated to travel through space voluntary.

Spatial: A real environment from anywhere in 3-dimensional space.

Virtual Reality: A device that presents the user or users virtually created scenes that is both real or unreal with additional components for input and output of senses.

Z Generation: Represent people who were born in millennium age, meaning after 2000s.

Chapter 9
Digital Gamification in the Tourism Industry

Emrah Özkul
https://orcid.org/0000-0002-2204-1378
Kocaeli University, Turkey

Emre Uygun
https://orcid.org/0000-0002-2204-1378
Kocaeli University, Turkey

Selen Levent
https://orcid.org/0000-0001-7980-4332
Kocaeli University, Turkey

ABSTRACT

In today's world where digital technology has become an indispensable part of human life, the use of digital platforms allows people to play various types of games to not only meet their personal needs, but also to keep people's minds away from the strenuous work tempo of daily life. The game is an action that allows people to socialize and gain different experiences and to have a pleasant time. It is limited by certain fixed rules for a purpose, including the outcome of winning and losing as a result of these rules. This action led to the emergence of a concept called gamification along with digitalizing technology. Gamification is the use of game elements, game design, and mechanics in the game environment that enable users to socialize and have fun in non-game environments. This chapter defines the concepts of game, gamification, and gamification in tourism, and examines the digital gamification applications in the tourism industry with certain parameters. Authors emphasize the relationship between tourism and gamification.

INTRODUCTION

Gamification activities make the process of introducing, marketing and teaching an idea, product or a

DOI: 10.4018/978-1-7998-1989-9.ch009

service interactive, enabling people to enjoy this process. The main purpose of gamification is to ensure user's commitment to businesses by contributing to the creation of the impression that they are part of the story designed to attract the user's attention through the right content and stories. Gamification is a model that makes it possible to apply by adjusting regarding to the characteristics of the sectors in which the companies operate. Although it has been functional for many years in the fields of marketing, services and finance, it is frequently seen in the tourism industry mainly in recent years. The confirmation of this statement can be seen especially from the use and creation of gamification designs in various forms by airline companies, tourism agencies and accommodation enterprises.

Companies in tourism industry (airline companies, tourism agencies, accommodation businesses, etc.), have started to transfer their reservation systems to digital platforms. The mobile applications used by the companies that continue to provide services on these digital platforms have enabled the touristic consumers to access to the businesses and services these companies offer without the limitations of the time and the place. Companies use gamification models to provide a more convenient and fun way to reach their services for consumers. Allowing tourist consumers to collect points for each flight, accommodation choice or mobile check-in transaction through mobile application and earning various gifts that they can use in later transactions can be mentioned as one of the exemplary gamification models that businesses have applied.

The use of gamification model in tourism is not limited to touristic organizations. Gamification models are also utilized for the promotion and marketing of touristic destinations. Especially, augmented reality (AR) and virtual reality (VR) applications are used to create a gamification model by transferring digital environments with various animations and modeling the historical, cultural and natural features in the destinations can help the tourists to fully experience the destination. To comprehend the idea better, a tourist visiting Berlin Wall visualizing the state of the wall before it was demolished on the screen of their smartphone, having instant access to information and videos about the story off the wall can be given as the example of the digitalization of destinations.

In this study; literature on game, gamification, gamification theories, elements and models, player's types, tourism and game, gamification in tourism marketing have been reviewed. As a result of the literature review, the relationship between tourism and gamification has been explained and sample gamification models in tourism industry have been included. In the last part of the study, based on literature review and sample gamification applications, a new model has been designed and suggestions for digital gamification studies in tourism have been presented.

BACKGROUND

Game

The concept of Game is a leisure activity that has taken place in our daily lives from the beginning of humanity to the present day and will continue as long as human beings exist. According to the information obtained from the archaeological excavations the people of Ancient Egypt, approximately 5000 years ago, played a game called "Senet" (Piccione, 1980). In another excavation, it was found that people living in Central Asia developed a game using bones of certain animals in 800 BC (Lovett et al., 1901: 280). Based on the findings of various excavations made through the history, it can be understood that games emerge in a different way in every culture and society (Akbulut, 2018: 96). Since the concept of play is an ongoing process from past to present, it makes it difficult to make a full definition in the literature. According to the definition made by Avedon and Sutton-Smith (1971), a game is a voluntary activity that is limited by certain rules to ensure an unequal outcome between the parties as a result of certain conflicts from equal conditions offered to players. In another definition, the game is a voluntary entertainment activity performed outside the real world (Caillois, 2001), where an artificial conflict is concluded with a measurable result (Salen & Zimmerman, 2004: 80).

The increase in digitalization has brought the traditional game culture to digital environments and has enabled the digitalization of games (Özkan, 2018: 13). With computers, smartphones and other smart devices, the increase in the use of the internet network makes it easier for people to access digital games (Kunduracıoğlu, 2018: 13). Digital games make the game more immersive and fun by creating a real-time interaction experience with other players (Gentes, Guyot-Mbodji & Demeure, 2010). The fact that digital games are more attractive to individuals than traditional games increases the number of digital game players. In an international survey in United Stated of America, it has been determined that the number of people playing digital games continuously has a higher ratio in countries like Germany (66%), Mexico (57%), Russia (53%), England (52%) comparing to the other countries (Herger, 2014). According to an international study made by Superdata research company, one out of every three people in the world (approximately 2.5 billion people) play free digital games on computers and mobile platforms, and the companies have earned approximately $ 82 billion in revenue (Superdata, 2017).

Game Culture From Traditional to Digital

In the world we live in, with the rapid change of technology in the last 3 decades, countless changes have happened in our work life, communication and games (Mclaughlin, 2012). With these changes, the number of academic studies on popular culture and digital games has increased in the last fifteen years (Seaborn & Fels, 2015: 14). Digital games quickly replacing traditional games have become an indispensable element of life (Mayra, 2008). In the traditional game culture, the children being the only ones playing games is the dominant idea throughout the society, whereas nowadays, it is seen that individuals in every age group of the society spend their free time by playing digital games (Kunduracıoğlu, 2018: 13). According to a study that was led in the United States, it is concluded that 70% of the gaming population is 18 years old and older and the average age for the gamers is thirty-four (Entertainment Software Association, 2018). With this research, it can be accepted that there is current concrete evidence that traditional game culture has changed with technology over time.

The report prepared by Entertainment Software Association (2018) includes findings on which platforms, people in America play digital games. Accordingly, it can be said that people who want to play digital games, mostly use their personal computers to access the games. The digital games that are played via computers create a higher sense of pleasure and relief on people compared to traditional games (Özkan, 2018:13). The fact that people who play digital games see themselves as part of the games they play, and that traditional games are replaced by digital games, is an indication of the digitalization of the game culture (Binark & Bayraktutan, 2011: 3).

Gamification

Gamification concept has been created with the combination of the words "game" and "add/adding" (ication) (Lim, Kim & Kim, 2018:23). This impression was used first in 2002 by computer program-

Figure 1. Gamification between Serious Game, Toys and Playful Design
Source: Deterding, Dixon, Khaled, & Nacle, 2011

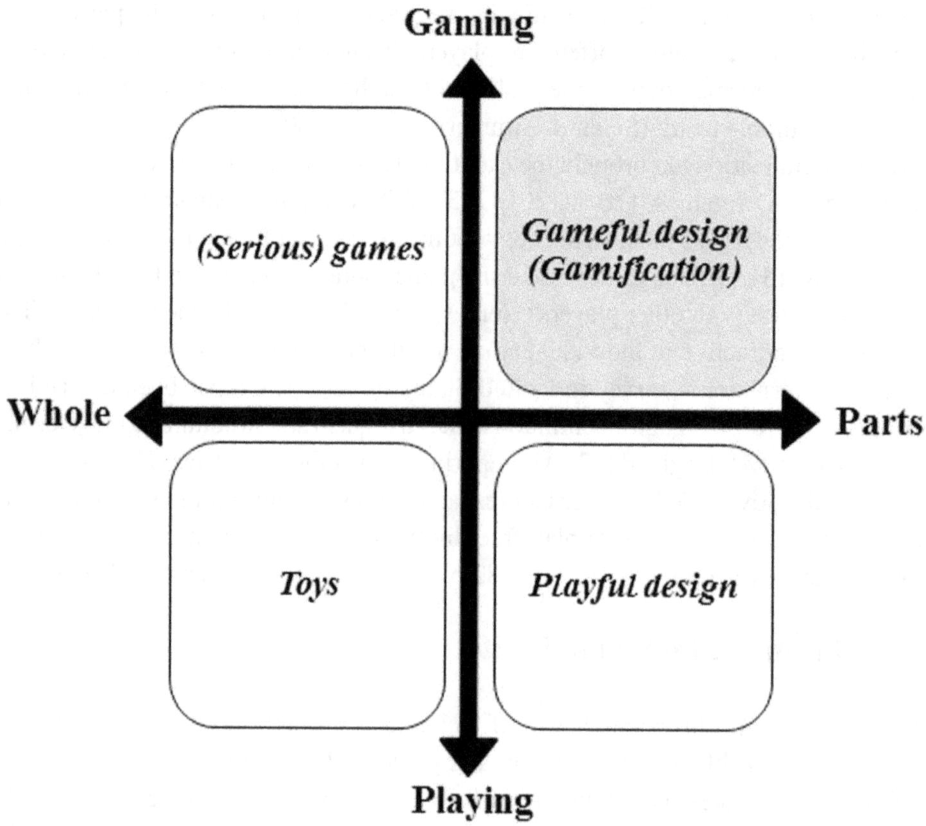

mer Nick Pelling (Marczewski, 2013: 3; Bayraktar, 2014: 5). Although the first definition of academic gamification was seen in the literature in 2008, this concept had not become popular until 2010 (Deterding et al., 2011: 9; Xu et al., 2014: 525). There are many definitions of the concept of gamification in the literature. Deterding et al. (2011) defines gamification as the use of game elements in non-game environments by making games more interesting, increasing the motivation of players. According to another definition, gamification is defined as communicating in order to solve the problems of users by using game and game mechanics (Zimmermann & Cunningham, 2011) and the use of game elements in non-game environments (Werbach, 2015).

Although nowadays some statements like productivity games, surveillance entertainment, fun ware, game layer and applied gaming are used for the gamification concept, the term "gamification" is commonly used in literature (Deterding et al., 2011). The relationship with the gamification concept and similar concepts are shown in Figure 1. The concepts indicated in the Figure 1 create the game concept altogether.

While the serious game concept shown in the figure expresses the usage of games with a serious aim, the toy concept expresses the simplification of rules of games with the aim of spending time in an enjoyable way. The 'gameful design' and 'playful design' concepts in the figure are the concepts that are evaluated in detail in literature. The difference in the roots of these two concepts is the main division separating the concepts of 'play' and 'game'. Calliois (2001) states that 'play' originates from the word 'Paidia', which constitutes an arbitrary combination of behaviors that are freer, improvised, and 'game' originates from the word 'Ludus' which indicates the existence of a competitive struggle for purposes based on certain regular rules. Due to this scope, the graceful design concept can be expressed as a combative game design created with regular rules with specific aims, and playful design concept can be expressed as game designs that make leisure times more entertaining and addictive for the gamer. Gamification can differ from other concepts in the literature due to its use of game elements without having to incorporate all of the other expressions in the figure (Matallaoui, Hanner & Zarnekow, 2017: 6).

Gamification Theories

Many theories related to the scope, practicality, development of the gamification and the motivation of gamers as a part of it have been asserted by experts along with the emergence of the gamification concept.

Flow Theory

The flow theory, created by Csikszentmihalyi (1990), states that on condition that the balance between the competition among the players and the size of the prize which can be obtained with a sense of competition is in parallel, it can make players happy by creating an element of motivation in the them. In the theory, gamers can gain experience by being motivated to achieve in game goals. Csikszentmihalyi (1990) has mentioned that this experience is possible with the combination of a balance between the personal skills of the players and the difficulty of the game tasks.

According to Nakamura and Csikszentmihalyi (2009) flow experience consists of the combination of six elements, the elements that make up the flow experience are expressed as follows:

- The intense and focused concentration of a person to a point during the experience
- The combination of behavioral awareness and behavior of the person
- Loss of self-awareness
- A sense of self-control over an action or activity
- Change in one's perception of temporal experience
- Experience of the situation or activities as a rewarding activity in itself.

The main purpose of using flow theory in games is to increase the intrinsic motivation of the players and to enable them to enjoy themselves by playing games (Kasa & Hassan, 2015). Figure 2 shows the emotions that players feel in the flow theory, depending on the variation between the abilities of the players and the difficulty levels of the tasks they face. It was stated that the player's ability to play increased with the difficulty level of the game, players played the games in a more focused manner and as a result they were happier. In the gamification efforts, it is possible to make the gamification activity successful if the masses of the players 'flow' so that they can be happy by focusing.

Self Determination Theory

Deci, Connell & Ryan (1989: 580) describes self-determination theory as having a sense of choice by choosing behavioral preferences that meet the innate psychological needs of individuals and enable them

Figure 2. Flow Theory
Source: Csikszentmihalyi, 1990

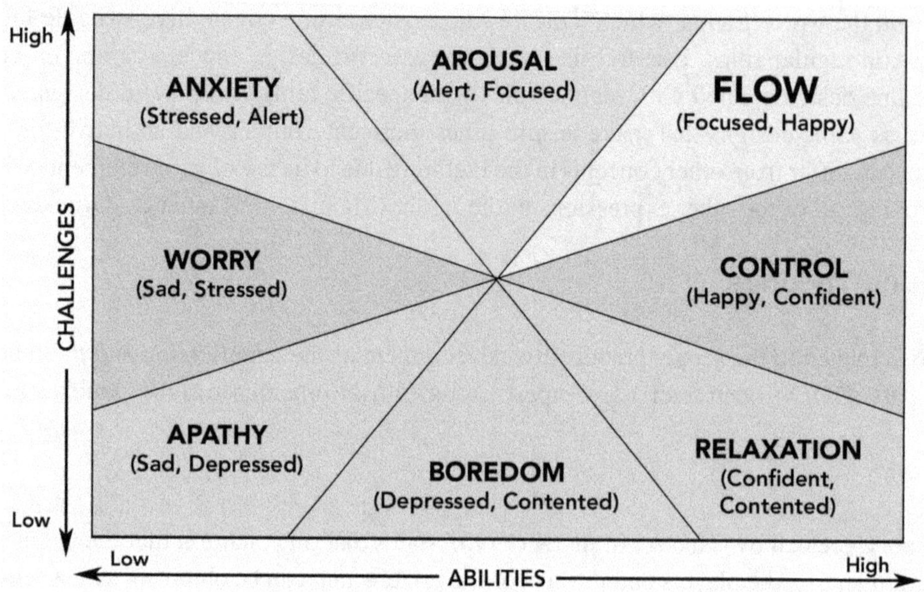

to create their own intrinsic motivation. Intrinsic motivation is defined as the motivation element that the individuals create within themselves, in order to achieve a satisfactory result for a specific purpose, regardless of external factors (Özkan, 2018: 27; Kunduracıoğlu, 2018: 32).

Ryan and Deci (2000: 68) emphasized that three personal needs, innate ‘‘competence, relatedness and autonomy”, are important in choosing the appropriate motivation among the internal and external motivations. These three personal needs can be seen in the Figure 3. Competence is the ability of individuals to be able to control their behaviors in a way that affects their surroundings (Kowal & Fortier, 1999). Competence in games can be seen as; completed tasks and gained points. Relatedness is one's need to establish a relationship with others to interact, to share with the need to socialize (Deci, Connell & Ryan, 1989). Relation inside games can be seen as; clans/teams chat rooms and inter-player messages. Autonomy is the desire of the person to act in accordance with his / her own control and wishes without being influenced by others, and not being completely independent (Andersen, Chen & Carter, 2000). Autonomy inside the games can be seen as; images, user characters (avatars), personalized options.

Hook Theory

In order to provide a continuity of the behaviors by directing people to certain actions, a four-stage theory called “hook” was developed by Nir Eyal. In the scope of this staged theory developed, a habit can be formed on users with specific features (Eyal & Hoover, 2014). Four stages of hook theory are shown in Figure 4.

These stages are briefly explained below (Eyal & Hoover, 2014; Özkan, 2018: 23):

1. **The Trigger**: It is defined as the elements that trigger the person to carry out the actions with the expectation of a reward. These triggering elements which enable the person to act are divided

Figure 3. Self Determination Theory
Source: Ryan & Deci, 2000

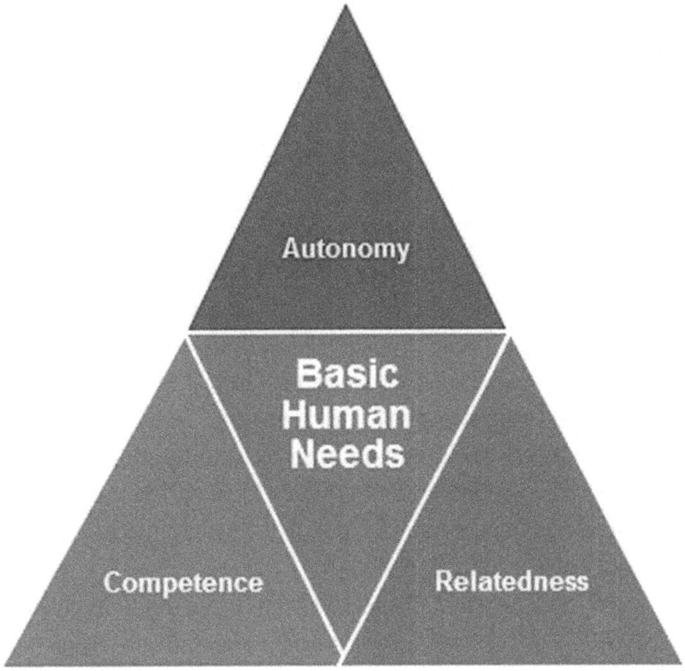

Figure 4. Hook theory
Source: Eyal & Hoover, 2014

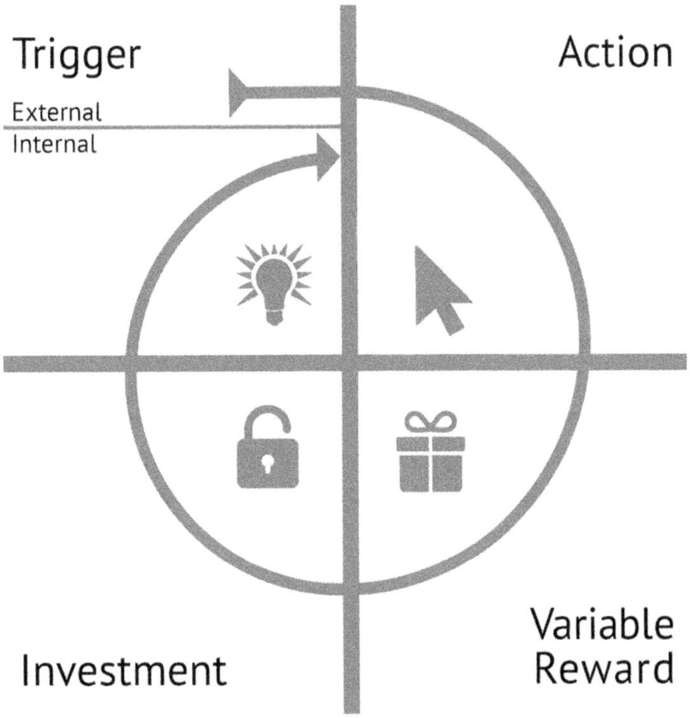

into two as internal and external triggers. While the internal triggering element is an automated motivation that exists in the minds of individuals coherent with a goal, the external triggering element consists of instructions and commands that trigger people. With various rewards, companies wanting to create a sense of expectation in the users, utilize different triggers. The creation of a clicking urge in users with the link to a website of the companies or the icons of mobile applications on smart phones can be shown as examples of the triggers that companies use to activate users.

2. **Action:** It may be insufficient for the users to be stimulated by internal and external triggers for a specific target with the expectation of a reward. In this context, users need to take a behavioral action in order to achieve the goals. Users who wish to benefit from promotion coupons or discount codes sent by companies can use the corporate websites and mobile applications of the companies can be evaluated in this context.
3. **Variable Rewards**: It is one of the most important tools that companies prefer to create brand loyalty and ensure that they become long-lasting consumers. The companies should meet the expectations of users that they have created with triggers for their purposes. Because, the expectation of a reward leads to an increase in the levels of dopamine hormones in the brain, thus making people happy. Therefore, the rewarded user can become a happy user and a loyal consumer. Chances of people winning prizes from scratch cards, lottery or other lottery applications organized by companies can be given as examples of variable prizes.
4. **Investment**: The aim of the investment phase, which is the final stage of the hook theory, is to ensure the loyalty of people by encouraging to act again with different triggers and to continue their loyalty to the company. The investment phase is also defined as the process in which users who are committed to the company make time and personal effort to influence the people in their social circles ensure that they become new users of the company.

Fogg Behavior Theory

Another theory asserted by Professor BJ Fogg in 2009 is called 'Fogg Behavior Theory'. This theory includes three elements which are motivation, ability and triggers that determine the behaviors of people (Fogg, 2009). Fogg (2009) expressing the factors that affect the behaviors that people will show for certain purposes explains these three elements as follows:

1. **Motivation**: These are the factors that lead to a certain behavior. There is a parallel relationship between the existing motivation of the person and the actions towards the goal. Individuals with low motivation are less likely to take actions.
2. **Ability**: These are the capabilities and potential that individuals have in order to achieve their goals. The element of motivation influences the effort of the person by enabling them to reveal their existing abilities to reach the target goal. Ability, which is directly related to motivation, may increase or decrease depending on the motivation of the person.
3. **Triggers**: It is the actual factor that triggers the behavior of a person directed at the target. Although the person has sufficient motivation and ability to achieve the goal, if there is no triggering element, the behavior does not occur. Triggers can appear in multiple ways. With the external triggers such as the corporate SMS messages, advertisements or announcements, there can also be internal triggers such as the feeling of hunger, the need for feeling satisfaction and the like. Evaluating Fogg's theory of behavior in the context of gamification. Wu (2017) describes the application of Fogg's

theory of behavior in gamification as the interest in games could be sustained by motivating the players' game points, badges, achievements towards their game tasks and making the games more attractive with the right prizes for the players.

First Gamification Examples

Although studies on gamification have been found in the 2000s, it is a concept that took its place in the literature in 2010. It is seen that the concept of gamification has been used in some fields prior to its definition in the literature and gamification efforts have been made with the applications in some fields. Gamification efforts by companies began in 1910 by Kellogg's, a cereal brand. The Kellogg's brand presented the book 'Funny Junhleland Moving-Pictures' as a reward to consumers who brought two cereal boxes. Following this implementation of the Kellogg's brand, Cracker Jack, a popcorn producer, began to place various surprise toys in each corn box produced in 1912 (Mccormick, 2013).

Although these cases are not accepted as a modern gamification application, giving toys to consumers as a reward has been a basis to the existence of gamification (Lloyd, 2014). The postage stamp manufacturer S&H Green Stamps, which has been active in the US since 1896 to 1970s, has encouraged people to collect postage stamps to increase the use of it. In this context, the company showed that postage stamps and postage stamp books produced by the company in 1975 can be used by people for entertainment purposes rather than solely on messaging cards (Lowbrow, 2014; Bayraktar, 2014: 15).

Gamification Elements

It is possible to make the people behave by making use of the gamification elements in the process of creating loyalty by directing people to specific behaviors and entertaining (Zichermann & Linder, 2010). In this context, although the concept of gamification is considered as a stand-alone concept, it is actually a combination of multiple elements. As one of these elements lacking or being absent may cause the collapse of gamification, gamification should be prepared correctly for the purpose. The elements that create gamification have been evaluated under three branches as mechanics, dynamics and components by. The three elements that make up gamification as a basis to a rising pyramid, these components form the foundation and there are mechanics on it with the dynamics on top (Werbach & Hunter, 2015).

Figure 5. The game element hierarchy
Source: Werbach & Hunter, 2015

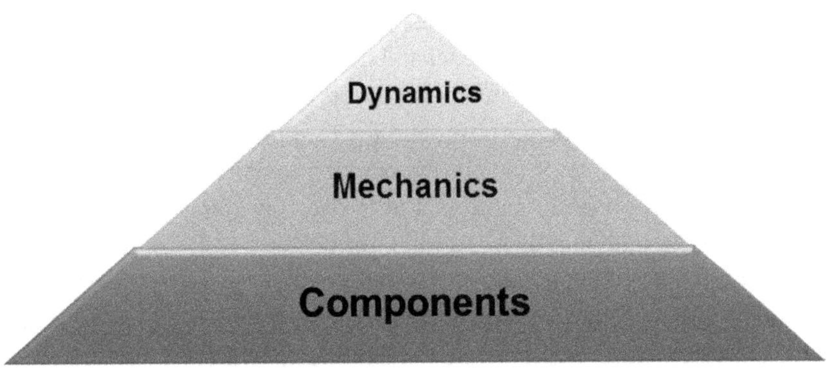

1. **Dynamics**: Among the components of gamification, the element of dynamics is at the top of the gamification pyramid. In the gamification that will be made, it is the element that directs the design of the structure that does not visibly exist but is intended to be created (Werbach & Hunter, 2015; Öztürk, 2015: 9; Tunga & İnceoğlu, 2016: 273) The dynamic element is divided into five sub-elements (Werbach & Hunter, 2012: 78; Werbach & Hunter, 2015; Matallaoui, Hanner & Zarnekow, 2017: 10). These sub-elements:
    - **Constraints**: The limitation of the behaviors that the players can exhibit and implement within the game in the designed games. It is the area that the players are free to move within the game with defined restrictions.
    - **Emotions**: The theme of the game is that the characters, music and stories create emotions like excitement, happiness, joy, surprise and anger on the players. Emotions are a set of emotional feelings that can affect player's enjoyment and motivation games and enable them to continue playing.
    - **Narratives**: Fictional stories that get the players to continue the games by creating a perception of reality. The consistency between the fictional story and the game makes the time spent in the game more effective.
    - **Progression**: It is the situation they face during gamers time in the game where players can unlock new tasks by allowing them to level up their in-game characters with various tasks and rewards. These situations can motivate players to make progress in-game quests by ensuring their dependence on games.
    - **Relationships**: These are the social interactions that gamers can establish with other players in the game. It is the element that provides socializing of gamers by providing the existence of feelings of friendship, competition and empathy with in-game tasks with team friends or rivals, in-game exchange and daily chats.
2. **Mechanics**: It is the element that takes place at the second stage of the pyramid that creates gamification and enables the gamers to interact with games. Mechanics are the fundamental rules of the game that has a supplementary feature of dynamic elements with increasing the motivations and loyalties of gamers to games (Werbach & Hunter, 2012: 79; Werbach & Hunter, 2015; Tunga & İnceoğlu, 2016: 273; Kunduracıoğlu, 2018: 21). The sub elements that constitute the mechanic element are classified into ten (Werbach & Hunter, 2012: 79; Karaarslan & Altuntaş, 2016: 437; Özkan, 2018: 36). These sub elements are;
    - **Challenges**: They are the challenges that gamers show towards the duties they face in order to reach in-game duties and rival gamers. These challenges are the hardships like solving a problem of gamers, finding or collecting the objects in order to gain the game.
    - **Chance**: It is the coincidental element that gamers can experience in the game such as excess of rival, duty numbers or hardship level and the reward number that they gain.
    - **Competition**: It is the situation in which players face the challenge of opposing players or computer players for in-game purposes and the result of winning or losing.
    - **Cooperation**: It is the state of the players working together to help other players to play the game in order to achieve an in-game task or to overcome a task.
    - **Feedback**: It is the state that the player is informed as a result of the system's evaluation of the behaviors of the players in the game. With the success statistics, systematic reports, current updates, screen notifications conveyed to the player by the game system, the players are informed about the game and motivated to continue playing.

- **Resource Acquisition**: It is the process of collecting the resources in the game in order to develop and strengthen the characteristics of the characters that the players have in order to complete the game tasks successfully and to reach new tasks.
- **Rewards:** These are in-game objects that are presented to the players by the game in return for the efforts of the players when the in-game missions are completed.
- **Transactions**: It is the commercial activity that players engage with other players who play the game in order to have objects that they do not have to complete the tasks they face. Swap behavior with other players in order to supply the missing objects and providing certain virtual/real money objects to the players by the virtual market in the game can be evaluated within the scope of in-game shopping.
- **Turns**: This is a limited period of time for players to collect resources and complete tasks to create equal opportunities. The player's required moves are made to each player in order to make their choices in order and the gameplay is presented equally.
- **Win States**: It is the win or lose situation that players will face depending on whether they have succeeded in the tasks they have completed individually or as a team in order to accomplish in-game goals.

3. **Components**: Components that make up gamification are located at the bottom of the gamification pyramid. The components that add a complementary feature to dynamic and mechanic elements are the elements that was taken into consideration with the interaction that gamers make with the envisaged game (Werbach & Hunter, 2012: 80; Werbach & Hunter, 2015; Tunga & İnceoğlu, 2016: 273; Karaarslan & Altuntaş, 2016: 437; Kunduracıoğlu, 2018: 22). Werbach and Hunter (2012: 71) have gathered the components of the games into fifteen sub-elements within the scope of gamification. These are;
    - **Points**: It is the numerical equivalent of the efforts of the players in the tasks they encounter in the process of achieving in-game goals. This numerical response to the player by the game is an indication of how close the targets are as a result of players finding in-game objects and completing tasks. The need for players to score more points in order to improve the strength and characteristics of game characters by achieving success in the designed games can lead to increase in-game competition (Sobocinski, 2017: 136), encouraging players to show more active behavior in games.
    - **Badges**: It is a visual representation of the in-game achievements of the players with the tasks they have completed and the points they have gained. The badges earned by the players are a symbol of their mastery in playing tasks against their opponents. The need of proving their mastery by collecting more badges than competing players is one of the factors that contribute to motivating player's in-game.
    - **Leaderboards**: They are the tables that the points, objects and badges of the players and the in-game statistical status in which the players rank in comparison to the other players who play the game. The leaderboard helps a player to see himself and his opponents to see how well he is in winning the game, which can motivate players more than others to achieve success in the game.
    - **Achievements**: These are the prizes given to reach the next quests as a result of the successful completion of each of the tasks within the designed in-game goal.

- **Avatars**: They are virtual game characters that represent each player in the games played. Avatars are visual elements that can be personalized with various objects in the game to reflect the personal characters of the players.
- **Boss Fights**: In in-game quests, players are challenged with high-level opponents to achieve a rare object or subsequent quests compared to their opponents.
- **Collections**: The collection or gathering of subjects or badges in the designed game by the gamer.
- **Combat**: Short-time battles with other players or computer controlled players during in-game missions.
- **Unlocking Content**: Unlocking and accessing the task elements, locked objects, maps as a result of completing certain tasks and collecting in-game objects.
- **Gifting**: These are the gifts that are given with specific time spaces to the gamers who play the game regularly by the designed game system. It is also called in-game gifting when players share the resources they have acquired in the game with each other.
- **Levels**: These are the experience points given to players in order to achieve other tasks at the end of each game task.
- **Quests**: Pre-determined challenges for achieving in-game rewards or goals.
- **Social Graphs**: It is a graph that shows the features of the players in their friend lists and their interaction with these players by connecting to the social networks in the game.
- **Teams**: These are the gamer communities that gamers ensemble to help each other with a mutual goal or a task.
- **Virtual Goods**: These are all the virtual objects that players use as they have certain meanings and functions within the game.

Gamification Models

There are different studies in the literature regarding the effective use of gamification and various models created in these studies. The most important and the most remarkable studies among the examined are the D6 model of Werbach and Hunter and the Octalysis model developed by Yu- Kai Chou that is known mostly in gamification sector (Yılmaz, 2017: 216).

Model of D6

In order to create the gamification design, it is necessary to create a large system in its infrastructure. It is important that each stage and step to be performed within this system is examined and planned in detail so that the objectives of the model can be expressed clearly. According to Werbach and Hunter, in order to establish an effective gamification design, six basic parameters must first be determined. Based on these parameters, Werbach and Hunter call their gamification model 'D6 Gamification Design Model'. The six basic parameters indicated according to this model are given in Figure 6 (Werbach & Hunter, 2015: 86).

Setting Goals: The first thing to be done in gamification is to determine the business goals. Therefore, the first stage of the model is to set a business objective. Business objectives are related to specific performance objectives within the gamification design. Some examples like creating brand loyalty, increasing efficiency of the workers and consumer loyalty can be given as business objectives (Werbach

Figure 6. D6 gamification design
Source: Werbach & Hunter, 2015

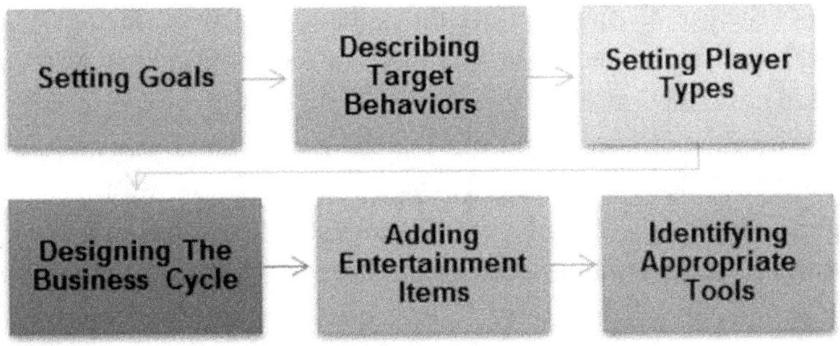

& Hunter, 2015: 86). Such objectives are referred to as SMART targets. The initials of the word SMART are created from the codes of the creation of the objectives and the word expansions; Specific (accurate, clear, open), Measurable (can be expressed with unit of measure that are accepted as measurable and standard), Attainable (visible and accessible), Realistic (Realistic and appropriate to company resources), Time-bound (Periodical, starting and ending within a certain time period) (Yılmaz, 2017: 86).

Describing Target Behaviors: The targeted behavior should be tangible and specific. After the desired target behaviors are listed, various metrics are designed to turn behaviors into measurable results. The applications that are used for gamification work with software applications. So, besides the seen actions the activities behind the scenes are converted to certain numbers to create feedback process. Scores are preferred as the most useful tools to measure behavioral changes. It is possible to monitor changes in target behaviors with a scoring system that can quantitatively measure any progress and change that will occur (Werbach & Hunter, 2015: 90).

Setting Player Types: Yılmaz (2017: 87) states that the most common mistake that is made during gamification design is that designs are designed for monotype players assuming that all players have the same characteristics and this design also is expected to have an effect on other types of players. The most common and biggest mistake is to think that gamification techniques are suitable for all player types and to design accordingly. Werbach and Hunter (2012: 93) suggest further elaboration of the segmentation of player types in order to manage the design process better and make it effective. The most effective way to understand the target audience better is to classify player types according to their distinct character traits. The best-known model for determining player types is the Bartle Player Model. In this model, player types are divided into 4; as achievers, explorers, socializers and killers. While achievers aim to raise levels or earn badges, the explorers want to discover new contents. Socializers want to make new friends, while the killers want to impose what they want and defeat other players (Werbach & Hunter, 2012: 92). Approximately 80 per cent of the player types are socializers, 50 per cent are explorers, 40 per cent are achievers and 20 per cent are killers (Zichermann & Cunningham, 2011: 23).

Designing the Business Cycle: Games always have a beginning and an end. Each end must have another beginning and games must be developed in a way which such a cycle exists. Thus, players can be kept in the system for a longer time. Social media applications are the most convenient and preferred systems for planning activity cycles. For example, while sharing photos on Facebook, one of the social media applications, tagging other people in the photo contributes to the formation of a new cycle. Other people also interact with specific actions. The cycle thus continues and contributes to the formation of new cycles. There are two types of operating cycles. These are loyalty cycles and progress steps. The loyalty cycle consists of 3 steps: motivation, action and feedback. In addition, the loyalty cycle is a micro-plane depiction of the overall picture of what the player does, why he does it, and how the system responds to it. This cycle constitutes the basic operations of the gamification system and progress steps are introduced when the next steps cannot provide users with different experiences. The progress steps reflect the change in player experience when playing the game. The loyalty cycle and progress steps form the basis of gamification design. The loyalty cycle gives the player awareness of the goals and the game, while the progressive steps give the player a journey into his world (Werbach & Hunter, 2015: 94).

Adding Entertainment Items: While creating gamification designs, there are some questions that the creators of the design should ask themselves. These questions are: Do players participate in the system voluntarily? If there is no external prize offered, will players still be able to continue playing? If the answer to these questions is no, the elements that can make game design more fun should be considered and worked on. Thus, players will participate voluntarily in games that become more fun. There are 4 different types of entertainment factor in gamification design. These are hard entertainment, easy entertainment, variable situations and social entertainment. Hard fun is the kind of fun that gives players the feeling of overcoming something by forcing them, while easy fun is the kind of fun that allows the player to discharge without forcing him too much. In changing situations, players recognize different characters and experience new experiences, while interacting with other players with social entertainment is defined as a competitive environment (Özkan, 2018: 45).

Identifying Appropriate Tools: After completing the other stages precisely, the last stage is to use the right tools. At this stage, the right tools to use are decided, and the decision-making phase and action process begins. It is recommended that this stage should be performed by many experts rather than by a single person. In order to implement gamification design correctly (Werbach & Hunter, 2012: 103);

- In the gamification process, the business objectives of the project are formed by experts.
- To have a good understanding of the target audience and to have the basic knowledge of psychology to guide this audience to the desired behavior.
- Choosing a game designer with sufficient technical knowledge.
- Finding a competent person to analyze and interpret the data provided by gamification system
- It is recommended to have technology experts to ensure the applicability of the project.

Octalysis Model

According to Yu Kai Chou, gamification is a design process that is based on human motivation Many other systems are based on the assumption of that the work is done in a short time and the employees will only do the work they need to do. Human-centered design reveals that people in a system have insecurities, emotions and reasons for whether they want to do certain things. Therefore, they take into account that they optimize their motivation, emotions and loyalty. In this context, Yu Kai Chou considers gamification as a human-oriented design and states that there are 8 main effects of gamification. Chou states that these basic elements are located in different positions according to left and right brain functions of human (Chou, 2016: 7).

Figure 7. Left Brain vs Right Brain Core Drives
Source: Chou & Yu-Kai, 2016

According to the factors in the octagon of Chou, by ascribing a meaning indicates that the player feels that he is chosen for this job and he is doing a great job, and he assigns a meaning to the task. Even if they do not have any financial gain during the game, their inner satisfaction that they feel enables them to concentrate on the game. The empowerment factor is related to the completion of the task. Unless there is a struggle in the game, awards or badges have no significant impact on players. There is an important bound between the concepts of success and struggle. It is not always sufficient for individuals to reveal their creativity in activity by itself. In addition, individuals will want to see the result of their creativity. Therefore, different ways should be provided to players through feedback. The feeling of ownership is related to player's sense of belonging to the game. The more individuals feel belonging to the game, the easier it will be to identify with the game. The social pressure factor is related to the relations of the player with other players. When strong relations are established, people's loyalty to the game increase. The scarcity effect, which is a basic human impetus, is related to the desire of a person to want because he does not have that thing. In order to have the treasures, badges and rewards that he does not possess, he stays in the game and spends time. Unpredictability, not having an idea of what he can encounter in the next step of the game, will arouse more curiosity and loyalty to the game. The last factor, avoidance, refers to the state of eschewal and avoidance of any scenario that is negative in the game (Chou, 2016: 9; Özkan, 2018: 48). Yu Kai Chou grouped these factors according to left and right brain functions. The left brain part of the octagon in Figure 7 is related to extrinsic motivation, while the right brain part is related to intrinsic motivation. Chou considered the negative motivation at the bottom of the octagonal scheme he designed as the black hat (the dark side of gamification) and the positive motivation at the top as the white hat. It emphasizes that success can be achieved if both parts of the gamification design scheme are used proportionately (Chou, 2016: 9; Şahin & Samur, 2017: 7).

Player Types

Player types are one of the most important factors that influence players during the game. Each of the players in the game behaves differently. Therefore, one player may like the game while the other player may not like the game at all. There are different factors in determining the game preference of the people. Their own personalities and styles are important in individual's game preference (Yee, 2006: 773).

The model that Bartle created in his study is shown in the figure 8. In his study, he examined multi-player games and determined four basic player types. In this study, expressed with X and Y axes, while the game world is seen on the right part of the X-axis, on the left side is the players. In the Y axis, there is interaction at the bottom and the game is at the top part. While the player that is near to top part in the Y-axis plays a game-focused game, the player that is near to below part plays an interaction-focused game. While a player who is close to the left part of X-axis plays a game focused on other players, the focal point begins to form the game world as he moves to the right. These player types located between X and Y axes are; achievers, explorers, killers, socializers (Bartle, 1996; Kunduracıoğlu, 2018: 23):

- **Achievers**: This player type is game-focused and tries to achieve in-game goals. The achievements of the game are gamers' priorities such as collecting badges, collecting points and gaining experience. Although, they feel the need of socializing in a duty that they cannot do on their own. Scouting is a tool for them in order to obtain new resources and information. They prefer the games that they can gain a lot of achievements and badges rather than game type.
- **Explorers**: This player type is interaction-focused and tries to explore new ways and methods. The mystery in the game and the urge of wonder that they feel is more important than earning

Figure 8. Player Types
Source: Bartle, 1996

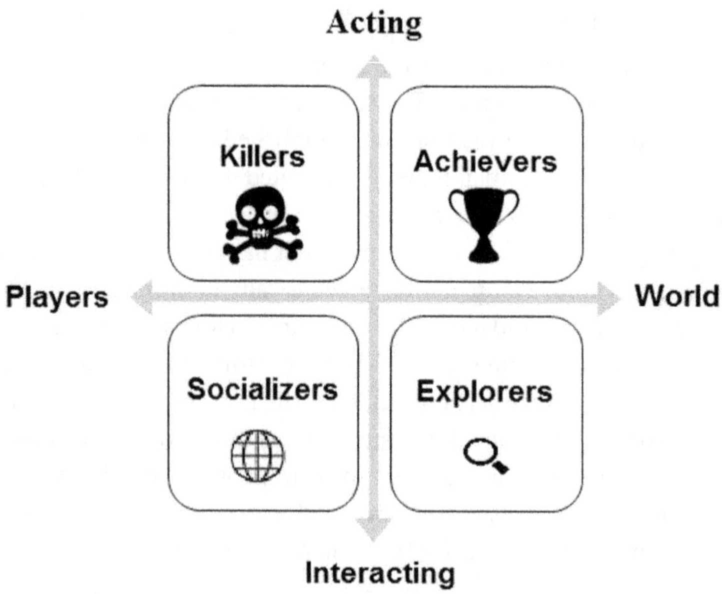

points. They do actions like killing in order to collect point just to enable them to continue the game and make new discoveries. These type players mostly prefer adventure games.

- **Killers**: For these player types that are focused on other players, in other words, their rivals, the only aim is to beat them. One of the main goals is to ensure that other players lose and prevent them from winning. What is important for them is not to win but to defeat their opponent. Therefore, games without losers do not attract their attention. They usually prefer race and battle games.
- **Socializers**: For this interaction-oriented player type, the game is a tool for socializing. They do not have aims like defeating opponents, collecting points or making discoveries. The thing that is truly important for them is to have a good time with other people. These types of players prefer easy-to-play, low-competition games. An example of this is Rummikub, which is a multi-user game where they can have a chat.

Although player types are covered in 4 main headings it is not possible to make certain distinctions. Every player can be a little achiever, a little explorer and a little social. While some players may mainly tend to a single type, while others may have several. It can be easily understood with observation during the game which player type the players belong to, from the type of game they play, their in-game goals and their behaviors during the game.

Bartle's classification is based on the game, but gamification is not exactly the same with the game as mentioned. Therefore, it is seen that Bartle's typology needs to be revised in the field of gamification. With Marczewski's work, more detailed and specific gamification types have emerged for gamified systems. In addition, contrary to the most common beliefs, apart from rewards such as money, motivating autonomy, mastery and goal factors have been shown to motivate users more in the game. He considered motivation 3.0 as the root of the typology of player types as it is shown in model 9. However, using the term player suggests that external motivation is important in gamification (Sever et al., 2015: 193).

Figure 9. Gamification user types
Source: Marczewski,2015

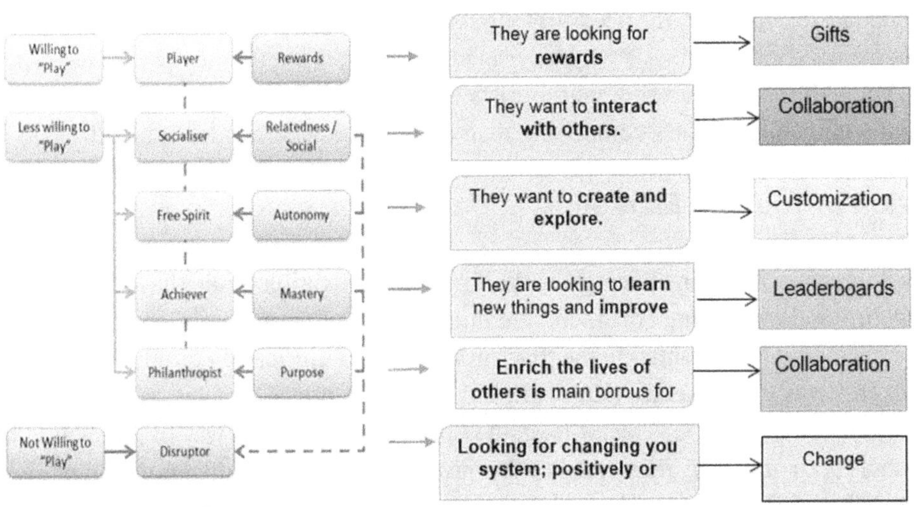

MAIN FOCUS OF THE CHAPTER

Tourism and Game

The use of games in tourism marketing can potentially offer great marketing opportunities. Tourism is an experience industry that can be created and personalized. In order to develop these types of experiences, some technology tools such as social media, smartphones and games are needed. Among the first companies to benefit from technological developments are the travel industries (Buhalis & Law, 2008). The current use of games by the tourism industry is divided into two major groups as social media and location-based platforms (Xu et al., 2014);

1. Social media, such as Facebook, based social games used to attract potential customers, create a target or company image, and promote brand awareness.
2. Location-based mobile games are informative games that provide tourists with more detailed information about travel destinations (Waltz & Ballagas, 2007). Furthermore, it is a very rich source of information for tourists in a tourism centre thanks to images, sounds and emotions that cannot be continuously fully simulated with computers (Linaza et al., 2014: 498). Most of the existing tourism destination games are based on the game principles of the classic treasure hunt. For example; The Amazing City Game was developed in Trondheim, Norway, in order to encourage tourists to take part in a quiz tour by solving tasks at different locations (Wu & Wang, 2011).

However, these gaming applications are not always successful. Some tourists are not interested in games and sometimes destitute of fun and facts about touristic places. When Çeltek's (2010) study was examined, it was concluded that the games other than VeGame and Geocaching did not provide any information about the country and the destination. However, this result is limited to the games examined in the study. Not all games made in the field of tourism can be generalized in this context. Game developers with technical backgrounds may sometimes not be fully aware of the tourist's need to play and motivation. That's why it should be designed with a user-centered approach to game design (Yovcheva et al., 2014: 15). The game experiences of tourist gamers can be benefited for the design of these games. Although games have theories on player motivation, they cannot address tourists in particular. Therefore, they are inadequate to determine the needs of tourists. While designing these games, you should look for answers of some questions, such as ''why tourists play games, when do they play, what types of games they prefer''. In addition, considering that tourists have limited time, the games that will be designed should be less ambiguous and less compelling than other games (Fernandes et al., 2013).

Gamification in Tourism Marketing

With the development of digital technology, marketing communication studies have begun to be reshaped. Mobile marketing and socializing are among the future trends of marketing. Gamification offers opportunities as a new marketing potential in the tourism industry. Gamification provides virtual experiences to individuals in terms of marketing (Xu et al., 2014). Games are about pleasure and there is a strong impact of marketing the pleasure. Therefore, games can offer a powerful way of interacting in a fun and useful way. It has been proved by many studies that promotion of tourism marketing is an important issue (Huang et al., 2013: 492). The use of the immersive virtual environment, which is one of the cutting-edge approaches for the promotion of tourism sites, will contribute to the successful interaction with the target audience (Bogdanovych et al., 2007). The use of gamification in the tourism industry offers a variety of opportunities such as tour marketing and new generation awareness, attracting potential customers, improving tourist's on-site experiences and increasing interaction (Hay, 2008: 346). In gamification, game design can be based on real environments such as tourist attractions (such as Thailand and China). Gamification activities present an informative and entertaining atmosphere, providing potential visitors a variety of experiences. Games are often offered with virtual reality or AR 3-D technology, providing an immersive and engaging experience of virtual and real targets (Huang et al., 2013: 495).

The tourism industry offers multidimensional and versatile experiences (Kim, Ritchie & McCormick, 2012: 15). There are different types of emotions, such as hope, fear and excitement, which arise from the pleasure of playing and the desire to continue playing to challenge their own abilities. With these emotion types, the participation and the dependence on playing play an important role in the marketing of tourism organizations and destinations. As a result of the studies, it was revealed that games are the main forms of entertainment for mobile and social generations born after 1996 and it was concluded that future marketing activities should be developed gamification oriented (Xu et al., 2014). Gamification is approached in three parts according to behaviors of buying based on the place and importance of gamification in tourism experience. These are evaluated as; before the purchasing and consuming of tourism experience / during its consumption/ after purchasing and consuming of tourism experience (Sigala, 2015):

Firstly, when the purchasing/consumption of tourism experience part is evaluated, given the tourists' desire to travel, the presentation of travel information is an important consideration in determining the needs of tourists. Therefore, at this stage, tourism companies are responsible for providing information to travelers in order to reveal and trigger their travel needs, curiosities and preferences, and to ensure that customers know what products and services are available to customers. Lufthansa developed ''www.snapshottraveler.gr'' which is an interactive online entertainment game that allows its users to virtually experience three destinations in Canada, Thailand and Brazil. While navigating the destinations virtually in this game, users are asked to search and photograph the animals they find. Users learn about the destination's tourist resources, the interesting places, existing travelling experiences and options by exploring and interacting with virtual words. As a result, they find the opportunity of compete for performance achievement by winning prizes and badges. While the game is linked to social media (Facebook and Twitter), in addition to its impact on external motivation, other game elements are also used to provide user's motivation and experience. Thus, it provides an opportunity to decide whether they prefer these destinations or not, having the knowledge and experience about the destination before going to that destination.

While utilizing gamification during the purchasing/consumption of tourism experience that is the second part, the aim of companies and destinations is to encourage tourists to increase their consumption levels and increase their product / service use while creating pleasant and memorable experiences. At this stage, gamification is utilized to implement crowd-sourced effects and motivation. For example; a game was developed that allows tourists in India to guess where they are, giving players points and various rewards for right answers (http://www.ixigo.com/yoindia). There are millions of photos of the places in the game. Players have the opportunity to learn and explore their location with these photos. For their goal of becoming a Traveling God, they share the badge they earn and the points they get via Facebook.

In the last part, it takes place after the purchase/consumption of tourism experience. Gamification applications in the tourism industry aim to help create customer databases and increase customer loyalty.

The JetBlue gamification application was developed with the loyalty program (TrueBlue) to improve customer loyalty and connection with businesses and to build a customer database. TrueBlue members that register on JetBlue's Facebook page earn 25 TrueBlue points when they check in with their mobile device at a JetBlue airport. They also earn Jetblue badges for each interaction they create by tagging their friends. Jetblue, a location-based social media application, is also known as Go Places. Apart from the Jetblue badges and points, they can take advantage of status privileges such as free services - updates and earn loyalty points by checking Jetblue locations.

Gamified Restaurant Experience

McDonald's, which is one of the Fast-Food chains, benefits from gamification. McDonald's uses mobile phones to let users get free food from the nearest restaurant and interact with digital billboards. This game called Pick N'Play had a great success in Sweden. In order to be sure that gamers are on the playing area, they only need to use their GPS and do not need to download an application. The goal of McDonald's is to increase sales with this application. Consumers are happy to get free burgers. Also, McDonald's have started a campaign in Canada asking customers to create a burger as a movie star. The best stories are shown in a film, using the restaurant chain as an ad. Participants receive free cinema tickets as awards for their participation (Digital Tourism Think Tank, 2013).

Gamification at the Airlines

In the early 1980s, American Airlines launched specific programs to increase brand loyalty. The prevalence of smartphones, the big data that are obtained and new forms of communication determine our travel style and preferences and will continue to do so in the future. Airline companies such as KLM and Virgin America continue to renew their long-term loyalty programs by integrating social networks. KLM developed the Meet & Seat feature to help passengers find out who they are and connect through social networks like Facebook, Google+ or LinkedIn. When passengers share their personal information with other customers, they are allowed to interact with other passengers. Virgin America developed an application using social networking to deliver an experience above the clouds with the 'seat-to-seat delivery' feature. The application works on the aircraft's built-in touch screen entertainment system. It offers opportunities like interacting with other passengers, communicating, sending drinks/food (Digital Tourism Think Tank, 2013).

Gamified Virtual Travel Experience

The potential tourists that are hesitant about their destinations, tourism-based games were developed that provides the opportunity of experiencing world tourism destinations even from their homes easily. For example, as one of the largest online travel agencies in the world, by Expedia.com is developed a travelling application in 2012 that is travel in 100 days around the world. The game was developed to increase Expedia's loyalty awareness and to connect with customers. Players learn more about destinations around the world and earn points. Users can redeem their points for travel bookings (Digital Tourism Think Tank, 2013).

Figure 10. Virgin America seat-to-seat delivery feature
Source: Retrieved from https://edition.cnn.com/travel/article/virgin-seat-to-seat-service/index.html

Methodology

Digital gamification in tourism is the activity of introducing and marketing enterprises or tourist destinations in the tourism industry to tourists through mobile applications (Xu et al., 2014). Gamification, which is considered as a new approach for the tourism industry, offers tourists the opportunity to explore and experience new places with fun. In this section, the concept of gamification and the elements that make up the concept of gamification are explained and their relationship. Descriptive research is a research aiming to describe a subject or activity of interest with a description. Descriptive research, in which no cause and effect relationship is sought between the subjects, can have a single aim or can have both descriptive and explanatory purposes. If the only purpose of the research is to describe a subject, the reasercher examines them observing the facts in detail and depicts the subject without developing any theoretical problems. Qualitative research methods such as case studies, ethnographic studies or similar qualitative data collection techniques are used in the studies conducted with this purpose. (Lin, 1976). Table 1 (adapted from Çeltek, 2010) was created to examine exemplary gamification practices used in the tourism industry. Table 1, which was formed by using secondary data obtained by using purposive sampling technique, was examined by using document analysis technique. Purposive sampling technique is the selection of a source that is suitable for the purpose of the study from a source of information with a certain consciousness (Tongco, 2007). The document analysis technique is the process of evaluating the materials in printed and electronic media in a systematic way (Corbin & Strauss, 2008). In the creation of Table 1, (game, gamification, gamification in tourism, mobile applications) keywords and articles published in national and international journals; papers presented / published in congresses and symposiums; national and international master theses; related books and websites; 'Google Play Store and

Apple Store' mobile applications have been utilized. Within the scope of this study, which is limited to tourism, a total of twenty-five gamification practices have been reached. These accessible gamification applications are evaluated separately in terms of the benefits they provide in terms of tourism sector applied, the platform played, the stage in which tourism is applied, self-determination theory, Hook theory, Fogg behavior theory, player types and tourism marketing.

Findings

In line with the information obtained from the articles, papers and books examined at national and international level, gamification practices used in the tourism industry have been reached through the document analysis technique. Table 1 was created by researching the gamification applications in tourism through web sites and by accessing the gamification applications in tourism from secondary data sources. As a result, the achieved gamification practices have been examined by using the parameters in Table 1. When gamification practices in tourism are evaluated in terms of the area they are applied in tourism industry, it is observed that the games are mostly designed for destinations / cities, while they are applied in transportation enterprises, museums, food and beverage enterprises, accommodation enterprises and travel agencies respectively. When the games are examined in terms of the platform that they are played, it is seen that location-based mobile applications are more than social media-based games. While it has been determined that 25 games reached during the stage of gamification in tourism are applied during the journey, a certain number of games can be applied before and after the travel. Also, as a result of findings, it was found that a significant number of games can be applied before travelling, during travelling and after travelling as in three stages. According to the theory of self-determination, when the characteristics of the games are examined, competence is seen in games handled outside of Air British Airways / Trivia Quizzes and Games, while the games that are few in number have relevance and autonomy. According to the Hook theory, it is observed that the most intense features in the games are the triggering feature, and that the action and variable rewards exist in some games, while the investment feature is found in very few games. When the Fogg Behavior Theory is considered, it is seen that even though motivation and trigger features are found in numerous games, ability feature exists in a few games. In the features of player type determination, it is seen that achiever and socializer types are target market numerously; explorers are the target market only in specific games. VR Games Zone Game is found to be the only game that determines the killers' type, which is one of the player types, as a target market. Also, it is seen that some games in the table do not appeal to a single type player; they have set more than one player type as a target market. When the benefits of gamification in tourism to tourism marketing are examined; while it has been observed that city/country promotion, sharing on social media/sharing with friends, giving information about the business, giving discount/promotion and product information take place in numerous games, the opportunity to purchase products in the game and interactive communication features with other players take place in only a few games.

Ghosts Games Wartburg

''Ghost Games'' mobile application can be considered as an example of digital gamification efforts in tourism. The mobile application company called Ghosts Games can provide an experience to the tourists visiting the region by making use of gamification and augmented reality applications of the works in

Table 1. Gamification Applications in Tourism Industry

	Foundation Name or Game Name	Pizza Hut / Beanbag Blitz	EpicMix	4foods	Starbucks	Tripventure	Drallo	Efteling	Ghost Game / Wartburg Castle	TravelPlot Porto	Virtual Romans Leicester	Tourism Ireland	British Airways /Trivia Quizzes and Games	Virgin Games	VR Games Zone	Ixigo Game	JetBlue	Pick'in Play	Expedia	American Airlines / Travel Games	Turkish Airlines	Foursquare / Swarm	Stray Boots	Pocket Ranger	My Marriott Hotel	La Quinta's Play & Stay game
The Tourism Sector That It is Applied	Food Beverage Foundation	X		X	X													X								
	Destination / Town		X			X	X		X	X	X	X										X	X	X		
	Accomodation Foundation																								X	X
	Travelling Agency															X			X							
	Transportation Business												X			X	X		X	X	X					
	Museum							X	X	X	X				X											
The Platform That The Game Is Played	The Games That Are Based On Social Media		X	X									X	X		X	X			X	X	X		X	X	X
	Mobil Games Based On Place	X	X		X	X	X	X	X	X	X	X		X	X	X	X	X	X		X	X	X	X		
The Stage That Gamification Is Applied In Tourism	Before Travelling	X	X	X									X	X		X	X								X	X
	During Travelling	X	X	X	X	X	X	X	X	X	X	X	X	X	X	X	X	X	X	X	X	X	X	X	X	X
	After Travelling	X	X							X			X	X		X	X	X							X	X
Game Features According To Self Determination Theory	Sufficiency	X	X	X	X	X	X	X	X				X	X	X	X	X	X	X	X	X	X		X	X	X
	Related		X											X						X			X	X		
	Autonomy														X			X				X		X	X	X

continued on next page

Table 1 Continued

	Foundation Name or Game Name	Pizza Hut / Beanbag Blitz	EpicMix	4foods	Starbucks	Tripventure	Drallo	Efteling	Ghost Game / Wartburg Castle	TravelPlot Porto	Virtual Romans Leicester	Tourism Ireland	British Airways /Trivia Quizzes and Games	Virgin Games	VR Games Zone	Ixigo Game	JetBlue	Pick'in Play	Expedia	American Airlines / Travel Games	Turkish Airlines	Foursquare / Swarm	Stray Boots	Pocket Ranger	My Marriott Hotel	La Quinta's Play & Stay game
Game Features According To Hook Theory	Trigger	X	X	X	X		X	X	X	X	X	X		X	X	X	X	X	X	X	X	X	X	X	X	X
	Action	X			X		X							X		X	X		X	X	X				X	X
	Changeable Rewards	X	X		X		X	X						X			X	X	X	X		X				X
	Investment		X		X	X	X							X												
Game Features According To Fogg Behaviour Theory	Motivation	X	X	X	X	X	X		X	X	X			X	X		X	X	X	X	X	X	X	X	X	X
	Ability		X			X							X	X	X	X								X		
	Triggers	X	X	X	X			X				X		X	X	X	X	X	X	X	X	X			X	X
The Targeting Features of Player Types	Achievers	X	X	X	X	X	X	X		X			X	X		X	X	X	X	X	X	X		X	X	X
	Explorers	X				X	X		X	X	X	X				X			X				X	X		
	Killers														X											
	Socializers	X	X		X		X			X				X	X	X	X	X	X	X	X	X	X	X	X	X

continued on next page

Table 1 Continued

	Foundation Name or Game Name	Pizza Hut / Beanbag Blitz	EpicMix	4foods	Starbucks	Tripventure	Drallo	Efteling	Ghost Game / Wartburg Castle	TravelPlot Porto	Virtual Romans Leicester	Tourism Ireland	British Airways /Trivia Quizzes and Games	Virgin Games	VR Games Zone	Ixigo Game	JetBlue	Pick'in Play	Expedia	American Airlines / Travel Games	Turkish Airlines	Foursquare / Swarm	Stray Boots	Pocket Ranger	My Marriott Hotel	La Quinta's Play & Stay game
The Benefit Features That Are Provided In Terms of Marketing Tourism	Sharing on Social Media / Sharing With Friends		X	X	X		X			X		X	X	X		X	X		X	X	X	X	X	X	X	X
	The Opportunity of Purchasing in Game				X			X						X			X			X				X	X	X
	Information of City / Country	X	X			X	X	X	X	X	X	X	X			X	X		X		X	X	X	X	X	
	Giving Promotion / Sale	X	X		X		X	X						X		X	X	X	X	X	X	X			X	X
	Giving Information About Foundation	X	X	X	X	X	X	X	X		X		X	X	X	X	X	X		X	X	X	X	X	X	X
	Giving Information About The Product	X	X	X	X		X	X	X					X	X	X	X	X		X	X	X		X	X	X
	Interactive Communication With Other Players		X							X				X	X		X	X		X		X	X	X		

certain destinations. It takes tourists to a journey through the history using different techniques, objects and the information about the daily lives of the people living in that destination (Falke, 2012).

The castle of Wartburg is located in Eisenach which is a part of the Thuringa state of Germany. This medieval castle, which has a thousand-year history, has an important place in German history besides being the first German castle in the UNESCO World Heritage List (Wartburg Castle, 2019). Different than other museums, this castle functions as a museum and conducts a gamification activity in which the visiting tourists can interact. ''Ghost Game Wartburg'' helps tourists explore the castle by providing interaction with the virtual ghosts living in the castle. The ghosts created by gamification, meet the tourists that use the application on their smartphones (Falke, 2012). This application, suitable for the use of all age groups, can make the tourists satisfied especially the ones who seek cultural pleasure by making sightseeing fun and entertaining.

A Gamification Model for Tourism

A gamification model was created based on the conceptual studies and sample applications regarding the use of digital gamification activities in Turkish tourism. Istanbul, which is a destination known throughout the world and visited every year by millions of local and foreign tourists, was preferred while fictionalizing the gamification model that has been created. The model illustrated in the figure 12 was created by benefiting from D6 gamification model that had been created by Werbach and Hunter (2015). The game was developed with an aim to contribute to an unforgettable destination experience by the tourists that use the application will satisfy especially the tourists included in the "explorers" group since the

Figure 11. Ghost Game Wartburg
Source: Falke, 2012

sense of wonder is one of the principles that the tourists in this group look for in a game. The sense of wonder can create an urge to visit the places that have never been explored. In the created model, the entertainment element that will make the tourists continue the gamification game consists of the type of hard entertainment in which the players are challenged to overcome some difficulties.

It is possible to prepare a mobile application interactive with social media applications. Using the application with GPS locations on, tourists visiting Istanbul will have access to historical and cultural information of the tourist spots in their area or in the nearest neighborhood and they are supposed to play games such as question-answer and puzzle to reach more information or any potential interesting information about an important person or story belonging to that area. The application includes popular destinations such as Topkapı Palace, Dolmabahçe Palace, Basilica Cistern, Galata Tower, Maiden's Tower, Egyptian Bazaar, Grand Bazaar, Blue Mosque, Hagia Sophia Museum, Chora Museum visited by local and foreign tourists from all ages.

Figure 12. Created gamification model

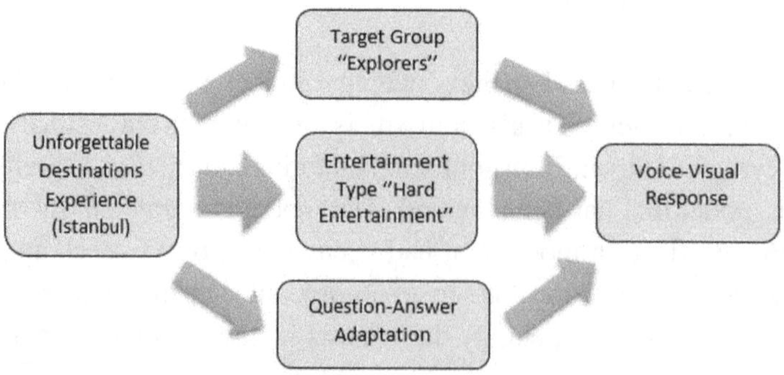

Figure 13. Example application question

Tourists who visit these and similar locations via the application will receive a screen similar to the one shown in the Figure 13 on their phones if they participate in the question and answer game in accordance with the notifications received. On the screen, they can reach questions and answers with written and audible warnings in English or Turkish language according to their preference. Buttons that provide access to an audible question, answer time for the question, and access to the next question are displayed on the screen.

The user who chooses an answer from the options, will receive a second screen depending on the answer being true or false. The second screen that users will see takes place in the Figure 14. Depending on the answer, a person from the history and present, who lived at the tourist spot about which the question was asked, will provide users a vocal or visual explanation related to the answer. There will also be internet links on the page where users can reach more information about the touristic destination. If they answer every one of the questions belonging to each touristic destination and share the photograph of the touristic destination on social media accounts via the application, users will have the chance to win specific badges. That these badges indicate the fact that the destination has been visited and the very user has answered more questions correctly than the other users, provides a competition among the tourists.

In theory, introduction of the cultural and historical sites in Istanbul through gamification with the help of the model created may especially be preferred by the explorer tourists using this game model. Therefore, the target market of the designed game consists of the users in explorer tourists group.

Figure 14. Example application answer

SOLUTIONS AND RECOMMENDATIONS

Digital gamification in tourism facilitates the move of the touristic business operations to the mobile platforms and strengthens the corporate brand image perceived by tourists providing faster, informative and entertaining products/services for tourists. The tourists who have experienced the gamification activities that tourism companies apply via digital platforms may have an increasing loyalty and want more products/services or might feel an urge to spend more money providing that they are motivated by being satisfied with the presented product or service by. Digital gamification activities for destinations are directly proportional to the success of turning the natural, cultural and historical beauties of the destination into an interactive entertainment activity due to its influence on tourist's destination preferences, duration of the stay, expenditures and their tendency to visit the destination again. Main aims of digital gamification applications in the tourism industry are:

- To satisfy tourists by creating an interactive experience during their holiday.
- Providing entertaining contents and services to targeted tourist group on the internet through digital platforms.
- Outclassing the rival companies and creating a higher preferability by the tourists.

The responsible ministries or municipal authorities of the destinations wishing to bring more tourists to the region by introducing and marketing the destinations can benefit from the following suggestions:

- Creating correct informational contents and game designs for the target tourist group expected to visit the destination.
- In the tourist spots, the scores obtained from the gamification activities should be usable in the social facilities such as the restaurants and museums in the region by the tourists.

- It is important to increase the recognition of destinations on social media platforms through tourists via digital gamification applications and to reach more masses by encouraging electronic word of mouth marketing activities.

FUTURE RESEARCH DIRECTIONS

This study has been restricted with academic studies reached by benefiting from the purposive sampling technique and digital gamification applications in the tourism industry. Therefore; although the findings have a feature of representing the universe, they do not represent the entire universe. A gamification model has been developed benefiting from document analysis techniques and literature scanning aimed at digital gamification applications to a destination in the existing study. In the future studies, this model can be improved and applied on a destination. In order to learn the expectations and satisfaction of tourists about this gamification model to be created for the destination, a study on digital gamification applications can be performed using semi-structured interview or survey techniques in the future tourism.

CONCLUSION

Gamification is a concept whose use by companies has been continuously increasing. Because of the fact that individual's needs have been met on digital platforms through technological devices such as smartphones, tablets and computers in a more efficient way in terms of time and cost comparing to their common daily life, it has been inevitable for the companies to use digital gamification applications. Businesses, who want to satisfy their expectations and needs by satisfying the consumers, aim to increase the consumer's loyalty increasing their motivation by making the process of providing the right products / services to the right target audience faster and more entertaining via gamification applications on digital platforms.

Accordingly, the enterprises within the tourism industry think that with the gamification activities they launch on various digital platforms, tourists may increase their perceptions and motivations about the enterprises or destinations and increase their demands for more products or services. The literature on digital gamification activities supports this opinion. Digital gameplay applications used in the tourism industry, especially in combination with augmented reality (AR) and virtual reality (VR) applications may enable tourists to interact more about the company and the destination, to gain experience, to be satisfied and helps them to share their real or virtual experiences which they had with their friends.

As a result of this study, it is seen that comparing to enterprises in the tourism industry, digital gamification activities are mostly used as a contribution to promote and market the destinations. It has been found that digital games prepared for destinations are designed to be location-based mobile applications in general and the usage stages of gamification applications developed for the tourism industry are not only limited to the service purchasing stage, but also designed to include the pre- and post-service stages. When the target markets of digital games developed for the tourism industry are examined, it is noteworthy that the players in the achievers group and the players in the socializers group are observed to be the main target. Besides promoting cities and countries in particular; gamification applications utilized in the tourism industry contribute to tourism marketing through various factors such as providing information about businesses, providing discounts and promotional activities for enterprises.

REFERENCES

Akbulut, D. (2018). *The use of "Gamification" approach in public relations*. International Symposium on Communication in the Digital Age Proceeding E-Book. Oct. 18-19, Mersin, Turkey. Academic Press.

Andersen, S., Chen, S., & Carter, C. (2000). Fundamental Human Needs: Making Social Cognition Relevant. *Psychological Inquiry*, *11*(4), 269–276.

Avedeon, E. M., & Sutton Smith, B. (1971). *The study of games*. New York: John Wiley & Sons.

Bartle, R. (1996). Hearts, clubs, diamonds, spades: Players who suit MUDs. *Journal of MUD Research*, *1*(1).

Bayraktar, Ö. (2014). *Gamification as a communication model* (Master dissertation). University of Maltepe, Turkey.

Binark, M., & Bayraktutan, G. (2011). ''E-status'' of participation: The virtual World of youth. Aslı Telli Aydemir (Ed.), Digital game culture map players: Digital player's habitus and career derivatives (pp. 303-330). Turkey: Alternative Information Publishing.

Bogdanovych, A., Esteva, N., Gu, M., Simoff, S., Maher, M. L., & Smith, G. (2007). The role of online travel agents in the experience economy. In *Proceedings of the 14th international conference on information technology in tourism ENTER*, Ljubljana, Slovenia. Academic Press.

Buhalis, D., & Law, R. (2008). Progress in information technology and tourism management: 20 years on and 10 years after the Internet—The state of eTourism research. *Tourism Management*, *29*(4), 609–623. doi:10.1016/j.tourman.2008.01.005

Caillois, R. (2001). *Man, play, and games*. Chicago, IL: University of Illinois Press.

CastleW. (2019). Retrieved from https://www.wartburg.de/en/wartburg-castle.html

Çeltek, E. (2010). Mobile advergames in tourism marketing. *Journal of Vacation Marketing*, *16*(4), 267–281. doi:10.1177/1356766710380882

Chou, Y.-K. (2016). *Actionable gamification beyond points, badges, and leaderboards*. Japan: Octalysis Media.

CNN. (2019). Retrieved from https://edition.cnn.com/travel/article/virgin-seat-to-seat-service/index.html

Corbin, J., & Strauss, A. (2008). *Basics of qualitative research: Techniques and procedures for developing grounded theory* (3rd ed.). Thousand Oaks, CA: Sage. doi:10.4135/9781452230153

Csikszentmihalyi, M. (1990). *Flow: The psychology of optimal experience*. New York: Harper Perennial.

Deci, E. L., Connell, J. P., & Ryan, R. M. (1989). Self determination in a work organization. *The Journal of Applied Psychology*, *74*(4), 580–590. doi:10.1037/0021-9010.74.4.580

Deterding, S., Dixon, D., Khaled, R., & Nacle, L. (2011). From game design elements to gamefulness: Defining "Gamification". In *Proceedings of the 15th International Academic MindTrek Conference on Envisioning Future Media Environments*, Tampere, Finland. 10.1145/2181037.2181040

Digital Tourism Think Tank. (2013). Retrieved from https://www.thinkdigital.travel/tag/2013/

Entertainment Software Association. (2018). *Essential facts about the computer and video game industry.* Retrieved from www.theesa.com/wp-content/uploads/2018/05/EF2018_FINAL.pdf

Eyal, N., & Hoover, R. (2014). *Hooked: How to build habit forming products.* Wood Dale, Ilinois: Penguin Group.

FalkeC. (2012). Retrieved from http://www.uni-weimar.de/kunst-und-gestaltung/wiki/images/IFD_mobile-culture_2.01_ChristopherFalke_GhostsGames.pdf

Fernandes, R. P. A., Almeida, J. E., & Rosseti, R. J. F. (2013). A collaborative tourist system using serious games. *Advances in Intelligent Systems and Computing*, *206*, 725–734. doi:10.1007/978-3-642-36981-0_67

Fogg, B. (2009). A behavior model for persuasive design. In *Proceedings of the 4th International Conference of Persuasive Technology*, Claremont, CA. 10.1145/1541948.1541999

Gentes, A., Guyot Mbodji, A., & Demeure, I. (2010). Gaming on the move: Urban experience as a new paradigm for mobile pervasive game design. *Multimedia Systems*, *16*(1), 43–55. doi:10.100700530-009-0172-2

Hay, B. (2008). Fantasy Tourism and Second Life. In S. Richardson, L. Fredline, A. Patiar, & M. Ternel (Eds.), *CAUTHE 2008: Tourism and hospitality research, training, and practice; "Where the 'Bloody hell' are we?"* (pp. 345–348). Retrieved from https://search.informit.com.au/documentSumary;dn=970132315896676;res=IELBUS

Herger, M. (2014). *Enterprise gamification engaging people by letting them have fun.* USA: CreateSpace Independent Publishing Platform.

Huang, Y., Backman, S. J., Backman, K. F., & Moore, D. (2013). Exploring user acceptance of 3d virtual worlds in travel and tourism marketing. *Tourism Management*, *36*, 490–501. doi:10.1016/j.tourman.2012.09.009

Ixigo. (2019). Retrieved from https://www.ixigo.com/%20yoindia

Karaarslan, M. H., & Altuntaş, B. (2016). Investigation of the selected marketing cases in Turkey within the concept of gamification. *Mehmet Akif Ersoy University Journal of Social Sciences Institute*, *8*(17), 433–447.

Kasa, M., & Hassan, Z. (2015). The role of flow between burnout and organizational citizenship behavior (OCB) among hotel employees in Malaysia. *Procedia: Social and Behavioral Sciences*, *211*, 199–206. doi:10.1016/j.sbspro.2015.11.084

Kim, H., Ritchie, J. R., & Mccormick, B. (2012). Development of a scale to measure memorable tourism experiences. *Journal of Travel Research*, *51*(1), 12–25. doi:10.1177/0047287510385467

Kowal, J., & Fortier, M. S. (1999). Motivational determinants of flow: Contributions from self-determination theory. *The Journal of Social Psychology*, *139*(3), 355–368. doi:10.1080/00224549909598391

Kunduracıoğlu, İ. (2018). *A content analysis on gamification concept* (Master dissertation). University of Balıkesir, Turkey.

Lim, C., Kim, M., & Kim, S. B. (2018, January). *A Q & A application using the concept gamification.* ISER 98th International Conference, January 8-9, 2018, Osaka, Japan. Academic Press.

Lin, N. (1976). *Foundations of social research.* USA: McGraw-Hill.

Linaza, M. T., Gutierrez, A., & Garcia, A. (2014). Pervasive augmented reality games to experience tourism destinations. In Z. Xiang, & I. Tussyadiah (Eds.), *Information and Communication Technologies in Tourism* (pp. 497–509). Switzerland: Springer International Publishing.

Lloyd, V. (2014). *A brief history of gamification.* Retrieved from https://www.thehrdirector.com/features/gamification/a-brief-history-of-gamification/

Lovett, E., & Longworth Dames, M. (1901). The ancient and modern game of astragals. *Folklore, 12*(3), 280–293. doi:10.1080/0015587X.1901.9719634

Lowbrow, Y. (2014). *Livin' the dream with green stamps: A 1975 catalog.* Retrieved from https://flashbak.com/livin-the-dream-with-green-stamps-a-1975-catalog-26187/

Marczewski, A. (2013). *A simple introduction tips, advice and thoughts on gamification, United State of America.* Lulu Press.

Marczewski, M. (2015). *Gamification user types.* Retrieved from http://www.gamified.uk/user-types/

Matallaoui, A., Hanner, N., & Zarnekow, R. (2017). Introduction to gamification: Foundation and underlying theories. In S. Stieglitz, C. Lattemann, S. Robra-Bissantz, R. Zarnekow, & T. Brockmann (Eds.), Gamification using game elements in serious contexts (pp. 3-18) Switzerland: Springer International Publishing.

Mayra, F. (2008). *Play in the mobile Internet: Towards contextual gaming.* Paper presented at Internet Research 9.0 Conference, October 15-18, Copenhagen, Denmark.

Mccormick, T. (2013). *Gamification: A short history.* Retrieved from https://foreignpolicy.com/2013/06/24/gamification-a-short-history/

Mclaughlin, D. (2012). *Game on! How video games and game design principles can help to make adult education more engaging for the learner* (Master dissertation). University of Ireland Maynooth, Ireland.

Nakamura, J., & Csikszentmihalyi, M. (2009). The concept of flow. In C. R. Snyder, & S. J. Lopez (Eds.), *Oxford handbook of positive psychology* (pp. 89–105). USA: Oxford University Press.

Özkan, Ç. (2018). *Gamification use in the context of corporate communication strategies: A research on the banking industry* (Master dissertation). University of Istanbul, Turkey.

Öztürk, S. (2015). *Using gamification in effective team planning and team activity* (Master dissertation). University of Bahçeşehir, Turkey.

Piccione, P. A. (1980). In search of the meaning of Senet. *Archaeological Institute of America, 33*(4), 55–58.

Ryan, R. M., & Deci, E. L. (2000). Self-determination theory and the facilitation of intrinsic motivation, social development, and well-being. *The American Psychologist*, *55*(1), 68–78. doi:10.1037/0003-066X.55.1.68 PMID:11392867

Şahin, M., & Samur, Y. (2017). Instructional method of digital age: Gamification. *Journal of Ege Education Technologies*, *1*(1), 1–27.

Salen, K., & Zimmerman, E. (2004). *Rules of play: Gaming Design Fundamentals*. Cambridge, MA: MIT Press.

Seaborn, K., & Fels, D. I. (2015). Gamification in theory and action: A survey. *International Journal of Human-Computer Studies*, *74*, 14–31. doi:10.1016/j.ijhcs.2014.09.006

Sever, N. S., Sever, G. N., & Kuhzady, S. (2015). The evaluation of potentials of gamification in tourism marketing communication. *International Journal of Academic Research in Business and Social Sciences*, *5*(10), 188–202. doi:10.6007/IJARBSS/v5-i10/1867

Sigala, M. (2015). Applying gamification and assessing its effectiveness in a tourism context: Behavioral and psychological outcomes of the TripAdvisor's gamification users. *Asia Pacific Journal of Information Systems*, *25*(1), 179–210. doi:10.14329/apjis.2015.25.1.179

Sobocinski, M. (2017). I gamified my courses and I hate that. *World Journal of Science, Technology, and Sustainable Development*, *14*(2/3), 135–142.

Superdata Games & Interactive Media Intelligence. (2017). *2017 year in review, digital games, and interactive media*. Retrieved from https://www.superdataresearch.com/market-data/market-brief-year-in-review/

Tongco, M. D. C. (2007). Purposive sampling as a tool for informant selection. *Ethnobotany Research and Applications*, *5*, 147–158. doi:10.17348/era.5.0.147-158

Tunga, Y., & İnceoğlu, M. M. (2016). Gamification design. In *Proceedings of the 3rd International Conference on New Trends in Education. April 26-29, 2016.*

Vardarlıer, P., & İnan, K. (2017). Gamification model proposal for the improvement of sales personnel performance. *Journal of Behavior at Work*, *2*(1), 8–19.

Waltz, S. P., & Ballagas, R. (2007). *Pervasive persuasive: A rhetorical design approach to a location-based spell-casting game for tourists in proceedings of situated play*. DiGRA 2007 - The 3rd International Digital Games Research Conference. Sept. 24-28, 2007, Tokyo.

Werbach, K. (2015). *Gamification online via Wharton University of Pennsylvania*. Retrieved from https://www.coursera.org/learn/gamification

Werbach, K., & Hunter, D. (2012). *For the win how game thinking can revolutionize your business*. Pennsylvania: Wharton Digital Press.

Werbach, K., & Hunter, D. (2015). *The gamification toolkit dynamics, mechanics and components for the win*. Pennsylvania: Wharton Digital Press.

Wu, B., & Wang, A. I. (2011). A pervasive game to know your city better. In *Proceedings of the 2011 IEEE International Games Innovation Conference*. Washington, DC: IEEE Computer Society. 10.1109/IGIC.2011.6115111

Wu, M. (2017). *The magic potion of game dynamics*. Retrieved from https://community.khoros.com/t5/Science-of-Social-Blog/The-Magic-Potion-of-Game-Dynamics/ba-p/19260

Xu, F., Weber, J., & Buhalis, D. (2014). Gamification in tourism. *Information and Communication Technologies in Tourism, 2014*, 525–537.

Yen, B. T. H., Mulley, C., & Burke, M. (2019). Gamification in transport interventions: Another way to improve travel behavioral change. *Cities (London, England)*, *85*, 140–149. doi:10.1016/j.cities.2018.09.002

Yılmaz, E. A. (2017). *Gamification*. Istanbul, Turkey: Abaküs Publishing.

Yovcheva, Z., Buhalis, D., Gatzidis, C., & Van Elzakker, C. (2014). Empirical evaluation of smartphone augmented reality browsers in an urban tourism destination context. *International Journal of Mobile Human Computer Interaction*, *6*(2), 10–31. doi:10.4018/ijmhci.2014040102

Zichermann, G., & Cunningham, C. (2011). *Gamification by design: Implementing game mechanics in web and mobile apps*. Canada: O'Reilly Media.

Zichermann, G., & Linder, J. (2010). *Game-Based marketing inspire customer loyalty though rewards, challenges and contests*. Hoboken, NJ: John Wiley & Sons.

ADDITIONAL READING

Garcia, A., Linaza, M. T., Gutierrez, A., & Garcia, E. (2019). Gamified mobile experiences: Smart technologies for tourism destinations. *Tourism Review*, *74*(1), 30–49. doi:10.1108/TR-08-2017-0131

Mantouka, E. G., Barmpounakis, E. N., Milioti, C. P., & Vlahogianni, E. I. (2018). Gamification in mobile applications: The case of airports. *Journal of Intelligent Transport Systems*, *23*(5), 417–426. doi:10.1080/15472450.2018.1473157

Negruşa, A. L., Toader, V., Sofica, A., Tutunea, M. F., & Rus, R. V. (2015). Exploring gamification techniques and applications for sustainable tourism. *Sustainability*, *7*(8), 11160–11189. doi:10.3390u70811160

Pamfilie, R., Vasilcovschi, A., & Bumbac, R. (2016). A new possible way of promoting tourist packages: Gamification. *Cactus Tourism Journal*, *13*(1), 5–10.

Robson, K., Plangger, K., Kietzmann, J. H., McCarthy, I., & Pitt, L. (2015). Is it all a game? Understanding the principles of gamification. *Business Horizons*, *58*(4), 411–420. doi:10.1016/j.bushor.2015.03.006

Sigala, M. (2015). The application and impact on gamification funware on trip planning and experiences: The case of TripAdvisor's funware. *Electronic Markets*, *25*(3), 189–209. doi:10.100712525-014-0179-1

Swacha, J., & Ittermann, R. (2017). Enhancing the tourist attraction visiting process with gamification: Key concepts. *Engineering Management in Production and Services*, *9*(4), 59–66. doi:10.1515/emj-2017-0031

KEY TERMS AND DEFINITONS

Achievers: A name given to players whose goal is to collect points, badges and prizes in the games and the players who enjoy collecting them while playing individually or in groups depending on the tasks they encounter.

Components: They are the elements that increase the game experience and pleasure as a result of the interaction of players via points, badges, tasks, virtual objects in the games that they play.

Dynamics: They are the basic fictional and invisible features during the gamification process that make up a game and increase the quality and practicability of the game such as game tasks and game story. Creating Dynamics wrongly or faultily can harm the loyalties of gamers to the game by causing the collapse of the game in terms of fiction.

Flow Theory: It is a theory stating that the size of the prize that can be obtained from the game depending on the level of the games and the player's abilities have an effect on the gamer's motivation.

Game: It is an activity that includes rules for a purpose and provides people with the opportunity to enjoy themselves and socialize by taking them away from the tiredness of daily hustle and bustle.

Gamification: The activities that make the process of introducing a product, service or idea to the people in order to be marketed by making use of game elements to make it fun.

Mechanics: They are the In-game elements that motivate players to spend more time in games by encouraging them to engage in more activities.

Section 2
Industry 4.0 and Customer Experience

Chapter 10
Technological Developments:
Industry 4.0 and Its Effect on the Tourism Sector

Hande Mutlu Ozturk
https://orcid.org/0000-0002-4404-0106
Pamukkale University, Turkey

ABSTRACT

Technological developments in recent years have been affecting the lives of people and societies more rapidly than in the past. Developments in the field of communication, robotics, transportation, etc. are called the 4th Industrial Revolution or Industry 4.0 in the industrial sector. Technological developments have created great changes in the services and industrial sectors. Industry 4.0 has also led to changes in the transformation of the tourism sector and is likely to occur in future processes. This chapter examines the impact of Industry 4.0 on the tourism sector.

INTRODUCTION

Technological developments have been increasing in recent years. At the end of the 18th century, the advancements in technology have been advancing rapidly since the 1st Industrial Revolution, which started with the use of steam engines. Developments in the field of communication and progress in informatics have affected the change in all fields.

The first three of the previous industrial revolutions are called mechanization, use of high electrical energy and automation and electronics, respectively (Lasi et al., 2014). Today, economies are turning to the fourth industrial revolution, defined by the use of cyber systems, smart factories and innovations in the service sector (Shamim et al., 2016, Lee et al., 2014). Industry 4.0 can also be defined as a subcomponent of digital transformation in existing businesses and processes (Porter and Heppelmann, 2014).

Depending on the changes in technology and digitalization; both the product and the production methods vary. This change is called industry 4.0. The 4th industrial revolution, called as Industry 4.0,

DOI: 10.4018/978-1-7998-1989-9.ch010

is the process in which the use of computers and automation in the industry. With Industry 4.0, use of automatic machine also called robots has been at the industry.

Peceny et al. (2019) discussed Tourism 4.0, a new tourism paradigm that emerged to reveal the potential for innovation in the entire tourism sector. In the study, it is assumed that technologies that provide important opportunities from Industry 4.0 such as Internet of Things, Big Data, Chained Disabilities, Artificial Intelligence, Virtual Reality and Augmented Reality can be applied to tourism sector. It is believed that by creating a common ecosystem involving local residents, local authority, tourists, service providers and government, an enriched tourism experience can be created in both the physical and digital worlds.

In their study, Shamim et al (2017) examined Industry 4.0 issues in the service sector and the situations in the hospitality industry. The challenges of Industry 4.0 require continuous innovation and learning, depending on people and the capabilities of the business. Appropriate management approaches play a vital role in the development of dynamic capabilities and in an effective learning and innovation environment. In the study, proposes a framework of management practices that can support innovation and the learning environment in an organization was investigated.

BACKGROUND

Industry 4.0

Depending on the technological developments, the current approaches in the tourism sector needs to change. In this issue, definitions such as Industry 4.0 or Tourism 4.0 have been made. The aim of Industry 4.0 or Tourism 4.0 in tourism is to reduce the negative effects of tourism, to see the effects of the use of technology in the tourism sector and to develop cooperation models in partners. Some researchers also define Tourism 4.0 or Industry 4.0 in the tourism industry as smart or intelligent tourism, and this is being discussed by many researchers. (Buhalis & Amaranggana, 2013, Gretzel et al., 2015a, Hunter et al., 2015, Gretzel et al., 2015b, Geissbauer et al., 2014, Schwab, 2016, Verevka, 2018, Goncharova and Bezdenezhnykh, 2018, Lebedev et al., 2018). Concepts and tools provided by intelligent tourism is a phenomenon that defines, extends and integrates Information and Communication Technology (ICT) into the tourism experience.

Factors that led to the fourth industrial revolution called Industry 4.0; the spread of robots, the Internet of Things, artificial intelligence, sensors, cognitive technologies, nanotechnology, services of the Internet, quantum informatics, wearable technologies, augmented reality, intelligent signalling, intelligent robots, big data, new generation technologies such as 3D and smart networks. Industry 4.0 technologies have started to change business environments and lifestyles by rapidly using them in business life, communication and education. Since the tourism industry is a dynamic industry that is rapidly adapting to innovations and technologies, Industry 4.0 technologies have quickly found application in the tourism industry. With the use of Industry 4.0 technologies for tourism purposes, 'smart tourism' concepts emerged. Topsakal et al. (2018) discussed why some people, and therefore tourists, have previously used industrial revolutions than other people or tourists, and what the causes are for it.

Mil and Dirican (2018) have focused on the effects of technological developments on the tourism sector and presented a large literature review on the subject. Icten and Bal (2017) examined the virtual and augmented reality and their application examples in fields such as education, art, traffic, engineering and tourism. Ilhan and Celtek (2016) carried out a detailed study on the effects of augmented reality on mobile marketing and tourism. In this study, details of how technological developments can be used in finding directions, examining two or three dimensional product images with mobile devices, getting information about travel, and visual trips in museums are presented.

Artificial intelligence technology reduces errors in applications and processes and makes hotel managers to make more accurate decisions about tourist demand and supply. Thus, a better marketing strategy can be implemented and financial management and human resources planning can be performed (Claveria et al., 2015). With the Internet of Things and sensors installed in / around the hotel even in the cities, make it is easier to access data on the availability of facilities, tourist location, weather, road conditions, traffic situation and airport traffic. While this information does not directly affect the experience of tourists, it affects the overall impression and satisfaction of tourists (Jin et al., 2014).

Ay (2009), in his study, discussed the reflections of the developments in information technologies and internet on travel agencies and online travel agency. In this study, the literature is searched on the subject and the subject is examined by making use of secondary data and sectoral applications. Ay (2009) indicated that the rapid developments in the field of information and communication technologies in recent years increase the tendencies towards distribution systems, electronic data transfer and globalization of communication systems. In the tourism industry, whoever adapted to the technological developments grow in a short time. Data banks, smart cards, automation and communication networks in production increase economies and competitiveness. In the tourism industry, internet applications and their relations with distribution systems have recently come to the forefront. Other technological developments (voice recognition systems, data mining, artificial intelligence applications, virtual reality or geographic information systems, etc.) are expected to be much more effective in the exchange of tourist products and distribution channels. With the widespread use of computers and the internet, travel agencies have started to provide their services via web pages. The person who lives in any part of the world can access the web page of the travel agency in any tourist area of the world via internet and make his / her holiday or reservations. All these developments have brought to the fore the issue of ensuring shopping and data security.

Kıroğlu (2012) examined the concept of internet use and electronic tourism in her study. In this study, possible developments of e-tourism have been discussed by taking into consideration current applications and future expectations. In the research, the process of people's orientation towards e-commerce is examined; The changes in the volume of electronic commerce over time and the structure of the user profile are examined. E-commerce applications in the tourism sector were examined and SWOT analysis of e-tourism was conducted. In the study, taking into consideration the benefits of electronic commerce in tourism sector, it is predicted that agency will end and the future potential of e-tourism is considered and the deficiencies of the current applications and solution suggestions are indicated.

Kapiki (2012) analysed the current and future trends affecting the tourism and hospitality industry, including globalisation, guests' safety and security, the importance of offering outstanding services, the new technologies that enhance competitiveness, the population ageing that impacts directly on tourist demand and the correlation between price and value. She gives detail on the title of "current trends in hospitality and tourism" about globalization, safety and security, diversity, service, technology, demographic changes and price-value. She indicated that important parameters for the hospitality industry in the future would be as;

- Green and eco-lodgings.

- Development of mega hotels (multi-purpose facilities with casino, shops, theatre, theme park, etc.).
- More boutique hotels.
- Intelligent hotels with advanced technology using the guest's virtual fingerprint in order to perform all the operations (check-in, charges, check-out, etc.).
- Increasing employee salaries in order to retain the existing staff.
- More emphasis on the internet and technology.
- Guests' virtual and physical social networks will be the best distribution channels.

Dubey (2016) examined the effects of new technologies on tourism. In this study, the effects of web-based applications such as digital technologies, social media, internet and cloud technologies on the preferences and experiences of tourists were investigated. Interactivity, augmented and virtual reality, space tourism was discussed in the study.

The main objective in the use of Industry 4.0 in the tourism sector is to create an interactive platform based on the latest technology infrastructure. One of the main objectives of the Industry 4.0 for tourism sector is to provide services and products in a sustainable way, accessible to everyone at any time.

The applications should be of a type that will facilitate the integration of the tourist to the destination, improve the quality of the experience and provide customized products and services. Practices should facilitate the integration of tourists to the destination, improve the quality of experience and provide customized products and services. In other words, it can be possible to develop the sector by providing the application of high technology products in tourism services in Industry 4.0.

MAIN FOCUS OF THE CHAPTER

Today, use of automation systems, computers or communication systems have affected not only the industry or production, but also all areas of life and all sectors. Industry 4.0 will provide the highest quality standards for all engineering, management, production, operations and logistics processes, as well as great flexibility and robustness to all sectors. With the possibilities of technological advances, various variables such as costs, availability, use of resources and market demand can be easily optimized in real time. In countries this paradigm shift will allow the two strategic sectors to converge or disintegrate with each. This provides opportunities for all sectors.

Industry 4.0 and its impact on the manufacturing sector have been studied in detail in many studies. However, there is very limited work in the service sector and it also faces the challenges of mass customization, digital development, intelligent work environment and efficient supply chain.

Like all sectors, tourism sector is also affected by technological developments. Especially communication systems, internet of things, cloud and cognitive computing systems, web based software systems, mobile systems, internet and mobile banking, developments in national and international money transfer affect tourism sector and increase tourism mobility all over the world (see Figure 1).

With the introduction of Industry 4.0, it was possible to access information easily, especially by using the infrastructure on the internet, making reservations easier and making choices by filtering the preferences according to requests and budget easily by using the internet infrastructure. This situation has helped to increase tourism mobility. The development of mobile communication and mobile internet has also had a positive impact on the development of tourism.

Figure 1. Industry 4.0 for tourism

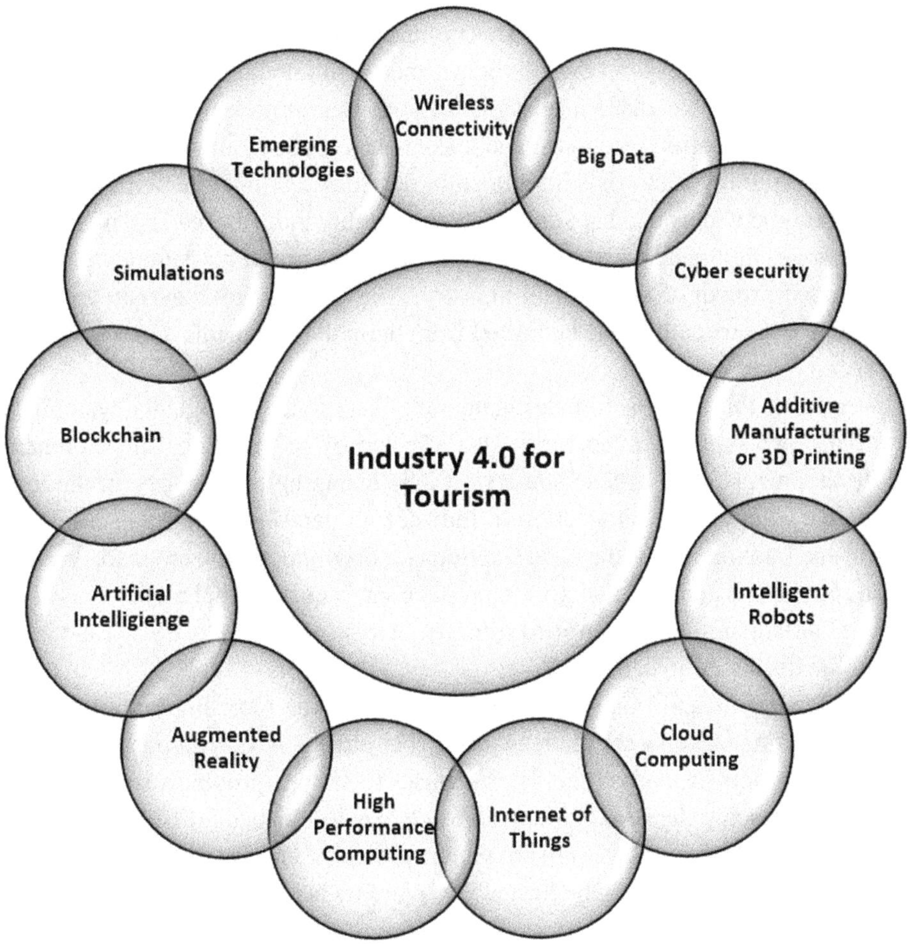

Some of the developments in the industry affect directly tourism and others indirectly. The developments in the transportation sector and the change in comfort and speed in airway, railway, highway and sea transportation have revived the tourism sector. Development in communication systems and social media make it easier to access information and arouse curiosity and desire to see.

Advances in technology offer new products and new inventions. Since many people now arrange the travel by themselves using the computer infrastructure, companies are also developing new booking methods according to this new situation. Nowadays there are countless different ways of booking for consumers, but before digitalization there was only one option: to use travel agency. Mobile phones have also increased the options. For example, online travel guides also offer the consumer some options and advantages; simple updates, links to specific information, search functions, bookmarks or feedback, and reviews of other travellers. All this both increases and facilitates the desire of the traveller to travel. While the older generation prefers to spend their holidays in a travel agency, the younger generation prefers to use online booking portals. Therefore, more and more hotels focus on online promotion. Unlike large hotels or large hotel chains, small hotels tend to rely on online sales systems and try to sell through large portal systems.

Today, new media culture has developed. This culture is differentiating in every aspect of life and has changed the way consumers spend their holidays and accommodations. At the moment, the widespread use of mobile devices has made it possible for potential tourists or consumers to communicate with each other, share information resources, and make reservations in all situations and environments. A large part of the hospitality sector has integrated into digitalization and new media culture. Mobile applications have become widely used not only in the hospitality industry but also in the travel industry. Last minute deals are easily tracked and experienced. Attractive accommodation and culinary delights contribute to the rapid growth of the tourism industry. At the same time, tourism areas that have not been used accessible until recently, have benefited from these developments and started to take a share in the tourism sector.

The main purpose of this study is to investigate the effects of technological developments on tourism sector. Like all sectors, tourism sector is also affected by technological developments. Today's technological developments, also called Industry 4.0, are changing the services in the tourism sector and the way they are offered. The application of Industry 4.0 leads to changes in many areas such as hospitality, travel and booking. Nowadays, reservations for accommodation are made via the internet or mobile applications. Robots are used in hotels to produce services or virtual reality is used in marketing in tourism. Smart tourism has been mentioned recently. In this section, Industry 4.0 and its applications in tourism sector are discussed in detail.

Technological developments bring some problems. One of the most important of these problems is data security. Protecting systems and data against cyber-attacks is very important. The replacement software systems and robots instead of people can causes to serious problems such as unemployment. As a result of robots replacing the worker, the workforce may lose its value. This change in the tourism sector may adversely affect the employees in the sector. Unmanned hotels where robots work, machines cook and software are serving can only be turned into places where accommodation is made and there is no social life.

INDUSTRY 4.0 AND TRAVEL

4th industrial revolution also affects travel systems. Compared to the past, today, mobility is constantly increasing. Today, more people travel than in the past. In 2030, considering that there will be 1 billion more people traveling in addition to today, and therefore transportation infrastructure will have to be transformed to meet this increase.

As a result of improvements in the internet and communication, transport and accommodation can be offered as a package. Using the infrastructure of mobile systems, a transportation system will be developed that will take you from your location to the hotel where you will be staying and then bring you back to where you live.

Nowadays, in the transportation and hospitality industry, most of the operations cannot be performed automatically but are performed manually. In the future, it seems that most of them will be done automatically. In this case, in the future, machines that we often call robots will interact with the customer. In addition to providing communication, mobile phones will be a device that confirms location sharing, likes, or dislikes, contributing to their preferences for subsequent travel.

Figure 2. Smart city and transportation

Communication or mobile phone is a modern need nowadays. People shares their travel experience using devices, channels, and back-end technology systems. Travel in an Industry 4.0 world where machines interact and react wisely to the physical environment will be different. Industry 4.0 will also make change on travel systems, and travel and transport will integrate (see Figure 2).

Travel agency is in a radical change in itself and the definitions, practices and management approaches in this field also change in line with global trends (Yolal, 2003). One of the changes caused by globalization in the tourism industry is that businesses start to operate on a global scale. They also have to compete on international platforms. Companies operating globally see the world as their field of activity, develop global strategies and maintain their presence in the global market. Global powers change the structures of enterprises and eliminate the boundaries restricting enterprises (Yılmaz and Yılmaz, 2005). The effects of information technologies on travel agencies are manifested in the production, marketing, distribution and management of tourism products for both the public and private sectors. Industry 4.0 increase the managerial efficiency and productivity in travel enterprises, while enabling the management of the business to adapt to the new business environment and benefit from new opportunities (Buhalis, 1998).

Digitalization offers an exciting opportunity for the aviation, travel and tourism ecosystem. In the next decade it has the potential to have a value of about $ 1 trillion in the industry and other sectors. The travel ecosystem changes with blurring boundaries and changing roles in the industrial landscape (Weinelt and Moavenzadeh, 2017).

Weinelt and Moavenzadeh (2017) were indicated that in the next decade (2016 - 2025), digitalization in aviation, travel and tourism is expected to be:

- Can generate $ 305 billion value for the sector with increased profitability
- Migration from traditional players to new opponents worth $ 100 billion
- Achieve $ 700 billion worth of benefits for customers and the wider community through cost and time savings for consumers with fewer environmental footprints, improved security and security

- As a result of the clear shift of existing jobs in the sector, which are expected to be partially offset by the creation of next generation skilled jobs within and outside the travel ecosystem

Digitalization will have a positive environmental impact, contributing to the footprint of a more sustainable sector through innovation in production, smart assets and efficient resource utilization. For customers, personal impact is expected to be important as travel becomes a seamless, more quality experience. The greatest social impact could be the impact of digital transformation on the travel labour force, which by 2025 could represent one in every 11 jobs worldwide (Weinelt and Moavenzadeh, 2017). Intelligent automation will change the nature of some travel jobs and completely eliminate others. However, digitally activated growth will create new employment opportunities that can eliminate the automation of existing roles, especially as strong growth is anticipated for the sector.

The aviation, travel and tourism industries are at the forefront of digital innovation. However, when industry and technology trends are analyzed, it can be predicted that there will be greater changes in the future. The transportation sector has been one of the early applications of digital technologies and platforms with other name Industry 4.0. On the other hand, in emerging markets, the demand for travel driven by a growing middle class emerges. The middle class plays a leading role in the use of digital technology. In the future, it can be expected that digital technologies will develop more and demand will increase. The travel ecosystem has helped shape customer expectations for on-demand and appropriate services through digital innovation, both within and across industry boundaries (Weinelt and Moavenzadeh, 2017). The next step is to change the way they work so that they can capture the opportunities that digital transformation offers, and to expand its use in areas where the digital sector has not yet been implemented.

INDUSTRY 4.0 AND HOSPITALITY

The concept of hospitality has changed drastically over the years. Online booking platforms have changed a lot, differentiated and greatly increased their market share. As one of the most important results of Industry 4.0, digitization has found application in hospitality sector and has created dynamic changes in accommodation.

Hospitality as a subdivision of tourism is a fundamental part of the indoor and outdoor entertainment market. Consistent tourism demand allows the hospitality industry to anticipate demand and identify opportunities to increase consumer spending by creating a wave of secondary financial impact' (Robinson, et al., 2013).

The hospitality industry has always contributed to the development of the imagination of travellers. The competitive environment in the hospitality industry pushes the tourism sector to find new and effective solutions. One of the main trends in this area is the ability to offer new services and develop innovative applications for the development of the hospitality industry. The applicability and competitiveness of all kinds of innovations in hotel and restaurant establishments is essential (Dzhandzhugazova et al., 2015).

In the last decade, there has been a significant increase in both the use of social media and the overall development of new technologies worldwide. The tourism and hospitality industry shows an impressive development due to technological developments. Industry 4.0 has started to find application in tourism sector. The increase in the use of social media, easier communication and evaluation of customers after the hotel accommodation, makes it easier to find new customers for accommodation. Technological developments offer new opportunities for the tourism and hospitality industry. The hospitality sector sees technological developments as a way to increase profits. However, one of the most important challenges that service providers face is how to integrate into social media marketing and to verify whether it is profitable to use their business (Benea, 2014).

Industrial problems, special mentality of consumers, cultural and national differences of customers benefiting from hotel services are among the factors that affect the application and development of hospitality (Pine et al., 1999). The innovative trend in the hospitality industry is reflected in the successful implementation of innovations in the hotel (Zaitseva, 2013). With the widespread application and combinations of new information, new services, products and technologies can be produced. If a hotel is not involved in any innovation process, its performance will eventually decline and may lose competitiveness (Ilyenkova and Kuznetsov, 2009).

Technology is changing the way travellers interact with brands before, during and after the journey. Away from the spirit of hospitality as a human-centred industry, the adoption of new technologies allows hospitality businesses to offer greater personalization and better service (Dzhandzhugazova et al. 2016). In their study, Shamim et al (2017) examined the applications of Industry 4.0 in the service sector and the situations in the hospitality industry. All the challenges of Industry 4.0 require continuous innovation and learning that depend on people and the capabilities of the business. Appropriate management approaches can play a vital role in developing dynamic capabilities and in an effective learning and innovation environment. Shamim et al (2017) proposes a framework of management practices that can support innovation and the learning environment in an organization, and therefore recommends the adoption of Industry 4.0, which will enable the adoption of technology, such as digital enhancements and the implementation of cyber physical systems (CPS).

The Hospitality Service Providers (HSP), use the Internet of Things (IoT) technology to provide competitive advantage in the market. In the accommodation sector; IoT, sensors, actuators, identification labels, mobile devices, etc. devices can communicate directly or indirectly between the traveller and the accommodation facilities.

The paradigm of IoT can provide to the HSP different ways for interacting with guests and collecting real-time data. When HSP systems are used, guest behaviors and preferences can be measured with very high accuracy. IoT also helps to the HSP for improving the efficiency of multiple departments (Figure 3).

The entertainment and hospitality industry is one of the most important elements of the global economy. In recent years, the widespread use of new technologies in the tourism sector, like all sectors, has affected the way in which services are provided and received. Kansakar et al. (2019) investigated how new technologies and guest experiences are currently being used in the hospitality industry and how they have changed the hospitality service platform. In this study, it is predicted that the Internet of Things (IoT) technology will affect hospitality services in the future.

Verevka (2019) examined the role of digital innovations in the hospitality industry and discussed the impact of the digital revolution on business management. In this study, the conceptual aspects of Industry 4.0 are discussed and the current situation and the development of the digital transformation of the industry are analyzed. Pilot projects have been researched to implement Industry 4.0 in the hotel and restaurant business. Based on the researches, the basic conditions for achieving successful digital transformation and reducing the risks of digital innovation in the hospitality industry have been determined.

Figure 3. Industry 4.0 for hospitality

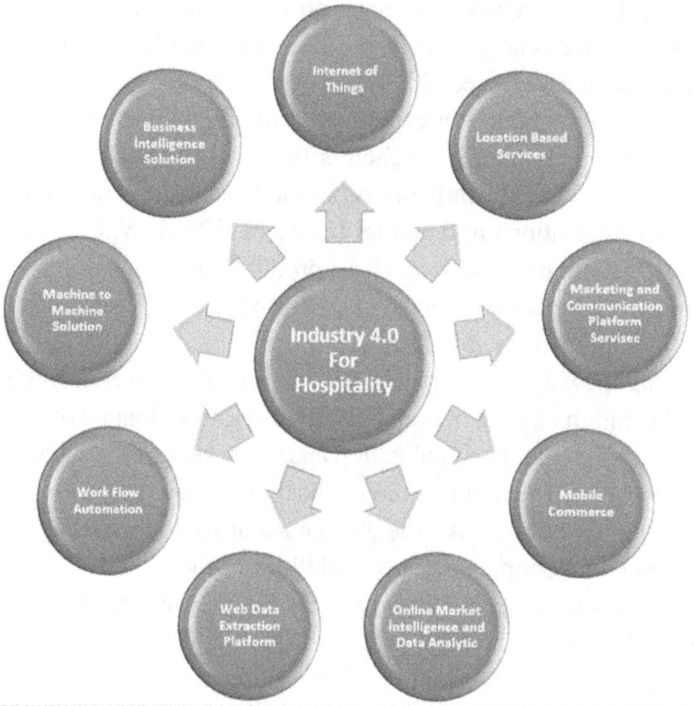

INDUSTRY 4.0 AND GASTRONOMY

Food is one of the most important key elements of identity and culture of a nation, as well as its symbols, history, discourses and myths (Smith, 1995). Studies have shown that there are strong historical links between food and tourism that unite them as a political force. In addition, food and agriculture have traditionally been strong economic sectors in the formulation of public policies and strategies in all societies. Agricultural systems, food products and outputs in the countries have been the most important inputs of the tourism sector and also sometimes have been among the symbols and rituals for the countries.

The concept of gastronomy emerged in line with the need for nutrition. The concept of gastronomy, which mostly used as the science of eating and drinking, has been influenced, changed and evolved according to conditions throughout history. Gastronomy and Gastronomic Culture is one of the most effective parameters for choosing a destination for the guest. One of the reasons for visiting a destination is the culinary culture of that area (Zagralı and Akbaba, 2015); Çevik and Saçılık, 2011). There is a broad consensus that gastronomy plays an important role in enhancing the pleasure and enjoyment of the guest. The food and drink culture of the local people varies in each destination and it can turn into a kind of entertainment by attracting the attention for the guests to discover what the local people eat and drink (Baytok et al., 2001).

The concept of food is one of the basic elements of the tourism sector and constitutes an important aspect of the accommodation process. Food Tourism is a concept that includes the socio-cultural characteristics of food. There are a limited number of international researches on gastronomic tourism (Hall & Sharples, 2003). It is thought that the globalization phenomenon, which is gaining momentum

Figure 4. 3D printing and Gastronomy application

nowadays, has started to make the world a market that offers the same or similar products (Toksöz and Aras, 2016). Gastronomy Tourism enables the realization of alternative tourism types. This requires the protection of socio-cultural heritage. Gastronomy is also important for preserving historical and cultural heritage and transferring it to future generations (Hall et al. 2003).

In their studies, Gunes et al (2018) discussed the historical development and current status of gastronomy. They examined the effects of advances in digital technology on gastronomy. It is stated that globalization and technological developments have an effect on food, beverage and gastronomy. It has been reported that the use and development of robots, 3D printers or cooking technologies to have an impact on gastronomy.

With the development and spread of 3D printers, it becomes also possible to apply them to the field of gastronomy. Thus, it is possible to produce food in very difficult and complex figures. Considering that gastronomy appeals to the taste and the eye, with technological advances and the use of robots, visual foods that cannot be made today can be cooked using 3D printers (see Figure 4). Depending on the development of robots, it will be also possible to produce food by robots in the future. Depending on the development of automation and cooking techniques, foods as same taste and similar appearance can be cooked. Automation can also prevent time loss. Temperature and time parameters can be easily set in the cooking units. This may prevent under- or over-cooking of food.

INDUSTRY 4.0 AND SMART TOURISM

'Smart' is generally used to describe sensors, big data, open data, new connection methods, and information exchange. The concept of smart develops depending on technological improvements and affects economic and social developments. Höjer and Wangel (2015) argue that individual technological developments are not so important. In fact, they argue that the rational creator has synchronization and harmonious use of interconnections of different technologies.

In some cases, the concept of 'smart' is simplified and used by politicians to direct societies and sell technological solutions. It is also used as to provide free wireless internet or to develop mobile applications. It should be noted, however, that internet or mobile systems are only infrastructure, not 'smart'. Using these infrastructures and technologies, data collection, management, processing and conversion to new data for a particular purpose are very important concepts. Although all of these are new approaches to the concept of smart tourism, they are not fully capable of explaining smart tourism.

In the context of tourism, smart is used to describe a complex connection of all intelligent systems used in everyday life. In the tourism sector, there is an incredible institutional support for switching to smart tourism applications. In some cases, there are also internal and external pressures in the sector to realize smart tourism. Efforts have been made to advance smart tourism applications around the world in regions and countries where the tourism sector is vibrant. Many governments provide financial and technological support for the establishment of technological infrastructure that supports smart tourism and endeavors to establish infrastructure.

Becoming 'smart' is widely acknowledged in the need for a more customer-oriented approach in order to develop the experiences, needs and preferences of tourists and to evaluate their needs to achieve better satisfaction (Correia et al., 2013, Prayag et al., 2013). Tourists are more interested in experiences from a wide range of products and services linked to the nature, history, gastronomy and culture of the region they visit. For this reason, tourism operators are required to provide 'all-inclusive' package solutions that can include a large number of products and services in order to increase the satisfaction of tourists (Buhalis & Law, 2008).

Gretzel et al. (2015a) identified smart tourism in their studies, addressing the current smart tourism trends and studied the technological developments and their effects on enterprises. They discussed the expectations and disadvantages of smart tourism. In their study, Neuhofer et al. (2013) focused on tourism experiences that develop technology as an integrative conceptual framework to define today's tourism as a combination of experiences, co-creation and technology.

The concept of smart tourism developed with technological developments. The concept of Industry 4.0 also prepared the ground for the emergence of smart tourism. The development of internet infrastructure, mobile applications and smart city projects also contributed to the spread of smart tourism and smart tourism destinations (see Figure 5). The focus of smart tourism is the development of innovation structures and the competitiveness of the tourism sector. Thus, it is possible to use resources less while providing efficiency. With the use of smart data, it will be possible to enrich tourism by using existing data and developing new ones. Smart end-user applications have been developed to transform the experiences of tourists into positive ones.

Smart Tourism has focused on the use of advanced technologies to transform data into efficient new business models by using and evaluating data collected through physical infrastructures and social connections. The objective of Smart Tourism is to analyze tourism products, services, venues and experiences for a wide stakeholder community and to create services that meet the requirements. By using new technological developments, it is possible to develop innovative and entrepreneurial enterprises and sectoral connections and create productive spaces.

Major changes have been experienced in the technological fields of tourism in recent years. As a result of e-tourism, global integration and global distribution and central reservation systems have been developed (Buhalis 2003; Werthner and Ricci 2004).

The spread of social media has accelerated the sharing of information in the tourism sector (Sigala et al. 2012). The use of the Internet and mobile internet has facilitated the knowledge of tourism sector, consumer expectations, mobility and demands of tourism consumers (Buhalis and Law 2008; Wang et al. 2012).

However, smart tourism has made progress especially with the use of digital systems and the physical and governance infrastructure of the tourism sector. New smart systems have started to be implemented in tourism systems and there has definitely been a significant progress in the evolution of ICT in the tourism sector (Gretzel, 2011).

The introduction and expansion of social media and the internet has helped to develop the dynamic infrastructure connections and technological networks necessary to achieve the Smart Tourism Goals. The development of social media is considered to be a useful and rich source of information in the development of tourism (Miah et al., 2016).

Big Data is one of the most fundamental data defining the 'knowledge economy' and stands out as an emerging field of research for researchers and practitioners (De Mauro et al., 2016, Erickson and Rothberg, 2014, Laney, 2001). Del Vecchio et al. (2018), in their study, aims to show how giant Social Big Data that can be obtained by tourists can improve the value creation process for Smart Tourism Goals. Bernabeu et al. (2016), taking into account the capacity of destinations, discussed the Big Data technology used in tourism planning and management for the new Smart Tourism Destination (STD) approach. Big data technology, using advanced technological infrastructure, produce large amounts of data. The HTA approach and Big Data Technology (BDT) can be handled together so that some of the most typical cases applicable to the tourism sector can be analyzed. With the use of this technology, strengths and weaknesses in companies and tourist destinations can be identified from a wider perspective. By cooperating with the enterprises in the tourism sector and analyzing the perspectives of the entrepreneurs and experts, a healthier and sustainable growth of the tourism sector can be achieved.

Given the high dependence on information and communication technologies (ICT), the information intensity of tourism is not surprising (Law et al. 2014; Koo et al. 2015; Werthner and Klein 1999; Benckendorff et al. 2014). Due to the development of the internet, there has been a major development in the area, also known as e-tourism. Innovations have begun to be applied easily to the sector and the sector and users (tourists or travelers) have benefited from the opportunities offered by technology. Sharing information easily has a positive impact on both the tourism sector and the user.

ICT applications are available for tourists in different cities around the world. Some cities offer to the tourist some physical infrastructure. For example, in Barcelona, tourists are informed about the arrival time and route of the buses. There is also USB ports for charging mobile phones at bus stops. The city of Brisbane has recently installed over 100 pointers at points of interest to communicate information to tourists via mobile application, if within a certain radius of their location. Amsterdam uses beacons to allow tourist signs to translate them into different languages, and Amsterdam ArenA is testing sensors for better crowd management. Seoul is investing heavily in providing tourists with free wireless internet and smartphones. Jeju Island in South Korea has declared it a smart tourist destination using innovative technology for content presentation (Gretzel et al., 2015b).

Figure 5. Components of smart tourism

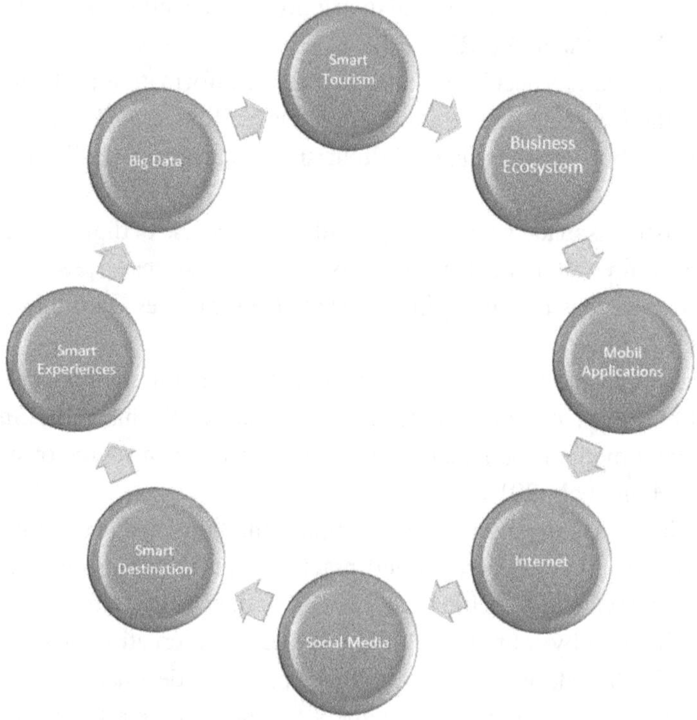

INDUSTRY 4.0 AND SHOPPING

Shopping is a very important economic activity. It is also used as a leisure tool to meet a wide range of social and psychological needs (Howard, 2007, Jones, 1999, Wakefield and Baker, 1998).

Shopping, which is also considered as a great leisure activity, has become an important element in tourism (Law and Au, 2000). All over the world, shopping is no longer just a means of meeting needs. It varies depending on the differences in social, cultural and economic trends. It is used as a method to meet the created demands. (Bellenger and Korgaonkar, 1980; Michalkó and Timothy, 2001; Timothy, 2005). With the changing social structures and the impact of globalization, tourists do not do shopping only in order to meet their needs. Tourists buy clothes, souvenirs, art and crafts to remember the countries they visit. Thus, shopping becomes a motivating factor for travel and becomes an important part of the tourism experience (Timothy and Butler, 1995; Timothy, 2005). In addition, shopping is considered to be one of the important parameters of the tourism sector and is part of their distillation (Goeldner et al. 2000).

Tosun et al. (2007) examined tourists' satisfaction with local shopping culture, staff service quality, product value and reliability, physical characteristics of stores, payment methods and other shopping and store features. The different levels of satisfaction of the tourists were investigated with different store and shopping features in the study. Based on the results of the research, it is suggested that providing a higher level of shopping experience for tourists and increasing the contribution of shopping to the regional economy should be supported by various financial and educational tools to continue the production and sale of authentic handicrafts and souvenirs.

Shopping acts as a factor that influences and motivates tourists' travel decisions and consumer behaviour, such as visiting, having fun, experiencing and getting to know the local culture (Fodness, 1994). Law and Au (2000) describe shopping as follows; the consumption of purchased goods is a source of pleasure and satisfaction and a driving force for tourism.

Shopping is one of the oldest and most common travel related things. In most cases, lack of shopping for tourists is seen as a lack of holiday experience. (Keown, 1989; Turner and Reisinger, 2001). Mostly, tourists buy goods or souvenirs not only for themselves, but also for relatives and friends (Anderson and Littrell, 1995; Kim and Littrell, 2001).

Shopping, which is one of the most important tourism activities, contributes economically to the visited countries because money is just spent for fun (Jackson, 1996; Asgary et al., 1997).

However, the effects of shopping are not limited to the local, regional and national economy. When tourists return to their country, products purchased by tourists can help them develop a positive image for the visited country. They may share feeling with their friends and relatives. People in general and tourists in particular tend to share their experiences through the photos, videos and items they buy.

Therefore, tourism can be used as an alternative economic growth strategy. It is also used as a political tool for the development of industry and its promotion to international markets (Richter, 1989; Hall, 1994; Tosun, 2001; Tosun and Fyall, 2005).

Depending on technological changes, the shopping style also changes. Shopping centres are becoming more and more widespread. In shopping centres, many products and services are offered together. Tourists visit the natural and historical places in the countries they visit, as well as shopping malls. In the countries visited, technological products are marketed as well as food, clothing and handicrafts.

SOLUTIONS AND RECOMMENDATIONS

It can be foreseen that technological application to the tourism industry will continue especially in tourism-intensive cities and countries. Depending on technological developments, tourists' expectations are changing. Tourists want to get easy information, find their way easily and easily go to wherever they want, while traveling comfortably in the countries they visit. At the same time, the ability to use mobile phones and access to the Internet easily has become a priority. Smart tourism starts with smart city infrastructure. However, tourists will need to be provided with information to improve their satisfaction by processing data in the cities. For this reason, it is very important to develop mobile applications that contain information about cities and provide them to visitors. Smart tourism should be approached as a whole. Hotels, restaurants, local governments, central governments, data providers, transportation sector are all part of this system. It is not possible for companies operating in the tourism sector to provide smart tourism.

By using technological applications in the tourism sector, guests can be provided with excitement, efficiency and comfort. Recreational industrial robots have been used in tourism for a long time mainly because of their efficiency and reliability. Robots with voice and face, at the reception, check-in and carrying goods are available at the hotels.

Robots appear almost at all levels of the tourism industry. The main drivers of its widespread use are: productivity, accessibility and service enhancement. With the use of artificial intelligence in the tourism sector, human-robots who learn more easily began to serve. With the development of technology and the increase in application areas to tourism, it can be seen that more robots are used in the entertainment sector.

Virtual Reality (VR) has entered daily life as a technology that reflects real-life situations and can be used to create interactive scenarios. With VR, complex situations can be simplified, simulated and recreated imaginatively. Virtual reality is also widely used in the tourism sector and hospitality. VR has emerged as a very important concept in the hospitality industry and has begun to be used. Guests can learn and experience the hotel's physical environment before visiting the hotel. The increased accessibility of virtual reality creates new marketing and brand opportunities in the hospitality industry. It is possible to reach customers with modern methods by using virtual reality. VR has begun to replace traditional brochures. Using virtual reality methods, resorts or hotels can provide more immersive experiences such as VR presentations, providing 360° videos or more personalized tours. Virtual reality has found its use as a marketing tool.

Virtual reality for companies in the field of tourism can be seen as a tool to promote. From a user's point of view, virtual reality can be a tool to experience new exciting things. With VR, virtual environments can be created for tourism planning. Using VR, tourists can experience an artificial and limitless environment. Considering the virtual reality in the tourism sector, one should be aware that a person - customer or tourist - has a virtual experience from a touristic point of view (Guttentag, 2010). This is due to the fact that virtual reality is an interactive and immersive world (Mazuryk and Gervautz, 1996). Virtual technology can provide an almost unlimited experience for users (Guttentag, 2010). It should be noted that, depending on the type of virtual environment, it has different effects on all five senses of the human body (Gruber 2015).

Technological developments and chat habits made via social media have enabled the development of chat applications in hotel technology. Chatbot is the chat interface and with this application, customers can interact with a person or artificial intelligence and contact the accommodation facility. The chat industry has great potential in the hotel industry as users spend a long time in messaging applications today. Using chat interfaces or applications, guests can make reservations at hotels, access hotel services, be aware of daily activities in the building, learn about breakfast or mealtimes, and forward any room problems to the housekeepers. Changing conditions and advances in technology both create a change in hospitality and offer new opportunities.

FUTURE RESEARCH DIRECTIONS

Industry 4.0 has a positive role in solving a variety of problems related to data management and other technological issues. The future of the hospitality industry will be shaped by current developments with the Internet of Things (IoT) technology.

The concept of Industry 4.0 also began to change the way we shop. Mobile or online shopping is becoming more common. With the impact of globalization, it became possible to purchase every product from anywhere in the world. Therefore, local products and crafts began to attract more tourists. One of the biggest problems for tourists is the return of purchased products. Due to the widespread use of internet shopping, there has been an incredible growth in the logistics sector. Serving the products purchased by tourists to their doors when they return home will increase the desire for shopping.

With the next development, by using data exchange on technological infrastructure, using a single image of the travellers and synchronizing data between all systems, taking into account previous experiences and decisions, preferences can be made for him according to his previous wishes. At this point, with the creation of new technologies such as robotics, the so-called Smart Travel systems can be developed.

Depending on technological advances, robots may be expected to provide services at airports, trains station or bus stations during the travel. As a result of Industry 4.0, in the future, travel technology can be expected to become smarter. In the future, as systems that process, measure and evaluate data evolve, it will be possible to transform the travel system into smarter systems.

Depending on the technological improvements, the concepts of smart buildings or smart houses are also developing. When the guests stay in a place, the comfort requests they desire and adjust in the environment they live in can be used in the next accommodation via mobile infrastructure or internet of objects by using the mobile big data 'infrastructure. This situation will increase the satisfaction of the guests regarding to the facility as well as increase the efficiency of the facilities.

In general, the hospitality industry should create a mutually beneficial platform between guests and HSP, with new technological advances and facilitating partnership. The platform should enhance HSP's operational management efficiency while providing an exceptional travel experience. In addition, after the evolving revolution of IoT technology, future potential hospitality services will also develop.

CONCLUSION

The tourism sector is growing rapidly. Tourism is an important sector in terms of the promotion of countries, normalization of relations between countries and the promotion of cultures. Tourism sector is also changing due to technological developments. Technological developments change the expectations of individuals and societies. The expansion of the Internet and mobile applications contributed to the growth of the tourism sector. People want to visit the countries and regions they see.

The spread of social media also increases the desire to travel. People have a desire to visit places that friends or famous people share on social media. Traveling habits or travel patterns also change. The reason for this rapid change is due to technological developments.

In this study, the effects of technological developments on tourism sector are examined. Technological change is called Industry 4.0. Although the concept of Industry 4.0 was initially used for the production sector, it is now being adopted for all sectors. With the digitalization, developments in communication infrastructure, and the use of mobile systems and internet infrastructure, tourism sector is one of the sectors positively affected by Industry 4.0.

REFERENCES

Anderson, L. F., & Littrell, M. A. (1995). Souvenir-purchase behavior of women tourists. *Annals of Tourism Research*, *22*(2), 328–348. doi:10.1016/0160-7383(94)00080-8

Asgary, N., De Los Santos, G., Vincent, V., & Davila, V. (1997). The determinants of expenditures by Mexican visitors to the border cities of Texas. *Tourism Economics*, *3*(4), 319–328. doi:10.1177/135481669700300402

Ay, L. (2009). Bilgi Teknolojisindeki Gelişmelerin Seyahat Acentalarına Yansımaları. *Online Seyahat Acentacılığı. Sosyal Ekonomik Araştırmalar Dergisi*, *9*(17), 117–136.

Baytok, A., Emren, A., Gürel, N., Dalkıranoğlu, A., & Töre, H. ve Güney H. (2001). Afyonkarahisar Mutfağı. Ankara: Uyum Ajans.

Bellenger, D. N., & Korgaonkar, P. K. (1980). Profiling the recreational shopper. *Journal of Retailing*, *56*(3), 77–92.

Benckendorff, P., Sheldon, P., & Fesenmaier, D. R. (2014). *Tourism information technology*. Oxford, UK: CAB International. doi:10.1079/9781780641850.0000

Benea, I. A. (2014). *Influences of social media on the tourism and hospitality industry*. Vienna, Austria: Modul Vienna University.

Bernabeu, M. A. C., Mazón, J. N., Giner, D., & Baidal, J. I. (2016) Big Data and Smart Tourism Destinations: Challenges and opportunities from an industry perspective. In *School of Hospitality and Tourism Management Conference*, University of Surrey, UK.

Buhalis, D. (1998). Statejik Use of Information in the Tourism Industry. *Tourism Management*, *19*(5).

Buhalis, D. (2003). eTourism: Information technology for strategic tourism management. Pearson Education.

Buhalis, D., & Amaranggana, A. (2013). Smart tourism destinations. Information and Communication Technologies in Tourism 2014 (pp. 553-564). Dublin, Ireland: Springer International Publishing.

Buhalis, D., & Law, R. (2008). Progress in information technology and tourism management: 20 years on and 10 years after the Internet—The state of eTourism research. *Tourism Management*, *29*(4), 609–623. doi:10.1016/j.tourman.2008.01.005

Çevik, S., & Saçılık, M. (2011). Destinasyonun Rekabet Avantajı Elde Etmesinde Gastronomi Turizminin Rolü: Erdek Örneği. 12. *Ulusal Turizm Kongresi Bildiriler Kitabı*, *12*, 503–515.

Claveria, O., Monte, E., & Torra, S. (2015). A new forecasting approach for the hospitality industry. *International Journal of Contemporary Hospitality Management*, *27*(7), 1520–1538. doi:10.1108/IJCHM-06-2014-0286

Correia, A., Kozak, M., & Ferradeira, J. (2013). From tourist motivations to tourist satisfaction. *International Journal of Culture, Tourism and Hospitality Research*, *7*(4), 411–424. doi:10.1108/IJCTHR-05-2012-0022

De Mauro, A., Greco, M., & Grimaldi, M. (2016). A formal definition of Big Data based on its essential features. *Library Review*, *65*(3), 122–135. doi:10.1108/LR-06-2015-0061

Del Vecchio, P., Mele, G., Ndou, V., & Secundo, G. (2018). Creating value from social big data: Implications for smart tourism destinations. *Information Processing & Management*, *54*(5), 847–860. doi:10.1016/j.ipm.2017.10.006

Dubey, A. K. (2016). Future Technology and Service Industry: A Case study of Travel and Tourism Industry. *Global Journal of Enterprise Information System*, *8*(3).

Dzhandzhugazova, E. A., Blinova, E. A., Orlova, L. N., & Romanova, M. M. (2016). Innovations in hospitality industry. *International Journal of Environmental and Science Education*, *11*(17), 10387–10400.

Dzhandzhugazova, E. A., Zaitseva, N. A., Larionova, A. A., & Pervunin, S. N. (2015). The Russian Hotel Market: condition and development under the crisis. *Mediterranean Journal of Social Sciences, 6*(3 S5), 289.

Erickson, S., & Rothberg, H. (n.d.). Big Data and Knowledge Management: Establishing a Conceptual Foundation. Electronic Journal of Knowledge Management, 12(2), 108–116.

Fodness, D. (1994). Measuring tourist motivation. *Annals of Tourism Research, 21*(3), 555–581. doi:10.1016/0160-7383(94)90120-1

Geissbauer, R., Schrauf, S., & Koch, V. (2014) Industry 4.0 – Opportunities and Challenges of the Industrial Internet assessment, PricewaterhouseCoopers. Retrieved from https://www.pwc.nl/en/assets/documents/pwcindustrie-4–0.pdf

Goeldner, C. R., Ritchie, J. R. B., & McIntosh, R. W. (2000). *Tourism: Principles, Practice, Philosophies*. New York: Wiley.

Goncharova, N. L., & Bezdenezhnykh, T. I. (2018) Employing the Elderly in the Service Sector in Conditions of Electronic and Fourth Innovation and Technology Revolution: Industry 4.0. *Proc. of the 31st Int. Business Information Management Association Conf., IBIMA 2018 - Innovation Management and Education Excellence through Vision 2020, IBIMA* 2018, pp. 2330 –36.

Gretzel, U. (2011). Intelligent systems in tourism: A social science perspective. *Annals of Tourism Research, 38*(3), 757–779. doi:10.1016/j.annals.2011.04.014

Gretzel, U., Sigala, M., Xiang, Z., & Koo, C. (2015a). Smart tourism: Foundations and developments. *Electronic Markets, 25*(3), 179–188. doi:10.100712525-015-0196-8

Gretzel, U., Werthner, H., Koo, C., & Lamsfus, C. (2015b). Conceptual foundations for understanding smart tourism ecosystems. *Computers in Human Behavior, 50*, 558–563. doi:10.1016/j.chb.2015.03.043

Güneş, E., Bayram, Ş. B., Özkan, M., & Nizamlıoğlu, H. F. (2018). Gastronomy Four Zero (4.0). *International Journal of Environmental Pollution and Environmental Modelling, 1*(3), 77–84.

Guttentag, D. A. (2010). Virtual reality: Applications and implications for tourism. *Tourism Management, 31*(5), 637–651. doi:10.1016/j.tourman.2009.07.003

Hall, C. M. (1994). *Tourism and politics: policy, power and place*. John Wiley & Sons.

Hall, M., & Sharples, L. (2003). *The consumption of experiences or the experience of consumption? An introduction to the tourism of taste in Food Tourism Around the World Development, management and markets*. Burlington, VT: Butterworth-Heinemann an Imprint of Elsevier Linacre.

Höjer, M., & Wangel, J. (2015). Smart sustainable cities: definition and challenges. In *ICT innovations for sustainability* (pp. 333–349). Cham, Switzerland: Springer. doi:10.1007/978-3-319-09228-7_20

Howard, E. (2007). New shopping centres: Is leisure the answer? *International Journal of Retail & Distribution Management, 35*(8), 661–672. doi:10.1108/09590550710758649

Hunter, W. C., Chung, N., Gretzel, U., & Koo, C. (2015). Constructivist research in smart tourism. *Asia Pacific Journal of Information Systems, 25*(1), 105–120. doi:10.14329/apjis.2015.25.1.105

Icten, T., & Bal, G. (2017). Artırılmış gerçeklik üzerine son gelişmelerin ve uygulamaların incelenmesi. *Gazi Üniversitesi Fen Bilimleri Dergisi Part C: Tasarım ve Teknoloji*, *5*(2), 111–136.

İlhan, İ., & Çeltek, E. (2016). Mobil Pazarlama: Turizmde Artırılmış Gerçeklik Kullanımı. *Gaziantep University Journal of Social Sciences*, *15*(2), 581–599. doi:10.21547/jss.256721

Ilyenkova, S. D., & Kuznetsov, V. I. (2009). *Innovation Management*. Moscow, Russia: Eurasian Open Institute.

Jackson, K. T. (1996). All the world's a mall: Reflections on the social and economic consequences of the American shopping center. *The American Historical Review*, *101*(4), 1111–1121. doi:10.2307/2169636

Jin, J., Gubbi, J., Marusic, S., & Palaniswami, M. (2014). An information framework for creating a smart city through internet of things. *IEEE Internet of Things Journal, 1*(2), 112-121.

Jones, M. A. (1999). Entertaining shopping experiences: An exploratory investigation. *Journal of Retailing and Consumer Services*, *6*(3), 129–139. doi:10.1016/S0969-6989(98)00028-9

Kansakar, P., Munir, A., & Shabani, N. (2019). Technology in the Hospitality Industry: Prospects and Challenges. *IEEE Consumer Electronics Magazine*, *8*(3), 60–65. doi:10.1109/MCE.2019.2892245

Kapiki, S. (2012). Current and Future Trends in Tourism and Hospitality: The Case of Greece. *International Journal of Economic Practices and Theories, 2*(1). Available at https://ssrn.com/abstract=2150562

Keown, C. F. (1989). A model of tourists' propensity to buy: The case of Japanese visitors to Hawaii. *Journal of Travel Research*, *27*(3), 31–34. doi:10.1177/004728758902700306

Kim, S., & Littrell, M. A. (2001). Souvenir buying intentions for self versus others. *Annals of Tourism Research*, *28*(3), 638–657. doi:10.1016/S0160-7383(00)00064-5

Kıroğlu, P. (2012). Elektronik turizm analizi, uygulamaları ve beklentiler (Doctoral dissertation, İstanbul Kültür Üniversitesi/Sosyal Bilimler Enstitüsü/İktisat Anabilim Dalı/Yönetim Ekonomisi Bilim Dalı).

Laney, D. (2001). 3D data management: Controlling data volume, velocity and variety. *META group research note, 6*(70), 1.

Lasi, H., Fettke, P., Kemper, H. G., Feld, T., & Hoffmann, M. (2014). Industry 4.0. *Business & Information Systems Engineering*, *6*(4), 239–242. doi:10.100712599-014-0334-4

Law, R., & Au, N. (2000). Relationship modeling in tourism shopping: A decision rules induction approach. *Tourism Management*, *21*(3), 241–249. doi:10.1016/S0261-5177(99)00056-4

Law, R., Buhalis, D., & Cobanoglu, C. (2014). Progress on information and communication technologies in hospitality and tourism. *International Journal of Contemporary Hospitality Management*, *26*(5), 727–750. doi:10.1108/IJCHM-08-2013-0367

Lebedev, O. T., Mokeeva, T. V., & Rodionov, D. G. (2018) Matrix Structures of Science and Technology Innovations Development and Implementation Trajectory. In *Proc. of the 31st Int. Business Information Management Association Conf., IBIMA 2018 - Innovation Management and Education Excellence through Vision 2020, IBIMA 2018,* pp. 1759 –68.

Lee, J., Kao, H. A., & Yang, S. (2014). Service innovation and smart analytics for industry 4.0 and big data environment. *Procedia Cirp*, *16*, 3–8. doi:10.1016/j.procir.2014.02.001

Mazuryk, T., & Gervautz, M. (1996). Virtual Reality-History, Applications, Technology, and Future. Retrieved from https://www.cg.tuwien.ac.at/research/publications/1996/mazuryk-1996-VRH

Miah, S. J., Vu, H. Q., Gammack, J., & McGrath, M. (2017). A big data analytics method for tourist behaviour analysis. *Information & Management*, *54*(6), 771–785. doi:10.1016/j.im.2016.11.011

Michalko, G., & Timothy, D. J. (2001). Cross-border shopping in Hungary: Causes and effects. *Visions in Leisure and Business*, *20*(1), 2.

Mil, B., & Dirican, C. (2018). Endüstri 4.0 Teknolojileri ve Turizme Etkileri. *Disiplinlerarası Akademik Turizm Dergisi*, *3*(1), 1–9.

Neuhofer, B., Buhalis, D., & Ladkin, A. (2014). A typology of technology-enhanced tourism experiences. *International Journal of Tourism Research*, *16*(4), 340–350. doi:10.1002/jtr.1958

Peceny, U. S., Urbančič, J., Mokorel, S., Kuralt, V., & Ilijaš, T. (2019). Tourism 4.0: Challenges in Marketing a Paradigm Shift. In Consumer Behavior and Marketing. IntechOpen.

Pine, B. J., Pine, J., & Gilmore, J. H. (1999). *The experience economy: work is theatre & every business a stage. Harvard Business Press*. doi:10.5860/choice.37-2254

Porter, M. E., & Heppelmann, J. E. (2014). How smart, connected products are transforming competition. *Harvard Business Review*, *92*(11), 64–88.

Prayag, G., Hosany, S., & Odeh, K. (2013). The role of tourists' emotional experiences and satisfaction in understanding behavioral intentions. *Journal of Destination Marketing & Management*, *2*(2), 118–127. doi:10.1016/j.jdmm.2013.05.001

Richter, L. K. (1989). *The politics of tourism in Asia*. University of Hawaii Press.

Robinson, P., Lück, M., & Smith, S. (2013). *Tourism* (1st ed.). Wallingford, UK: CABI. doi:10.1079/9781780642970.0000

Schwab, K. (2016) The Fourth Industrial Revolution: what it means, how to respond, World Economic Forum. Available at https://www.weforum.org/agenda/2 016/01/the-fourth-industrialrevolution- what-it-means-and-howto- respond/

Shamim, S., Cang, S., Yu, H., & Li, Y. (2016). Management approaches for Industry 4.0: A human resource management perspective. In *2016 IEEE Congress on Evolutionary Computation (CEC)* (pp. 5309-5316). IEEE. 10.1109/CEC.2016.7748365

Shamim, S., Cang, S., Yu, H., & Li, Y. (2017). Examining the feasibilities of Industry 4.0 for the hospitality sector with the lens of management practice. *Energies*, *10*(4), 499. doi:10.3390/en10040499

Sigala, M., Christou, E., & Gretzel, U. (Eds.). (2012). *Social media in travel, tourism, and hospitality: Theory, practice, and cases*. Ashgate Publishing.

Smith, A. D. (1995). Gastronomy or geology? The role of nationalism in the reconstruction of nations. *Nations and Nationalism*, *1*(1), 3–23. doi:10.1111/j.1354-5078.1995.00003.x

Timothy, D. J. (2005). *Shopping tourism, retailing and leisure*. Clevedon, UK: Channel View Publications. doi:10.21832/9781873150610

Timothy, D. J., & Butler, R. W. (1995). Cross-border shopping: A North American perspective. *Annals of Tourism Research*, *22*(1), 16–34. doi:10.1016/0160-7383(94)00052-T

Toksöz, D., & Aras, S. (2016). Turistlerin seyahat motivasyonlarında yöresel mutfağın rolü. *Journal of Tourism and Gastronomy Studies*, *4*(1), 174–189.

Topsakal, Y., Yüzbaşıoğlu, N., Çelik, P., & Bahar, M. (2018). Turizm 4.0-Turist 5.0: İnsan Devriminin Neden Endüstri Devrimlerinden Bir Numara Önde Olduğuna İlişkin Bakış. *Journal of Tourism Intelligence and Smartness*, *1*(2), 1–11.

Tosun, C. (2001). Challenges of sustainable tourism development in the developing world: The case of Turkey. *Tourism Management*, *22*(3), 289–303. doi:10.1016/S0261-5177(00)00060-1

Tosun, C., & Fyall, A. (2005). Making tourism sustainable: prospects and pitfalls. *Environmentalism in Turkey: Between democracy and development*, 249-262.

Tosun, C., Temizkan, S. P., Timothy, D. J., & Fyall, A. (2007). Tourist shopping experiences and satisfaction. *International Journal of Tourism Research*, *9*(2), 87–102. doi:10.1002/jtr.595

Turner, L. W., & Reisinger, Y. (2001). Shopping satisfaction for domestic tourists. *Journal of Retailing and Consumer Services*, *8*(1), 15–27. doi:10.1016/S0969-6989(00)00005-9

Verevka, T. (2018). Key performance indicators of high-tech enterprises. IV International Scientific Conference "The Convergence of Digital and Physical Worlds: Technological, Economic, and Social Challenges" (CC-TESC2018). In SHS Web of Conferences (Vol. 44, p. 00077). EDP Sciences. 10.1051hsconf/20184400077

Verevka, T. V. (2019). Development of Industry 4.0 in the Hotel and Restaurant Business, *IBIMA. Business Review (Federal Reserve Bank of Philadelphia)*. doi:10.5171/2019.324071

Wakefield, K. L., & Baker, J. (1998). Excitement at the mall: Determinants and effects on shopping response. *Journal of Retailing*, *74*(4), 515–539. doi:10.1016/S0022-4359(99)80106-7

Wang, D., Park, S., & Fesenmaier, D. (2012). The role of smartphones in mediating the tourism experience. *Journal of Travel Research*, *51*(4), 371–387. doi:10.1177/0047287511426341

Weinelt, B., & Moavenzadeh, J. (2017). Digital Transformation Initiative: Aviation, Travel and Tourism Industry. In Geneva, Switzerland: World Economic Forum.

Werthner, H., & Klein, S. (1999). *Information technology and tourism: a challenging relationship*. Vienna, Austria: Springer. doi:10.1007/978-3-7091-6363-4

Werthner, H., & Ricci, F. (2004). E-Commerce and Tourism. *Communications of the ACM*, *47*(12), 101–105. doi:10.1145/1035134.1035141

Yılmaz, D., & Yılmaz, S. (2005). Küreselleşme ve Bilgi Teknolojilerinin Turizm Endüstrisine, Etkileri. Retrieved from www.bilgiyönetimi.com

Yolal, M. (2003). *Türkiyede'ki Küçük ve Orta Ölçekli Konaklama İşletmelerinde Bilgi Teknolojileri Kullanımı*. Eskişehir: Anadolu Üniversitesi Yayınları.

Zagralı, E., & Akbaba, A. (2015). Turistlerin Destinasyon Seçiminde Yöresel Yemeklerin Rolü: İzmir Yarımadası'nı Ziyaret Eden Turistlerin Görüşleri Üzerine Bir Araştırma. *Journal of Yaşar University*, *10*(40), 6633–6644.

Zaitseva, N. A. (2013). *Management in service industry: tourism and hospitality*. Moscow, Russia: Academia Publishing House.

ADDITIONAL READING

Azmi, A., Nur'Hidayah Che Ahmad, K. K., Abdullah, D., & Zubir, H. A. (2018). Industry 4.0: Teaching Preferences, Perceptions, and Challenges among Tourism and Hospitality Academicians. *Journal of Academic Research in Business and Social Sciences*, *8*(15), 350–365.

Dohale, V., & Kumar, S. (2018). A Review of Literature of Industry 4.0. In National Convention of IIIE and International Conference.

Gerd, J. H. (2019). Industry 4.0: A supply chain innovation perspective. *International Journal of Production Research*. doi:10.1080/00207543.2019.1641642

Güneş, E., Bayram, Ş. B., Özkan, M., & Nizamlıoğlu, H. F. (2018). Gastronomy Four Zero (4.0). *International Journal of Environmental Pollution and Environmental Modelling*, *1*(3), 77–84.

Holjevac, I. A. (2003). A vision of tourism and the hotel industry in the 21st century. *International Journal of Hospitality Management*, *22*(2), 129–134. doi:10.1016/S0278-4319(03)00021-5

Ivanović, S., Mijolica, V., & Roblek, V. (2016). A holistic approach to innovations in tourism. *Proceedings of ICESoS*, *2016*, 367–380.

Mil, B., & Dirican, C. (2018). Industry 4.0 technologies and its effects on tourism economics. *Journal of Multidisciplinary Academic Tourism*, *3*(1), 1–9. doi:10.31822/jomat.347736

Minaudo, M. (2018, October). New Technologies Applied to Tourism 4.0. In *Global Conference on Business, Hospitality, and Tourism Research (GLOSEARCH 2018).*

Oztemel, E., & Gursev, S. (2018). Literature review of Industry 4.0 and related technologies. *Journal of Intelligent Manufacturing*. doi:10.100710845-018-1433-8

Papathanassis, A. (Ed.). (2011). *The long tail of tourism: Holiday niches and their impact on mainstream tourism*. Springer Science & Business Media. doi:10.1007/978-3-8349-6231-7

Pindžo, R., & Brjaktarović, L. (2018). Digital Transformation of Tourism. *In TISC-Tourism International Scientific Conference Vrnjačka Banja*. 3 (1), pp. 340-355.

Rio, D., & Nunes, L. M. (2012). Monitoring and evaluation tool for tourism destinations. *Tourism Management Perspectives*, *4*, 64–66. doi:10.1016/j.tmp.2012.04.002

Trappey, A. J., Trappey, C. V., Govindarajan, U. H., Chuang, A. C., & Sun, J. J. (2017). A review of essential standards and patent landscapes for the Internet of Things: A key enabler for Industry 4.0. *Advanced Engineering Informatics*, *33*, 208–229. doi:10.1016/j.aei.2016.11.007

Wallace, S., & Riley, S. (2015). [an industry perspective. *Journal of Tourism Futures*.]. *Tourism (Zagreb)*, ▪▪▪, 2025.

Zhong, R. Y., Xu, X., Klotz, E., & Newman, S. T. (2017). Intelligent manufacturing in the context of industry 4.0: A review. *Engineering*, *3*(5), 616–630. doi:10.1016/J.ENG.2017.05.015

Ziyadin, S., Litvishko, O., Dubrova, M., Smagulova, G., & Suyunchaliyeva, M. (2019). Diversification Tourism in the Conditions of the Digitalization [IJCIET]. *International Journal of Civil Engineering and Technology*, *10*(02), 1055–1070.

KEY TERMS and DEFINITIONS

3D Printer: A machine allowing the creation of a physical object from a three-dimensional digital model, typically by laying down many thin layers of a material in succession.

Industry 4.0: The 4th industrial revolution, called as Industry 4.0, is the process in which the use of computers and automation in the industry.

IoT: The growing network of devices that can connect, communicate and transfer data between one another.

Robot: A robot is a machine—especially one programmable by a computer— capable of carrying out a complex series of actions automatically.

Smart tourism: Smart Tourism has focused on the use of advanced technologies to transform data into efficient new business models by using and evaluating data collected through physical infrastructures and social connections.

Tourism 4.0: The aim of Tourism 4.0 in tourism is to reduce the negative effects of tourism, to see the effects of the use of technology in the tourism sector and to develop cooperation models in partners.

Virtual Reality: Virtual reality (VR) is a simulated experience that can be similar to or completely different from the real world.

Chapter 11
Expansion of Technology Utilization Through Tourism 4.0 in Slovenia

Jurij Urbančič
Arctur d.o.o., Slovenia

Alenka Vavroš
Arctur d.o.o., Slovenia

Vesna Kuralt
Arctur d.o.o., Slovenia

Simon Mokorel
Arctur d.o.o., Slovenia

Hrvoje Ratkajec
Arctur d.o.o., Slovenia

Urška Starc Peceny
Arctur d.o.o., Slovenia

Matevž Straus
Arctur d.o.o., Slovenia

Tomi Ilijaš
Arctur d.o.o., Slovenia

ABSTRACT

With more and more people traveling worldwide (the number doubling in the last 20 years), tourist destinations are now more than ever trying to maintain and enhance their competitiveness in the global market. In this regard, novel business models combined with state-of-the-art technology can play a crucial role in not only satisfying the increasing tourism demand but also ensuring a sustainable growth to avoid the deteriorating effects on both the social and natural habitat. For these reasons, the Slovenian government included tourism as one of the priority areas for investment. Hence, the Tourism 4.0 initiative was launched to unite tourism stakeholders with high-tech companies and unify their scattered ideas, experiences, knowledge, and expertise. The ambition is to position Slovenia as a top destination for sustainable tourism with high economic value. This chapter introduces the basic concepts behind Tourism 4.0 and how it relates to technologies for an enhanced tourism experience.

DOI: 10.4018/978-1-7998-1989-9.ch011

INTRODUCTION

Within this proposed book chapter, the authors present the largest government-funded research and development project in the field of Slovenian tourism, called Tourism 4.0 (T4.0) and launched in September 2018. The name itself originates from the fourth and ongoing industrial revolution known as Industry 4.0 (Vogel-Heuser & Hess, 2016), which is characterized by the cyber-physical systems (CPS) manufacturing, consisting of a heterogeneous data exchange (Lu Y., 2016). If Industry 4.0 aims to achieve higher added-value products and services though operational efficiency and the automation of the production process by utilizing the modern technologies, Tourism 4.0 is aimed at doing the same for the tourism sector.

The motivation behind the project is that, unfortunately, a gap between the tourism industry and the use of emerging technologies still exists. Some of the obstacles for a greater technology implementation are the lack of sufficient knowledge, tools and strategies. These issues are already being addressed within the context of Smart Tourism (Gretzel, Sigala, Xiang, & Koo, 2015), an initiative aimed at applying technological innovations to improving the efficiency and sustainability of the tourism sector, while at the same time enhancing the local residents' quality of life. There are three main components associated with Smart Tourism:

1. Smart Destination –a tourist destination where the state-of-the-art technology is integrated in the destination's own infrastructure. It applies the same principles of accessibility, efficiency, sustainability and quality of life as Smart Cities, only in this case they are extended to include tourists (in addition to local residents) (Buhalis & Amaranggana, 2013).
2. Smart Experience – the convergence of technology and tourism experience (Hunter, Chung, Gretzel, & Koo, 2015). One of the requirements is a technological platform for instant tourism-related information exchange between the stakeholders, thereby also creating a large amount of data in this process – known as Big Data (Buhalis & Amaranggana, 2015). Through this data, destinations can provide personalized services and products to tourists in order to enhance their tourism experience. In this context, the tourists themselves become active participants in the creation of their own experience by sharing the information about their preferences and past experiences.
3. Smart Business - a complex business ecosystem which includes a strong collaboration between the public and the private sector (Buhalis, 2000), defined by the digitalization of business activities and by the rapid responses to the changes in the market. As a result of the data sharing between all stakeholders, consumers are also included in the product/service creation process as an important source of knowledge for innovation (Foss, Laursen, & Pedersen, 2011).

Recognizing the potential of the Industry 4.0 technologies by merging them with the concept of Smart Tourism, the T4.0 project ambition is to create a benchmark for the transformation of today's Slovenian tourism industry into an innovation-driven economy for the benefit of all participants. The main goal is the establishment of a Collaborative Platform that will facilitate the exchanges between all of the stakeholders in the tourism ecosystem. The platform will also enable the collection, exchange and analysis of data for the needs of strategic activities such as marketing, resource allocation, energy consumption and tourist dispersion, all of which are aimed at improving the tourist experience and minimizing the negative impact on the local environment. To this aim, the project partners have a set of defined technologies for the improvement of business processes and activities in the tourism sector.

BACKGROUND

The first forms of tourism were known already in ancient Egypt, whereas the ancient Greeks and Romans devoted considerable attention to traveling for pleasure and even developed a concept associated with what we understand as vacations (Rabotić, 2014; Korstanje & Busby, 2010). Modern tourism, according to many historians (Zuelow, 2015), started to develop at the end of the eighteenth century and is recognized as one of the most important commercial activities today. It accounted for generating 10% of the total employment in 2018 and represented a share of 10.4% in the global economy GDP in 2018 (WTTC, 2018).

However, it has long been reported (Forster, 1964; Young, 1973; Pizam, 1978; Sunlu, 2003) that host locations can also experience several negative consequences, resulting in a degradation of both the natural and social environment. At the destination level, examples of such negative consequences include the following:

i.) Overconsumption of water, where in some countries the tourist sector represents more than 10% of the domestic water consumption (or as high as 40% in the case of Mauritius) (Gössling et al., 2012); and

ii) Skyrocketing of rental prices, like in the case of Barcelona, where the price of residential apartments can reach vacation apartments price level (Martín Martín, Guaita Martínez, & Salinas Fernández, 2018).

At the global level, tourism-related activities already account for 8% of the global greenhouse gas emissions (Lenzen, et al., 2018), representing a significant factor, associated with climate change. Similarly, the negative effects on the cultural heritage and its role in society have been associated with over-tourism and banalizing or "disnifying" tourism (Ashworth & Tunbridge, 1990), which have led to a 'multiplicity of standardized attractions that reduce the uniqueness of urban identities even while claims of uniqueness are more intense (Zukin, 1998).

At the same time, when properly implemented, tourism has the potential of driving positive social transformation (Tovar & Lockwood, 2008), bringing economic benefits (Frechtling, 2000) and resulting in sustainable development (McCool & Moisey, 2001). The latter is recognized by the United Nations Organisation in their recommendations on implementation of the 17 Sustainable Development Goals (SDGs) (World Tourism Organization and United Nations Development Programme, 2018). In this regard, the use of new technologies plays an important role in facilitating a more environmentally, socially, culturally and economically sustainable interaction between visitors, local communities, tourist landscapes and the cultural heritage, e.g. the so-called "intelligent entertainment" (Mencarelli & Pulh, 2012) and "edutainment" (Addis, 2005).

Before moving on to explain the main focus of this chapter and how new technologies can improve every stakeholder's experience in tourism, the authors present the characteristics of Slovenia as a destination with a great potential for improving its current state of tourism. Slovenia is a relatively small country with a population of around 2 million and covering an area of 20 273 km^2, located at the crossroads between the central and Southeastern Europe at the intersection of three globally recognizable regions (the Alps, the Mediterranean and the Pannonian plains). After gaining its independence from Yugoslavia in 1991, the country faced a sharp decline in tourism and lost almost 75% of foreign visitors (Sirse & Mihalic, 1999). However, in the last 20 years a number of measures were implemented in order to boost

Figure 1. A division of Slovenia into 4 macro destinations

the Slovenian tourism and make it internationally competitive. Among the actions taken is the establishment of a system of macro-destinations by the Slovenian Tourist Board, based on geographically defined units and with similar strategic elements (e.g. main products and services). These destinations, shown in Figure 1 are: the Alpine Slovenia, the Mediterranean Slovenia, the Thermal Pannonian Slovenia and Central Slovenia with the capital Ljubljana. Two more important drivers of Slovenia's tourist brand and its tourist offer are also the natural environment and the natural heritage (especially the outdoor activities, the natural sights and the rich gastronomy). The cultural heritage (comprising more than 30.000 immovable cultural heritage monuments), on the other hand, still presents immense opportunities for further interpretation.

As a result of past efforts, there are record numbers of tourist visitors today, going hand in hand with a notable global interest in Slovenia as a destination. In 2018, more than 5.9 million tourist arrivals and 15.6 million overnight stays were recorded in Slovenia. Most of the visitors come from abroad, with Germany, Italy, Austria, the Netherlands and Croatia topping the list. Visitors from these countries combined accounted for more than 30% of the total visitors (see Table 1). This represents an increase of 8% in arrivals and 10% of overnight stays, as compared to 2017, making this the fifth consecutive year of growth in tourist visits. In the same year, tourism accounted for 12.3% of the total gross domestic product and 12.8% of all employment (representing around 110 700 jobs). With this above-average growth continuing in 2019 (see Total Slovenia News, 2019), which is forecast to continue in the near future, the Slovenian tourism industry is becoming one of the most important economic sectors in the country.

However, despite the aforementioned achievements, some pressing issues remain open, and continue to hinder further progress of tourism. Among the most visible are the shortage of workforce and the low wages. This is illustrated by Figure 2, depicting the average salary of hotel employees in the past 4 years. The pay in this sector accounts for only around 80% of the national average salary. When inspecting the employees in holiday homes, the situation is even worse - their average gross salary is below 70% of the national average salary. As a consequence, employers are experiencing great setbacks when looking for trained and competent staff, mainly due to the harsh working conditions and low salaries.

Table 1. Number of foreign tourist arrivals and overnight stays in 2018. The data is courtesy of the SURS—Statistical Office of the Republic of Slovenia

Country of origin	Tourist arrivals		Tourist overnight stays	
	Number	Percent of total	Number	Percent of total
Germany	506 081	8.6	1 362 214	8.7
Italy	598 825	10.1	1 334 059	8.5
Austria	381 709	6.5	1 011 135	6.5
Netherlands	185 257	3.1	612 710	3.9
Croatia	218 896	3.7	527 118	3.3
other countries	2534371	42.9	6 328 774	40.5

In order to tackle these issues, the partnership for Strategic Development and Innovative Partnership in Tourism (SRIPT) was launched by the Tourism and Hospitality Chamber of Slovenia with the objective to include tourism into the Slovenian Smart Specialization Strategy (S4) (published by the Government of the Republic of Slovenia, 2015). The Smart specialization focuses on areas with growth potential. It highlights the fields in which Slovenia has the critical mass of knowledge, capacities and competences to reach a recognizable position within the global markets, making such sectors investment priorities.

Figure 2. Comparison between the national average gross salary and salary of employees in hotels or similar accommodations and holiday homes or similar accommodations. Data courtesy of Statistical Office of Slovenia.

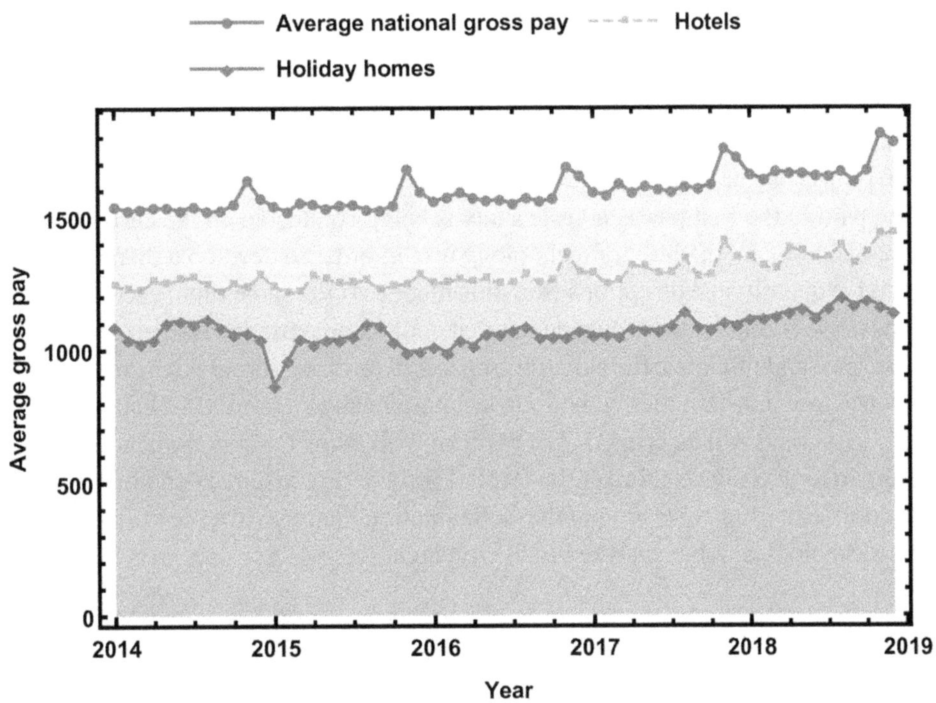

Among the key SRIPT performance indicators defined for the period until 2023 are the following:

1. Increasing the added value in tourism by 15%;
2. Increasing the revenue under the heading of export travel by 4-6% annually;
3. Improving the energy efficiency of tourist facilities by 20%.

Tourism 4.0 follows the SRIPT goals, directly addressing two key priority fields: information-based marketing and technological solutions. The project initiative was launched by the high-tech company Arctur, which identified the lack of readiness to embrace the use of the key enabling technologies in the tourism sector, especially among small and medium enterprises. Consequently, the Partnership for Tourism 4.0 was established to strengthen the collaboration between various stakeholders willing to participate in the research and in the further development of tourism. The project brings together business companies, governance bodies and top researchers in tourism and IT technology. The evolving partnership already boasts a number of prominent members, such as the Slovenian Ministry of Economic Development and Technology as well as the Ministry of Public Administration, the Tourism and Hospitality Chamber of Slovenia, the Association of Towns and Municipalities of Slovenia, the University of Ljubljana, the University of Maribor, the University of Primorska and many other partners.

MAIN FOCUS OF THE CHAPTER

The primary goal of the project is to establish a collaborative ecosystem that is focused on merging both the physical and the digital space into a personalized experience. At the same time, it is also aimed at promoting sustainable and socially conscious aspects of tourism. For this reason, the Tourism 4.0 ecosystem is defined under the assumption of connecting all stakeholders, which includes the local community, the various levels of government, the tourist service providers and the tourists themselves, while at the same time promoting their collaboration (see Figure 3). The existing business activities and processes are being identified and analyzed within this collaboration in order to encourage more innovation in tourism and to foster strategic thinking.

Tourism 4.0 project can be roughly divided into two main phases, both of which are further explained in the following part. In the first phase, a basic study is being conducted on the current state of the Slovenian tourism, including on the level of technology use, in order to develop a theoretical model of the ideal tourism and to quantify the deviations from this model. At the same time, a technological research on selected innovative technologies is carried out that could help optimize business processes, improve the tourist experience and minimize the negative impacts of tourism. Based on the results obtained from the aforementioned research activities, a modern, technologically-supported Collaboration Platform for all stakeholders in tourism will be created. The platform will allow for communication, monitoring and analytically supported decision-making at the level of providers, consumers and destinations. The second phase includes demonstrative testing of the Collaboration Platform in a controlled real-environment along with the technological tools analyzed in the research.

Figure 3. Tourism 4.0 ecosystem of collaboration between all stakeholders in tourism

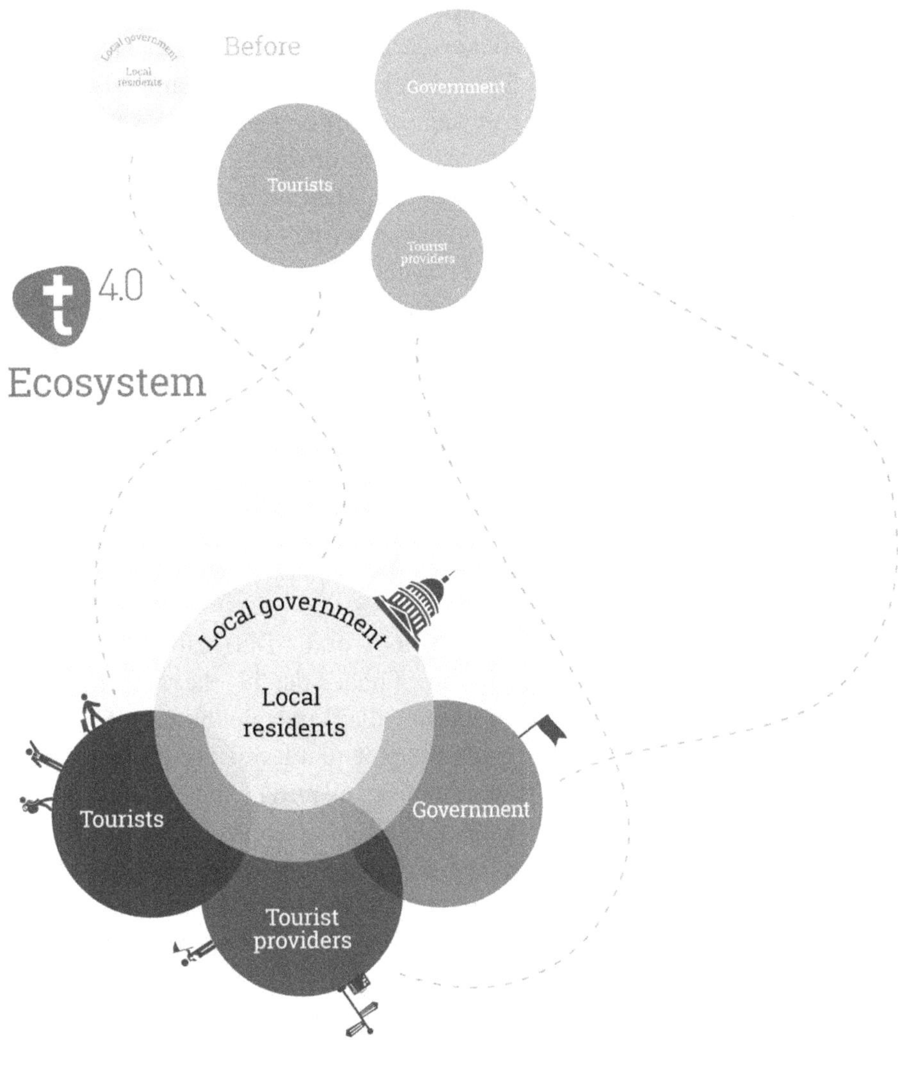

Basic Research Phase

Tourism and technological advancement have been inter-connected for a long time (Poon, 1993; Buhalis & Law, 2008). Among the major milestones in this process, one could note the introduction of the Computer Reservations Systems (CRSs) in the 1970s, the Global Distribution Systems (GDSs) in the 1980s and the spread of Internet in the 1990s. Because tourism is a part of the global economy today, tourist destinations have to be, now more than ever, innovative as regards the use of new business models and state-of-the-art technology. To achieve this, first and foremost, the adequate information on the state-of-the-art and future trends is necessary.

Through the basic research phase, the aim of the project is to develop and introduce a comprehensive ecosystem, based on the theoretical model named Tourism Impact Model (TIM). TIM encompasses a system of motivation and rewards for positive behavior, i.e. behavior with positive consequences on the social, environmental and economic aspects of the tourism sector. The latter encompasses positive impacts on tourism as regards local inhabitants, tourists and other stakeholders. The model is based and sustained with the data gathered with the help of key enabling technologies, as known from the Industry 4.0. These technologies will be used initially to develop the tools that support the management of tourist flows in the context of the local tourist carrying capacity and will be consequently providing the instruments for potentially preventing the occurrence of over-tourism. In parallel, bibliographic research is being conducted on selected technologies for establishing their present and potential use in tourism.

This chapter section is dedicated to providing an explanation of both TIM and the technologies selected for a detailed study. The results will allow us to compare the ideal tourism ecosystem based on the literature review with the existing state of tourism and show us how the key enabling technologies could be used to manage the transition from the existing tourism ecosystem closer to an ideal one. The outcome will be presented in a research report, summarizing the results and suggesting a set of recommendations on how to improve the overall state of tourism. Also based on the above-mentioned research, the development of a prototype version of the Collaboration Platform will begin, encompassing steps from user environment design to generic technology development and technology integration.

Tourism Impact Model

The Tourism Impact Model (TIM) represents one of the tools of the T4.0 platform. It is designed to support the digital transformation of processes with the aim of understanding the respective micro location's existing state and future predictions according to the so-called strategic thinking 4.0, which is focused on raising sustainable tourism to its maximum possible potential level. The greatest value of TIM for users is the planned integration of the quick response mechanism. The model combines a study of the methodology of the carrying capacity with the technological components, including data from different sources, to provide recommendations for respective micro locations. These recommendations would be more in-depth than mere policies and adjusted according to the specific characteristics of each micro location, respectively, rather than staying general. The overall approach to the development of the TIM tool follows the stages described below.

The first stage is the identification and a critical review of the existing methodologies for measuring the carrying capacity at tourist destinations and of the models for the visitors' management. There are several tourist-carrying-capacity models, both applied and theoretical. For example, the maximum carrying capacity concept could be formalized by applying the Fisher and Krutilla model (Fisher & Krutilla, 1972) or the Canestrelli and Costa's model (Canestrelli & Costa, 1991). The selection and development of the most suitable method will take into account the local and regional specificities (i.e. macro destinations in Slovenian tourism) and the interconnection of indicators (e.g. for benchmarking), as well as their dynamics. Monitoring the indicators dynamics is important, since the carrying capacity is a category that sets the scale for measuring environmental stresses on one hand. The dynamic monitoring, on the other hand, makes it possible to assess the degree of tourism-related pressures within a given period and opens up space for timely action.

In the second stage, the methodology will be developed for measuring the tourist capacity of the various types of territory of stakeholders, respectively. An analysis of the strategic tourist flows and the niche tourist flows in each type of the stakeholders' territory is planned, in order to design the model for measuring the carrying capacity of destinations and to cater for the selection of the relevant digital tools for the data harvesting and analysis. There are a great number of potential indicators for monitoring the carrying capacity, as well as for drafting recommendations for the macro and micro destinations (in the case of Slovenian municipalities) on how to properly collect and process the required data. When measuring the values of these indicators, we are primarily interested in two major aspects: what is the impact of tourism activities at a specific micro location and what are the degrees of collaboration between different stakeholders at the same micro location. Thematically, the identified indicators are categorized into six groups:

1. Economic indicators (growth, income, accommodation, employment, local crops and livestock products, real estate, seasonality);
2. Social indicators (impact on local communities, safety and health, adjustments for people with disabilities, cultural heritage, concentration of demand and supply, meeting the needs of tourists and local residents);
3. Environmental indicators (air quality, consumption of drinking water, disposal and treatment of waste water, waste management, traffic-related data such as transport-related pollution, energy consumption, environmental assessments);
4. Infrastructure indicators (cycling route network, charging stations for electric cars, road network congestion, public transportation, parking lot occupancy);
5. Political-civilian indicators (participation of local residents in local government decision making, NGO support, collaboration between neighboring locations);
6. Collaboration indicators (although interwoven in all other aspects above, a special focus and care is dedicated to the understanding and evaluation of the natural environment and to the level of collaboration between different stakeholder).

Through the selection of indicators, the project aims to determine the tourist carrying capacity in all of the indicator groups by applying the Multi-Attribute Decision Making (MADM) theory, using the open-source DEXi visualization and the analytical platform (Bohanec, 2008). The methodology of MADM DEXi is based on the decomposition of a large complex problem into sub problems, which are smaller, less complex and easier to solve than the overall problem. The situation is presented graphically by a hierarchical tree of attributes. Each alternative is evaluated by each attribute and subsequently formulated again by a single value. The aggregation of values (the utility function) is defined by a set of simple "if-then" rules, which are, compared to the linearly weighted sum, more suitable for expressing nonlinearities in decision knowledge and are easier to understand. The hierarchical tree of attributes presented as a set of decision rules (utility functions) that are defined by experts, represents the knowledge base.

The TIM will enable each respective location to determine, with the help of experts, the threshold values of its carrying capacity in terms of the basic indicators like air quality, water supply, waste disposal, tourist accommodation, traffic load, etc. In this way, the destination can obtain a detailed assessment of the status and a comparison to the referential and the regional average values. Each destination will be able to see what their advantages, opportunities and risks are. Depending on the availability of the data, the system will enable for real-time impact assessments, thus providing decision makers with a reliable foundation for strategic decision-making and adopting adequate measures in the processes of digitalization. The result in the form of visual representations will help them to grasp their level of preparedness and empower by providing the guidelines for further actions. Through the analysis of the data, provided from selected sources, TIM can also generate tailored reports for the respective destinations. Such reports can include the visualization in the form of a "traffic light", depicting in this way the existing state as

Figure 4. Key enabling technologies with the potential for use in the tourism sector

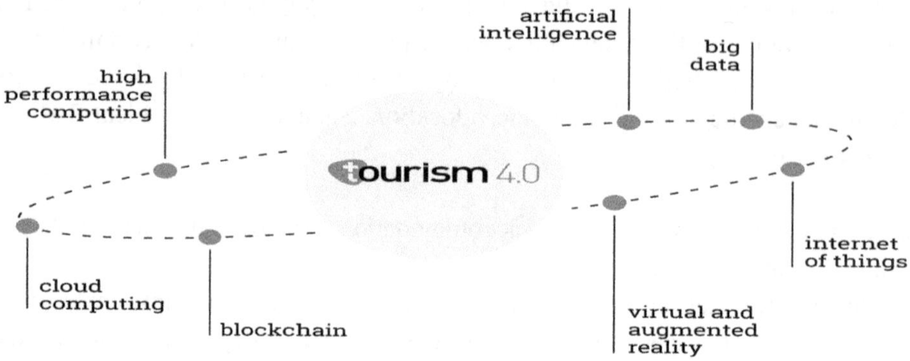

compared to the critical limit values. In such visual representations, areas that are below the predefined critical values are marked in green, whereas yellow indicates an approaching warning and red appears when the values are above the critical benchmark.

What the TIM strives to unlock is the innovation potential of the advanced technologies, originating in the Industry 4.0. TIM aims to do that with the goal of tackling the challenges of over-tourism and allowing for insights into the degree to which tourism in a respective location still has a positive impact on the ground. In contrast to other similar initiatives from all over the world, we are trying to achieve more sustainability in the tourism sector. This is encouraged by a bottom up approach, encouraging collaboration among all stakeholders in the tourism ecosystem, among the actors who are and always will be co-creating the tourist experience. The results of our activities help destinations, be it municipalities or any selected location of any size, to understand their limits and act strategically to create destinations that people want to visit and live in.

Technology Research

The technology research in the scope of Tourism 4.0 is a bibliographic study conducted on pre-defined technologies, such as blockchain, the internet of things, big data, high performance & cloud computing, artificial intelligence and virtual & augmented reality, as shown in Figure 4. It is aimed at gathering the required technical descriptions and identifying the (potential) uses in the tourism sector. Advanced technological solutions are of particular importance for a sustainable use of resources within the accommodation facilities and for the optimization of business processes, such as:

- the reduction of costs;
- offering higher quality services;
- the reduction of time spent on individual activities in the course of the business process.

In addition to being included in the recommendations for both tourism service providers and policy makers, these technologies will be integrated into the Collaborative Platform to provide a new cooperative system for all tourist stakeholders. The respective technologies and their use in tourism within the context of Tourism 4.0. are described below.

Blockchain Technology

The first blockchain was introduced by Satoshi Nakamoto in 2008 as the supporting technology for Bitcoin, a peer-to-peer electronic cash system (Nakamoto, 2008). While there is no universally valid definition of blockchain, one can provide the following definition based on the peer-reviewed literature (Seebacher & Schüritz, 2017): "*A blockchain is a distributed database, which is shared among and agreed upon a peer-to-peer network. It consists of a linked sequence of blocks, holding timestamped transactions that are secured by public-key cryptography and verified by the network community. Once an element is appended to the blockchain, it can not be altered, turning a blockchain into an immutable record of past activity*".

The blockchain technology presents a wide range of possibilities, and can be implemented into any service provided, thus gaining attention of many different industries. Tourism is among these, with a special interest pertaining to the areas of booking, reservation and payment systems (Önder & Treiblmaier, 2018). However, the real potential of this technology lies in building reliable, flexible and stable systems. Such systems tend to rely on a large number of transactions between the users, where any interaction between the users is considered a transaction, whichis not limited solely to financial transactions. Therefore, the most important factors to be considered when dealing with smart distributed systems are the transaction costs, the transaction times and the permeability of transactions.

The blockchain technology is an ideal match for systems and platforms that would profit from the fact that no single person or company can exercise control over them. As such, two specific use cases are being developed to be included into the Collaboration Platform later:

- *A Digital Passport* for tourists, where the blockchain technology can be utilized as a data storage solution. With the Digital Passport, tourists are able to build their own personal profile based on the shared information, preferences and interests, with the purpose of receiving recommendations for trips, activities or products. All of the user data or their hashes can be stored in a public or private blockchain. The main advantage of this is that no one owns or controls the database and the access to the data in this way. In a "traditional" system, if the database owner ceases to exist or decides that they are no longer interested in maintaining the database, other connected products can be left nonfunctional. In distributed systems, however, this kind of risk is averted, because a consortium owns the database, both legally and through a technical implementation of the blockchain technology. Blockchains can also be used to store information which users share through their profiles. The benefits include the fact that a clear and transparent system is put in place, which allows for backtracking of profile sharing and data access. Should any privacy issues occur, there is a record for every time a user has shared their data with an agent.
- *A Collaboration Impact Token*_(Peceny, Urbančič, Mokorel, Kuralt, & Ilijaš, 2019). The idea of the Collaboration Impact Token (CIT) is to provide a reward system for tourists. These tokens are a blockchain-based crypto currency, awarded to tourists as incentives for the behavior which is beneficial to the local community and to the environment. The emphasis is on sustainability, e.g. consuming less energy or using public transport. In turn, these tokens can be used for purchasing other tourist goods. Their most important feature is the assignment of a dynamic value, dependent on the location and a given time, i.e. a higher value at the touristic periphery or during low tourist season. The CIT will be stored in a digital personal user wallet, which can be integrated into Digital Passport.

Internet of Things

The IoT or Internet of Things represents all of the inter-connected devices that communicate with each other through the Internet. There can be more than one communication mode, ranging from the communication between devices and people to communicating among the devices themselves. A larger number of uniquely determined devices can be connected into the IoT, which, with the help of sensors, transmit the various data to the Internet, thus making it available for other users. The technology breakthrough in IoT (along with smart cities, Big Data, and Cloud computing) has contributed to designing the notion of smart tourism (Kaur & Kaur, 2016). The greatest impacts of the IoT technology are expected in the fields related to personalization (e.g. the use of tablet computers or smart TVs in hotel rooms for adjusting the room temperature, the television control, the lights, etc.), in accessing real-time data (e.g. notification of the change of flight directly to one's mobile device), in simplifying certain tasks, in easier management of customer relations, for improving the tourist experience or in simplifying maintenance tasks (e.g. instantaneous error notification) (Dave, 2018).

For Tourism 4.0, the IoT can present an effective tool to monitor and reduce the negative impacts of tourism, notably through installing the sensors for monitoring the energy and water consumption or the waste production. It can also offer an upgrade of the already existing services, such as virtual bulletin boards, by displaying context-sensitive ads in accordance with a generalized Digital Passport profile of tourists who are located near a respective location.

Big Data

Big data refers to a large amount of data that can be processed, analyzed and used to support decision-making (Hung, 2016). While there is no single definition of big data, the definitions usually include the characteristics, pertaining to data volume (the quantity of data), velocity (the rapid data collection and analysis) and variety (the various types of data from different sources). Big data has already proven useful for marketing purposes (Elgendy & Elragal, 2014; Xu, Frankwick, & Ramirez, 2016) and in the fields of the biomedical research and healthcare (Luo, Wu, Gopukumar, & Zhao, 2016; Raghupathi & Raghupathi, 2014). In tourism (Li, Xu, Tang, Wang, & Li, 2018), big data is being used, too, notably to study the tourist behavior (Miah, Vu, Gammack, & McGrath, 2017). Big data can be effectively used to facilitate exploring the popular tourism destinations (i.e. where to travel) (Lee, Cai, & Lee, 2014), the searching for effective travel routes (Kurashima, Iwata, Irie, & Fujimura, 2013; Lu, Wang, Yang, Pang, & Zhang, 2010; Okuyama & Yanai, 2013) and for identifying the best timeframe to visit a specific destination.

Tourists produce a huge amount of data on the global scale, which can provide valuable insights into their practices and attitudes. This data is collected and "owned" by a range of actors, public and private, for their own business and administrative purposes, such as billing. However, after the primary use this data often remains un-shared, thus limiting the potential for cross-analysis and business intelligence that could provide a boost for the tourism sector. With the appropriate data collection and analysis, complex Big Data analytics can be applied within the context of Tourism 4.0. This would allow for a deeper insight into the tourism sector, as compared to more traditional methods of surveys and interviews with focus groups. This in turn can encourage more sustainable strategies and business models in order to tackle the negative externalities presented by modern tourism, while at the same time improving the experience of both returning and new visitors. A special focus is put into developing a tool for monitoring and visualizing the tourist mobility patterns through various sources, such as reports of mobile phone activities, traffic counters and street cameras. By combining the (appropriately anonymized) mobility data in tourism with other data available (i.e. accommodation capacities, resources available, population density, traffic pollution, etc.), trends in tourism throughout seasons can be monitored, in order to identify the destinations (both local and regional) where peaks are unsustainable, thus identifying the possible strategies for a diversification as well as new marketing approaches. Through this process, a diversification of the tourist flows can be promoted or a more effective distribution can be planned over time, to minimize the negative impacts of the existing peaks and – most importantly – to identify the areas for improvements and innovation.

High Performance Computing and Cloud Computing

A high-performance computing (HPC) system is a tool used to tackle problems that require substantial computing resources or a length of time and therefore cannot be processed using personal computers due to their limitations. Normally the HPC services are provided by supercomputers that can be accessed remotely, using the network. Traditionally, they have been used in scientific and engineering fields, which require large quantities of data to be processed, e.g. in weather forecasting, steel casting simulations and in aerodynamic research. Recent uses include projects in the artistic and creative industries, which are due to the rapid advances in the area of 3D technologies.

Cloud computing refers to the practice of using a network of remote servers hosted on the Internet to manage, store and process data, instead of using a local server or a personal computer. These services are broadly divided into three categories: Infrastructure-as-a-Service (IaaS), Platform-as-a-Service (PaaS) and Software-as-a-Service (SaaS). The most relevant advantage of using cloud computing is the overall reduction of fixed costs and the possibility of shifting those costs into operative costs depending on the demand (Etro, 2009). This facilitates the market entry for smaller entities like SMEs, startups or public institutions, since it removes the need to purchase and maintain expensive equipment. Also, the expanded use of the Internet of Things has led to the production of large amounts of data that requires storage, processing and appropriate access, with cloud computing providing an appropriate dynamic and flexible support infrastructure.

Cloud computing offers many advantages over other distributed platforms, lately including HPC as a Service (HPCaaS), which aims to provide scalability of resources, cost-effectiveness and no-maintenance for end users (Sha, et al., 2019).

While certain case studies on the uses of HPC and Cloud computing in the tourism sector can be found, they are mostly limited to big players like Airbnb (AirbnbEng, 2010), Red Lion Hotels Corporation (AWS Case Study: Red Lion Hotels, 2019) and SETTOUR (AWS Case Study: SETTour, 2019). On the other hand, there are many examples of success stories of SMEs using HPC that have led to business benefits and could also be applied to tourism. For example:

- In the field of urban planning, building-performance was modelled in order to create more sustainable buildings (HPC-Cloud-based urban planning, 2019). Large simulations were performed quickly and effectively through Cloud-based HPC, which significantly decreased the operation time and substantially increased the number of buildings per simulation. The same principle can be applied when designing new or improving existing tourist accommodations.
- For predicting air quality, Cloud-based HPC services have been used to investigate the air-quality in large urban areas, cities and towns (HPC-Cloud-based prediction of air quality, 2019). Running simulations by using HPCaaS helps to increase the number of scenarios which can be simulated

realistically in a given time and to reduce the computational time needed for simulations. The outcomes of using this system include shorter times for simulations and cost reductions. In tourism, this can be used to monitor the increase in pollution, caused by tourism-related activities in near-real-time, and consequently to adapt appropriate reduction scenarios.

- HPCaaS has been used in environmental consulting for the design of industrial plants and public infrastructure by using simulations for modeling the emissions and water consumption, in order to estimate their impact on the environment (Cloud-based environmental modelling, 2019). Tourism encompasses the need for adequate infrastructure, and a similar method can be used to improve the planning of tourism-related facilities.

The whole Tourism 4.0 Collaborative Platform will be developed on an HPC infrastructure, consisting of the general computing infrastructure, the GPU core processors and a dedicated memory partition, coupled with a highly fault tolerant parallel storage. All this is available through Arctur, a leading HPCaaS provider in the region.

Artificial Intelligence

When discussing Artificial intelligence (AI), people often refer to the simulations of human intelligence processes by computer systems and other machines. These processes include, but are not limited to learning, reasoning and self-correction. AI has been applied to a number of different fields, including organic chemistry (Lindsay, Buchanan, Feigenbaum, & Lederberg, 1980), healthcare (Hanson & Marshall, 2001), photovoltaic (Mellit & Kalogirou, 2008), finance (Bahrammirzaee, 2010) and many others. Considerable attention has been given to the application of AI to tourism forecasting (Cho, 2003; Yu & Schwartz, 2006). It has been proven that the AI-based forecasting methods like the neural network model, the fuzzy time-series theory and the hybrid grey theory can surpass the traditional forecasting methods (Wang, 2004).

It can be concluded, therefore, that the use of forecasting with the help of AI in Tourism 4.0 lies in the ability to predict long- and short-term demands of tourists, in order to support decision-making regarding resource allocation, investment planning and economic policy directions.

Virtual and Augmented Reality

Virtual reality (VR) is a three-dimensional, interactive computer-generated environment that allows the user to experience both real and unreal events (Perry Hobson & Williams, 1995). There are three key elements that define a VR experience (Cruz-Neira, Sandin, DeFanti, Kenyon, & Hart, 1992; Williams & Hobson, 1995):

1. **Visualization Components:** These components include the 3D vision, the clarity and linearity of vision as well as the ability to look around a particular point, all of which is achievable by using a virtual reality headset.
2. **Immersion:** A key factor of a VR experience is the perception by the VR user of actually being present in a virtual world, i.e. their isolation from the real world. This is achievable through panoramic surroundings and the surrounding acoustics, by the reaction of images to the movements of the user and through physical representations of objects.

3. **Interactivity:** This is the control of participants over the experience. The combination of many kinetic effects helps to provide additional realism to the whole VR experience by tricking the human brain into believing the illusion.

Since the tourism sector is aiming to offer different types of experiences to guests, the VR is quickly gaining importance for the sector. Some of the immediate uses for VR in tourism are (Guttentag, 2010): in planning and management, in marketing, entertainment and education, as well as relating to accessibility and preservation of heritage.

Augmented reality (AR) is a visualization technique that superimposes computer-generated data (video, graphics, text and other multimedia formats) on real-world images that are captured by the camera (Kounavis, Kasimati, & Zamani, 2012). In this way the AR can provide an augmentation of a person's view of reality by enhancing the perception of the surroundings with the help of a computer or a mobile device. Thanks to smartphones, the AR technology has become available in one small portable device and accessible to an even wider public, including as a tool to improve the tourist experience. For example, by pointing the smartphone with an AR application towards a particular object or location, the user can see additional information on top of the real-world camera view (Yovcheva, Buhalis, & Gatzidis, 2012). Hence, the AR can assist tourists with important information regarding attractions, events or destinations.

For Tourism 4.0, VR & AR are particularly important in view of designing the innovative digital services that offer enriched touristic experience, like visiting locations with difficult accessibility or experiencing digital depictions of historic sites. Users would also be able to create their own personalized virtual excursions, contributing to the personalization of tourism in this way.

Experimental Testing Phase

The next phase of the Tourism 4.0 project is the T4.0 Living Lab Slovenia, which covers the setup of an experimental environment, enabling a demonstration of the technological tools described in the previous section and their potential for use in tourism. This project connects Slovenian tourism stakeholders and high-tech companies for the first time in history, in order to identify and unify the dispersed ideas, experiences, knowledge and expertise, along with linking them into a collaborative relationship. For the purpose of testing, validation and demonstration of product innovations, the T4.0 Living Lab Slovenia project will establish a testing ground near the town of Postojna, (see Figure 5), one of Slovenia's most important tourism destinations due to its long-standing history of cave tourism (Gunn, 2004; Šebela, 2019).

The goal of this phase of the project is to further develop and test the comprehensive business and innovative tools developed by Tourism 4.0 in a controlled environment. This will facilitate the tourist economy to reach higher levels of interaction between the guests and the providers of tourism services and products. The products and services presented in the Living Lab will enable tourists to have a completely new personalized experience, while companies will achieve greater business excellence and have better business outcomes, due to the fact that their tactical and strategic business decisions can now be empirically supported.

The partners of this project will develop innovative solutions in line with the recommendations provided through the preceding research phase. All users (tourists, tourism service providers and public institutions) will be able to get acquainted with several interconnected technological modules. This phase also provides an opportunity to test the Collaboration Platform with the support of the tools, developed by using the technologies stemming from Industry 4.0, the result of which will be implemented into the

Figure 5. The location of the T4.0 Living Lab Slovenia, near the town of Postojna

final market-ready business product. Following additional elements for the comprehensive ecosystem will also be developed:

- An interactive interface to facilitate communication between tourists and the tourism service providers, which will enable connectivity with all enabling tools, user applications and databases on the Collaboration Platform.
- Applicative solutions in five focus areas (the development of which will be carried out by providers of tourism services with the participation of partners, active in the technological field):
    1. Developing smart systems and services to optimize resource use. The project will explore and test technologies for reducing the consumption of energy in tourism facilities;
    2. Developing smart systems and services to improve the tourism experience;
    3. Additional enabling tools (implemented by technology partners);
    4. A digital tool for measuring and understanding the behavior of the guest in different contexts, which will be the basis for the development of new business decisions and solutions;
    5. A web-based tool, together with a reservation system for managing and integrating tourism offers in destinations, with the aim of achieving the best possible experience for the guest and raising business opportunities in the field of tourism. The latter leads to an increase in the value added of tourism, to more job opportunities in tourism and the rise of business opportunities in tourism.

The testing, the validation and the demonstration of the technology, the applications (the products and the services) and the Collaboration Platform itself will be carried out in an experimental center consisting of 20 high-tech demonstration units with a central space and the supporting infrastructure. During the duration of the project, the authors expect to have a minimum of 1000 tourists and a minimum of 50 tourist stakeholders participating in this innovative environment.

Innovative Potential of the Project

The Collaboration Platform represents a completely new basis for the development of a comprehensive innovative environment in the field of tourism. For a successful implementation it should include knowledge about:

- The field of tourism, i.e. the definition of cooperative roles of the tourism providers, the tourist, the local community and the national government. Among other things, it needs to define the models of interaction, and define the tourist both as a co-creator and as a user of the new products in tourism. It should be defined what tourism providers and tourists can acquire using the latest technologies, without forgetting one of the most important topics - the safety of personal data.
- The field of ICT technologies, i.e. the definition and application of technological standards in the tourism ecosystem. This includes the development of integration platforms IoT, blockchain, data and text mining using artificial intelligence and others.

The central innovation of the T4.0 Living Lab Slovenia project is the establishment of a research environment, allowing for development and innovation, facilitating the direct interaction between tourists and other stakeholders. In this environment, the tourist appears in three roles: as a guest of a particular destination, as a temporary resident of the local environment and as an innovator contributing to the reduction of negative effects of tourism activities. In terms of the technological achievements, this environment consists of advanced applications that act as an interface between the tourism providers and the tourists. As mentioned above, these applications are the first tools of their kind in the world, with which tourists are able to communicate directly, interactively and safely with the tourism providers, in order to improve their user experience. Since it is crucial that an interactive cooperation is established, a transparent and trusted environment must be created, which will encourage the development of new tourism solutions using industry-wide enabling technologies for an enriched tourism offer.

Non-technological innovations are marked by the transformation of the tourism providers and tourists into groups of people that use modern digital technologies for new tourist experiences and practice sustainable behavior. This cooperation requires new business models and the rethinking of processes by tourism providers, as well as new forms of tourist behavior. The innovation for companies means a significant leap towards a more personalized offer for the tourist, improved practical uses of enabling technologies in order to increase the satisfaction of the tourists and of other stakeholders involved in tourist experiences, as well as to achieve better business results.

SOLUTIONS AND RECOMMENDATIONS

The results of the basic research are the assessment of the level of technology-use in the field of tourism and the preparation of a set of recommendations to tourist service providers, destination managers and public institutions, including the use of technology for the optimization of business processes and activities. These will be set out in the form of recommendations, made publicly available to tourist associations, the Chamber of Commerce and Industry of Slovenia and to the Association of Municipalities and Towns of Slovenia. The project proposed also results in an innovative tourist platform for the active collaboration of tourism stakeholders, with the tourism providers and the tourists as the key actors.

Through the experimental phase T4.0 Living Lab Slovenia, the project will also provide both a demonstrative and an educational environment for tourist workers, students, managers and policy makers, too. By piloting the tools provided within the Tourism 4.0 project it will be possible to leverage the tourist potential by promoting synergies in a range of tourism "niches" (including as regards heritage, cultural tourism, wine-related activities, city- and, rural tourism, to name but a few), while simultaneously being able to scale-up the cross-national and regional benefits of tourism. Overall, the testing phase will provide an opportunity to:

- Test, verify and expand the potentials of data collection for support to the tourism sector;
- Discuss and develop concrete analytical tools to be used by all stakeholders in the sector;
- Foster a greater involvement of the data owner and identify further business opportunities;
- Set the foundations for the incubation and financing of business ideas (SMEs-based) in the region.

With the project, the authors want to demonstrate the true value of the key enabling technologies of Industry 4.0 for the tourism sector, in terms of business excellence as well as in terms of sustainability. Their introduction in tourism is expected to lead to high quality monitoring of the business processes for empirically supported strategic planning, a higher level of responsible behavior from the tourists, increasing the level of personalization of tourist offers and to an overall increase in the value added of Slovenian tourism. The actions planned, the lessons learnt and the recommendations will also allow to upscale the benefit of this pilot initiative EU-wide, notably by advancing sustainable tourism innovations across Europe.

FUTURE RESEARCH DIRECTIONS

In the light of the predicted future increase in the tourist activity, it is essential that the tourism supply is constantly innovative and sustainable at the same time. The establishment of the Tourism 4.0 ecosystem could address both of these issues by providing an environment that stimulates all stakeholders to exercise more informed behavior. This innovative environment is independent of any geographic location and thus implementable worldwide. Currently, plans are underway to establish large international consortiums in the Alpine, the Danube and the Black Sea macro regions, where the tools developed through Tourism 4.0 will be implemented. Through the guidelines and recommendations provided, the results of the targeted analysis are to be translated into strategic guidelines and policies. However, the aim is beyond general, as the policy-specific guidance will be offered for a successful implementation and the use of tools to help manage specific destinations move towards sustainable development.

Last but not least, Tourism 4.0 puts a special emphasis on the use of the smart technology applications for the heritage sector and the cultural tourism in the form of digitisation (transforming the real-life cultural heritage into digital formats) and digitalisation (using new business processes for managing and presenting the digitised heritage). This is primarily aimed at developing new points of interest, as well as heritage-inspired digital branding, marketing and post-tourism creative narratives, and equally at the use in management, monitoring and citizen participation. Hence, the proposed chapter will discuss the potential of Tourism 4.0 for the value chain – from "creation", production/publishing, dissemination/ trade, exhibition/transmission, promotion, as well as use and re-use. It will include an overview of the technologies available and possible, the case studies and business models in the heritage sector, as well as an in-depth discussion of the required skill-sets for a successful implementation.

CONCLUSION

Through this book chapter, the project Tourism 4.0 is presented, which began in September 2018 and is still underway. Tourism 4.0 strives to continuously optimize the interaction between the tourist and the tourism providers, with the aim of responsibly and innovatively meeting their respective needs, while including the local population and government bodies into the tourism process. Although initially lunched in Slovenia, the project is already expanding across the national borders.

The authors introduce the two main phases of the project: the initial research and the experimental phase. In the research phase, the consortium partners aim at gaining an in-depth understanding of the present state in the Slovenian tourism with the aim of providing future improvements of tourist satisfaction levels, increasing the value added for service providers and minimizing the negative impacts of tourism on the local community. For this reason, a great effort is put into the study of emerging technologies. In this part, a definition of an ideal tourism ecosystem based on the stakeholder collaboration is being developed, accompanied by a description of the role of technology in reaching the desired goals. The experimental phase is dedicated to setting up a testing environment where the technologies can be implemented in a controlled real environment, in order to demonstrate the characteristics of sustainable and personalized tourism.

To conclude, the Tourism 4.0 project aims to unlock the innovation potential in the whole tourism industry by combining the tourism resources with the available technology, to leverage the existing local and distributed attempts, the ready-made solutions, the development proposals and the initiatives which are running at the levels of tourist destinations. The project not only intends to stimulate the exchange of best practices and solutions between the stakeholders in tourism, but more importantly, wishes to inspire innovation, sharing and collaboration between various entities, thus creating a unique innovation environment. Tourism 4.0 also understands that technological infrastructure and smart tourism solutions can only act as enablers towards more sustainable tourist communities and as such encourage the evolution towards symbiotic destinations.

ACKNOWLEDGMENT

The work for this chapter was carried out within the framework of the research project Tourism 4.0—enriched tourist experience (OP20.03536), co-funded by the Slovenian Ministry of Education, Science and Sport and the European Regional Development Fund. The leader of the project consortium is the Arctur Company and includes top experts from the Faculty of Tourism Studies—Turistica (University of Primorska), the Faculty of Tourism (University of Maribor) and the Faculty of Computer and Information Science (University of Ljubljana).

The authors would like to acknowledge the support received from the Association of Municipalities and Towns of Slovenia (Skupnost občin Slovenije—SOS).

The authors would also like to thank Tjaša Zornik for providing the graphic designs for this work.

REFERENCES

Addis, M. (2005). New technologies and cultural consumption–edutainment is born! *European Journal of Marketing*, *39*(7/8), 729–736. doi:10.1108/03090560510601734

AirbnbEng. (2010, Nov. 15). *MySQL in the cloud at Airbnb*. Retrieved from https://medium.com/airbnb-engineering/mysql-in-the-cloud-at-airbnb-336e5666bc94#.llrxogduu

Ashworth, G. J., & Tunbridge, J. E. (1990). *The tourist-historic city Belhaven. AWS Case Study: Red Lion Hotels*. (2019, Aug. 13). Retrieved from https://aws.amazon.com/solutions/case-studies/red-lion/

AWS Case Study: SETTour. (2019, Aug. 13). Retrieved from https://aws.amazon.com/solutions/case-studies/settour/

Bahrammirzaee, A. (2010). A comparative survey of artificial intelligence applications in finance: Artificial neural networks, expert system and hybrid intelligent systems. *Neural Computing & Applications*, *19*(8), 1165–1195. doi:10.100700521-010-0362-z

Bitcoin Transaction Fees. (2019, June 15). Retrieved from https://bitcoinfees.info/

Bohanec, M. (2008). *DEXi: Program for Multi-Attribute Decision Making User's Manual*. Ljubljana, Slovenia: Institut Jozef Stefan.

Buhalis, D. (2000). Marketing the competitive destination of the future. *Tourism Management*, *21*(1), 97–116. doi:10.1016/S0261-5177(99)00095-3

Buhalis, D., & Amaranggana, A. (2013). Smart Tourism Destinations. In *Information and Communication Technologies in Tourism 2014* (pp. 553–564). Cham, Switzerland: Springer. doi:10.1007/978-3-319-03973-2_40

Buhalis, D., & Amaranggana, A. (2015). Smart Tourism Destinations Enhancing Tourism Experience Through Personalisation of Services. In *Information and communication technologies in tourism 2015* (pp. 377–389). Cham, Switzerland: Springer. doi:10.1007/978-3-319-14343-9_28

Buhalis, D., & Law, R. (2008). Progress in information technology and tourism management: 20 years on and 10 years after the Internet—The state of eTourism research. *Tourism Management*, *29*(4), 609–623. doi:10.1016/j.tourman.2008.01.005

Canestrelli, E., & Costa, P. (1991). Tourist carrying capacity: A fuzzy approach. *Annals of Tourism Research*, *18*(2), 295–311. doi:10.1016/0160-7383(91)90010-9

Cho, V. (2003). A comparison of three different approaches to tourist arrival forecasting. *Tourism Management*, *24*(3), 323–330. doi:10.1016/S0261-5177(02)00068-7

Cloud-based environmental modelling. (2019). Retrieved from FORTISSIMO: https://www.fortissimo-project.eu/experiments/514

Cruz-Neira, C., Sandin, D., DeFanti, T., Kenyon, R., & Hart, J. (1992). The CAVE: Audio visual experience automatic virtual environment. *Communications of the ACM*, *35*(6), 64–73. doi:10.1145/129888.129892

Dave, N. (2018). *digitaldoughnut*. Retrieved from https://www.digitaldoughnut.com/articles/2018/january/ways-in-which-iot-is-shaping-the-future-of-travel

Elgendy, N., & Elragal, A. (2014). Big data analytics: a literature review paper. *Industrial Conference on Data Mining* (pp. 214–227). Springer. 10.1007/978-3-319-08976-8_16

Etro, F. (2009). The economic impact of cloud computing on business creation, employment and output in Europe. *Review of Business and Economics*, *54*(2), 179–208.

Fisher, A., & Krutilla, J. (1972). Determination of optimal capacity of resource-based recreation facilities. *Natural Resources Journal*, *12*, 417.

Forster, J. (1964). The Sociological Consequences of Tourism. *International Journal of Comparative Sociology*, *5*(2), 217–227. doi:10.1177/002071526400500208

Foss, N. J., Laursen, K., & Pedersen, T. (2011). Linking Customer Interaction and Innovation: The Mediating Role of New Organizational Practices. *Organization Science*, *22*(4), 980–999. doi:10.1287/orsc.1100.0584

Frechtling, D. C. (2000). Assessing the Impacts of Travel and Tourism-Measuring Economic Benefits. *International Library of Critical Writings in Economics*, *121*, 9–27.

Gössling, S., Peeters, P., Hall, C., Ceron, J.-P., Dubois, G., Lehmann, L., & Scott, D. (2012). Tourism and water use: Supply, demand, and security. An international review. *Tourism Management*, *33*(1), 1–15. doi:10.1016/j.tourman.2011.03.015

Government of the Republic of Slovenia. (2015). *Slovenia's Smart Specialisation Strategy*. Retrieved from http://www.svrk.gov.si/fileadmin/svrk.gov.si/pageuploads/Dokumenti_za_objavo_na_vstopni_strani/S4_document_2015_ENG.pdf

Gretzel, U., Sigala, M., Xiang, Z., & Koo, C. (2015). Smart tourism: Foundations and developments. *Electronic Markets*, *25*(3), 179–188. doi:10.100712525-015-0196-8

Gunn, J. (2004). *Encyclopedia of caves and karst science*. Taylor & Francis. doi:10.4324/9780203483855

Guttentag, D. (2010). Virtual reality: Applications and implications for tourism. *Tourism Management*, *31*(5), 637–651. doi:10.1016/j.tourman.2009.07.003

Hanson, C. III, & Marshall, B. (2001). Artificial intelligence applications in the intensive care unit. *Critical Care Medicine*, *29*(2), 427–435. doi:10.1097/00003246-200102000-00038 PMID:11269246

HPC-Cloud-based prediction of air quality. (2019). Retrieved from https://www.fortissimo-project.eu/experiments/410

HPC-Cloud-based urban planning. (2019). Retrieved from https://www.fortissimo-project.eu/experiments/406

Hung, P. (2016). *Big data applications and use cases*. Berlin, Germany: Springer. doi:10.1007/978-3-319-30146-4

Hunter, W. C., Chung, N., Gretzel, U., & Koo, C. (2015). Constructivist research in smart tourism. *Asia Pacific Journal of Information Systems*, *25*(1), 105–120. doi:10.14329/apjis.2015.25.1.105

Kaur, K., & Kaur, R. (2016). Internet of things to promote tourism: An insight into smart tourism. *International Journal of Recent Trends in Engineering & Research*, *2*(4), 357–361.

Korstanje, M., & Busby, G. (2010). Understanding the bible as the roots of physical displacement: The origin of tourism. *Ereview of Tourism Research*, *8*(3).

Kounavis, C., Kasimati, A., & Zamani, E. (2012). Enhancing the tourism experience through mobile augmented reality: Challenges and prospects. *International Journal of Engineering Business Management*, *4*, 10. doi:10.5772/51644

Kurashima, T., Iwata, T., Irie, G., & Fujimura, K. (2013). Travel route recommendation using geotagged photos. *Knowledge and Information Systems*, *37*(1), 37–60. doi:10.100710115-012-0580-z

Lee, I., Cai, G., & Lee, K. (2014). Exploration of geo-tagged photos through data mining approaches. *xpert Systems with Applications, 41(2)*, 397-405.

Lenzen, M., Sun, Y.-Y., Faturay, F., Ting, Y.-P., Geschke, A., & Malik, A. (2018). The carbon footprint of global tourism. *Nature Climate Change*, *8*(6), 522–528. doi:10.103841558-018-0141-x

Li, J., Xu, L., Tang, L., Wang, S., & Li, L. (2018). Big data in tourism research: A literature review. *Tourism Management*, *68*, 301–323. doi:10.1016/j.tourman.2018.03.009

Lindsay, A., Downs, D., & Lunn, K. (2003). Business processes—Attempts to find a definition. *Information and Software Technology*, *45*(15), 1015–1019. doi:10.1016/S0950-5849(03)00129-0

Lindsay, R., Buchanan, B., Feigenbaum, E., & Lederberg, J. (1980). *Applications of artificial intelligence for organic chemistry*. McGraw-Hill.

Lu, X., Wang, C., Yang, J.-M., Pang, Y., & Zhang, L. (2010). Photo2trip: generating travel routes from geo-tagged photos for trip planning. *In Proceedings of the 18th ACM international conference on Multimedia* (pp. 143-152). Florence, Italy: ACM. 10.1145/1873951.1873972

Lu, Y. (2016). Industry 4.0: A survey on technologies, applications and open research issues. *Journal of Industrial Information Integration*, *6*, 1–10. doi:10.1016/j.jii.2017.04.005

Luo, J., Wu, M., Gopukumar, D., & Zhao, Y. (2016). Big data application in biomedical research and health care: a literature review. *Biomedical informatics insights, 8, BII-S31559.*

Martín Martín, J., Guaita Martínez, J., & Salinas Fernández, J. (2018). An Analysis of the Factors behind the Citizen's Attitude of Rejection towards Tourism in a Context of Overtourism and Economic Dependence on This Activity. *Sustainability*, *10*(8), 2851. doi:10.3390u10082851

McCool, S., & Moisey, R. (2001). *Tourism, recreation, and sustainability: Linking culture and the environment*. Wallingford, CT: Cabi. doi:10.1079/9780851995052.0001

Mellit, A., & Kalogirou, S. (2008). Artificial intelligence techniques for photovoltaic applications: A review. *Progress in Energy and Combustion Science*, *34*(5), 574–632. doi:10.1016/j.pecs.2008.01.001

Mencarelli, R., & Pulh, M. (2012). Museoparks and re-enchantment of the museum visits: An approach centred on visual ethnology. *Qualitative Market Research, 15*(2), 148–164. doi:10.1108/13522751211215877

Miah, S., Vu, H., Gammack, J., & McGrath, M. (2017). A big data analytics method for tourist behaviour analysis. *Information & Management, 54*(6), 771–785. doi:10.1016/j.im.2016.11.011

Nakamoto, S. (2008). *Bitcoin: A Peer-to-Peer Electronic Cash System (White Paper).* Retrieved from https://bitcoin.org/bitcoin.pdf

Okuyama, K., & Yanai, K. (2013). A Travel Planning System Based on Travel Trajectories Extracted from a Large Number of Geotagged Photos on the Web. In *The era of interactive media* (pp. 657–670). New York: Springer. doi:10.1007/978-1-4614-3501-3_54

Önder, I., & Treiblmaier, H. (2018). Blockchain and tourism: Three research propositions. *Annals of Tourism Research, 72*(C), 180–182. doi:10.1016/j.annals.2018.03.005

Peceny, U., Urbančič, J., Mokorel, S., Kuralt, V., & Ilijaš, T. (2019). Tourism 4.0: Challenges in Marketing a Paradigm Shift. In Consumer Behavior and Marketing. IntechOpen.

Perry Hobson, J., & Williams, A. (1995). Virtual reality: A new horizon for the tourism industry. *Journal of Vacation Marketing, 1*(2), 124–135. doi:10.1177/135676679500100202

Pizam, A. (1978). Tourism's impacts: The social costs to the destination community as perceived by its residents. *Journal of Travel Research, 16*(4), 8–12. doi:10.1177/004728757801600402

Poon, A. (1993). *Tourism, technology and competitive strategies.* Wallingford, CT: CABI.

Rabotić, B. (2014). Special-purpose travel in ancient times: 'Tourism' before tourism? *Turističko poslovanje, (14)*, 5-17.

Raghupathi, W., & Raghupathi, V. (2014). Big data analytics in healthcare: promise and potential. *Health information science and systems, 2*(1), 3.

Šebela, S. (2019). Postojna—Planina Cave System, Slovenia. In Encyclopedia of caves (pp. 812-821). Academic Press.

Seebacher, S., & Schüritz, R. (2017). Blockchain technology as an enabler of service systems: A structured literature review. In *Proceedings 8th International Conference on Exploring Service Science, IESS 1.7* (pp. 12-23). Cham, Switzerland: Springer. 10.1007/978-3-319-56925-3_2

Sha, C., Zhang, J., An, L., Zhang, Y., Wang, Z., Ilijaš, T., . . . Ji, Q. (2019). Facilitating HPC Operation and Administration via Cloud. *Supercomputing Frontiers and Innovations, 6*(1).

Sirse, J., & Mihalic, T. (1999). Slovenian tourism and tourism policy: A case study, 54(3). *Tourism Review*, 34–47.

Sunlu, U. (2003). Environmental impacts of tourism. In *Proceedings Conference on the Relationships between Global Trades and Local Resources in the Mediterranean Region*, (pp. 263-270). Academic Press.

Total Slovenia News. (2019, July 31). Retrieved from https://www.total-slovenia-news.com/travel/4215-strong-first-half-for-slovenian-tourism-in-2019

Tovar, C., & Lockwood, M. (2008). Social impacts of tourism: An Australian regional case study. *International Journal of Tourism Research, 10*(4), 365–378. doi:10.1002/jtr.667

Vogel-Heuser, B., & Hess, D. (2016). Guest editorial industry 4.0–prerequisites and visions. *IEEE Transactions on Automation Science and Engineering, 13*(2), 411–413. doi:10.1109/TASE.2016.2523639

Wagar, J. A. (1964). The carrying capacity of wild lands for recreation. *Forest Science, 10*(suppl_2), a0001-24.

Wang, C.-H. (2004). Predicting tourism demand using fuzzy time series and hybrid grey theory. *Tourism Management, 25*(3), 367–374.

Williams, P., & Hobson, J. (1995). Virtual reality and tourism: Fact or fantasy? *Tourism Management, 16*(6), 423–427. doi:10.1016/0261-5177(95)00050-X

World Tourism Organization and United Nations Development Programme. (2018). *Tourism and the Sustainable Development Goals – Journey to 2030*. Madrid, Spain: UNWTO.

WTTC. (2018). Retrieved from World Travel & Tourism Council: https://www.wttc.org/-/media/files/reports/economic-impact-research/regions-2019/world2019.pdf

Xu, Z., Frankwick, G., & Ramirez, E. (2016). Effects of big data analytics and traditional marketing analytics on new product success: A knowledge fusion perspective. *Journal of Business Research, 69*(5), 1562–1566. doi:10.1016/j.jbusres.2015.10.017

Young, G. (1973). *Tourism: Blessing or blight?*

Yovcheva, Z., Buhalis, D., & Gatzidis, C. (2012). Smartphone augmented reality applications for tourism. [ertr]. *Ereview of Tourism Research, 10*(2), 63–66.

Yu, G., & Schwartz, Z. (2006). Forecasting short time-series tourism demand with artificial intelligence models. *Journal of Travel Research, 45*(2), 194–203. doi:10.1177/0047287506291594

Zuelow, E. (2015). *A history of modern tourism*. London, UK: Macmillan International Higher Education.

Zukin, S. (1998). Urban lifestyles: Diversity and standardisation in spaces of consumption. *Urban Studies (Edinburgh, Scotland), 35*(5-6), 825–839. doi:10.1080/0042098984574

KEY TERMS AND DEFINITIONS

Artificial Intelligence: The ability of computers to imitate human intelligence and give the impression that the computer has learned human-like habits and patterns. This can be achieved through complex algorithms, machine learning technologies and behavioral patterns.

Augmented Reality: Technology that upgrades the image of the real world with additional computer-generated information or virtual effects, such as images and sounds.

Big Data: Volumes of data that are too large and too complex to be processed by traditional data-processing application software.

Blockchain: A chain of blocks containing data that is bundled together. This database is shared across a network of computers (so-called distributed ledger network). Each data block links to the previous block in the blockchain through a cryptographic hash of the previous block, a timestamp, and transaction data. The blockchain only allows data to be written, and once that data has been accepted by the network, it cannot be changed.

Cloud Computing: Cloud computing refers to the practice of using a network of remote servers, hosted on the Internet to manage, store and process data instead of using a local server or a personal computer.

HPC: A high-performance computing (HPC) system is a tool used to tackle problems that require more computing resources or time than they can obtain on the personal computers, available to the respective users to address them.

Industry 4.0: The name given to the fourth industrial revolution that includes the trend of automation and optimization of the manufacturing processes using smart autonomous systems.

Internet of Things: Internet of Things or IoT represents all connected devices that communicate with each other through the Internet.

Tourism 4.0: The trend of applying the tools and concepts of Industry 4.0 to the tourism sector, in order to create a personalized travelling experience and a more sustainable tourism.

Virtual Reality: Computer technology that uses realistic images, sounds and other sensory experiences, created with specific software in order to mimic a real or imaginary environment and to simulate the user's physical presence.

Chapter 12
Customer Experience in the Restaurant Industry:
Use of Smart Technologies

Ilhami Tuncer
https://orcid.org/0000-0002-6883-4461
Karamanoglu Mehmetbey University, Turkey

ABSTRACT

One of the most important points of consideration that will ensure the sustainability and profitability of businesses in the digitized world market is the experience offered to customers. Experience in business preferences, especially of the digitized customers, is quite specific. This chapter discussed the contribution of smart technologies to customer experience for restaurants, and emphasized its significance. Moreover, the subjects of artificial intelligence, smart technology, and QR code were addressed based on customer experience. The contribution offered to customers was emphasized by giving examples of smart technology applications used in restaurant businesses to improve customer experience. The chapter will contribute theoretically to the subject which has not been adequately studied in the literature.

INTRODUCTION

Undoubtedly, technology has contributed greatly to the development of businesses, their gaining awareness and attaining a modern structure. In the process to date, technology has had a limited level of interaction with businesses, although it has evolved through many phases of development. But, today, because of the high level of interaction and integration of technology with businesses, we are witnessing major changes in the business world. It is observed that the reflections of especially the development of technologies such as artificial intelligence and augmented reality in businesses is a harbinger of radical changes in the business world. It can be said that the impact of these smart technologies on businesses and the acceptance process of businesses is at an initial level. However, while smart technologies are more active in some industries, they have not yet been integrated into some sectors.

DOI: 10.4018/978-1-7998-1989-9.ch012

Today, smart technologies are broadly used in marketing as well as in many different areas. The digitization of consumers has emerged as a result of the digitization of marketing. This is because businesses can experience effective interaction with their customers in digital environments. Many businesses invest in digital environments by seeing the opportunities. Digital media has resulted in a proliferation of increased technology use in marketing activities.

An increase in the use of smart devices such as smartphones and tablet computers has been one of the most important reason to encourage smart technologies to be integrated into marketing. The fact that especially smartphones with the Internet can be accessed without any place or time limitations has been the most important factor allowing the use of smart technologies. Many businesses can reach and contact their customers by integrating smart technologies into smartphones. The tendency for new-generation customers to interact with digital technologies, in particular, supports this process. Therefore, businesses' investment in and integration with smart technologies will give them significant advantages in their sustainability and profitability.

Smart technologies are used in many areas of marketing, as well as playing a major role in improving customer experience. Many businesses rely on smart technologies to deliver a diverse and also better experience to their customers. Many businesses can be said to have started to use especially artificial intelligence and augmented reality technologies and such technologies affect customer experience. In particular, devices and robots equipped with these technologies provide consumers with a different experience, as well as providing great conveniences. The use of these technologies, their adoption by businesses and consumers regardless of their being new, and the recognition of the clear benefit of the experience offered by them have increased the spread of the technologies.

The service sector is a very favorable sector for the use of smart technologies due to the heavy labor in the sector. Innovative developments are experienced in terms of serving consumers by the integration of smart technologies into robots and smart devices. The innovations brought by smart technologies are also observed quite intensely in restaurant businesses. Today, many restaurant businesses invest in smart technologies to differentiate from competitors, deliver better service, and provide a good customer experience. They employ many smart robots as substitutions for waiters. Therefore, smart technologies will be able to identify the future of restaurant businesses.

The use of smart technologies in customer experience has become a very important subject. This thesis appears to be supported by the increasing share of the service sector in the world and the competition experienced in the service sector. There is a considerable shortage of academic studies in this area. It was aimed through this study to contribute to the enrichment of the literature and to guide marketing managers. In this study, it was aimed to discuss the impact of smart technologies on customer experience in restaurants and to provide current examples from the industry regarding the use of smart technologies. To that end, first, customer experience and then the subject of smart technologies were scrutinized and supported by applications and examples available in the industry.

BACKGROUND

Customer Experience

In today's business world where competition is maximized, the necessity to be customer centered has become very important for businesses. Therefore, customer experience has become very important for each and every business. Grewal (2009) has argued that customer experience is the most important business strategy in which both customers and businesses gain. This is because a good customer experience creates value, and influences satisfaction, image, differentiation and loyalty (Jain, Aagja, & Bagdare, 2017). Therefore, the most distinctive feature in terms of customers is the experience offered by businesses. It is very important that businesses have a holistic communication with their customers so that they can best reflect the experience offered to the customers. Therefore, the experience offered to customers during the customer–business interaction will reveal the level of distinction of a business. The level of distinction of a business will lead to the competitive advantage.

Although customer experience is defined in different ways in the service and marketing literature, they complement each other (Helkkula, 2011). For example, Zomerdijk and Voss (2010) have suggested that customer experience is a holistic concept of a combination of several aspects of products or services offered by a business. Customers give cognitive, emotional, sensory and social reactions based on the interactions they experience with a business at various touchpoints (Lemon & Verhoef, 2016; Verhoef et al., 2009). These subjective and multidimensional intrinsic reactions constitute the customer experience (Meyer & Schwager, 2007). Similarly, Homburg, Jozić and Kuehnl (2017) has defined customer experience as a set of cognitive, sensory, emotional, relational, and behavioral responses to the brand or business on the journey through the touchpoints of interactions with the business before, during and after a purchase. It is understood from the definitions that customer experience is conceptually based on the interaction between the customer and the business (Gentile, Spiller, & Noci, 2007). Customer experience has also been conceptualized in progressive terms. Progressive customer experience has been labeled pre-contact, during contact, post-contact, and service failure. Touchpoints have to be mapped in a holistic and catchy way in order for progressive customer experience to create a brand and customer value (Sultan, 2018).

The impact of customer experience on factors that would provide competitive advantage, such as customer loyalty and satisfaction, has driven business managements to adopt the idea of customer experience management. Many researchers (Lemon & Verhoef, 2016; Macdonald, Kleinaltenkamp, & Wilson, 2016) agree that customer experience creates value through activities and actions at multiple touchpoints. Therefore, businesses have to innovatively design touchpoints and renovate them according to developments, for a good customer experience. The design of touchpoints should be tailored especially to the customer perspective rather than the business perspective (Lemon & Verhoef, 2016). The main goal of this action of the business is to create long-term customer loyalty (Homburg et al., 2017).

Customer journey and touchpoints are the most important building blocks that make up customer experience. Customer journey is a visual representation of touchpoints that are accompanied by the customer's emotional signs and is one of the best techniques used to visualize service design (Halvorsrud, Kvale, & Følstad, 2016; Segelström, 2013). Customer journey is a temporal process that starts before the customer's purchase and continues until after the purchase. In this process, the customer interacts and communicates with the business directly or indirectly at several points. And these points form customer–business multiple contact touchpoints. Touchpoints are verbal or non-verbal events that an individual perceives about a business (Duncan & Moriarty, 2006). Therefore, a customer's experience of a business consists of an overall reaction to all of these events. The increase in marketing communication channels has resulted in large increases in customer journey touchpoints. According to Voorhees et al (2017), touchpoints can be designed in many ways, such as online platforms, physical environments, or personal interactions. Examples for touchpoints include printed catalogs, kiosks, telephones, digital technologies, stores, emails, staff, billboards, and mobile platforms. This makes it difficult for businesses to design a multichannel customer experience (Verhoef, Kannan, & Inman, 2015). Therefore, the overall customer experience provided by a business that seamlessly designs and implements both online and offline touchpoints will be strong (Lemon & Verhoef, 2016).

The growing customer interest in technology has also changed expectations of businesses. The digital revolution has transformed the information and communication technologies beyond their classic roles and gave them a strategic role for businesses. The competitiveness of any business that can benefit from digital technologies can be very high. This is because the business gains the ability of agility to deal with the opportunities and threats of the market. Especially the fact that the multi-channel structure has begun to dominate businesses reveals the importance of digital technologies. Many businesses adopt innovative digital technologies such as artificial intelligence, Internet of things, blockchain, augmented reality, and business process automation. This is because digital technologies are changing customer behavior and expectations, the structure of businesses, and the boundaries of people's business zone (Lemon, 2016). Digital technologies can answer customer questions and give them advise without the limitation of time and space.

Digital technologies are the most important factor causing an increase in the number of touchpoints for customer journey. Customers' interest in business touchpoints has been enhanced by the emerging science and technology, enrichment of media channels, emergence of innovative business and revenue models, and the hyperlinked world (Wind & Hays, 2016). While one-way communication has dominated the touchpoints of traditional customer journey, mutual interaction has begun to dominate each touchpoint of the digitized customer journey. Many tools of interaction at touchpoints enable businesses to collect important big data such as customer reviews and textual data (Tirunillai & Tellis, 2014). Such big data can provide businesses with insights into the touchpoints causing problems on the customer journey (McColl-Kennedy, Zaki, Lemon, Urmetzer, & Neely, 2019). Therefore, it will contribute to the customer experience that businesses create digital substructures on customer touchpoints existing in each station of the customer journey.

As a reflection of digital transformation in the future, it is predictable that the majority of businesses will communicate and interact with their customers through multiple channels. This thesis is supported by the integration and relationship of the rising generation of customers with technology. Consequently, it will be a great advantage for businesses to attach importance to digital touchpoints during the process of designing customer experiences. Customers' interest in using smart technologies, in particular, will enhance their communication and interaction with touchpoints. What makes technology "smart" is its being integrated into an interactive device that can access the Internet (Foroudi, Gupta, Sivarajah, & Broderick, 2018). Industry 4.0 has popularized the use of new generation technologies such as artificial intelligence, Internet of things, augmented reality, and virtual reality. Many businesses use these technologies at customer touchpoints to improve customer experience and be able to make a difference. Bodhani (Bodhani, 2013) has determined that many major retailers use smart technologies to improve sales, and to enhance customer relations and experience. Moreover, smart technologies can improve customers' emotional interaction at customer touchpoints with their experiential aspects and extremely realistic and interactive interface designs (Pantano & Timmermans, 2014). The benefits of smart technologies to businesses are not limited to these. For example, smart technologies strive to create information about customers, to generate specific data, and to obtain, manage, and deliver the data as a value while determining which technology is more useful for the business, and choosing and delivering it (Foroudi et al., 2018). This means that smart technologies can be used to answer questions such as how the customer journey should be designed and managed successfully, which channels will be more useful at touchpoints, and many other similar questions.

Artificial Intelligence

While the use of artificial intelligence technologies has increased in every sector today, it has also increased in the food and beverage service sector. Like other business lines, restaurants also aim to beat the competition in a profitable way using smart technologies to maximize customer experience in a positive manner. In the most fundamental sense, artificial intelligence means programs, algorithms, systems or machines that demonstrate intelligence. It is typically used to specify a set of tools that can enhance the intelligence of a product, service, or solution (Shankar, 2018). Artificial intelligence is also a fast-growing field. From banking to law, from food and beverage services to tourism, the use of artificial intelligence technologies has increased in the entire service sector. Artificial intelligence has only been effective in certain areas in the past, but now it makes human life easier altogether and has an active role in decision-making processes. It would be wrong to investigate the concept of artificial intelligence under a single heading. Artificial intelligence gathers around two main concepts in general terms:

- Machine Learning
- Deep Learning

Machine Learning

Machine Learning is a methodological paradigm that makes inferences from existing data using mathematical and statistical methods and estimates an unknown by using these inferences (Aladag, 2015). Machine learning focuses on forecasts by creating an infrastructure using previously recorded features. And, it benefits the user by making these predictions meaningful. Thus, it provides prediction models for personal marketing, spam filtering, fraud detection and network security, and consequently predictable outcomes for subsequent processes. When these predictions are made, multiple different models of prediction emerge. These (Birbil, 2018; Castañón, 2019; Ş. Şeker, 2013; Şengöz, 2017) are as follows:

Decision Trees: Decision trees make observations for specific activities and determine the optimal algorithm for the output that is targeted.

K-Means Clustering: It makes a prediction by basing a limited amount of data into similar structures and using a grouping process.

Neural Networks: Neural networks use training data to learn and handle future data, and to identify multiple variables regardless of whether they are dependent or independent.

Reinforcement Learning: This learning model includes repeated attempts by previously prepared functions. The steps that produce logical outputs are rewarded, and the existing algorithm continues until the most logical process. Negative consequences are punished and not used in subsequent steps.

Deep Learning

Deep learning is an important and up-to-date subfield of artificial intelligence that is very popular nowadays. Deep learning methods have been developed mainly based on the studies of artificial neural

Table 1. Deep Learning Method and Feature

Deep Learning Method	Features
• Deep Neural Networks	• Successfully implemented in many fields
• Deep Autoencoders	• An unsupervised data learning method that does not require marked data for learning
• Deep Belief Networks	• They are used in a supervised and unsupervised manner. However, the duration of the training can be very slow.
• Deep Boltzmann Machine	• Its difference from deep belief networks is that it has non-directional connections between all layers except for the top two layers. Therefore, it is more costly compared to deep belief networks in terms of calculation.
• Recurrent Neural Networks	• They are successful in determining sequential patterns. Therefore, their sub-types, such as Long-Short-Term Memory (LSTM), provide successful results in areas such as natural language processing.
• Convolutional Neural Networks	• They usually give successful results in the field of computerized vision and on visual data, even though they require a large amount of marked data. They have also been successfully applied to problems in the field of natural language. processing.

Source: (Ravì et al., 2017; A. Şeker, Diri, & Balık, 2017)

networks (ANN). However, unlike these studies, they are based on a greater number of hidden neurons and layers (Ravì vd, 2017). Deep learning methods consist of 6 concepts in general (Ravì et al., 2017).

1. Deep Neural Networks (DNN)
2. Deep Autoencoders
3. Deep Belief Networks (DBN)
4. Deep Boltzmann Machine (DBM)
5. Recurrent Neural Networks (RNN)
6. Convolutional Neural Networks (CNN)

There are many learning methods, as can be seen from the table. The output obtained from artificial intelligence along with machine learning and deep learning is used in various service sectors. From among these ideas, the artificial intelligence systems used in restaurants are often utilized to provide convenience to customers and internal operations. They make the operation faster and more efficient, as well as uplifting the customer experience and improving the service quality and profitability of the restaurant.

Various projects have been developed to use artificial intelligence in restaurants, to improve the customer experience positively. For example, a system, using an algorithm developed through artificial intelligence, makes a classic phone call to a restaurant which does not exist in the digital environment, instead of the customer's having to waste time talking to the restaurant. By this means, the customer's time saving reaches the maximum level. Additionally, the customer's data are saved in the restaurants included in the digital system. When the customer wants to come back to these restaurants, the customers are presented with a recommendation on an appropriate day and time according to their previous habits. The menus that the customer has always ordered and their prices are offered via the automatic phone application as a quote. In this way, customers will not have to repeat orders in the same menu if they want to, and will be able to order the same menu faster. At the same time, if they do not want to, orders similar to the same menu can be filtered by the algorithm and presented as a second menu. The next algorithm is prepared by evaluating the orders made by customers. At the same time, for first-time customers, menu options that have been previously ordered more frequently can also be offered in the first place. Thus, artificial intelligence standardizes many people's pleasure and experience by clustering them according to the majority, and offers these experiences to other first-time customers.

Dishes can be cooked and served based on data that have previously been defined in restaurants by combining the artificial intelligence algorithm and machine automation. Food items placed in compartments in a machine after being equipped with various options are presented to customers for them to decide. Moreover, chefs who manage background kitchen operations in restaurants can make a selection in their own kitchens and cook dishes without touching them by using artificial intelligence pans or cooking utensils. For example, as a first scenario, the customer places an order manually, and the first stage involves the chef's — who manages the operation in the background kitchen — being notified of the order. In the subsequent process, the chef prepares the dish by entering the command that he desires into the machine equipped with artificial intelligence according to the order given. The dish —variances of which are determined, such as temperature, time and taste, and ingredients of which are measured by a pre-taught algorithm — is cooked/prepared and served with the help of artificial intelligence. The customer experience in this situation creates a technology-integrated restaurant perception with a positive marketing tactic for firms, and minimizes user errors, yielding a cost and time benefit for the company. The success rate is higher in standardized fast food chains as there are no specific menus, whereas in restaurants offering special menus, it is more difficult to maximize customer experience. This is because, in such restaurants, customers are better able to choose ingredients of a meal according to their own tastes, while more algorithm components are required for the relevant artificial intelligence. In this scenario, more loops and instructions should be supplied to the artificial intelligence algorithm in order to determine how much the dish will be cooked or how much more of the current proportional ingredient there will be. As a second scenario, after the customer orders food via the application on the device, the data sent to the artificial intelligence machine automation are organized into a meaningful structure, and the order is prepared directly. Thus, the order of the classical sort is eliminated, and an entirely digital order process and a satisfaction process are created.

Augmented Reality

The AR concept is one of the most discussed subjects in the digital environment. The concept of augmented reality, used by businesses as a customer experience and marketing tool, also plays an educational role on the part of many institutions. Institutions such as the Military Academy and Police Academy of the United States provide a realistic conflict environment to their staff with the help of AR goggles. Companies use AR goggles with the aim to have their customers use a product or service again and to ensure that their customers have a positive feeling after they purchase a service or product. Moreover, businesses use AR goggles as a way of marketing for a customer mass that will create a new demand pool. Although the concept of AR is often considered in a similar sense with virtual reality, the two concepts differ. This conceptual confusion can be eliminated if the concepts of "reality" and "virtual" are discussed before the AR and virtual reality distinction.

To put it simply, reality expresses the existing physical world, and we turn to the people and objects around us with whom we live, with all the senses every day. This reality is what causes most of the commercial transactions initially, despite the developments in all commercial activities and information and communication technologies (Farshid, Paschen, Eriksson, & Kietzmann, 2018).

The menu held in the hand, the space, staff, and utensils used for food are the reality for a customer who buys service from a restaurant. Such reality is not a purely artificial or semi-artificial mixture, but a reality whose existence from the beginning of history to the present time is accepted by all authorities. Online stores and restaurants equipped with AR are completely outside this category, whereas the production and service factor of the classic restaurant concept are included in this category.

Virtual reality (VR) refers to the complete, three-dimensional virtual representations of the real world or of the objects within. For example, AutoCAD software enables architects, engineers, and design specialists to draw real 3d sketches before they make changes to buildings (Farshid et al., 2018).

The greatest feature of virtual reality is that when an individual or a subject being studied enters the virtual reality world, his connection to the current physical environment is reduced to almost nothing. In individuals using virtual reality goggles, brain operations take place in the virtual environment while the body is in touch with the real world. An individual visiting a restaurant within software image graphics is not actually in the restaurant. He gains experience by being in the environment only with the help of the graphical software. This marketing experience, as a marketing tool, has not yet been launched in restaurant businesses for potential customers who have never previously been in a certain restaurant before and have not observed service, speed and other similar features of that restaurant. However, if it is launched, a new marketing channel will emerge. The ability to set up virtual reality in this marketing channel for customers who want to be in the environment set up by companies in booth centers in specific community areas (squares, shopping malls away from restaurants, walkways, etc.), and experience and test the event virtually will create a new channel in terms of marketing. In this case analysis, such restaurant chains will be able to provide potential customers with a choice of momentary food and beverage product samples in their hands, as well as a trip or a seating option in the restaurant. In this way, the customers will be given the opportunity to experience advertising and marketing as well as a potential customer experience.

Fundamentally, augmented reality is similar to virtual reality. However, unlike virtual reality, augmented reality allows the user to see the real world with virtual objects superimposed on top of the real world or in combination with virtual objects. Therefore, augmented reality completes reality instead of completely changing reality, unlike virtual reality (Kipper & Rampolla, 2012).

In augmented reality, an individual does not sever the connection with the real world completely or at the highest level. The user integrates objects into his own world and observes them graphically in his life. In virtual reality, an individual greatly reduces access to his physical world and operates in a space equipped with different graphical software. In augmented reality, however, the individual integrates the virtual world into his own world. By using a phone application, information can be obtained about various games, shopping products (testing whether shoes fit to a foot size that was registered beforehand, testing whether a dress fits a body that was registered beforehand), and restaurant product images— with their rates of calories, vitamins and mineral content.

Other important AR applications developed in the last 10 years include information displays embedded in vehicle windshields. Such displays are also known as "Heads Up Displays — HUDs." Thanks to these displays, important signaling information can be displayed at the windshield level of the vehicle without requiring any glasses. Through the developments in micro-electronics, micro-mechanics and micro-optics technologies in the 2000s, sensors and computers have become smaller and more functional. Thanks to these advances, smartphones have emerged, becoming the new human-computer interaction medium. Similarly, wearable devices have begun to become more intelligent and gradually reduced to the shape of glasses. The first example of this was Google Glass, which was produced by Google (Ürey & Akşit, 2018).

Currently, augmented reality applications in restaurants are not widely available around the world. Often restaurants that are newly in business offer the opportunity to use such applications. Businesses that operate in the classical restaurant sector and in food and beverage services offer little opportunity or do not provide such an opportunity at all. The reason for this is that the marketing concept of old-fashioned restaurant businesses and the marketing concept of new restaurant businesses differ according to the new world system. The understanding in the newly opened restaurant businesses is to offer the customer many extra experiences — in addition to providing food and beverage services to the customer — and to consider it to be a positive experience and marketing tool of the business entirely for the customer. The experience of the customer is targeted from the customer's selection of items to order, to the point of payment, and the process after the customer leaves the restaurant is also targeted for this experience to spread it to new people through a one-to-one advertising strategy.

Before the customer gets his order, if the customer points his phone to the menu through tools such as devices and goggles equipped with augmented reality, the customer can be provided with virtual information on the contents and rates of calories, fats, proteins, carbohydrates, vitamins, and so forth. In this way, the customer will be able to learn how many calories and nutrients the food to be consumed have with the help of a pre-designed augmented reality system. It is possible to extend this scenario: It is possible to project the data of which products are consumed more from the menu in front of the customer, and thus, to project the content consumed within that week or percentages of special offers of that week. The projection of the actual dimensions of food and beverages in the menu content to the table will also be possible through the augmented reality system.

Customers with a desire to know how a product in a menu is prepared and what stages it undergoes will yield a positive or negative customer experience in terms of hygiene. The vast majority of past restaurant scandals consist of speculation in the sense of health-related information and lower the brand value of restaurants. An online or offline video showing the preparation and cooking stages of a product and under what conditions it is prepared can be presented to the customers. The customer can learn instantly and online which stages the product goes through during the preparation process. This will be accomplished via an Internet connection through the application or device. In the offline scenario, when the customer points a device with an integrated application to the menu, a previously prepared video will be projected, thus providing a positive customer experience and marketing benefit.

In addition, for customers with allergic reactions to certain spices or substances, spice rates and allergic chemical substance rates can be viewed through applications or devices without prompting service personnel. In this way, customers will be able to make their selections in a healthy way.

QR Code Systems

The QR code is a new medium that provides a bridge between traditional communication tools and digital technologies, while it is also referred to as a communication technology. It is placed on all kinds of printable surfaces, such as traditional books, magazines and newspapers, and transforms traditional communication tools into hybrid communication tools (Aktaş, 2014).

What makes QR codes more successful than other codes is their ability to transmit much more data in a small and still space. With this capability, they store a high amount of data. Although a QR code

Figure 1. QR code systems characteristics
Source: (Coşkun, 2012)

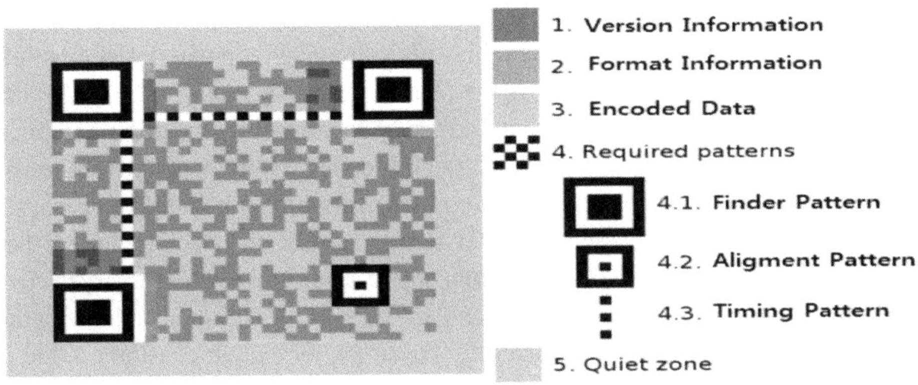

is actually a structure consisting of dots, each dot has the characteristic of a door that opens up to a meaningful data pool. This is a feature that creates a difference.

- **Version Information:** Identifies the version of the QR code. Each code's data storage capacity is different, and their versions increase as the data they store increases. Moreover, the codes have different error correction levels.
- **Format Information:** This includes error correction levels with a length of 15 bits and a mask pattern indicator. The scanner first reads the information here.
- **Encoded Data:** The data of the QR code is storage in this area. The data is stored by translating the binary numbers "0" and "1" into black and white cells. This data area is read by the scanner.
- **Required Pattern**
    - **Finder Pattern:** By using this structure, which is located in three corners of the code, the direction, size and angle of the symbol are determined. It ensures that the code is read from any angle (360 degrees) at high speed.
    - **Aligment Pattern:** This template is very effective in correcting the deformities in the QR code. If there are deformations in the QR code, the scanner understands the error and performs the necessary adjustment.
    - **Timing Pattern:** It is a structure with a black and white pattern between the identification templates. The central coordinate information of each cell is determined by using this pattern. It corrects the central coordinate of the cell when the code is bent or when an error occurs in cells. It does this in horizontal and vertical directions.
- **Quiet Zone:** The blank space surrounding the QR code where there is no data. Nothing is written or printed in this field. It must have a width of at least 4 modules (each dot is 1 module). This blank space allows the code to be read in a hassle-free way (Çoşkun, 2012).

As data capacity, it provides numerical only code (a maximum of 7,089 characters), alphanumeric (a maximum of 4,296 characters), binary (a maximum of 2,953 characters), and Kanji-Kana (a maximum of 1,817 characters) functions. It has a very rich data capacity in terms of character count. The QR code has shown widespread prevalence in various sectors as areas of use. This technology — used for promotion, pharmaceutical packaging, e-mail and SMS services, automotive sector, cinema industry and food business sector — is also used as a marketing and sales instrument.

In restaurant businesses, QR code technology is at a popular level in terms of menu pricing. When customers who enter a restaurant are seated at tables, they are given a menu defined with a QR code. Customers perform a scan on the QR code with the help of the applications in the smartphone system. Within the bounds of options, they can order meals directly by contacting the order acceptance section, without the assistance of the attendant. To do this, the QR code system must have an online interface. Moreover, customers can complete the remote order placement process without being physically in the restaurant by using the online ordering service of the restaurant, QR code technology on the website or the QR code interface integrated into the food ordering applications. It is also necessary to create a virtual wallet that integrates the payment systems of the bank and combine the virtual wallet with the QR code.

Many restaurant businesses organize special offers over QR for existing customers and prospective customers. Restaurant businesses offer a variety of menu prices in applications with a QR code infrastructure to increase sales consumption figures for special offers of the week. The goal here is to come up with a tactical strategy to maximize the consumer's sense of curiosity. Moreover, the restaurant businesses that desire to measure the customer experience and analyze feedback, such as complaints and appreciations, accumulate in their own internal pool the feedback from customers with the QR code they share with the customers after an order. Customers will thus not be obliged to enter their information to create a record of complaint or appreciation, which takes a long time to enter otherwise. Instead, they quickly identify their information with the QR code that was previously defined. Thus, they can create a record of satisfaction or dissatisfaction.

Customers are provided with information accompanied by a QR code when they are served, in order for them to get transparent information about the products that restaurants serve and to get information on calories, minerals and other ingredients (proteins, carbohydrates, and so forth). When the customer scans the QR code system that is brought with the product, he can see the stages through which the materials that make up the product were collected, the stages through which the materials were combined, and the stages after which the materials were served by using smart applications.

The concept of making payments by using the QR code system in restaurants is a fast-growing behavior across the globe. The QR code payment system, which is widely used in Europe and the USA, has started to spread in China thanks to Alipay. The need for tourists — coming to restaurants around the world — to make a quicker and safer payment has also been effective in this case, in addition to the technological reasons of its widespread expansion. Due to the credit card copying and theft events, the QR code provides a more secure payment through virtual wallet accounts. As a result of their research, Lou et al. (2017) have stated that the QR payment system can create a positive experience for tourists. In that study conducted on 247 tourists, it was found that the tourists preferred the payment system via QR, which was satisfactory to them (Lou, L., Tian Z., Koh, J., 2017).

The QR code, which was invented in Japan in 1994 by Denso Wawe, has begun to be preferred as a safe and fast payment method worldwide, creating a butterfly effect. This technology, which was designed for monitoring vehicles during production in the first place, was later used as payment systems and has been used in many useful functions such as payment instruments in restaurants. We can list these benefits as the proliferation of brand demand, the advertising activities that we describe as face to face advertisement, the rise in brand awareness, the creation of a data base system for customers, the increase of sales figures, and the extreme minimization of marketing costs compared to those of the classical marketing concept.

MAIN FOCUS OF THE CHAPTER

Examples of Smart Technology Implementation in Customer Experience at Food and Beverage Businesses

Drive-Thru Menus/Mobile Apps

To deliver personalized customer experience, McDonalds acquired the Dynamic Yield technology company for about 300 million dollars. A digital menu developed by McDonalds has been built into the interior and exterior of restaurants. This menu has a technology equipped with machine learning. It offers personalized menu service to customers. It continuously updates menus according to the weather, traffic and time of the day on that day. If the weather is hot, it can offer additional food such as ice cream in addition to the recommended menu. It also collects data about the menus purchased by the customer with the license plate recognition system and recommends items in the subsequent purchase menus based on the past data. It also shares the trend menu with its customers by using data from branches in nearby locations. Moreover, if young people around the branch have basketball or football games, it adds to the digital menus items that will accelerate teens' metabolisms (Tiffany, 2019). This technology used in Canada improved sales by approximately 3% to 3.5%. It improves customer experience through the mobile application used by McDonalds on smart phones. It offers many advantageous prices in addition to offering mobile payment and mobile ordering. It offers products that complement the customer's products with artificial intelligence technology. Personalized real-time marketing activities can be implemented by registering when and how often customers log into the mobile application. Customers using mobile applications in Japan have purchased 35% more sales through the recommendation of the application (Marr, 2018).

Say2eat's Facebook Messenger Ordering Technology

With this technology, placing an order is as easy as sending a message to a friend without even downloading any mobile application or registering in a system. A chatbot system is created by integrating artificial intelligence technologies to the Facebook Messenger application. An order can be placed directly from Facebook messenger by calling any business using the Say2eat technology. Because this technology gets integrated to Facebook with 6 billion active users as well as providing customers with an excellent experience, sales are also growing. Without the hassle of other tools, it can take orders of customers and solve their problems through chatbots 7/24 without any time limits. Other than Facebook, orders can also be placed through message applications such as Twitter, WhatsApp, SMS and Google Home. One of the most important features of it is that chatbots are integrated with POS. That is, orders placed by customers directly pass to the POS system. It increases the satisfaction and experience level of customers due to the short delivery time of the order as much as it benefits the business (Say2eat, 2017). One of the restaurant chains that uses this application is Domino's Pizza.

Figure 2. Facebook messenger ordering application example
Source: https://medium.com/

Pepper Robot

Developed by a Japanese technology firm using the artificial intelligence technology, Pepper is able to serve as a waitress in the restaurant. Pepper, which can communicate with customers, can make food recommendations, provide information about the content of dishes, serve dishes and accept credit card payments. Moreover, it is used to book tables and provide information to customers when the seats are ready. It can also offer customers personalized promotional products. Pizza Hut and many restaurants in Tokyo employ Pepper as an employee. Although such robots cannot offer a warm hospitality, it is no doubt that they offer a different experience that will excite customers (Eriksen, 2018).

Facial Recognition Technology

Particularly in China, many restaurants accept payment by a facial recognition system with artificial intelligence technology. It is observed that its use is more common among businesses offering self-service. Another technology offered to customers' experience with the facial recognition system is food recommendations. When the customer enters the restaurant, the kiosk introduces his face. The smart technology then offers product recommendations based on the age and physical characteristics of the customer. KFC implements this technology in its branches in China. These features shorten the customer's waiting period as well as offering a different experience to the customer. Another feature is implemented in restaurants offering the table d'hôte system. The customer selects and receives his meals. Then, the amount to pay is displayed on the screen when the table is placed on the device with the face recognition system.

Kabaq AR Menu

The use of AR technology is becoming increasingly widespread in restaurants to increase customer experience. The Kabaq mobile application, which hosts AR technology, was developed by a New York-based company. This mobile application offers the possibility to see the three-dimensional and realistic version of the food you choose when you place the food on your empty plate. Before ordering the food, you can see its actual size, the ingredients in it and how it will be served. Moreover, it can be seen how the orders are packaged to be delivered to homes and workplaces, and custom-designed packaging is also an option (Morozova, 2018).

Domino's Ad Interaction

Domino's Pizza uses AR technology for advertising in order to interact with its customer. Customers can scan outdoor advertisements by using their phone's camera and see the menu offers — available in three dimensions with augmented reality technology — specific to them through the Blipar App (Aaronallen, 2013). Moreover, the customer can also purchase items on Snapchat with AR technology by following offers specific to them. This interaction creates a desire to purchase on the customer's part. Domino's Pizza has increased its sales by 10% by making the customer experience seamless and memorable by using this technology (Cameron, 2018). AR technology focuses on menus, as in this example.

Inamo AR Table

Tables in a restaurant in the UK have been equipped with a three-dimensional technology to interact with customers and offer them a different experience. At these tables with AR technology, customers can play games, change the color and picture of the table, place orders, look at the menus in three dimensions through AR and learn about the local environment. This application has been welcomed by many customers and received many awards. Such applications improve sales and bring about a memorable customer experience. Its providing information about the local environment contributes to tourism.

QR Code-Mobil Ordering

One of the smart technologies that facilitate quick ordering is the QR code technology. There is no need for a waiter to order when you have this technology. One can download the required application to smartphones or tablet computers and place an order directly by pointing the device to the QR code on the menu on the table. Thus, service can be provided in a short period of time without having to wait for the waiter. For example, one can easily order through the application by downloading the Quick Order application to a tablet computer or smartphone. This application can be put into service in hotels and restaurants.

One can also easily place orders through the Barcode Menu application. The customer can access the digital menu by pointing his device to the barcode on the table. He selects the menu and confirms it. The approved menu is passed to the automation of the restaurant and then directed to the kitchen for preparation. Moreover, by using this application, mobile payments can be made, customer-specific special offers may be organized, and comments about the restaurant can be made. More information about the contents of the food can be learned by integrating QR codes on the items exhibited in the restaurants.

SOLUTIONS AND RECOMMENDATIONS

Smart technologies are now one of the biggest factors in the digital world that enable enterprises to exist, sustain themselves and ensure profitability in the future. This is because the ability to maximize customer experience now depends on smart technologies based on this kind of innovation. This factor is estimated to significantly affect especially the preferences of new generation customers.

The ability of restaurant businesses to invest especially in smart technologies and adapt to current developments will make significant contributions to the improvement of customer experience and make their profitability sustainable. Employing smart robots to reduce costs, in particular, will enhance the competitiveness of restaurants. On the other hand, the fact that robots offer a different experience will improve customer satisfaction. Today, many customers want every service to be instant and easy to reach and turn to restaurants that provide these opportunities. For this reason, the integration of order, payment and other services of assistance into smart applications will give a very important advantage to restaurant businesses in terms of customer experience.

The development of AR technologies compatible with smartphones in restaurant businesses is also a very important issue. Especially the widespread adoption of the menu AR application service will accelerate the time required for customers to make decisions. This is because the customer will be able imagine what he or she wants to order in the menu with the help of the AR application service. Therefore, this smart technology will provide a different experience as well as customer convenience. The inclusion of AR applications in different service processes by restaurant businesses will improve the level of satisfaction and loyalty of customers. Smart devices, such as tablets and phones, are indispensable smart technologies. The fulfillment of such infrastructure tools by restaurant businesses in the medium term will ensure the widespread use of smart technologies. This will make restaurants more innovative, speeding up the transition of potential customers to being loyal customers. In summary, the future of customer experience in restaurant businesses will be established by the importance given to and investments made in smart technologies. For that reason, it is very important for restaurant managers to attach importance to these issues in the planning of the future.

FUTURE RESEARCH DIRECTIONS

This study was limited to examples of smart technologies used in restaurant business. Content analysis studies or empirical studies can be carried out in the future to find out whether the features of smart technologies used are satisfactory to customers or whether they are sufficient for consumers. In addition to artificial intelligence and augmented reality, smart technology examples used in restaurant businesses can be investigated. Studies can be carried out to find out on whether the use of artificial intelligence and smart technologies can impact customer satisfaction and loyalty. Customer experience and smart technologies can also be studied in different sectors.

CONCLUSION

Smart technologies, expected to have a significant impact on the future of the world economy, have already caused radical changes in many sectors and areas. Especially the emergence of smart devices and the tendency of consumer audiences to use digital technologies are the most important factors influencing and triggering this change. The relationship between smart technologies and customer experience, which is a fairly new topic in the literature and needs to be enriched, was scrutinized and discussed in this study. The current importance of customer experience was emphasized and the role of smart technologies in this experience was scrutinized. There are many smart technologies involved. But, in this study, artificial intelligence and augmented reality technologies widely used in the field of customer experience were addressed, and their contributions to customer experience were discussed. While explaining these issues, the importance and use of these technologies especially for the restaurant businesses were addressed. In the last section of the paper, examples of smart technology applications used in many parts of the world to increase customer experience were presented. It was observed that these applications facilitated the habits of consumers and also provided great savings to the businesses. Especially because restaurants are a service sector, production and consumption take place at the same time. Therefore, restaurants are the places where the customer–business interaction occurs the most. For this reason, smart technologies have many advantages and opportunities to deliver a good customer experience in this area where interaction is intense.

REFERENCES

Aladag, E. (2015). Makine Öğrenmesi Nedir? Retrieved from https://www.emrealadag.com/makine-ogrenmesi-nedir.html

Birbil, I. (2018). Tahmin ve Çıkarım. Retrieved from http://www.veridefteri.com/2018/04/04/tahmin-ve-cikarim-makine-%0Aogrenmesi-yapay-ogrenme/

Bodhani, A. (2013). Getting a purchase on AR. *Engineering & Technology*, *8*(4), 46–49. doi:10.1049/et.2013.0408

Cameron, N. (2018). Domino's debuts augmented reality pizza ordering. Retrieved from www.cmo.com.au/article/649324/domino-debuts-augmented-reality-pizza-ordering/

Castañón, J. (2019). 10 Machine Learning Methods that Every Data Scientist Should Know.

Coşkun, A. (2012). QR Kod Nedir? Retrieved from http://aylincsknn.blogspot.com/2012/08/qr-kod-cagn-barkodu-olarak-anlan-qr_11.html

Duncan, T., & Moriarty, S. (2006). How integrated marketing communication's 'touchpoints' can operationalize the service-dominant logic. *The Service-Dominant Logic of Marketing: Dialog, Debate, and Directions*, 236–249.

Eriksen, K. (2018). Benefits of Artificial Intelligence in the Restaurant Industry. Retrieved from https://www.deputy.com/blog/benefits-of-artificial-intelligence-in-the-restaurant-industry

Farshid, M., Paschen, J., Eriksson, T., & Kietzmann, J. (2018). Go boldly! Explore augmented reality (AR), virtual reality (VR), and mixed reality (MR) for business. *Business Horizons*. doi:10.1016/j.bushor.2018.05.009

Foroudi, P., Gupta, S., Sivarajah, U., & Broderick, A. (2018). Investigating the effects of smart technology on customer dynamics and customer experience. *Computers in Human Behavior*, *80*, 271–282. doi:10.1016/j.chb.2017.11.014

Gentile, C., Spiller, N., & Noci, G. (2007). How to sustain the customer experience: An overview of experience components that co-create value with the customer. *European Management Journal, 25*(5), 395–410. doi:10.1016/j.emj.2007.08.005

Grewal, D. (2009). *Enhancing the Retail Customer Experience*. Elsevier.

Halvorsrud, R., Kvale, K., & Følstad, A. (2016). Improving service quality through customer journey analysis. *Journal of Service Theory and Practice, 26*(6), 840–867. doi:10.1108/JSTP-05-2015-0111

Helkkula, A. (2011). Characterising the concept of service experience. *Journal of Service Management, 22*(3), 367–389. doi:10.1108/09564231111136872

Homburg, C., Jozić, D., & Kuehnl, C. (2017). Customer experience management: Toward implementing an evolving marketing concept. *Journal of the Academy of Marketing Science, 45*(3), 377–401. doi:10.100711747-015-0460-7

Jain, R., Aagja, J., & Bagdare, S. (2017). Customer experience – a review and research agenda. *Journal of Service Theory and Practice, 27*(3), 642–662. doi:10.1108/JSTP-03-2015-0064

Kipper, G., & Rampolla, J. (2012). *Augmented Reality: An Emerging Technologies Guide to AR.*

Lemon, K. N. (2016). The art of creating attractive consumer experiences at the right time: Skills marketers will need to survive and thrive. *GfK Marketing Intelligence Review, 8*(2), 44–49. doi:10.1515/gfkmir-2016-0015

Lemon, K. N., & Verhoef, P. C. (2016). Understanding customer experience throughout the customer journey. *Journal of Marketing, 80*(6), 69–96. doi:10.1509/jm.15.0420

Macdonald, E. K., Kleinaltenkamp, M., & Wilson, H. N. (2016). How business customers judge solutions: Solution quality and value in use. *Journal of Marketing, 80*(3), 96–120. doi:10.1509/jm.15.0109

McColl-Kennedy, J. R., Zaki, M., Lemon, K. N., Urmetzer, F., & Neely, A. (2019). Gaining customer experience insights that matter. *Journal of Service Research, 22*(1), 8–26. doi:10.1177/1094670518812182

Meyer, C., & Schwager, A. (2007). Understanding customer experience. *Harvard Business Review, 85*(2), 116. PMID:17345685

Morozova, A. (2018). How Much Does It Cost to Build an AR App for Food Menu like KabaQNo Title. Retrieved from https://jasoren.com/how-much-does-it-cost-to-build-an-ar-app-for-food-menu-like-kabaq/

Pantano, E., & Timmermans, H. (2014). What is smart for retailing? *Procedia Environmental Sciences, 22*, 101–107. doi:10.1016/j.proenv.2014.11.010

Ravì, D., Wong, C., Deligianni, F., Berthelot, M., Andreu-Perez, J., Lo, B., & Yang, G. (2017). Deep Learning for Health Informatics. *IEEE Journal of Biomedical and Health Informatics, 21*(1), 4–21. doi:10.1109/JBHI.2016.2636665 PMID:28055930

Say2eat. (2017). Top 5 Reasons to Partner with Say2eat's Modernized Food Ordering Technology. Retrieved from https://medium.com/@say2eat/top-5-reasons-to-partner-with-say2eats-modernized-food-ordering-technology-4d5a3b70e6c1

Segelström, F. (2013). *Stakeholder Engagement for Service Design: How service designers identify and communicate insights*. Linköping University Electronic Press. doi:10.3384/diss.diva-97320

Şeker, A., Diri, B., & Balık, H. H. (2017). Derin Öğrenme Yöntemleri ve Uygulamaları Hakkında Bir İnceleme. [GMBD]. *Gazi Mühendislik Bilimleri Dergisi*, *3*(3), 47–64.

Şeker, Ş. (2013). Sınıflandırma. Retrieved June 2, 2019, from http://bilgisayarkavramlari.sadievrenseker.com/2013/03/31/siniflandirma-classification/

Şengöz, N. (2017). Yapay Sinir Ağları. Retrieved June 3, 2019, from http://www.derinogrenme.com/2017/03/04/yapay-sinir-%0Aaglari/

Shankar, V. (2018). How Artificial Intelligence (AI) is Reshaping Retailing. *Journal of Retailing*, *94*(4), vi–xi. doi:10.1016/S0022-4359(18)30076-9

Sultan, A. J. (2018). Orchestrating service brand touchpoints and the effects on relational outcomes. *Journal of Services Marketing*, *32*(6), 777–788. doi:10.1108/JSM-12-2016-0413

Tiffany, K. (2019). McDonald's new drive-thru menus will change based on the weather, traffic, and time of day. Retrieved from www.vox.com/the-goods/2019/3/27/18284045/mcdonalds-tech-startup-artificial-intelligence-menu-drive-thru

Tirunillai, S., & Tellis, G. J. (2014). Mining marketing meaning from online chatter: Strategic brand analysis of big data using latent dirichlet allocation. *JMR, Journal of Marketing Research*, *51*(4), 463–479. doi:10.1509/jmr.12.0106

Ürey, A., & Akşit, K. (2018). Sanal Gerçeklik ve Artırılmış Gerçeklik Nedir? Retrieved May 20, 2019, from https://sarkac.org/2018/03/sanal-gerceklik-artirilmis-gerceklik-nedir/

Verhoef, P. C., Kannan, P. K., & Inman, J. J. (2015). From multi-channel retailing to omni-channel retailing: Introduction to the special issue on multi-channel retailing. *Journal of Retailing*, *91*(2), 174–181. doi:10.1016/j.jretai.2015.02.005

Verhoef, P. C., Lemon, K. N., Parasuraman, A., Roggeveen, A., Tsiros, M., & Schlesinger, L. A. (2009). Customer experience creation: Determinants, dynamics, and management strategies. *Journal of Retailing*, *85*(1), 31–41. doi:10.1016/j.jretai.2008.11.001

Voorhees, C. M., Fombelle, P. W., Gregoire, Y., Bone, S., Gustafsson, A., Sousa, R., & Walkowiak, T. (2017). Service encounters, experiences, and the customer journey: Defining the field and a call to expand our lens. *Journal of Business Research*, *79*, 269–280. doi:10.1016/j.jbusres.2017.04.014

Wind, Y. J., & Hays, C. F. (2016). *Beyond advertising: creating value through all customer touchpoints*. John Wiley & Sons.

Zomerdijk, L. G., & Voss, C. A. (2010). Service design for experience-centric services. *Journal of Service Research*, *13*(1), 67–82. doi:10.1177/1094670509351960

ADDITIONAL READING

Akaka, M. A., & Schau, H. J. (2019). Value creation in consumption journeys: Recursive reflexivity and practice continuity. *Journal of the Academy of Marketing Science*, *47*(3), 499–515. doi:10.100711747-019-00628-y

Femenia Serra, F., & Neuhofer, B. (2018). Smart tourism experiences: conceptualisation, key dimensions and research agenda.

Flavián, C., Ibáñez-Sánchez, S., & Orús, C. (2019). The impact of virtual, augmented and mixed reality technologies on the customer experience. *Journal of Business Research*, *100*, 547–560. doi:10.1016/j.jbusres.2018.10.050

Hamilton, R., & Price, L. L. (2019). Consumer journeys: Developing consumer-based strategy. *Journal of the Academy of Marketing Science*, *47*(2), 187–191. doi:10.100711747-019-00636-y

Kuehnl, C., Jozic, D., & Homburg, C. (2019). Effective customer journey design: Consumers' conception, measurement, and consequences. *Journal of the Academy of Marketing Science*, *47*(3), 551–568. doi:10.100711747-018-00625-7

Scholz, J., & Duffy, K. (2018). We ARe at home: How augmented reality reshapes mobile marketing and consumer-brand relationships. *Journal of Retailing and Consumer Services*, *44*, 11–23. doi:10.1016/j.jretconser.2018.05.004

Wirtz, J., Patterson, P. G., Kunz, W. H., Gruber, T., Lu, V. N., Paluch, S., & Martins, A. (2018). Brave new world: Service robots in the frontline. *Journal of Service Management*, *29*(5), 907–931. doi:10.1108/JOSM-04-2018-0119

Wozniak, T., Schaffner, D., Stanoevska-Slabeva, K., & Lenz-Kesekamp, V. (2018). Psychological antecedents of mobile consumer behaviour and implications for customer journeys in tourism. *Information Technology & Tourism*, *18*(1-4), 85–112. doi:10.100740558-017-0101-8

KEY TERMS AND DEFINITIONS

Artificial Intelligence: It is a set of software and hardware systems that have many abilities such as human behaviors, digital logic, motion, speech and sound perception.

Augmented Reality: It is called superimposing virtual objects on real images by using the object recognition feature of the devices.

Experience: It refers to the totality of knowledge, skills and attitudes that a person gains through direct perceptions and activities.

Machine Learning: It is a method paradigm that makes inferences from existing data using mathematical and statistical methods and makes predictions about what is unknown with these inferences.

Touchpoint: This is each of the moments of communication or interaction between customers and the company or its products.

Virtual: It is something that has no place in reality, does not exist in reality, but is designed in the mind.

Chapter 13
Electronic Customer Relationship Management in Tourism

Cenk Murat Koçoğlu
Karabük University, Turkey

Merve Yıldırım Kalem
Karabük University, Turkey

ABSTRACT

Tourism companies, like other companies producing goods and/or services, practice different strategies to achieve competitive advantage and satisfy the needs and wants of customers/guests. The most important recent strategy is electronic customer relationship management (e-CRM). In today's competitive environment, a successfully implemented e-CRM enhances a customer centric culture that is key to tourism growth and competitive advantage. For tourism companies, e-CRM succeeds only if it is totally integrated strategically and operationally in the business and in the information system of the company. This chapter focuses on e-CRM concept, its importance in tourism, dimensions of e-CRM, major challenges of e-CRM and its opportunities for companies, its implementation in tourism, e-CRM Technologies used in tourism, e-CRM success factors, and the reasons of e-CRM failures.

INTRODUCTION

In a globalised world, the competition between enterprises has increased both in nationally and internationally. In order to gain a competitive advantage, businesses have begun to use modern marketing tools. In this context, a customer-oriented concept which is called customer relationship management (CRM) that adopts the demands and needs of customers has emerged. Customer Relationship Management (CRM) has become one of the most important marketing strategies of the late 20th and early 21st centuries. CRM is used in various fields including tourism industry for developing customer relationship and sustain the enterprise's existence in the competitive environment.

DOI: 10.4018/978-1-7998-1989-9.ch013

Internet-based technology has been growing in business-to-consumer and business-to-business environments. From a customer point of view, Internet-based services decrease the cost of searching; facilitate the selection of service providers, provide control over products or services and increase the avaliability (Anderson &Srinivasan, 2003).

By the rapid development in the Internet technologies, Electronic Customer Relation Management (e-CRM) has come into prominence in the last decade. The new technologies including the Internet, e-mails, mobile phones and digital televisions facilitate the CRM activities and provide its cost-effective use. E-CRM is defined as part of an enterprise's comprehensive business and marketing strategy that enabled through the use of the Internet (Feinberg & Kadam, 2002). In today's complex and dynamic competitive market, organizations aim is to create long-term relationships with their customers and increase their profits though e-CRM. Tourism industry is also adopting e-CRM strategies.

Many research works carried out so far about e-CRM in various fields including tourism. Some of these are as follows: Ahmad and Chowdhury (2008) used multiple case study strategy about customers' perception of value from e-CRM features on airline electronic ticketing websites. Caber (2010) assessed the influence of website quality on satisfaction and trust in tour operator-travel agency electronic customer relationship managementin her study. GilaniNia, Balaei and Niyari (2013) assessed the effect of e-CRM system to gain a competitive advantage in the hotel industry in the city Sarein in their study. Dubihlela and Khosa (2014) have studied the impact relationship between effective e-CRM implementation, customer loyalty, customer retention and customer profitability. The results of their study provides information for hotel managers to increase customer loyalty, customer retention and customer profitability. Erkiah and Ladkoo (2018) aimed to analyze the effectiveness of e-CRM as a strategic tool in the hotelindustry in Mauritius.

This chapter provides a better understanding of the concept of CRM and e-CRM and the importance of e-CRM in tourism. This chapter focuses on dimensions, implementation, benefits of e-CRM and perceived barriers to adoption of e-CRM Technologies in tourism. Morever, e-CRM Technologies used in tourism industry will be presented in this chapter. Later on, e-CRM success factors in tourism and the reasons of e-CRM failures will be presented. Last section of the study will be conclusion and recommendation section. In this study, recommendations will be made for the improvement of e-CRM for the marketers operating in tourism industry.

BACKGROUND

E-CRM

Due to the competitive business environment, Customer Relationship Management (CRM) is one of the most important academic and practical topic in marketing. Customer Relationship Management (CRM) has become the leading business strategy in today's dynamic and competitive business environment. There are several definitions made on CRM. "A strategy and process of acquiring, retaining, and partnering with selective customers to create superior value for the organization and the customer. It involves the integration of marketing, sales, customer service, and the supply-chain functions of the enterprise to reach efficiency and effective customer value" is one definition claimed by Parvatiyar and Sheth (2001). Croteau and Li (2003) defined CRM as "a customer focused business strategy which aims to enhance customer satisfaction and customer loyalty to offering a more responsive and customized service to each customer". As stated by Kotler and Keller (2006), CRM is the process of managing detailed information about each customer and creating stronger relationship with them to raise customer loyalty. Another definition for CRM is"the combination of marketing efforts, business processes and technology that will let the organizations to know and distinguish their customers from various aspects" (Chang, 2010).

By the rapid development in the Internet technologies, e-CRM has come into prominence in the last decade. The new technologies including the Internet, e-mails, mobile phones and digital televisions facilitate the CRM activities and provide its cost-effective use. Electronic communication has created many opportunities to establish a more appropriate and friendly relationship with customers (Güleri, 2000). The cost-effectiveness and velocity of the Internet make many businesses turn to online operations around customer needs to increase customer loyalty, customer retention and customer profitability. Enterprises have found new ways to meet with customers through developments on the Internet. The use of the Internet has further enabled marketing activities. Through the use of the Internet and Information Technologies, businesses have redesigned their customer relationship management (CRM) to meet customer needs and desires (Garrido-Moreno & Padilla-Meléndez, 2011). Information Technology (IT) requiresusing computer hardware, software and telecommunication devices. CRM has reachedimportant developments and developed its applications with the use of information technology (IT) to integrate sales and services aswell as promote a customers' loyalty program (Chen, Lin & Yang, 2011). New method of customer relationship focuses on web-based interactions between organizations and customers (Kourtesopoulou, Papaioannou & Kriemadis, 2011) has been formed, named electronic customer relationship management. Electronic customer relationship management (e-CRM) differs from CRMin the manner of its the implementation method (Chen, Lin and Yang, 2011). E-CRM develops CRM tools by integrating new electronic channel technologies (Pan & Lee, 2003).

The e-CRM is defined by Feinberg and Kadam, (2002) as part of company's business and marketing strategy which enabled through the use of the Internet. Lee-Kelley, Gilbert and Mannicom (2003) defined e-CRM as " the marketing activities, and tools delivered over the Internet technologies such as websites and e-mail, data-capture, data warehousing, data mining with the purpose of buildingand improving long lasting customer relationships". Malik and Kumar (2013) and Muro, Magutu and Getembe (2013) claimed that e-CRM is a strategic technology-oriented model which combines traditional CRM applications with electronic business market place applications. E-CRM has gained great importance in enterprises operating in tourism industry. Due to being a service-oriented industry, a great emphasis is given to customer satisfaction. In order to create long term relationship with customers, to increase customer satisfaction and loyalty, e- CRM can be used as a marketing tool.

Importance of Electronic Customer Relationship Management in Tourism

In today's competitive business environment, organizations should pay attention to the relationship of their own customers. Particularly for service-based industries such as hospitality, tourism and transport industries, managing customer relationship is essential because of the inherent intangibility and heterogeneouscharacteristics of service industry (Cheng, Chen, and Chang, 2008). By using information and communication technology such as web sites and e-mails, businesses operating in tourism sector are trying to become close to their customers so they can build long-term customer relationship. Hotels can receive feedback from guests such as complaints. In these terms, giving value to knowledge management and implementation of an e-CRM system is essential for achieving a successful business performance. The World Tourism Organization (WTO) argues that "in order to be successful, organizations should identifyconsumer needs and reach potential customers with comprehensive, customized and current information" (Buhalis, 1998). Dhillon, Joshi, & Verma (2012) stated that hotel managers should be well-informed about each customer's details from their needs to tastes in order to be in a position to deliver and meet the expected requirement of their customers. Businesses such as hotels, airlines, boat charters, tour operators, and other tourism related enterprises should use e-CRM to promote their services to achieve the desired profits by using the latest Technologies.

Technological trends in hospitality industry are improving day by day. With the advent of information communication technology, the way customers interact and seek information has changed (Ip, Leung, & Law, 2010). Traditional hotel reservation system has changed with the help of the Internet technology. Technological improvements give a chance to potential customers to read comments about hotels, compare hotel rates and rooms.

To provide a better service, hotels developed some e-CRM practices that enable and empower guests to take part in the hotel service process value chain by getting involved in theservice production, delivery, consumption and postconsumption processes (Sigala, 2006). Guests have a chance to create their profile online (e.g. vegetarian, double pillows, dietary requirementsetc.) To illustrate, Wyndham Hotel has a programme called ByRequest that provides selection of pillow, preferable newspaper delivery. Such e-CRM practice plays an important role in tourism for customer satisfaction and for developing long-term customer relationship

Furthermore, hotels and travel agencies are using online marketing platforms to increase bookings. With the help of the developments in computer graphics and web images, images of hotels and destinations can be displayed. By this way, guests can use virtual tours to view the interior and exterior of the hotel better. It has become easy for them to be informed about destinations, hotel rooms and rates. Potential guests will have the chance to experience their stay before arriving at the hotel.

Airline companies create relationships with business partners in order to contribute to the success of e-CRM goals. They can monitor their relationships with theircustomers by knowing the level of customers' loyalty based on the e-CRM systems (Will, 2009). For example, The STAR Alliance network the first global airline alliance offers customers worldwide reach and an unforgettable travel experience. Member passengers of the Star Alliance airline frequent flyer programmes can earn miles on all flights across the Star Alliance network. Star Alliance Silver and Star Alliance Gold status provide other benefits such as priority boarding and worldwide lounge access that passengers become entitled towhenthey reach a higher status in an individual airline frequent flyer programme (Buttle, 2009).

Dimensions of Electronic Customer Relationship Management

Ab Hamid and McGrath (2005) proposed twelve dimensions of e-CRM which are information quality, ease of navigation, consumer service quality, fulfilment, integrated marketing channels, online community, rewards, personalization level, site security, value-added services, perceived trust and price attractiveness.

1. **Information Quality:** Information quality is important for potential customers to find what they are searching for. This leads to an easier decision-making process. It is also important that the information presented on websites should beup-to-date and comprehensible.
2. **Ease of Navigation:** Many travel enterprises, particularly hotels, should have creative, eye-pleasing, practical and functional websites. Ease of navigation indicates the content layout and ease of use.

3. **Consumer Service Quality:** Enterprises operating in travel and hospitality industry are trying to improve the quality of service that they providein order to create customer loyalty. Service quality is the key factor for success in service industries mainly in tourism. Guests constantly demand for continous communicationwith hotel managers. The service that they receive is very important for their revisit intention.
4. **Fulfilment:** The order fulfilment processis delivering the right service at the right moment and answeringwhat customer demands.
5. **Integrated Marketing Channels:** This dimension refers to the integration of the virtual and physical marketing operations of an enterprise. Managing customer relationships in both channelsis very important for hotel managers.
6. **Online Community:** Technological developments have made customers more active and enabled them to share their comments about destinations, hotels airline companies or any other tourism enterprise. Customers' exhange information and get online help from their online members.
7. **Rewards:** Reward programs provide membership privileges for returning consumers. Rosenbaum, Ostrom, & Kuntze (2005) stated that a sense of exclusive membership makes members feel happy about being part of a group of guests or clients who receive special services from the enterprises and have stronger connections with the organization.
8. **Personalization**: E-CRM technologies allow customers to customize their own corporate web pages. Customers of the travel enterprise www.lastminute.com can customize their home page (mylastminute.com) that stores their preferences (www.lastminute.com).

With the help of e-CRM practices, guests' knowledge can be collected for creating personalized services such as online personalization of a hotel room. Moreover, e-CRM technologies enableguests to make online bookings and transactions easier as well as being able to customise their hotel experience. Guests can personalize room service features to their profile or preferences. This will lead to guests' satisfaction. E-CRMenables automatic sending of emails highly personalized for each guest. The following examples include online reservation confirmation emails, pre-arrival emails (messages which remind guests about their reservation, thank you emails and guest satisfaction surveys).

9. **Website Security:** Detailed personal information and credit card information are necessary for online booking. This requires the security of customers' information and the security of transactions. Enterprises face the challenge of privacy concerns. Although customers are willing to exchange their personal information for discounts, room upgrades and free nights, they also worry about identity theft.
10. **Value-Added Services:** Nysveen and Lexhagen (2002) and Nysveen (2003) defined value-added services as "Website applications that help customers access different forms of information about the service provided".In hospitality and tourism sector, the Internet technology provides value added and customised services to be delivered. Internet technology allows travel search, comparison, and booking.
11. **Perceived Trust**: Online transactions (including online booking) can be associated with perceived risk. For the web security, there must be tools of trust such as digital certificatessigned by a trusted third party and secure servers.

12. **Price Attractiveness**: E-tailers (Retailers that primarily use the Internet for services provided) such as Priceline and Lastminute allow consumers compare prices of hotel rooms, tours and activities, rental cars, flights and many other travel experiences.

MAIN FOCUS OF THE CHAPTER

The Benefits of Electronic Customer Relationship Management in Tourism

The implementation of e-CRM will involve changes in operations of enterprises and result in an improvement in its performance and long-term, sustainable competitiveness.

An example of an enterprise that benefits from e-CRM can be given as Pegasus Airlines. It is a low-cost airline company operating in Turkey. Pegasus Airlines is benefiting form using e-CRM technologies. The enterprise is continually working to meet passenger expectations. For ticket pricing, a system is complemented by electronic ticket policy whereby passengers receive ticket information via e-mail and/or SMS. Moreover, Pegasus Airlines has developed a credit/loyalty card that offers passengers various benefits such as insurance rate reductions. The enterprise receives awards and recognition for initiatives in website development. Pegasus Airlines has social network relations with its customers. To illustrate, the company offers customers special competition through its Twitter account. Pegasus Airlines has a Facebook page completes with a game entitling customers have chance to win free flights tickets. Furthermore, the enterprise developed a campaign called "mobile phone fly" whereby customers can receive discounts. Pegasus Airlines aims to build sustainable relationship with its customers placing emphasis on social Networks, e-CRM activities and digital Technologies. The enterprise is aware of the fact that customers are the most important aspect of any service industry (Kotler &Armstrong, 2012).

E-CRM provides various benefits; some of them can be given as:

1. **Increased Customer Loyalty:** To gain a competitive advantage, hotel top management should add value to their services on hotel websites. Providing features such as reservation modifications, tracking transaction history, web page personalization will result in better customer service and increase customer loyalty (Hamid, Cheng, & Akhir, 2011).
2. **Efficient Marketing:** E-CRM allows analyzing customer data from various point of views to know which marketing campaign element is suitable for sales and profitability (Greenberg, 2001). E-CRM provides data base of historical records of interactions with each guest. E-CRM applications enhance sales and profitability by customer profiling and e-mail marketing.
3. **Ease of Doing Operations:** The workflow of an organization can be improved with the integration of e-CRM applications with back-office systems. Integrating customer data into a single database which enablesusers to share the information (Scullin, Allora, Lloyd, & Fjermestad, 2002) using e-mail across departmentshelps information distribution (Awad, 2000).

To illustrate, Marriott International set up a ''Codification System'' to convert what all hotel employees informed about Hotel day-to-day operations and standard operational procedures. Moreover, Marriott International created a reward system for employees who shared, created and mobilized new knowledge relevant to the Hotel (Gupta &Govindarajan, 2000). Marriott International believes that if employees are satisfied while doing their work, they will provide the best service to the guests. This will increase customer retention rates.

4. **Efficiency and Cost Reduction:** e-CRM Technologies simplify the process of managing transactions. E-CRM has significantly reduced the cost of data storage. Some examples of cost reduction in internal operations include automated routing, automated reporting and using auto responses. Using e-mails across departments reduces cost of doing business. The application of e-CRM in tourism industry will reduce internal costs and marketing costs.
5. **Greater Customer Support:** E-CRM provides a customer support with improved response time through auto responses, live connections, e-mails and multi-language support. Using chat enables enterprises to answer more than one customer at once. For phone and email support, enterprises can only respond to a single customer at once, but with live chat, they respond to many customers at a time.

The Perceived Barriers to Adoption of Electronic Customer Relationship Management Technonogies

Although E-CRM has several advantages, there are some costs and risks. In the highly competitive environment, enterprises are trying to gain strategic advantage through their customer relations, but there are several factors that make enterprises successful or unsuccessful after implementing E-CRM strategies.

E-CRM implementation is expensive and time consuming (Rahimi, Nadda, &Wang, 2015), enterprises have to overcome the perceived barriers to adoption of e-CRM Technologies. Factors listed below are considered as common reasons for e-CRM failure:

1. *Lack of skilful human resources* (Almotairi,2009): Untrained employees will not be able to provide personalized services to customers.
2. *Absence of management support* (Freeland, 2002; Almotairi, 2009)
3. *Absence of integration with other systems* (Ward, 2001; Amstel, Eijk, Haasdijk,& Kuilman, 2000) Lack of well defined strategies and objectives (Chen &Popovich, 2003; Greenberg, 2002): Company strategies should follow current trends in the tourist market in order to communicate with tourists. The incomprehension of the e-CRM system can be one of the key reasons for failure.
4. *Absence of cost-benefit analysis for e-CRM implementation* (Goodhue, Wixom, &Watson, 2002)
5. *Absence of appropriate IT infrastructure* (Chen &Chen, 2004): Transactions involving the use of word processor, spreadsheet, powerpoint and various hotel softwares for specific needs make hotel operations easier. Moreover, guests may expect to use computers for their business purposes and the Internet purposes. Therefore the right ICT equipments and facilities are important to meet customer needs and to makethem satisfied (Asabere &Doku, 2013).
6. *Poor customer-centric orientation* (Almotairi, 2009): In his book *Why CRM Doesn't Work: How to Win by Letting Customers Manage the Relationship* (Newell & Seth, 2003) Newell discussed how many CRM projectsget failed because the idea of customer-centricity contradicts with the way many enterprises work.
7. *Lack of intimacy with the customer* (Kincaid, 2003): Intimacy between existing and/or potential customer does not exist online environment. Online environment may seem unfamiliar and confus-

ing for many people. Some people want a real person to answer their questions. Human aspects of face-to-face interaction will always appreciated.

8. *Lack of trust* (Almotairi, 2009): Trust is an essential factor for e-CRM. For a business transaction, trust must exist between company and customer. From a customer point of view, lack of trust in terms of safety and security of information provided online may act as a barrier for e-CRM. Although online shopping has grown rapidly, many people are worried about sharing their personal information over the Internet. In a survey, 75 percent of participants claimed that they do not want expressing personal and credit card information over the Internet (www.searchengineland.com). In the absence of trust, it becomes very hard to get customers to share the data that is needed for the creation of an effective e-CRM strategy. Therefore, privacy policies and guarantees are crucial issues for building trust.
9. *Lack of financial support* (Constantelou, 2002): Financial investment for the introduction of e-CRM Technologies is a barrier for tourism related companies usually operating with low margins.

Electronic Customer Relationship Management Implementation in Tourism

The use of Internet Technologies is rapidly increasing in all areas of the enterprises and this situation provides various benefits to institutions and individuals. The development of web-based technology encourages enterprises to start e-CRM implementation. (Pan and Lee, 2003). Before starting an e-CRM Project, it is very important to understand and analyze the organization's culture, objectives, responsibilities and its strategies. An organizational culture is a pattern of shared values and beliefs which help people understand organizational functioning and so provide them with the norms for behavior in an organization (Hurcombe, 1998). It is the key element for managing an e-CRM strategy implementation. A customer-orientated organization culture makes enterprises more precise to customer needs. The strategy of a customer-oriented enterprise should also be a customer-focused strategy. Moreover, defining where the organization is, where it wants to go and where it is actually going are also very important (Chalmeta, 2006).

Once the organization's objectives, responsibilities and strategies are analyzed, training of employees should occur before the implementation (Grover, 2011). For example, sending employees (who require access to the new e-CRM system) to training facilities or bringing in an on-site consultant can be training options. Training should be an ongoing activity as systems must continuously change. E-CRM strategy has to include training programs for employees.

Tourism organizations are implementing e-CRM practices to be ahead of the expectations of tourists' and to gain competitive advantage in tourism market. E-CRM analysts develop *data warehouses* and *data mining* techniques which are the most popular systems for providing e-CRM capabilities (Kotler &Armstrong, 2012).A data warehouse is a company-wide electronic data base that provides managers with up-to-date and historical decision support information not easy to access or present in traditional operational data stores (Hendawi & El-Shishiny, 2008). A data warehouse is a computerised data base system whereby the gathered information about customers from different operational systems, call centers and similar sources are pulled together into an accessible location. Enterprises use data warehouses for the purpose of developing long relationship withthe loyal customers. Once the data warehouse brings the data together, tourism enterprises can use data mining techniques to select interesting information about customers (Kotler &Armstrong, 2012). Capturing the right data from identified sources is important for successful implementation of e-CRM.

With e-CRM implementation, organizations often reach positive outcomes, such as positive word of mouth, referrals and loyalty (Kim &Cha, 2002). The need for effective e-CRM implementation is becoming important in the electronic business environment since customer satisfaction and loyalty are hard to establish in competitive nature of the hotel industry market (Cherapanukorn, 2017). As a consequence, understanding the dynamics of e-CRM and assessing its impact on organizations' outcomes are very important, mostly in service industries (Kımıloğlu &Zaralı, 2008).

Within the hotel industry, Sigala, Lockwood & Jones, (2001) highlighted the need tointegrate yield management (YM), customer databases, corporate and distributionsystems in order to maximise yield per each customer during his/her business lifetime, per distribution channel and/or per hotel chain or properties-network. To illustrate, LeMeridien integrated their yield management and e-CRM systems for personalisingtheir rates, services and promotions to their frequent customers. Additionally, Wells, William & Choobineh (1999) explained how electronic networks enable the distribution of guestdatabases across Ritz Carlton hotel properties allowing employees to accommodate andtreat individual guests based on preferences from their previous stays.

Moreover, in order to successfully compete and achieve some online processes (e.g. online personalization of a hotel room), e-CRM practices should be effectively integrated with offline practices (e.g. disseminate information to housekeeping for making necessary changes). There should be an integration of e-CRM and offline hotel operations (Sigala, 2006).

Electronic Customer Relationship Management Technologies Used in Tourism

Nowadays, more and more people are using the Internet to plan their travel. People are using search engines to be informed about destinations. Choosing the right hardware, software and systems from the range of technology options can be a challenging task for many businesses. Whatever the option or the combination of options is chosen, the main principle is that the tecnological infrastructure should create a control center, integrating different customer data into customer interactions that create superior customer experiences (Payne & Frow, 2005). E-CRM technologies have to be integrated with legacy systems.

E-CRM Technologies used in tourism are listed below:

Website Features

Along with the developments in information technologies, the widespread use of the Internet has also affected the method of doing business. Enterprises have started to use organizational websites by using information technologies to protect and increase their market share. Nowadays, customers prefer to engage in hotels, airline companies, boat charters, tour operators and many other tourism related businesses by visiting their websites. They search for service information and make the buying decision. The quality of the interaction between customers and service providers during this level is crucial because customers judge the services during this level. Therefore, an enterprise's website features are very important for achieving success.

Tourism enterprises should take advantage of the Internet techonology and organizational websites in their advertising and promotion activities. Thus, sustainable competitive advantage can be achieved and promotion and marketing activities of tourism products can be made more effectively.

The website features of a typical e-CRM organization are identified and classified into three areas as; general website features, electronic trade features and after sales support features (Collier &Meyer, 1998; Feinberg, R. A., Kadam, R., Hokama, L., & Kim, I., 2002).

General Website Features

An organizational website should contain an introduction page where visitors can get the information about how to use the website more efficiently.

Moreover, website personalization is very important for an organization that uses e-CRM Technologies. Websites that have the personalization feature allow people to filter the content. E-CRM help users to personalize on their first visit, so for the next entry, users can see only his/her own preferences. Website personalization plays an important role in creating loyal customers and managing customer relations. Background color selection, adjusting screen resolution, turning on and off the music (if available) are examples of personalization features. Guests can personalize their profile according to their preferences and needs. They have the chance to design their hotel experince in advance (e.g. ByRequest programme Wyndham provides an opportunity for selection of pillow, welcome drink, preferable newspapers delivery, etc.)

General website features consist of various ways for customers to contact the organization such as sending an e-mail, using fax, calling toll-free numbers. Or, when a customer pushes the call-back button in the website, leave user comments, recommedations and requests to electronic bulletin board supported by website, then, hotel or travel agency has to get back to the customer.

Furthermore, when the visitor searches on key words, he/she wants to reach quickly for the specific information on website that he/she is looking for.

Membership feature is very important in e-CRM. Membership benefits make members feel happy of being part of a group of customers that receive special treatment and services provided by the enterprise (Rosenbaum et al., 2005). Membership allows customers to request a password. With this password he/she can surf on password protected web pages within the website. For example, The Museum of Modern Art (MoMA) in New York is operating a customer database, a membership program and an awardwinning website. MoMA offers membership benefits for each category of membership. Furthermore, MoMA has introduced barcoded swipe cards, which register data on attendances – how often, to what event, cost, etc. Members have chance to enjoy exclusive benefits and get invitations to events, including special discounts at MoMA Stores (www.moma.org).

If the visitor wants to be informed about the services, special offers, discounts, etc., he/she can receive automated e-mails from the enterprise. This feature helps enterprises to know their potential customers. Additionaly, a site map and a site tour features are also important.

Another feature of an organizational website is visitors' interaction with each other. The web chat feature allows a visitor to chat withanother visitor or customer representatives.Consumers also use web sites such as GetSatisfaction.com, TripAdvisor.com, Complaints.com to ask questions, know about complaints and offer suggestions. In such virtual communities, customers advise other people on restaurant, hotels, tour operators, features and services.

Electronic Trade Features

Nowadays, e-commerce is the most frequently used form of trade. The reason is that both the cost of advertising and transportation is cheaper and it is easy to serve many users at the same time. In addition, it provides 24/7 shopping opportunities. All this features have led many businesses to turn to e-commerce.

E-trade features of an e-CRM website are essential for businesses because these features help enterprises to change shopping customers into loyal customers. One of these features is online purchasing. With the occurrence of marketing in the Internet environment, restruction and evaluation of the marketing mix elements (product, price, place and promotion) has become obligatory and benefits of marketing on the Internet have brought a new shopping environment for customers (Özgüven, 2011).

The Internet as a new marketing and sales channel, offers a different shopping environment to consumers. They can buy all sort of tourism products such as museum tickets, flight tickets, hotel rooms, rental cars etc. online. Online bookings are often much faster and cheaper when compared to traditional offline bookings. Having a website integrated with online reservation technology available 24/7 allows customers to purchase services at anytime they want. Hotels, restaurants, airline companies, tour operators and boat charters can increase reservations with online reservation system. Nowadays, more and more people are turning to their mobile phones to search for travel options and online bookings. For this reason, it is very important for hotels and airline companies to optimize their websites for mobile usage and mobile transactions. Many airline companies and hotels use online reservation applications. In the online environment, customers can purchase services by comparing with competitors' prices and getting recommendations and ideas fromother people. Moreover, websites should enable customers to choose payment method that they prefer such as credit card or pay later (in the hotel).

Once they book their flight or hotel room, they can customize their service online before ordering if this feature is enabled in the website. For instance, Pegasus offers customer service options such as a discount to passengers on international flights who order their in-flight meals online in advance or choosing online from a range of refreshments and catering options offered by Pegasus Flying Café.

The purchase can be seen online in an e-CRM web site. Purchase conditions are shipping policies, return policies, warranty, guarantee, information about taxes and fees and other organization commitments.

Many tourism products or destinations can be viewed before purchase. Tourists have the chance to receive online service information. Moreover, the external Linkage feature is very important for a website. This feature helps a customer easily be linked to the external links of complementary products from other enterprises. Many airline companies' websites have this feature. Customers are linked to car rental websites in order to rent a car online, they are linked to hotel websites and other websites that they can learn about the activities in the destination to where they will travel.

The ease of shopping, care and service are a combination of an e-CRM website features aiming to increase customer service, care and online transactions (Sigala, 2006).

After Sales Support Features

From a consumer's perspective, in the post-sales stage, a consumer assesses the service or travel experience as satisfying or not. From the enterprise perspective, the enterprise wants to know whether the customer is satisfied with the service he/she had received, his/her revisit intention andthe customer will tell others about his/her experiences with the service.

For tourism related businesses to have an e-CRM organizational website with post-purchase features enhance customer satisfaction. These businesses should be aware of the fact that the importance of the Internet during the stage of post-purchase (e.g., complaints, suggestions, feedback).

Frequent asked questions (FAQs) which is an important post-sales support feature act as a self-help for customers looking for answers to their questions. For example, in Turkish Airlines Website there is FAQs secition where passengers can find answers about their questions about check-in, excess baggage, fare rules, free baggage, cabin baggage, visa and travel requirements etc. (www.turkishairlines.com)

But the Internet has somehow changed this phenomenon.
Complaining through Internet is faster, easier, and has larger scale of audience.
(Harrison-Walker, 2011).

Customers complain when their expectations are under-performed degree that falls outside their zone of tolerance, or when they think that they have been treated unfairly (Buttle, 2009). Some research results show that 60 (Andreasen, 2000) percent of consumers and 70 (Goodman, 1999) percent of customers who are not satisfied do nothing. In addition, research shows that enterprises do not receive any information from 96 percent of customers who are not satisfied with their products and services (Plymire, 1991). Instead of complaining to hotel management, customers choose not to come again or telling about their negative travel experiences to their family and friends. However, this situation has somehow changed with the Internet. Enterprises are aware of the fact that one single complaint of a customer may affect hundreds of potential customers. Tourism-related businesses are aware of the fact that the internet has changed tourists' behavior and as a result, in some websites, there is a specific area that allows customers to write their complaints about their service experience.

Electronic Customer Relationship Management Success Factors

In today's competitive market place, for being successful in highly competitive tourism market, e-CRM success is very important. Some of the most important e-CRM success factors can be ranked (figure 4) in below section (Almotairi, 2009):

1. **Top Management Team Support:** It plays an important role in e-CRM implementation and financing e-CRM Technologies. Board members support is a key requirement for successful e-CRM.
2. **A Clearly Defined e-CRM Strategy:** A successful e-CRM strategy should focus on the improvement of customer satisfaction, efficient processesin transactions, brand image, effective database management and latest technology usage (Kımıloğlu & Zaralı, 2008). A well-defined e-CRM strategy should becompatible with the enterprise's vision and mision. An enterprise should focus on a customer-oriented approach for e-CRM success.
3. **Cultural Change:** Tourism-related organizations should be ready for technological inovations. There might be a need for a cultural change to implement the latest marketing trends and develop a customer-oriented organization. A friendly culture for knowledge sharing is very important. CRM success needs to realize its objectives and organizations should develop a culture where all staffs are encouraged to share and learn from the new work structure and information (Alexander, 2004).
4. **Information Technologies infrastructure:** E-CRM need to be considered not only as a technological solution but also as a strategy and IT solution.Supporting e-CRM with the right Technologies is

very important. An example from hospitality business is explained below how well they managed IT infrastructure:

With the emergence of database marketing, Radisson Hotels and Resorts is aware of the fact that they need to take a different marketing approach. New customer touch points like a website and customer loyalty programs are some examples of this marketing trends.After analyzing marketing trends, Radisson had successful infrastructural efforts. First successful effort was to move all Radisson franchisees to a common suite of applications for reservations, operations, and guest tracking (from 1994 to 1997). Moreover, there was an enterprise data model and data dictionary development in 1996. Furthermore, Radisson included the central reservation system in 1999. By the end of the year 1999, Radisson integrated data infrastructure. After that, "Look to Book" online loyalty program an incentive program aimed at travel agents was introduced. It enables the marketing department to create promotions targeted at travel agents, encouraging them to book their clients in Radisson hotels. The standardized reservation system plays an important role in this program (Goodhue et al., 2002).

5. **Inter-Departmental Integration**: E-CRM needs all departments involvement. The importance of interdepartmental corporation need to be considered in e-CRM success. Especially, marketing and IT mostly work together for guest transactions and marketing campaigns.
6. **Skillful Staff:** Employees who requires access to the new e-CRM system should be well trained. Training is an ongoing activity whichis crucial in the implementation process. Employees should be trained before the implementation of the new e-CRM system. They should be encouraged to exchange their knowledge. Examples for training include sending employees to training facilities or bringing in an on-site consultant (Grover, 2011).

Moreover, since the idea of creating ideal guest experiences only through e-CRM systems is usually impossible, tourism organizations need to create intelligent-response teams consisting of skillful employees. For dealing with guests' needs and desires, employees will be required to have social skills such as communications and personality identification (Olsen &Connolly, 2000). Showing empathy and responding are very important for delivering excellent experience in interaction with guests.

7. **Key Information on Customers:** To increase business profitabilits and guest loyalty, hotels should give importance to the implementation of e-CRM strategies which target to seek, gather and store the right information on customers, validate and distributeit throughout the organisation. Then the information is used for creating customized, unique guests' experiences (Sigala & Connolly, 2004; Olsen & Connolly, 2000; Siguaw & Enz, 1999).

Top Management Support / Commitment: The role of board level is essential in backing the CRM implementation process and securing required amount of financing for putting CRM projects into action.

- **Defined CRM Strategy:** A clear definition of the CRM strategy and alignment of this strategy to the company's strategy would facilitate the transition of changing work structure and environment toward customer-centric approach Top Management Support / Commitment: The role of board level is essential in backing the CRM implementation process and securing required amount of financing for putting CRM projects into action.

The Reasons of Electronic Customer Relationship Management Failures

Although e-CRM systems are implemented with the purpose of revenue growth and providing a better customer service, many organizations get failed because of some main reasons. The main reasons for tourism enterprises end in failure can be:

1. **Improper Use of Information Technologies:** An e-CRM strategy might fail if the information technology (IT) is not used properly. For this reason, appropriate use and adoption of hardware and software in e-CRM activities is one of the greatest opportunities in the hospitality business that can lower a failure implementation in e-CRM strategy (Sin, Tse, & Yim, 2005).
2. **Incorrectly Defined Strategy:** For quickly implementing e-CRM, enterprises usually start investing on e-CRM Project before developing a clear strategy that dictates how e-CRM will address marketing, sales, and services issues. To justify the financial support required to successfully implement an e-CRM system, top management should identify and define a business strategy to see positive returns for their investment (Scullin et al., 2002). E-CRM requires a combination of interdepartmentally clarified strategies, programs and processes (Kale, 2004).
3. **Mismatch Between an Organization and the Vendor's Software:** A vendor should sell a productflexible enough to match the organization's best practices. A single software solution cannot meet all e-CRM requirements equally well. For this reason, each organization must find a solution that best concerns with the critical customer-facing functions (Scullin et al., 2002).
4. **Rejection by Final User Employees:** happens when business functions are retooled. In order a new e-CRM system not to end in failure, new processes needed should be developed with the knowledge, help and acceptance of the end users (Scullin et al., 2002).
5. **Orgnizational Change:** Underestimating the organizational changes involved in becoming an organization with customer-centricity approachis one of the majorchallenge of the implemental failure of e-CRM. Organizational changes are needed in the adoption of new e-CRM system (Kale, 2004).

An example can be presented from Radisson Hotels and Resorts that have faced some data quality problems arising from poor data entry practices at franchises. Improper practices of franchises, such as appending clerk initials to a reservation code, or accepting defaults, such as using the hotel's own address in guest records were common. Consequently, director of CRM at Radisson Hotels and Resorts has recognized the need to change the corporate mindset which has not been easy because some employees are willing and able to make the shift, while others are not. Therefore, IT had to train final users in franchises about data entry practices (Goodhue et al., 2002).

6. **Size of Project:** Some e-CRM implementations condemned to fail if their scope is too broad. Beginning with a pilot e-CRM implementation will reduce the risk of failure and it gives the enterprise the opportunity to evaluate the positives and negatives of the pilot as the enterprise plans for a larger scale implementation (Scullin et al., 2002).
7. **Considering e-CRM as a One-Time Event:** Enterprises sometimes underestimate that e-CRM is as an evolutionary development for their organization that will require repetitions to implement. E-CRM is not a one-time event, strategy should be refined, and plans should be developed to achieve new e-CRM goals.

SOLUTIONS AND RECOMMENDATIONS

With globalization, competition has increased rapidly. In this environment, in order to meet the competitive challenges, enterprises need to focus on customer relations, customer satisfaction and customer loyalty. In todays changing marketing environment, it is very crucial for enterprises to improve and maintain mutually beneficial relationships with stakeholders and customers, especially in the tourism sector. The Internet technology has changed how enterprises manage relationships with their customers. Through the use of Information Technologies, enterprises have redesigned their CRM. Enterprises implement e-CRM to better understand their customers, to create and manage long relationships with them. E-CRM is an effective marketing approach that focuses on web-based interactions between enterprises and their customers. With the widespread use of the Internet, tourism enterprises give more importance to undertake marketing activities over the Internet.

Tourism enterprises need to consider the internet as a strategic tool in e-CRM. It is recommended that enterprises operating in tourism and hospitality industry should keep up to date with new Informaiton Technologies. To offer better services to their customers/guests, hotels and other tourism enterprises have to take advantage of new technologies in their operations and marketing activities. Implementing e-CRM strategies help them to satisfy and retain customers and also to increase their profitability. Tourism is a dynamic and competitive industry that requires tourism enterprises to be innovative and creative with their campaigns, they have to aim to stick customers' /guests' minds and encourage customer interaction. Tourism organizations have to be aware of the fact that technology can only act as an enabler in managing customer relationship successfully, to be successful in e-CRM there must be involvement of people and processes. Concerning a tourism organization, with regard to the corporate culture, the enterprise should be customer-centric to provide the best service (Ivanovska, 2008). The importance of social media requires tourism enterprises to turn to mobile advertisements and other customer acquisition campaigns. Enterprises can benefit from a variety of Search Engine Optimization (seo) methods to increase customer loyalty. When they do seo studies for their websites, they will be able to get high rankings in the search engines andthey can meet customer expectations with their websites. As a result, the customer retention rate will increase so that their profit will increase. As a conclusion, it is recommended that tourism organizations should implement e-CRM strategies/tools to satisfy and retain customers, create loyal customers and also increase profits.

FUTURE RESEARCH DIRECTIONS

With the development of the Internet and the continuous improvement of technology, the concept of e-CRM in the tourism industry needs further consideration. Future researchs can bemade by collecting data through interviews with CRM managers and hotel customers in order to understand how the Internet is used as a marketing tool in e-CRM in the tourism industry. Most of dissertations made on the subject of CRM. However, there are limited dissertations in the field of e-CRM in tourism. Researcher can make further emprical studies on this subject. Other areas which are closely connected to this subject such as Management Information Systems (MIS) and E- tourism could be studied.

CONCLUSION

E-CRM Technologies used by tourism organizations are e-mail, instant messaging and websites. General website features, e-trade website features and post-sales website features are the website features of a typical e-CRM organization (Collier & Meyer, 1998; Feinberg, R. A., Kadam, R., Hokama, L., & Kim, I., 2002). Tourism organizations offer highly personalized content on their websites. Customers/guests can interact with other people (other customers or customer representatives) over the web chat. Moreover, there are websites that customers can make comments about their experiences about hotels, restaurants, destinations and many other tourism related businesses. They enter these websites to ask questions, make complaints and offer suggestions about destinations, restaurants, tour operators and/or hotels. Since all kinds of tourism products such as museum tickets, flight tickets are sold online, tourism enterprises have also started to sell services on their websites. Online bookings are much faster and cheaper. Many tourism enterprises offer customization feature. Once a customer book his/her flight or hotel room, he/she can customize the service online before ordering. Shopping convenience, customer service, care and quality are the features that an e-CRM website should include. (Sigala, 2006).

Implementation of e-CRM is very important for tourism organizations to be ahead of the expectations of tourists' and to gain competitive advantage in tourism market. The interactivity of Internet technology and e-CRM technologies brings advantages and challenges for tourism enterprises aiming to build customer relationships. Some of the main advantages are ease of doing operations, increased customer loyalty, efficient marketing and cost reduction. However, in order to succesfully implement e-CRM, there have to be some necessities such as top management support, a well-defined strategy, appropriate IT infrastructure and trained employees. Although e-CRM systems are implemented with the aim of revenue growth, many compaines condemned to fail because of some main reasons such asimproper use of IT, incorrectly defined strategy, employees rejections of new e-CRM system and lack of organizational changes.

REFERENCES

Ab Hamid, N. R., & McGrath, G. M. (2005). The Diffusion of Internet's Interactivity on E-tail Web Sites: A Customer Relationship Model. *Communications of the International Information Management Association*, *2*, 45–70.

Ahmad, I., & Chowdhury, A. R. (2008). *Electronic Customer Relationship Management (eCRM) - customers' perception of value from eCRM features on airline e-ticketing Websites.* (Master thesis). Available from Diva-portal student theses database. (1020631)

Almotairi, M. (2009). A Framework for Successful CRM Implementation. In *Proceedings of the European and Mediterranean Conference on Information Systems 2009.* İzmir, Turkey.

Amstel, P., Eijk, P., Haasdijk, E., & Kuilman, D. (2000). An interchange format for cross-media personalized publishing. *Computer Networks*, *33*(1-6), 179–195. doi:10.1016/S1389-1286(00)00049-9

Anderson, R. E., & Srinivasan, S. S. (2003). E-Satisfaction and E-Loyalty: A contingency framework. *Psychology and Marketing*, *20*(2), 123–138. doi:10.1002/mar.10063

Andreassen, T. W. (2000). Antecedents to satisfaction with service recovery. *European Journal of Marketing*, *34*(1/2), 156–175. doi:10.1108/03090560010306269

Asabere, N. Y., & Doku, V. (2013). Measuring Customer Relationship Management (CRM) in the Hospitality Industry of Some Selected Hotels in Accra, Ghana: The Role of Information and Communication Technologies (ICTs). [IJAIEM]. *International Journal of Application or Innovation in Engineering & Management*, *2*(3), 19–28.

Awad, E. M. (2000). The structure of e-commerce in the banking industry. In *Proceedings of the 2000 ACM SIGCPR Conference on Computer Personnel Research.* Chicago, IL. New York: ACM. 10.1145/333334.333375

Buhalis, D. (1998). Strategic use of information technologies in the tourism industry. *Tourism Management*, *19*(5), 409–421. doi:10.1016/S0261-5177(98)00038-7

Buttle, F. (2009). *Customer Relationship Management Concepts and Technologies* (2nd ed.). Burlington, VT: Elsevier.

Caber, M. (2010). *Tur Operatörü-Seyahat Acentesi Elektronik Müşteri İlişkileri Yönetiminde Web Sitesi Kalitesinin Memnuniyet Ve Güven Üzerine Etkisi.* (Doctoral dissertation). Available from Yükseköğretim Kurulu Ulusal Tez Merkezi. (250541)

Chalmeta, R. (2006). Methodology for customer relationship management. *Journal of Systems and Software*, *79*(7), 1015–1024. doi:10.1016/j.jss.2005.10.018

Chang, C. K. (2010). Acceptability of an asynchronous learning forum on mobile devices. *Behaviour & Information Technology*, *29*(1), 23–33. doi:10.1080/01449290701806337

Chen, I. J., & Popovich, K. (2003). Understanding customer relationship management (CRM): People, process and technology. *Business Process Management Journal*, *9*(5), 672–688. doi:10.1108/14637150310496758

Chen, Q., & Chen, H. (2004). Exploring the success factors of eCRM strategies in practice. *Journal of Database Marketing and Customer Strategy Management*, *11*(4), 333–343. doi:10.1057/palgrave.dbm.3240232

Chen, R. H., Lin, R. J., & Yang, P. C. (2011). The relationships between eCRM, innovation, and customer value-An empirical study. In *Proceedings 2011 IEEE International Summer Conference of Asia Pacific Business Innovation and Technology Management (APBITM 2011).* Dalian, China: IEEE Press. 10.1109/APBITM.2011.5996343

Cheng, J., Chen, F., & Chang, Y. (2008). Airline relationship quality: An examination of Taiwanese passengers. *Tourism Management*, *29*(3), 487–499. doi:10.1016/j.tourman.2007.05.015

Cherapanukorn, V. (2017). Development of eCRM Success: A Case Study of Hotel Industry. *International Journal of Trade, Economics, and Finance*, *8*(2), 90–95.

Collier, D. A., & Meyer, S. M. (1998). A service positioning matrix. *International Journal of Operations & Production Management*, *18*(12), 1223–1244. doi:10.1108/01443579810236647

Constantelou, A. (2002). Emerging Trends in Customer Relation Management Using ICT: the Travel Industry, Issue Report N. 22, STAR (SocioEconomic Trends Assessment for the DigitalRevolution) Retrieved May 10, 2019, from http://www.cerna.ensmp.fr/Enseignement/CoursModelesEcoNumerique/TD/4-Constantelou.pd

Croteau, A. M., & Li, P. (2003). Critical Success Factor at CRM Technological Initiative. *Canadian Journal of Administrative Sciences*, *20*(1), 21–30. doi:10.1111/j.1936-4490.2003.tb00303.x

Dhillon, J. S., Joshi, M., & Verma, R. (2012). Indian Hospitality Industry: Moving Towards Customer Oriented Information System (COIS). *Journal of Business Management & Social Sciences Research (JBM&SSR)*, *1*(1), 58-69.

Dubihlela, J., & Khosa, P. M. (2014). Impact of e-CRM Implementation on Customer Loyalty, Customer Retention and Customer Profitability for Hoteliers along the Vaal Meander of South Africa. *Mediterranean Journal of Social Sciences*, *5*(16), 175–183.

Erkiah, S., & Ladkoo, A. D. (2018). A Quantitative Study about Assessing the Effectiveness of Electronic Customer Relationship Management: A Case of Two Hotels in Mauritius. *International Journal of Social and Business Sciences*, *12*(10), 1438–1446.

Feinberg, R., & Kadam, R. (2002). E-CRM Web Service Attributes as Determinants of Customer Satisfaction with Retail Web Sites. *International Journal of Service Industry Management*, *13*(5), 432–451. doi:10.1108/09564230210447922

Feinberg, R. A., Kadam, R., Hokama, L., & Kim, I. (2002). The state of electronic customer relationship management in retailing. *International Journal of Retail & Distribution Management*, *30*(10), 470–481. doi:10.1108/09590550210445344

Freeland, J. (2002). *The Ultimate CRM Handbook: Strategies and concepts for building enduring customer loyalty and profitability*. New York: McGraw-Hill.

Garrido-Moreno, A., & Padilla-Meléndez, A. (2011). Analyzing the impact of knowledge management on CRM success: The mediating effects of organizational factors. *International Journal of Information Management*, *31*(5), 437–444. doi:10.1016/j.ijinfomgt.2011.01.002

GilaniNia, S., Balaei, S., & Niyari, F. S. (2013). Impact of ECRM on the Create Competitive Advantage in Hotels of Sarein. *Arabian Journal of Business and Management Review*, *3*(4), 77–85.

Goodhue, D. L., Wixom, B. H., & Watson, H. J. (2002). Realizing business benefits through CRM: Hitting the right target in the right way. *MIS Quarterly Executive*, *1*(2), 79–94.

Goodman, J. (1999). Basic facts on customer complaint behavior and the impact of service on the bottom line. *Competitive Advantage*, *8*(1), 1–5.

Greenberg, P. (2001). Capturing and Keeping Customers in Internet Real Time. *The Mcgraw-Hill Companies*. Retrieved May 10, 2019, from https://www.amazon.com/CRM-Speed-Light-Capturing-Customers/dp/0072127821

Greenberg, P. (2002). *CRM at the Speed of Light*. New York: Mcgraw-Hill Companies.

Grover, D. (2011). Effective Customer Relationship Management through e-CRM. *Viewpoint, 2*(1), 27–38.

Güleri, T. (2000). CRM throughout the enterprise. *Call center solutions, 18*(12), 44-49.

Gupta, A., & Govindarajan, V. (2000). Knowledge Management's Social Dimension: Lessons From Nucor Steel. *Sloan Management Review, 42*(1), 71–80.

Hamid, N. R., Cheng, A. Y., & Akhir, R. M. (2011). Dimensions of E-CRM: An Empirical Study on Hotels' Web Sites. Retrieved May 2, 2019, from IBIMA Publishing: http://www.ibimapublishing.com/journals/JSAR/2011/820820/820820.pdf

Hendawi, A. M., & El-Shishiny, H. (2008). A Data Warehouse Prototype for the Tourism Industry: A Case Study from Egypt. Retrieved May 9, 2019, from https://www.academia.edu/17324396/A_Data_Warehouse_Prototype_for_the_Tourism_Industry_A_Case_Study_from_Egypt

Hurcombe, J. (1998). *Developing strategic customers and key accounts*. Bedford, PA: Policy Publications.

Ip, C., Leung, R., & Law, R. (2010). Progress and Development of Information and Communication Tehnologies in Hospitality. *International Journal of Contemporary Hospitality Management, 23*(4), 533–551. doi:10.1108/09596111111130029

Ivanovska, L. P. (2008). E-CRM in the Tourism Sector. *Revista de Turism, 5*, 14–19.

Kale, S. H. (2004). CRM failure and the seven deadly sins. *Marketing Management, 13*(5), 42–46.

Kim, W. G., & Cha, Y. (2002). Antecedents and consequences of relationship quality in hotel industry. *International Journal of Hospitality Management, 21*(4), 321–338. doi:10.1016/S0278-4319(02)00011-7

Kımıloğlu, H., & Zaralı, H. (2008). What signifies success in e-CRM? *Marketing Intelligence & Planning, 27*(2), 246–267. doi:10.1108/02634500910945011

Kincaid, J. W. (2003). *Customer Relationship Management: Getting it Right!* New Jersey: Prentice-Hall.

Kotler, P., & Armstrong, G. (2012). *Principles of Marketing* (14th ed.). Harlow, UK: Pearson Education Limited.

Kotler, P., & Keller, K. (2006). *Marketing Management* (12th ed.). New Jersey: Pearson Prentice Hall.

Kourtesopoulou, A., Papaioannou, A., & Kriemadis, T. (Eds.). 2011. Exploring E-crm application: A case study in outdoor recreation sector. In *Proceedings of the 19th International Congress on Physical Education and Sport, ICPES.* Komotini, Greece. Academic Press.

Kumar, N., & Malik, K. (2013). E-CRM golden opportunity for banks. *International Journal of Multidisciplinary and Academic Research, 2*, 1–10.

Lastminute.com. Retrieved April 5, 2019 from https://www.lastminute.com

Lee-Kelley, L., Gilbert, D., & Mannicom, R. (2003). How e-CRM can enhance customer loyalty. *Marketing Intelligence & Planning, 21*(4), 239–248. doi:10.1108/02634500310480121

Marjani, M., Nasaruddin, F., Gani, A., Karim, A., Hashem, I. A. T., Sıddıqa, A., & Yaqoob, I. (2017). Big IoT Data Analytics: Architecture, Opportunities, and Open Research Challenges. *IEEE Access: Practical Innovations, Open Solutions, 5*, 5247–5261. doi:10.1109/ACCESS.2017.2689040

Muro, M., Magutu, P., & Getembe, K. (2013). The strategic benefits and challenges in the use of customer relationship management system among commercial banks in Kenya. *European Scientific Journal, 9*, 327–349.

Newell, F., & Seth, G. (2003). *Why CRM Doesn't Work: How to Win By Letting Customers Manage the Relationship*. Princeton, NJ: Bloomberg Press.

Nysveen, H. (2003). The importance of risk-reducing value-added services in online environments: An exploratory study on various segments of leisure tourism. *Information Technology & Tourism, 6*(2), 113–127. doi:10.3727/109830503773048228

Nysveen, H., & Lexhagen, M. (2002). Reduction of perceived risk through on-line value added services offered on tourism businesses websites. In K. W. Wöber, A. J. Frew, & M. Hitz (Eds.), *Proceedings of the ENTER 2002 Conference*. Wien, Austria: Springer-Verlag.

Olsen, M. D., & Connolly, D. J. (2000). Experience-based travel. How technology is changing the hospitality industry. *The Cornell Hotel and Restaurant Administration Quarterly, 41*(1), 30–40. doi:10.1177/001088040004100121

Özgüven, N. (2011). Tüketicilerin Online Alışverişe Karşı Tutumları ile Demografik Özellikleri Arasındaki İlişki Analizi. *Karamanoğlu Mehmetbey Üniversitesi Sosyal ve Ekonomik Araştırmalar Dergisi, 13*(21), 47–54.

Pan, S. L., & Lee, J. N. (2003). Using e-CRMfor a Unified View of the Customer. *Communications of the ACM, 46*(4), 95–99. doi:10.1145/641205.641212

Parvatiyar, A., & Sheth, J. N. (2001). Customer relationship management: Emerging practice, process, and discipline. *Journal of Economic and Social Research, 3*(2), 01-34.

Payne, A., & Frow, P. (2005). A Strategic Frameworkfor Customer Relationship Management. *Journal of Marketing, 69*(4), 167–176. doi:10.1509/jmkg.2005.69.4.167

Plymire, J. (1991). Complaints as opportunities. *Journal of Consumer Marketing, 8*(2), 39–43. doi:10.1108/07363769110035009

Rahimi, R., Nadda, V., & Wang, H. (2015). CRM in Tourism. In R. Nilanjan (Ed.), *Emerging Innovative Marketing Strategies in the Tourism Industry* (pp. 16–43). Hershey, PA: IGI Global. doi:10.4018/978-1-4666-8699-1.ch002

Rosenbaum, M., Ostrom, A., & Kuntze, R. (2005). Loyalty programs and a sense of community. *Journal of Services Marketing, 19*(4), 222–233. doi:10.1108/08876040510605253

Scullin, S., Allora, J., Lloyd, G. O., & Fjermestad, J. (2002). Electronic customer relationship management: benefits, considerations, pitfalls and trends. In *Proceedings of the IS One World Conference, 3-5*. Las Vegas, NV. Academic Press.

Sigala, M. (2006). E-Customer Relationship Management in the hotel sector: Guests' perceptions of perceived e-service quality levels. *Turizam*, *54*(4), 333–344.

Sigala, M. (2006). Culture: The Software of Customer Relationship Management. *Journal of Marketing Communications*, *12*(3), 203–223. doi:10.1080/13527260600811787

Sigala, M., & Connolly, D. (2004). In search of the next big thing: IT issues and trends facing the hospitality industry: A Review of the Sixth Annual Pan-EuropeanHospitality Technology Exhibition and Conference (EURHOTEC 2001). International Hotel and Restaurant Association, Feb. 19–21, 2001. Palais Des Congress, Paris, France. *Tourism Management*, *25*(6), 807–809. doi:10.1016/j.tourman.2003.08.009

Sigala, M., Lockwood, A., & Jones, P. (2001). Gaining advantage from yield management: Strategic implementation in the rapidly developing world of IT. *International Journal of Contemporary Hospitality Management*, *17*(3), 364–377. doi:10.1108/09596110110403956

Siguaw, J. A., & Enz, C. A. (1999). Best practices in marketing. *The Cornell Hotel and Restaurant Administration Quarterly*, *40*(5), 31–43. doi:10.1177/001088049904000507

Sin, L. Y. M., Tse, A. C. B., & Yim, F. H. K. (2005). CRM: Conceptualization and scale development. *European Journal of Marketing*, *39*(11/12), 1264–1290. doi:10.1108/03090560510623253

Sterling, G. (2008). Pew: Americans Increasingly Shop Online But Still Fear Identity Theft. Retrieved April 12, 2019, from https://searchengineland.com/pew-americans-increasingly-shop-online-but-still-fear-identity-theft-13366

The Museum of Modern Art MOMA. Retrieved May 9, 2019 from https://www.moma.org/support/membership/

Turkish Airlines. Retrieved May 2, 2019 from https://www.turkishairlines.com/en-tr/any-questions/

Ward, H. (2001). CRM initiatives fail due to lack of integration. *ComputerWeekly.com.* Retrieved May 1, 2019 from https://www.computerweekly.com/news/2240042772/CRM-initiatives-fail-due-to-lack-of-integration

Wells, J. D., William, L. F., & Choobineh, J. (1999). Managing information technology for one-to-one-customer interaction. *Information & Management*, *35*(1), 53–62. doi:10.1016/S0378-7206(98)00076-7

Will, D. (2009). Information Technology's Strategic Impact on the Air Travel Service Industry. *Journal of Information & Management for American Airlines*, *15*(2), 24–56.

World Tourism Organization (WTO). (1998). *Guidelines for the Transfer of New Technologies in the Field of Tourism*. Madrid, Spain.

ADDITIONAL READING

Al-Khouri, A. M. (2012). Customer Relationship Management: Proposed Framework from a Government Perspective. *Journal of Management and Strategy*, *3*(4), 34–54. doi:10.5430/jms.v3n4p34

Ivanovska, L. P. (2008). E-CRM in the Tourism Sector. *Revista de Turism, 5*, 14–19.

Kotler, P., & Armstrong, G. (2012). *Principles of Marketing* (14th ed., pp. 26–58). Harlow, England: Pearson Education Limited.

Kotler, P., & Armstrong, G. (2012). *Principles of Marketing* (14th ed., pp. 120–153). Harlow, England: Pearson Education Limited.

Lindgreen, A., Palmer, R., Vanhamme, J., & Wouters, J. (2005). A relationship-management assessment tool: Questioning, identifying,and prioritizing critical aspects of customer relationships. *Industrial Marketing Management, 35*(1), 57–71. doi:10.1016/j.indmarman.2005.08.008

Marjani, M., Nasaruddin, F., Gani, A., Karim, A., Hashem, I. A. T., Sıddıqua, A., & Yaqoob, I. (2017). Big IoT Data Analytics: Architecture, Opportunities, and OpenResearch Challenges. *IEEE Access: Practical Innovations, Open Solutions, 5*, 5247–5261. doi:10.1109/ACCESS.2017.2689040

Tuzunkan, D. (2018). Customer Relationship Management In Business- to-Business Marketing: Example of Tourism Sector. *Geo Journal of Tourism and Geosites, 22*(2), 329–338.

Zikmund, W. G., McLeod, R. Jr, & Gilbert, F. W. (2003). *Customer relationshipmanagement: Integrating marketing strategy and information technology*. Hoboken, New York: John Wiley & Sons.

KEY TERMS AND DEFINITIONS

Business-to-Business: Transactions between two businesses.

Business-to-Consumer: Transactions between business and an individual consumerthe end-user of a product or service.

Cost-Benefit Analysis: A technique used to compare the total costs of a project with its benefits. This analysis provides the identification of the net cost or benefit associated with the project. Cost-Benefit Analysis enables companies to analyze decisions or determine a value for intangibles.

Customer Acquisition: The process of gaining new customers, persuading them to buy a company's goods or services.

Customer-Centricity: A way of doing business in a way that keeping customer satisfaction above all else. Customer-centricity aims to keep the customer satisfied and provide a positive customer experience, both during and after the sale. By doing this, companies aim to create loyal customers and increase their business profits.

Customer Lifetime Value: The financial measurement of customers relationships for a business. Since companies do not know how long each relation will be, they make an prediction and define customer lifetime value.

Customer Retention: A marketing strategy aimed at increasing the repurchases of a company's products and services by existing customers.

Seo: SEO stands for Search Engine Optimization, itmakes certain websites more attractive to search engines than other websites listed under the same keywords.

Supply-Chain: A network of all the individuals, suppliers, resources and technology involved in producing a product and distributing it to its final buyer.

Chapter 14
In Which Department(s) the Robots Might Be Employed:
Evidence From Turkish University Students

Seden Doğan
https://orcid.org/0000-0001-8547-7702
Ondokuz Mayis University, Turkey

ABSTRACT

The development of service robotics comes along with the development of industrial robotics. Service robots are designed to perform professional job tasks as well as for service users in areas of everyday life. One of the fields for service robots' application is medicine. This chapter determines the opinions of the university students who have tourism education towards tasks that may be performed by the robots. The research question was "in which hotel department the robots might be employed". In total, 396 valid surveys were collected from the students, Mann Whitney U and Kruskal Wallis-H Tests were conducted to test the hypotheses. It was found that gender, department, and grade are significantly effective on the students' opinions towards tasks that might be performed by service robots.

INTRODUCTION

Digital technologies are transforming many industries and creating new challenges that require understanding. These may be stated as speed of change, cultural transformation, defining the skills for the future, outdated regulations and financing of digital and physical systems. TÜBİTAK (The Scientific and Technological Research Council of Turkey) (2016) states that by year 2018 the number of robots to be used in the industry will be approximately 3 million and the number of interconnected devices will increase to 29 billion, by year 2025 economic impact created by industrial robots will be annually between 0,6 and 1,2 trillion USD and by year 2030 digital technologies will have strong effects on productivity, income distribution and environment. On the other hand, it is foreseen that only in Europe investments worth 140 billion Euros will be made until 2020 (Çelik, Güleryüz, & Özköse, 2018:87).

DOI: 10.4018/978-1-7998-1989-9.ch014

The International Federation of Robotics (IFR) forecasts that sales of service robots will grow between 20 and 25 per cent by 2020 (Windsor, 2018).

Aim of this chapter to explore the opinions of the university students who have tourism education, towards tasks may be performed by the robots. The research question was "in which hotel department the robots might be employed".

BACKGROUND

Robots

The term "robot" which was derived from "robota" that means "subordinate labor" in Slav languages, was mentioned for the first time by the Czech playwright Karel Capek in his play "Rossum's Universal Robots (R.U.R.) in 1920 (Siciliano & Khatib, 2019:4; Murphy, Gretzel, & Pesonen, 2019; Hambling, 2018:3; Murphy, Hofacker, & Gretzel, 2017:105). The term "robot" was used to describe its "artificial people". These narratives link the creation of robots similar to humans in shape and size, able to communicate and be communicated with in familiar human ways, with the easy replacement of human workers in human-tailored working environments (Sandry, 2015:336). After that, in 1940 Isaac Asimov who is a Russian science-fiction writer refered the ethics of the interaction between robots and humans and the well-known three laws in his novel name I, Robot.

Developed in six months by the Hungarian noble Wolfgang von Kempelen upon the order of the Empress of Austria Maria Theresia in 1769, the chess instrument, which was exhibited for the empress for the first time in 1770 and destroyed by a fire in the Chinese Museum in 1854, was named as Mechanical Turk because of the robot-player sitting on and wearing Turkish clothes of the period. It consisted of a Turkish figure wearing a mustache and Ottoman clothes sitting in front of a wheeled cabin that is 120 cm long, 105 cm wide and 60 cm high maple tree and with a chessboard.

When the front cover was opened many large and small lifts, pulleys and other mechanical spaces could be seen in the cabinet of Mechanical Turk. When Mechanical Turk, that work by winding started

Figure 1. Mechanical Turk
Source: Şahin, 2016

to play chess with the volunteer opponent, was laying his eyes on the chessboard, shaking his head from time to time and moving the chess pieces with his hands. Having shaken his head three times to express that his move was over, Mechanical Turk was able to answer the questions from the spectators after the match in German language, although in a few words, by combining the letters on a special tray next to the chessboard. For example, when making the winner move of the game, Mechanical Turk could say "Schach" literally means "Checkmate". There was a second chessboard inside the cabin, which helped the operator follow the game. Under the main chessboard played by the Mechanical Turk, there was a spring-like mechanism under each square and a magnet under each piece. Thanks to this system, a real chess player in the cabin (firstly Baron Kempelen himself, then his assistant Anthon) was able to track which piece was playing to which position and to move the Mechanical Turk by using the special mechanism that submits his moves to the main board.

After the death of Kempelen in 1804, Mechanical Turk traveled from hand to hand and reached Johann Maelzel. Mechanical Turk defeated Napoleon Bonaparte in 1809 and continued his chess victories in France and England. In 1820, he made a match with Charles Babbage also known as the father of the computer. John Mitchell, the doctor and surgeon, was the new owner of Mechanical Turk, which was auctioned after Maelzel died. Mitchell founded a club and began showing the secrets of Mechanical Turk to the club members for a fee. At first the show seemed to gain success but he donated the automat to the Chinese Museum in the United States in 1854 due to his failure as a showman (Şahin, 2016; Chip Online, 2014; Satranç Dünyası, 2009; Köksal, 2004). In recent history, Hollywood and popular culture in the 1950s brought the concept of robotics and inspiring robotics to a wider audience. In 1956, Unimation, the first company to produce a robot, was founded (Ivanov et al., 2019).

Robots are virtual or mechanical objects that are used to facilitate daily activities. Developments in artificial intelligence studies lead to questioning how the process of robots replacing manpower will proceed (Hwang, Park, & Hwang, 2013:459). The first industry robot was tested within the automobile sector in 1961. Later, in the early 1980s, Japanese industry first began to implement this technology on a large scale in their factories, acquiring strategic competitiveness by decreasing costs and increasing the quality of their products. As to industrial robots, such automatically controlled, reprogrammable, and multipurpose manipulator machines were increasingly employed until 2000s in fields as diverse as refined petroleum products and nuclear fuel, textile and leather goods, communication and motor vehicles, agriculture and food products, and more. Finally in 2010s, self-driving cars showed up (Pagallo et al., 2018:4).

Since those days, the concept of robot has rapidly developed and there are a wide variety of robots with different features. Non-humanoid robots with little or no social skills have been very successful in both the industry and domestic sectors. Interaction between human and robot consists of three phase (Li et al., 2019:369):

1- The robot proactively detects a potentially interested user who seems to want to interact with it.
2- The robot understands what kind of intention that the user would like to express.
3- The robot provides services or responses for the user according to the comprehension of interactive intention from the user.

Bartneck and Forlizzi (2004) define a social robot as *"an autonomous or semi autonomous robot that interacts and communicates with humans by following the behavioral norms expected by the people with whom the robot is intended to interact"*. The design of social robots is either focused on human-like

Figure 2. Kismet
Source: https://robots.ieee.org/robots/kismet/

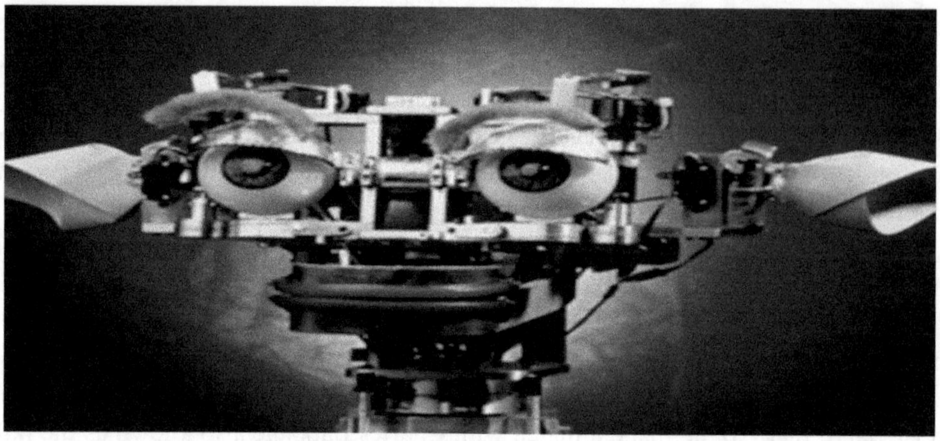

appearance or human-like manners. Kismet (Fig.1), which is the first social robot with anthropomorphic features, was designed to mirror the empathy and emotion of the interacting humans (Bhaumik, 2018:197). It has a cartoonish face, speak with a squeaky baby voice, and can always make people smile.

Robota is a sophisticated educational toy robot designed to build human-robot social interactions with children with motor and cognitive disabilities. It is 45 cm high, with motors to drive its legs, arms and head. Robota is being used in different experiments that investigate the use of toy robots for normal children and for children with disabilities (Billard, 2003:260).

Maggie, which is a social robot developed at RoboticsLab, has an artistic design of a 1.35 meters tall girl-like doll (Salichs et al, 2006).

As humans like to interact with each other, look for human characteristics such as human face and voice, creation of robots that are directly inspired by human capabilities and/or selectively imitate aspects of human form and behavior, is the main focus of robotic companies. Humanoids may be in a variety of shapes and sizes, from complete human-size legged robots to isolated robotic heads with human-like sensing and expression (Siciliano & Khatib, 2019:6). Humanoid robots may be used to assist and communicate with people in homes, working places, public spaces, hospitals, and tourist attractions.

Pepper, which is 1,2 m high and has a touch screen on its chest, is a humanoid robot released by Softbank Robotics and specifically designed for social interaction, that has been deployed in a variety of public environments (De Jong et al., 2018:883; Suddrey, Jacobson, & Ward, 2018).

As robots become increasingly present in, attitudes towards them vary. According to a research conducted by Eurobarometer, the ruslts showed that younger males, those living in urban centres, and more educated respondents are more favourable, generally, of using robots to care for the elderly (Hudson, Orviska, & Hunady, 2017:199). Reich-Stieber and Eyssel (2015) found that gender, age, and commitment to technology play a role upon a person's attitude towards education robots. Pochwatko et al. (2015) found that respondents' gender and the appearance of robots strongly correlated with attitudes towards robots. Ivanov, Webster and Garenko (2018) found that young Russian adults are quite supportive of the introduction of service robots into the hotel industry.

Figure 3. Maggie
Source: ROS Robots, 2019

Figure 4. Pepper
Source: Bauter Engel, 2018

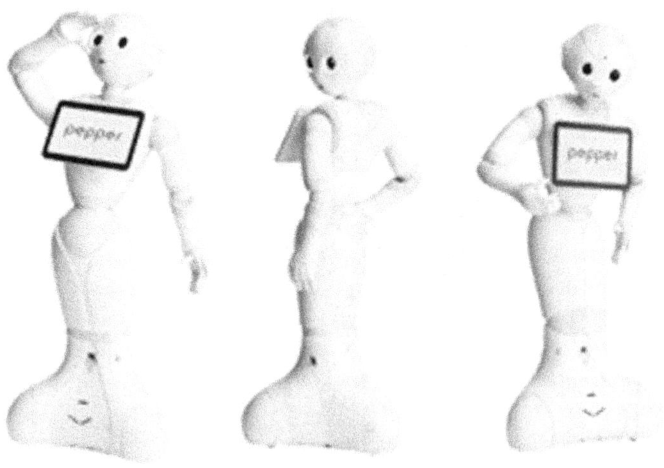

Service Robots in The Hospitality Industry

Fraunhofer Institute for Manufacturing Engineering and Automation (Fraunhofer IPA) defined service robot as *"a freely programmable mobile device carrying out services either partially or fully automatically. Services are activities that do not contribute to the direct industrial manufacture of goods, but to the performance of services for humans and institutions"* (Decker, Fischer, & Ott, 2017:349).

IFR defines service robots as *"a service robot is a robot which operates semi- or fully autonomously to perform services useful to the well-being of humans and equipment, excluding manufacturing operation"* (Pieska et al.,2012:191). In 1995, IFR and the United Nations Economic Commission for Europe (UNECE) adopted a preliminary system for service robots classification according to categories and the ways of interaction with them. The service robots are first divided into two groups according to their role in the community (for personal/domestic use and service for professional purposes) (Karabegovic et al., 2015:186).

The ultimate goal of a service robot is to serve people by providing information or helping for instance at home, in hospitals, hotels or industrial environments (Rodriquez-Lizundia et al., 2015:83). As they perform their tasks in the same environment as humans, the service robots should have the same abilities with people. They should be able to recognize faces, gestures, signs, objects, speech and atmosphere. Successful realization of set tasks results in bypassing obstacles without collision and destruction in the shortest possible time and distance. They should communicate with people on the basis of emotion (Karabegovic & Dolecek, 2017:10).

In the tourism industry, the service robots which need to work in dynamic environments in close interaction with humans (Stückler & Behnke, 2010) perform tasks such as welcoming guests, taking

Figure 5. Eva
Source: Yell Robot, 2019

orders, preparing food, room service, room cleaning, check-in / check-out, providing information about the business or the city (Ivanov & Webster, 2017:75; Hospitality Technology, 2017). For hotels, efficiency of activities performed by staff is measured by the time needed to execute them; the less time, the less expensive labour cost would be. Hence, investment in robot labour is often less expensive than paying humans. However, the adoption of service robots changes the nature of service experience as some service encounters are redefined by human-robot interaction. Different from industrial robots, the success of service robots depends on the satisfaction of the users (Tussyadiah & Park, 2018).

Researchers and practitioners are currently testing and applying robotics across various contexts in tourism and hospitality. Here below, some examples are presented.

Eva, who was launched in Februay, is a wedding photographer. It is 1,52m high and designed to move freely in a pre-mapped area via its infrared sensors. It can select guests with facial recognition. People who want to take photos can arrange photo preferences from Eva's 23.8-inch touchscreen (Yell Robot, 2019).

The EU-funded Spencer has been deployed at Schiphol Airport in a pilot scheme designed to aid passengers through the airport terminal. On November 30th in 2015, Schiphol Airport in Amsterdam welcomed Spencer that interacts with passengers guiding them through the terminal building in multiple languages. The human-robot interaction technology has been developed by Örebro University in Sweden who collaborated with researchers and businesses in five different countries including KLM, the end-user in the consortium. It aims to assist passenger transit and passenger flow avoiding bottlenecks and additional costs caused by wayfinding problems, delayed arrivals and language barriers (Sadler, 2015).

Operating in Nagasaki, Japan since 2015, Henn-na Hotel's front office department has robot personnel work as a receptionist and luggage carrier (Henn-na Hotel, 2019). Three robots speaking multiple languages (one in the form of a dinosaur) welcome guests, check-in, and record their requests, complaints

Figure 6. Spencer
Source: Sadler, 2015.

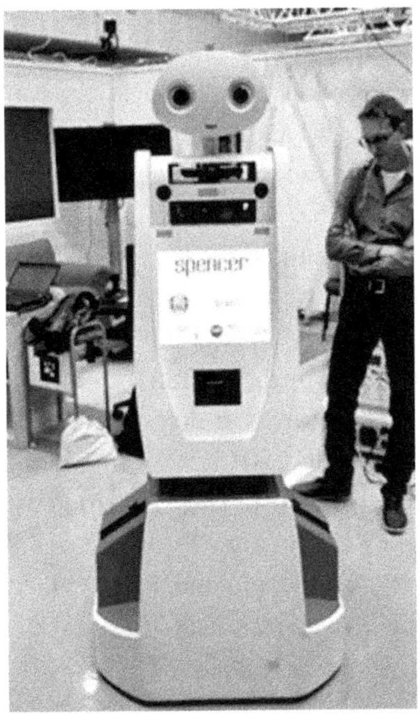

and demands. In the luggage room, a mechanism consisting of robotic arms stores the luggage. In terms of being the first hotel where robots work, in the Guinness World Record holding hotel a total of 186 robots are employed (Alexis, 2017: 211; Jarvis, 2016; Osawa et al., 2017: 219; Pierce, 2015).

IBM's robot Connie operates at Hilton McLean Tysons Corner, located in McLean, Virginia, United States, Connie, named after Hilton founder Conrad Hilton, is about 60 cm tall and has moving arms and legs. The robot provides information about the hotel and its surroundings to guests in reception desk (Ivanov, Webster, & Berezina, 2017: 1506; Trejos, 2016; Gagliordi, 2016).

Located in Cupertino, California, the Aloft Hotel employ a robot-servant called A.L.O., who assists human employees in the tasks such as taking room materials such as shampoo, soap, towels, bed linen to the room and taking the dirty materials to the laundry. It is dressed in uniform and has also a name badge (Ivanov, Webster, & Berezina, 2017: 1506; Trejos, 2014).

The Atlanta-based InterContinental hotel group launched a robot employee, Dash, at the Crown Plaza San Jose-Silicon Valley hotel in 2015. The Royal Caribbean Hotel has a waiter team consisting of robots in Bionic Bar section (Yalçın Kayıkçı & Kutluk Bozkurt, 2018: 58).

Advanced Telecommunications Research's humanoid robot Robovie acted as a museum guide at the Osaka Science Museum in Japan and autonomously led groups of school children around the museum and taught them about the exhibits (Kahn et al., 2012:303).

Operating in New York, Yotel provides transport of 300 pieces of luggage per day with the ABB IRB 6640 industrial robot, which they call Yobot. This allows guests to check in quickly (Social Tables, 2019).

Hotel EMC2 in Chicago has two robotic staff, Cleo and Leo. These two robots with name badges and uniforms are approximately three feet high and can be charged. Their tasks include delivering forgotten items and bringing extra towels, food or any other requested items to the rooms (Social Tables, 2019).

Valerie, which is the product of a collaboration between the Robotics Institute and the School of Drama at Carnegie Mellon, is considered the world's first robot receptionist with a personality, university officials said Wednesday. The blonde robo-ceptionist interacts with people by talking about her boss, her psychiatrist and her dream of being a lounge star (Gockley et al., 2005:1339)

Robutler consists of various bottles and glasses that are freely arranged on a table. Its task is to prepare and offer drinks. The system integrates a mobile platform, an arm optimized for light weight, and an articulated dexterous four-finger hand with basic navigation skills, real-time scene analysis by stereo vision, compliant motion control, and an intuitive man-machine interface. It comprises many of the key features required to realize the vision of a robotic servant that acts and interacts in the human living environment (Hillenbrand et al.,2005).

Ada is a waitress robot which is 1,5m high and 36kg weight has an artificial intelligence developed by Akın Robotics in Turkey. It is able to taking order, serving food and drink, walking and talking (Akın Robotics, 2019).

The effectiveness of the tasks performed by the staff in the hotels is measured in due time. The shorter the time is, the lower the labor cost. Therefore, the investment in the service robot will be lower than the investment in human (Tussyadiah & Park, 2018:309). In addition hotels may provide services that they cannot provide previously due to high cost. They provide customers with a different experience and create added value by saving money (Bowen & Morosan, 2018:729). In 2016, Travelzoo conducted a survey with 6211 people from Brazil, Canada, China, France, Germany, Japan, Spain, UK and USA. It has been concluded that robots will have an advantage in terms of providing indefinite service in different languages.

Figure 7. Henn-na Hotel Robots
Source: Rajesh, 2015.

Figure 8. Connie
Source: Gaudin, 2016.

Figure 9. A.L.O.
Source: Crook, 2014.

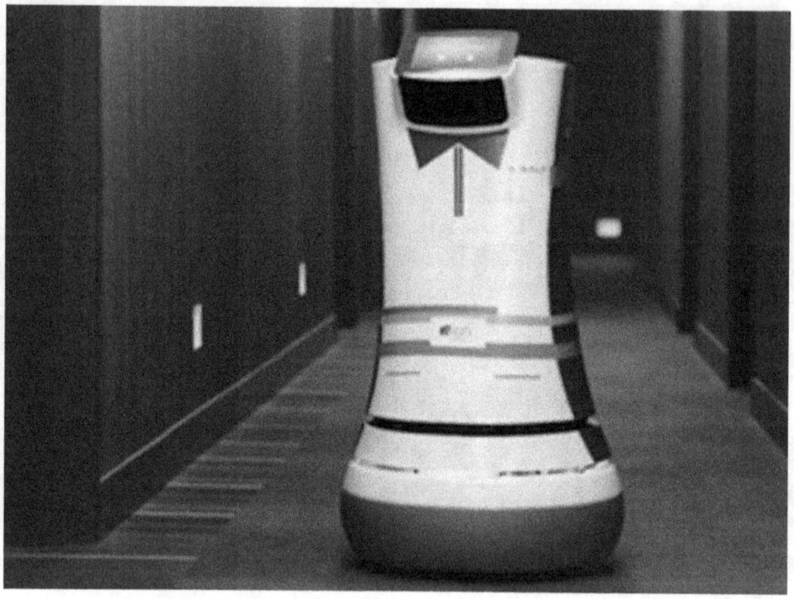

Figure 10. Robovie II
Source: Moumita, 2010.

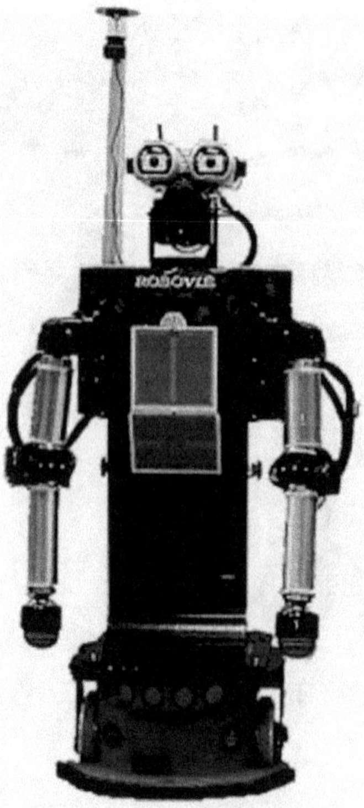

Figure 11. Yobot
Source: Katz, 2011.

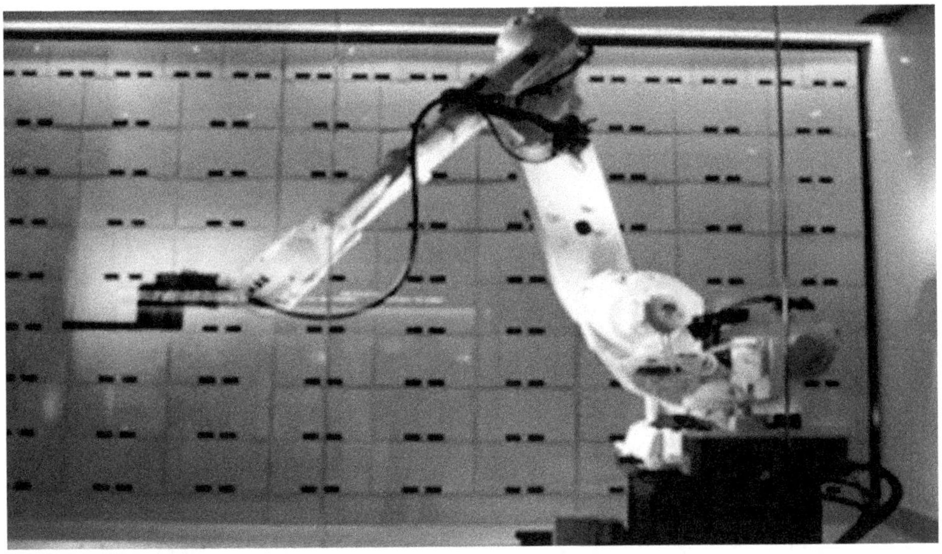

It is stated that robots will be an important asset in the tourism industry by 2030. On the other hand, it is foreseen that robots will create many new jobs aside from causing unemployment. Increasing interest in the enterprises where robots work will necessitate increasing the number of employees. In addition, the increase in the income level in all enterprises where robots work will create new employment opportunities (Bowen & Morosan, 2018:731; Hospitality Technology, 2017). That's why, it is important to understand the characteristics of robots that induce positive reactions from consumers in service settings such as hotels because the success of service robots depends on the satisfaction of the consumers (Tussyadiah & Park, 2018).

Figure 12. Cleo and Leo
Source: Whitley, 2018.

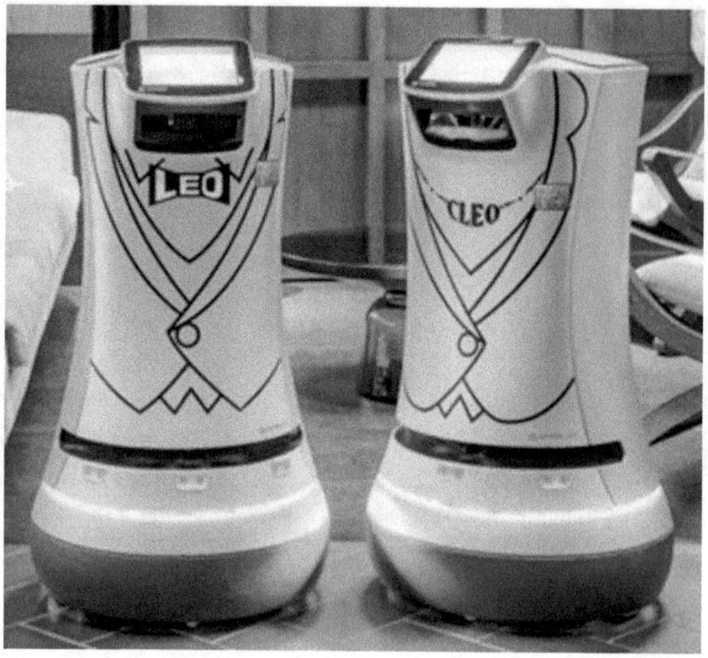

Figure 13. Valerie
Source: Lin, 2004.

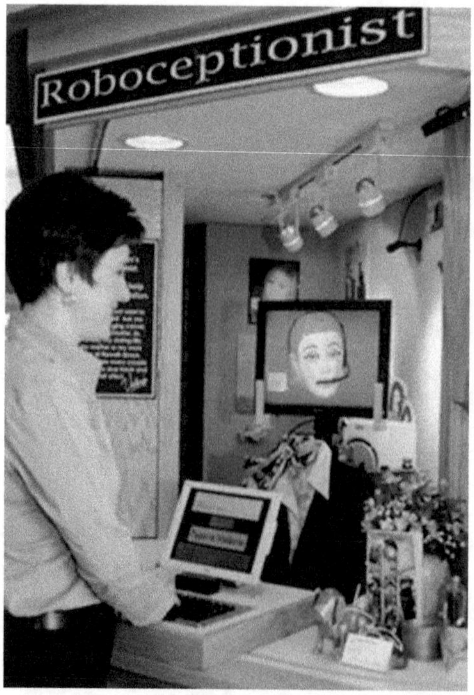

Figure 14. Robutler
Source: Hillenbrand et al., 2005.

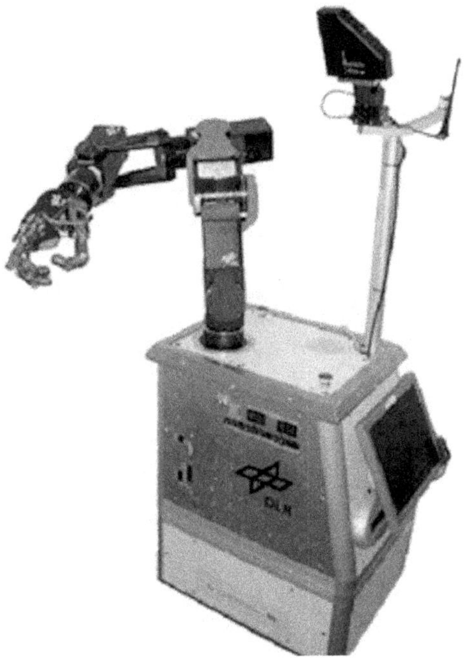

Academical Studies on Robots

It is seen important that robots taking place in tourism industry is discussed academically and consumers, employees, managers and other stakeholders in the industry identify their views and opinions about robots. Some academic studies on this subject are included in this section.

In 2019, Yu and Boyol Ngan investigated the gender and cultural differences of perceived human versus human-like robot smile in service. They found that male and female customers from different cultures perceive robot and human staff with different degrees of head tilt quite differently.

Lu, Cai and Gürsoy (2019) developed a Service Robots Integration Willingness Scale (SRIW) Scale with a survey of 1348 people. As a result of the construct validity and reliability tests applied to the 7-point Likert-type scale consisting of 6 dimensions (performance effectiveness, intrinsic motivation, anthropomorphism, social impact, status facilitation and emotions) and 36 expressions, it was determined which service areas the service robots can be used. These include airlines (check in, baggage handling, robotic flight attendant, robotic guide), hotel (check in, check out, robotic concierge, room service), restaurant (serving as a companion and usher, waiter and, cashier) and retail shops (to find the needs in the store and serve as a carrier).

Ivanov and Webster (2019) surveyed 1003 people and the participants agreed that service robots could perform at information desk and reservation office and also serve as cleaner and cashier.

Wirtz et al. (2018) presented the definition and key attributes of service robots with ethical issues about robot-delivered services at individual, market and societal level.

In 2018, Ivanov, Webster and Garenko conducted a survey with 260 Russian participants and determined that the participants who have more supportive views against service robots in hotels are male and

Figure 15. Ada
Source: Akın Robotics, 2019.

it was concluded that robots can take part in baggage handling, towel / bed linen delivery, information submission and payment processes.

In a survey conducted with 393 Iranian participants in 2018, Ivanov, Webster and Seyyedi unexpectedly found that women exhibited a more positive attitude towards robots, and participants under 30 were surprisingly more skeptical towards robots.

Sun Tung and Au conducted content analysis of 329 consumer reviews written on TripAdvisor, Agoda, Yelp and Booking.com websites for the four robotic hotels in 2018, suggesting that robotic services have an impact on human perceptions, allowing guests to proactively seek opportunities to interact and communicate with robots.

In 2017 Ivanov, Webster and Berezina evaluated the current and potential adoption of service automation and robots by hotels, restaurants, airports, theme and amusement parks, meeting and events, car rentals, travel agencies and tourist information centers, museums and art galleries. They indicated that economic efficiency, customer expectations and company's competitiveness will be determiner about which service processes may or have to be robotised.

Kuo, Chen and Tseng (2017) made a SWOT analysis with the data extracted with expert panel and semi-structured interview techniques. As a result, it is concluded that the strength of service robots is the chance of promoting them with the curiosity aroused in consumers and the weakness of service robots as the lack of skill in system integration.

In 2017, Sun Tung and Law aimed to present the potential directions for advancing theoretical, methodological and managerial implications for tourism experience research in human-robot interaction.

Pinillos et al. in their experimental study in 2016, an experimentally designed service robot, Sacarino, was commissioned as a bellboy at the Novotel Hotel in Valladolid, Spain. They received feedback from the guests and identified the necessary issues to improve the functionality of the robot.

In their experimental study in 2015, Rodriguez-Lizundia et al. determined that the hotel guests who interacted with an experimentally designed service robot, Sacarino, acting as a bellboy at the Novotel Hotel in Valladolid, Spain, tend to keep the robot at arm's length. At this point, it was stated that a more active looking robot would be attractive for long term interaction and a passive looking robot would create shorter interaction.

METHODOLOGY

Research Question

The research question was determined as "According to students who study in tourism, in which department(s) the robots might be employed?".

Data Collection

Data were collected during the period of January-May 2018. An online questionnaire was developed and distributed via social media platforms and Whatsapp groups. Targeted group was university students who study tourism related departments such as Tourism Management, Culinary, Catering and Tourism & Travel Services. Main reason of choosing these students is that they receive technology related lectures. The final sample included 396 students aged between 17 and 48 years old. The sample's characteristics are presented in Table 1.

Questionnaire was developed based on Ivanov et al.'s (2018) studies. It took appr. 10 minutes to complete the online survey. The questionnaire consisted of 40 individual questions in two parts. The first part was dedicated to the demographic characteristics of students. The second part delved into the students' opinions about the acceptability of various activities in different departments that robots could perform in a hotel.

As seen in Table 1, %54 of students who answered the questionnaire are male, %52 of them are 20 years old and younger, %58 of them are undergraduate student and %43 of them are 1st grade student.

Second part of the questionnaire was delevoped to investigate the students' opinion about the acceptability of various activities in different departments that robots could perform in a hotel. There are 36 statements in the part. The result of mean rate is shown in Table 2.

As seen in Table 2, the highest mean rate (3.91) belongs to *carrying luggage* and lowest mean rate (2.22) belongs to *baby sitting*. In general, all jobs performing in Front Office and Housekeeping Departments, some jobs performing in Food & Beverage Department and some additional services were found acceptable for service robots by the students. The results are supportive some of the previous studies. According to Ivanov et al.'s (2018) study, Russian young consumers think that robots may deliver some items to the guest room, provide information about hotel or destination and process payment. Another study landed up that according to Iranian tourists cleaning, delivering items, providing information, tak-

Table 1. Students' demographic characteristics

Characteristic		n	%
Gender	Male	215	54,3
	Female	181	45,7
Age	18 and younger	64	16,2
	19	65	16,4
	20	76	19,2
	21	57	14,4
	22	50	12,6
	23	36	9,1
	24 and older	48	12,1
Education	Associate Degree	231	58,3
	Undergraduate	165	41,7
Department	Tourism Management (undergraduate) (TM-g)	165	28,3
	Culinary (Associate degree) (C-u)	104	17,8
	Catering (Associate degree) (Ca-u)	58	9,9
	Tourism Management (Associate degree) (TM-u)	40	6,9
	Tourism and Travel Services (Associate degree) (TTS-u)	29	5,0
Grade	1st	171	43,2
	2nd	98	24,7
	3rd	26	6,6
	4th	101	25,5
TOTAL		**396**	**100,0**

ing an order or processing payment are the most appropriate functions for robots (Ivanov et al., 2018). According to new research conducted by Ivanov and Webster (2019), the most commonly approved of usage of robots is perceived to be information provision, housekeeping activities and processing bookings, payments and documents. The robots were found to be suitable to carry luggage, deliver items, clean the common areas and collect garbage for 1000 respondents.

Testing the Hypotheses

The Kolmogorov-Smirnov Z-test showed that the distribution of answers was statistically different from normal for all questions (all Z values were significant at $p < 0.01$). That is why the non-parametric Mann-Whitney U test and the Kruskal-Wallis H were used in data analysis. Specifically, it was used Mann-Whitney U test to identify the impact of gender and graduation on students' answers, while the Kruskal-Wallis H was used for evaluating the influence of department and grade towards service robots in general on students' answers.

As it can be seen in Table 3, there is a difference of opinion between male and female students about the delivering ready laundry to the room and performing calculations in accounting by service robots. Female students have more positive opinion about delivering ready laundry to the room but have more

In Which Department(s) the Robots Might Be Employed

Table 2. Directions of robots application in hotels

Direction of Robot Application in Hotels		Mean	Std. Dev.
Front Office	Carrying luggage	3,91	1,044
	Concierge services (taxi, ordering ticket etc.)	3,65	1,084
	Providing information about hotel facilities	3,57	1,122
	Providing information about destination	3,51	1,158
	Check-in / Check-out process	3,30	1,168
	Processing credit/debit card payment	3,24	1,153
	Guiding/Excorting to the room	3,20	1,179
	Processing cash payment	3,12	1,133
	Welcoming a guest	2,98	1,196
	Receiving general complaints/requests	2,89	1,217
Housekeeping	Cleaning common areas in the hotel	3.78	1.121
	Delivering ready laundry to the room	3.72	1.045
	Cleaning the room	3.68	1.122
	Taking orders for laundry	3.62	1.052
	Delivering requests about the room (towel, pillow etc.)	3.58	1.108
	Receiving complaints/requests about the room	3.40	1.164
Food & Beverage	Cleaning the tables	3.63	1.098
	Washing the dishes in the stewarding area	3.60	1.133
	Delivering food and drink to the room	3.22	1.214
	Taking order in the restaurant	3.08	1.153
	Taking order in the bar	3.02	1.196
	Serving food/drinks in the pool/beach area	3.01	1.232
	Serving food in the restaurant	3.01	1.183
	Guiding/Escorting to the table in the restaurant	2.99	1.269
	Welcoming a guest in the restaurant	2.99	1.269
	Preparing drink order in the bar	2.95	1.261
	Serving drink in the bar	2.94	1.213
	Preparing food order in the kitchen	2.86	1.245
Additional Services	Delivering/collecting towels in the pool/beach area	3.48	1.202
	Gardening services	3.39	1.118
	Promoting the hotel in exhibitions	3.19	1.270
	Calculation in the accounting	3.13	1.318
	Animation	2.98	1.288
	Guarding/Security	2.88	1.360
	Massage in SPA	2.38	1.230
	Baby sitting	2.22	1.207

Table 3. Result of Mann Whitney U Test (Gender)

Direction of Robot Application in Hotels	Gender	Mean Rank	Z	p
	Male	188.39		
Delivering ready laundry to the room	Female	210.51	-2.067	0.039
	Male	209.53		
Calculation in the accounting	Female	185.40	-2.144	0.032

negative opinion about calculations in accounting than male students. It should make further detailed studies to examine why there is difference between males and females.

As it can be seen in Table 4, the opinions of the undergraduate and associate degree students about serving food in the restaurant, serving drinks in the bar, preparing food in the kitchen, preparing beverage orders in the bar and serving food and beverages in the pool / beach area are carried out by service robots differ. The undergraduate students have more positive opinion than the students who have associate degree. This may be because of educational and working experience of the students. Undergraduate students study four years and that's why, they receive more lectures that contains new trends and development. They also have chance to have more working experiences in the industry during their educational life and these experiences may help to improve their point of views to technological innovations like robotics.

As shown in Table 5, Kruskal Wallis H Test revealed differences in the opinions of the students studying in different departments about delivering request about the room, delivering food and drink to the room, preparing food orders in the kitchen, preparing drink orders in the bar, cleaning the tables, gardening services and delivering/collecting towels in the pool/beach area performing by the service robots. As a result of Mann Whitney U Test, the students of the four-year Tourism Management and the students of the two-year Tourism and Travel Services Program students have more positive opinions than the other students. Students at culinary program and catering program receive lectures that less

Table 4. Result of Mann Whitney U Test (Educational Degree)

Direction of Robot Application in Hotels	Graduation	Mean Rank	Z	p
	Associate degree	188.00		
Serving food in the restaurant	Undergraduate	213.20	-2.222	0.026
	Associate degree	188.90		
Serving drink in the bar	Undergraduate	211.94	-2.030	0.042
	Associate degree	182.08		
Preparing food order in the kitchen	Undergraduate	221.49	-3.467	0.001
	Associate degree	184.74		
Preparing drink order in the bar	Undergraduate	217.77	-2.905	0.004
	Associate degree	188.30		
Serving food/drinks in the pool/beach area	Undergraduate	212.78	-2.156	0.031

Table 5. Result of Kruskal Wallis H Test (Department)

Direction of Robot Application in Hotels	Chi Square	p	Mann Whitney U			
			Z	p	Diff.	Mean Rank
			-2.040	0.041	C-u / TM-u	76.63 / 61.78
Delivering requests about the	9.823	0.044	-2.215	0.027	TM-g / TM-u	107.23 / 85.56
room (towel etc.)			-2.748	0.006	TTS-u / TM-u	42.19 / 29.79
			-2.228	0.026	C-u / Ca-u	87.47 / 70.79
			-2.489	0.013	TM-g / Ca-u	118.17 / 94.45
Delivering food and drink to the	11.587	0.021	-2.486	0.013	TTS-u / Ca-u	53.31 / 39.34
room			-2.162	0.031	TTS-u / TM-u	40.95 / 30.69
			-3.503	0.000	TM-g / C-u	147.81 / 114.67
Preparing food order in the	14.417	0.002	-2.145	0.032	TTS-u / C-u	80.17 / 63.33
kitchen			-2.867	0.004	TM-g / TM-u	108.64 / 79.75
			-2.178	0.029	TTS-u / TM-u	41.00 / 30.65
			-2.540	0.011	TM-g / C-u	144.28 / 120.27
Preparing drink order in the bar	11.584	0.021	-2.012	0.044	TM-g / Ca-u	117.02 / 97.72
			-2.867	0.015	TM-g / TM-u	107.87 / 82.90
			-2.084	0.037	TTS-u / C-u	78.95 / 63.67
Cleaning the tables	10.152	0.038	-2.389	0.017	TM-g / Ca-u	117.79 / 107.65
			-2.310	0.021	TTS-u / Ca-u	52.55 / 39.72
			-2.954	0.003	TM-g / TM-u	108.73 / 79.36
Gardening services	10.708	0.030	-2.650	0.008	TTS-u / TM-u	42.19 / 29.79
			-3.714	0.000	TTS-u / C-u	89.45 / 60.74
			-3.023	0.003	TTS-u / Ca-u	55.14 / 38.43
Delivering/collecting towels in	15.223	0.004	-2.767	0.006	TTS-u / TM-u	42.45 / 29.60
the pool/beach area			-3.139	0.002	TTS-u / TM-g	125.95 / 92.50

related to new trends or technological innovations. That's why, most probably they have idea clue about robotics in hospitality industry.

As shown in Table 5, Kruskal Wallis H Test revealed differences in the opinions of the students studying in different grades about check-in/check out process, concierge services and processing credit/debit card payment. As a result of the Mann Whitney U Test, the students in the 4th grade who generally have more positive opinion than the students in the other grades. This result is acceptable because the 4th grade students have the highest level of education and more experiences than the other student. That's why, their perception about robotics in hospitality industry may be more positive than the others.

Table 6. Result of Kruskal Wallis H Test (Grade)

Direction of Robot Application in Hotels	Chi Square	p	Mann Whitney U			
			Z	p	Diff.	Mean Rank
Check-in / Check-out process	13.625	0.003	-3.202 -2.676 -3.571	0.001 0.007 0.000	1st / 2nd 2nd / 3rd 4th / 3rd	137.53 / 130.58 66.79 / 46.35 69.61 / 42.19
Concierge services (taxi, ordering ticket etc.)	8.290	0.040	-2.288 -2.821	0.022 0.005	1st / 2nd 4th / 3rd	138.73 / 128.49 68.33 / 47.17
Processing credit/debit card payment	13.330	0.004	-2.255 -2.870	0.024 0.004	1st / 2nd 4th / 2nd	140.76 / 124.95 111.11 / 88.55
			-3.085	0.002	4th / 3rd	68.88 / 45.04

SOLUTIONS AND RECOMMENDATIONS

Robots are here and they will stay. Today, industrial robots are mainly used in the production process in manufacturing, medicine, military, automotive and agriculture industries. It is predicted that a third of jobs that exist today could be taken by Smart Technology, Artificial Intelligence, Robotics, and Algorithms (STARA) by 2025 (Brougham & Haar, 2018:239). Robots will sooner or later change our daily lives, such as: assistants, servants, helpers, friends, assistant to surgeons in medical operations, intervention in hazardous environments of any search or rescue, in agriculture and forestry, cleaning, digging, dangerous transportation, construction, and demolition (Karabegovic & Dolecek, 2017:11).

Even though the concept of a robot is not old, usage of them in travel, hospitality and tourism industry is relatively new. Most probably, as this industry is labor-intensive and human-driven, employing robots is regarded as irrational. But we have to face the reality that robots are surfacing in the industry and becoming research topics in academia.

Cleaning the swimming pool or restaurant at the hotel may be done by industrial robots. That kind of robots may also deliver luggage to the rooms or vacuum the carpets. Humanoid robots can easily provide guidance to customers at the hotel reception; welcome them at the entrance of the lobby or restaurant. It is possible to see a humanoid robot at the lobby bar while washing the glasses or making a cocktail. The increase of using robots in the travel, hospitality, and tourism bring lots of questions. One of these questions is which hotel departments are more suitable for robot employment. This also was the question of this research. It focuses on the appropriateness of robots in hotels. There are only a few researches that evaluated people's opinion about robots in hotels.

This chapter contributed to the body of knowledge by identifying the activities which (potential) tourists consider as appropriate to be delivered by robots. University students having tourism education in Turkey were asked to answer the research question which was "in which hotel department the robots might be employed". In total, 396 valid surveys were collected from the students. According to

Mann Whitney U and Kruskal Wallis-h Tests, female and male students have different opinions about delivering ready laundry to the room but have more negative opinion about calculations in accounting by service robots. The department and grade of the students make also difference in the opinions towards tasks perfomed by the service robots. The results are both surprising and not surprising. It is surprising because it was expected more differences on opinions towards tasks perform by the service robots between male and female students. It is not surprising because the Gen Y (Millenials) and Gen X are technology experts, accept new technologies and don't resist. Born in the late 80s and 90s, Gen Y (Millenials) met internet when they were children. They accept as first digital natives. They connect with the digital world through play, enjoyment and desire (Csobanka, 2016:66; Pendergast, 2010:6). They spend a considerable amount of time contributing, sharing, searching for and consuming content on the net (Del Chiappa & Fotiadis, 2018:323).Gen Z is the first generation to be surrounded by the world of technology since their first days. Technology is integrated into almost all areas of their lives and using it on a daily basis has been the norm since their youngest age. When 92% of Gen Z members go online daily and 24% are online almost constantly, technology is an inseparable part of their lives (Larkin, Jancourt, & Hendrix, 2018; Lenhart, 2015).

According to the survey results, the activities performing in Front Office and Housekeeping Departments, some jobs performing in Food & Beverage Department and some additional services such as gardening services, promoting hotel in exhibitions seem to be easiest departments to be automated. The tasks in these departments were found that customers might perceive as most appropriate for robotisation. On the other hand, not all activities such as baby sitting and massaging can and should be automated. But it can be said that robot-delivered activities in hotels will be welcome by the customers. The results supported the researches conducted by Ivanov and Webster (2019_a; 2019_b) and Frey and Osborne (2013).

According to the survey results, there is significant difference between male and female students' attitude towards tasks may be performed by robots. In their research, Ivanov et al. (2018) and Pochwatko et al. (2015) found gender significance on attitude towards robots. There are also significant differences between undergraduate students and studesnt having associate degree. Undergraduate students have more positive opinions towards tasks may be performed by robots.

There are several limitations of this paper. First, sample size contains only students. Second, because of lack of time, only 369 students were reached to invite the research. Third, none of the students have received a lecture that contains concept of robot or interacted with a service robot, therefore their opinions are based on only personal feelings and work experiences.

As there is not enough academic research regarding service robots in hospitality industry, the results of this research cannot be evaluated sufficiently. In Turkey, there is just one research about robotics in tourism which contains only literature review (İbiş, 2019). So, even though this research has some limitations, the results are expected to be clarified the attitude towards robots in hospitality industry.

FUTURE RESEARCH DIRECTIONS

Future researches may focus on the effect of gender, generation, national culture, work experience, job title and working place of respondents on their acceptance of robots in different areas in the hospitality industry. It is also important to determine consumers', employees', managers' and other stakeholders' opinions about robots in the industry. By conducting in-depth interviews with these stakeholders, the advantages and disadvantages of the employment of robots in touristic enterprises can be determined.

CONCLUSION

Robots have arrived and they will not go back. The hospitality industry has to face this reality and accept them by understanding the advantages and disadvantages of them. There should also be lectures at tourism schools to introduce robots and robotic applications in the hospitality industry. If the employees learn to cooperate and accept to work with the robots, this will help to increase the efficiency of hotel operations and hotel guests' satisfaction.

REFERENCES

Akın Robotics. (2019). *Waitress Robot Ada.* Retrieved from https://akinrobotics.com/tr/akinsoft_garson_robot_ada.php

Alexis, P. (2017). R-Tourism: Introducing the Potential Impact of Robotics and Service Automation in Tourism. *Ovidius University Annals. Economic Sciences Series*, *XVII*(1), 211–216.

Bartneck, C., & Forlizzi, J. (2004). A Design-Centred Framework for Social Human-Robot Interaction, *2004 IEEE International Workshop on Robot and Human Interactive Communication*, September 20-22, Kurashiki, Japan. 10.1109/ROMAN.2004.1374827

Bauter Engel, J. (2018). *SoftBank Taps Affective to Boost Pepper Robot's Emotional IQ*. Retrieved from https://xconomy.com/boston/2018/08/28/softbank-taps-affectiva-to-boost-pepper-robots-emotional-iq/

Bhaumik, A. (2018). *From AI to Robotics: Mobile, Social, and Sentinent Robots*. Boca Raton, FL: CRC Press. doi:10.1201/9781315372549

Billard, A. (2003). Robota: Clever Toy and Educational Tool. *Robotics and Autonomous Systems*, *42*(3-4), 259–269. doi:10.1016/S0921-8890(02)00380-9

Bowen, J., & Morosan, C. (2018). Beware Hospitality Industry: The Robots are Coming. *Worldwide Hospitality and Tourism Themes*, *10*(6), 726–733. doi:10.1108/WHATT-07-2018-0045

Brougham, D., & Haar, J. (2018). Smart Technology, Artificial Intelligence, Robotics, and Algorithms (STARA): Employees' Perceptions of our Future Workplace. *Journal of Management & Organization*, *24*(2), 239–257. doi:10.1017/jmo.2016.55

Çelik, K., Güleryüz, S., & Özköse, H. (2018). 4. Endüstri Devrimine Kuramsal Bakış. *Avrasya Sosyal ve Ekonomi Araştırmaları Dergisi*, *5*(9), 86–95.

Chip Online. (2014). *Unutulmayacak Mekanik Türk Efsanesi*. Retrieved from https://www.chip.com.tr/haber/unutulmayacak-mekanik-turk-efsanesi_48558.html

Crook, J. (2014). *Starwood Introduces Robotic Butler at Aloft Hotel in Cupertino*. Retrieved from https://techcrunch.com/2014/08/13/starwood-introduces-robotic-butlers-at-aloft-hotel-in-palo-alto/

Csobanka, Z. E. (2016). The Z Generation. *Acta Technologica Dubnicae*, *6*(2), 63–76. doi:10.1515/atd-2016-0012

De Jong, M., Zhang, K., Roth, A. M., Rhodes, T., Schmucker, R., Zhou, C., . . . Veloso, M. (2018). Towards a Robust Interactive and Learning Social Robot, *AAMAS 2018,* July 10-15, Stockholm, Sweden.

Decker, M., Fischer, M., & Ott, I. (2017). Service Robotics and Human Labor: A first technology assessment of substitution and cooperation. *Robotics and Autonomous Systems, 87*, 348–354. doi:10.1016/j.robot.2016.09.017

Del Chiappa, G., & Fotiadis, A. (2018). Generation Y Perspective of Hotel Disintermediation and User-Generated Content. In D. Gursoy (Ed.), *The Routledge Handbook of Hospitality Marketing*. New York: Taylor & Francis Group.

Frey, C. B., & Osborne, M. A. (2013). *The Future of Employment: How Susceptible are Jobs to Computerisaton?* Retrieved from https://www.oxfordmartin.ox.ac.uk/downloads/academic/The_Future_of_Employment.pdf

Gagliordi, N. (2016). *This Watson-Powered Robot Concierge is Rethinking the Hotel Industry.* Retrieved from https://www.zdnet.com/article/this-watson-powered-robot-concierge-is-rethinking-the-hotel-industry/

Gaudin, S. (2016). *Meet Connie, Hilton's Smart Robot Concierge.* Retrieved from https://www.computerworld.com/article/3042401/meet-connie-hilton-s-smart-robot-concierge.html

Gockley, R., Bruce, A., Forlizzi, J., Michalowski, M., Mundell, A., Rosenthal, S., ... Wang, S. (2005). Designing Robots for Long-Term Social Interaction, *2005 IEEE/RSJ International Conference on Intelligent Robots and Systems*, August 2-6, Alberta, Canada. 10.1109/IROS.2005.1545303

Hambling, D. (2018). *We: Robot, The Robots that already Rule our World.* London, UK: Aurum Press.

Hillenbrand, U., Ott, C., Brunner, B., Borst, C., & Hirzinger, G. (2005). *Towards Service Robots for the Human Environment: The Robutler.* Retrieved from https://www.dlr.de/rm/Portaldata/52/Resources/Roboter_und_Systeme/Robutler/abstract.pdf

Hospitality Technology. (2017). *Robots in Hospitality: Five Trends on the Horizon.* Retrieved from https://hospitalitytech.com/robots-hospitality-five-trends-horizon

Hudson, J., Orviska, M., & Hunady, J. (2017). People's Attitudes to Robots in Caring for the Elderly. *International Journal of Social Robotics, 9*(2), 199–210. doi:10.100712369-016-0384-5

Hwang, J., Park, T., & Hwang, W. (2013). The Effects of Overall Robot Shape on the Emotions Invoked in Users and the Perceived Personalities of Robot. *Applied Economics, 44*, 459–471. PMID:23157974

İbiş, S. (2019). Turizm Endüstrisinde Robotlaşma. *Türk Turizm Araştırmaları Dergisi, 3*(3), 403–420. doi:10.26677/TR1010.2019.169

Ivanov, S., Gretzel, U., Berezina, K., Sigala, M., & Webster, C. (2019). Progress on Robotics in Hospitality and Tourism: A Review of the Literature. *Journal of Hospitality and Tourism Technology.* Retrieved from https://www.emeraldinsight.com/doi/pdfplus/10.1108/JHTT-08-2018-0087

Ivanov, S., & Webster, C. (2017). Designing Robot-Friendly Hospitality Facilities. *Scientific Conference: Tourism. Innovations. Strategies*, October 13-14, Bourgas, Bulgaria. Academic Press.

Ivanov, S., & Webster, C. (2019a). Perceived Appropriateness and Intention to Use Service Robots in Tourism. 2019 Information and Communication Technologies in Tourism, January 30-February 1, Nicosia, Cyprus.

Ivanov, S., & Webster, C. (2019b). What should Robots do? A Comparative Analysis of Industry Professionals, Educators and Tourists. 2019 Information and Communication Technologies in Tourism, January 30-February 1, Nicosia, Cyprus. Academic Press.

Ivanov, S., Webster, C., & Berezina, K. (2017). Adoption of Robots and Service Automation by Tourism and Hospitality Companies. *Revista Tourismo & Desenvolvimento, 27/28*, 1501–1517.

Ivanov, S., Webster, C., & Garekno, A. (2018). Young Russian adults' attitudes towards the Potential Use of Robots in Hotels. *Technology in Society, 55*, 24–32. doi:10.1016/j.techsoc.2018.06.004

Ivanov, S., Webster, C., & Seyyedi, P. (2018). Consumers' Attitudes towards the Introduction of Robots in Accommodation Establishments. *Tourism: An International Interdisciplinary Journal, 66*(3), 302–317.

Jarvis, H. (2016). *Robot-Run Hotel Gets Guinness Approval*. Retrieved from https://standbynordic.com/robot-run-hotel-gets-guinness-approval/

Kahn, P. H., Kanda, T., Ishiguro, H., Freier, N. G., Severson, R. L., Gill, B. T., ... Shen, S. (2012). "Robovie, You'll Have to Go into the Closet Now": Children's Social and Moral Relationships With a Humanoid Robot. *Developmental Psychology, 48*(2), 303–314. doi:10.1037/a0027033 PMID:22369338

Karabegovic, I., & Dolecek, V. (2017). The Role of Service Robots and Robotic Systems in the Treatment of Patients in Medical Institutions. In M. Hadzikadic, & S. Avdakovic (Eds.), *Advanced Technologies, Systems, and Applications – Lecture Notes in Networks and Systems*. Switzerland: Springer International Publishing AG. doi:10.1007/978-3-319-47295-9_2

Karabegovic, I., Karabegovic, E., Mahmic, M., & Husak, E. (2015). The Application of Service Robots for Logistics in Manufacturing Processes. *Advances in Production Engineering & Management, 10*(4), 185–194. doi:10.14743/apem2015.4.201

Katz, L. (2011). *NY hotel books Yobot the luggage-handling robot*. Retrieved from https://www.cnet.com/news/ny-hotel-books-yobot-the-luggage-handling-robot/

Köksal, G. (2004). *Napolyonu Mat Eden Türk*. Retrieved from https://bianet.org/bianet/print/31804-napolyonu-mat-eden-turk

Kuo, C.-M., Chen, L.-C., & Tseng, C.-Y. (2017). Investigating an Innovative Service with Hospitality Robots. *International Journal of Contemporary Hospitality Management, 29*(5), 1305–1321. doi:10.1108/IJCHM-08-2015-0414

Larkin, C. M., Jancourt, M., & Hendrix, W. H. (2018). The Generation Z World: Shifts in Urban Design, Architecture and the Corporate Workplace. *Corporate Real Estate Journal., 7*(3), 230–242.

Lenhart, A. (2015). *Teens, Social Media, & Technology. WWW document*, Retrieved from http://www.pewinternet.org/2015/04/09/teens-social-media-technology-2015/

Li, K., Sun, S., Zhao, X., Wu, J., & Tan, M. (2019). Inferring User Intent to Interact with a Public Service Robot Using Bimodal Information Analysis. *Advanced Robotics*, *33*(7-8), 369–387. doi:10.1080/01691864.2019.1599727

Lin, J. (2004). *University Unveils Chatty "Roboreceptionist"*. Retrieved from http://www.nbcnews.com/id/4306856/ns/technology_and_science-science/t/university-unveils-chatty-roboceptionist/#.XTstbegzbIU

Lu, L., Cai, R., & Gürsoy, D. (2019). Developing and Validating a Service Robot Integration Willingness Scale. *International Journal of Hospitality Management*, *80*, 36–51. doi:10.1016/j.ijhm.2019.01.005

Moumita. (2010). *Robovie II Helps You with your Grocery Shopping*. Retrieved from http://gadgether.com/robovie-ii-grocery-shopping/

Murphy, J., Gretzel, U., & Pesonen, J. (2019). Marketing Robot Services in Hospitality and Tourism: The Role of Anthropomorphism. *Journal of Travel & Tourism Marketing*. Retrieved from https://www.tandfonline.com/doi/pdf/10.1080/10548408.2019.1571983?needAccess=true

Murphy, J., Hofacker, C., & Gretzel, U. (2017). Dawning of the Age of Robots in Hospitality and Tourism: Challenges for Teaching and Research. *European Journal of Tourism Research*, *15*, 104–111.

Osawa, H., Akiya, N., Koyama, T., Ema, A., Kanzaki, N., Ichise, R., . . . Kubo, A. (2017). What is Real Risk and Benefit on Work with Robots? From the Analysis of a Robot Hotel, *HRI2017 Companion of the 2017 ACM/IEEE International Conference on Human-Robot Interaction*, March 6-9, Vienna, Austria. 10.1145/3029798.3038312

Pagallo, U., Corrales, M., Fenwick, M., & Forgo, N. (2018). The Rise of Robotics & AI: Technological Advances & Normative Dilemmas. In M. Corrales, M. Fenwick, & N. Forgo (Eds.), *Robotics, AI and the Future of Law*. Singapore: Springer Nature. doi:10.1007/978-981-13-2874-9_1

Pendergast, D. (2010). Getting to Know the Y Generation. In P. Benckendorff, G. Moscardo, & D. Pendergast (Eds.), *Tourism and Generation Y*. Oxford, UK: CABI.

Pierce, A. (2015). *A Hotel Staffed by Robots*. Retrieved from http://www.technologytoday.us/columnPDF/A_Hotel_Staffed_by_Robots.pdf

Pieska, S., Luimula, M., Jauhiainen, J., & Spiz, V. (2012). Social Service Robots in Public and Private Environments, In D. Biolek, K. Volkov, & K. M. Ng (Eds.), Recent Researches in Circuits, Systems, Multimedia, and Automatic Control, Wisconsin: WSEAS Press.

Pinillos, R., Marcos, S., Feliz, R., Zalama, E., & Gomez-Garcia-Bermejo, J. (2016). Long-Term Assessment of a Service Robot in a Hotel Environment. *Robotics and Autonomous Systems*, *79*, 40–57. doi:10.1016/j.robot.2016.01.014

Pochwatko, G., Giger, J.-C., Rozanska-Walzcuk, M., Swidrak, J., Kukielka, K., Mozaryn, J., & Piçarra, N. (2015). Polish Version of the Negative Attitude Toward Robots Scale (NARS-PL). *Journal of Automation, Mobile Robotics, & Intelligent Systems*, *9*(3), 65–72.

Rajesh, M. (2015). *Inside Japan's First Robot-Staffed Hotel*. Retrieved from https://www.theguardian.com/travel/2015/aug/14/japan-henn-na-hotel-staffed-by-robots

Reich-Stiebert, N., & Eyssel, F. (2015). Learning with Educational Companion Robots? Toward Attitudes on Education Robots, Predictors Of Attitudes, and Application Potentials for Education Robots. *International Journal of Social Robotics*, *7*(5), 875–888. doi:10.100712369-015-0308-9

RobotsR. O. S. (2019). *Maggie*. Retrieved from https://robots.ros.org/maggie/

Robots. (2019). *Kismet*. Retrieved from https://robots.ieee.org/robots/kismet/

Rodriguez-Lizundia, E., Marcos, S., Zalama, E., Gomez-Garcia-Bermejo, J., & Gordaliza, A. (2015). A bellboy Robot: Study of the Effects of Robot Behaviour on User Engagement and Comfort. *International Journal of Human-Computer Studies*, *82*, 83–95. doi:10.1016/j.ijhcs.2015.06.001

Sadler, K. (2015). *SPENCER Robot to Assist Passengers at Schiphol Airport*. Retrieved from https://www.internationalairportreview.com/news/21360/spencer-robot-assists-passengers-schiphol-airport/

Şahin, E. (2016). *18. Yüzyılda Satranç Oynadığı Herkesi Madara Eden Türk Robot: Mekanik Türk*. Retrieved from https://www.webtekno.com/internet/mekanik-turk-h15195.html, Accessed on May 10, 2019.

Salichs, M. A., Barber, R., Khamis, A. M., Malfaz, M., Gorostiza, J. F., Pacheco, R., ... Garcia, D. (2006). Maggie: A Robotic Platform for Human-Robot Social Interaction. In *Proceedings 2006 IEEE Conference on Robotics, Automation, and Mechatronics*, June 1-3, Bangkok, Thailand. 10.1109/RAMECH.2006.252754

Sandry, E. (2015). Re-evaluating the Form and Communication of Social Robots. *International Journal of Social Robotics*, *7*(3), 335–346. doi:10.100712369-014-0278-3

Satranç Dünyası. (2009). *İlk Satranç Oynayan Makine (Türk)*. Retrieved from http://www.satranc.net/lk-satranc-oynayan-makinaturk/

Siciliano, B., & Khatib, O. (2019). Humanoid Robots: Historical Perspective, Overview, and Scope. In A. Goswami, & P. Vadakkepat (Eds.), Humanoid Robotics: A Reference. The Netherlands: Springer.

Social Tables. (2019). *6 Hotel Brands Leading with the Way with Using Robot Technology*. Retrieved from https://www.socialtables.com/blog/hospitality-technology/hotel-brands-robot/

Sprenger, M., & Mettler, T. (2015). Service Robots. *Business & Information Systems Engineering*, *57*(4), 271–277. doi:10.100712599-015-0389-x

Suddrey, G., Jacobson, A., & Ward, B. (2018). *Enabling a Pepper Robot to provide Automated and Interactive Tours of a Robotics Laboratory*. Retrieved from https://arxiv.org/pdf/1804.03288.pdf

Sun Tung, V. W., & Au, N. (2018). Exploring Customer Experiences with Robotics in Hospitality. *International Journal of Contemporary Hospitality Management*, *30*(7), 2680–2697. doi:10.1108/IJCHM-06-2017-0322

Sun Tung, V. W., & Law, R. (2017). The Potential for Tourism and Hospitality Experience Research in Human-Robot Interactions. *International Journal of Contemporary Hospitality Management*, *29*(10), 2498–2513. doi:10.1108/IJCHM-09-2016-0520

Travelzoo. (2016). Travellers Expect Robots on Their Holidays by 2020. Retrieved from https://press.travelzoo.com/robophiles--robophobes--britons-divided-over-use-of-robots-in-travel/

Trejos, N. (2016). *Introducing Connie, Hilton's New Robot Concierge*. Retrieved from https://www.usatoday.com/story/travel/roadwarriorvoices/2016/03/09/introducing-connie-hiltons-new-robot-concierge/81525924/

TÜBİTAK. (2016). *Yeni Sanayi Devrimi Akıllı Üretim Sistemleri Teknoloji Yol Haritası*. Retrieved from http://www.tubitak.gov.tr/sites/default/files/akilli_uretim_sistemleri_tyh_v27aralik2016.pdf

Tussyadiah, I. P., & Park, S. (2018). Consumer Evaluation of Hotel Service Robots. In B. Stangl, & J. Pesonen (Eds.), *Information and Communication Technologies in Tourism 2018*. Cham, Switzerland: Springer. doi:10.1007/978-3-319-72923-7_24

Whitley, D. (2018). *Chicago's Hotel EMC2 employs robot butlers to deliver room service to guests*. Retrieved from https://www.thesun.co.uk/uncategorized/7946799/chicagos-hotel-emc2-employs-robot-butlers-to-deliver-room-service-to-guests/

Windsor, M. (2018). *What is the future of service robots*? Retrieved from https://www.eenewspower.com/news/what-future-service-robots/page/0/1

Wirtz, J., Patterson, P. G., Kunz, W. H., Gruber, T., Lu, W. N., Paluch, S., & Martins, A. (2018). Brave New World: Service Robots in the Frontline. *Journal of Service Management*, *29*(5), 907–931. doi:10.1108/JOSM-04-2018-0119

Yalçın Kayıkçı, M., & Kutluk Bozkurt, A. (2018). Dijital Çağda Z & Alpha Kuşağı, Yapay Zeka Uygulamaları & Turizme Yansımaları. *Tekirdağ Namık Kemal Üniversitesi Sosyal Bilimler Enstitüsü Dergisi: Sosyal Bilimler Metinleri*, *3*(1), 54–64.

Yell Robot. (2019). Robot Wedding Photographer can Recognize Guests. Retrieved from https://yellrobot.com/robot-wedding-photographer-eva/. Accessed on June 19, 2019.

Yu, C.-E., & Boyol Ngan, H. F. (2019). The Power of Head Tilts: Gender and Cultural Differences of Perceived Human vs Human-Like Robot Smile in Service. *Tourism Review*, *74*(3), 428–442. doi:10.1108/TR-07-2018-0097

ADDITIONAL READING

Coeckelbergh, M. (2010). Robot Rights? Towards a Social-Relational Justification of Moral Consideration. *Ethics and Information Technology*, *12*(3), 209–221. doi:10.100710676-010-9235-5

Ford, M. (2015). *Rise of the Robots*. New York: Basic Books.

Jörling, M., Böhm, R., & Paluch, S. (2019). Service Robots: Drivers of Perceived Responsibility for Service Outcomes. *Journal of Service Research*, *22*(4), 404–420. doi:10.1177/1094670519842334

Kanda, T., & Ishiguro, H. (2013). *Human-Root Interaction: Social Robotics*. New York: CRC Press.

Lin, P., Abney, K., & Bekey, G. (2011). Robot Ethics: Mapping the Issues for a Mechanized World. *Artificial Intelligence*, *175*(5-6), 942–949. doi:10.1016/j.artint.2010.11.026

Parisi, D. (2014). *Future Robots*. Philadelphia: John Benjamins Publishing Company. doi:10.1075/ais.7

Robertson, J. (2014). Human Rights vs. Robot Rights: Forecasts from Japan. *Critical Asian Studies*, *46*(4), 571–598. doi:10.1080/14672715.2014.960707

Stock, R. M., & Merkle, M. (2017). A Service Robot Acceptance Model: User Acceptance of Humanoid Robots During Service Encounters, *1st Workshop on Emotion Awareness for Pervasive Computing qith Mobile and Wearable Devices*, 13th-17th March, Hawaii, The USA. 10.1109/PERCOMW.2017.7917585

Stock, R. M., & Merkle, M. (2018). Can Humanoid Service Robots Perform Better Than Service Employees? A Comparison of Innovative Behavior Cues, *51st Hawaii International Conference on System Sciences*, 3rd-6th January, Hawaii, The USA. 10.24251/HICSS.2018.133

Van Pinxteren, M. M. M., Wetzels, R. W. H., Rüger, J., Pluymaekers, M., & Wetzels, M. (2019). Trust in Humanoid Robots: Implications for Services Marketing, *Journal of Services Marketing*, doi/. doi:10.1108/JSM-01-2018-0045

KEY TERMS AND DEFINITIONS

Henn-na Hotel: A hotel that is operated by only robots.

HRI (Human-Robot Interaction): Interaction between social or service robots and humans.

Mechanical Turk: A mechanical chess player.

Robot: An automatic, electrically powered device that can be used to perform a specific job or various tasks when programmed, used in place of human in hazardous work in factories.

Robotic Applications in Hotels: Usage of robotic Technologies such as social or service robots and chatbots in hotels.

Service Robot: A robot that assist, help and support people for their own environment.

Social Robot: A robot that can manage and interact with human and other autonomous systems within the assigned social rules and the tasks assigned to it.

Chapter 15
Smart Hotels and Technological Applications

Ali Dalgic
Isparta University of Applied Sciences, Turkey

Kemal Birdir
https://orcid.org/0000-0003-1353-3618
Mersin University, Turkey

ABSTRACT

Technological developments and demographic changes have started to encourage the hotel businesses to use automation systems and adapt the concept of smart hotel. Smart hotels use state-of-the-art technology to provide customers with different and more technology-oriented experiences. A robot works as receptionist, concierge, bellboy, and housekeeper in such hotels. These applications are observed to minimize service failures and save on costs. For instance, chatbots created in smart hotels by means of artificial intelligence respond to customers' problems and demands quickly and accurately. Virtual reality applications used in the marketing of smart hotels make it possible for guests to take a virtual tour of the cultural heritage sites of the destinations. Internet of things (IoT) is another technology that is often used in smart hotels to offer customers a more comfortable holiday and get instant data. Likewise, in smart hotels, self-service software technology is often used and enables customers to perform many tasks such as check-in / out, getting hotel room keys, etc.

INTRODUCTION

Changes in micro and macro environment have necessitated innovations in all the businesses. It is extremely difficult for companies to embrace macro scale changes (demographic, socio-cultural, technological and economic changes etc.) and adopt these changes to their businesses in a short span of time. Therefore, it is important to follow trends and use new products and services to adopt to these very fast and technology-driven changes. It is possible to say that hotel businesses are using more automa-

DOI: 10.4018/978-1-7998-1989-9.ch015

tion systems and some hotel businesses have already started to use smart hotel concepts in almost all areas(Dalgıç and Birdir, 2017: 317).

Buhalis and Leung (2018) stated that "smartness refers to the integration of network of organizations and smart features that engage in interoperable and interconnect systems to simplify and automate daily activities and do add value throughout the ecosystem for all stakeholders". Hotel businesses have integrated robots into their business processes to take advantage of artificial intelligence in order to attract customers and find solutions to customer problems. In hotels, robots started to perform the tasks that human beings would do, and they now work as a receptionist, concierge, bellboy, and housekeeper. In addition, Chatbots with artificial intelligence were created and customers' requests and demands can now be answered 24 hours a day (Muthumanicam, 2018). Hotel businesses also use IoT to offer a more comfortable holiday experience to their customers and create a giant network among the objects at the hotel. In the light of internal and external data taken from the objects, hotel businesses provide a better holiday environment for their customers while creating a huge database for themselves (Kansakar, Munir and Shabani, 2017: 1; Buhalis and Leung, 2018: 41).

The technological uses of hotels are not limited to robots, artificial intelligence and the IoT. In recent years, many hotel businesses have started to use virtual reality applications within the hotels in order to provide a better experience and to promote their hotels. The virtual reality application enables potential customers to visit the hotel before booking their stay (Beck and Egger, 2018: 3-4). Customers who purchase the service may have the opportunity to travel around the hotel and the region's important cultural heritage with virtual reality applications without leaving the hotel and their rooms (Terry, 2016). Self-service software is another technological application that is frequently used in the hotel businesses. With the help of self-service technologies, customers can perform check-in / out, pick up hotel room keys and view their account / folio without any support from hotel employees (Rosenbaum and Wong, 2015: 1863).

Robots used in smart hotels, artificial intelligence, IoT, virtual reality and self-service technologies create many advantages for hotel companies. They reduce the costs, create a positive brand image, provide targeted marketing opportunities, enable customers to experience different technologies and create competitive advantage against other businesses. On the other hand, such technologies offers customers an opportunity for different experiences during their stay, quick solutions to their problems, and instant access to information both inside and outside the hotel. This chapter aims to present robots, artificial intelligence, IoT, virtual reality and self-service technologies and help readers discover different intelligent technologies used in such hotels and the innovations offered to customers.

BACKGROUND

Smart Hotels

Nowadays, the number of people who want to stay in luxury and more comfortable hotel rooms is increasing. Most of these people want to be in control of the service they are receiving during their stay. The managers who are aware of this trend have introduced the smart hotel room as a new product to serve the needs of this trend (Royal-hospitality, 2016). As a striking example for instance, all rooms at the Wynn Las Vegas Hotel are smartly designed using high- tech applications. Customers have the chance to control the room's amenities and available services by verbal commands. Some of these applications are turning on the light, turning off the curtains, turning on the television, and lowering the room temperature, etc. (Ting, 2017).In such technology driven hotels, the smart concept is evident from the moment one enters to the hotel. It is no longer a dream for a robot to meet and serve you, call you by your name and offer you a refreshment as you enter the hotel door. At Hilton Worldwide, a concierge named Connie uses cognitive technology to communicate and interact with customers (Terry, 2016). In addition, smart hotels started to employ robots as receptionists. Henn-na Hotel in Japan for example has three robot receptionists that use AI technology to provide the services and welcome that any receptionist supposes to offer to guest. Although the appearances of these receptionists are different (dinosaur, human, small robot, a famous celebrity, etc.), each of them knows dozens of languages and responds to customers' problems in a much efficient and service-oriented approach. In smart hotels, there are many technological devices such as automatic check-in and eye scanning systems (Soegjobs, 2016); the internet facilities are endless, and the highest level of technology is being used to meet the needs and wants of hotel guests. Many services are easily available on the internet (meal orders, special requests, room reservations and cancellations, direct payment, etc.). Powerful wi-fi connection is also provided on the beaches of the hotels so that customers can communicate their requests via their telephone, tablet or laptop (Conroy, 2016).

The hotel enterprises, known as a labour-intensive sector, is slowly adapting to technological developments in order to provide fast, accurate, satisfactory and cost efficient service to their guests. Robots in hotels do the tasks that people can do and can work as a receptionist, concierge, bellboy, housekeeper and almost all the roles a human employee can perform at a hotel. This means that guests are offered different experiences while the costs are reduced, and business processes are easily facilitated for hotels (Smart Hotel Technology Guide, 2018). Virtual technology is another technology used by hotel businesses that want to offer different experiences to their guests. Smart hotels can offer virtual and real-time marketing activities, and the opportunity to visit the hotel's facilities and cultural heritage sites whilst the guests are in their rooms (Dalgıç and Birdir, 2017: 317). A further technology used in smart hotels is artificial intelligence technology. Chatbots created with artificial intelligence technology can respond to guests' requests 24 hours a day. This technology can help hotel businesses to increase customer satisfaction by providing quick and on-time real service (Muthumanicam, 2018). On the other hand, the IoT technology provides data transfer and analysis of the items in the hotel through a network. This technology provides customers with a comfortable holiday and creates big data (Buhalis and Leung, 2018: 41). Self-service software technology used in smart hotels enable guests to do many services such as check-in / check-out, payments and change of booking (Rosenbaum and Wong, 2015: 1863).

Smart hotels offer the customer a high level of technology experience as well as many advantages for hotel managers. With smart technology, energy and water consumption is less than other hotels. In addition to maximum comfort in rooms, security is at the top level. These increase the service quality (Petrevska, Cingoski and Gelev, 2016: 202). The most striking problem in smart hotels is that automation may not be preferred by all customers as it reduces the interaction and communication with humans while they are staying at the hotel. Moreover, the increase in automation in smart hotels will also cause redundancy and there will be fewer employment opportunities in tourism sector.

MAIN FOCUS OF THE CHAPTER

Technological Applications for Smart Hotels

Robotics

The interdisciplinary work of computer and engineering sciences made robots to become capable of autonomous mobile, to develop the ability to perform given tasks and to provide a data-driven perspective while performing their tasks (Smart Hotel Technology Guide, 2018). The use of robots is becoming more widespread in the hospitality sector. In the accommodation businesses, receptionist, concierge, bellboy, housekeeper services are now being replaced by robots (QR Code 1 and QR Code 2). On the other hand, robots are also used in all other service areas such as information provision and guidance at train stations, tour guides in museums, surveillance at airports and other public areas, cleaning robots at airports, and even dancing robots for events etc. (Ivanov and Webster, 2019a). International Federation of Robotics (IFR, 2016) defines a service robot as "a robot that performs useful tasks for humans or equipment excluding industrial automation application".There are three basic levels of service robots in the hotel business. The first level is the hardware and mechanical structure including sensors and actuators. At this level, the robot can consist of two main parts; a mobile base for navigating through the hotel and an anthropomorphic body to interact with people and hotel guests. The second level covers the control architecture of the robot. Architectural design is difficult because it consists of many elements such as different software components, programming techniques and modelling approaches, navigation techniques, interfaces, communication and recognition systems, and cognitive modelling. The third level is the application level. At this level, the services to be provided by the robot are defined (Pinillos, Marcos, Feliz, Zalama and Gómez-García-Bermejo, 2016: 41-42). Some of the important services that can be defined at this level are giving information about the hotel facilities, providing on-line information from the internet, giving event information, taxi call, breakfast control, videoconference service to the hotel desk and accompanying guests (Pinillos, Marcos, Feliz, Zalama and Gómez-García-Bermejo, 2016: 43).

Robots are used in helping customers at reception, handing suitcases to guest rooms and collecting suitcases, storing and bringing suitcases in luggage rooms, taking food orders to the rooms, baking eggs and pancakes during breakfast (Smart Hotel Techology Guide, 2018; Ivanov and Webster, 2019b). The Jen Shangri-La hotel uses robots to send items to the rooms. Henn-na Hotel in Japan is the first robotic hotel to offer fully automated customer services to its customers and has around 80 robots (Alexis, 2017). Robots work in many different positions such as concierge robots, cleaning robots, and arm robots carrying luggage. (Osawa, Ema, Hattori, Akiya, Kanzaki, Kubo, Koyama and Ichise, 2017: 241). There is a concierge robot that carries luggage of customers and greets them at Hilton Hotel (Buhalis and Leung, 2018: 44). At the Aloft Hotel in Starwood, Cupertino, there is an avid robotic butler called Boltr that provides guests with facilities such as toothbrushes and towels (Crook, 2014). The robot Sacarino works as a bellboy in a hotel in Valladolid, performs many jobs and undertakes the following tasks (Pinillos, Marcos, Feliz, Zalama and Gómez-García-Bermejo, 2016: 43);

- It provides visual and audio information about hotel facilities, dining times and restaurant services. Guests can also get information about the city, shopping centres and museums. The interaction between Sacarino and guests is being carried out by sound and touch screen. In addition to written and verbal information on the screen, guests are informed through images and maps.

- Sacarino provides weather forecasts, daily news, fun information (such as jokes and proverbs) and has more than 5000 jokes and proverbs databases.
- Sacariono also provides information about facilities including the activities such as conferences to be held at hotel. Sacarino conveys details such as event description, meeting room and program to the guests verbally and on screen.
- Sacarino arranges a taxi for guests. To avoid bad practices, guests fill out a form via the Sacarino web page.
- Sacarino can check guests for breakfast, waits at the door of the restaurant to ask guests their room number through laser scanner.
- Sacarino provides video conference service. If it cannot find an answer to a guest's question, it allows guests to connect hotel desk and have a video call via skype.
- Sacarino can accompany guests. When any guest wants to go somewhere within the hotel, Sacarino asks that the guest to follow him.

BellBot, developed in Spain, can provide the following services to satisfy the hotel guests (López, Pérez, Zalama and Gómez-García-Bermejo, 2013: 2);

- It can give commands to elevators and escalators in transportation of people. This robot, which senses the arrival of people, can send commands to the central system for the arrival of the elevators.
- BellBot can be used to control smoke detectors and control alarm systems. When one of the smoke detectors is activated, BellBot can control the scene.
- It can perform different tasks through sensors. For example, the activation of a sensor in the newspaper box can start the delivery of a newspaper task.

The use of robots in hotel enterprises has many advantages. First of all, hotel businesses can reduce labour costs by using robots (Ivanov and Webster, 2018: 192). Besides, it is possible to increase the efficiency by facilitating the business processes with robots. The robots used in hotel businesses can be seen as an innovation for customers and this may increase the rate of preference of hotel. It may be possible to increase hotel brand awareness through innovative marketing. One of the biggest advantages of robots for hotel businesses is that they minimize the problems that may arise in terms of communication and overcome the language barrier (Smart Hotel Technology Guide, 2018). There are some disadvantages

Figure 1. Robots in hotels

QR Code 1.A Dinosaur Robot Receptionist

QR Code 2.Robot Assistant in Lobby

related to the use of robots in hotels. Hotel customers not belonging to generation Y and Z may wish to communicate and interact with people and robots may cause dissatisfaction for them. In addition, the robots cannot perform service compensation as a result of service failures, so hotels may lose customers. Robots can lead to unemployment and cause qualified workers to leave from the sector.

Artificial Intelligence

Artificial intelligence is a field of computer science that enables machines to work and behave like humans (Smart Hotel Techology Guide, 2018). Basically, artificial intelligence refers to computers or machines that traditionally perform tasks that require cognitive function (Madakam, Ramaswamy and Tripathi, 2015). Artificial intelligence can be used in hotel business, customer service, sales processes and perform the tasks that people can do in hotel businesses. Hotel owners can benefit from this situation to save money, eliminate human errors and provide superior service (Revfine, 2019a).All existing and future hotels will need an ecosystem and an acquiring, contextualizing, processing and analyzing platform to attract customers to their hotels and create a good customer experience. Such a system is possible with artificial intelligence. Artificial intelligence can establish a system to help transfer all the data obtained from hotel departments. Bookings, transactions, satisfaction surveys and third-party providers, and data obtained from hotel personnel, are all valuable sources for hotels to interact and provide better and timely service to their guests. These data can be processed with artificial intelligence (AI), to communicate with customers, to provide customized services and to create promotional offers (Muthumanicam, 2018) and to analyze data. The Dorchester Collection chain of hotels collects data about their business through surveys and online comments and draws conclusions about the overall performance (Revfine, 2019a).

Artificial intelligence in hotel businesses is often used to respond to customer needs and to meet their needs. In hotels, there are several applications that provide online chat services to respond questions and requests, created via artificial intelligence (Li, Bonn and Ye, 2019: 173). Chatbots are used to answer

Figure 2. Chatbot

QR Code 3.Chatbot

customer questions 24 hours a day on the web sites of hotels or social media platforms(Lasek and Jessa, 2013). This is invaluable in terms of ensuring constant interaction and communication for hotel businesses (Muthumanicam, 2018). Guests at the Las Vegas Cosmopolitan Hotel are given a card by the receptionist upon check-in. This card contains information about Chatbot (QR Code 3) called Rose. Guests can get in touch with Rose to get information about hotel facilities and event times(Tkaczyk, 2017).

Hotel businesses can use artificial intelligence to receive requests, put guests in touch with the relevant department, respond to online reviews, make customized recommendations, and perform targeted digital marketing with customized information (Smart Hotel Technology Guide, 2018). The use of artificial intelligence brings advantages such as data-driven marketing, upsell and cross-sell activities, continually evolving segmentation of guest from all markets and backgrounds, and develop super-enhanced guest profiles (Muthumanicam, 2018) to better understand and serve their needs and wants. In addition, artificial intelligence can reduce queues, create facilities for guests, provide a better guest experience and increase profitability with highly focus and targeted marketing (Smart Hotel Technology Guide, 2018).

Virtual Reality (VR)

Virtual reality can be defined as the virtual world in the computer environment that enables individuals to get real experiences related to their needs and wants, trips and excursions, destinations to visit and various activities in the targeted destination (Desai, Desai, Aimera and Mehta, 2014). Virtual reality systems can be categorized in non, semi- or fully-immersive. The less the user can perceive from the real world on the outside, the more immersive the system. Non-immersive systems are defined as desktop-based virtual reality systems and are known for their low cost and ease of installation and use. Semi-immersive systems offer a multi-user experience that provides 3D sound equipped with large projection screens. Fully-immersive systems isolate the user from the real world and often offer a single user experience (Gutiérrez, Vexo and Thalmann, 2008; Dörner, Jung, Grimm, Broll and Göbel, 2013; Beck, Rainoldi and Egger, 2019). In addition, virtual reality systems have two different structures as synthetic and captured virtual worlds. In the first years of virtual reality technology, virtual worlds were made with synthetic computer. This trend has changed in recent years and 360-degree images and videos are being used in the real world and mounted on virtual reality systems (La Valle, 2016; Wei, 2019).

In the hospitality industry, information and promotions were mostly advertised through brochures, catalogues and 2-dimensional (2D) web pages. These materials used in promotional activities have been replaced with 3-dimensional form in recent years. Instead of 2D media, virtual reality technology offers sensory experiences and multi-sensory modes and media content. Potential customers who want to get detailed information about the accommodation facility through virtual reality can experience the hotel, the details of the rooms, the range and services of amenities and more (Huang, Backman, Backman and Chang, 2016: 116; Beck and Egger, 2018: 3-4). Virtual experiences enable a person to get experiential knowledge in an interactive multimedia environment as well as rich, perceptual and cognitive knowledge (Beck and Egger, 2018: 7; Tussyadiah, Wang, Jung and Dieck, 2018).

Virtual reality applications in smart hotels are also increasing in recent years (QR Code 4). Virtual reality offers its customers the opportunity to visit all the places they want with a pair of glasses from the hotel room (Soegjobs, 2016). The Marriott Hotels Group has piloted the virtual reality program to enable customers to navigate different exotic areas. The customers will be able to experience a lot of places using virtual reality glasses and programmes (Terry, 2016). In the following years, virtual reality technology is likely to become widespread. It will be one of the most desired features in hotel businesses

Figure 3. Virtual reality technology in smart hotels

QR Code 4.A Businessman in the Hotel Room

as customers will be able to visit different locations without paying extra fees (tours, guides, food, beverages, etc.). Another advantage is security. Customers will be able to navigate the unknown destinations without security concerns. Although virtual reality has positive advantages in terms of customers and hotel management, it may have economic negative effects for some other stakeholders. Customers experiencing the places through virtual reality will not buy any kind of services such as tours, guidance, food & beverage, souvenirs or any other goods and services. Therefore, stakeholders and tourism destinations can be economically affected and may lose jobs and money (Dalgıç and Birdir, 2017: 317).

Virtual reality technology has a lot of benefits for customers and accommodation facilities. Hotel managers can inform customers via virtual reality system and use this system for marketing purposes. Virtual reality technology helps gaining the trust of potential customers and increase sales. It creates easier sales to customers abroad. In addition, it will create a sense of curiosity about the destination and thus provides cross selling opportunities. The virtual reality system offers customers the opportunity to experiment before they book their rooms. After the purchase, it creates a destination experience and offers the opportunity to visit the cultural heritage sites in the destination (Smart Hotel Techology Guide, 2018).

Augmented Reality (AR)

AR is a technology in which the interaction between the objects is created by combining the virtual and the real world. As a result of the integration of the objects into the real world, this technology enables synthesis real and virtual image with this technology (Chung, Han and Joun, 2015: 589).AR applications can be classified into "marker-less" and "marker-based" (Jung, Chung and Leue, 2015: 76). Marker-based AR "requires specific labels to register the position of 3D objects on the real-world image" (Cheng and Tsai, 2013: 451). Marker-based AR technology uses predefined and detectable computer vision techniques (Jung, Chung and Leue, 2015: 76).On the other hand, marker-less AR applications do not require code. They can detect certain features from area-based GPS locations and can be used in outdoor environments (QR Code 5)(Patkar, Singh and Birje, 2013: 66).

Recently, AR technology has been used frequently by hotel businesses especially in marketing activities. Potential customers can get real images and written information on many issues such as hotel, hotel's destination, and hotel rooms using their smartphone, tablet or cameras (İlhan and Çeltek, 2016: 582).

In addition to promotional and marketing activities, hotel businesses can use AR technology for their guests to have a fun holiday. For instance, The Hub Hotel has hanged maps on the hotel room walls for its customers. Customers staying in hotel rooms can get information about local attractions when they point their smartphone to the points on the map and enjoy a comfortable and enjoyable stay. Likewise, Holiday Inn has placed signs for guests. Guests holding their smartphones to these signs encounter realistic virtual depictions of celebrities. A similar application has been developed by Best Western Hotel. This hotel placed signs at certain points for guests' children, and children who read these signs via smartphone or tablet see themselves with Disney characters (Revfine, 2019b).

Internet of Things (IoT)

The IoT can be specified as a giant network of inter-connected objects. Data exchange and analysis are provided through this network. Hotel businesses can use this data to introduce new business ideas and benefit from different opportunities to make more money and even gain new customers (Smart Hotel Technology Guide, 2018). IoT is the interconnection where daily physical devices, such as sensors, actuators, identification tags, mobile devices can communicate directly or indirectly with each other via local networks or the Internet (Kansakar, Munir and Shabani, 2017: 1; Cavada, Elahi, Massimo, Maule, Not, Ricci and Venturini, 2018: 351). IoT can be separated into three layers. First layer is *smart systems* which comprises physical objects augmented with sensing and communications capabilities to form smart objects. Data acquisition is achieved within this layer. Another layer is *connectivity*. This layer comprises the wired and wireless networking infrastructure that connects the various systems. The primary function of the connectivity layer is data transmission. The last layer is *analytics or sense-making*. This layer is involved in the processing of data (sense-making) to yield higher order information. This information can be used to actuate other IoT objects (Chuah, 2014: 216). Massive sensory information storage, computing and processing is the core of the IoT. Data from individual objects can be collected from internal and external environments and be analyzed by the intelligent system. Human intervention is minimal (Alsaadi and Tubaishat, 2015). Intelligent lighting solutions use data from sensors to control lighting fixtures in the room. Smart thermostats utilize the climate (such as temperature and humidity) information inside the room for heating and ventilation. Both applications are important for lowering energy consumption and providing better comfort (Chuah, 2014: 218). Energy saving creates advantages

Figure 4. Augmented reality technology

QR Code 5.AR Information Technology on Smartphone Screen

for IoT hotel in terms of effective waste and green management, and recycle and reuse programs. IoT technology can be used in many parts of the hotel business. Table 1 provides detailed information on the use and functions of IoT technology in hotel businesses.

Marriott International uses items that can collect data and interact with each other in hotel rooms. This system improves the comfort of the guests by collecting data such as room temperature and luminance. In addition, external information such as the weather and traffic congestion in the hotel location can be obtained through the IoT (Buhalis and Leung, 2018: 44). For instance, Starwood Hotels and Resorts' daylight harvesting is an energy-saving program that automatically saves energy-saving LED lighting according to natural light entering the hotel room, and it improves indoor lighting consistency (Kansakar, Munir and Shabani, 2017: 3). Hotels such as Mondrian SoHo, The Plaza and The Marlin have placed tablets in their rooms to check in-room functions. The Peninsula Hotels have developed proprietary in-room tablets for its guests to benefit from services such as order room service, message the concierge, transportation services, and the possibility to choose the movies they want to see in their hotel rooms. (Kansakar, Munir and Shabani, 2017: 2). Hilton and Starwood hotels offer guests automatic check-in and keyless entry via mobile apps (Kansakar, Munir and Shabani, 2017: 2).

The IoT can also help hotel businesses collect data such as marketing statistics (economic, political and environmental data). This data can be used in decision making processes of hotels by creating and processing big data. All objects in the hotel are connected to each other by a network of objects and they play an active role in the creation of big data (Borgia, 2014: 3; Martins, Ramos, Rodrigues, Cardoso, Lam and Serra, 2015; Buhalis and Leung, 2018: 41). Generally, hotel businesses can use IoT to inform employees about location and to welcome guests before arriving at the hotel. IoT can also be used to personalise way-finding via sensing of mobile app presence, activate lift access via sensing of mobile app presence, utilise mobile devices as electronic keys, deliver targeted personalized marketing promotions, and prepare real-time reports for repair of defective items. IoT has benefits such as creating better guest experience, increasing revenue with target marketing, streamlining business processes, and helping overcome inconveniences caused by faulty assets in guestrooms (Smart Hotel Technology Guide, 2018).

Cloud Computing

Today's businesses use information technology strategies that include cloud computing (Oliveira, Thomas and Espadanal, 2014: 497). Cloud computing is a technology that provides flexible, economic and convenient access to information from anywhere, beyond the classical information technology infrastructure (QR Code 6) (Garrison, Kim and Wakefield, 2012). Cloud Computing can be defined as information and communication services on common resources. It can be scaled according to needs and it is ready for immediate use and easy to manage. In other words, cloud computing is a general concept that expresses information and use communication technology services (Armutlu and Akçay, 2013: 24).The services that cloud computing offers to its users are available in 3 models: IaaS (platform as a service), PaaS (platform as a service) and SaaS (software as a service) (Velte, Velte and Elsenpeter, 2010). IaaS stands for infrastructure services such as servers, data center space. Instead of owning such infrastructures, hotel businesses can increase the capacity of these infrastructures at any time. PaaS, on the other hand, is an information system that enables hotel businesses to create applications and collaborate on projects without purchasing and maintaining any infrastructure, servers, or data center space. Guests of the hotel facilities can access this platform as long as they have internet connections. Finally, SaaS allows hotel businesses to quickly access existing or new software. In this way, hotel companies can easily access the

Table 1. IoT location and functions

IoT Location	Type of IoT/sensor	Function
Guest Room	Movement sensor	According to the presence and location of the guest in the room, the ambiance can be adjusted with energy management systems. This may include lighting and head adjustments and water management.
	Voice sensor	Curtains, lighting, room temperature conditions can be set by the guest with voice activation systems that control in-room devices.
	Temperature sensor	Room temperature measurement devices can be used to keep guests in a more comfortable environment.
	Door lock	Mobile applications can be used as a room key for the door lock system.
	Wearable sensor	Monitor guest health situation during their workout and provide detailed information about their health status.
Restaurant & Lobby	Location sensor	Guest locations can be identified, and guests can receive a welcome message and invitations to various events.
Hotel facilities	Availability beacon	Delivery availability notifications to hotel guest can be managed.
Warehouse	Inventory tag	Detects the profile and position of objects, helps to determine the expiration date and stock level. Can also help stock orders and re-stocking goods and materials.
Building	Temperature sensor	It helps to measure the outside temperature and adjusts the energy management systems.
	Light sensor	Detects sunlight and helps to adjust the lighting system and brightness.
Roadside	Traffic sensor	It helps to control the parking space and traffic situation.
Social Network	Content sensor	It helps to examine social networks related to the hotel and give feedback to managers.

Source: Buhalis and Leung, (2018: 48)

Figure 5. Cloud computing flowchart

QR Code 6.Cloud Computing Flowchart

software they need (Morch, 2018). Cloud computing has many advantages such as enabling lower-cost computing, reducing IT infrastructure and maintenance costs, upgrading business software quickly, offering unlimited storage capacity, increasing data security, easier group collaboration, universal access to documents, and providing access via one device (Miller, 2008: 24-28).

Self-Service Software Technology

Traditionally, the experience of the hotel guests was based on communicating / interacting with hotel staff, the hotel's services, the comfort of the hotel room, and meeting the expectations of the guests. Thanks to technological developments, guests are now able to use self-service software (Schaap, 2018). Self-service software is a technological product that provides electronic support without the need for interaction with a service representative. Such software provides users with control over their personal experiences. Such services can be realized through kiosks or mobile phones (Kaushik, Agrawal and Rahman, 2015; Smart Hotel Technology Guide, 2018). Examples of such self-service software technologies include self-check-in kiosks (QR Code 7 and QR Code 8), mobile check-in applications, automatic teller machine, and automatic vending machines(Liu and Hung, 2019: 225). Self-service software technologies include check-in / out, collecting hotel room keys, and viewing account / folio. It also includes services such as maps, information about the hotel, coupons for hotel activities, making reservation changes and, room selection. (Rosenbaum and Wong, 2015: 1863).

Resort World Genting hotel in Malaysia uses kiosks that provide service to guests. Guests who come to these kiosks firstly register their identity cards or passports. Then, they can check their reservations. During this process, it is possible to perform some changes in the reservation (type of room, number of nights, etc.). When the reservation has been confirmed, check-in takes place and payment can be made to the kiosks (credit card or cash). Immediately after payment, guests can also receive electronic room keys (Selfservicesmartsolutions, 2018). There are many benefits of this system such as cost saving, improving efficiency, improving hotel image, and enhancing customer experience. However, this system also has disadvantages. For instance, it may decrease customer-to-hotel communication and fail to satisfy customers' needs (Liu and Hung, 2019: 229-231).

Overall, self-service software technology offers many opportunities for both businesses and customers. It collects customer information, their accommodation preferences, stores credit card information, reduces customer waiting time and provides personalized marketing promotion. This technology offers

Figure 6. Self-service software technology

QR Code 7.Self Check-in Kiosks

QR Code 8.Customers Use Self-ordering Kiosks

customers advantages such as ID identification, generating mobile key cards, checking the functions in the room, making requests directly from the department (F & B, housekeeping etc.), providing real-time information about hotel facilities and ease of payment (Kim, Christodoulidou and Brewer, 2012; Smart Hotel Technology Guide, 2018).

SOLUTIONS AND RECOMMENDATIONS

Tourism enterprises continue to develop different implementations and polices to keep up with complex and dynamic changes. They prioritize technological applications to increase the satisfaction of today's ever demanding customers, provide customer-oriented fast service, lower operating costs, make more profit and develop competitive advantage. In some tourism enterprises, technological applications have been carried out at an advanced level and the concept of smart hotel has been introduced to their systems. In smart hotels, it is possible to see robots in many departments, artificial intelligence on hotel websites, virtual reality in hotel promotion efforts, internet of thing applications in rooms and self-service software technology applications in hotel entrances. Application of similar technologies is expected to increase as the technology advances. Therefore, following suggestions could be proposed for tourism enterprises:

- First, hotels should provide necessary infrastructure before proceeding with technological applications. They should also investigate the potential benefits and harms of the technological applications to be used.
- Service robots, which can be seen in almost all smart hotels, can bring business advantages such as reducing labor costs, timely execution of business processes and engaging with potential customers. However, the issue that companies should pay attention to is to provide continuous control. In addition, the recovery of service failures caused by the robots should be addressed by the hotel managers. The robots will be inadequate in terms of service recovery, so hotel managers will need to work on service recovery to avoid customer complaints.
- In hotels, it is important to meet the demands of the guests quickly and to produce quick solutions to their problems. Hotel managers, who are aware of this, use artificial intelligence applications in hotel rooms. These applications can respond directly to customers' requests. However, some problems may arise. One of the biggest problems is the misunderstanding of customers' requests and a wrong response from the technology. It is important that hotel managers constantly assign employees who will check how the requests have been responded and take action in case of any misunderstanding and miscommunication.
- In recent years, hotel enterprises have been using virtual reality applications to support their marketing activities. These applications make it possible to present the hotel rooms, hotel units, the surrounding area and the hotel destination as a real experience to potential customers. The fact that these applications may not always reflect the real circumstances and facilities of hotels to potential customers may create poor image for hotels. In this case, it is important to present the reality. Another important issue is that customers can experience the monuments and historic places around the hotel with virtual reality application from the rooms they stay. Some hotels offer this app as a service to their customers. It should be noted that hotels offering this service may have a negative impact on local residents, travel agencies and tour guides due to loss of business.

- One of the most important problems for smart hotels or hotels using technological applications is customers with different demographic characteristics. Not all customers may want to use technology-related applications. Therefore, it would not be wrong to argue that hotels cannot be successful only with technological applications, and they will always need human touch for different customers and different services. Lastly, under all circumstances and technological developments hotel managers should continue to hold onto to the qualified employees in tourism industry which will continue to be a human oriented business.

FUTURE RESEARCH DIRECTIONS

Technological applications are more common than before in tourism enterprises. Therefore, it is important to study the benefits hotels gain through the use of technological applications. The disadvantages of the technological applications should also be explored. Furthermore, research that will evaluate the opinions of customers in smart hotels will be important for hotel managers. Thus, hotel managers will be able to see the applications they need to develop for better serving their guests. Finally, it is important to investigate what service failures the robots have made and what service recoveries will be available in order to prevent customer losses in hotels.

CONCLUSION

Emerging technology and demographic change have started to change labour-intensive characteristics of hotel enterprises. Hotel businesses benefit from technological developments and the concept of smart hotel is becoming more common. Smart hotels benefit from technological software and applications, connect the existing items within the hotel with a network and operate robots like humans. Robots can be seen at the hotel entrances and receptions of smart hotels and they work as receptionists, concierge, bellboy, and housekeeper, which in turn reduce the costs of the businesses. In addition to robots, virtual reality applications are widely used in smart hotels. Virtual reality is a practice that creates a competitive advantage by using potential customers for promotion and information. Customers can choose to visit the hotel's facilities and rooms prior to purchase. In addition, customers use virtual reality to experience cultural heritage destinations before traveling. One of the technological applications used by hotel businesses in recent years is AR. Through AR, hotel businesses strengthen their promotional and marketing activities and ensure that their guests have an enjoyable stay.

Another technology used in smart hotels is IoT. IoT is a gigantic network that allows data exchange within the hotel. Smart hotels benefit from the internet of things to enhance comfort in the rooms and to create big data. Smart hotels also use cloud computing technology to reduce information technology costs and access data quickly and reliably. With the help of cloud computing, hotel businesses can develop an advance and unrestricted data storage infrastructure. Smart hotels also benefit from artificial intelligence in order to find a quick solution to customer problems and fulfil their demands 24 hours a day. Chatbots, created through artificial intelligence, work for the satisfaction of customers and they can immediately forward requests to relevant departments. A further application that is frequently used in smart hotels is self-service software technology. With this technology, customers can perform many operations such as check-in / out, receiving hotel room keys, and viewing their account / folio.

All technological applications used in smart hotels are designed to increase customer satisfaction. Robots may reduce the human source service failures and provide customers with different experiences. Chatbots, created with artificial intelligence, are designed to find solutions to customers' problems and to increase their satisfaction by fulfilling their wishes. Similarly, the IoT is used to ensure that customers spend their holidays comfortably in their rooms. With this technology, it is aimed to provide guests with different holiday experiences. All these technologies have advantages such as easing business processes, increasing productivity, saving costs and collecting customer data.

REFERENCES

Alexis, P. (2017). R-Tourism: Introducing the potential impact of robotics and service automation in tourism. *Ovidius University Annals. Series Economic Sciences*, *17*(1), 211–216.

Alsaadi, E., & Tubaishat, A. (2015). Internet of things: Features, challenges, and vulnerabilities. *Int. J. Adv. Comput. Sci. Inf. Technol*, *4*(1), 1–13.

Armutlu, H., & Akçay, M. (2013).Bulut bilişimin bireysel kullanımı için örnek bir uygulama. *Akademik Bilişim Konferansı*, 23-25.

Beck, J., & Egger, R. (2018). Emotionalise me: Self-reporting and arousal measurements in virtual tourism environments. In *Information and Communication Technologies in Tourism 2018* (pp. 3–15). Cham, Switzerland: Springer. doi:10.1007/978-3-319-72923-7_1

Beck, J., Rainoldi, M., & Egger, R. (2019). Virtual reality in tourism: A state-of-the-art review. *Tourism Review*, *74*(3), 586–612. doi:10.1108/TR-03-2017-0049

Borgia, E. (2014). The internet of things vision: Key features, applications and open issues. *Computer Communications*, *54*, 1–31. doi:10.1016/j.comcom.2014.09.008

Buhalis, D., & Leung, R. (2018). Smart hospitality—Interconnectivity and interoperability towards an ecosystem. *International Journal of Hospitality Management*, *71*, 41–50. doi:10.1016/j.ijhm.2017.11.011

Cavada, D., Elahi, M., Massimo, D., Maule, S., Not, E., Ricci, F., & Venturini, A. (2018). Tangible tourism with the internet of things. In *Information and Communication Technologies in Tourism 2018* (pp. 349–361). Cham, Switzerland: Springer. doi:10.1007/978-3-319-72923-7_27

Cheng, K. H., & Tsai, C. C. (2013). Affordances of augmented reality in science learning: Suggestions for future research. *Journal of Science Education and Technology*, *22*(4), 449–462. doi:10.100710956-012-9405-9

Chuah, J. W. (2014). The internet of things: An overview and new perspectives in systems design. In *Proceedings 2014 International Symposium on Integrated Circuits (ISIC)*. Singapore: IEEE Computer Society. pp. 216–219. 10.1109/ISICIR.2014.7029576

Chung, N., Han, H., & Joun, Y. (2015). Tourists' intention to visit a destination: The role of augmented reality (AR) application for a heritage site. *Computers in Human Behavior*, *50*, 588–599. doi:10.1016/j.chb.2015.02.068

Conroy, J. (2016). Hotel trends for 2016. Retrieved from http://www.telegraph.co.uk/travel/hotels/articles/Hotel-trends-for-2016/

Crook, J. (2014). Starwood introduces robotic butlers at Aloft hotel in Cupertino. https://techcrunch.com/2014/08/13/starwood-introduces-robotic-butlers-at-aloft-hotel-inpalo-alto/

Dalgıç, A., & Birdir, K. (2017). Otel işletmeciliğinde yeni trendler. In Y. Akgündüz (Ed.), *Otel İşletmeciliği* (pp. 311–328). Paradigma Akademi.

Desai, P. R., Desai, P. N., Ajmera, K. D., & Mehta, K. (2014). A review paper on oculus rift-a virtual. [IJETT]. *Int. J. Eng. Trends Technol, 13*(4), 175–179. doi:10.14445/22315381/IJETT-V13P237

Dörner, R., Jung, B., Grimm, P., Broll, W., & Göbel, M. (2013). Einleitung. In R. Dörner, W. Broll, P. Grimm, & B. Jung (Eds.), *Virtual und Augmented Reality (VR/AR)* (pp. 1–32). Berlin, Germany: Springer. doi:10.1007/978-3-642-28903-3_1

Garrison, G., Kim, S., & Wakefield, R. L. (2012). Success factors for deploying cloud computing. *Communications of the ACM, 55*(9), 62–68. doi:10.1145/2330667.2330685

Gutiérrez, M. A. A., Vexo, F., & Thalmann, D. (2008). *Stepping into virtual reality*. London, UK: Springer. doi:10.1007/978-1-84800-117-6

Huang, Y. C., Backman, K. F., Backman, S. J., & Chang, L. L. (2016). Exploring the implications of virtual reality technology in tourism marketing: An integrated research framework. *International Journal of Tourism Research, 18*(2), 116–128. doi:10.1002/jtr.2038

IDS. (2014). A hotel system for the future. Retrieved from https://sites.google.com/a/ids.software.ucv.ro/smart-hotel/about#section1

IFR. (2016). Service robots. Retrieved from www.ifr.org/service-robots/

İlhan, İ., & Çeltek, E. (2016). Mobile marketing: Usage of augmented reality in tourism. *Gaziantep University Journal of Social Sciences, 15*(2), 581–599. doi:10.21547/jss.256721

International Organization for Standardization. (2012). ISO 8373:2012(en) Robots and robotic devices—vocabulary. Retrieved from https://www.iso.org/obp/ui/#iso:std:iso:8373:ed-2:v1:en:term:2.2

Ivanov, S., & Webster, C. (2018). Adoption of robots, artificial intelligence and service automation by travel, tourism and hospitality companies. In V. Marinov, M. Vodenska, & M. Assenova (Eds.), *Traditions and Innovations in Contemporary Tourism* (pp. 190–203).

Ivanov, S., & Webster, C. (2019a). What should robots do? A comparative analysis of industry professionals, educators, and tourists. In *Information and Communication Technologies in Tourism 2019* (pp. 249–262). Cham, Switzerland: Springer. doi:10.1007/978-3-030-05940-8_20

Ivanov, S., & Webster, C. (2019b). Perceived appropriateness and intention to use service robots in tourism. In Pesonen, J. & Neidhardt, J. (Eds.) *Information and Communication Technologies in Tourism 2019, Proceedings of the International Conference in Nicosia, Cyprus*, pp. 237-248. 10.1007/978-3-030-05940-8_19

Jung, T., Chung, N., & Leue, M. C. (2015). The determinants of recommendations to use augmented reality technologies: The case of a Korean theme park. *Tourism Management*, *49*, 75–86. doi:10.1016/j.tourman.2015.02.013

Kansakar, P., Munir, A., & Shabani, N. (2017). Technology in hospitality industry: Prospects and challenges. *arXiv preprint arXiv:1709.00105*.

Kaushik, A. K., Agrawal, A. K., & Rahman, Z. (2015). Tourist behaviour towards self-service hotel technology adoption: Trust and subjective norm as key antecedents. *Tourism Management Perspectives*, *16*, 278–289. doi:10.1016/j.tmp.2015.09.002

Kim, J., Christodoulidou, N., & Brewer, P. (2012). Impact of individual differences and consumers' readiness on likelihood of using self-service technologies at hospitality settings. *Journal of Hospitality & Tourism Research (Washington, D.C.)*, *36*(1), 85–114. doi:10.1177/1096348011407311

La Valle, S. M. (2016, Forthcoming). *Virtual Reality. Draft October 31*. Cambridge, UK: University Press.

Lasek, M., & Jessa, S. (2013). Chatbots for customer service on hotels' websites. *Information Systems Management*, *2*(2), 146–158.

Li, J. J., Bonn, M. A., & Ye, B. H. (2019). Hotel employee's artificial intelligence and robotics awareness and its impact on turnover intention: The moderating roles of perceived organizational support and competitive psychological climate. *Tourism Management*, *73*, 172–181. doi:10.1016/j.tourman.2019.02.006

Liu, C., & Hung, K. (2019). Understanding Self-service Technology in Hotels in China: Technology Affordances and Constraints. In *Information and Communication Technologies in Tourism 2019* (pp. 225–236). Cham, Switzerland: Springer. doi:10.1007/978-3-030-05940-8_18

López, J., Pérez, D., Zalama, E., & Gómez-García-Bermejo, J. (2013). Bellbot-a hotel assistant system using mobile robots. *International Journal of Advanced Robotic Systems*, *10*(1), 40. doi:10.5772/54954

Madakam, S., Ramaswamy, R., & Tripathi, S. (2015). Internet of Things (IoT): A literature review. *Journal of Computer and Communications*, *3*(05), 164–173. doi:10.4236/jcc.2015.35021

Martins, D., Ramos, C., Rodrigues, J. M., Cardoso, P., Lam, R., & Serra, F. (2015). Challenges in building a big data warehouse applied to the hotel business intelligence. In *6th Int. Conf. on Applied Informatics and Computing Theory (AICT'15), in Recent Research in Applied Informatics*. 110-117.

Miller, M. (2008). *Cloud computing: Web-based applications that change the way you work and collaborate online*. Que publishing.

Morch, A. (2018). How cloud computing will revolutionize modern hotel marketing and tap into new uncontested markets. Retrieved from https://aremorch.com/how-cloud-computing-will-revolutionize-modern-hotel-marketing-and-tap-into-new-uncontested-markets/

Muthumanicam, S. K. (2018). The evolution of artificial intelligence (AI) in hotel industry. Retrieved from https://chatbotnewsdaily.com/the-evolution-of-artificial-intelligence-ai-in-hotel-industry-23298d1e8dd4

Oliveira, T., Thomas, M., & Espadanal, M. (2014). Assessing the determinants of cloud computing adoption: An analysis of the manufacturing and services sectors. *Information & Management, 51*(5), 497–510. doi:10.1016/j.im.2014.03.006

Osawa, H., Ema, A., Hattori, H., Akiya, N., Kanzaki, N., Kubo, A., . . . Ichise, R. (2017, March). What is Real Risk and Benefit on Work with Robots?: From the Analysis of a Robot Hotel. In *Proceedings of the Companion of the 2017 ACM/IEEE International Conference on Human-Robot Interaction (pp. 241-242)*. ACM.

Patkar, R. S., Singh, S. P., & Birje, S. V. (2013). Marker based augmented reality using Android os. [IJARCSSE]. *International Journal of Advanced Research in Computer Science and Software Engineering, 3*(5).

Petrevska, B., Cingoski, V., & Gelev, S. (2016). From smart rooms to smart hotels. *Zbornik radova sa XXI međunarodnog naučno-stručnog skupa Informacione tehnologije-sadašnjost i budućnost, Žabljak.* Feb. 29-March 5, 2016, 21, 201-204.

Pinillos, R., Marcos, S., Feliz, R., Zalama, E., & Gómez-García-Bermejo, J. (2016). Long-term assessment of a service robot in a hotel environment. *Robotics and Autonomous Systems, 79*, 40–57. doi:10.1016/j.robot.2016.01.014

Revfine. (2019a). How to use artificial intelligence in the hospitality industry. Retrieved from https://www.revfine.com/artificial-intelligence-hospitality-industry/

Revfine. (2019b). How augmented reality is transforming the hospitality industry. Retrieved from https://www.revfine.com/augmented-reality-hospitality-industry/

Rosenbaum, M. S., & Wong, I. A. (2015). If you install it, will they use it? Understanding why hospitality customers take "technological pauses" from self-service technology. *Journal of Business Research, 68*(9), 1862–1868. doi:10.1016/j.jbusres.2015.01.014

Royal-hospitality. (2016). Smart hotel. Smart return. Retrieved from royal hospitality.net/images/lighting/smart_hotel.pdf.

Schaap, J. (2018). How Self-Service Is Replacing Traditional Guest Service In The Ideal Guest Experience. Retrieved from https://www.hotel-online.com/press_releases/release/how-self-service-is-replacing-traditional-guest-service-in-the-ideal-guest

Self service smart solutions. (2018). Otellerin self service kiosk kullanması için 5 neden. Retrieved from https://selfservicesmartsolutions.com/makaleler/blogss/otellerin-self-service-kiosk kullanmasi-icin-5/

Smart Hotel Technology Guide. (2018). Using technology to navigate the guest experience journey. https://sha.org.sg/userfiles/ckeditor/Files/Smart%20Hotel%20Technology%20Guide%202018.pdf

Soegjobs. (2016). What are the top trends in hospitality industry in 2016? Retrieved from https://soegjobs.com/2016/05/21/top-trends-hospitality-industry 2016/#globalhportal

Terry, L. (2016). 6 Mega-Trends in hotel technology. Retrieved from http://hospitalitytechnology.edgl.com/news/6-Mega-Trends-in-Hotel Technology105033

Ting, D. (2017). 10 hotel trends that will shape guest experience in 2017. Retrieved from https://skift.com/2017/01/03/10-hotel-trends-that-will-shape-guest-experience-in-2017/

Tkaczyk, C. (2017). Meet Rose, the sassy new bot concierge in Las Vegas. Retrieved from https://www.travelandleisure.com/hotels-resorts/cosmopolitan-las-vegas

Tussyadiah, I. P., Wang, D., Jung, T. H., & tom Dieck, M. C. (2018). Virtual reality, presence, and attitude change: Empirical evidence from tourism. *Tourism Management*, *66*, 140–154. doi:10.1016/j.tourman.2017.12.003

Velte, A. T., Velte, T. J., Elsenpeter, R. C., & Elsenpeter, R. C. (2010). *Cloud computing: A practical approach* (p. 44). New York: McGraw-Hill.

Wei, W. (2019). Research progress on virtual reality (VR) and augmented reality (AR) in tourism and hospitality: A critical review of publications from 2000 to 2018. *Journal of Hospitality and Tourism Technology*.

ADDITIONAL READING

Belias, D., & Varelas, S. (2019). To be or not to be? Which is the case with robots in the hotel industry? In *Strategic Innovative Marketing and Tourism* (pp. 935–941). Cham: Springer. doi:10.1007/978-3-030-12453-3_108

Brougham, D., & Haar, J. (2018). Smart technology, artificial intelligence, robotics, and algorithms (STARA): Employees' perceptions of our future workplace. *Journal of Management & Organization*, *24*(2), 239–257. doi:10.1017/jmo.2016.55

Çeltek, E. (2019). Advantages of Augmented Reality, Virtual Reality, QR Code, Near Field Communication, Geo-Tagging, Geo-Fencing, and Geo-Targeting for Marketing Tourism. In Smart Marketing With the Internet of Things (pp. 94-113). IGI Global.

Grieco, L. A., Rizzo, A., Colucci, S., Sicari, S., Piro, G., Di Paola, D., & Boggia, G. (2014). IoT-aided robotics applications: Technological implications, target domains and open issues. *Computer Communications*, *54*, 32–47. doi:10.1016/j.comcom.2014.07.013

Guo, Y., Liu, H., & Chai, Y. (2014). The embedding convergence of smart cities and tourism internet of things in China: An advance perspective. [AHTR]. *Advances in Hospitality and Tourism Research*, *2*(1), 54–69.

Jaremen, D. E., Jędrasiak, M., & Rapacz, A. (2016). The concept of smart hotels as an innovation on the hospitality industry market - case study of PURO hotel in Wrocław. *Economic Problems of Tourism*, *36*(4), 65–75.

Jung, S., Kim, J., & Farrish, J. (2014). In-room technology trends and their implications for enhancing guest experiences and revenue. *Journal of Hospitality and Tourism Technology*, *5*(3), 210–228. doi:10.1108/JHTT-11-2013-0035

Kokkinou, A., & Cranage, D. A. (2013). Using self-service technology to reduce customer waiting times. *International Journal of Hospitality Management*, *33*, 435–445. doi:10.1016/j.ijhm.2012.11.003

Lee, S. A. (2018). m-servicescape: Effects of the hotel mobile app servicescape preferences on customer response. *Journal of Hospitality and Tourism Technology*, *9*(2), 172–187. doi:10.1108/JHTT-08-2017-0066

Tung, V. W. S., & Au, N. (2018). Exploring customer experiences with robotics in hospitality. *International Journal of Contemporary Hospitality Management*, *30*(7), 2680–2697. doi:10.1108/IJCHM-06-2017-0322

Wu, H. C., & Cheng, C. C. (2018). Relationships between technology attachment, experiential relationship quality, experiential risk and experiential sharing intentions in a smart hotel. *Journal of Hospitality and Tourism Management*, *37*, 42–58. doi:10.1016/j.jhtm.2018.09.003

Yüncü, D. (2015). Relationships between perceptions of virtual destination environment, satisfaction and loyalty. *Journal of Hospitality and Tourism Technology*, *6*(2), 160–173. doi:10.1108/JHTT-07-2014-0021

KEY TERMS AND DEFINITIONS

Artificial Intelligence: is an area of computer science that helps machines to sense, comprehend and learn. It also helps machines to work and react like humans do. AI can also be a mean to recognise patterns and relationships in data sets with help of machine learning (Smart Hotel Technology Guide, 2018).

Big Data: is huge data sets that may be analysed computationally to discover patterns, trends, and associations, especially relating to human behaviour and interactions.

Chatbot: is a advance computer program specifically designed to initiate conversation with human users (at our case, hotel guests), especially over the Internet.

Robots: can be defined as "actuated mechanism programmable in two or more axes with a degree of autonomy, moving within its environment, to perform intended tasks" (International Organization for Standardization, 2012).

Self-Service Software: is an electronic support without need for interaction with a service provider. In such an environment, users are in control of their own service experince thanks to intuitive capabilities of such software. The technology offers many possibilities for applications via kiosks or mobile phones (Smart Hotel Technology Guide, 2018).

Smart Hotel: is a new development that heavily benefits from technology to replace registering procedure to a hotel, opening doors, changing lights and powering on the plugs with the press of a buton, voice commands and maybe thought control (using smartphones). There are limitless opportunities to apply smart Technologies to make the hotels smarter including creating a virtual alter-ego of a guest so that their device can be used to communicate with the hotel server and thus be given access to room controllers (IDS, 2014).

The Internet of Things (IoT): is inter-connected objects of a giant network ranging from smartphones and corridor signs to room thermostats. This broad network enables the exchange and analysis of data. This data can help hotels to discover new business insights and see and act to new opportunities (Smart Hotel Technology Guide, 2018).

Virtual Reality: is an imaginary 3D environment which immerses users via the use of a VR headset. Hotel guests can "experience" an artificial hotel, move around in it and even interact with virtual features and services of a hotel (Smart Hotel Technology Guide, 2018).

Section 3
Smart Tourism

Chapter 16
Smart Applications in Tourism

Cemal Ince
https://orcid.org/0000-0003-2005-7090
Gaziosmanpaşa University, Turkey

Gülmira Samatova
Kyrgyz-Turkish Manas University, Kyrgyzstan

ABSTRACT

Technology process that started with the reservation provided many services in accommodation establishments such as online room selection, room colour selection, air conditioning adjustment, temperature and light control, virtual scenery creation, and online service. Another area that is affected by technology is food and beverage services. Food and beverage businesses also benefit from technology while serving their guests. Many services are provided through technological revolutions such as online reservation, online service, smart tables, robots, and so on. Today, technology also affects museums and museum services positively. Virtual reality museums offer visitors the opportunity to visit their works online. Online ticket sales also save visitors time. Credit cards and ticket sales provide economic convenience to the guests. With the help of technology, heat and light adjustments can be made in museums and technology helps to protect the works exhibited in these museums.

INTRODUCTION

One of the factors that play an important role in the improvement of tourism is technological developments. Without technological advances today people would not be able to make an economic, comfortable journey in such a short time. Hospitality companies could not offer such quality service to their guests in a short time and quality. The hotel rooms would not have been so clean. The museums could not bring such works together and present them to their visitors. Food and beverage establishments would not be able to prepare and offer their guests such a wide range of food and beverages satisfyingly, economically in a healthy way and in a short time. Establishments offering recreation and entertainment services would not be able to offer businesses, entertainment and leisure services so easy, economic and of good quality. People who with travelling would not be able to cross country borders so easily and would have

DOI: 10.4018/978-1-7998-1989-9.ch016

to deal with bureaucracy. These explanations show that technology made a positive contribution on all components of tourism and became an important factor in the development of tourism. Stipanuk (1993) states that technology plays six main roles in tourism. These roles are creative, protective, developer, focal point, vehicle and destructive roles. Five of these roles are positive and one of them is negative which is a destructive role.

Law, Leung, Buhalis (2009) examined the evolution of information technologies in terms of consumers, businesses and technology. In this context, they stated that information technologies become more critical for tourism and hospitality businesses to provide competitive advantage and to manage their distribution and marketing on a global scale.

Xiang et al. (2015) in their study about travel, information technology and consumer behaviour stated that in general, the adoption of the Internet has reached a level of saturation in travel and some traditional channels such as online travel agencies (OTAs) continue to dominate travel planning. They found that traditional internet use for travel planning is common in all customer segments, and that social media use is common among Y generation travelers. That is why, it is stated that traditional internet usage will continue in online travelers but for groups such as generation y, large enterprises are obliged to search for new sources of information and transaction channels.

Gretzel et al. (2015) in their study about the tendency in present smart tourism applications revealed the technologic and commercial foundations of smart tourism. The key point of smart tourism emphasizes the adaptation of information and communication technologies to the physical system. Smart tourism is not a component of a destination but rather a social phenomenon generated from the integration of information and communication technologies with tourism experience (Hunter et al., 2015). Smart tourism experience is particularly personalization of technology-mediated tourism experiences, context awareness and real-time experiences. Within this scope, smart tourism consists of an intelligent information layer aimed at collecting data, an intelligent transformation system that supports interconnection (a Web-based system), and an intelligent processing system that enables data analysis, visualization and intelligent use (Tu and Liu, 2014).

BACKGROUND

The relationship between tourism and technology has started to attract the attention of scientists especially after 1980s. Since the 1980s, information communication technologies (ICT) has been playing an important role in the global development and transformation of tourism (Poon, 1993; Sheldon, 1997; Porter, 2001). Many scientists working in the field of tourism examined the relationship between tourism and technology from various perspectives.

Buhalis ve Law (2008) in their study titled "Progress in information technology and tourism management: 20 years on and 10 years after the Internet—The state of Tourism research" examined the effect of developments in information technologies on tourism. Again Buhalis (1998) in his study titled "Strategic Use of Information in the Tourism Industry" examined technology usage strategies in tourism industry. Buhalis and O'Connor (2005) in their study titled "Revolutionary Information Communication Technology in Tourism" researched information communication technology in tourism. Poon (1993) in his study titled "Tourism, technology and competition strategies" examined tourism technology relation in terms of competition strategies. Sheldon (1997) in his work titled "Tourism information technology" made an examination regarding the effect of information Technologies (IT) on all sectors of tourism.

Pesonen ve Horster (2012), examined the use of near-field communication (NFC), a wireless technology covering mobile devices in their study titled "Tourism in Near Field Communication Stipanuk (1993) in his study titled "Tourism and technology: interactions and applications" examined the relationship between tourism technology and examples of tourism industry in many countries mainly the USA and New Zeland. Huang, Backman, S. Backman (2016) in their study titled "Exploring the Implications of Virtual Reality Technology in Tourism Marketing: An Integrated Research Framework" examined the Internet and other technological innovations that change the structure of the tourism industry, the perception and marketing of tourism destinations. Brown, Chalmers (2003) in their study titled "Tourism and mobile technology" made a study that shows the expectations of the city tourists from the tourism technology and offered suggestions to the enterprises in this direction. Baines (1998), in his study titled "Technology and tourism" examined the impact of technology on the services offered and the quality of business in the industry. Inkpen (1998) in his study titled "Information technology for travel and tourism" he analyzed the global distribution system (GDS) in tourism and examined distribution systems such as global distribution systems, hotel distribution system, tour reservation system and travel agency systems.

Frew (2000) in his study titled "A critical analysis of tourism information technology research" he criticized the focus of studies on the relationship between tourism and technology, tourism information systems in general, electronic distribution and commerce and various web-based applications and argued that the areas of quality assurance, mobile access and education should also be examined within the framework of tourism technology.

Law, Leung, Buhalis (2009) in their study titled "Information technology applications in hospitality and tourism: a review of publications from 2005 to 2007", they examined the evolution of information Technologies by analyzing their findings in the category of consumers, technologies and businesses.

O'Connor and Murphy (2004) in their study titled "Research on information technology in the hospitality industry" examined the researches on the use of technology in the hospitality industry. Researches focused on the effects of the internet on distribution, pricing and consumer interactions. As a result of the research, it was obtained that information technology may have undesirable consequences for the tourism industry that the predictions are often incorrect and do not yield the expected results.

TOURISM AND TECHNOLOGY

Although tourism is defined as a labor-intensive sector, technological developments are important milestones in the journey of tourism in time. The development of tourism started with the development of travel vehicles such as ship, road and railway transportation. Technology takes a very important place in the development of tourism in time. In providing the desired quality in the journey from the places where the tourist is constantly living, to the touristic destinations and the place where they live continuously, technology, travel, accommodation, food-beverage and entertainment / recreation services are among the main factors.

Since technology affects the nature, type and quality of the services provided in accommodation establishments, travel establishments, food and beverage companies, entertainment and recreation businesses, it also affects the quality of tourism destinations directly. The innovation activities of tourism enterprises to provide their guests with better quality and high performance services have been directly affected by the global effects of modern technology. The use of technology in tourism that began in the 1970s with computerized reservation (CRS) applications, grew rapidly with the global distribution system (GDS) in the late 1980s, the Internet in the 1990s and mobile communication technologies in the 2000s (Werther et. al, 2015; Yalçınkaya et. al, 2018). However today with the wide usage of data mining, the internet of objects, cloud computing, intelligent technology and artificial intelligence are increasing the level of both service quality and level of competition in tourism (Buhalis and Amaranggana, 2013). Especially the development of artificial intelligence technology has revealed the concept of smart city. Smart city applications have led to the introduction of tourism destinations with smart applications. Technological development has been moved to the upper level with the internet automation in tourism enterprises. With the cloud technology, guests can experience the world of beach and air conditioning in their rooms. Guests can give their orders directly with mobile communication tools and make travel reservations via mobile vehicles (In the United States, in 2007 54% of travel bookings were made online) (Kang, 2011).

At the same time, with internet and communication technologies modern tourists can obtain all kinds of information such as their preferred destinations, travel vehicles, tourism activities, local food and drinks, recreation activities, culture and beliefs of local people and so on with one key starting from planning and deciding process of their trips. This gives tourists time, economy, motivation and hedonic feelings.

In tourism technology is consisted of two parts (Baydeniz, 2019):

a) Information technologies: Computer, communication tools, operating system, application software and multi-media technologies.
b) Communication technologies: Communication factors such as smart phone, internet, social media etc.

Although information technologies and communication technologies function different, they are in close relation with each other. Since communication technologies provide verbal, written and visual communication with every part of the world, computer technologies provide benefit from time, economy, quality and competitiveness by developing business processes and service products (Benli and Kızgın, 2002).

As a result of the meeting of tourism with technology, the concept of smart tourism has come out. Smart tourism are processes in which modern tourism products supported by information and communication technologies and smart tourism, are presented to the guests by various tourism businesses. Smart tourism is a system that uses intelligent technology to create, manage and present technological touristic experiences. Smart tourism focuses on both experiences of public and tourists as smart destinations, smart travel, smart accommodation, smart gastronomic and recreational activities and aims to benefit from the sustainability technology in tourism (Gretzel et al., 2015:180).

In the tourism sector technology is not only limited with internet because its usage area is gradually expanding. For example booking offices will completely disappear in the near future. Turkish Airlines that is operating in the travel industry has removed the reservation offices and makes check-in transactions with cards. This system can also be used for accommodation management in the near future and online booking and check-in can be done online by the guest. In this way, guests can go directly to their rooms without waiting in the reception. Moreover, tourists will have the opportunity to perform these operations from their countries before travelling.

Mobile communication tools are also used to serve guests in modern hotel businesses. Now with software installed on mobile devices, many services such as bellboy, technical service, room service purchasing from other units etc. can be purchased directly without needing to communicate with employees. Nowadays the applications such as heat and light control of rooms can be made by the guests with mobile communication tools, forms the current examples of the widespread use of technology in the tourism sector. In addition with biochemistry system guests can easily enter their rooms with sound, fingerprint or retinal scanning doors.

Travel agencies and tour operators have again started to use technology intensively. During travelling services such as computer games, television, listening to the radio, etc. are offered them for having a good time.

MAIN FOCUS OF THE CHAPTER

Information technology which has developed greatly since 1980s, affected the tourism sector with numerous applications and also affected the service understanding of the sector.

The process that started with the use of computers for reservation purposes, provided a great transformation with virtual reality, cloud applications and mobile applications that became widespread in every field of tourism. The main service providers such as accommodation enterprises, travel enterprises, food and beverage enterprises, recreation and animation enterprises, gift businesses, agricultural enterprises, etc. in tourism sector as well as ancillary service enterprises are also affected by information communication practices. These enterprises want to benefit from tourism to reach the target groups in various categories, increase them, provide the best service, provide competitive advantage, minimize costs and benefit from information communication technologies for the efficiency and profitability of the enterprise.

In various countries of the world, tourism enterprises benefit from many applications such as mobile applications, online check-in / Check-out transactions, smart technologies, cloud applications, virtual reality applications, augmented reality applications. Information and communication technologies continue to develop. These developments affect people, their social and economic habitats and their level of life. This affects the expectations of the target groups from tourism. This directly affects the future projections of the tourism enterprises that want to meet the expectations of the tourists.

For this reason, the target point of this study is to emphasize the current smart technologies used in various fields of tourism to meet the expectations of tourists as well as give examples of applications in various areas of tourism industry in the World. In addition, make analysis regarding the future of smart technology in the tourism sector.

In the study the effects of technology on the development of tourism; smart tourism, smart destinations, smart hotels and other tourism enterprises under the technology titles with the help of concrete examples were tried to be examined.

SMART APPLICATIONS IN TOURISM

Technological developments that started with the use of extensive information technology concept in 1991 by Mark Weiser (Gökalp and Eren, 2016, 282) caused the occurrence of concepts such as smart cities, smart homes and smart hotels in recent years. In tourism sector the usage of common information technology applications involves the operations of multiple computer systems connected to each other in the physical environment. For working of this system, tablets, iPads, dashboards, integrated computers, sensors, wireless networks and systems that provide open data, are required.

Figure 1. Smart tourism scheme
Source: Zang et.al., 2012; Çelik and Topsakal,2017.

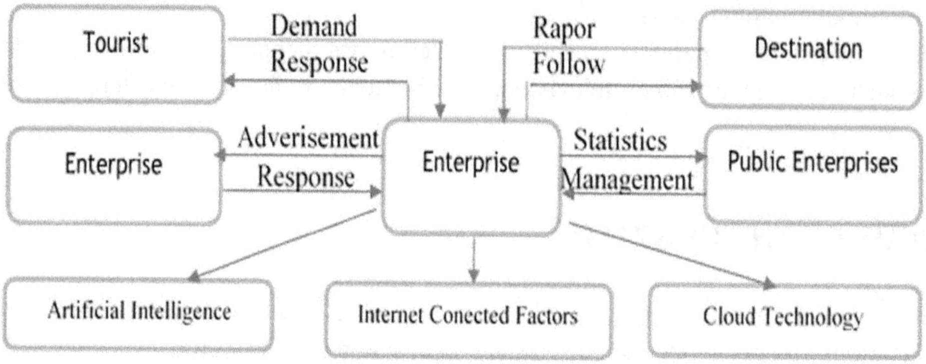

Intelligent tourism expresses smarten up of touristic services, management, marketing, entertainment, food and beverage and travel activities with the usage of oil information technologies. Smart applications in tourism offers important conveniences for tourists in destinations and tourism enterprises where they are used. Smart tourism, in cooperation with smart cities, is a system that focuses on increasing the quality of touristic life of tourists and benefits from information technologies that are widely used for this purpose. Intelligent tourism was firstly mentioned by G. Philips in 2000 as a long-term, sustainable and complex approach to develop, diversify and commercialize tourism businesses and products (Liv d., 2017). In 2015 it was defined by the World Tourism Organization (UNWTO) by integrating it with the concepts of information communication technologies (ICT) and smart destinations (Yalçınkaya et.al., 2018:37). Smart tourism is an expression of a transition from product orientation to service orientation (Wang and Yunpeng, 2013: 59; Gretzel et.al, 2015: 183).Intelligent tourism practices do not only affect the quality of service in tourism enterprises. Intelligent tourism reveals with the economic developments in various sectors (online or credit card payment, smart shopping, payment facilities, etc.), environmental initiatives, energy saving and renewable energy sources, the quality of social life, quality increase in public services (traffic services, meteorology services etc., recording of tourist movements, etc.) in company (Romao and Neuts, 2017).

Smart tourism has been developed as a system. According to Zhang et al. (2012), intelligent tourism consists of five elements such as information processing centre, tourists, government agencies, natural areas and services (enterprises). The request of searching for tickets, purchase, reservation, social communication, destination, employment and information about the services provided are sent to information processing centre by the tourists. The information processing center provides online and written and visual information to these demands. At the same time data processing centers also perform the functions of recording and storing statistical data. These statistical data are transferred to the official institutions and related enterprises to determine the tourism policies of the following periods. Tourism establishments (travel agencies, accommodation establishments, food and beverage companies, recreation and entertainment establishments) are willing to provide data from the information processing centers towards tourists in advertising, promotion and sales efforts (Zhang et al., 2012; Çelik and Topsakal, 2017:154).

Table 1. Tourism applications in smart tourism destinations

Tourism Applications in Smart Tourism Destinations	Subsidiar Functions	Destination Compounds	SmartTourism Destination Dimensions
Provides visitors to experience the digital recreation of touristic areas and time travel (Chillon 2012) Augmented Reality (AR)	Interpretation	Attraction	Smart People, Smart Mobility
Vehicle tracking system that provides real- time transportation network information and distribution to end-user devices (Arup 2010)	Planning	Accessibility	Smart Life, Smart Mobility
The hotel should be able to form energy demand based on environmental management and make energy audits to make them (Metric Stream 2013).	Sustainability	Facilities	Smart Environment
A multi-lingual application that offers a variety of services such as an electronic travel guide as well as number of packages available for tourists (Jordan 2011)	Guidance	Accessible Packages	Smart People, Smart Mobility
NFC tags and QR codes (GSMA 2012) to access information about nearby places of interest through mobile devices.	Estimate (Close) Marketing	Activities	Smart Mobility
A Complaint Management System supported by various ICT channels such as SMS or mobile applications that can direct complaints of tourists directly to the appropriate officers (Metric Stream 2013)	Feedback	Subsidiary Equipment	Smart Life

Source: Buhalis and Amaranggana, 2013:559.

Smart Destinations

The European Union which thinks that smart tourism is inevitable in the ice zone surrounded by all areas of technology, has developed the smart tourism capital project in order to manage the synergy created by tourism in EU cities and environments, strengthen the innovative applications, increase the attractiveness of destinations and enrich the economic growth and business areas. With the European Smart Tourism Capital project, it is aimed to form a basis for sharing best practices among European cities, cooperate and establish new partnerships.

In order to become a European Tourism Capital, it is necessary to be at high level in the categories of sustainability, digitalization, cultural heritage and innovation as well as being a role model for developing smart tourism destinations (etourism.org, 2019). Helsinki and Lyon which were found suitable in terms of these criterions, succeeded to become the European Tourism Capital of 2019. In light of these explanations, smart destinations have become one of the most important tourism issues that is supported for the European Union.

In order to develop smart destinations in tourism, three sources such as cloud communication technology, internet connected elements and internet service systems for end users, are needed (Zang et.al., 2012). Cloud communication technology provides access to web platforms with internet networks as well as advantage to businesses (Etro, 2009). On the other hand, internet-connected elements express defining, finding, managing and monitoring smart tools (Mingjun et.al., 2012). Internet service systems are various virtual reality applications that provide speed, aesthetics and convenience to tourists in destinations, hotels, travel, recreation and entertainment services with the help of cloud computing and internet connected elements (Table 1).

Table 2. Characteristics of smart tourism destinations

Tourism Organizations	It acts as a smart hub that coordinates all related information and allows users to access real-time information easily It enables digitalization of basic business processes It optimizes their energy use It make connections with local communities, tourists and government in creating tourism experience It responds to needs of customers based on corporate agility, rapid decision-making and timely predictions. It provides precise targeting and personalized service
Public Organizations	It establishes information management supporting open data It regulates data privacy. It establish public-private partnership
Local Collaborators (Local people)	It provides continuous connection It must be empowered and creative enough It is a technology enthusiast It makes citizen journalism It actively takes place in the development of intelligent heritage / e- culture
Tourists	They are well connected and well informed They are active critics and buzz marketers Request high personalized service They are connected to both social and technology They dynamically discuss through social media They form common touristic experience They contribute to the content of the service They use end-user devices at multiple touch points
Environment	It has interconnection with the internet of things. The presence of cloud computing services is required It has an innovative ecosystem. It is equipped with sensor networks throughout the environment It combines digital information and social contexts that will increase geophysical reality. It has workable social platforms

Source: Buhalis and Amaranggana, 2013:560

Smart tourism destinations have certain characteristics. Tourism organizations in smart tourism governments, tourists, local people and the environment are equipped with information communication facilities provided by smart technology. A common cooperation should be established between them. Table 2 shows the characteristics of smart tourism destinations.

Tourism organizations which are the elements of smart destinations have to undertake a variety of tasks for the operation of public organizations, local collaborators, tourists and smart environment, destination. The characteristics of smart destination items are described in Table 2. Tourism organizations serve as a hub for the future tourist destination of tourism organizations and prepare all the data they need. The public provides data confidentiality, networking for data transfer and data transfer between the public and the private sector. Tourists who prefer to come to smart destination reflect individual characteristics that can use the technology very well, await personalized service from tourism organizations, share their satisfaction / dissatisfaction very quickly and create common tourism experiences with social media tools. On the other hand the environment has time, energy saving, innovative ecosystem characteristics. These elements which have different characteristics, are connected to each other by smart networks and are always part of the intelligent destination in continuous communication.

Smart Applications in Hotels

Smart cities, smart destinations, and as a result, smart tourism businesses are tourism factors which are complementary elements of a chain. At the core of the smart hotel business, there are technological applications that are sensitive to the environment, increase the quality of life of tourists and provide better quality service for hotels (Jaremen, Jędrasiak, and Rapacz, 2016). Factors that play a role in the development of smart hotel management:

- Environment-friendly enterprises are preferred by tourists,
- Energy conservation is desired to be done by reducing enery use,
- Obtaining competitive advantage by providing better quality service to guests,
- Providing sustainability characteristics to hotel management,
- Providing saving by recycling,
- Increasing productivity in businesses,
- Providing personalized service to guests,
- Taking the expenses of hotels under control,
- Providing guests to communicate with more tourism destinations in time and in economic savings with virtual reality applications.

When the factors mentioned above are analyzed, it is seen that the basis of smart hotel applications is achieving the three purposes of society such as obtaining profit, maintaining the existence of the enterprise and being beneficial to the society. In this framework hotel investors developed smart hotel management for the reasons such as providing competitive advantage, maximizing profit, providing better service to the guests, increasing customer satisfaction, maximizing the life of the business and so on. Within this scope in smart hotel managements the services below were developed (Atay et al., 2019):

1. Smart telephone applications: Making online check-in, opening room doors with smart phones, making heat, light and ventilation settings in rooms, accessing hotel guide online, being able to benefit from high multimedia services in rooms, creating various concepts in rooms with cloud applications, paying bills online.
2. Personalized Services: Providing the possibility to determine services and fees according to the request of guests with online systems. Buying online services such as room services, food and beverage services, social services, SPA, Fitness etc. Buying online services etc. Purchasing services via tablets, phones.
3. Smart Card Services: Being able to benefit from hotel services with cash loaded cards.
4. Smart Order: Being able to order directly on the system with kitchen and restaurant employees and also providing the appropriate food and beverage service of their expectations by uploading the guest information to the tablet devices used by the employees.
5. Smart Room: Providing access to menus via virtual monitors, daily social and cultural activities, water, air conditioning and so on. Being able to adjust the temperature, monitoring the personal health status of the guests online, making the guests feel like on the shore with cloud applications.
6. Child Services: Providing services for parents to monitor their children in playgrounds, pools etc.
7. Following up performance: Being able to follow the instant performances of the employees with online systems and directing them in the working process.

8. Smart Follow-up Systems: Applications of following up hotel rooms, department and main warehouse stocks, mini bar and so on. online and tracking of cabinets and renovation of contents.

When smart hotel applications are examined, it is seen that the most common uses are online data flow, room technologies, applications to improve service quality, energy saving applications, waste control and follow-up applications, applications to reduce bureaucracy, practices to increase guest satisfaction, practices for pleasure, time saving and competitiveness applications, technological applications aimed at increasing the safety of guest, management and employees.

In smart technology applications, mobile key cards come to mind first. Starwood, Marriot and Hilton chain hotel businesses use mobile phone-operated smart card systems in their hotels. Wynn Las Vegas Amazon Alexa provide a smart voice command device that allows users to adjust the lighting, curtain, television on and off, and room temperature adjustment in their rooms (Balakrishnan, 2016). Cloud.7 Hotels that started its activities in Istanbul, offers its guests the opportunity to connect rooms with the latest technology, share experiences or isolate itself from the World completely. It provides comfortable bedrooms and pillows, rain showers with bluetooth speakers, cordless phones, 7/24 Whatsapp access to the staff of Cloud.7, keyless entry to the room and so on. applications, smart online check- in and check-out, lobby area without reception, 24-hour gym as well as allowing guests to set their own schedules. Also "Meet the Living" platform of Cloud.7 provides guests of Cloud.7 to meet with the inhabitants of the world with similar world views and learn about the city's favorite places and hidden corners (tourismcurrent.com, 2019).

Hilton Garden Inn Isparta has the property of being the first user of Hilton's hotels intelligent application of Digital Switch (Digital Key). With mobile-enabled room key technology, guests can make digital check-in, in room selection with intelligent technology guests can go directly to the room without losing time at reception. With digital key, guests can benefit from the Hilton Honors programme and can use the window or room lighting in their rooms easily and safely (turizmhaber.com,2019). In Poland Hotel Blow Up Hall 5050 operates does not use a number system in reception and rooms. Instead of a key, the hotel provides the phone to the guests to find and open the room. Instead of a key, the hotel provides iphone to the guests to find and open the room (blowuphall5050.com,2019). With the navigation in the smart phone, the suitable room is found and opened. The sensor in the room at the Loews Hotel 1000 in Seattle, USA, informs the cleaning staff for housekeeping and informs the cleaning staff. It warns the staff by alarming to collect and clean the room. The CitizenM Amsterdam hotel in the Netherlands offers guests a private computer. At the same time, guests can remotely control the air conditioning and light settings via tablets provided to them. With these tablets, alarms can be set and information about the hotel and its surroundings is offered (adonisotel.com, 2018).

One of the increasingly intelligent technologies in the tourism sector is augmented reality (AR) applications. In accommodation establishments AR applications can occur as directing guests to rooms, online check-in /check-out, cloud applications in rooms, customer relationship management applications, applications in SPA service areas etc. Marriot Hotels, Omni Hotels and Resorts, Premier Inn are some of the hotel managements that benefit from AG technologies. City Club Hotel also allows guests to change the room color to any color with AG applications (Özgüneş and Bozok, 2017).

Smart Applications in Food and Beverage Establishments

The most used area of smart applications in food and beverage business is marketing and promotion of products and enterprises. Information search in smart technology is used in evaluation of alternatives, mobile registration and payment issues (Güngör et al., 2017).

In the from food and beverage businesses, kiosk order taking, self-check-out kiosks, digital menus and digital navigation applications are used. Also, facial recognition systems started have been introduced. Face recognition systems are systems that recognize guests and analyze previous orders. KFC uses this system in China. The restaurant chain named Cali Burger and Fuel uses the same system in America (Broadway World, 2018). Again Taco Bell serves with self-order kiosks with All Access (full access) vision. Robot waiters serve in Naulo restaurant in Kathmandu, Nepal. You can order the waiters named Ginger by touching the screens and Ginger prepares these orders and serve them to the tables (Beraat Gökkuş, 2018). There is a restaurant located in Pakistan Multan using robot waiters (ekonomihaber7.com, 2017) and in the Chinese war there are restaurants using robot waiters (ntv.com.tr, 2013).

In the studies of Spence and Piqueras-Fiszman (2013) who examined the technological developments in different restaurants, predicted that the technology and the kitchen would merge completely within a few years and that maybe even a waiter could not be needed with the developing technology and also in the studies of Aksoy and Akbulut (2016) who examined the technological developments in different restaurants, they mention that Inamo Restaurant provides the best customer experience through technology. With a menu that offers Augmented Reality on the smart tables, the restaurant serves customers and learns about the food and drinks they will order as well as the customers can see what kind of food they will meet.

Ebony restaurant in Doubai uses smart dining tables. There is no need to wait for waiter with smart tables. Guest can browse the dining table through the smart table, visit the food gallery and order. The process of preparing and baking the ordered food can also be viewed on the smart desk. Again on the smart desk, news can be read, videos can be watched and social media shares can be made. A taxi service is available while leaving (morfikirler.com, 2018).

One of the smart technologies that can be used for food and beverage businesses is the cloud-based bill system. With this system you can order online, follow product stock status and prevent the order of products that are not in stock. Again, the use of smart technologies will gradually increase in technology of order. For example, food can be ordered and delivered with drones. In addition to this inventory tracking, personnel charts, smart pos systems and industry blogs will develop more.

The 3D restaurant of world, Food Ink Restaurant offers different experiences to its customers and the selected dishes are prepared and served with 3D technology (foodink. Io). In the near future it will not be wrong to say that food and beverage enterprises and similarly virtual reality applications in the kitchen will be able to come forefront in product marketing and technology and 'smart plates' will place on 'smart tables'.

Smart Applications in Travel Enterprises

Travel businesses are one of the areas where smart internet is widely used. The most common application here is the Internet of Things (IoT) application. The Internet of objects means connection ability of everyday objects to each other over a network. In modern technology, it does not matter how many physical objects are connected.

Figure 2. Cloud based bill system
Source: https://adisyo.com/index.html, A.D. 23.05.2019

One of the travel companies that use the Internet of Things mostly is the airline business. In airline management internet technology of objects is used in many points from baggage monitoring to cabin climate control. Internet applications provide convenience to businesses in many areas such as lost luggage, flight delays and customer service and play an important role in reducing customer complaints. With the usage of internet of things application, convenience in many areas such as lost luggage, flight delays and customer service as well as reduction in reducing customer complaints is provided.

Since travel is the main condition for occurrence of tourism, it is observed that there are various smart applications in the transportation sector that mediates the travel (Drummond, 2016). Many technology channels, traditional destinations and websites of agencies, personal blogs, interview sites and social networks came out as smart tourism technologies today. These technologies have directly affected the whole sector especially travel agencies and the guests who buy services (Huang et al., 2017). Smart technology that makes itself effective in booking and payment process at the beginning, is now changing the travel concept. Smart applications in the travel sector are used in finding the current location with internet and mobile communication devices, determining the location on the map, calculating the distance between the two cities, video searching and weather monitoring etc. (Jinendra et al., 2012).

Smart technology entered the travel sector in the 1950s with central reservation systems applications (Yüksek, 2013; Pemberton et al., 2001; Zhou, 2003) and continued with global distribution systems (Hançer and Ataman, 2006). However, internet started to be used in 1990s. Central reservation systems and global distribution systems provide datas such as travel, accommodation, entertainment services, tourist attraction areas and holiday packages to travelers who want to travel (Buhalis, 1998; Yüksek, 2013).

Smart technologies also have become irreplaceable applications of modern airports. For example, robotic technology can be used to detect hidden weapons in airports. By using robotic technology some of the brands produce luggage with smart tracking system. In addition, some airline security companies try to minimize the waiting minutes of passengers by using robots for fast pre-screening operations (renkliyolculuk.com, 2019). Artificial intelligence applications are used especially in customer service in travel industry. Some of the requests and questions of customers are answered quickly by chatterbots having artificial intelligence. In addition, these bots learn new things from the interactions with the customers and can improve themselves.

Travel companies are preparing applications and websites that are more suitable for mobile use in information technologies. Those who will travel with these applications can easily get all kinds of information about their travels and get instant answers to their questions. In Princess Cruise ships the equipment named Ocean Medallion tries to make optional food and drinks, interactive games, special entertainment and smart navigation applications more fun. The medallion makes navigation simpler, more personal and fun ever. The medallion protects digital identities of guests and allows them to communicate with thousands of readers on the ship and in the harbor. The on-off button helps you discover everything unproblematically, without the need for menu or charging. In addition using the medallion for payments, unlocking the room door and accelerating the landing, help the crew to see the information they can use to provide better service (Princes.com, 2019).

Qantas airline that operates in the field of air transport, started its first virtual reality application with Samsung in 2015. Qantas with virtual reality headsets, tries to provide its passengers having a pleasant time on their travels while collecting data about their passengers. By analyzing this data, it puts the applications that meet their personal expectations, into practice. Many airline companies that follow applications of Qantas have started to use virtual reality applications technology to increase their peripheral service sales by enabling their guests to see their seats and cabins in advance (GlobalData Technology, 2019). Virgin Atlantic airline manufactures Boeing 787 aircraft and cargo equipment connected to IoT. Aircraft with iod system the system works interconnected including the engine and landing gear and aims to resolve mechanical problems that may occur (Drummond, 2016).

On the other hand Air Asia airline company started smart route application in IoT field in cooperation with GE. Air Asia aims to save fuel in flight efficiency with this application. Delta airline uses IoT technology for baggage monitoring and control. Until the time of boarding, on the plane and at the baggage delivery place, it performs baggage tracking with delta mobile application. German Lufthansa airline uses the IoT application to eliminate the lost baggage incident and checks and monitors the baggage with the mobile application.

Abu Dhabi International Airport uses the smart travel application to reduce queues and facilitate the travel of passengers at airports. In this application passengers make check-in from kiosks, get boarding pass and deliver their baggage to automatic baggage system. With this system, passengers pass through passport control quickly with face recognition system. Face recognition and IoT technology started to be used in the International Istanbul Airport that entered into service in 2019. A new infrastructure based on cloud technology has been established at the new Istanbul airport that works with servers, storage systems, backup systems, cloud software and components, helpdesk, logging, monitoring and data center management systems.

In this system, internal and external customers can make check in from kiosks and pass passport control easily with face / retina recognition system. London Gatwick Airport is the first to use a broad range of cloud services for basic IT needs (globaldata.com, 2019).

With Jet Blue system communication and ticketing systems between IoT and airline systems becomes automatic. After the reservation, an automatic ticket and a seat are provided to the passenger 24 hours before departure without having to enter the application or website. The seat is based on the criteria set by the passenger when booking a ticket (Drummond, 2016).

Railway enterprises have started to benefit from smart technology applications. Especially with smart mobile applications, train tickets can be purchased without going to the offices. The United Kingdom announced that the paper ticket application will end in two years and the smart ticket application will be used. At the same time, the transreport mobile application was developed for trains. Accordingly, users have the opportunity to report problems they are not satisfied with such as broken toilet, irritating graffiti or very cold / hot train. The Passenger Assistant application also provides support to disabled people in need of assistance (Smart Rail World, 2018).

Luke Upton (2014) explains 10 innovations that will shape rail and subway transport systems:

- Virtual Ticketing Agents: Ticketing is the first area where smart technology is applied in railway transportation. Paper tickets are replaced by smart tickets in the world. The next development after the smart ticket application is the virtual agent applications that will manage ticket office, vending machine and call center services from a single point (kiosk). A passenger using a Virtual Agent can talk to a 'real-time' real-time person who offers an experience similar to a ticket window other than a video connection. The ticket agent spoken may be at a central ticket office.
- Li-Fi Applications; Wi-Fi applications are included in modern transportation. However, the new emerging Li-Fi applications will become the new address of the modern internet. Li-Fi transmits data using the spectrum of light and operates more efficiently than Wi-Fi. Since visible light is more than the radio spectrum, it can transmit data density more. Therefore, it will replace Wi-Fi applications in the future.
- Big Data and Railway; Railway enterprises have two main potentials with large data collection. These are wireless signaling that provides smarter communication on trains and allows operators to receive data from trains constantly, monitor the performance of trains more effectively and predict potential problems or maintenance performances. This may prevent trains from being out of service and disrupting travels.
- Travel without tickets; Transport for London (TfL) has announced that it will offer debit cards to metro passengers for ticketless travel in London. Again, EMV debit cards (Europay, MasterCard and Visa) and Barclaycard's PayTag labels, including the use of debit cards as a ticket-free solution on rail and subway are available. It has started to be used seasonal ticketing and smart card applications with the NFC (Near Field Communication) system installed on smartphones.
- High-speed and Super-Speed Trains; Rapidly spreading fast trains and super-fast trains will be used in international passenger transportation. Particularly in Japan, China and Europe big investments are made on speed trains. The production of super-fast trains having no collision risk, faster than airplane, 4 times faster than the bullet and will work with full solar energy and IoT-equipped named Hyperloop continues to be produced. At the same time these trains will work automatically without a machinist.
- Ergonomically Desgined Stations; In the rapidly growing economic use of railways, smart station designs will be increased to meet the expectations of passengers in order to transfer passengers from one station to another quickly and safely. For example, station designs are made that direct passengers from the right side of the escalators and where the tickets are prepared to the exit door. In addition, a system is designed to be used automatically in the station by other smart cards. An ergonomic station designed with big data and smart ticketing will provide a new and exciting experience for passengers.
- Railway Loyalty Schemes; Data from smart card technology and data storage systems which is an important factor for airlines, have become more important in the development of passenger loyalty

programs. The use of large data storage systems makes it easy to provide data in the creation of a customer loyalty scheme.

· Entertainment for Passengers; Modern railway management offers many options both entertain and inform passengers during travel. On the trains, videos are displayed on a fixed system, and films, TV programs, music, games and news are provided over the Wi-Fi for passengers to watch from their own devices. For example in Italy NTV that offers settled portal presents Italolive over Wi-Fi system that allows passengers to watch TV, movies, news and sports.
· 4G Mobile Internet technology has made a revolution in travelling for rail passengers. High-speed connectivity, increased video and music streaming speed, and the in-car passenger entertainment system provided a way for faster service. The prevalent use of 4G allows mobile ticketing and in-vehicle e-commerce to become widespread.
· Smart Applications; People are now accepting smart applications. In 2017, 268.69 billion people downloaded smart applications. The category of smart applications includes virtual personal assistants technology as well as ticket booking, travel organization, instant notification of conditions and schedule changes. There are also applications for instant measurement of the satisfaction of passengers during travel.

Smart Applications in Museums

The technology was firstly used in museums to hold inventory, display, lighting, temperature and light control. Since the 2000s, the concept of virtual museum as well as exhibiting collections on the internet has revealed (Yücel 2012). In addition to this, with the help of intelligent technology, virtual guidance services, virtual animations, sound and visual effects, the visitors started to get better quality services.

Smart technologies do not just bring innovations to tourists who visit museums. These technologies are used in museums for educational purposes. For this purpose in museums attractive areas full of simulators, headphones, kiosks, virtual reality applications, 3D printers, computer rooms with touch systems are prepared. With these systems, visual information is transmitted in museums for the new generation that is connected to technology. Digital devices such as mobile phones, i-pad and i-tules provide great convenience in introducing museums to the new generation (Erbay, 2017). Developing technology has assigned a new task to museums and museums have become an active learning areas (Keş and Akyürek, 2018). Various museums such as Gaziantep Zeugma Mosaic Museum, Corum Archeology Museum, Burdur Archeology Museum, Ankara State Painting and Sculpture Museum, Cleveland Museum of Art, MOSTI (Museum of Science and Technology of Islam in Saudi Arabia) provide better museum experience to its visitors by using interactive applications.

In recent years, information and communication technologies have been used in museums around the world for using the museums comfortably and economically, informing visitors better as well as increasing and measuring the user experience. Ceipidor et al. (2013) developed a multimedia mobile guide application aimed at researching near-field communication (NFC) technology and user experience (UX) in the tourist-cultural area. Angelaccio et al. (2012) developed a virtual NFC (Near Field Communication) based AmI information system that facilitates the entrance of tourists interested in cultural heritage sites without affecting the historical environment.

Fevgas et al. (2014) developed a virtual museum system called IMUSE that aims to increase the museum visits and make the visitors navigate freely within the museum and making the most of WEB, RDIFs and communication technologies.

Smart technology applications in museums are provided by applications such as audio-guide (Topkapi Palace museum, Los Angeles city museum, London Science Museum), head-set (London Science Museum, Natural History Museum, British Muesum), virtual assistant, mobile guide, virtual museum, guide books and land information system. Smart museum assistants are Bluetooth-connected systems that provide continuous and free communication. With these systems, visual and virtual access is provided to the museum content in the desired language and the possibility of applications to be kept in mind is increased. With smart museum assistant system visitors (Altaş 2018):

- Visitors will be able to access the location of the nearest museum,
- Will be able to see the museum on the map and reach the museum in a simple way with audio and visual directions,
- The website of the museum can be accessed,
- They can mark their favorite museums,
- They can share their experiences about the museum on their website,
- They can communicate with the museum with the registered phone number,
- They can list all activities in the museum and
- Access to all announcements from museums.

For a long time, audio guide technology has been used in the narration of museums. Audio guides that were started to serve as MP3 recordings, serve visitors with voice, video and language options with the effect of technological developments. With their properties, audio guides provide great convenience for hearing impaired visitors. Intelligent technologies that provide audio guide services are among the important exhibiting sources of world museums.

Kiosks that are aimed to inform visitors, touch tables and mobile sensors that are used to introduce works are smart technologies in museums. Besides this, smart tablets and phones are also used to introduce museums. Detailed information about the artifacts exhibited in museums can be obtained with phones with barcode application. By scanning the barcode on the work exhibited in the museum, all kinds of detailed information (written and visual) about the work can be transmitted to the visitors.

Internet applications are also highly used in modern museums the most common use of the internet is 360° virtual tours that is generally used as a source in advertising and archives. Visitors entering the website of museums have the opportunity to visit the works exhibited in the museums through virtual tours. With virtual tours, interest is attracted to museums and people are encouraged to visit the actual state of the museum. One of the smart technologies used in museums is the augmented virtual reality applications. Web 3D and virtual reality applications offer many advantages to museum visitors. In augmented virtual reality museums, virtual information (3-D objects, multimedia types including textual or imaginary information, etc.) provides visitors to be closer to objects and see more realistic views with live or recorded video frames (Styliani, Liarokapis, Kostas & Petros, 2009). With the help of computer interaction techniques, visitors can touch virtual objects and fully examine them through components or sensors (Coşkun, 2017). Virtual reality applications are used by many museums today. In addition, scientists are trying to develop new applications for museums equipped with intelligent technologies. For example, Damala et al. (2008) developed a museum guide based on AR technology at the Museum of Fine Arts in Rennes, France. Again Cianciaruloa (2015) used a AR project in a museum in Viggiano Italy, exhibiting products from old farms. In Turkey Sakıp Sabancı Museum, Maritime Museum, Bursa Clock Museum, Topkapı Palace Museum, Carpet Museum, SEKA Paper Museum, Museum of Natural

Burdur Kavakli Greek Church and Hatay Archeology Museum are using smart technology as LV systems (Sertalp, 2016). Museums such as Vincent Van Gogh Museum, Saint Climent de Taull Church, etc. redesign their original artworks with AR technologies and offer visitors the opportunity to examine their original designs (İlhan & Çeltek, 2016). Similarly museums such as the Louvre Museum, the British Museum, The Hermitage Museum, also offer guests the opportunity to examine the originals of the designs with AR.

Benefıts of Smart Applications in Tourism

Smart technologies can be used in all service areas of tourism sector. Smart technologies in tourism sector can easily be applied in hotel businesses, travel businesses, food and beverage businesses, recreation and animation businesses, museums and so on. Some of the benefits of these applications are listed below (Gökalp and Eren, 2016):

- It can make increase in the service quality of tourism enterprises: Smart phone applications, personalized services, smart room applications, child tracking systems, smart card, smart wristband, smart waiter and smart kiosk etc. applications can increase the quality of service. The following issues in the quality of service can be assisted:
- Reliability: Useful hotel equipment such as handheld devices, kiosks, access to in-room services by voice command and smart mini bar have positive effects on this dimension. As it reduces human error automatic provision of services instead of hotel staff positively affects reliability (Paryani et al., 2010).
- Pocessing Speed: Services such as online check-in and check-out, automatic concierge service, smart waiter services allow to respond the expectations of guests quickly and reduce the workload of the staff.
- Security: With smart ID cards accession of guests is allowed. With this service, you cannot have access to private areas of someone. Therefore, guests feel safe and comfortable.
- It can increase guest satisfaction: With the help of smart technologies such as augmented reality applications, online check-in and check-out operations, smart card, smart waiter applications, smart mina bar and cloud applications etc. guest satisfaction can be increased.
- Increased Profitability and Productivity: By using smart technologies in tourism sector, service quality of enterprises increases while saving time and cost elements can be achieved. With time savings, the solution accelerates and increases guest satisfaction in the business and operations of the guests. The workload on staff is reduced. This can result in a reduction in workload costs. Smart technology also reduces product inventory costs. As a result of this, productivity and profitability can be increased in the enterprises.
- Innovation: Smart technology means innovation. Smart technologies mean innovations in process, product, system and service. These innovative approaches attract the attention of guests. These innovations lead to positive experiences for the guests. Innovative practices can also lead to an increase in service quality.
- Increased Environmental Awareness: Smart technology provides facilities in subjects such as intelligent lighting, intelligent water management, intelligent garbage collection systems, garbage processing and recycling systems, intelligent irrigation, smart city planning and environmental pollution management and so on (Yavuz, 2019). As a result, environment-friendly tourism enterprises (green

hotels, clean sea, etc.) have started to spread. As tourism enterprises with environmental sensitivity are preferred by the guests particularly, this has been effective in the development of environmentally sensitive enterprises in tourism.

Problems of Smart Applications in Tourism

- Cost. The most obvious loss of smart technologies is temporary cost increases. Certain level of software, smart telephones, tablets, wireless networks, sensors, RFID tags and many other technologies such as small and sensitive devices and servers are needed for smart applications. In the sector where there are mixed applications such as tourism, the number of IT enterprises that will develop and implement smart applications and the increasing demand is increasing the costs of smart technology systems considerably.
- Privacy and Security. Although smart technologies have great benefits, they also carry risks. In tourism businesses, smart technologies are provided by integrated cloud technologies system. Cloud technologies include high privacy and security protection. When security and privacy are not given enough attention as this technology is set up, security gaps arise. Monitoring of guests and storing their information brings risks. In addition, shortages that may arise during the access control process may cause the information of the guests to be received by third parties.
- Induration. One of the negative effects of smart technologies in the social sense is the damage of human relations. Intelligent technologies are transforming the human-human relationship into a human-technology relationship. Even though intelligent technologies are particularly beneficial in terms of production-time and cost, they may lead to the relevance of monotonous, unilateral and robotic relationships in time. This responds to the technological needs of the guests while eliminating emotional (conscience, interaction, sharing, etc.) interactions. This may cause dissatisfaction in guests.
- Unemployment. The widespread use of smart technologies in the tourism sector also carries the risk of unemployment. Some of the work to be performed by the employees is carried out by information and communication technologies. As a result, the workload of the employees is reduced. In this case, the number of employees required by the enterprise decreases. This may lead to an increase in unemployment rates.
- Confusion. Smart technologies are innovative applications. This approach may cause confusion in employees. It may also cause problems for employees who are accustomed to old practices to adopt or become accustomed to new practices. This will cause conflicts of old-new understanding and confusion. Another confusion is the excessive focus of managers on technology.

Smart technology in tourism is now a reality. Smart technologies will increase the variety of tourism services, cause changing of services and will continue to be determinant in the future. As travel establishments, accommodation establishments, food enterprises etc. sometimes follow smart technologies, they sometimes will continue to provide better, better quality and more economic service to their guests by creating their own smart applications. Therefore, intelligent technology in tourism sector will continue to develop continuously.

SOLUTIONS AND RECOMMENDATIONS

Technology will continue to influence and shape the tourism sector at an increasing rate. What is important here is how much technology will play an important role in which points of tourism. What will be the limit of technology in tourism? Where and how long will the use of technology in tourism reach its saturation point? How long will technology continue to please tourists? How much longer will tourists accept technology? Will the satisfaction, acceptability and sustainability of the technology invasion be achieved in the relationship between tourism and technology?

The answers to the above questions will determine the extent and limits of the interaction between tourism and technology. It will also shape the future of the use of technology in tourism. Another important factor is what the dependence of generation Y, millennium generation will be on technology. Or what the degree and duration of technology affecting the generation y and millennium will be.

There are benefits of the use of technology in tourism for both tourists and tourism enterprises in terms of time, cost, competition, employment, satisfaction. Technology is an important factor in providing competitive advantage for enterprises. However, how much materialization and technological superiority can satisfy the emotions (mind, heart, conscience, etc.) and psychological expectations of tourists in abstract-satisfied tourism enterprises?

When receiving services in destinations how will tourists communicate and interact with technology that only serves but does not show any emotional characteristics? For example how will technology solve the psychological problems that a guest will face in the process of receiving services? How will technology share the joy of a guest? Or how will a guest protect his privacy from technology that must be hidden?

In the light of these explanations, technology in tourism;

1. Should be a factor that facilitates the services as well as saves time and cost for the guests and businesses.
2. Technology; may play a role in improving human relations in tourism enterprises but should not be decisive in inter-human relations.
3. In human oriented tourism sector technology cannot give services expected from human such as smile, friendship, conversation, discussion etc. Therefore, these services should not be expected from technology. Particularly from robot technology service, and so on. Inability to achieve the expected performance in the fields is an example of this situation.
4. Technology that is economic at the beginning can turn out to be high costs (loyalty, purchase, renewal, depreciation, etc.) after a while.
5. Another event is the necessity of adapting to the changes that may occur in the expectations of the target audiences, may necessitate the technological changes of the enterprises. This can be a cost-increasing factor as well as may cause difficulties in accessing new technologies.

In the light of these explanations, the points in relation between technology and tourism should be taken into consideration;

1. Technology is required in areas such as travel, accommodation, food and beverage and recreation.
2. Location, time and amount of this requirement depends on;
    a) Technological expectations of tourists from enterprises,

   b) Depends on the technological supply of tourism enterprises. So supply depends on the level of the relationship of demand.
3. The level of contribution of technology to the sustainability of tourism should be considered.
4. The effects of technology on the tourism environment (environmental pollution (water, waste, air) as well as its contribution to environmental protection (protection of green areas, cleanliness), impact on natural areas, etc.) should be considered.

It should not be forgotten that technology has both benefit and harm. When technology is used in place and quantity, it can make a very positive contribution to the experiences of tourists. But extreme technology and mechanization cannot help tourists seeking naturalness in the experience, in contrast it can lead to negative experiences.

FUTURE RESEARCH DIRECTIONS

Technology in tourism will always exist. Even the effectiveness of the technology will increase more. One of the main factors of this is the desire of investors to maximize profit. Tourism businesses will use two methods to maximize profits:

a) Minimizing costs (reducing fixed and variable costs, particularly personnel costs, equipment costs etc.),
b) Increasing customer satisfaction (providing quality, fast and economic services).

One of the tools used to carry out all these, is the application of information communication technologies in tourism.

Scientists are naturally searching the relationship between tourism and technology, the reasons why they are preferred and their competitive advantages. When researching the relationship between tourism and technology, scientists should investigate the social dimensions of the event, its sociological importance, its consequences, and the side effects of technology. Especially they should investigate the negative effects of technology on customer satisfaction in the short and long term because tourism is an event with abstract consequences. Therefore, abstract results may not always be explained by taking material elements such as technology.

The effects of technology on the buying, benefiting and after-behaviors of tourists should also be examined. It is also important for scientists to research the expectations of tourists from tourism technology in order to correctly locate the future of tourism technology relationship. Another issue that scientists need to research is how much technology can replace human beings in tourism service and how sustainable it will be.

CONCLUSION

As a result technology shows its existence and effect in all areas of human life for this period. It is inevitable that tourism should stay away from it. Tourism enterprises will always benefit from technology and its infrastructure (information and communication technologies) in order to provide the quality services expected by its customers. Instead, the amount and time will be decided by the level of expectations of customers from tourism businesses. The use of technology in tourism is a supply-demand relationship balance. Businesses that can achieve this balance will be able to meet expectations of customers and satisfy them. The enterprises that fail to achieve this balance will not be able to respond positively to the expectations of the customers.

One of the determining factors of tourism technology relationship is to follow the expectations of target and potential tourists from tourism. Investigating and analyzing the expectations of the target groups of tourism enterprises will be a determining factor in the level of technology that the enterprise will use. Finally;

As long as technology and informatics exist in human life, tourism enterprises will continue to use these technologies and to benefit from these technologies. The type, degree and usage of this will be determined by the wishes and expectations of customers.

REFERENCES

Adonisotel.com. (2018). 5 Applications from smart hotels you will say "wow" when you see. Retrieved from https://www.adonisotel.com/Destek/Otel-Yonetimi-Makaleler/gorunce-vay-can%C4%B1na-diyecek-siniz

Aksoy, M., & Akbulut, B. A. (2016). Restoranlardaki teknolojik yeniliklerin deneyim pazarlaması açısından değerlendirilmesi. In II Eurasia Tourism Congress, Konya. Retrieved from https://www. researchgate. net/publication/310994877

Altaş, A. (2018). Geographical information system applications utilized in museum in Turkey within the scope of the cultural heritage tourism: a case study of mobile application of museum asistance. In C. S., & R. N. (Eds.), GIS applications in the tourism and hospitality ındustry (pp. 42-60). Hershey, PA: IGI Global: Business Science References.

Baines, A. (1998). Technology and tourism. *Work Study*, *47*(5), 160–163. doi:10.1108/00438029810370492

Balakrishnan, A. (2016). Wynn Las Vegas to Add Amazon Alexa to all Hotel Rooms. Retrieved from http://www.cnbc.com/2016/12/14/wynn-las-vegas-to-add-amazon-alexa-to-all-hotel- rooms.html

Baydeniz, E. (2019). The Relation of tourism with technology. Retrieved from https://www.academia.edu/12421334/TUR%C4%B0ZM%C4%B0N_tEKNOLOJ%C4%B0_%C4%B0LE_%C4%B0L%C4%B0%C5%9EK%C4%B0S%C4%B0

Benli, T., & Kızgın, Y. (2002, April). A Field study on the importance of use of information technology in hospitality enterprises: The case of Muğla province. *Journal of Legislation*, *5*(52).

Blowuphall5050. (2019). Innovations. Retrieved from http://www.blowuphall5050.com/hotel/

Broadway World. (2018). Order with your face at philly restaurant: facial recognition software debuts at fuel. Retrieved from https://www.broadwayworld.com/philadelphia/article/Order-With-Your-Face-At-Philly-Restaurant-Facial-Recognition-Software-Debuts-At-Fuel-20180404

Buhalis, D. (1998). Strategic use of information in the tourism industry. *Tourism Management*, *19*(5), 409–421. doi:10.1016/S0261-5177(98)00038-7

Buhalis, D., & Amaranggana, A. (2013). Smart tourism destinations. In *Information and communication technologies in tourism 2014* (pp. 553–564). Cham, Switzerland: Springer. doi:10.1007/978-3-319-03973-2_40

Buhalis, D., & Law, R. (2008). Progress in information technology and tourism management: Status of etourism survey 20 years and 10 years after the internet. *Tourism Management, 29*(4), 609–623. doi:10.1016/j.tourman.2008.01.005

Ceipidor, U., Medaglia, C. M., Volpi, V., Moroni, A., Sposato, S., Carboni, M., & Caridi, A. (2013). NFC technology applied to touristic-cultural field: a case study on an Italian museum. In *Proceedings 5th International Workshop on Near Field Communication.* 10.1109/NFC.2013.6482445

Cianciarulo, D. (2015). From local traditions to "augmented reality". The MUVIG Museum of Viggiano (Italy). *Procedia: Social and Behavioral Sciences, 188*, 138–143. doi:10.1016/j.sbspro.2015.03.349

Coşkun, C. (2017). Augmented reality as an exhibition method. *Art and Design Magazine* (20), 61-75.

Damala, A., Cubaud, P., Bationo, A., Houlier, P., & Marchal, I. (2008). Bridging the gap between the digital and the physical: Design and evaluation of a mobile augmented reality guide for the museum visit. In *Proceedings of the 3rd International Conference on Digital Interactive Media in Entertainment and Arts (DIMEA '08)* (pp. 120-127). ACM, New York. 10.1145/1413634.1413660

De Avila, A. L. (2011). Tourism: What lies ahead?, World Tourism Organization (UNWTO) affiliate members AM-reports – Technology in Tourism (Vol. 1).

Drummond, M. (2019). 5 Great ways airlines are using the ınternet of things. Retrieved from https://w3.accelya.com/blog/5-great-ways-airlines-are-using-the-internet-of-things

Ekonomihaber7.com. (2017). The waiters of this restaurant are robot! Retrieved from http://ekonomi.haber7.com/ekonomi/haber/2373365-bu-restoranin-garsonlari-robot

Erbay, M. (2017). Education in museums with new generation Technologies. *National Education Magazine, 46*(214), 255–268.

Etro, F. (2009). The Economic impact of cloud computing on business creation, employment, and output in Europe. *Review Business and Economics, 54*(2), 179–208.

Frew, A. J. (2000). A critical analysis of tourism ınformation technology research. In *Information and Communication Technologies in Tourism* (pp. 39–52). Vienna, Austria: Springer.

GlobalData.com. (2019). Cloud computing in tourism - thematic research. Retrieved from https://www.globaldata.com/https://hot-topics.globaldata.com/reports/cloud-computing-in-tourism-thematic-research/

Gökalp, E., & Eren, E. (2016). Application of ıntelligent technologies in tourism and hospitality sector. In V. Tecim, C. Tarhan, & C. Aydın (Eds.), Smart Technology & Smart Management: Smart Technology & Intelligent Management, 278-287.

Gökkuş, B. (2018). First robot waiter restaurant of South Asia in service. Retrieved from https://tr.euronews.com/2018/12/09/guney-asya-nin-ilk-robot-garsonlu-restorani-hizmette

Gretzel, U., Sigala, M., Xiang, Z., & Koo, C. (2015). Smart tourism: Foundations and developments. *Electronic Markets, 25*(3), 179–188. doi:10.100712525-015-0196-8

Güngör, Y. M., Güngör, O., & Doğan, S. (2017). Evaluation of food and beverage businesses in Kuşadası from the perspective of SoLoMo marketing. *International Management Economics and Business Journal, 13*(3), 683–704.

Hançer, M., & Ataman, C. (2006). Use of communication technology in travel agencies and evaluation of web sites: The Case of Aegean Region, Dokuz Eylül University. *Journal of Social Sciences.*

Huang, C. D., Goo, J., Nam, K., & Yoo, C. W. (2017). Smart tourism technologies in travel planning: The role of exploration and exploitation. *Information & Management, 54*(6), 757–770. doi:10.1016/j.im.2016.11.010

Huang, Y. C., Backman, K. F., Backman, S. J., & Chang, L. L. (2016). Exploring the ımplications of virtual reality technology in tourism marketing: An ıntegrated research framework. *International Journal of Tourism Research, 18*(2), 116–128. doi:10.1002/jtr.2038

Inkpen, G. (1998). Information technology for travel and tourism (No. Ed. 2). Addison Wesley Longman Ltd.

Jaremen, D. E., Jędrasiak, M., & Rapacz, A. (2016). The Concept of smart hotels as an ınnovation on the hospitality ındustry market - case study of PURO Hotel in Wrocław. *Economic Problems of Tourism, 36*(4), 65–75.

Jinendra, D. R., Bhagyashri, J. R., Pranav, G. Y., Seema, V. U., & Parag, A. N. (2012). Smart travel guide: Application for android mobile. International Journal of Electronics [IJECSCSE]. *Communication and Soft Computing Science & Engineering, 2*, 115.

Kang, B. (2011). The Impact of channel knowledge on shopping orientations in consumer buying behaviour. *Anatolia- An International Journal of Tourism and Hospitality Research, 22*(1), 120-124.

Keş, Y., & Akyürek, A. B. (2018). Interactive museums for the new generation growing with technology. *IMU Journal of Art, Design, and Architecture 4*(2), 95-110.

Law, R., Leung, R., & Buhalis, D. (2009). Information technology applications in hospitality and tourism: A Review of publications from 2005 to 2007. *Journal of Travel & Tourism Marketing, 26*(5-6), 599–623. doi:10.1080/10548400903163160

Mingjun, W., Zhen, Y., Wei, Z., Xishang, D., Xiaofei, Y., Chenggang, S., ... & Jinghai, H. (2012, October). A research on experimental system for Internet of things major and application project. In *Proceedings 2012 3rd International Conference on System Science, Engineering Design and Manufacturing Informatization* (Vol. 1, pp. 261-263). IEEE.

Morfikirler.com. (2018). World's first smart restaurant opened in Dubai. Retrieved from https://morfikirler.com/dubaide-dunyanin-ilk-akilli-restorani-acildi/

Ntv.com.tr. (2013). Robot waiters are at your service. Retrieved from https://www.ntv.com.tr/galeri/yasam/robot-garsonlarhizmetinizde,9SpVjbRyZkyEQgok65ZnCw

O'Connor, P., & Murphy, J. (2004). Research on information technology in the hospitality industry. *International Journal of Hospitality Management, 23*(5), 473–484. doi:10.1016/j.ijhm.2004.10.002

Özgüneş, R. E., & Bozok, D. (2017). Virtual competitor of tourism sector (is that so?): Augmented reality, 1st International Sustainable Tourism Congress, November 23-25, 2017, Kastamonu-Turkey.

Pemberton, J. D., Stonehouse, G. H., & Barber, C. E. (2001). Competing with CRS-generated information in the airline industry. *The Journal of Strategic Information Systems, 1*(1), 59–76. doi:10.1016/S0963-8687(01)00042-7

Poon, A. (1993). *Tourism, Technology, and Competitive Strategies*. Oxford, UK: CAB International.

Porter, M. (2001). Strategy and the Internet. *Harvard Business Review, 79*(3), 63–78. PMID:11246925

Princes.com. (2019). Ocean medallion. Retrieved from https://www.princess.com/learn/faq_answer/pre_cruise/ocean_medallion.jsp

Renkliyolculuk.com. (2019). Technological developments in tourism. Retrieved from https://www.renkliyolculuk.com/2019-turizm-alanindaki-teknolojik-gelismeler/

Romão, J., & Neuts, B. (2017). Territorial capital, smart tourism specialization and sustainable regional development: Experiences from Europe. *Habitat International, 68*, 64–74. doi:10.1016/j.habitatint.2017.04.006

Sertalp, E. (2016). The Usage of augmented reality (ag) in tourism areas. In *Proceedings 21st Internet Conference in Turkey* (pp. 1-7). TED University, Ankara, Turkey.

Sheldon, P. (1997). *Tourism information technologies*. Oxford, UK: CAB.

Spence, C., & Piqueras-Fiszman, B. (2013). Technology At The Dining Table. *Flavour (London)*, 1–13.

Stipanuk, D. M. (1993). Tourism and technology: Interactions and applications. *Tourism Management, 14*(4), 267–278. doi:10.1016/0261-5177(93)90061-O

Styliani, S., Liarokapis, F., Kostas, L., & Petros, P. (2009). Virtual Museums, a Survey and Some Issues for Consideration. *Journal of Cultural Heritage, 10*(4), 520–528. doi:10.1016/j.culher.2009.03.003

Tu, Q., & Liu, A. (2014). Framework of smart tourism research and related progress in China. In *Proceedings International Conference on Management and Engineering (CME 2014)*, pp. 140-146, Pennsylvania: DEStech Publications.

Turizmgüncel.com. (2018). Akıllı ev konseptindeki ilk otelini İstanbulda açtı. Retrieved from https://www.turizmguncel.com/haber/akilli-ev-konseptindeki-ilk-otelini-istanbul'da-acti-15-otel-daha-geliyor-h27337.html

Turizmnews.com. (2018). Akıllı otel uygulaması Türkiye'de ilk kez Isparta'da. Retrieved from https://www.turizmnews.com/akilli-otel-uygulamasi-turkiye-de-ilk-kez-isparta-da/12773/

Upton, L. (2014). The top 10 innovations that are shaping the future of rail and metro. Retrieved from https://www.smartrailworld.com/it-and-wifi/the-top-10-innovations-that-are-shaping-the-future-of-rail-and-metro

Wang, D., Li, X., & Yunpeng, L. (2013). China's smart tourism destination initiative: A Taste of the service dominant logic. *Journal of Destination Marketing & Management, 2*(2), 59–61. doi:10.1016/j.jdmm.2013.05.004

Werther, H., Koo, C., & Gretzel, U. (2015). Special issue on smart tourism systems: Convergence of information technologies, business models, and experiences. *Computers in Human Behavior, 50*, 556–557. doi:10.1016/j.chb.2015.03.042

Xiang, Z., Magnini, V. P., & Fesenmaier, D. R. (2015). Information Technology and Consumer Behavior in Travel and Tourism: Insights from travel planning using the ınternet. *Journal of Retailing and Consumer Services, 22*, 244–249. doi:10.1016/j.jretconser.2014.08.005

Yalçınkaya, P., Atay, L., & Karakaş, E. (2018). Smart tourism applications. Gastroia: *Journal of Gastronomy and Travel Research,* Cilt 2. *Sayı, 2*, 34–52.

Yavuz, M. C. (2019). Smart destination: Tourism, ınnovation, entrepreneurship, challenges. *Journal of Tourism Theory and Research, 5*(2) Futourism, 203-211.

Yücel, D. (2012). *New Media Art and New Museum.* İstanbul, Turkey: İstanbul Kültür University Publications.

Yüksek, G. (2013). Development of ınformation technologies, travel enterprises and global distribution system example. *Journal of Internet Applications & Management Journal, 4*(1). eturizm.org. (2019). Helsinki ve Lyon, 2019 Avrupa Akıllı Turizm Başkentleri ödüllerini aldı. Retrieved from https://eturizm.org/helsinki-ve-lyon-2019-avrupa-akilli-turizm-baskentleri-odullerini- aldi/

Zhou, Z. (2003). *E-Commerce and ınformation technology in hospitality and tourism.* California: Thomson Delmar Learning.

ADDITIONAL READING

Boes, K., Buhalis, D., & Inversini, A. (2015). *Conceptualising smart tourism destination dimensions. ın ınformation and communication technologies in tourism 2015* (pp. 391–403). Cham: Springer.

Gretzel, U., Werthner, H., Koo, C., & Lamsfus, C. (2015). Conceptual foundations for understanding smart tourism ecosystems. *Computers in Human Behavior, 50*, 558–563. doi:10.1016/j.chb.2015.03.043

Katsoni, V., & Segarra-Ona, M. (2018). Smart tourism as a driver for culture and sustainability: Fifth International Conference IACuDiT, Springer Nature Switzerland AG.

Su, K., Li, J., & Fu, H. (2011, September). Smart City and the applications. In *2011 International Conference on Electronics, Communications and Control (ICECC)* (pp. 1028-1031). IEEE. 10.1109/ICECC.2011.6066743

Xiang, Z., & Fesenmaier, D. R. (2017). *Analytics in smart tourism design concepts and methods.* Springer. doi:10.1007/978-3-319-44263-1

KEY TERMS AND DEFINITIONS

Augmented Reality: It is the superimposition of virtual objects over real images using the object recognition feature of the devices.

Smart Application: The concept of smart technology internet is a system of internet, communication systems and networks, technological innovations, data management and processing systems and ecological management systems.

Smart Destination: Cloud communication, internet-connected things and end-user internet destinations.

Smart Hotel: Smart hotel applications have gained importance with the emergence of smart building designs. Smart building designs are developed for the development of smart living and smart working spaces by using sustainable technologies that affect the quality of life such as lighting and heating systems.

Smart Museum: Smart museum applications have gained importance with the emergence of smart building designs. They are smart applications for the development of intelligent workspaces in museums using sustainable technologies such as lighting, heating systems and augmented reality.

Smart Restaurant: Smart restaurant applications, service, food preparation, smart applications during meals smart applications for the development of smart workspaces using sustainable technologies such as lighting, heating systems and augmented reality.

Smart Tourism: A type of tourism supported by efforts in an estimation to collect physical infrastructure, social connections, state / organization resources and human mind data.

Chapter 17
Smart Tourism, Smart Cities, and Smart Destinations as Knowledge Management Tools

Halime Göktaş Kulualp
Karabuk University, Turkey

Ömer Sarı
Karabuk University, Turkey

ABSTRACT

The rapid increase in the population has caused problems in the correct use and management of resources in cities. Solutions to these problems have been sought based on knowledge management. In today's digital age, the concept of smartness of cities has been put forward together with the web-based applications. It has changed the expectations and needs of tourists and residents, especially in the tourism sector in the service sector. Smart tourism destinations, which are seen as a solution to the expectations and needs of the changing tourists and local people, bring holistic innovations covering all the stakeholders in the tourism ecosystem. In this direction, it is thought that knowledge management makes the right use of resources obligatory and contributes to sustainable tourism understanding. For this purpose, in the chapter, suggestions were made to provide maximum benefit from web-based projects that are carried out in qualitative direction of knowledge management and smart tourism destinations.

INTRODUCTION

Rapid population growth in metropolitan cities has caused problems with the correct use and management of resources. For this reason, especially in large cities, programs are needed to provide conveniences for the increasing number of people and their needs. In the digital world we live in, web-based solution proposals have been sought for these problems. With the widespread use of computers and the internet, the concept of smart city emerged when knowledge about cities was transformed into web-based and artificial intelligence formats in order to increase the efficiency of cities and facilitate human needs.

DOI: 10.4018/978-1-7998-1989-9.ch017

In the ongoing process, such practices have led to the rapid spread of smart cities and even radical changes in various sectors. This situation has changed the expectations and needs of tourists and residents in tourism, especially in the service sector. Smart tourism destinations that are expected as a solution to the changing expectations and needs of tourists and residents introduce holistic innovations that cover all stakeholders within the tourism ecosystem. In addition, it has made it imperative to adopt a protective approach to the use of environmental and socio-cultural resources, which is the capital of tourism. It contributes to sustainable tourism understanding by necessitating the correct use of resources. Therefore, the concept of smart destination has enabled the establishment of an integrated structure in which all stakeholders are involved. For this purpose, in the study, a qualitative assessment of the smart tourism destinations that have been examined in the international literature in the 1990s, but which have recently been discussed in Turkey, has been made.

It is seen that enterprises, which are an indispensable member of the social environment, use the knowledge obtained with the increasing speed of technology and the use of computers for the purpose of sustainable competitive advantage in the market through projects such as web design and software. Especially in the service sector, knowledge management applications have gained significant advantages against their competitors.

Tourism destinations showed great interest in the concept of smart city. In this context, tourism destination enterprises are trying to establish superiority in competition by increasing the satisfaction of tourists and increase market share as well as regional destination managers adopt the concept of smart tourism destination in order to provide their citizens with a better quality of life. In line with this knowledge, it is aimed in this study to conduct an extensive examination in the smart tourism destinations (Çelik and Topsakal, 2017:164) which are a new subject in Turkey in a qualitative way. Initially, it was aimed to examine knowledge management and to reveal the sectoral and regional importance of knowledge. Then, the concept of smartness was examined in qualitative terms and a judgment on what smart cities are like was made. Afterward, the compilation obtained was scrutinized with the concept of smart destination in a holistic way. Finally, an evaluation of the smart tourism destination for the tourism economy and the stakeholders in the ecosystem has been made.

BACKGROUND

Concept of Knowledge Management

For centuries, human beings have been able to guide the events they have lived with the experiences they gained and their innate abilities. Because they maintained their lives by establishing a causative relationship with the knowledge they have obtained from these experiences and abilities. Therefore, the use of knowledge in human life has become an indispensable actor. According to TDK (Turkish Language Association) (2018), the concept of knowledge or its components are defined as "the product of thought resulting from the study of human intelligence". It is possible to say that knowledge is a belief that is accepted by the person. It doesn't matter for a person whether knowledge is the absolute right. The person evaluates knowledge with internal consistency rather than objective reality (Lang, 2001: 45). Thanks to its existence and use, the knowledge has enabled the emergence of today's technology. In this context, knowledge, which enters every area of human life, is seen as an indispensable element (Tokcan, 2015:7).

Knowledge and management concepts can be difficult to perceive together at the first glance. Knowledge is an advanced personal concept aimed at comprehension. Management, on the other hand, refers to organizational processes that require teamwork for common purposes. It is widely accepted that knowledge is an important organizational resource that provides predominance in the market. Because of this reason, the management of knowledge is such an important subject that cannot be left up to chance (Güçlü and Sotirofski, 2006: 355).

The concept of knowledge management was first introduced to the management literature by Dr. Karl Wiig in 1986. In the 1990s, some consulting companies realized that they could gain a competitive advantage by sharing their knowledge. In 1994, large consulting companies started to provide systematic knowledge to their customers for the first time. In short, we can say that knowledge management is the result of changes experienced in the field of business in the 1990s (Yılmaz, 2009: 18).

Denning (2000: 114) states knowledge management as *"it includes multiple dimensions, including strategy, implementation communities, help desks, acquisition, storage and dissemination of knowledge, taxonomy, quality assurance, authentication procedures, budget incentives and knowledge measures."*

According to Lee and Lee (2006: 21), knowledge management is a practice to improve organizational performance through the process of obtaining knowledge, converting it into a useful form, applying the knowledge and using the knowledge systematically.

According to OECD (2003), "knowledge management" generally refers to an extensive collection of organizational practices for creating, promoting, and disseminating trade secrets known as know-how in business science within an organization or with the outside world. Fundamentally, it includes the following (OECD, 2003: 4):

- Organizational arrangements (decentralization of authority, the establishment of bureaucratic divisions, use of knowledge and communication technologies, etc.);
- Personnel development (guidance and training practices, mobility etc.) and management of skills;
- Transfer of competencies (databases of personnel qualifications, the outlines of good business practices, etc.);
- Administrative changes and incentives for personnel to share knowledge (performance of personnel
- Evaluation and promotion based on knowledge sharing, development of the role of managers, etc.)

It is possible to say that knowledge management is based on many disciplines such as economics, psychology and information management. Knowledge management constitutes an advantage for organizations. Departments of information management cover people, technology and processes and impacts business processes positively. The purpose of knowledge management is to capture what is required for a process and to support knowledge workers to share this knowledge. If an organization captures and disseminates knowledge, the benefits it will acquire will be greater (Güçlü and Sotirofski, 2006: 355-356). According to Ayaz and Göktaş Kulualp (2019), a well-established human resources system is required to support the sharing of knowledge of employees. In addition, having a strategic awareness in the process of human resources management will provide competitive advantage to the organization.

Importance of Knowledge Management in Tourism

In the digital age in which we are in, it is thought that it obliges enterprises to maintain their existence in the sustainable competitive system and to adapt to the information system brought by the age in order to keep profitability at maximum level and to use information most effectively in pragmatic ways (Önaçan, 2015: 20).

When the researches of the last twenty years are examined, it is thought that knowledge management has increased the positive and desired effects on the business. These effects are thought to provide enterprises with significant benefits such as opportunities and advantages in the competitive system. Therefore, it is thought that enterprises carry out more studies in the field of knowledge management and increase their investments in this field and support the production of innovative ideas in this field (Sarvary 1999; 4; Karakoçak, 2007; Choi et al., 2008; Yusof et al., 2012: 318; Liao and Wu, 2009; Zack et al., 2009; Zaied et al., 2012; Rašula et al., 2012; Torabi and El-Den, 2017; Xu et al., 2018).

Córdova and Gutiérrez (2018) in their research on the knowledge management system in service companies assume that the existence of an information cycle that include service offerings, including socialization, externalization, consolidation and internalization in strategic knowledge generation, as well as culture factor, identifying opportunities and offering the added value proposition, will improve the long-term survival and quality of life of enterprises.

In the tourism sector, where the service is predominant, it is thought that information management and its tools are becoming more important with the beginning of computerized processes as in other sectors. In the research conducted by Jia et al. (2012) in the field of knowledge management in tourism it is assumed that computer and web-based artificial intelligence systems can provide positive benefits to the business by producing appropriate strategies with the properties that can sort, store and share the information needed for tourism enterprises.

According to Lee (2008: 23), knowledge management can increase communication channels with employees, suppliers and customers in hospitality enterprises. In this context, it assumes that it strengthens its staff, coordination and creativity in order to ensure that the products they use are innovative. Yang and Wan (2004: 593), on the other hand, emphasize that knowledge management is a solution to the loss of information caused by a high turnover rate, especially in accommodation enterprises.

Tourism enterprises are considered to be the biggest obstacle to the implementation of knowledge management and specially to create costs in long-term projects. However, it also includes several obstacles. The first is expressed as a complete data-based commitment, which is thought to be a technology gap. This can mean the source of complex problems in poorly designed programs. Second, the program application gap occurs when there is unawareness about employees that they know the decisions and practices. Thus, this causes a closed or unnoticed stack of knowledge. The third originates from the mismatch between knowledge management and other elements of the organization. Finally, it is because information is still not understood during the transfer process and to carelessly designed projects for implementation. This indicates that the generation, transfer and use of the knowledge are informal and unstructured. Taking strategic decisions without being aware of what information is available could be the biggest problem of tourism (Cooper, 2006: 58-59). Such obstacles may be related to the nature of tourism. However, according to Tribe and Liburd (2016: 59), it is defined as *"The tourism information system not only reveals the main processes of theoretical progress, practical problem solving and its commitment to the real world, but also reveals the radical possibilities of ontological policies in tourism research."*

In this regard, the emergence of these opportunities allows tourism enterprises and destinations to be positioned differently in the tourism market. The benefits of smart technologies, known as an application of knowledge management, to tourism enterprises and destinations, are seen as innovative approaches.

Thanks to this innovation, the above-listed benefits are thought to provide the competitive advantage that tourism enterprises can sustain in the market (Gökalp and Eren, 2016).

MAIN FOCUS OF THE CHAPTER

Smart City Concept

In today's city planning, the infrastructure of the districts and buildings has been abandoned and digitalization, quality of life, integration, equality and the needs of citizens have been emphasized (Axelsson ve Granath, 2018:1). Historically, it has emerged in mid-1800s to define the self-governing and fertile cities of Western America (Yigitcanlar et al., 2018: 146). The concept of smart city, referring to sustainable growth, has later become a concept that has gained popularity since the last decade of the 21st Century. For the first time, the California Institute for Smart Communities has focused on designing a city knowledge technology application that can transform people living in the city into a smart society (Albino et al., 2015: 4). Definitions and theme codes that are broadly accepted in the literature regarding the concept of Smart City are shown in Table 1 below.

A definition and theme related to the smart city is examined, it is emphasized that the focus of human life instead of technology can be better quality by demonstrating an innovative approach to changing human needs with the help of technology. It is emphasized that smart city implementation needs the cooperative approach of the local government, non-governmental organizations and society (TBV, 2017; Caragliu and Del Bo, 2018:1).

In this context, it is thought that the concept of smart city means a system that allows citizens to meet their needs and expectations faster, healthier and livable in metropolises, which are becoming more and more crowded with each passing day (GSMA,2013). However, in order to protect citizens' rights, such as their personal rights, financial gains and trust, smart cities must implement sustainable models and policies with effective cooperation and provide a serious infrastructure (Anand and Navío-Marco, 2018:1).

The European Parliament emphasizes that there are some problems with the increase in cities. Environmental damage and energy consumption and management are among these problems. Therefore, an integrated system of information and communication technologies that will increase people's well-being and enable sustainable economic growth is proposed. In the methods created, as in Figure 1, it was formed in the axes of Smart Economy, Smart Mobility, Smart Governance, Smart Environment, Smart Life and Smart Citizen axes (European Parliament, 2014: 17-18).

Vanolo (2013: 887) explains these axes (2013: 887) as follows:

- Smart Economy: Innovative and entrepreneurial spirit, flexibility in the labor market, integration and transformation capability in international markets
- Smart Mobility: Local and trans local access, the presence of modern, sustainable and secure transport systems of ICTs.
- Smart Governance: Participation in decision-making processes, transparency in governance systems, presence of public services and accessibility, quality of political strategies.
- Smart Environment: The attractiveness of natural environmental conditions, lack of pollution and sustainable management of resources.

Table 1. Definition and theme of Smart City

Reference	Definition	Theme
Lara et al. (2016)	A society that systematically supports general prosperity for all its members and is resilient enough to become an increasingly better place to live, work and play in a proactive and sustainable manner.	Community, prosperity, sustainability, livability
Yiğitcanlar (2016)	It is the ideal form for building the sustainable cities of the 21st century in the event of a balanced and sustainable view of economic, social, environmental and institutional development.	Sustainability, productivity, management, community
Piro, Cianci, Grieco, Boggia and Camarda (2014)	A city that can be considered as a city environment supported by common ICT systems. To improve the overall quality of their lives, they can provide advanced and innovative services to citizens.	Technology, livability, policy
Alkandari, Alnasheet, and Alshaikhli (2012)	A city that uses an smart system characterized by the integration of infrastructure, capital, behavior, and interaction between cultures.	Technology, productivity, community, management
Lazaroiu and Roscia (2012)	A city in the future, it is a city model where technology serves people and a city that improves the quality of economic and social life.	Technology, welfare, livability, welfare
Schaffers et al. (2012)	A safe, environmentally friendly green and efficient urban center of the future with advanced infrastructures such as sensors, electronic devices and networks to promote sustainable economic growth and a high quality of life.	Technology, productivity, livability, sustainability
Caragliu et al. (2011)	Investments in human and social capital and traditional transportation and modern ICT infrastructure investments provide sustainable economic growth and high quality of life through participatory governance through smart management of natural resources.	Community, technology, livability, sustainability, management, policy, accessibility
González et al. (2011)	A series of next-generation services based on information and communication technologies and a public administration or authority that provides or targets the infrastructure.	Management, policy, technology
Hernandez-Munoz et al. (2011)	A city that represents an exceptionally rich ecosystem to encourage massive distribution of applications and services across a wide range of business sectors.	Technology, management
Nam and Pardo (2011)	A humane city with many opportunities to evaluate human potential and to lead a creative life.	Community, welfare, productivity
Zhao (2011)	A city that improves quality of life, including ecological, cultural, political, institutional, social and economic components, without leaving a burden to future generations.	Livability, management, sustainability, community, productivity
Belissent (2010)	A city that uses ICT to make critical infrastructure components and services - city management, education, health, public safety, real estate, transportation and public services- more consciously, interactively and efficiently.	Technology, accessibility, livability, governance
Eger (2009)	A special idea that uses ICT to re-invent and strengthen the role of local people, city governments, businesses and residents and the community in the new service economy, creating business locally and improving the quality of community life.	Community, management, technology, livability, productivity
	It is a special idea that local people, city governments, businesses and residents use ICT to re-invent and strengthen the role of the community in the new service economy, locally creating jobs and improving the quality of community life.	Community, management, technology, livability, productivity

continued on following page

Table 1 Continued

Reference	Definition	Theme
Rios (2008)	It is a city that inspires, shares culture, knowledge and life, a city that motivates its residents to create and evolve in their own lives, - an awe-inspiring city. A vessel for intelligence, but finally an incubator of the reinforced areas.	Community, livability, productivity
Giffinger et al. (2007)	A city built on the smart combination of the intelligence and activities of independent and conscious citizens, who determine themselves in the future, in the economy, in people, in governance, in mobility, in the environment and in life.	Community, management, accessibility, technology, efficiency, policy
Keklik (2004)	A city that actively embraces new technologies that aim to keep people's promises, access to services, and stay in touch with what's going on around us, in a simple and inexpensive way, and become a more open society.	Technology, community, accessibility, livability
Odendaal (2003)	A city that benefits from opportunities to promote the welfare and impact of information and communication technologies.	Technology, efficiency
Bowerman et al. (2000)	A city that monitors and combines all of its critical infrastructures, such as roads, bridges, tunnels, rails, airports, seaports, communications, water, energy, even large buildings, can better optimize and perform preventive maintenance. While monitoring activities and safety aspects it can maximize its services to its citizens.	Policy, management, accessibility, livability
Hall et al. (2000)	An urban center of the future has become a safe, environmentally friendly green and efficient structure because of its power, water, transportation, etc. All structures, sophisticated, integrated materials, sensors, electronics, and databases are interconnected with computerized systems consisting of monitoring and decision-making algorithms.	Sustainability, technology, management

Source: Adapted from "Lara, A. P., Da Costa, E. M., Furlani, T. Z., & Yigitcanlar, T. (2016). Smartness that matters: towards a comprehensive and human-centred characterisation of smart cities. Journal of Open Innovation: Technology, Market, and Complexity, 2(1), 8" and "Mora, L., Bolici, R., & Deakin, M. (2017). The first two decades of smart-city research: A bibliometric analysis. Journal of Urban Technology, 24(1), 3-27".

- Smart Life: Desired and measured standard of living in terms of the existence of cultural and educational services, the existence and accessibility of digital services, tourist destinations, social cohesion, healthy environment, people and housing safety
- Smart Citizen: Competence and quality level of human and social capital, flexibility, creativity, tolerance, cosmopolitanism and public participation.

For a city to be considered smart, the proper protection and management of the resources of the city must be ensured. In addition, the established systems bring solutions to these problems (Uçar et al., 2017:1793). The factors that constitute the infrastructure of smart cities are listed below (http://www.akillisehirler.org, 2018):

- Adequate water supply,
- Adequate electricity supply,
- Sanitation, solid waste management,
- Active city mobility and public transport,
- Affordable food and beverage facilities,

Figure 1. Smart City components
Source: Innovation and Change Guide for Smart Cities, Retrieved from http://www.akillisehirler.org/

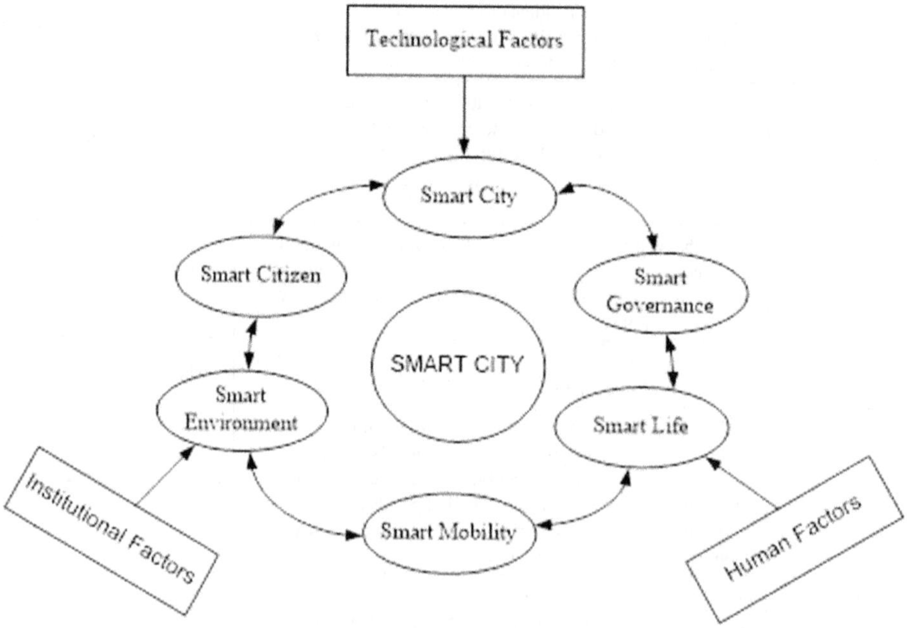

- Robust IT infrastructure, digitalization
- Good governance, e-Government, e-Municipality, participatory management,
- Sustainable resource and environmental management
- Security of citizens
- Health and education services.

A city can transform into a smart city only after a certain process. This process can only be achieved through innovative ideas of high-tech systems. Smart urbanization in Turkey has started to be applied in the 2000s on the basis of the problems experienced in large cities. In this context, it is seen that policies on the rationalization of cities have been created in the 10th Development Plan, in National Science and Technology Policies, which are the documents of Annual Programs and relevant strategy papers and 2003-2023 Strategy Document-Vision 2023. With the title "First Smart City in Turkey", Karaman is seen to be involved in environmental cleaning, traffic, education etc. by using smart technologies (Uçar et al., 2017: 1792-1793).

In this context, smart city applications in Turkey are given in Table 2 (Akgül, 2013).

As shown in Table 2 above, smart city applications in Turkey, the Eco-Tech residential area was established in Yalova as an IT Valley Project (Uçar et al., 2017:1793). It is also seen in Fatih, Kadikoy and Beyoğlu districts in Istanbul. In this context (Yılmaz, 2016:8-9) model implementations listed below were carried out.

- Smart Meters,
- EU Smart Space, Smart Buildings, Fatih Sports Complex,

- ISKI, drinking water distribution management,
- IBB, traffic control center
- Ispark, Smart Park pricing and automatic parks,
- Istanbul card: a service integrated with the same card in different transportation vehicles,
- Isbak, Smart Container, MOBESE and digital library

In Ankara, the following model implementations were carried out (Yılmaz, 2016:8):

- ASKİ, Administrative audit and data acquisition system,
- EGO, EGO Mobile, Smart Stop, Car Passenger Information and Camera Systems,
- Smart Buildings, Industrial Park,
- Automated Fire Command Center.

In the following processes, it is observed that these have increased rapidly in cities such as Antalya, Kars and Konya. In the report published by a research company, Frost & Sullivan, it is thought that 26 cities will be included in the scope of smart city by 2025. In addition, it is seen in the report that the cities of Istanbul, Bursa, Ankara, Eskişehir, Izmir, Denizli, Antalya and Adana are evaluated within the scope of smart city (Uçar et al., 2017:1794).

In this direction, City Digital Maps, Smart Signalization, Geographic Information System, Smart Meter Systems, Transportation Information System, High Speed Internet Infrastructure, Information Kiosks, Citizen Address and Information System, Wireless Internet Infrastructure 3G – 4G GSM Technology, Waste Management System, Smart Building Architecture, Biological, Mechanical and Rail Systems are carried out throughout Turkey (Alkan, 2015: 76).

As a result, smart city urbanization can be said to improve the quality of life of citizens in the following areas (Yılmaz, 2015: 12):

- Increasing the efficiency of urban systems especially in transportation and energy fields,
- Improvement in living areas (decrease in air and noise pollution, etc.),
- Improvement in services provided to citizens,
- Increasing the competitiveness of cities.

Smart Tourism Destinations

Recently, the internet and technology have allowed tourists to change behavior patterns. This situation is not limited to hotel flight bookings of tourists; they access all the visual and audial data about the destination's characteristics and use them in the destination decision process. This allowed destinations to develop in terms of technology. Therefore, this has laid the groundwork for the emergence of the popular concept "smart tourism" (Kim et al., 2017:362).

Smart tourism is based on the adoption of technologies that collect large amounts of data to create new values, such as smart devices, sensors, social media and mobile technologies. This type of information-based perspective includes information and requests of tourism and tourists Del Vecchio e al.,2018:848; Li et al.,2017:300). Therefore, to add smartness to tourism means that information communication technologies and destinations can be exchanged quickly in the decision-making process by strengthening the communication between stakeholders (Buhalis and Amaranggana,2015:378). The World Tourism

Table 3. Evolution of tourism destination concepts

Traditional Tourism Destination Concept	Systemic Approach to Tourism Destination	Smart Tourism Destination
This includes the gathering of places and services of interest; the lack of cooperation in destination and the role of tourists in destination.	Interaction between tourists, service providers and local people in destination, continuous contact with the macro environment and non-linear links between stakeholders are mentioned within this approach.	Combining digital and real fields, public-private consumer cooperation, government involvement, creative and sophisticated people, value creation and personalized services are among the characteristics of smart destination.

Source: Jovicic, D. Z. (2017). From the traditional understanding of tourism destination to the smart tourism destination. Current Issues in Tourism, p.5. Retrieved from https://doi.org/10.1080/13683500.2017.1313203.

Organization (UNWTO, 2017), on the other hand, refers to the social importance of smart tourism destination as follows: *"Smart destinations" are key to sustainable development and contribute not only to the developments in the tourism sector but also to the developments in societies [...]"*.

Smart destinations are seen to be part of the tourism industry and the community living in the region. Therefore, it is seen that smart destinations have become an element of smart tourism (Yalçınkaya et al.,2018:39). Reviewing these smart destinations according to their development process, Jovicic (2017:5) has evaluated them in three stages.

Tourists will not hesitate to keep the data the destination service providers need as long as they trust the providers. Thus, destination service companies will design alternative smart applications that will best meet the needs and expectations of tourists with these data obtained. In fact, they can make applications more fun by gamification and increase the benefits the tourists obtain (Femenia and Neuhofer,2018:16). Buhalis and Amaranggana, (2013:560) present which characteristics of smart tourism are for stakeholders by adapting the characteristics of smart tourism destinations from Hedlund (2012) in the following Table 4.

The characteristic features of the stakeholders in smart tourism destinations are outlined in the table above. By Buhalis and Amaranggana (2015), a model of how a smart tourism activity can contribute to the tourism experience was developed. In this context, when the Figure is examined, it is observed that the Internet and web-based applications increase tourist experience positively. In fact, when the steps are examined, it is seen that it starts in a smart destination and passes towards a smart tourism destination. This situation not only increases the competitiveness and quality of service in tourism destination enterprises but also is thought to benefit the individual satisfaction of tourists by increasing the quality of the tourist experience. Figure 2 shows how smart tourism destination can enhance tourist experience.

With this model, Buhalis and Amarangana (2015:387) listed the characteristics of smart tourism destinations as follows. It is emphasized that it may be possible by, the first, to support the decision-making process by giving all the real information to the tourist before the trip, the second, to enable advanced access networks and real-time feedback networks to discover personalized services during travel, and finally, to use feedback networks to review the experiences after the trip and to gain prospective experience.

In line with this, the "No Homes Without Internet" project was carried out in Turkey, facilitating the public services for the locals, in order to easily access information about Kars province. In addition, a mobile system has been established with the "Kars mobile" project, which aims to introduce the culture and history of Kars (Uçar et al., 2017:1794).

Topsakal (2018), in his study, recommends the installation of mobile systems to facilitate the holidays of disabled individuals. In this context, he has proposed the Turkey 4all project, which is thought

Table 4. Characteristics of smart tourism destinations

No.	Stakeholders	Output Features
1	Tourism organizations	▪ Serves as a smart central function that coordinates all relevant information and makes it easy for users to access real-time information ▪ Digitizing basic business processes ▪ Optimizing energy use ▪ Collaborating with local communities, tourists and government to create the tourism experience together ▪ Provides organizational agility, rapid decision-making and the need to respond to customer needs based on timely analysis ▪ Provides precise targeting and personalized service
2	Governments	▪ Information management that supports data disclosure ▪ Organizes data privacy ▪ Establishment of Public-Private Partnership
3	Local Residents/Local Communities	▪ Establishes constant communication ▪ Provides savings creation and empowerment ▪ Brings an understanding of technology ▪ Provides citizen journalism ▪ Participates actively in the development of smart heritage / E-culture
4	Tourists	▪ Makes them very well connected and informed ▪ Creates active critics and whisper marketers ▪ Very personalized service demand arises ▪ Gathers both socially and technologically ▪ Provides the opportunity to discuss dynamically in social media ▪ Allows to create experience together ▪ Contributes to the content ▪ Uses end-user devices at common use points
5	Environment	▪ Allows entities to connect to each other via the internet. ▪ Presence of cluster information services ▪ Formation of the innovative ecosystem ▪ Sensor networks in the environment ▪ Combines digital information and social contexts to enhance Geophysical reality ▪ Creates interoperable social platforms

Source: Buhalis, D., & Amaranggana, A. (2013). Smart Tourism Destinations. Information and Communication Technologies in Tourism, p. 560. Retrieved from https://doi.org/10.1007/978-3-319-03973-2_40.

to enable individuals with disabilities to enjoy their holidays in Turkey. Therefore, it is thought that it represents an integrated system that is nested in the properties of the target group in places such as airports, museums, food and beverage enterprises, etc. where tourism is carried out.

In the project titled "myAntalya", which started to serve tourists in April 2013 in Antalya, enterprises are reached in 20 categories in English, German and Russian languages. On request, this project offers access to social media accounts of these enterprises. The services and topics given in the scope of this project are listed in Table 5.

Çelik and Topsakal (2017), in their research, examined the city of Antalya within the scope of smart tourism destination. In this research, it is considered that electronic traffic control system, smart health service, free Wi-Fi, free SMS system, unmanned service points, smart public transportation system, smart environment and smart energy systems should be installed for the smart city. In addition, since Antalya is a tourism city, smart applications have been started to provide better service to tourists. For this purpose, kiosks and mobile applications that will provide interactive information in a number of languages have been initiated. These projects are considered to take place in the initial stage.

Figure 2. Contribution of smart tourism destination to tourism experience
Source: Çelik, P., & Topsakal, (2017). Akıllı turizm destinasyonları: Antalya destinasyonunun akıllı turizm uygulamalarının incelenmesi. Seyahat ve Otel İşletmeciliği Dergisi, 14(3), p. 156.

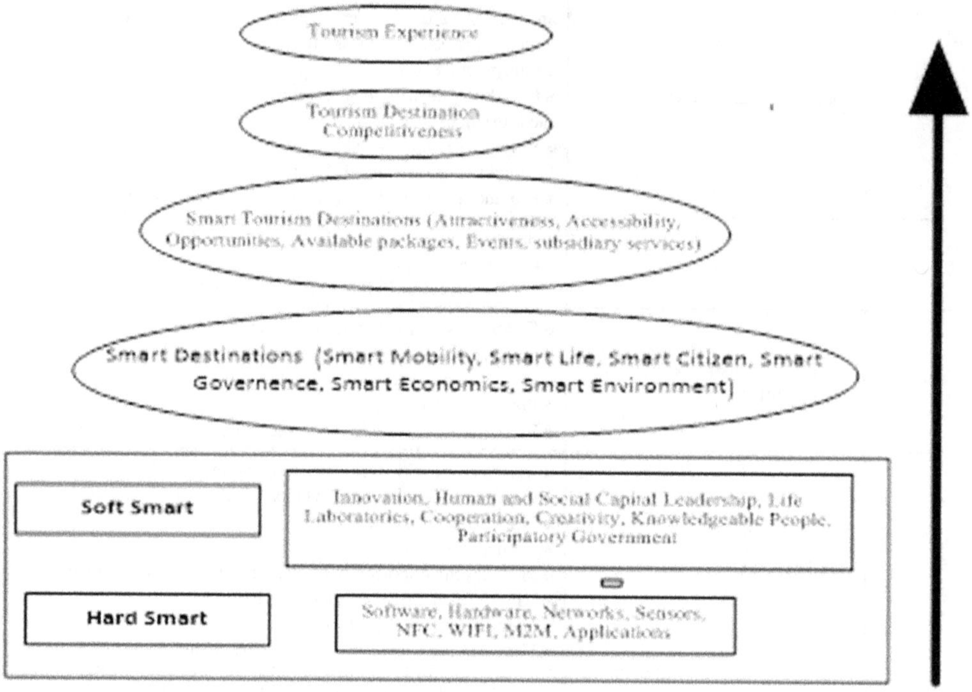

SOLUTIONS AND RECOMMENDATIONS

Smart tourism destination systems will adopt a protective approach to nature and the environment, which is the capital of tourism, in case of its preparation by taking into consideration the carrying capacity. In this context, through the effective use of these resources, it will enable significant positive developments in the regional and national tourism economy. It will provide the locals with a better quality of life and allow them to enrich their tourist experiences. It will also provide a different competitive understanding for destination businesses with its innovative aspects.

Göktaş Kulualp (2016) mentioned talent maturity models in the establishment of an information management system that will enable the development of organizations. These capability models are road maps that show where institutions are in the area of information management. Therefore, local administrations can provide their development by applying these knowledge management maturity models.

If the mobile applications implemented for the smart destination model are gamified, they can be transformed into a more enjoyable state and they will increase the frequency of use. This is thought to contribute to the increase in advertising revenues from the application as well as the promotion of destinations. Therefore, it is assumed that in the short term, advertising can come out of the cost item and turn into income item, and in the long term, it can provide competitive advantages in the marketing of destinations.

Table 5. Scope of "MyAntalya" project

Offered Service	Sub Headings
Accommodation	Apart Hotel, Boutique Hotel, Camping Area, Ski Resort, 5-Star Hotel, 3-Star Hotel, 2-Star Hotel, Special Category Hotel, Pension, Holiday Village
Touristic	Bar-Pub, World Cuisine Restaurants, Turkish Restaurants, Home Cooking Restaurants, Fast-Food, Bagel Sellers, Pitta Restaurants, Cafes, Pastry, Starbucks
Culture-Art	Museums, Art Galleries, Culture And Art Centers
Transportation	Motorway Rest Areas, Suburban, Sea Transportation, Airport, Marina, Car Rental, Bus Terminal, Taxi, Cable Car, Tramway
Fun	Bowling, Amusement Parks, Beach, Cinema, Theatre Play Cafes
Social	Park-Grove Areas, Social Facilities
Sports	Horse Riding, Ice Skating, Golf Courses, Sports Centers, Sports Courses, Stadiums
Health	Hospitals, Pharmacies
Shopping	Shopping Malls, Supermarkets, Bookshops, Sporting Goods, Stationery, Liqueur Store
Security	Gendarmerie, Police, Traffic, Fire Department, Constabulary
Education	Elementary School, Higher Education, Student Dormitories, Libraries
Business-Trade	Fair-Congress Centers, Business Centers, Industrial Centers, Winery
Public-State	Courthouse, Municipality, Marriage Department, governorship, consulates, Mukhtar offices, notary, Post Office, Land Registry Office, Türk Telekom, governorship, electricity-natural gas, National Education Directorate, SGK Directorate, Tax Office
Shipment-Logistics	Carrier, UPS, TNT, PTT, MNG Kargo, Yurtiçi Kargo, Fedex, DHL, Aras Cargo
Settlement	Neighborhoods, Districts, Sites, Co-Operatives, Lodgments
Religious	Sanctuaries
Car park	Indoor Car Parks, Outdoor Car Parks
Gas Station	BP, Opet, Petrol Ofisi, Shell, Total, Türk Petrol, Akpet, Alpet, Bpet, Starpet, Sunpet

Source: Çelik, P., & Topsakal, (2017). Akıllı turizm destinasyonları: Antalya destinasyonunun akıllı turizm uygulamalarının incelenmesi. Seyahat ve Otel İşletmeciliği Dergisi, 14(3), p. 160.

In recent years, it has been seen that web applications that offer gifts to consumers have become the focus of attention. In this context, web-based projects that will be put into practice regarding smart tourism destinations are thought to increase the benefit of the project to be carried out if local gifts are presented in order to attract the attention of the participants.

Smart tourism destinations will take on the pro-active role in regard to the destruction of resources by means of unplanned perseveration of tourism in the global sense. Therefore, it will adopt the concept of sustainable tourism that does not allow the resources that are concerned today and in the future to be destroyed by misuse.

FUTURE RESEARCH DIRECTIONS

With the increase in computer usage, it is seen that the process of accessing information from the information stage has been transferred to the computer in the desired time. With the strategic use of the information obtained, it is foreseen that local administrations gain an advantage over the competition.

In this respect, it is possible to measure how much the use of information obtained by the local governments from the visitors in the future will affect the recognition of the city. It is thought that this finding will contribute to the determination of benefits obtained from local governments.

It is thought that quantitative measurement of the effect of the intelligent applications examined in the literature will contribute to the validity of future web applications. In this context, it is recommended to measure the return of smart applications in figures in future studies.

It is thought that providing a competitive advantage in the tourism sector, which is seen as a human sector, is more difficult than the industrial sector. The competition system is a decisive criterion for maintaining its existence in local governments. For this reason, the importance of information management practices in tourism sector in the future research can be emphasized and the importance of information management applications can be emphasized.

The concept of smart tourism destination, which has just begun to be examined in Turkey, has been discussed in qualitative terms in this study. In future studies, the examination of the importance of smart tourism destinations among the reasons for the choice of destination for the tourists will be useful for both marketing of the touristic product and for the development of smart tourism. Because, if smart destinations take their place among the preference reasons, it is expected that the projects and applications on this subject show an increase.

CONCLUSION

First, the World Environment and Development Commission based on sustainable development in 1987, based on the Brundtland Report "today's needs, meeting the needs of future generations to meet the development without compromising the idea" to be implemented on a global scale to implement applications (T. C. Ministry of Foreign Affairs, 2019).

In the next process, the rapid increase in today's cities brings about the correct use of resources and management issues. Therefore, digital projects prepared to prevent the problems that can be experienced have been started to be implemented. As a result of this situation, the concept of smart cities was put forward. In this context, it has made the tourism cities where the population is concentrated to be the focus of attention.

Smart tourism destinations will be possible through the use of all the resources within the tourism ecosystems together and by following a policy that creates added value to all stakeholders. In order to provide more benefits from coordinated and integrated computing systems, destination managers are obliged to consider all aspects of the smartness.

In the development of regional tourism, the share of local governments, which benefit from the development models, is of great importance. In this context, local governments' participation and support will be key in achieving the success of smart tourism applications.

REFERENCES

Akgül, M. K. (2013). Kentlerin E-dönüşümü: akıllı kentler. *Kalkınmada Anahtar Verimlilik Dergisi, 25*(291). Retrieved from http://anahtar.sanayi.gov.tr, Date of Access: 12.10.2018.

Albino, V., Berardi, U., & Dangelico, R. M. (2015). Smart cities: Definitions, dimensions, performance, and initiatives. *Journal of Urban Technology*, *22*(1), 3–21. doi:10.1080/10630732.2014.942092

Alkan, T. (2017). Akıllı kentler ya da 21. Yüzyıl şehirleri. Retrieved from http://www.bilisimdergisi.org, Date of Access: 29.11.2018.

Anand, P. B., & Navío-Marco, J. (2018). Governance and economics of smart cities: opportunities and challenges, Telecommunications Policy, Available online October 11, 2018 In Press, Corrected Proof, /. doi:10.1016/j.telpol.2018.10.001

Axelsson, K., & Granath, M. (2018). Stakeholders' stake and relation to smartness in smart city development: Insights from a Swedish city planning project. *Government Information Quarterly*, *35*(4), 693–702. doi:10.1016/j.giq.2018.09.001

Ayaz, N., & Göktaş Kulualp, H. (2019). Strategic role of human resource management in tourism enterprises (pp. 293-308) in (R. Efe, I. Koleva, M. Öztürk, & R. Arabacı (Eds.), Recent Advances in Social Sciences, Chapter Eighteen, 1st Edition, Cambridge Scholars Publishing.

Buhalis, D., & Amaranggana, A. (2013). Smart Tourism Destinations. Information and Communication Technologies in Tourism, 553-564. . doi:10.1007/978-3-319-03973-2_40

Buhalis, D., & Amaranggana, A. (2015, February). Smart tourism destinations enhancing tourism experience through personalisation of services. Paper presented at the meeting of Information and Communication Technologies in Tourism 2015: *Proceedings of the International Conference in Lugano, Switzerland*, pp. 377-389, 10.1007/978-3-319-14343-9_28

Caragliu, A., & Del Bo, C. F. (2018). Smart innovative cities: The impact of smart city policies on urban innovation. *Technological Forecasting and Social Change*. doi:10.1016/j.techfore.2018.07.022

Çelik, P., & Topsakal, Y. (2017). Akıllı turizm destinasyonları: Antalya destinasyonunun akıllı turizm uygulamalarının incelenmesi. *Seyahat ve Otel İşletmeciliği Dergisi*, *14*(3), 149–166. doi:10.24010oid.369951

Choi, B., Poon, S. K., & Davis, J. G. (2008). Effects of knowledge management strategy on organizational performance: A complementarity theory-based approach. *Omega*, *36*(2), 235–251. doi:10.1016/j.omega.2006.06.007

Cooper, C. (2006). Knowledge management and tourism. *Annals of Tourism Research*, *33*(1), 47–64. doi:10.1016/j.annals.2005.04.005

Córdova, F. M., & Gutiérrez, F. A. (2018). Knowledge management system in service companies. *Procedia Computer Science*, *139*, 392–400. doi:10.1016/j.procs.2018.10.275

Del Vecchio, P., Mele, G., Ndou, V., & Secundo, G. (2018). Creating value from social big data: Implications for smart tourism destinations. *Information Processing & Management*, *54*(5), 847–860. doi:10.1016/j.ipm.2017.10.006

Denning, S. (2000). *The springboard: how storytelling ignites action in knowledge era organizations*. Oxford, UK: Butterworth-Heinemann.

European Parliament, Mapping Smart Cities in the EU. (2014). Retrieved from http://www.europarl.europa.eu/RegData/etudes/etudes/join/2014/507480/IPOL-ITRE_ET(2014)507480_EN.pdf

Femenia-Serra, F., Neuhofer, B., & Ivars-Baidal, J. A. (2018). Towards a conceptualisation of smart tourists and their role within the smart destination scenario. *Service Industries Journal*, 1–25.

Gökalp, E., & Eren, P. E. (2016). Akıllı Teknolojilerin Turizm ve Otelcilik Sektöründe Uygulanması, Date of Access: 04.11.2018. Retrieved from www.researchgate.net/profile/Ebru_Goekalp/publication/309619452_

Göktaş Kulualp, H. (2016). Bilgi yönetimi olgunluk modellerine göre işletmelerin bilgi yönetimi düzeyleri: genel bir bakış. Akademik Bakış Uluslararası Hakemli Sosyal Bilimler Dergisi, (56), 69-91.

GSMA. (2013). Guide to Smart Cities the Opportunity for Mobile Operators. Retrieved from https://www.gsma.com/iot/wp-content/uploads/2013/02/cl_sc_guide_wp_02_131.pdf, Date of Access: 04/11/2018

Güçlü, N., & Sotirofski, K. (2006). Bilgi yönetimi. *Türk Eğitim Bilimleri Dergisi*, *4*(4), 351–373.

Jia, Z., Shi, Y., Jia, Y., & Li, D. (2012). A framework of knowledge management systems for tourism crisis management. *Procedia Engineering*, *29*, 138–143. doi:10.1016/j.proeng.2011.12.683

Jovicic, D. Z. (2017). From the traditional understanding of tourism destination to the smart tourism destination. *Current Issues in Tourism*, 1–7. doi:10.1080/13683500.2017.1313203

Karakoçak, K. (2007). *Bilgi yönetimi ve verimliliğe etkisi: Türkiye Büyük Millet Meclisi uygulaması*. Ankara, Turkey: Ankara Üniversitesi Sosyal Bilimler Enstitüsü Yayımlanmış Doktora Tezi.

Kim, K., Park, O., Yun, S., & Yunc, H. (2017). What makes tourists feel negatively about tourism destinations? Application of hybrid text mining methodology to smart destination management. *Technological Forecasting and Social Change*, *123*, 362–369. doi:10.1016/j.techfore.2017.01.001

Lang, J. C. (2001). Managerial concerns in knowledge management. *Journal of Knowledge Management*, *5*(1), 43–59. doi:10.1108/13673270110384392

Lara, A. P., Da Costa, E. M., Furlani, T. Z., & Yigitcanlar, T. (2016). Smartness that matters: Towards a comprehensive and human-centred characterisation of smart cities. *Journal of Open Innovation: Technology, Market, and Complexity*, *2*(1), 8. doi:10.118640852-016-0034-z

Lee, M. (2008). *A qualitative case study approach to define and ıdentify perceived challenges of knowledge management for casino hotel ındustry, Graduate College, University of Nevada Las Vegas William F.* Harrah College of Hotel Administration. Doktora Tezi.

Lee, Y.-C., & Lee, S.-K. (2006). Capabilities, processes, and performance of knowledge management: A structural approach. *Human Factors and Ergonomics in Manufacturing*, *17*(1), 21–41. doi:10.1002/hfm.20065

Li, Y., Hu, C., Huang, C., & Duan, L. (2017). The concept of smart tourism in the context of tourism information services. *Tourism Management*, *58*, 293–300. doi:10.1016/j.tourman.2016.03.014

Liao, S. H., & Wu, C. C. (2009). Knowledge management and ınnovation: the mediating effects of organizational learning. In *Proceedings IEEE International Conference on Industrial Engineering and Engineering Management*, pp. 1850-1854. IEEE.

Mora, L., Bolici, R., & Deakin, M. (2017). The first two decades of smart-city research: A bibliometric analysis. *Journal of Urban Technology*, *24*(1), 3–27. doi:10.1080/10630732.2017.1285123

OECD. (2003). Introduction and Draft Results of the Survey of Knowledge Management Practices in Ministries/Departments/Agencies of Central Government. Retrieved from www.oecd.org/officialdocuments/publicdisplaydocumentpdf/?cote=GOV/PUMA/HRM(2003)2&docLanguage=En

Önaçan, M. B. K. (2015). *Organizasyonlar için bilgi yönetimi çerçevesi ve bilgi yönetim sistemi mimarisi önerisi: dOBLYN (Doküman ve bilgi yönetimi)*. Ankara Üniversitesi Sosyal Bilimler Enstitüsü İşletme A.B.D., Doktora Tezi.

Rašula, J., Vukšić, V. B., & Štemberger, M. I. (2012). The impact of knowledge management on organisational performance. *EBR*, *14*(2), 147–168.

Sarvary, M. (1999). Knowledge management and competition in the consulting industry. *California Management Review*, *41*(2), 95–107. doi:10.2307/41165988

T. C. Ministry of Foreign Affairs. (2019) Retrieved from http://www.mfa.gov.tr/dunya-surdurulebilir-kalkinma-zirvesi_johannesburg_-26-agustos---4-eylul-2002_.tr.mfa. Date of access: 23/02/2019.

TDK. (2018). Retrieved from www.tdk.gov.tr/index.php?option=com_gts&arama=gts&guid=TDK.GTS.5bdc1f11cbf6d6.64662412, Date of access 02.11.2018.

Tokcan, H. (2015). *Bilginin üretimi ve kullanımı açısından bilgi yönetimi: üniversitede akademik yöneticilerin bilgi yönetimi algıları üzerine bir uygulama, Doktora Tezi*. T.C. Kırıkkale Üniversitesi Sosyal Bilimler Enstitüsü İşletme Anabilim Dalı, Kırıkkale.

Topsakal, Y. (2018). Akıllı Turizm Kapsamında Engelli Dostu Mobil Hizmetler: Türkiye 4.0 İçin Öneriler. *Journal of Tourism Intelligence and Smartness*, *1*(1), 1–13.

Torabi, F., & El-Den, J. (2017, November). The impact of knowledge management on organizational productivity: a case study on koosar bank of ıran. Paper presented at the meeting of 4th Information Systems International Conference 2017 (ISICO 2017), Bali, Indonesia, *Procedia Computer Science, 124,* 300-310.

Tribe, J., & Liburd, J. J. (2016). The tourism knowledge system. *Annals of Tourism Research*, *57*, 44–61. doi:10.1016/j.annals.2015.11.011

Türkiye Bilişim Vakfı. (2017). 2016 Çalışma Raporu. Retrieved from http://www.tbv.org.tr/core/uploads/page/document/1123_24031710643.pdf, Date of Access 04/11/2018.

Uçar, A., Şemşit, S., & Negiz, N. (2017). Avrupa Birliği Akıllı Kent Uygulamaları Ve Türkiye'deki Yansımaları. Suleyman Demirel University Journal of Faculty of Economics & Administrative Sciences, 22.

UNWTO. (2017). Innovation, technology and sustainability - pillars of smart destinations. Erişim Tarihi: 05.11.2018. Retrieved from http://media.unwto.org/press-release/2017-02-21/innovation-technology-and-sustainability-pillars-smart-destinations

Vanolo, A. (2014). Smartmentality: The Smart City as Disciplinary Strategy. *Urban Studies (Edinburgh, Scotland)*, *51*(5), 883–898. doi:10.1177/0042098013494427

Xu, J., Hou, Q., Niu, C., Wang, Y., & Xie, Y. (2018). Process optimization of the University-Industry-Research collaborative innovation from the perspective of knowledge management. *Cognitive Systems Research*, *52*, 995–1003. doi:10.1016/j.cogsys.2018.09.020

Yalçınkaya, P., Lütfi, A. T. A. Y., & Karakaş, E. Akıllı Turizm Uygulamaları. (2018). Gastroia. *Journal Of Gastronomy And Travel Research*, *2*(2), 34–52.

Yang, J., & Wan, C.-S. (2004). Advancing organizational effectiveness and knowledge management implementation. *Tourism Management*, *25*(5), 593–601. doi:10.1016/j.tourman.2003.08.002

Yigitcanlar, T., Kamruzzaman, M., Buys, L., Ioppolo, G., Sabatini-Marques, J., da Costa, E. M., & Yun, J. J. (2018). Understanding 'smart cities': Intertwining development drivers with desired outcomes in a multidimensional framework. *Cities (London, England)*, *81*, 145–160. doi:10.1016/j.cities.2018.04.003

Yılmaz, E. (2016). Yeni bir hikâye konya akıllı şehir olabilir mi? Konya Ticaret Odası Araştırma Raporu. Retrieved from http://www.kto.org.tr/d/file/yeni-bir-hikaye-konya-akilli-sehir-olabilir-mi---emre-yilmaz.pdf

Yılmaz, İ. A. (2009). *Ağırlama işletmeleri yöneticilerinin bilgiye ve bilgi yönetimine yönelik yaklaşımları: Antalya'daki beş yıldızlı otel işletmeleri üzerinde bir uygulama*. Mersin: Mersin Üniversitesi Sosyal Bilimler Enstitüsü Turizm İşletmeciliği ABD, Yayınlanmamış Yüksek Lisans Tezi.

Yılmaz, Ö. (2015). Akıllı kentler ve bilgi toplumu stratejisi, T. C. Kalkınma Bakanlığı, Information Society Strategy and Action Plan (2015-2018) and Smart Cities. Retrieved from https://www.csb.gov.tr/db/cbsgunu/webmenu/webmenu15780.pdf, Erişim Tarihi: 17.11.2018.

Yusof, M. N., Bakar, A. H., & Tufail, M. A. (2012) Knowledge Management and Growth Performance: Construction industry perspective, DOI: , Conference: Sustaining the World with Better Structures & Construction Practice (APSEC-ICCER 2012), At Surabaya, Indonesia. Retrieved from www.researchgate.net/profile/Mohamad_Nizam_Yusof/publication/268331671_Knowledge_Management_and_Growth_Performance_Construction_industry_perspective/links/546894700cf20dedafd0d168.pdf doi:10.13140/2.1.3118.0482

Zack, M., McKeen, J., & Singh, S. (2009). Knowledge management and organizational performance: An exploratory analysis. *Journal of Knowledge Management*, *13*(6), 392–409. doi:10.1108/13673270910997088

Zaied, A. N. H., Hussein, G. S., & Hassan, M. M. (2012). The role of knowledge management in enhancing organizational performance, I. J. *Information Engineering and Electronic Business*, *4*(5), 27–35. doi:10.5815/ijieeb.2012.05.04

ADDITIONAL READING

Ardito, L., Ferraris, A., Petruzzelli, A. M., Bresciani, S., & Del Giudice, M. (2019). The role of universities in the knowledge management of smart city projects. *Technological Forecasting and Social Change*, *142*, 312–321. doi:10.1016/j.techfore.2018.07.030

Bakıcı, T., Almirall, E., & Wareham, J. (2013). A smart city initiative: The case of Barcelona. *Journal of the Knowledge Economy*, *4*(2), 135–148. doi:10.100713132-012-0084-9

Chourabi, H., Gil-Garcia, J. R., Pardo, T., Nam, T., Mellouli, S., Scholl, H. J., ... Nahon, K. (2012). Understanding Smart Cities: An Integrative Framework. *45th Hawaii International Conference on System Sciences*, Hawaii.

Gretzel, U., Sigala, M., Xiang, Z., & Koo, C. (2015). Smart tourism: Foundations and developments. *Electronic Markets*, *25*(3), 179–188. doi:10.100712525-015-0196-8

Komninos, N., Pallot, M., & Schaffers, H. (2013). Special issue on smart cities and the future internet in Europe. *Journal of the Knowledge Economy*, *4*(2), 119–134. doi:10.100713132-012-0083-x

Mangır, F. (2016). Smart city: strategies for local governments: the case of Konya in Turkey. *Selçuk Üniversitesi Sosyal Bilimler Meslek Yüksekokulu Dergisi*, *41*, 17–36.

Negre, E., Rosenthal-Sabroux, C., & Gascó, M. (2015). A knowledge-based conceptual vision of the smart city. In 2015 *48th Hawaii International Conference on System Sciences*, pp. 2317-2325. 10.1109/HICSS.2015.279

Stewart, R. A., Willis, R., Giurco, D., Panuwatwanich, K., & Capati, G. (2010). Web-based knowledge management system: Linking smart metering to the future of urban water planning. *Australian Plants*, *47*(2), 66–74. doi:10.1080/07293681003767769

KEY TERMS AND DEFINITIONS

Knowledge Management: Knowledge management is to be aware of the information in order to make maximum use of the process, to identify, implement, disseminate, develop and store the aims of information.

Smart City: The smart city is the name given to an environment where technology is integrated into the city.

Smart Destination: Smart destination is the integrated concept in which the principles are applied to the new city profile and not only in the citizens but also in the exchange process.

Smart Environment: Smart environment is a concept that emerged in the early 1990s where city residents are continuously interacting with objects and sensors seamlessly to better their lives.

Smart Public Transportation System: Connected solutions for shared passenger transport services such as buses, trains and ferries. Includes applications for connected vehicles and related infrastructure, such as passenger information, ticketing & payment systems, cloud & analytics services as well as traffic management & control.

Smart Tourism: Smart tourism is physical tourism, social connections, state/organization resources and tourism supported by efforts in a destination for gathering data from the human mind.

Talent Maturity Models: A model lies in improving the processes of an institution. Describes which elements are missing on the basis of process areas or which elements should be present.

Chapter 18
Smart Municipalities in Tourism

Nuretting Ayaz
https://orcid.org/0000-0003-2117-2015
Karabük University, Turkey

Bayram Akay
Burdur Mehmet Akif Ersoy University, Turkey

ABSTRACT

Tourism is a fragmented and information-oriented sector covering tour operators, travel agencies, hotel sales representatives, associations, meeting offices, transportation, car rental, airlines, cruise, souvenirs, restaurants, hotels/motels, and entertainment. In the tourism sector, technology is an important requirement for the interconnection of all these fragmented organizations, activities, and elements, and for the satisfaction of tourists. However, the effective use of technology in a destination requires the leadership of a leader. This chapter approaches the smart municipality applications oriented to tourism sector with the example of Turkey, a well-known destination around the world, from the perspective of municipalities, which is regarded as a democratic leader at local level and provides a general framework regarding the studies on an innovative, environment-friendly, solution-oriented, and more livable urban life. Suggestions are provided for smart technologies to improve the quality of life of local citizens and to the satisfaction of tourists.

INTRODUCTION

Natural, cultural, historical and artificial attractions, societies that protect their authenticity, a tourism consciousness that is a result of the harmony between the local cultural characteristics and the tourism, interest of the national, regional and local authorities in tourism and achievement of the international standards for transportation, accommodation and food and beverage sectors are among the resources of a tourism destination (EC, 1999: 18). The expectation from local authorities regarding the tourism sector include funds for the development of tourism, investments in infrastructure, reconstruction, care and cleaning services for touristic attractions, education of tourism employees and local people and professional support (Wilson et al., 2001: 134). Local governments are highly important for providing rapid

DOI: 10.4018/978-1-7998-1989-9.ch018

services in an integrated way. Besides, within the context of the notion of "think global, act locally", local governments are expected to be responsible for sustainability, support for development, raising awareness and developing local partnership (Leslie and Hughes, 1997: 145).

Among the regional and local authorities, municipalities are the main contact points for the local resident. Municipalities are highly important authorities since their responsibilities and functions of the municipalities significantly influence the daily lives of the inhabitants (Nadaroğlu, 1978: 272-313).

The concept of "smart city", which have been constructed because of the developments in communication and information technologies, has been considered as an element of the municipalities since the early 2000s. Parallel to this, the concept of "smart tourism destination" has been formed. Smart tourism destination depends on the information and communication technologies of a smart city, including, internet, mobile communication, cloud services, artificial intelligence technologies (Yüzbaşıoğlu et al., 2018: 708). Technologies and databases in a smart destination are used for communication and flow of information between the stakeholders (Başer and Doğan, 2018).

Tourism is a fragmented and information-related sector that is composed of various enterprises, including, tour operators, travel agents, hotel's sales representatives, unions, transportation firms, car rentals, airlines, cruise, souvenir shops, hotels, restaurants and entertainment firms (Buhalis and Laws 2001). Technology is an important factor that connects these fragmented enterprises in order to provide high-quality experiences for visitors. In addition, technology is an important necessity for tourism destinations to sustain both tourist and destination safety (Koçoğlu and Acar, 2019). So, a leader is required for the effective use of information and communication technologies and a strategic management approach in a tourism destination (Ayaz and Kulualp, 2019).

This study deals with the smart municipality services in Turkey from the perspectives of the municipalities. We review the existing studies in order to provide a general framework to maintain an innovative, environment-friendly, solution-oriented and habitable urban life. In order to improve the quality of life of the local inhabitants and the visitors, we also provide suggestions for the integration of the technologies, systems, services, and capabilities to an organic network, which is open to new developments.

BACKGROUND

Smart Tourism

Smart tourism consists of integrated efforts to collect data from physical infrastructure, social connections, public and private businesses and individuals, and to use innovative technologies to find innovative methods. Smart tourism is based on the ability to intelligently store, process, merge, analyze and use collected data, instead of collecting large amounts of data (Gretzel et al., 2015). In smart tourism, which can be characterized by innovation and facilitation, it is essential to improve the quality of life of local residents and tourists in destinations using high levels of technology and interfaces (Khan, et al., 2017).

In smart cities where natural resources are used more efficiently and wisely, where cost and energy saving practices are present, environmental pollution is low, carbon emission is low, quality of life and participation of people in city decisions are high (Male, 2017: 60). Tourism has emerged as a subset of the smart city concept, which aims to provide tourists with solutions that meet specific travel-related needs (Khan, et al., 2017).

Table 1. Smart solutions by disruptive technologies and social innovations

Cloud - on-demand access to rich application functionality and computing power allow for exponential scalable solutions in a "pay-per-use" mode.	**Blockchain** - algorithm that facilitates registering transactions in an indisputable way without the use of an intermediary or a central administration.
Mobile - ubiquitous mobile internet, powerful mobile devices and apps with intuitive interfaces act as front end for many innovative solutions.	**Co-creation** - an initiative that brings together parties to jointly create a mutual valued outcome. In the context of smart cities, these are citizens, private companies, knowledge institutes and public organizations that form an ecosystem to solve societal problems in a creative way.
Social media & Digital platforms – use digital technology to connect people in new and powerful ways forming the basis of new collectives and communities.	**Crowd sourcing** - the process of obtaining needed services, ideas, or content by soliciting contributions from a large group of people, and especially from an online community, rather than from traditional employees or suppliers.
Big data - large volumes of highly volatile and rich data (structured data, sensor data, audio, video, social media) form the lifeblood of smart solutions.	**Sharing economy** - using information technology as an enabler to bring together supply and demand in new ways. Use digital platforms to share and reuse excess capacity in a variety of goods and services.
Artificial intelligence - cognitive systems combine machine learning with the ability to interact via natural language and create insight from data.	**Gamification** - the use of game thinking and game design elements (competition, levels, leaderboards, badges, etc.) in non-game contexts. The objective of gamification is to stimulate people to change their behavior. In a desired way.
Internet of things - physical objects equipped with advanced sensors and connectivity transform into smart objects and generate a plethora of data.	**Self organization** - a process where overall order and coordination arises out of the local interactions between people in an initially disordered system. It is spontaneous and not necessarily controlled by any auxiliary agent. The resulting organization is wholly distributed and typically robust.
Robotics and drones – social robots and drones replace human labor on a large scale, not only for routine work but increasingly for providing services.	**Renewable energy** – New sources of energy like photo voltaic cells and energy from wind and water are an alternative to large scale generation based on fossil fuel.
3D printing - additive manufacturing allows for efficient production of unique products where and when they are needed.	

Source: www.deloitte.com.

The first priority issue in establishing good relations with tourists in promoting and presenting the products in a destination to tourists and having a good reputation (Koçoğlu and Kıycı, 2019) has recently become prominent in smart technologies used in destinations. Technologies that facilitate the needs of tourists throughout their travels have become an important reason for choice.

Smart City

The concept of "Smart City" covers all efforts towards innovative, environment-friendly, solution-oriented, more livable urban life. The smart city application, which first started in Barcelona, includes operations center, smart library, smart intersection, traffic forecasting, showing alternative routes to drivers, free high-speed internet in city squares and parks, security where services are conducted from a single point (Yılmaz, 2016: 4). Smart Cities aim to improve the quality of life for both citizens and tourists. To achieve this goal, it requires integration of technologies, systems, services and capabilities into a flexible and open access organic network for future developments (Vecchio et al., 2018: 848). Smart cities have four basic features. These are (Steel & Roe, 2017: 97b);

- Usage of existing technology for citizens and the city,
- Usage of digital data to improve the quality of work and life of citizens,

- Integrating information technologies into the city,
- Applying good practice throughout the region to promote innovation, learning and knowledge transfer.

The smart city is a global term used to describe modern societies influenced by sustainable economic, social and environmental values. This term is built around building sustainable communities, using resources efficiently, meeting the needs of city residents, and economic growth. These cities are expected to include not only information and communication technologies, but also modern transport technologies, logistics services, urban traffic and intelligent systems that increase residents' mobility, security, green, efficient and sustainable energy resources (www.smart-cities.eu).

Smart cities are cities that use ICT for cost and energy savings, better service delivery and quality of life in the context of efficient and efficient use of resources. The main themes for these cities are; smart living, smart building & home, smart transportation, smart energy (renewable generation & storage, ami, pqm, plm, oms), smart water management, smart waste management (recycling of waste, residual management, recovery of waste organics & energy), smart education (e-education), open data, urban mobility, smart governance, smart medical facility (e-medical), smart communications, smart networks, environmental awareness (i.e. changing weather conditions; human defined changes), smart crisis management and metrics and performance indicators (https://unctad.org). The prominent application tools for these themes are listed as follows (www.segittur.es):

Sensors: These are sensors that detect air pollution, warn relevant institutions, detect temperature, humidity and CO_2 levels in buildings, vibrations in bridges and historical artifacts, perform image scans to prevent armed attackers, detect the level of dangerous gases, monitor energy usage, measure noise levels, monitor traffic intensity and that are located at critical points of the city, providing support for the improvement of the quality of life and taking precautions by warning the relevant institutions in the elimination of city problems.

Intelligent applications in transportation are traffic monitoring systems, electronic payment systems, and smart stops.

Intelligent applications in the field of energy are smart street lighting systems and smart grids and meters used in electricity distribution.

Smart applications in the field of water are electronic payment systems, intelligent water meters and demand management and automatic water quality monitoring.

The role of the cities has become more important and all the sustainability issues have relevant urban dimensions. By utilizing ICT options environmental, social and economic sustainability issues can be handled and solved at higher level, however, this cannot be done without the capability for radical/disruptive innovation. The radical innovation vision of Industry 4.0 is a response to the global competitiveness challenges, which also requires a supporting infrastructural and social environment having urban dimensions as well. The concept of smart cities is an answer for the environmental, economic and social sustainability issues generated by the previously listed challenges. Smarter cities are utilizing ICT technologies and in order to do so they need to innovate (Nick et al., 2018: 53).

Smart Municipality

Municipalities, as important and indispensable units of public administration, are important institutions in the efficient implementation of a large number of public services in accordance with the preferences of local people and institutionalizing democracy (Ulusoy and Akdemir, 2009: 260). As elected representatives, they are important technical units at the level of public interest, public service, efficiency, effectiveness and accountability at the local level (Elliot, 1997: 153).

It is a known fact that municipalities provide quickness and convenience in providing many services to the public. However, it is important for local governments to perform these functions more easily and to reduce the burden of the central administration. EU countries believe that the development of local government will increase integration. For this purpose, it attaches more importance to its local administrations and tries to provide more autonomous structures for them to assume an integrative role (Ministry of Interior Affairs General Directorate of Local Administrations, 1998: 11).

With technological advances, many municipalities have started to emerge as institutions with the opportunity to develop sustainable cities and communities for the future. New technologies such as Big Data, IoT, real-time analytics, digitization and sharing economics started to take their place at the center of the concept of "smart city" in terms of energy, water, mobility, health, buildings and public services. However, cooperation networks between cities are developing around "smart innovation" and "smart city" programs. Many municipalities are working to adapt to these technological developments and involve society. The Academy collaborates with residents, businesses and the municipality to create intelligent, sustainable solutions called "Quadruple Spiral" that benefits the whole community. In this process, it is an important requirement for smart municipalism that many players are willing to participate and have a sense of working together in different fields and thoughts.

When a "smart city" program is created, municipalities should lead as a process initiator and accelerator leader. Municipalities need to provide opportunities to provide more sustainable solutions to the inhabitants, the growth of local businesses, and closer cooperation between academy and society.

Within the scope of smart municipal services, two main implementation tools, "Local Agenda 21" and "Eco-municipality" can be put forward in order for municipalities to carry out sustainable activities. However, these tools should be used in the framework of *"corporate social responsibility approach"* in terms of infrastructure regulation, protection of natural and cultural environment, promotion of the regions, developing suitable products for the regions, directing the local people and auditing the services.

In the corporate social responsibility approach, it is essential that "an organization pursues a working strategy and policy in accordance with the economic and legal conditions, business ethics, and the expectations of individuals in and around the organization" and that the resources of the institution are used for the benefit of the society. In this context, prominent applications are voluntary business practices and investments that support community welfare improvement, health and safety, environmental protection and social issues. These works and practices can be realized in two dimensions as "volunteering" and "philanthropy" (Eren, 2000: 99; Vural and Coşkun, 2011: 63). As the organizations, which achieves to be prominent in the corporate social responsibility approach that are suggested to develop "social dialogue" and work in cooperation with different parts of society in various formats and levels in resolving the common problems that concern the society (Ateş, 2005: 46; Gül, 2011: 92). Corporate social responsibility (CSR) provides a stakeholder relationship strategy that brings competitive advantage and more financial benefit to the organizations (Gürlek and Tuna, 2019: 195). Organizations could gain sustainable competitive advantage by satisfying various demands of stakeholders with the help of CSR (Gürlek and Tuna, 2018).

Social responsibility approach in tourism is associated with four dimensions; namely customers, employees, environment, and society. In customer dimension, it is suggested to provide quality services, honesty and clear information; in employees dimension, it is suggested to provide training opportunities during and after hiring, create a suitable environment for health and safety, provide equal employment opportunities for female and male employees; in the environment dimension; it is suggested not to overuse the natural resources such as water, to use energy efficiently and in social dimension, it is envisaged to govern other cultural differences such as other value systems, religious beliefs, lifestyles and traditions with the dimensions of philanthropy and association (Argandoña, 2010: 21-25).

Local Agenda 21

Following the economic crisis in the 1970s, the concept of "local" began to replace the concept of "region" (Eraydın, 2004: 126). With this change process, local governments are now expected to develop behaviors in order to manage and improve their own performances within the scope of the notion of "think globally, act locally", combine with sustainable development goals and actions and policies of local government, focus on awareness and training activities, involve the public in consultation and the public, make partnership with groups in the society and to measure, monitor and report on the process of sustainability (Leslie and Hughes, 1997: 145). In the realization of these behaviors; Local Agenda 21 (LA-21) is proposed as a model which is shaped around the concept of "governance" and the concept of governance which is based on multi-actor and social partnerships based on criteria such as transparency, accountability, participation, work harmony, relevance and effectiveness (Öztürk and Ayaz, 2010).

LA-21, which came to the agenda with the United Nations Conference on Environment and Development in Rio de Janeiro in 1992 and described with the phrase "humanity is at a historical turning point", aims to achieve a balance between development and environment and is considered as an action plan that has the highest level of global reconciliation and political commitments for shifting the concept of sustainable development into life. LA-21 is shown as a solution to all the problems of human beings and the ongoing deterioration in the world, especially from environmental problems, from the population explosion to international inequality (Emrealp, 2005: 13; Ökmen and Görmez, 2009: 109-110).

Within the scope of LA-21, it is recommended to municipalities within the local government units to establish city councils that will determine their problems and priorities with other organizations, especially civil society. As the city council represents all dynamics in the city, it is desirable that the city council adopts the basic principles of protection of rights and law, sustainable development, environmental sensitivity, social assistance and solidarity, transparency, accountability, participation, and decentralization. In particular, municipalities are expected to play a facilitating role at every stage of the LA-21 process, to internalize LA-21 activities (to place LA-21 understanding at every level of their administration) and to prepare their plans and programs in line with LA-21 efforts (Emrealp, 2005: 39-40). It is believed that LA-21 implementations, which are foreseen to be carried out under the leadership of local administrations, can play an important role in tourism and social development with a planned, democratic and cooperative approach that includes community participation in sustainable development, environment and development decisions (Jackson and Morpeth, 2000: 119). However, at the center of sustainable tourism, LA-21 practices are also seen as a tool for identifying the predominant actions in the establishment systems and procedures of government offices, local administrations, national tourist offices and professional chambers and merging them in the decision-making process for tourism. The main activities in the context of tourism through LA-21 applications are summarized as follows (Middleton and Hawkins 1998: 104):

- Assessing the existing regulatory, economic and voluntary framework capacity to bring sustainable tourism,
- Evaluating economic, social, cultural and environmental effects of tourism sector activities,
- Designing new tourism products at the center of sustainability as an integral part of the sustainable tourism development process,
- Facilitate the exchange of knowledge, skills and technology related to sustainable tourism between developed and developing countries,
- Creating local awareness through courses and trainings,
- Planning for sustainable tourism development,
- Ensuring the participation of all sectors of society,
- Measuring the success of sustainable development at the local level,
- Developing partnerships for sustainable development.

Despite all these predictions, tourism cannot provide the necessary support for LA-21 applications. Jackson and Morpeth (2000) reported that little progress has been made in the field of tourism, even where LA-21 applications are best developed. In a study conducted by Leslie and Hughes (1997) in the UK, causes of failure in LA-21 applications are listed as local governments having little benefit from their implementation, the financial resource failures required for LA-21, the concentration of LA-21 practices around the environmental strategy and the abstraction of other stakeholder partners. In many countries, local authorities do not have much involvement in tourism and therefore have little experience in planning, developing and managing tourism. It is believed that this has changed in recent years and that local authorities play a greater role in tourism (UNEP, 2003: 4).

Eco-Municipality

Eco-Municipality is a model of municipalism that was launched in Sweden in 1983. Climate change in Sweden has been instrumental in the emergence of this model. Eco-Municipality is a vision of environmentalist municipality with strategic and technical goals, where participation is prominent and envisages the development of projects in line with sustainable development goals. The environmental efforts of LA-21 and city councils, the initiatives of the municipalities to ensure the EMAS (Eco-Management and Audit Plan) and ISO (International Standard Organization) standards, and the efforts to get the green certificates of the buildings create the strategic dimension of eco-municipality, while conservative and sustainable green transformation activities in renewable energy, recycling and development plans constitute the technical dimension thereof (Bostancı, 2011: 251).

The eco-municipality model, which aims for a democratic and participatory development process in the long term, takes its name from the words economy and ecology. It adopts a model based on the desire to develop an ecologically, economically and socially healthy society. The most important components of the eco-municipal model are widespread community awareness, a common language of sustainability and integrated community participation (www.glc.org). Within the scope of developing an economically and ecologically sustainable framework and receiving the support of the whole community, an eco-municipality is expected to develop activities in the following areas (www.stevenspoint.com):

- Social awareness and education,
- Development and land use,

- Green building (buildings suitable for the climate and land structure, with renewable energy sources, using natural and non-waste materials) and energy,
- Waste and recycling,
- Water and wastewater,
- Food,
- Transportation,
- Parks and tourism.

Within the scope of tourism mission, eco-municipality aims to create and maintain a desirable natural environment for the host community and visitors. To this end, it encourages the host community and visitors to conserve natural resources. It aims to develop sustainable tourism through partnerships or voluntary activities to further adopt environmental sustainability practices to government and tourism businesses. It pioneers social cooperation in the management process of parks and natural habitats, organizing parks for recreational, recreation and health needs of the host community and visitors, preserving natural habitats and special species through economic cooperation (http://www.stevenspoint.com).

Eco-municipality activities are directly related to tourism, which considers natural environment and natural people as the main theme. If municipalities incorporate the eco-municipality model into their own bodies, they can achieve effective results in sustainable tourism management and make positive contributions to the tourism development of their regions.

Municipal Services in Turkey

The local government system in Turkey consists of provincial special administration, metropolitan municipality, municipality, local administration unions and villages. There is no legal hierarchical relationship between these five levels, nor is there a cooperative relationship. Within these units, municipalities are defined as "a public legal entity with administrative and financial autonomy, established to meet the local needs of the local residents and formed by the electorate of the decision-making body, the voters". According to the Municipality Law No. 5393, the duties and responsibilities of the municipalities include a wide range such as urban infrastructure such as reconstruction, water, sewage and transportation; geographic and urban information systems; environment and environmental health, cleaning and solid waste; police, fire, emergency, rescue and ambulance; urban traffic; burials and cemeteries; afforestation, parks and green areas; housing; culture and arts, tourism and promotion, youth and sports; social work and assistance, marriage, occupation and skills; economy and trade development services, opening protection houses for women and children, opening pre-school education institutions, construction and maintenance and repair of all state-owned school buildings, meeting all kinds of tools, equipment and materials, opening and operating all kinds of health-related facilities; to ensure the protection and maintenance and repair of cultural and natural assets, historical texture and the places and functions that are important in terms of city history, to rebuild the ones that cannot be protected in accordance with the original, to provide material to the students, amateur sports clubs and to provide the necessary support, all kinds of amateur sports events and giving awards to the athletes who have achieved outstanding success in national and international competitions by the decision of the city council.

In the scale of Turkey, it is also recommended to focus on efforts to develop tourism at the local level. In the SWOT analysis study, prepared based on the tourism trends around the world and in Turkey titled *"Turkey as a Tourism Destination"*, the statement "the failure of the public, namely local administration, to pay sufficient attention to the tourism" stands out. In the Ninth Development Plan, it is requested to "focus on the establishment of a development environment based on local dynamics and internal potential, increasing institutional capacity at the local level and accelerating rural development" (DPT, 2006: 91).

MAIN FOCUS OF THE CHAPTER

Municipalities and Tourism in Turkey

The "locality" is at the center of the production and consumption in tourism in terms of coverage of specific features. The promotion and marketing of its services, the capitalization of certain characteristics of certain places, the identification of actions leading to tourist consumption, and the promotion and continuity between policies and planning in a tourism destination that stands out with international information and human movements (Cawley et al., 2002: 66) and all destinations and the environment, including businesses, should be taken into account and local tourism needs to be addressed before the other institutions in the destination (http://www.southhams.gov.uk).

Despite these expectations, local governments that have an important role in the development of tourism, tourism policy and planning could not draw the necessary attention on them (Hall, 1994: 152). Nevertheless, it is an undisputed fact that local governments are the best decision-making institutions for the benefit of tourism policies to be implemented in their regions (Jeffries, 2001: 47). However, tourism stands out as a management activity in which local governments are very eager to attract tourists to their regions in order to revive their economy during the recession periods (Elliot, 1997: 138). Increasingly competitive budgetary pressures, in particular, significantly undermine the ability of local authorities to participate in informal services such as tourism. Therefore, it is emphasized that local governments should address priority tourism issues and functions that cannot be fulfilled by any other institution (http://led.co.za). The functions of local governments that can address tourism are listed as follows (Long, 1994: 21):

- Partnership with the private sector, other organizations and other local authorities,
- Involving indigenous people,
- Assertive promotional activities,
- Quality information that can be easily found,
- Promoting special events,
- Developing sustainable tourism,
- Protecting the environment,
- Ensuring that employees receive appropriate wages,
- Consulting with the advisors in matters requiring expertise,
- Preparing training programs for employees,
- Strengthening the infrastructure of centers of attraction,
- Establishing a connection between centers of attraction.

Many governments around the world are looking for ways to integrate sustainability into tourism sectors at national and regional levels. In this process, local administrations can deeply affect the success of the local tourism sector. They can play a role in preserving the values to which the future of the

Table 2. Technical support matrix for tourism

Technical Support Providers	Leadership education	The market research	Planning	In-kind Partnerships	Finding Funds
Local / regional government agencies / councils			X	X	X
Regional, state and national tourism offices and agencies		X		X	X
Private sector tourism industry representatives		X		X	
University tourism centers / academic programs	X	X	X		X
Consultants for the tourism industry	X		X	X	X
Growth / development programs	X		X		
Regional, state and national associations of art and history			X	X	X
Regional, state and national natural resource and wildlife associations			X	X	X

Source: Nuckolls and Long, 2009: 8.

destination depends. In particular, the involvement of local administration plays a critical role in view of the impacts of functions such as land use and environmental regulation on the sector. In the process of significant interaction between visitors and tourism businesses as well as the interaction between the environment and the host community, the contribution of local governments cannot be ignored. As one of the key actors that can assist in sustainable tourism, the local government can generally be involved in the following activities (South Australian Tourism Commission and Local Government Association, 2006: 7-8):

- Strategic planning,
- Marketing and promotion,
- Data collecting,
- Sponsorship,
- Economic development,
- Visitor services,
- Public land management,
- Capacity building for society.

Within the scope of Tourism Technical Support Matrix prepared by Nuckolls and Long (2009: 8), local administrations are listed as important technical support providers in planning, in-kind partnerships and funding.

If local tourism development is to be successful, all stakeholders in the community should be involved in the planning and decision-making process. As the product of tourism renews itself and grows, these issues will inevitably come back to the society in different ways. Society should be able to manage its relationship with the tourism development process as well as the product itself. For those who make up the product and have an emotional connection with it, it is often difficult to understand that not everyone in the community has the same feelings for it. Therefore, the psychological and social anxiety generated by the tourism product should be identified and discussed as part of the process of revealing ongoing values. It should be understood that some of these values may be confronted with the threat of tourism product and there may be situations where steps should be taken to eliminate this threat (George et al., 2009: 246).

Given the fact that tourism is an economic activity that has deep ties to a particular region and local variables, central governments are not always appropriate to manage the complex structure and economic and social environment of the tourism sector. In many cases, for some reason, local governments seem to be in a better position to tackle tourism-related problems. They have direct knowledge of the region's strengths and weaknesses and are more aware of the opportunities and challenges that the business environment faces. They can intervene in time to important changes or problems affecting the community or people living in it (Bertucci, 2002: 4).

Tourism is among the last priorities of local administrations that have many responsibilities at the local level. However, following the 1990s, more local governments are taking decisions to encourage tourism development as a way to stimulate the local economy and increase revenues. In order to make successful and useful decisions on tourism, attention is paid to the sensitivity of the local people and the environment. It is recommended that the policies developed at the local level be implemented and that the support of the indigenous people is gained. It is expected from the public authorities to manage the effects of tourism in a way that will not be harmful and to develop the region in economic and social terms. In particular, informing and educating local politicians and the public about tourism issues, including the advantages and disadvantages of national tourism management plans and tourism development are among the important tasks of local administrators (Elliot, 1997: 137). The key role of local authority in tourism has become more noticeable in recent years. The reasons for this are related to the following five situations (UNEP, 2003: 8):

- The fact that tourism is a sector where the consumer is brought to service and not the other way around, and the complex structure of the tourism sector,
- A better understanding of the fact that tourists select not only individual products but also destinations,
- The dispersed structure of the sector, the majority of which consists of small-scale enterprises,
- Recognizing that, in terms of sustainability, the state authorities should play a regulatory role by not only leaving the tourism development in destinations to market rules,
- They are orientations towards administrative centralization which increases the response power of local authorities.

In the SWOT analysis study, prepared based on the tourism trends around the world and in Turkey titled "Turkey as a Tourism Destination", the statement "the failure of the public, namely local administration, to pay sufficient attention to the tourism" stands out (TUBITAK, 2003: 24). It is believed that tourism in a country will be meaningful with the internalization by the central government, local administrations, NGOs, professional organizations and the public. Especially, considering that tourism activities take place in a region, local administrations are seen as prominent organizations not only for their own people but also for the tourists visiting the region. It is pointed out that local administrations should be prepared rather than implementing plans and policies for the protection and optimal use of tourism resources in the region. In addition, local administrations are accepted as important institutions that can adopt tourism and receive their support to citizens and professional associations in their regions (Baş et al., 2007: 1-12).

How to expand tourism in a country's economy and how to secure it as a means of sustainable growth and development are important questions. In this context, it is accepted that a healthy management environment can play a very important role in making tourism a successful activity for sustainable development. In this context (Bertucci, 2002:3); it is suggested to develop;

- Effective and transparent political institutions that are accountable, allow political participation and representation, and respond to the needs and demands of citizens,
- Consistent and stable policies and a regulatory framework that address all aspects of tourism in a holistic way,
- Partnerships between the government, the private sector and civil society for tourism, which affects many aspects of society's economic and social life.

Municipalities are seen as advantageous institutions in local governments in terms of determining social needs of local people including the above principles accurately and in meeting these needs directly in terms of developing social policies and services for the public compared to central administrations (Seyyar, 2008: 30). In this context, in relation to tourism, the relationship between municipality and tourism needs to be evaluated in more detail, unlike other local governments.

Among the local administrations, municipalities have the strongest relationship with tourism. Municipal administrations are closely related to tourism with many duties and responsibilities. Activities such as cleaning and environmental protection services, infrastructure services, municipal police services, firefighting services, zoning plans, control of buildings, transportation services, consumer protection services are closely related not only to local people but also to tourists coming to the region. It is clear that the duties of the municipalities for the purpose of service are closely related to the tourism sector, and when taken into consideration, that the municipalities cannot be excluded. In this context, municipalities should accelerate their activities in tourism (Dönmez, 1996; Baş et al., 2007).

The role of municipalities in tourism can be mainly grouped in two themes; namely, providing quality physical infrastructure for destinations and ensuring tourism cooperation and ensuring the destination's needs. Within the scope of a quality physical infrastructure; street cleaning, street lighting, transportation, parks and gardens, coast protection and coast cleaning, water security, protection of natural environment, construction and planning control, environmental health, public toilets, maintenance and repair of sidewalks, parking lots, security, signposts, empty time activities and artistic activities can be seen as priority physical infrastructure works for destinations. They can gather tourism services under one roof as a part of tourism cooperation. They can prepare local development plans, local transportation plans, local cultural strategy plans and Local Agenda 21 (LA-21) action plans to ensure the sustainability of tourism. They can increase local awareness by supporting local culture, people, products and environment with their democratic base. They can pool resources in line with common goals and local ownership. They can directly transfer the needs of local sectors to the higher boards and central management units. They can bring people together through honest brokerage roles and form partnerships. They can work effectively to promote their region effectively in the tourism market (www.southhams.gov.uk).

Municipalities can carry out important activities in the field of tourism through cooperation with local tourism associations, especially hospitality. For local people, they can organize meetings and conferences on issues such as house, street and avenue cleaning, protection of antiquities, hospitality to foreign and local tourists, not selling at high prices and not disturbing tourists. They can ensure that public toilets, fountains and water resources are kept clean, and that tourist and natural beauties and historical monuments are preserved. They can benefit from the experts and their ideas for the studies within the scope of tourism while preparing the zoning plans. They can create simple plans (main road departures, car parking spaces, museums, monuments, squares, camps, grove and tourism offices) that show important places at the city entrances and offices for tourists to consult. They can play an important role in arranging city squares in an aesthetic view, specifying parking places, arranging the ways of selling goods and vehicles in the parking spaces, putting trash cans and ensuring cleanliness. They can support the promotion of the historical and cultural riches of their regions within the scope of the promotion activity which is an important factor in the development of tourism. They can lead the organization of local folklore performances, festivals, fairs, exhibitions. They can contribute to the production of various souvenirs within the scope of touristic crafts (Tortop, 1988: 7).

In the literature, the activities that municipalities can carry out within the scope of environment, reconstruction, recreation, culture, art, sports, promotion and supervision aimed at tourism and which may affect tourism are summarized as follows (Ağaoğlu, 1992; Ünal, 1992; Yeter, 1993; Long, 1994; Reed,1997; Wilson et al., 2001; Andriotis, 2002; Azaklı, 2004; Baş et al., 2007; http://www.southhams.gov.uk):

- Making planning and construction audits for tourism purposes,
- Carrying out activities to protect the environment and prevent pollution,
- Making zoning planning and landscaping in a way to protect the tourist attraction,
- Creating recreation areas for tourism (water sides, beaches, recreational facilities, motels, camping, children and youth camps, health and treatment facilities, hiking and sports fields, small recreation areas with small agricultural production, etc.)
- Improving urban ecology,
- Following rational land-use policies in rural areas,
- Eliminating urban deficiencies, disturbances and distortions,
- Bringing local building and architectural features to the forefront,
- Establishing and operating entertainment facilities such as cinema, theater, etc. issuing documents and inspections to established facilities,
- Organizing local, national and international cultural and tourism activities,
- Establishing and operating museums,
- Making accommodation facilities such as hotels, motels, pensions, camping,
- Supervising touristic facilities and enterprises,
- Issuing operating certificates for accommodation or food and beverage facilities established by private entrepreneurs,
- Protecting cultural assets of touristic value,
- Transforming their cultural identities into attraction by restoring cultural works in a balance of protection and use,
- Opening a cultural and tourism office,
- Providing guidance services,
- Establishing fairs, exhibitions, fairs, organizing festivals,
- Assisting marketing of touristic products (analysis, planning and implementation).
- Organizing seminars and meetings to raise awareness of local people on tourism,
- Opening various vocational courses and schools in order to train staff in the tourism sector,
- Educating municipal employees about tourism.

Table 3. Tourism-related activities of municipalities in Turkey

Marketing Activities	• Efforts for identification and analysis of local tourism market, • Retention activities for the promotion of tourism markets.
Cultural Activities	• Establishing and operating municipal cinemas, theaters, • Creating reading rooms for local people, • Establishing and operating museums, • Organizing local, national and international cultural activities, • Making expropriations to protect cultural assets, • Preserving local historical and cultural assets.
Environmental Activities	• Infrastructure services suitable for tourism potential (water, sewerage, electricity network, road, parking, cold storage, toilet), • Efforts to improve environmental protection (garbage collection, pest control, afforestation), • Zoning regulations compliant with environment.
Audit Activities	• Planning and construction supervision, • Inspections for environmental protection and prevention of pollution, • Cleaning, order, service and price controls in the facility and its establishments.

Source: Adapted from Kılıç, 2006: 83-88.

Kılıç (2006: 83-88) showed the municipalities as more effective institutions than the Ministry of Culture and Tourism in revealing and evaluating the tourism potential of the town they represent, and suggested that local promotional activities should be carried out by the municipalities in that region or their opinions would be taken. Tourism-related activities of the municipalities in Turkey have been associated with activities shown in Table 22 as four dimensions, namely marketing activities, cultural activities, environmental activities, and audit activities. He argued that the municipalities can contribute to the development of tourism if the municipalities fulfill their responsibilities under a certain priority. Meanwhile, Torlak (1999: 97) stated that the municipalities' tourism marketing activities were limited to advertising and public relations activities and that they could not benefit from other opportunities that the tourism market could create.

Tourism Related Smart Municipality Applications in Turkey

Most metropolitan municipalities in Turkey offer smart municipal services that benefit tourism. These practices are explained below under the headings (Elmastaş, 2019; Google play store, 2019).

Smart City Mobile Application: Kayseri Metropolitan Municipality's corporate mobile application which name is a Smart City Kayseri is attracted attention. Users of this device with Android and iOS operating system can get information about transportation, city information system, mobile map, and culture and art events by downloading the mobile application. With the City Information System in the Mobile Application, the nearest hospitals, pharmacies, gas stations, historical monuments, free wifi areas, notaries, ATMs, mosques, schools, parking lots, bicycle stops, taxi stops can be reached. Additionally, information on how to get to places such as streets, buildings and districts is provided. With the mobile application, time and location information of the cultural and art events organized by the municipality can be easily accessed. Adana Metropolitan Municipality's corporate mobile application "Adana Smart City" can be downloaded from Google play store. Users can get information about transportation, city information system, mobile map, culture and art events by downloading this mobile application. Another example is Bursa Metropolitan Municipality's corporate mobile application which name is "Bursa in Mobile phone". It can be downloaded also from Google play store. Users can get information about transportation, city information system, mobile map, and culture and art events by downloading the mobile application. Konya Metropolitan Municipality's corporate mobile application "Konya"is one of the most well-known examples. Users can get information about smart transportation system, where's my bus, smart bike stops, city information system, mobile map, and culture and art events by downloading the mobile application. In Turkey most of the Metropolitan Municipalities have smart city mobile applications which give touristic information about the city, these applications can be downloaded from google play. Some of these applications are: İZUM (İzmir Transportation Center), Gaziantep BB (Gaziantep Metropolitan Municipalities mobile guide application), IBB Kültür (İstanbul Metropolitan Municipalities culture and art mobile application).

Smart Bicycle Application: Sakarya Smart Bicycle Application, implemented by Sakarya Metropolitan Municipality, provides services in 15 different points. With the Sakarya Smart Bike System, citizens can rent bicycles through the stations established in 15 different points in the city by paying it with both member card and credit card. Like this, İstanbul Metropolitan Municipality's corporate mobile application "isbike" can be used in İstanbul. User can get information about the system through the application. This application has the following function: İt can list the bike stations from the nearest point or from the farthest point to the location. On the map, user can see a list of all stations, the number of bikes available and the number of spaces available. User can get directions to the bike station that he/she chose. If the users are not member of the application can subscribe via their phone. Through the credit card, user can rent bicycles.

Smart Junction Application: Dynamic Junction Control System was established in the 14 intersections where traffic density was determined and coordinated green wave application and was initiated to ensure traffic fluidity at 7 intersections as a result of feasibility studies conducted by Afyonkarahisar Municipality. Dynamic junction control system reduces the waiting times in traffic by calculating vehicle density at intersections by using cameras and applying longer green light time for intense directions. Variable signal duration saves gasoline and reduces the amount of carbon monoxide released into nature. . Coordinated green wave application was implemented in the 7 intersections. by providing uninterrupted flow, the stop-and-go ratio is reduced in traffic.

Smart Stop Application: Muğla Metropolitan Municipality has launched the smart stop application that brings convenience to transportation. In the first stage, Menteşe, Bodrum, Milas, Marmaris and Fethiye districts have established smart stops in various regions. LED screens are located at the stops. In these screens, the vehicles approaching the station at that moment can be seen in detail. Which bus will be here in a few minutes? Where will the bus go which is coming? The smart stop application can answer these questions. With the GPS system, instant information can be received between the stop and the bus.

Smart Signs Application: Smart signs application developed by Kayseri Metropolitan Municipality provides information about parking lots. Smart signs are placed at 11 points of the city. Drivers can easily access parking services with smart signposts showing the occupancy rates of multi-storey car parks. Thus, drivers have the opportunity to follow the occupancy information of the parking lots.

Augmented Reality Application: İstanbul Metropolitan Municipality's corporate mobile application "IBB AR" can be downloaded from Google play store. This application is designed to be used in Istanbul Metropolitan Municipality Augmented Reality Historical Places Flash Cards. Tourist can find information about the purchase of the cards from the Istanbul Metropolitan Municipality official web site or at https://www.medya.istanbul/. Also, Mardin Metropolitan Municipality's corporate mobile application "Mardin AR" can be downloaded from Google play store. This application was prepared to be used in "Mardin AR Cards". Users can run it by keeping their phone on the sample images on the link https://atfstore.com/mardin/.

Smart Museum Application: İstanbul Metropolitan Municipality's corporate mobile application "Panorama1453" can be downloaded from Google play store. The application has 3 language options: Turkish, English and Arabic. This application provides audio and detailed information about the artifacts in the museum. Like this, İstanbul Metropolitan Municipality's corporate mobile application "Miniatürk" can be downloaded from Google play store. This application is for introducing of Miniatürk. The application gives detailed information about history of monuments in Miniatürk. Application has many foreign language options.

In addition to these topics, some examples of smart municipal services are as follows: the virtual reality technology in Turkey destination offers the opportunity of 3D touring to tourists in Red Yard, Zeus, Athene and Asklepios temples in Bergama antique city in order to help planning for tourists, to provide them with tour information and to improve their travel experience. In order to provide information to the tourists coming to the city and to introduce the city, Mardin Metropolitan Municipality has established an Information Culture and Promotion Center in Mardin province and augmented reality application under the name of "Mardin Ar" was put into operation.

SOLUTIONS AND RECOMMENDATIONS

Smart municipalities establish free internet, electronic traffic control system, smart public transportation system, smart tourism points, unmanned service points, smart health service, barrier-free SMS system, smart environment and smart energy systems in order to become a smart city in transportation, health, security and municipal services. In particular, it leads the way in which big data is generated based on IoT technology (each object somehow accessing the Internet and communicating with other devices). In addition, Augmented Reality (AR) technology provides a vivid image of a physical, real-world environment enhanced by computer-generated input such as information, audio, video, graphics, or GPS data, and provides a natural stereo image of a three-dimensional environment with Virtual Reality (VR) technology. is also used quite often. Following the concepts such as smart signaling, smart stop, smart intersection, smart ticket and transportation; water, energy and meteorological applications are the prominent applications in smart municipalism (Memiş, 2019). Tourism in the context of this application is supported by the following activities may generate significant outcomes in the development and sustainability of tourism locally:

- Trainings can be given on the importance and benefits of smart tourism.
- In order to integrate the current tourism technologies into the physical tourism infrastructure of the city, smart tourism initiative program(s) can be developed starting with local tourism enterprises.
- The development of smart tourism at micro (hotels, restaurants, etc.) and macro (city infrastructure etc.) levels can be encouraged throughout the city.
- Satisfaction tracking and analysis of current and potential tourists can be done.
- By creating spaces closed to vehicle traffic, the satisfaction and safety of traveling tourists can be increased.
- Smart junctions and sensors that determine the number of vehicles can be used.

- Emergency vehicles such as ambulance and fire brigade can also be provided with the GPS tracking.
- Integration of bicycles in public transport can be supported.
- Smart bike stops can be installed.
- Access cards can be combined and standardized.
- With the smart stops, the bus driver can be informed in advance of the disabled or elderly passengers at the stop.
- With the Smart City Project, wireless access, irrigation, lighting and traffic lights can be controlled as needed.
- Blood pressure and blood measurement system, smart phone device and medical coaching service can be provided to people in need of care.
- Smart street lighting can save up to 60-70 percent energy.
- With the smart irrigation project, the moisture status of the soil can be monitored remotely by sensors placed in parks and gardens.
- With the mobile applications to be downloaded to mobile phones, multilingual tourist guides, quick response codes (QR) and ICT usage can be improved.

FUTURE RESEARCH DIRECTIONS

Municipalities are institutions that carry social responsibility in the environmental and social dimension. They are important technical units that can direct and supervise the local community in the prevention of overuse of natural resources, energy efficiency and waste management. They are democratic representatives of the local community who can succeed in managing value, belief, tradition and cultural differences. Therefore, it is important for municipalities to develop social dialogue within the scope of social responsibility in the tourism dimension in terms of localization of tourism services in their regions. In particular, in the perspective of *"Local Agenda 21"* and *"Eco-municipality"* tools, municipalities are important autonomous institutions that can support their region in tourism. In this scope, it is needed scientific research that can support smart municipality functions in the perspective of *"Local Agenda 21"* and *"Eco-municipality"* tools. Municipalities should consider of "Local Agenda 21" and "Eco-municipality" when developing smart technologies in tourism destinations. Future research should focus on smart technologies related to eco municipalism.

CONCLUSION

Local governments are as important organizations as national governments. Because they represent the people. Above all, they have the opportunity to deal with people directly and influence them wherever they are. They are also important because they play an active role in the effective and efficient implementation of decisions and plans formulated by statesmen at the center. In this context, they are very important in terms of ensuring local people's support (Elliot, 1997: 138).

The tasks expected from the municipalities, which are considered as an important unit of local governments within the understanding of local development, in terms of economic development, social sharing and inclusiveness, political participation and cultural pluralism are listed as follows (Göymen, 2004: 25):

- Providing infrastructure support especially for the economic development of the region,
- Working to make the region a center of attraction for investments,
- Supporting local investors to create new business areas,
- Preparing the conditions of sustainable development,
- Contributing to the development of social capital in the region,
- Preparing the conditions for the region to be a "learning region",
- Encouraging cooperation, common approach, knowledge-skill sharing and common movement towards economic and social development in the region,
- Communicating with the world; facilitating the follow-up of universal developments in the local stakeholders and encouraging horizontal relations abroad.

In addition, municipalities are emphasized in functionality with their ability to make better evaluations about the desires and wishes of local people and to make quick and effective decisions (Tortop, 2008: 17). Ensuring social security at the local level is also associated with important tasks such as public health, security management, crime reduction, disaster preparedness and sustainable development (Strömgren and Andersson, 2010: 288). It is stated that the effectiveness of the municipalities, which are considered as the most functional management units within the local government units, can come to the forefront with the creation of new resources, better use of the existing resources, product and service development, application of modern and innovative management approaches, and managerial knowledge and experience structure (Kutlar et al., 2011: 5-26).

Tourism is a fragmented and information-oriented sector covering tour operators, travel agencies, hotel sales representatives, associations, meeting offices, transportation, car rental, airlines, cruise, souvenirs, restaurants, hotels/motels and entertainment (Buhalis and Laws, 2001). Smart tourism or smart city is based on information and communication technology consisting of internet of objects, mobile communication, cloud computing and artificial intelligence technology (Yüzbaşıoğlu et al., 2018: 708). It unites the tourism sector by linking all these organizations, assets, activities and elements. In smart tourism, it is essential to provide tourists with a high level of tourism experience by fulfilling the needs and demands of tourists more efficiently. In addition, smart tourism destinations or smart cities offer solutions to the city people, infrastructure and facility managers as well as tourists (Başer and Doğan, 2018).

Turkey is a destination known worldwide international tourist arrivals and tourism revenues. According to 2018 report of UNWTO (World Tourism Organization), Turkey ranks at 6th place in international tourist arrivals and 10th place in tourism revenues. In this sense, it is a destination that has to follow the innovations that may arise in tourism.

REFERENCES

Ağaoğlu, O. K. (1992). Yerel yönetimler ve turizm. *Anatolia Turizm, Çevre ve Kültür Dergisi*, *3*(4), 18–23.

Andriotis, K. (2002). Local authorities in Crete and the development of tourism. *Journal of Tourism Studies*, *13*(2), 53–62.

Argandoña, A. (2010). Corporate social responsibility in the tourism industry: Some lessons from the Spanish experience. Retrieved from www.iese.edu/research/pdfs/DI-0844-E.pdf

Ateş, M. (2005). Avrupa'da sosyal diyalogun kuramsal yapısı: AB ekonomik ve sosyal komitesi. *Ankara Avrupa Çalışmaları Dergisi*, *5*(1), 45–66.

Ayaz, N., & Kulualp, H. G. (2019). Strategic role of human resource management in tourism enterprises. In R. Efe, I. Koleva, M. Öztürk, & R. Arabacı (Eds.), *Recent Advances in Social Sciences* (pp. 275–289). Newcastle, UK: Cambridge Scholars Publishing.

Azaklı, S., & Özgür, H. (2004). *Belediyelerin turizme yönelik ve turizmi etkileyen hizmetleri. A. Yüksel ve M. Hançer (Editörler). Turizm İlkeler ve Yönetim* (pp. 121–142). Ankara, Turkey: Turhan Kitabevi.

Baş, M., Kılıç, B., & Güçer, E. (2007). Türkiye'de yerel yönetimler ve turizm. *Mevzuat Dergisi*, *9*(119), 1–12.

Başer, G., & Doğan, O. (2018). Future Trends for Smart Tourism Destinations: Case of Antalya. In *Proceedings 1st International Congress of New Generation and New Trends in Tourism* pp: 694-697, (November 1-3, 2018), Sakarya, Turkey.

Bertucci, G. (2002). Strengthening local governance in tourism-driven economies. In statement prepared for the International Colloquium on Regional Governance and Sustainable Development in Tourism-driven Economies Strengthening. Cancun, Mexico, Feb. 20-22.

Bostancı, S. H. (2011). Kent konseylerinin Eko-Belediyecilik vizyonuna katkıları. In E. B. Keskin (Ed.), Kent Konseyleri Sempozyum Bildiri Kitabı (pp. 250-264), Bursa, Turkey: Bursa Kent Konseyi Bilimsel Yayınlar Dizisi-1.

Buhalis, D., & Law, E. (2001). Tourism Distribution. *Channels. Continuum Journal*, *12*(2), 207–220.

Cawley, M., Gaffey, S., & Gillmor, D. A. (2002). Localization and global reach in rural tourism: Irish evidence. *Tourist Studies*, *2*(1), 63–86. doi:10.1177/1468797602002001097

Çelik, P., & Topsakal, Y. (2017b). Smart Tourism Destinations: Review of Smart Tourism Applications of Antalya Destination. *Journal of Travel and Hospitality Management*, *14*(3), 149–166.

Demirezen, B. (2019). Artırılmış gerçeklik ve sanal gerçeklik teknolojisinin turizm sektöründe kullanılabilirliği üzerine bir literatür taraması. *Uluslararası Global Turizm Araştırmaları Dergisi*, *3*(1), 1–26.

Dönmez, M. (1996). *Mahalli İdarelerin Teşkilat Yapısı, Organları ve Görevleri*. Ankara, Turkey: Mahalli İdareler Derneği Yayınları.

DPT. (2006). *Dokuzcu Kalkınma Planı*. Ankara, Turkey: DPT Yayınları.

EC. (1999). *Towards Quality Rural Tourism*. Brussels, Belgium: Enterprise Directorate General Tourism Unit.

Elliot, J. (1997). *Politics of Tourism*. London, UK: Routledge.

Elmastaş, L. (2019) Antalya Akıllı Şehir Mobil Uygulaması, 05.08.2019. Retrieved from https://rayhaber.com/2019/03/antalya-akilli-sehir-mobil-uygulamasi/

Emrealp, S. (2005). Yerel gündem 21 uygulamalarına yönelik kolaylaştırıcı bilgiler elkitabı. İstanbul, Turkey: IULA-EMME (UCLG-MEWA) Yayını.

Eren, E. (2000). *İşletmelerde Stratejik Yönetim ve İşletme Politikası*. İstanbul, Turkey: Beta Basım Yayım.

Erkek, S. (2017). Akıllı Şehircilik anlayışı ve belediyelerin inovatif uygulamaları. *Medeniyet ve Toplum Dergisi*, *1*(1), 55–72.

George, E. W., Mair, H., & Reid, D. G. (2009). *Rural Tourism Development Localism and Cultural Change*. Bristol, UK: Channel View Publications.

Göymen, K. (2004). Yerel kalkınma önderi ve paydaşı olarak belediyeler. Yerel Kalkınmada Belediyelerin Rolü (pp. 21-30), İstanbul, Turkey: Pendik Belediyesi Kültür Yayınları 21.

Gretzel, U., Sigala, M., Xiang, Z., & Koo, C. (2015). Smart tourism: Foundations and developments. *Electronic Markets*, *25*(3), 179–188. doi:10.100712525-015-0196-8

Gül, M. (2011). Yerel yönetimlerde sosyal sorumluluk faaliyetlerinin seçmen davranışlarına etkileri: Kayseri Büyükşehir Belediyesi örneği. (Unpublished Master's Thesis), Erciyes Üniversitesi Sosyal Bilimler Enstitüsü, Kayseri.

Gürlek, M., & Tuna, M. (2018). Reinforcing competitive advantage through green organizational culture and green innovation. *Service Industries Journal*, *38*(7-8), 467–491. doi:10.1080/02642069.2017.1402889

Gürlek, M., & Tuna, M. (2019). Corporate social responsibility and work engagement: Evidence from the hotel industry. *Tourism Management Perspectives*, *31*, 195–208. doi:10.1016/j.tmp.2019.05.004

Hall, C. M. (1994). *Tourism and Politics, Policy, Power, and Place*. Chichester, UK: John Wiley & Sons.

http://led.co.za/sites/led.co.za/files/The_Role_of_Local_Authorities_in_Tourism.pdf

https://unctad.org/meetings/en/Presentation/CSTD_2015_ppt11_Latvia_en.pdf

https://www2.deloitte.com/content/dam/Deloitte/tr/Documents/public-sector/deloitte-nl-ps-smart-cities-report.pdf

https://www.segittur.es/opencms/export/sites/segitur/.content/galerias/descargas/documentos-en/Smart-Destination.pdf

http://www.glc.org/announce/06/pdf/Schnook-small.pdf

http://www.smart-cities.eu/download/smart_cities_final_report.pdf

http://www.southhams.gov.uk/administration/committee_agendas/prosperity_pdg/23Jun05/Tourism%20and%20Marketing%20Strategy%20Review_Appendix%20A_23.06.05.pdf

http://www.stevenspoint.com/index.aspx?NID=620

İçişleri Bakanlığı Mahalli İdareler Genel Müdürlüğü. (1998). *Yerel Yönetimlerde Halk Katılımı*. Ankara, Turkey.

Işıkçı, Y. (2002). *Türk turizminde altyapı ve çevre, II. Turizm Şurası Bildirileri* (pp. 267–273). Ankara, Turkey: T.C. Turizm Bakanlığı Yayınları.

Jackson, G., & Morpeth, N. (2000). Local Agenda 21 Reclaiming Community Ownership in Tourism or Stalled Process? In R. Greg, & D. Hall (Eds.), Tourism and Sustainable Community Development (pp. 119-135), London, UK: Routledge.

Jeffries, D. (2001). *Governments and Tourism. (First published).* Oxford, UK: Butterworth- Heinemann.

Khan, M. S., Woo, M., Nam, K., & Chathoth, P. K. (2017). Smart city and smart tourism: A case of Dubai. *Sustainability*, *9*(12), 1–24. doi:10.3390u9122279

Kılıç, G. (2002). Belediyelerin bölge turizmine olan etkileri. (Unpublished Master's Thesis), Sakarya Üniversitesi Sosyal Bilimler Enstitüsü, Sakarya.

Kılıç, G. (2006). Belediyelerin turizme ilişkin faaliyetlerini değerlendirmeye yönelik bir uygulama. *Çağdaş Yerel Yönetimler Dergisi, 15*(3), 79-110.

Koçoğlu, C. M., & Acar, A. (2019). Safety perception in touristic destinations: the example of Istanbul province. In R. Efe, I. Koleva, M. Öztürk, & R. Arabacı (Eds.), *Recent Advances in Social Sciences* (pp. 305–319). Newcastle, UK: Cambridge Scholars Publishing.

Koçoğlu, C. M., & Kıycı, Ş. (2019). The evaluation level of foreign tourists for five-star hotels in Belek in terms of corporate reputation. *Journal of Tourism Theory and Research*, *5*(2), 112–122. doi:10.24288/jttr.523396

Kutlar, A., Yüksel, F., & Bakırcı, F. (2011). Türkiye'de Belediyelerin Ekonomik Etkinliği ve Etkinliğe Etki Eden faktörler Üzerine Bir Araştırma. Retrieved from http://www.tbb.gov.tr/onlineyayinlar

Leslie, D., & Hughes, G. (1997). Agenda 21, local authorities and tourism in the UK. *Managing Leisure*, *2*(3), 143–154. doi:10.1080/13606719708559316

Long, J. (1994). Local authority tourism strategies-a British appraisal. *Journal of Tourism Studies*, *5*(2), 17–23.

Memiş, L. (2018). Smart technologies and reflections on local public issues in Turkey: A review of metropolitan municipalities. *AVRASYA Uluslararası Araştırmalar Dergisi*, *6*(14), 141–163. doi:10.33692/avrasyad.510640

Middleton, V. T. C., & Hawkins, R. (1998). *Sustainable Tourism: A Marketing Perspective*. London, UK: Butterworth-Heinemann.

Nadaroğlu, H. (1978). *Mahalli İdareler Felsefesi Ekonomisi Uygulaması. (Birinci Baskı).* İstanbul, Turkey: Sermet Matbaası.

Nick, G., Pongrácz, F., & Radács, E. (2018). Interpretation of Disruptive Innovation in the Era of Smart Cities of the Fourth Industrial Revolution. *Deturope*, *10*(1), 53–70.

Nuckolls, J., & Long, P. (2009). *Organizing resources for tourism development in rural areas. Center for Sustainable Tourism East Carolina University*. Retrieved from www.sustainabletourism.org

Ökmen, M. (2009). *Sürekli değişme ve gelişme sürecinde kamu yönetimi ve yerel yönetimler. K. Görmez ve M. Ökmen (Editörler). Yerel Yönetimlerin Güncel Sorunları: Küresel, Bölgesel ve Yerel Perspektifler* (pp. 5–41). Ankara, Turkey: Beta Basım A. Ş.

Öztürk, Y., & Ayaz, N. (2010). *Turizm olgusunda Yerel Gündem 21 uygulamaları üzerine bir araştırma. O. E. Çolakoğlu (Editör). 11. Ulusal Turizm Kongresi* (pp. 324–333). Ankara, Turkey: Detay Yayıncılık.

play.google.com. (2019). 05.08.2019 Retrieved from https://play.google.com/store/search?q=b%C3%BCy%C3%BCk%C5%9Fehir%20belediyesi&c=apps

Reed, M. G. (1997). Power relations and community-based tourism planning. *Annals of Tourism Research, 24*(3), 566–591. doi:10.1016/S0160-7383(97)00023-6

Seyyar, A. (2008). Yerel siyasetin gelişiminde sosyal politikaların önemi. Retrieved from www.yerel-siyaset.com

South Australian Tourism Commission and Local Government Association. (2006). Local government's engagement in tourism. Final Report, Retrieved from http://www.tourism.sa.gov.au/webfiles/tourismpolicy/LGEIT_Final_Report

Strömgren, M., & Andersson, R. (2010). The usage of safety management tools in Swedish municipalities. *Safety Science, 48*(3), 288–295. doi:10.1016/j.ssci.2009.11.009

Torlak, Ö. (1999). Belediye hizmetlerinde pazarlama. *Çağdaş Yerel Yönetimler Dergisi, Ocak (1),* 96-114.

Tortop, N. (1988). Turizm ve belediyeler. *Amme İdaresi Dergisi, Haziran (2),* 3-10.

Tortop, N. (2008). Yerel yönetimlerin gelişmesinde turizmin önemi. *İller ve Belediyeler Dergisi, (721),* 17-18.

TÜBİTAK. (2003). *Vizyon 2023 Ulaştırma ve Turizm Paneli Raporu.* Retrieved from http://www.tubitak.gov.tr/tubitak_content_files/vizyon2023/ut/utp_son_surum.pdf

Ünal, T. (1992). Yerel yönetimler ve turizm. *Anatolia Turizm, Çevre ve Kültür Dergisi*, (25-26), 5-7.

UNEP. (2003). Tourism and Local Agenda 21: The Role of Local Authorities in Sustainable Tourism. First edition, United Nations publication.

Vecchio, M., Mele, G., Ndou, V., & Secundo, G. (2018). Creating value from Social Big Data: Implications for Smart Tourism Destinations. *Information Processing & Management, 54*(5), 847–860. doi:10.1016/j.ipm.2017.10.006

Vural, Z. B. A., & Coşkun, G. (2011). Kurumsal sosyal sorumluluk ve etik. *Gümüşhane Üniversitesi İletişim Fakültesi Elektronik Dergisi, (1),* 61-87.

Wilson, S., Fesenmaier, D. R., Fesenmaier, J., & van Es, J. C. (2001). Factors for success in rural tourism development. *Journal of Travel Research, 40*(2), 132–138. doi:10.1177/004728750104000203

Yeter, E. (1993). Yerel yönetimler açısından turizmde denetim boyutu. *Anatolia Turizm, Çevre ve Kültür Dergisi, Mart-Nisan (2),* 25-27.

Yılmaz, E. (2016). *Yeni Bir Hikaye Konya Akıllı Şehir Olabilir mi?* Retrieved from http://www.kto.org.tr/yeni-bir-hikaye-konya-akilli-sehir-olabilir-mi

Yüzbaşıoğlu, N., Çelik, P., Topsakal, Y., & Bahar, M. (2018). Endüstri 4.0 ve Akıllı Turizm: Antalya Destinasyonu Akıllı Turist Rehberi Uygulama Geliştirilmesi, Innovation and Global Issues in Social Sciences III (pp. 16-27.), April 26-29, Antalya-Turkey.

ADDITIONAL READING

Gretzel, U. (2018). From smart destinations to smart tourism regions. *Investigaciones Regionales-Journal of Regional Research*, (42),171-184.

Pierce, P., & Andersson, B. (2017). Challenges with smart cities initiatives: A municipal decision makers' perspective. *Proceedings of the 50th Hawaii International Conference on System Sciences*. Hawaii, USA, January 4-7. 10.24251/HICSS.2017.339

Snow, C. C., Håkonsson, D. D., & Obel, B. (2017). A smart city is a collaborative community. *California Management Review*, *59*(1), 92–108. doi:10.1177/0008125616683954

KEY TERMS AND DEFINITIONS

Eco-municipality: Eco-Municipality is a vision of environmentalist municipality with strategic and technical goals, where participation is prominent and envisages the development of projects in line with sustainable development goals.

Local Agenda 21: LA-21, which came to the agenda with the United Nations Conference on Environment and Development in Rio de Janeiro in 1992 and described with the phrase "humanity is at a historical turning point", aims to achieve a balance between development and environment and is considered as an action plan that has the highest level of global reconciliation and political commitments for shifting the concept of sustainable development into life.

Smart Cities: Collaborative communities which is sustaining integrated systems facilitating efficiency, citizen engagement, and providing the life quality.

Smart Municipality: It is a municipality which can be used new and smart technologies for their locals and tourists.

Smart Tourism: It relies on new technologies such as ICT, mobile communication, cloud computing, artificial intelligence, and virtual reality in order to provide tourist sufficient in the destinations.

Chapter 19
A New Concept in Tourism:
Smart Tourism Destinations

Gamze Coban
https://orcid.org/0000-0001-8628-7222
Faculty of Tourism, Nevsehir Haci Bektas Veli University, Turkey

Şule Aydın
https://orcid.org/0000-0002-8760-643X
Faculty of Tourism, Nevsehir Haci Bektas Veli University, Turkey

ABSTRACT

This chapter provides an insight into the topic of smart destinations. By adopting smart technologies, tourism destinations, as well as cities, gain more opportunities to offer better quality of life for residents and visitors. Smart cities aim to improve resource management, sustainability, and living conditions in urban environments by utilizing ICTs. The concept of smart tourism destination, deriving from smart city, refers to the use of technology in tourism destinations to increase the service quality and tourist satisfaction by focusing on tourists' expectations. In other words, smart tourism destinations aim to integrate technology into the destination for these purposes. This chapter presents the concepts of smart city, smart tourism, and smart destination. The emergence of smart city and smart destination concepts and the issues and challenges they might face are discussed. In addition to some future research directions, a brief discussion on potential controversies is presented.

INTRODUCTION

Management of a city or a destination has always been a complex task. Although it is hard to claim that technology made this task totally easy, the advancements in technology are clearly providing the managers and governors with more tools for a successful city/destination management. On the other hand, the challenges are getting more complicated. Cities are facing many challenges due to the rapid increase in urban population (Buhalis and Amaranggana, 2014). More than half of the world population lives in cities which ultimately results in increasing pressure on infrastructure (Falconer and Mitchell,

DOI: 10.4018/978-1-7998-1989-9.ch019

2012). Pollution, traffic congestion, and many other infrastructural problems created by this rapid increase (Kim and Han, 2012) are not easy to solve only with conventional methods such as building wider roads and pipelines. This trend in world population, i.e. increase in urban population, is estimated to continue (United Nations Population Fund, 2007). The problems related to increasing urban population threaten economic and environmental sustainability of cities (Neirotti, De Marco, Cagliano, Mangano, and Scorrano, 2014). Since conventional solutions are not that effective anymore, 'smart' solutions and knowledge and smarter use of current infrastructure and resources gain even more significance. Fortunately, technology could provide many opportunities to manage these issues. According to Komninos, Pallot, and Schaffers (2013), innovational internet-based services could help to solve these issues by meeting contemporary demands.

'Smart' might have many meanings but in terms of urban planning and management, it means a city environment that is integrated with contemporary technologies, especially information and communication technologies (ICTs) which also include data management, artificial intelligence, and etc. According to Boes, Buhalis, and Inversini (2015), a smart city's ultimate goal is to increase the residents' quality of life by using these technologies. Smart cities also aim to achieve resource optimization and sustainability through these technologies (Gretzel, Sigala, Xiang, and Koo, 2015a). As well as Boes et al. (2015), Gretzel et al. (2015a) do not ignore that 'smart' has also become a fashionable marketing word. In smart cities, smart technologies are used in order to manage and develop all city services (Washburn et al., 2010). Thus, smart cities offer many services to their residents and find solutions to urban problems with the help of information and communication technologies. In this chapter, the concepts of smart city and smart destination are discussed. Smart tourism and the components of smart city are also mentioned in this context. Another discussion is made on the issues on smart cities and smart destinations. Finally, some recommendations for future research are provided.

BACKGROUND

Smart Cities, Smart Tourism, and Smart Destinations

The term smart usually refers to the use of advanced ICTs in many different fields. The term of smart city is thought to be first appeared in 1990s (Söderström, Paasche, and Klauser, 2014). 'Smart' is highly associated with technology (Gretzel et al., 2015a) but adopting appropriate technologies does not make a destination smart unless appropriate policies that are mature enough to refer to contemporary issues such as privacy are applied (Lehr, 2018). The main goal is to increase quality of life by using these technologies and ensure sustainability with better resource optimization and governance (Gretzel et al., 2015a). Advanced ICT technologies provide the cities and destinations with appropriate tools for these purposes. Not surprisingly, smartness started to be applied after the recent developments in ICTs. As well as businesses, many cities adopted smart technologies in order to increase the quality of life, manage or overcome some infrastructural challenges, and operate more competitively and sustainably as a destination in the global marketplace. According to Bakıcı, Almirall, and Wareham (2013), smart cities use the ICTs in order to enhance residents' quality of life and provide sustainable development. Bakıcı et al. (2013, p. 139) note that a smart city, in the case of Barcelona, "… implies a high-tech intensive and an advanced city that connects people, information and city elements using new technologies in order to create a sustainable, greener city, competitive and innovative commerce and a recuperating life quality with a straightforward administration and a good maintenance system". Nam and Pardo (2011) argue that smart cities integrate the information into the physical infrastructure so that they gain many advantages such as efficiency, environmental benefits, better decision making and resource allocation, and collaboration. Hall et al. (2000) also note the importance of integrating the infrastructure for some similar reasons and they also mention the main benefit, i.e. in their terms 'maximizing services to the citizens'. Smart cities are those which focus on people and sustainability, adopt a transparent and participative management style with high contemporary ethical standards, and use advanced technological capabilities in order to increase quality of life in the city in line with these principals.

Advancements in related technologies and smartness have also some reflections on tourism. ICTs started to be used in order to develop innovative products, increase service quality and tourist satisfaction, and improve competitiveness in tourism industry. As the competition became fiercer, tourism businesses and destinations had to reconsider their strategies. Technological advancements have provided useful tools for better business and destination management. Technology is smart destinations' tool for meeting the visitors' and residents' demands (Lamsfus and Alzua-Sorzabal, 2013). All these resulted in the emergence of knowledge-based cities and destinations. Emergence of smart destinations is linked to smart cities but there is not a clear distinction between a smart destination and a smart city yet. However, smart destinations should also focus on visitors whereas smart cities' primary concern is the residents.

Smart destinations should ensure that stakeholders have access to a system that enables the instant exchange of tourism information (Buhalis and Amaranggana, 2015). Smart tourism destinations collect data from many resources such as physical infrastructure, individuals, organizations, and etc. and use their technological capabilities in order to provide enriched experiences in a sustainable and efficient way (Gretzel et al., 2015a). Data collection and exchange are critical issues in smart settings since these are expected to have a major impact on the effectiveness of a smart system. The more advanced data collection and exchange possibilities a city/destination possesses, the more room to adopt smart applications it gains. Despite being a prerequisite for smartness, technology itself is not sufficient to become a smart city/destination. It could be argued that a participative management approach is crucial and technology is, though very useful, just a tool for achieving some goals such as sustainable development. The following parts of this chapter discuss the issues in more detail.

MAIN FOCUS OF THE CHAPTER

Smart City

Managing a city is clearly a complex task due to the responsibility of providing many services to many people. Cities have limited resources and they need to provide those services and develop the city at the same time with these resources. Moreover, they cannot ignore the environment and sustainability in the contemporary context. The tasks of city governors are not limited to these and Meijer and Bolivar (2015) mention the management of ethnical, religious, and socio-economic diversity and culture. In addition to many other efforts, managing all these issues also requires investment in innovational technologies. The term 'smart' would not emerge, at least in its contemporary context, without the technological advancements. Clearly, technology lies at the heart of smartness (Dameri, 2013; Buhalis and Amaranggana, 2014). Those cities which successfully establish ICT infrastructures become more competitive (Caragliu, Del Bo, and Nijkamp, 2011; Bakıcı et al., 2013) and gain the other benefits of smartness. In order to be defined as 'smart', a city should be using ICT infrastructure, namely software systems, network infrastructure, and etc., to provide services such as transportation, utilities, and etc. (Washburn et al., 2010). As the complexity of cities increases, ICT tools become more important since they are used for coordinating all those services and informing and engaging the residents (Buhalis and Amaranggana, 2014). Adoption of smart technologies creates more business opportunities and provides sustainable development while increasing the residents' quality of life and enabling wiser management of resources (Bakıcı et al., 2013). As ICT tools and applications advance, they are embedded into the physical spaces in urban areas (Schaffers et al., 2011; Komninos et al., 2013). Since physical space and digital world could become connected through both the availability and smart use of infrastructure, integration and connectivity, as wells as innovation and competitiveness, are the key terms for smart cities (Gretzel et al., 2015a). There is not a commonly accepted clear definition of smart city yet. On the other hand, it is not a totally blurred term, either. The following two definitions are sufficient to clarify the framework of the concept.

a smart city is a well defined geographical area, in which high technologies such as ICT, logistic, energy production ... cooperate to create benefits for citizens in terms of well being, inclusion and participation, environmental quality, intelligent development; it is governed by a well defined pool of subjects, able to state the rules and policy for the city government and development. (Dameri, 2013, p. 2549)

According to Caragliu et al. (2011, p. 50) a city can be defined as a smart city "when investments in human and social capital and traditional (transport) and modern (ICT) communication infrastructure fuel sustainable economic growth and a high quality of life, with a wise management of natural resources, through participatory governance". Therefore, a smart city is a city or a geographical area that aims to enhance the residents' quality of life, makes necessary investments in human and social capital and conventional and contemporary infrastructure, and secures the sustainability of natural and other resources.

Possession of the technological capabilities is not enough to become a smart city unless these technologies are used for enhancing the quality of life. Forrester emphasizes the necessity of integrating smart technologies into education, healthcare, transportation, and other services in smart cities (as cited by Washburn et al., 2010). All these efforts and applications in smart cities result in creating a knowledge society (Bakıcı et al., 2013). As well as technology, human capital and education are also important for smart cities (Caragliu et al., 2011). Human capital and social infrastructure are considered as key issues for urban performance by Caragliu et al. (2011). The importance of education is also emphasized by Giffinger et al. (2007). In addition, interconnection of technological infrastructure makes the whole system smart by these systems' ability to collect data and share them with each other which also helps to make better operational decisions (Harrison et al., 2010). In the similar vein, Gretzel, Werthner, Koo, and Lamsfus (2015b) emphasize the use of advanced ICTs for the optimization of resource production and consumption. According to Bakıcı et al. (2013), smart cities should focus on infrastructure, human capital, and information, which they call three main pillars.

It is arguable that smart cities gain the 'smartness' through the use of internet of things, cloud computing, and general ICTs. Data are collected from many resources such as transportation facilities, hospitals, water systems, power grids, and etc. and shared through the internet (Su, Li, and Fu, 2011). This virtual city infrastructure then can be used for better decision making (Harrison et al., 2010). The integration is not limited to machines but residents are also an important part of cocreation process (Bakıcı et al.,

2013). Thus, residents of smart cities should be included in and they should contribute to and cooperate with the process.

Smart city concept deals with economy, people, environment, and governance. Cohen (2012) developed a useful tool called 'Smart City Wheel' which consists of 6 components, namely smart environment, smart mobility, smart economy, smart living, smart people, and smart governance. The Smart City Wheel can be useful in developing strategies for smart cities and monitoring their progress. The components reflect the criteria for being a smart city and they could be used for estimating the level of smartness. Smart environment includes all the efforts and management of physical conditions that aim to protect and improve the environment. Natural conditions, resource management, and environmental protection are the keys for a smart environment (Giffinger et al., 2007) as well as smart buildings and urban planning (Cohen, 2012). Smart mobility refers to technologies and systems that facilitate mobility (Cohen, 2012) which also include local and international accessibility (Giffinger et al., 2007). Smart economy is about innovational and competitive economy with national and international integration (Giffinger et al., 2007; Cohen, 2012). Smart living refers mostly to quality of life as well as touristic attractivity whereas smart people component includes the social structure which is open and inclusive (Giffinger et al., 2007; Cohen, 2012). On the management side, this openness is referred in smart governance component where the administration is participative and services for residents are also included in this component (Giffinger et al., 2007; Cohen, 2012). The importance of people is also emphasized by Collins, Paquet, Roy, and Wilson (2002) in the form of smart community term by which they refer to a society with locus of control and initiative. In other words, the society participates in the decision making collectively thanks to the overall smart system. Education, innovation, networks, residents with appropriate mindsets, and welfare are all about being smart (Schaffers et al., 2011) and these are all embedded in the smart city wheel.

Smart Tourism

The emergence of smart cities, thus destinations, has also some reflections on tourism. Smart tourism is a new tourism type in which smart technologies are used. The components of or criteria for smart cities are also valid for smart tourism. These factors such as resource use, environmental awareness, sustainability, people with appropriate mindsets, participative administration, smart mobility, use of ICTs are among the criteria for smart tourism and tourists.

As well as residents, tourists benefit the services in a destination and become a part of the society or social structure at least for a while. Thus, their welfare is as important as the residents'. Moreover, an open and integrated society and economy should welcome tourists and tourism. The important point is that tourism should be practiced in the same smart way. For example, the tourists should have uninterrupted access to transportation, travel and destination information and they should have a say in the services they receive.

Smart tourism is also considered as a part of smart world and tourists' needs are better focused, service quality is enhanced, and newer services are offered thanks to the technologies used in a smart setting (Huang, Yuan and Shi, 2012). Smart tourism is supported by ICTs and consists of many layers (Gretzel et al., 2015a). Once achieved, a smart setting is expected to bring some positive outcomes for tourism industry. Although they are interconnected, according to Zhu, Zhang, and Li (2014), there are three ways through which smart tourism has a positive impact on tourism industry. The first one is about the centralization of many fragmented information so that tourists can conveniently receive information and services. The second way is that smart technologies make analyzing tourists' demands easier. The third way is that smart technologies make information processing times shorter.

Smart destinations have the central role in smart tourism. Destination is the place where tourism experience mainly occurs. It does not mean that other components of tourism should be neglected but the destination has a major role in tourism experience. Thus, smart destination concept can be regarded as an integrated part of smart tourism.

Smart Destination

It can be claimed that a smart city is also a smart destination and the possible distinction between these two can be considered to be a little bit blurred (Khomsi and Bédard, 2017). According to Gretzel et al. (2015a), smart destinations are those which apply smart principles to urban or rural destinations and consider the tourists as well as residents.

The concept of smart destination is derived from the concept of smart city (Buhalis and Amaranggana, 2014). As mentioned above, smart destinations take the tourists into consideration as well as residents. All the advantages that are aimed to be provided to the society are also considered for the visitors. Moreover, in a smart concept, visitors' needs are addressed before (such as providing various information that helps decision making and planning), during (such as helping the tourist in the destination by providing information and personalized service), and after (such as providing the tourists with the opportunities to give feedback) their visit to the destination according to Buhalis and Amaranggana (2014; 2015). Not surprisingly, all these efforts for tourists' welfare are closely related to technological advancements. As also mentioned by Jovicic (2017), the developments in ICTs result in some changes in destinations by making them smart rather than conventional. In a smart destination, ICTs should be integrated into physical infrastructure (Gretzel et al., 2015a). Smart destinations are those tourism areas that utilize technological infrastructure in order to enhance tourist experience and benefit other tourism stakeholders such as the local community, businesses, and administrative bodies. Similarly, Boes et al. (2015) emphasize that smart destinations utilize technology in order to provide benefits to tourists, organizations, and the destination itself.

A smart destination refers to an area that promotes the residents' and visitors' quality of life where smart technologies are integrated into the areas' infrastructure. The effectiveness and quality of all services in the destination such as education, healthcare, transportation, and other public services are improved through the use of technology.

Technologies and concepts like cloud computing and internet of things are important factors for the development of smart destinations (Wang, Li, and Li, 2013). Cloud computing enables convenient data access through a network and this eliminates the need for installing a program into a hard drive (Jovicic, 2017). For example, a tour guide system can be used without being installed, thus more conveniently (Wang et al., 2013). This way, a traveler has more opportunity to access data and she/he can plan and realize her/his trips independently and more flexibly. As observed in this example, exchange of and access to data are critical factors both of which have become more convenient thanks to the advancements in ICTs. In a similar sense, Gretzel et al. (2015a) claim that ICT is the key for smart tourism.

Another important factor for smart destinations' development is the internet of things which is also expected to have many major social and economic impacts. The term 'Internet of things', in a broad sense, refers to the data collecting and sharing ability of non-humans (computers, devices, and etc.) from and with any 'thing'. According to Ashton (2009), internet of things can create big changes since it does not depend on human-entered data. Thus, internet of things and other similar technologies and systems could help the destinations become smarter since they make collecting, sharing, and accessing many data on destinations easier. Jovicic (2017) emphasizes that smart destinations are knowledge-based. According to the author, they use ICTs in a way that information and knowledge can be exchanged instantly. The use of ICTs in the proper way could solve many problems of urban living (Neirotti et al., 2014) and tourism destinations. The advantages gained through ICTs, such as the provision and use of real time data, benefit both residents and visitors (Buhalis and Amaranggana, 2014). Better understanding of visitors' expectations, interconnected stakeholders, easier exchange of tourism related information (Buhalis and Amaranggana, 2015), and better business and marketing practices (Wang et al., 2013) are among the expected benefits of smart destinations or smart tourism. Thanks to the technologies such as location-based and proximity-awareness systems, marketing messages can be targeted at the relevant audience in a specific location (Buhalis and Amaranggana, 2015). This could also increase the efficient and effective use of resources which ultimately benefits the businesses, destination governing bodies, and residents and visitors. Smart destinations benefit from technological environments, advanced feedback opportunities, and a more dynamic network of stakeholders (Buhalis and Amaranggana, 2014). Therefore, thanks to higher levels of ICTs, thus advanced exchange of data/information, these destinations become more responsive to tourists' demands, find solutions to the current and potential problems in a shorter period of time, and ultimately increase the quality of service.

Some destinations have already begun to practice many smart applications at various levels. For example, in Barcelona, visitors and residents can find touristic information and arrival times on bus shelters and they can also charge their mobile devices through the USB ports installed into these shelters (Gretzel et al., 2015a). The same city has a bicycle share/rental system called 'bicing'. This service is also used for data collection through its service application and these data are used for improving the share system and optimizing vehicle transit (Barcelona City Website, 2019a). Cycling infrastructure is also aimed to be improved not only for the sake of bicyclists but also for other vehicles and pedestrians (Barcelona City Website, 2019b). The city focuses on solar energy with the aim of enhancing residents' quality of life. For example, they have built Solar Thermal Ordinance infrastructure which benefits homes as well as commercial and public service buildings (Zygiaris, 2013). The city also focuses on the use of electric vehicles in transportation for a better environment and minimum level of CO_2 emissions. They have built 300 free charging points for electric vehicles in addition to 180 charging points in municipal car parks (Rahyaputra, Muna, and Rizal, 2016). Another smart step in transportation is the plan to fit smart driving systems into 5000 public and private fleet vehicles from 2019 through 2021 (Barcelona City Website, 2019c). This project is expected to reduce traffic accidents and the data collected through these vehicles are to be used for further enhancement of safer mobility. Barcelona's smart city concept also includes smart education. For example, the Smart City Campus Project aims at some cooperative efforts to benefit businesses, educational institutions, and people (Rahyaputra et al., 2016).

Stockholm is planning to quit fossil fuel use by 2050 and the city is installing power supply systems called 'smart grids' for local energy production in houses. The main goal is to minimize CO_2 emissions and increase the sustainability in energy production and use. The city's district heating system, on the other hand, has already reduced the emissions (Stockholms Stad, 2019). Stockholm aims to offer an accessible and reliable transportation system to the residents. In addition to the efforts on improving public transportation systems such as extending the metro line to some suburban areas, the city's 'The Walkable City' plan focuses on safe, quality, and integrated paths so that the city becomes even more pedestrian friendly (Stockholms Stad, 2018). Alternative fuels and renewable energy sources already play an important role for the transportation and there are higher goals (Riva Sanseverino, Riva Sanseverino, Vaccaro, Macaione, and Anello, 2017).

The city of Amsterdam aims to reduce CO_2 emissions and enhance sustainability with 'Climate Street' project (Amsterdam Smart City, 2019a). Amsterdam, while also aiming to use renewable sources more, aims to reduce energy consumption via some digital technologies some of which are capable of providing feedback on the consumption (Zygiaris, 2013). The feedback provided to stakeholders informs them on how to reduce their energy consumption. 'The Climate Street' and 'West Orange' are the major projects that aim at encouraging people and organizations to save energy (Angelidou, 2016). The city's 'GridFriends' project aims to implement a smart grid for clean and shared distribution of energy (Amsterdam Smart City, 2019b).

Masdar City aims to offer highest quality of life while leaving the lowest environmental footprint. Renewable energy production and sustainable resource use are the keys. The city is already using solar and wind power. The buildings are designed to consume less energy and water and they are constructed with more environmentally friendly materials. Each building must meet some certain high criteria for sustainability (Masdar, 2019a). Masdar, the company, has also signed a joint agreement for developing the UAE's first waste-to-energy power plant (Masdar, 2019b). This could be expected to increase the sustainability in the city as well as contributing to the sustainability in the region. Personal Rapid Transit system, introduced in 2010, includes automated electric vehicles that run along specific corridors. As also stated by Masdar (2019c), this system combines the privacy advantages of a taxi service with the environmental performance of public transportation systems. Masdar City supports innovative technologies and projects such as transforming the vaporized water to fresh water with a solar-powered device (Air-to-Water), transforming the plastic waste material into fuel (Plastic-to-Fuel), a solar-powered hydrogen fueling station project (Solar Powered Hydrogen Fueling Station), some sustainable farming projects specifically for the region, and etc. (Masdar, 2019d). In Masdar City, there are also recreational facilities for both residents and visitors (Masdar, 2019e).

London has a cycling action plan and aims to improve cycling infrastructure which is expected to help to eliminate some problems such as poor air quality and congestion (Reid, 2018). The city aims for 80% of residents' transportation to be made by bicycles, public transportation, or on foot by 2041 and the entire transportation system of the city is aimed to be zero-emission by 2050; and there are already some concrete changes such as the use of hybrid buses (London, 2019a). In order to increase the air quality in the city, they introduced the Ultra Low Emission Zone (London, 2019b). There are also scrappage schemes for small businesses and low-income residents to encourage them to switch to cleaner vehicles (London, 2019c). They also aim to reduce waste and increase recycling with their policies based on London Environment Strategy (London, 2019d).

All these efforts address some common issues such as environmental protection, traffic problems, sustainability, public health, and a better community. Use of smart technologies could help to increase the effectiveness of these sorts of efforts and once achieved, the developments are expected to gain momentum since the data gathered and exchanged as a result of the use of these technologies are likely to lead to better decision making. Simple or complex, all current efforts on becoming smart destinations are important since they can be regarded as the preliminary steps of smart transformation.

Smart destination is frequently used to characterize a destination that fosters learning, technological advancement, innovation, and creativity. As Nam and Pardo (2011) state, creativity, knowledge, and learning are the key factors for a smart destination. Technological advancements could enrich tourist experience in the destinations and increase destination competitiveness. The ICTs which play a major role in creating knowledge and enhancing economic development are at the core of the success of smart cities/destinations. A destination which adopts a smart city concept also adopts a strategical mission which encourages innovation, science, and creativity in order to create a developing knowledge community.

Issues and Challenges

Although there is great potential for smart destinations and they could bring many advantages, there are also some challenges. First, it requires quite a lot of investment, i.e. financial resources, to build and improve technological infrastructure. Moreover, there is also need for specialists in smart (and related) technologies. Another challenge is that changing the infrastructure and integrating it with the rest of the world or the country may take a very long time and necessitate other investments. It should also be kept in mind that residents and visitors should adapt to these technologies socially. This social change might bring many other challenges such as an enormous generation gap, alienated people through a fast-changing society, and etc. Ethical issues are also among the factors to be considered. Although the destinations are trying to address ethical considerations such as privacy, the debate might continue for a long time mainly due to data collection possibilities from everywhere, everything, and everybody. In addition, while data collection possibilities are quite advanced in an ideal smart destination, the accuracy of these data might not be perfect. In other words, reliability of the collected data is open to discussion unless the aforementioned technologies reach to a certain level of maturity. Another discussion could be raised about data protection. Once the data are collected, securing the available data will be the next challenge. Although similar issues are relevant today, it could be argued that they might get even more controversial in a smarter future.

In a tourism destination, or in a city, there are various stakeholders. These stakeholders have different interests. In a smart destination concept, there should be a participative management style. However, even if this ideal way of management is adopted, stakeholders' interests might remain different from each other and more importantly, sometimes clash. Adopting a participative decision-making procedure and achieving to address every stakeholders' needs and wishes might not always be possible at the same time. Thus, some stakeholders might gain more power than others and if these stakeholders are not those previously neglected in the conventional way of decision-making, a true democratic way of management might still be missed by smart destinations. In other words, power imbalance issues are among the factors to be considered while changing the way of management for the sake of smarter destinations.

Some professions, thus jobs, are likely to be eliminated due to the advancements in technology. For example, mobile phone applications are already becoming alternatives to tour guides, travel agencies, and etc. These developments are already affecting the job market. On the other hand, some new jobs are created due to the same developments. Technology related jobs are likely to be in demand. Expected skills for tourism employees might change. For example, foreign language skills might not be that important anymore if translation applications reach to a certain level. In addition, tourism industry which is conventionally considered as labor-intensive might even lose this characteristic.

Lastly, becoming a smart destination is more challenging for those regions that do not possess a good technological infrastructure or financial resources to make the necessary investment for this unless there is a kind of global incentive system to support them.

SOLUTIONS AND RECOMMENDATIONS

More advanced systems should be developed to collect data accurately and secure the collected data since the data are the key for smartness. It is also important to train specialists in various technological areas. Smart destinations are not all about technical experts but all individuals and community. For this reason, continuous learning should be encouraged and the formal education should be revised if not revolutionized. Continuous learning is also helpful for keeping people employable in a changing job market.

In order to minimize generation gap, senior citizens should be focused. Some special projects might be helpful to overcome this issue. When it comes to the destinations with less technological infrastructure and financial resources, they should prepare long term strategic plans if they are to become smart destinations. On the other hand, failure to become a smart destination might not be a total disadvantage. If many destinations become smart, then those conventional destinations might be perceived as authentic.

It is difficult to meet the demands of or reconcile stakeholders with clashing interests. However, participative management, if conducted appropriately for a reasonable time, might lead to a different culture. Although clashing interests might continue to coexist, solving the issues and finding common solutions might become easier when a more democratic culture is achieved.

FUTURE RESEARCH DIRECTIONS

Since many changes are expected in smart destinations, future researchers might primarily focus on an ideal education system in smart cities. A proper education might help to avoid many potential challenges such as the aforementioned social changes and employment challenges. Moreover, since changes are also likely to create resistance, there might be people who tend to reject the smart concept and its reflections on their lifestyle. This might be another interesting domain of research for those from various disciplines.

Becoming a smart destination, or improving a current one, requires quite a lot of investment. This, in turn, might increase the prices of some products (or taxes or necessary payments in any other form) due to an expected return on these investments. Thus, researchers might analyze the price/cost trends in destinations which aim to become or are already smart.

The form of services provided are likely to change as a result of the use of smart technologies. These changes might have various impacts both on employees and consumers in the service industries including tourism. These impacts and related issues could be other topics for the future research projects or conceptual debates.

Lastly, since newer technologies are to be used as the destinations become smarter, there is the risk of creating some other problems similar to those eliminated. For example, increasing use of wireless devices, increasing production of many other products such as sensors, solar panels, and etc. using natural resources might result in some environmental or public health issues. It is important to consider these issues and get smarter in the use and creation of newer technologies in advance. Future researchers should therefore focus on the long-term impacts of smart technologies on environment and public health and smart technology projects that have the minimum level of potential impacts.

CONCLUSION

Population growth, rapid consumption of limited natural resources, and CO_2 emission and greenhouse impact issues are all among the contemporary discussions due to their potential and actual major impacts that could even risk the survival of human beings and other creatures on Earth. In order to create a better environment and ensure sustainability, better solutions are being sought. Smart technologies enable cities and destinations to address these concerns to some extent and the concepts of 'smart city' and 'smart destination' have already been put into practice. Smart destinations and cities use ICTs in order to preserve the resources, enhance the quality of life for both residents and visitors, and ensure sustainability. Increasing use of smart technologies brings some changes to tourism industry. These technologies enable businesses and other institutions to better understand changing expectations while also providing them with more advanced managerial and operational tools and all these have some impacts on services provided and users' satisfaction. In addition, effective and ideal use of smart technologies in destinations requires cooperation and coordination among stakeholders which might ultimately change the nature of competition.

Despite many advantages gained through applying the smart concept to a destination, there are also some challenges such as the requirement of technological infrastructure, adequate human resource, investments, long-term planning, and participative management. Moreover, the possible changes are likely to have reflections on society some of which may result in social challenges such as generation gap, rapid change of culture, and etc. Ethical and technical issues on data collection and protection are also other issues to be considered. Although use of smart technologies also aims to eliminate some environmental issues and preserve the nature, they controversially have the potential to create some similar risks to some extent. Increasing production and use of some technological tools or systems such as sensors, solar panels, wireless communication systems, and etc. may put pressure again on natural resources, environment, and public health.

One of the human resource considerations is that labor intensive industries such as tourism might face some major changes. Employment patterns are likely to change as a result of smart applications that eliminate some jobs while creating some others. In addition to employment challenges, positive and negative social changes could also be expected.

Lastly, it is useful to note that social changes that are likely to occur as a result of the development of smart cities and destinations are to be monitored for a better future community. As the emergence of the smart destinations brought more opportunities to increase the quality of life, this notion should be preserved also for the future of smart destinations.

REFERENCES

Amsterdam Smart City. (2019a). *Climate street (Klimaatstraat)*. Retrieved from https://amsterdamsmartcity.com/projects/climate-street

Amsterdam Smart City. (2019b). *GridFriends*. Retrieved from https://amsterdamsmartcity.com/projects/grid-friends

Angelidou, M. (2016). Four European smart city strategies. *International Journal of Social Science Studies*, *4*(4), 18–30. doi:10.11114/ijsss.v4i4.1364

Ashton, K. (2009). That 'Internet of Things' thing. *RFID Journal*. Retrieved from http://www.itrco.jp/libraries/RFIDjournal-That%20Internet%20of%20Things%20Thing.pdf

Bakıcı, T., Almirall, E., & Wareham, J. (2013). A smart city initiative: The case of Barcelona. *Journal of the Knowledge Economy*, *4*(2), 135–148. doi:10.100713132-012-0084-9

Barcelona City Website. (2019a). *Barcelona Digital City, Bicing*. Retrieved from https://ajuntament.barcelona.cat/digital/en/digital-transformation/urban-technology/bicing

Barcelona City Website. (2019b). *Info Barcelona*. Retrieved from https://www.barcelona.cat/infobarcelona/en/tema/mobility-and-transport/approval-to-move-bike-lanes-from-pavements-onto-the-road-surface-in-av-diagonal_784832.html

Barcelona City Website. (2019c). *Smart vehicles to reduce accidents*. Retrieved from https://ajuntament.barcelona.cat/en/noticia/smart-vehicles-to-reduce-accidents_735098 Accessed on July 4, 2019.

Boes, K., Buhalis, D., & Inversini, A. (2015). Conceptualising smart tourism destination dimensions. In L. Tussyadiah, & A. Inversini (Eds), *Information and communication technologies in tourism 2015: Proceedings of the international conference in Lugano, Switzerland*, February 3-6, 2015 (pp. 391-403). Cham, Switzerland: Springer.

Buhalis, D. & Amaranggana, A. (2014). Smart tourism destinations. In Z. Xiang, & L. Tussyadiah (Eds.), *Information and communication technologies in tourism 2014: Proceedings of the international conference in Dublin, Ireland*, January 21-24, 2014 (pp. 553-564). Cham, Switzerland: Springer.

Buhalis, D. & Amaranggana, A. (2015). Smart tourism destinations enhancing tourism experience through personalisation of services. In L. Tussyadiah, & A. Inversini (Eds.), *Information and communication technologies in tourism 2015: Proceedings of the international conference in Lugano, Switzerland*, February 3-6, 2015 (pp. 377-389). Cham, Switzerland: Springer. 10.1007/978-3-319-14343-9_28

Caragliu, A., Del Bo, C., & Nijkamp, P. (2011). Smart cities in Europe. *Journal of Urban Technology*, *18*(2), 65–82. doi:10.1080/10630732.2011.601117

Cohen, B. (2012). *The smartest cities in the world 2015: Methodology*. Retrieved from https://www.fastcompany.com/3038818/the-smartest-cities-in-the-world-2015-methodology

Collins, B., Paquet, G., Roy, J., & Wilson, C. (2002). E-Governance and smart communities: A social learning challenge. In *Proceedings of SSHRC Knowledge Based Economy Conference* (pp. 1-23). Newfoundland, Canada. Academic Press.

Dameri, R. P. (2013). Searching for smart city definition: A comprehensive proposal. *International Journal of Computers and Technology*, *11*(5), 2544–2551. doi:10.24297/ijct.v11i5.1142

Falconer, G., & Mitchell, S. (2012). Smart city framework: A systematic process for enabling smart + connected communities, 1-11. Cisco Internet Business Solutions Group (IBSG). Retrieved from https://smartcitiescouncil.com/sites/default/files/public_resources/Smart%20city%20framework_a%20systematic%20process.pdf Accessed on July 10, 2019.

Giffinger, R., Fertner, C., Kramar, H., Kalasek, R., Pichler-Milanovic, N., & Meijers, E. (2007). *Smart cities - Ranking of European medium-sized cities.* Final Report, Vienna University of Technology. Retrieved from http://curis.ku.dk/ws/files/37640170/smart_cities_final_report.pdf

Gretzel, U., Sigala, M., Xiang, Z., & Koo, C. (2015a). Smart tourism: Foundations and developments. *Electronic Markets, 25*(3), 179–188. doi:10.100712525-015-0196-8

Gretzel, U., Werthner, H., Koo, C., & Lamsfus, C. (2015b). Conceptual foundations for understanding smart tourism ecosystems. *Computers in Human Behavior, 50*, 558–563. doi:10.1016/j.chb.2015.03.043

Hall, R. E., Bowerman, B., Braverman, J., Taylor, J., Todosow, H., & Von Wimmersperg, U. (2000, September). *The vision of a smart city.* Paper presented at the 2nd International Life Extension Technology Workshop, Paris, France.

Harrison, C., Eckman, B., Hamilton, R., Hartswick, P., Kalagnanam, J., Paraszczak, J., & Williams, P. (2010). Foundations for smarter cities. *IBM Journal of Research and Development, 54*(4), 1–16. doi:10.1147/JRD.2010.2048257

Huang, X., Yuan, J., & Shi, M. (2012). Condition and key issues analysis on the smarter tourism construction in China. In F. L. Wang, J. Lei, R. W. H. Lau, & J. Zhang (Eds.), *Multimedia and signal processing communications in computer and information science: Second international conference proceedings CMSP 2012* (Vol. 346, pp. 444-450), *Shanghai, China, December 7-9, 2012*. Heidelberg, Germany: Springer. 10.1007/978-3-642-35286-7_56

Jovicic, D. Z. (2017). From the traditional understanding of tourism destination to the smart tourism destination. *Current Issues in Tourism, 22*(3), 276–282. doi:10.1080/13683500.2017.1313203

Khomsi, M. R., & Bédard, F. B. (2017). From smart city to smart destination: The case of three Canadian cities. *ARA: Revista de Investigaciónen Turismo, 6*(2), 69–74.

Kim, H. M., & Han, S. S. (2012). City profile Seoul. *Cities (London, England), 29*(2), 142–154. doi:10.1016/j.cities.2011.02.003

Komninos, N., Pallot, M., & Schaffers, H. (2013). Special issue on smart cities and the future internet in Europe. *Journal of the Knowledge Economy, 4*(2), 119–134. doi:10.100713132-012-0083-x

Lamsfus, C., & Alzua-Sorzabal, A. (2013). Theoretical framework for a tourism internet of things: Smart destinations. *Journal of Tourism and Human Mobility, 2*, 15–21.

Lehr, T. (2018). Smart cities: Vision on-the-Ground. In S. McClellan, J. A. Jimenez, & G. Koutitas (Eds.), *Smart cities: Applications, technologies, standards, and driving factors* (pp. 3–17). Cham, Switzerland: Springer. doi:10.1007/978-3-319-59381-4_1

London (2019a). *Green transport.* Retrieved from https://www.london.gov.uk/what-we-do/transport/green-transport

London (2019b). *The Mayor's ultra-low emission Zone for London.* Retrieved from https://www.london.gov.uk/what-we-do/environment/pollution-and-air-quality/mayors-ultra-low-emission-zone-london

London (2019c). *Our air and our vehicles.* Retrieved from https://www.london.gov.uk/sites/default/files/shorthand/cleaner_vehicles/

London (2019d). *Waste policy.* Retrieved from https://www.london.gov.uk/what-we-do/environment/waste-and-recycling/waste-policy

Masdar (2019a). *Sustainability.* Retrieved from https://masdar.ae/en/masdar-city/the-city/sustainability

Masdar (2019b). *Waste-to-Energy.* Retrieved from https://masdar.ae/en/masdar-clean-energy/technologies/waste-to-energy

Masdar (2019c). *Mobility.* Retrieved from https://masdar.ae/en/masdar-city/the-city/mobility

Masdar (2019d). *Research & Development.* Retrieved from https://masdar.ae/en/masdar-city/the-city/randd

Masdar (2019e). *Recreation.* Retrieved from https://masdar.ae/en/masdar-city/the-city/recreation

Meijer, A., & Bolívar, M. P. R. (2015). Governing the smart city: A review of the literature on smart urban governance. *International Review of Administrative Sciences, 82*(2), 392–408. doi:10.1177/0020852314564308

Nam, T. & Pardo, T. A. (2011). Conceptualizing smart city with dimensions of technology, people, and institutions. In *Proceedings of 12th Annual International Digital Government Research Conference: Digital Government Innovation in Challenging Times* (pp. 282–291). USA: ACM. 10.1145/2037556.2037602

Neirotti, P., De Marco, A., Cagliano, A. C., Mangano, G., & Scorrano, F. (2014). Current trends in smart city initiatives: Some stylised facts. *Cities (London, England), 38*, 25–36. doi:10.1016/j.cities.2013.12.010

Rahyaputra, V., Muna, N. K., & Rizal, N. S. (2016). *Barcelona's smart city: The Frontrunner in digital transformation.* CfDS Case Study Series No. 4. Retrieved from http://cfds.fisipol.ugm.ac.id/uploads/files/posts/72/CFDS_CASESTUDIES_BARCELONA.pdf

Reid, C. (2018). *London launches 5-year plan to entice more people to fall in love with cycling.* Retrieved from https://www.forbes.com/sites/carltonreid/2018/12/17/london-launches-5-year-plan-to-make-more-people-fall-in-love-with-cycling/#6fb296e624fd

Riva Sanseverino, E., Riva Sanseverino, R., Vaccaro, V., Macaione, I., & Anello, E. (2017). Smart cities: Case studies. In E. Riva Sanseverino, R. Riva Sanseverino, & V. Vaccaro (Eds.), *Smart cities atlas: Western and eastern intelligent communities* (pp. 47–140). Cham, Switzerland: Springer. doi:10.1007/978-3-319-47361-1_3

Schaffers, H., Komninos, N., Pallot, M., Trousse, B., Nilsson, M., & Oliveira, A. (2011). Smart cities and the future internet: Towards cooperation frameworks for open innovation. In *The Future internet assembly* (pp. 431–446). Heidelberg, Germany: Springer; doi:10.1007/978-3-642-20898-0_31

Söderström, O., Paasche, T., & Klauser, F. (2014). Smart cities as corporate storytelling. *City, 18*(3), 307–320. doi:10.1080/13604813.2014.906716

Stockholms Stad. (2018). *Stockholm City Plan.* Retrieved from https://vaxer.stockholm/globalassets/tema/oversiktplan-ny_light/english_stockholm_city_plan.pdf

Stockholms Stad. (2019). *Stockholm, a sustainably growing city*. Retrieved from https://international.stockholm.se/globalassets/ovriga-bilder-och-filer/stockholm-a-sustainably-growing-city.pdf

Su, K., Li, J., & Fu, H. (2011). Smart city and the applications. In *Proceedings of International Conference on Electronics, Communications and Control (ICECC)* (pp. 1028-1031). Ningbo, China: IEE.

United Nations Population Fund. (2007). *State of world population 2007: Unleashing the potential of urban growth*. Retrieved from www.unfpa.org/swp/

Wang, D., Li, X., & Li, Y. (2013). China's "smart tourism destination" initiative: A taste of the service-dominant logic. *Journal of Destination Marketing*, *2*(2), 59–61. doi:10.1016/j.jdmm.2013.05.004

Washburn, D., Sindhu, U., Balaouras, S., Dines, R. A., Hayes, N., & Nelson, L. E. (2010). Helping CIOs understand "smart city" initiatives. Cambridge, MA: Forrester Research Inc. Report.

Zhu, W., Zhang, L., & Li, N. (2014). Challenges, function changing of government and enterprises in Chinese smart tourism. *Information and Communication Technologies in Tourism, 10.*

Zygiaris, S. (2013). Smart city reference model: Assisting planners to conceptualize the building of smart city innovation ecosystems. *Journal of the Knowledge Economy*, *4*(2), 217–231. doi:10.100713132-012-0089-4

ADDITIONAL READING

Albino, V., Berardi, U., & Dangelico, R. M. (2015). Smart cities: Definitions, dimensions, performance, and initiatives. *Journal of Urban Technology*, *22*(1), 3–21. doi:10.1080/10630732.2014.942092

Anthopoulos, L. G. (2015). Understanding the smart city domain: A literature review. In M. P. Rodríguez-Bolívar (Ed.), *Transforming city governments for successful smart cities* (pp. 9–21). Cham, Switzerland: Springer. doi:10.1007/978-3-319-03167-5_2

Cocchia, A. (2014). Smart and digital city: A systematic literature review. In R. P. Dameri & C. Rosenthal-Sabroux (Eds.), *Smart city: How to create public and economic value with high technology in urban space* (pp. 13–43). Cham, Switzerland: Springer.

Dameri, R. P., & Rosenthal-Sabroux, C. (Eds.). (2014). *Smart city: How to create public and economic value with high technology in urban space*. Cham, Switzerland: Springer.

Deakin, M. (Ed.). (2014). *Smart cities: Governing, modelling and analysing the transition*. Abingdon, England: Routledge.

Dustdar, S., Nastić, S., & Šćekić, O. (2017). *Smart Cities: The internet of things, people and systems*. Cham, Switzerland: Springer. doi:10.1007/978-3-319-60030-7

Gretzel, U. (2018). From smart destinations to smart tourism regions. *Journal of Regional Research*, *42*, 171–184.

Mahmood, Z. (Ed.). (2018). *Smart Cities: Development and governance frameworks*. Cham, Switzerland: Springer. doi:10.1007/978-3-319-76669-0

Maheswaran M. & Badidi E. (Ed.). (2018). Handbook of Smart Cities: Software services and cyber infrastructure. Cham, Switzerland: Springer.

KEY TERMS AND DEFINITIONS

Destination: Destination, or tourism destination, refers to an area or region which possesses tourism attractions and appropriate services.

ICTs: Standing for information and communication technologies, it refers to the technologies that facilitate obtaining, storing, processing, and exchanging data and information.

Knowledge Society: It refers to a society which focuses on the creation and examination of knowledge. The major factor of production in a knowledge society is knowledge.

Smart City: The term smart city refers to a city/town/area that adopts an interactive/participative management style and aims to enhance the residents' quality of life and the quality of services provided in the city/town/area thanks mainly to advanced ICTs. Other technologies are also important since they can serve the same purposes. A cleaner environment, minimum levels of CO_2 emissions and ideally zero-emission, effective and efficient use of resources, and ultimately sustainability are at the focus of a smart city's efforts.

Smart Destination: There is not a clear distinction between a smart city and a smart destination. It can be defined as a destination that adopts an interactive/participative management style and aims to enhance the residents' and tourists' quality of life by using ICTs and other technologies for data collection, storage, exchange, and processing. Smart destinations focus on the effective and efficient use of resources, a cleaner environment, and ultimately sustainability.

Smartness: Smartness refers to smart use of resources, smart and effective management, and a network that enables participative and interactive management. Both the approach to governance and the use of technologies play key roles in smartness.

Smart Tourism: A new way of practicing tourism that enables tourists to access the services and information regarding their tour more conveniently thanks to some advanced technologies and interactive/participative management. Smart tourism does not only deal with tourists but also with residents. Tourists' and residents' wishes and needs can be understood more accurately in a smart setting due to advanced technology and interactive/participative management.

Chapter 20
Tourist Experience in Smart City Destination of Jaipur (Rajasthan)

Lakhvinder Singh
Government College, Kaithal, India

Kamal Kumar
Pt. Chiranji Lal Sharma Government P. G. College, Karnal, India

ABSTRACT

Smart city concept is emerging as a new venture in the tourism sector especially in a country like India. This chapter analyzes the tourist experience in Jaipur city of Rajasthan which is emerging as a popular tourist destination taken over for development under smart city mission. The chapter adopted empirical approach, but to some extent it also used secondary data sources. The authors analyzed tourist experience by using smart attraction, smart mobility, smart amenities, smart ancillary, smart safety, and smart available packages as important dimensions of smart tourism destination. Regarding tourist experience, the chapter noticed smart available packages, smart safety, and smart attractions are providing a pleasant experience to tourists in contrast with remaining aspects of smart tourism destination. The result of the study provides implication for tourism marketer and policy maker to take appropriate steps to enhance the tourist experience by improving smart facilities & services.

INTRODUCTION

The increasing strength of urban population has elicited the growth of tourism throughout the world. Popular cities of the world are large multifunctional entities into which tourists can be effortlessly absorbed and thus largely becoming the major tourist hubs (Falconer & Mitchell, 2012). Hence, due to increased urbanization and demand for travel to cities has been resulted in to continuous enhancement of facilities and services for transformation of urban areas in to smart cities. Therefore, it is necessary to be smart in order to survive in the tourism industry as to facilitate the products, actions, processes and services in real-time, by engaging different stakeholders simultaneously to optimize the collective performance and competitiveness and generate solutions and value for all.

DOI: 10.4018/978-1-7998-1989-9.ch020

Although, this is very emerging concept but the coverage and consideration received by this is not sufficient, so, it is of paramount importance to go through the meaning of smart city. The meaning and definition of this emerging phenomenon is perceived differently by scholars. As far as, Indian scenario is concerned it is described as a place equipped with ample infrastructural and super structural facilities catering to the needs of citizens and visitors. The government of India also launched Smart City Campaign for promoting towns of the country in order fulfill the infrastructural requirement, improved living conditions, hygienic atmosphere and implementation of latest technology. The Government of India (Ministry of Urban Development, Government of India, 2015) through its campaign intends to develop model cities in the country and identified major facilities and components of increasing smartness in cities are:

1. Supplying sufficient water
2. Ensured Power availability
3. Cleanliness, disposal of litter
4. Improved connectivity and government transportation
5. Economic house availability
6. Strong information technology
7. Better government administration
8. Ecological sustainability
9. Safe & Secure living environment
10. Better healthcare and educational services.

This phenomenon is widely visible during the commencement of 20th century due to rapid urbanization& to find a durable solution to this, many experiments was carried out (Kumar, 2016) especially the smart city phenomena. With the rapid development of the Indian economy, the requirement of urban culture facilities has raised a demand of sustaining the growing population. The concept of smart city is a new phenomenon in developing countries like India. India is moving towards becoming the worlds' most emerging nation with growing urbanization from 37 million in 2015 to 590 million in 2030. By considering this scenario, the Indian Government put emphasis on urbanized areas of the country for development. As a leading step in this line, the smart city concept was introduced and launched in 2015 in India. The mission aims to overall development of urban cities through smart solutions such as growing economy, infrastructural enhancement, technological enhancement and sustainable development (Government of India, 2015).Rajasthan, one of the most coveted travel destinations in India and worldwide, is the land of glistening sandy deserts, multi-colored fabrics, dulcet folk songs and sprightly dances, ancient traditions and rich handicrafts. It is a state with less than 25% of its population living in urban areas. However, the rate of urbanization in the last decades has increased considerably and has witnessed a 29% growth. The identified smart cities in the states are very well connected with all the major transportations means, equipped with sufficient boarding and lodging facilities. As far as the attraction at selected smart cities is concerned, the capital city of Jaipur is a place you can plan for a weekend leisure trip from any point of the country. Jaipur featured with monuments, palaces, variety of foods, shopping avenues and so on. The smart city of Jaipur has huge potential and prospects for tourism and continuously moving towards receiving increasing tourists' footfalls both domestic tourists visits (DTV) as foreign tourist arrivals (FTA) as presented in the table 1 indicated that tourists in selected destinations are growing year by year which

Table 1. Tourist traffic in smart cities of Rajasthan

Year	Domestic Tourist Visits	Foreign Tourist Visits
2014	1170152	568234
2015	1201152	596756
2016	1544730	565978
2017	1702665	633990

Source: Department of Tourism Rajasthan (India)

indicted positive growth of tourism in the area. The domestic tourism registered an enhanced increase over the previous years in contrast to foreign tourist arrival in Jaipur.

Taking a note from this growing phenomenon i.e. smartness of cities, the focal point for managing tourist places leads to implementation of information & communication technology in the travel sector. This also directed towards increased use of digitalization for travel services among tourists as well as service providers (Buhalis & Law, 2008). This will be used to ensure the smartness of tourist destinations.

Some researchers such as Ivars, Solsona and Giner (2016) elaborate a replica of smart travel places with better facilities & services availability for present and potential visitors, betterment of locals through development of tourism and overall contribution towards fulfillment of sustainable development agenda. The smart city innovation at tourist sites proved as boon to increase pleasing capacity of the destinations towards tourist as well as also support stakeholders to apply concrete tourism planning (Neuhofer, Buhalis, & Ladkin, 2014). Guo, Liu, and Chai (2014) also ascertained smart destination as important component of smart city campaign. Hence, there is need to consider its importance and explore further in this area.

Various researchers have been working on how to inculcate smartness to tourism industry by creating a proper framework. As cities become increasingly competitive and complex, Information and Communications Technology (ICT) will coordinate all activities and services, leading to connected, better informed and engaged citizens. This make cities more accessible and enjoyable for both residents and visitors through interactive service interconnecting all local organizations to provide real-time services and use data centrally for better coordination.

The tourism at smart city destination has popular strategy to provide authentic and unique experience to tourists and over the few years concept of smart city destinations and tourism has become popular throughout the world. The parameters to compare the smartness of the cities were based on certain criteria such as safety, ecology, urban products, mobility, technology, connectivity, culture-heritage, infrastructure, availability of information technology and so on. All these components contributed towards the development of smart tourism at smart city destinations (Buhalis & Amaranggana, 2014).Thus, the development of Smart City could also encourage the formation of Smart Tourism Destinations. With technology being embedded within the destinations environment, it can enrich tourist experiences and enhance destinations competitiveness.

Although several empirical and conceptual coverage of smart city destinations undertaken by many authors, but very less number of studies area found focusing on experience of tourists at such destinations. So, in this connation, the present chapter intends to highlight the tourist experience at smart city destination of Jaipur.

The chapter analyzes the tourist experience at smart city destination of Jaipur in Rajasthan of India. The chapter mostly used primary data collected from the tourists in order to observe their experience on the basis of five point Likert scale (strongly agree-5 to strongly disagree-1) towards smart city destination characteristics and service components. The survey instrument used in the study has been framed in two different parts, first part covered questions related to socio-demographic profile of the tourists, second part contained statements related to experience gained by tourists' different features/characteristics of smart city destination in Jaipur. The study, in order to maintain statistical equality randomly takes the opinion of four hundred tourists in Jaipur. The collected data has been analyzed through suitable statistical techniques.

BACKGROUND

The developing countries put their efforts to drive the smart city tourism destinations in their frontline and concentrated their planning on development of tourism at smart city destinations (Hwang, Park, & Hunter, 2015).

Regarding smart destinations the most crucial emphasize is laid to sustainability and the wellbeing of locals, but from the consumer's perspective, a destination is said to be smart when it makes intensive use of the technological infrastructure in order to enhance the tourism experience of visitors by personalizing and making them aware of both local and tourism services and products available to them at the destination (Xian, Wang, O'Leary & Fesenmaier, 2015).

The smart city concept in tourism sectors aims at achieving success in tourism business by enhancing the experience of the tourists at the destination by providing them comfortable and convenient facilities and services (Boes, Buhalis & Inversini, 2015). Hence, tourism is considered as the industry which largely contributed towards the attainment of benefits suggested under smart city mission and smart city destinations contributed to enable demand and supply match for the tourist at the destination.

In relation to tourist experiences, Lee, Lee, Chung, and Koo (2017) have recently created a model to show the effect of smart technology in the tourist's perceived experience and in his/her happiness. Therefore, in order to ensure the success of a tourism destination its importance to ensure that human resources and innovation are collaborating at all the levels (Ritchie & Crouch, 2005). The smart city is required to maintain transportation, IT connectivity, water and waste management. New smart cities with multiple provisions can overcome the urban decay and a more balanced distribution could be attained. With improved facilities, the tourism can boost up and the tourist can experience the unknown places. The investment in tourism sector by the private agencies and government authorities can help to generate sensible revenue for the government. Smart cities not only serve the citizens who dwell in the city, but also accommodate those who visit.

Buhalis (2000) in this phenomena further elaborated the importance of facilitation of attractions, accessibility, amenities, available packages, activities and ancillary services (6 A's) at tourist destinations to add value to the touristic experience. In other words, Cohen (2012) has defined few indicators for each of the smartness dimensions, namely-smart governance that relates with aspect of transparency within governance systems through modernization of city administration by supporting data openness and public involvement; smart environment which is related to energy optimization that leads to sustainable management of available resources; smart mobility which referred to accessibility within the city as well as outside the city and availability of modern transportation systems; smart economy which is related to implementation of economic strategies based around digital technology; smart people which linked to the qualification level of city's human capital; and smart living which involve the quality of life which measured in terms of healthy environment, social cohesion, tourist attraction and availability of cultural and educational services. The ultimate aim is to utilize the system to enhance tourism experience and improve the effectiveness of resource management towards maximizing both destination competitiveness and consumer satisfaction while also demonstrate sustainability over an extended timeframe.

This concept of smart tourism destination should also perform smartness by implementing appropriate tourism applications within Smart Cities' components as defined by Cohen (2012). A range of smart services can be seen on below table which shows how Buhalis' (2000) 6A's Destination Components (representing destinations element) and Smart Tourism Destination Dimensions as derivate from Cohen's Smart City Dimensions(representing smartness element) could be combined and possibly generate tourism applications with each of its utility function to be implemented in smart tourism destination. Broadly foreseeable changes caused by smartness immersion in tourism destinations build characteristic of smart tourism destinations, though it triggers different outcome for each stakeholder.

Tourism authorities should ensure that any information generate from every development of new application should be made openly available subject to their commercial and legal agreement without unreasonable additional cost (Reischl 2013). There are two main information sources: information coming from the city resulting from sensors, city elements and Open Data; and (2) information coming from the citizens and visitors as digital footprint from their social media activities. Users could use this information to identify problems as well as customized potential solutions to overcome those problems.

Destinations' smartness contributed to increasing satisfaction level of visitors, smartly collection and disbursal of required information at destination, facilitation in appropriately distributing of local facilities and also provide integrated platform to service provider (Rong 2012). Through this technology, tourist can seek more information about the visited areas. The tourism planners also facilitated to understand the diverse demands of the tourists for formulation of suitable tourism packages (Haubensak, 2011).

The real sense of smart tourist destination is to focus on tourists' needs by combining the ICT with casual culture and tourist innovation industry in order to promote tourism service quality, improve tourism management and enlarge industry scale to a broader extent (Huang, Yuan, & Shi, 2012). For this reason, technology plays an increasingly important role in promoting tourist destinations, distributing and marketing tourism, and supporting tourists before and during their stay in the destination. The important thing is to provide the tourist with an unforgettable visit, but for this to happen it is essential to innovate the destination, to make it attractive and capable of generating emotions. The tourist does not choose a destination only because of the monuments, the beautiful landscapes, the culture or the gastronomy that it has. All this is important, but today's tourist wants to be surprised by the unknown of the destination and wants information about what to do, what to visit and how to get to a place by using technologies.

Based on this approach and the given importance of smart city destination, it is considered interesting and relevant to see if smart city destinations able to provide unique and satisfactory experience to tourists during their visit and stay at the destination. On looking its different side, the author detected that despite the importance of smart technology at smart city destination, very less or no study address the concept of tourists experience, particularly in study area so specific study has been conducted so far. Hence, there exists a gap in this field, the author in order to fill this gap, decided to conduct a meaningful study to fill this gap.

MAIN FOCUS OF THE CHAPTER

The present chapter confined to capital of Rajasthan, also known as pink city is known for its numerous attractions and one of the famous tourist destinations in the state. Jaipur gets its name from its founder Maharaja JaiSingh II (1693-1744) the great warrior and astronomer. The city is surrounded by the Nahargarh hills in the north and Jhalana in the east, which is a part of Aravalli hills ranges. To the south and the west of the city are also prevailing hillocks but they are isolated and discontinuous in formation. The southern end of the city is open to plain and stretches far and wide towards Sanganer and beyond. The climate of Jaipur is semi-Arid and characterized by high temperature, low rainfall and mild winter. Thus the January and June are the coldest and hottest months. Normal amount of rain can be observed during summer months. Jaipur featured with monuments, palaces, variety of foods, shopping avenues and so on. Some of the prominent tourist places in Jaipur are Hawa Mahal, Jantar Mantar (Observatory), Forts of Amber, Nahar Garh, Jai Garh, Sisodia Rani Garden, Albert Hall, City Palace, Birla Auditorium, Jawahar Circle, RajMandir Cinema, Chokhi Dhani Fun & Eating Joint, Revolving Restaurant, Rose Garden, GovindDev Ji Temple etc. First of all, this section presents the overview of different aspects of smart city destination of Jaipur followed by the analysis of tourists' socio-demographic features and their experiences at smart city of Jaipur.

Emergence of Jaipur as a Smart City

Jaipur city of Rajasthan has been identified under Smart City Mission. The Jaipur smart city aims at conserving the heritage of the city by maintaining and improving the existing infrastructure around heritage buildings. 'Heritage Walks' in the walled city have been made wherein following measures are being taken, underground cabling of overhead wires, necessary measures for improvement of facade of the buildings and shops, basic cleanliness and improvement in sanitation conditions. The smart city transformation of Jaipur is planned in two different phases covering area-based development projects (ABD) and pan city development projects (PCD). Through these phases of smart city development, it can be cherished by residents of Jaipur as well as can be seen and explored by travelers who specially visit Jaipur for its architectural beauty and rich culture. The Government of India (Ministry of Urban Development, Government of India, 2015) has taken several steps under this phase are explained below.

Smart Heritage and Tourism

Location based developmental aim in Jaipur is main agenda of smart city mission. The aims are at development of suitable life conditions to citizens with better ecological management. It also looks towards developing heritage and tourism in the area through adoption to smart technology. As well as the arena of connectivity and infrastructural facilities also considered in this area. Numerous initiatives have been taken under this scheme which includes:

- Customized reprocess of inheritance structures for improvement of tourist attraction to please the visitors with memorable experience.
- Aesthetic renovation of the town and increase its traditional image as pink city. This will helpful in rejuvenating the visitors during their stay in the area.

- The development of 24/7 markets for visitors under this theme is considered important and will definitely work as a game changer for overseas and domestic visitors in the city.
- Restoration of Raam Nivas Garden will contribute towards increase in tourist traffic and also force locals to recreate themselves by paying visit.
- Transformation of Gowind Jee temple, *City Palace*, *Janttar Manttar* through smart project results in increased attraction among present and potential visitors.
- Reformation and renaissance of historical wells (Baworis) will increase the nature based tourist locale in the area. This is also helpful to maintain ecology of area.
- Developing Tal-Katora Lake under this mission is a crucial step towards increasing attractions for visitors as well as improving the water management at the site.
- Reinstallation of art like painting, monuments, crafts, ivory, and pottery-initiated increase in tourists' visits to local market and contributed towards economic gains to the locals. This also helps to increase length of tourist stay at the destination.
- Karishna Circuit development under smart cities mission in Jaipur is identified as important step to promote tourism.

Smart Mobility

Overcrowding and traffic problem were usually seen in the Jaipur during tourism peak seasons. This also brings the issue of parking of vehicles at tourist sites. Improper walking paths and road condition were .also identified as major hurdles in operating an intelligent traffic management in the city. The implementation technology under smart city mission in Jaipur not only tackles such issues but also bring convenience to citizens of the area.

For promoting eco-friendly transport in the city, the concept of renting of bicycle is introduced in the city. Built-in transportation managing systems in city introduced for optimization of transportation services for both traveler and transport operators. The travelers can easily get latest updates regarding timetable and operators can easily track the seat booking status through digital devices as well as the can monitor the speed, location, efficiency of the vehicles. The use of technology also enables them to handle security through cameras and video recording.

The local authorities under smart mission also introduced general ticket management system to implement paper travelling tickets to use public transportation means. The similar technology is applied in the forts, palaces & museums in the city to issue smart gate passes. This initiative contributed towards attaining sustainable development in the city.

Managing traffic signals, signage display of important information regarding facilities, emergency services, amenities etc. through using smart technology in Jaipur ensure vehicle movement and enhanced experience among locals as well as visitors. Smart city mission also introduced the computer enabled vehicle parking management in Jaipur helps to crowd management at tourist sites.

Smart and Sustainable Civic Infrastructure

The development of intelligent infrastructural facilities transformed the ambience of the city in Jaipur. The assessment of qualitative content of the air with the help of technology helps to make the environment hygienic and clean. The sufficient electric supply, qualitativeness of drinking water availability also supported smart management of amenities and utilities in the area. Further modification in security measures through electronic cameras, monitoring of accidents, renovation of sanitation system, up gradation of technology usage in educational institutions, providing technology enabled clean & hygienic public utilities, wireless internet connection, will contributed towards assurance of quality living environment to civilians in the city. The local authorities initiated several projects in this arena.

The assessment and management of airworthiness in environment through this agenda is the important phase for achieving sustainability towards environment in the city for residents as well as visitors. The tour planner also predicts the polluted and fresh air in the city before recommending the area to the potential visitors. Moreover, the technology updates regarding temperature, climatic and weather conditions also helping environment management in the city.

Automated online water quality monitoring system is also implemented in Jaipur to its supplying system. Implementation of such efforts will provide clean drinking water to the residents as well as for present and potential visitors in the city.

Mobile application enabled street problems reporting is a unique feature of smart city project in Jaipur to highlight all the problems and miss happening. The introduction of intelligent system through digitalization helps to monitor the activities and any incidents. The government also has taken several steps to use this mission for managing safety & security of visitors at tourist sites. The safe environment at destination forced the visitors to stay longer in the city.

Smart city projects in Jaipur also contributed towards construction of modern and smart building. King Library of Jaipur is an ideal monument taken for renovation under smart city mission. The transformation of this library will ensure the number of more visitors to the area for educational and recreation purpose.

The display of directional signage and important information through digital display boards or panels in roads or local streets in the city is helpful for stranger and visitors to conveniently travel within the city. The application of smart technology also helps to control and manage the pollution level and improve the sustainability in the area.

Smart Waste Management

The administration and controlling of wastage and liter in the city is smartly managed by applying latest technology and intelligent device. This will helpful in achieving minimum wastage level through enhanced recycling of wastage, to monitor disposing litter through digital solutions. Accordingly, all this will enhance the clean and hygienic appeal of the city.

Technology enabled measures for reducing waste disposal was started under smart city project. It facilitates collection of wastage from doorstep of citizens and to enhance the recycling process at residential and tourist spots. Smartly managed intelligent applications and equipments were used and operated for hassle free examination of garbage management.

Sewage handling system set up in the city for increasing competency of the system to maximization of usage of water for reuse. This facilitates aiding the scarcity of soil dampen water resources. The introduction of these initiatives contributed towards fulfillment of Swacchh Bhaarat Project. The empanelment of recycling and garbage picking agencies is a main step in this regard to reduce the garbage from the city and availability of clean & hygienic residential areas and tourists spots.

Analysis of Tourist Experience in Smart City Destination Jaipur

This section focuses on empirical assessment of tourist experience towards different service elements and facilities in smart city of Jaipur. First of all, the chapter highlights some description of socio-demographic overview of tourists visited in the area. Thereafter, it highlights the tourist experience in the city.

In terms of demographic overview of tourists visited to Jaipur, the author observed several important facts and noticed that out of total tourists' respondents, 77 were male whereas 23 were female. It indicates that male respondents are more active in responding during field survey. In terms of age wise classification, 38 were in the age group of 21-40, followed by 27 percent in 41-60, 18 above 60 years and 17 up to the age of 20 years. It means that senior and younger tourist groups are less interested to visit area instead of middle age group tourists. According to marital status classification of the respondents which indicated that 65 percent were married and 35 unmarried respondents. According to place of residence, 63 were Indian or domestic visitors as compared to 37 international or foreign visitors. This shows that there is need to focus on marketing & promotion strategies to attract more foreign tourists in the area. Education wise distribution highlighted that 34 respondents were graduate followed by 26 with senior secondary, 27 with Post graduation or above, and 13 respondents were higher secondary education only. This depicted that the more than half of the respondents are in the highly educated group i.e. above graduation. On analyzing the occupation of the respondents, it came to notice that 31 servicepersons, 26 businesspersons, 24 involved in other occupations (such as retirees, farmers, freelancers etc.) and 19 were students. It shows that the every segment of the society visited the area. Income wise classification highlighted that 33 were in the group of 2-4 lacs followed by 28 percent up to 2 lacs, 22 percent in 4-6 lacs and remaining 17 percent in above 6 lacs group.

In terms of the travel behavior features of the respondents, 73 were repeat visitors in contrast to 27 first time visitors in Jaipur. The largest portion of respondents i.e. 42 used rail transportation to reach the destination as compared to 30 through self-driven vehicles of private taxi/bus services. The remaining 26 were travelled through air transportation. It shows that the study area has the easy accessibility through above means of transportation. According to the classification of accommodation used, 47 used to stay in hotels followed by 28 in guest houses, 17 in inns and remaining 8 used to stay at their friends or relatives' houses. This highlights that the selected destination different types of the accommodation to offer to tourists according to their taste & budget. In terms of length of the stay at the destination, 38 stay for more than four days at the destination followed by 29 who stay 2-4 days, 22 stays 1-2 days and remaining 11 were return from Jaipur on same day. For the sources of information used to search about the destination, 30 searches about the destination from word of mouth publicity, 25 used electronic media, 17 used printed media and 28 used to search through travel agents or tour operators. With regards to travel companion 38 percent visited with family members followed by 27 visited in friends group, 23 traveled in group tour of school/colleges or office tours and remaining 12 visited alone to Jaipur. On asking about the purpose of visit, 52 visited for leisure (holiday) purpose, 22 for seeking novel experience about culture & heritage, 17 visited for business purpose and 9 visited for the purpose of some education or research work. It support that more than half of the respondents visited Jaipur for purely tourism or holiday purposes which highlights the touristic nature of the place.

Table 2. Experience towards smart attraction

Variables of Smart Attraction Dimension	Mean
Historical monuments/forts/palaces/museums are offered through smart devices	4.04
Visitor Centre and modern building are personalized with geo-location, video and audio guides	2.74
Cinemas, live theatres, casinos, shopping malls applying ICTs in their operation and promotion	3.50
Attractive local tourism website	3.82
Digital tourist information boards at tourist sites	3.91
Online .information about hotels & restaurants	3.80
Online booking of attraction/entertainment tickets	3.90
Mobile enabled check in and checkout in hotel	3.92
Crowd management through ICT at tourist sites and major events	3.74

Tourist Experience towards Different Dimension of Smart City Destination

The tourists' experience towards different aspects of smart city destination identified on the basis of extensive literature review is illustrated through descriptive statistics.

The tourist experience with smart attraction in Jaipur is indicating positive responses towards most of the statements of this dimension. Tourists were found most satisfied with most of the initiative taken for improvement in attraction under smart city mission such as offering of cultural tourist sites through smart devices, mobile enabled check in and check out, digital information boards, online booking attraction/entertainment tickets, efficient management of local tourism website, management of crowd through ICT at sites and events with mean rating of all these statements above 3.50. The application of ICT through smart device in cinemas, live theaters, shopping malls, casinos is moderately experienced by tourists whereas tourists were found dissatisfied with poor ICT based management of visitor Centre and modern building through geo location, video and audio guides

Regarding experience of tourists with smart mobility in Jaipur, the author observed that display of digital signage and boards at sites display of information of road panels, local web pages and mobile application, considering visual impairments and video monitoring of road tunnels, metro and unsafe

Table 3. Mean perception towards smart mobility

Variables of Smart Mobility Dimension	Mean
Adequate public transport support for disabled and elderly tourists	2.86
Real-time traffic management system updated with optimal routes	3.00
Efficient management of intermodal transport	3.35
Efficient management of the parking area through mobile applications	3.43
Video monitoring in tunnels, metros and unsafe areas	3.88
Travel information is provided on road panels, local webpages and mobile applications, considering visual impairments	3.98
Use of digital signage and boards about places of interest	4.04
Efficient GPS & navigation system in public transportation means	3.51

Table 4. Mean Perception towards Smart Amenities

Variable of Smart Amenities Dimension	Mean
Digital display of public utilities	3.73
Provide geo-location of nearby 24/7 chemists, hospitals, police post and medication services	3.01
Smart lighting at tourists sites & market area	4.27
Smart environmental management systems (EMS) to the management of the natural amenities	3.65
Eco-regulation and sustainable management are suitably adopted in accommodation and tourist sites	3.01

areas were providing satisfactory experience to tourists. The adoption of GPS and navigation system in public transportation, parking management through mobile devices and intermodal transport management are moderately dealing with tourist experience. The real time traffic management with optimal route and availability of transport means for disabled and elder tourists were dissatisfactory experienced by tourists in Jaipur.

Tourist experience towards smart amenities in Jaipur highlighted that smart lighting at tourists sites at market area, digital display of public utilities and smart environment management system of natural amenities found satisfactory by tourists in comparison to geolocation based management of hospitals, chemists, police posts & medication services and adoption eco and sustainable management practices by hotels and tourist sites with moderate experience to tourists.

Tourists in Jaipur noticed that innovativeness and friendly behavior of local community, existence of smart banking and ATM services, safe & secure payment gateways for tourists, uninterrupted internet & Wi-Fi and mobile charging points availability providing satisfactory experience. The digital souvenir purchasing system observed as moderate and digital postal service support as well as application of digital customer relationship management system by hotels in Jaipur was observed dissatisfactory by tourists.

The available online information of police posts and their contact numbers, installation of CC TV cameras at important places, mobile application or portals for lost & found complaint and emergency mobile application for tourists were satisfactory experienced by tourists in contrast to digital security system in hotels for safety of luggage and belongings of tourists is moderately experienced by tourists.

Table 5. Mean perception towards smart ancillary

Variable of Smart Ancillary Dimension	Mean
Provide smart banking and ATM services	4.28
Use safe digital payment systems specific for tourists	4.27
Provide postal services for tourists through tourism websites or mobile applications	2.88
Innovative and friendly communities	4.36
Digital platform for purchasing souvenirs	3.49
Uninterrupted internet and Wi-Fi connectivity	4.10
Uninterrupted mobile network at destination	3.89
Availability of mobile charging stations	3.67
Hotels & restaurants in area applying digital customer relationship management systems(CRM) for efficiently manage their services	2.59

Table 6. Mean perception towards smart safety

Variables of Smart Safety Dimension	Mean
Online information about tourist police posts and their contact details	4.23
Apps & portals for lodging complaints of lost & found	3.82
Installation of CCTV cameras at tourist sites & important places	4.05
Digital security system in hotels for safety of luggage & baggage	3.03
Emergency Apps for tourists	3.78

The result of tourist experience with smart available packages highlights that the availability of mice tourism facilities with latest equipment & technical support, display of quick online link or access to third party by tourist authorities, efforts of local tourism enterprises for providing unique experience through co-creation and technical support, efficient management of transport services in packages and updated mobile application as well as websites were satisfactory available in Jaipur. However, local tourism marketer moderately applied destination management systems (DMS) in their operations, so it is providing moderate satisfaction experience to tourists in the area.

The overall assessment of tourist experience with smart city aspects in Jaipur provides an insight that the intervention of smart city mission in Jaipur proved as an added advantage to boost the level of services in order to extend unique experience to tourists. The local authorities successfully implement the smart city mission and considerably putting its efforts for improving the touristic image of the city among contemporary travelers as well as extend its enhanced level of local services and infrastructure to its citizens also. The smart city concept also helps to attain sustainable development through all around growth. The empirical analysis in the chapter although highlights pleasant tourist experience with smart destination aspects, but from a concluding overview, it noticed smart available packages, smart safety and smart attractions are sufficiently enabling tourist

Satisfaction followed by smart amenities, smart ancillary and smart mobility which area indicted a moderate level of experience to tourists and hence need some attention of stakeholders to attain a crucial role to enhance tourists' experience.

Table 7. Mean perception towards smart available package

Variable of Smart Available Package	Mean
MICE-tourism facilities (meetings, incentives, conferences, exhibitions) are available with latest equipment and technical support in hotels at destination	4.37
Local tourism authorities are providing quick online access to third-party sources such as activities, time-table, travel planning or event ticket reservations	3.93
Local tourism marketer applied destination management systems (DMS)to manage efficiently nature and adventure activities at the destination	3.51
Efficient management of transport services in the packages-saving energy, improving sustainability, avoid traffic jams, respecting parking areas	3.68
Implement up-to-date mobile applications and websites to offer available packages with on-line reservation	3.64
Tourism enterprises putting their sincere efforts for contributed delivering unique tourist experiences through co-creation with a high level of technological support	3.88

SOLUTIONS AND RECOMMENDATIONS

Digitization proves as key to improve communities and countries, maintain global competitiveness, increase GDP growth, foster innovation and create new jobs. Smart city concept is a powerful platform to realize this vision and today, the Jaipur is emerging as a role model for digital transformation under smart city mission. The present chapter highlights that Jaipur is a popular tourist destination and has significant cultural & heritage importance. Its appeal as a tourist destination could be developed by implementation of the following initiatives:

1. The local authorities need to focus on introducing safety and security solutions at major tourist's sites and busy areas in Jaipur city to monitor activity and movement in high traffic areas. This was a key factor towards improving the safety of specific profiles, such as female or solo travelers. It would also improve conditions for local residents and reduce crime rates in the city.
2. The solutions provided under smart city mission need to build into a single, easy-to-understand digital platform. It works by aggregating data from various sensors and data analytics to support a wide range of urban services.
3. Smart city mission in Jaipur need to put emphasis on smart Wi-Fi hotspots at selected tourist sites and major locations, so that tourists and residents alike could take advantage of this smart city feature and have access to the internet. Improved and interactive tourist experience.
4. Advanced IP-based Surveillance Solutions required installing at main locations to monitor their respective areas and record incidents that occur throughout the city. It also delivers live video feed to the Jaipur Police control room, leading to faster response times and improved success rates, allowing citizens and tourist alike to feel safer in Jaipur.
5. Environmental Sensors need to install at key locations that provide air quality status reports in real-time to handle pollution and problems caused by desert conditions when they arise.
6. In the case of city infrastructure, installed Parking Information, Remote Kiosks and Facility Management Services at select locations should be implemented to improve on facilities and parking under smart city project.
7. Smart Pods must be installed for the provision of government services, citizens now have highly secured access to customized content and information that doesn't compromise on the personal privacy of its users.
8. The concerned authorities ensure the introduction of smart toilets with bio-digester facilities at all tourist sites and important places in order to maintain the green environment.
9. Eco-friendly conveyance should be introduced for tourists to travel from the town to nearby attractions to safeguard the environment.
10. There is need to focus on adoption of new technologies through smart devices to attract new tourists to the destination.
11. Local tourism authorities must develop, update and maintain the social medial page covering major attractions and facilities in the area to connect to the potential and future tourists through social media network.
12. Digital panels or totems can be placed both indoors (tourist offices, shopping centers, museums and cultural centers, etc.) and outdoors (at routes stop points, stations and public transport stops, etc.) to offer 24/7 service for tourists.

13. Jaipur is frequently visited by foreign tourists, so, local tourism websites need to be developed and designed in multi languages for convenience of tourists of different nationalities.
14. The web portal together with the mobile applications should configure tour packages according to the tastes or characteristics of the visitors, propose routes in the destination to improve the experience of tourists.

FUTURE RESEARCH DIRECTIONS

The present study assess the tourist experience towards smart city aspects in Jaipur and observed that the smart city mission has made this vision a reality, and our city is benefitting in every aspect – from safety and security, to easy access to information, to overall improvement of image and stepping into the unique tourist destination.

In the future, many smart services can be added to the system like smart object sensing and recognition, service rating, tourist crowded estimation, hotel booking, ticket buying, restaurant reservation, car renting and many other services. The smart tourism services (STS) could be implemented in the culture & heritage tourism sites including museums, forts, palaces etc. as a pilot project in the state and for implantation in any part of the country. Moreover, inclusion of future research on different aspects such as smart destination image and its relation with decision of the travel will be an extension in the area of study.

CONCLUSION

The understanding of phenomena of smart cities and smart destination is influenced by developmental aim and availability of infrastructure in different areas. From a broader perspective this is described an area equipped with improved digital innovations, technology adoption, infrastructural facilities and better experience to visitors. Jaipur city of Rajasthan is among the important tourist destinations undertaken for development through smart city mission in the country. Therefore, it is the need of the hour today is to understand experience gained by tourist from different smart city concepts during their visit in the area. The present chapter observed that Jaipur has been designated as smart city by Indian government in the year 2015 in order to improve the overall position of the country and ensure the well being of its citizens. This particular initiative provided a boon to tourism industry to excel its growth and improve the image as global tourism destination. Due to this, Jaipur proved an added advantage in beautification of their attractions, improvement in mobility, tourist and general infrastructure enhancement, better communication & technology and improved utilities. The level of tourist services and facilities were improved due to its intervention with technology and smart city mission. The chapter identified important aspects of smart city destination in Jaipur as smart attraction, smart mobility, smart amenities, smart ancillary, smart safety and smart available packages. Thereafter, the chapter provides a comprehensive overview of tourist experience towards above dimensions of smart city destination in Jaipur. The empirical analysis tourists got satisfactory experience with most of the variable of above dimensions except few. The aspects such as smart available packages, smart safety, and smart attraction are providing better experience to tourists in comparison to smart amenities, smart ancillary and smart mobility. This indicates that tourist experience in Jaipur is enhanced due to initiation of smart city mission and merger of different facilities

& services with technology. The study from its findings provides implication to different stakeholders to maintain the competitiveness and marketing strategies at destination in order to ensure a prominent position in global tourism market.

REFERENCES

Boes, K., Buhalis, D., & Inversini, A. (2015). Conceptualizing Smart Tourism Destination Dimensions. In I. Tussyadiah, & A. Inversini (Eds.), *Information and Communication Technologies in Tourism* (pp. 391–403). Heidelberg, Germany: Springer.

Buhalis, D. (2000). Marketing the competitive destination of the future. *Tourism Management, 21*(1), 97–116. doi:10.1016/S0261-5177(99)00095-3

Buhalis, D., & Amaranggana, A. (2014). Smart Tourism Destinations. In Z. Xiang, & I. Tussyadiah (Eds.), *Information and Communication Technologies in Tourism* (pp. 553–564). Heidelberg, Germany: Springer.

Buhalis, D., & Law, R. (2008). Progress in information technology and tourism management: 20 years on and 10 years after the Internet, The state of eTourism research. *Tourism Management, 29*(4), 609–623. doi:10.1016/j.tourman.2008.01.005

Cohen, B. (2012). *Smart cities hub*. Retrieved July 22, 2013, from http://smartcitieshub.com/2012/11/11/smart-cities-ranking-methodology/

Department of Tourism Rajasthan. (2018). *Annual Progress Report 2017-18*. Retrieved from Rajasthan Tourism Department website: http://www.tourism.rajasthan.gov.in/content/dam/rajasthan-tourism/english/others/tourism-department-annual-progress-report-2017-18.pdf

Falconer, G., & Mitchell, S. (2012). *Smart city framework. Cisco Internet Business SolutionsGroup*. IBSG.

Government of India, Ministry of Urban Development. (2015). *Mission Statement and Guidelines*. Retrieved from http://164.100.161.224/upload/uploadfiles/files/SmartCity Guidelines(1).pdf

Guo, Y., Liu, H., & Chai, Y. (2014). The embedding convergence of smart cities and tourism internet of things in China: An advance perspective. *Advances in Hospitality and Tourism Research, 2*(1), 54–69.

Haubensak, O. (2011). *Smart cities and Internet of things*. Zurich, Switzerland: ETH.

Huang, X. K., Yuan, J. Z., & Shi, M. Y. (2012). Condition and key issues analysis on the smarter tourism construction in China. In *Multimedia and signal processing* (pp. 444–450). Berlin, Germany: Springer. doi:10.1007/978-3-642-35286-7_56

Hwang, J., Park, H. Y., & Hunter, W. C. (2015). Constructivism in smart tourism research: Seoul destination image. *Asia Pacific Journal of Information Systems, 25*(1), 163–178. doi:10.14329/apjis.2015.25.1.163

Ivars Baidal, J. A., Solsona Monzonís, F. J., & Giner Sánchez, D. (2016). Gestión turística y tecnologías de la información y la comunicación (TIC): El nuevo enfoque de los destinos inteligentes. *Documents d'Analisi Geografica, 62*(2), 327–346. doi:10.5565/rev/dag.285

Kumar, S. (2016). Developing India as Smart Tourism Destination-A Sap Lap Analysis. *South Asian Journal of Tourism and Heritage*, *9*(2), 124–136.

Lee, J., Lee, H., Chung, N., & Koo, C. (2017). An Integrative Model of the Pursuit of Happiness and the Role of Smart Tourism Technology: A Case of International Tourists in Seoul. In *Information and Communication Technologies in Tourism* (pp. 173–186). Cham: Springer.

Neuhofer, B., Buhalis, D., & Ladkin, A. (2014). A Typology of Technology-Enhanced Tourism Experiences: Technology-Enhanced Tourism Experiences. *International Journal of Tourism Research*, *16*(4), 340–350. doi:10.1002/jtr.1958

Reischl, C. (2013). *Data will drive smart cities' medical tourism success*. Denver, CO: Mercury Advisory Group.

Ritchie, J. R., & Crouch, G. I. (2005). *A model of destination competitiveness. Competitive destination: A sustainable tourism perspective*. Wallingford, CT: Cabi.

Rong, A. (2012). *China economic net*. Retrieved July 22, 2013 accessed from http://en.ce.cn/Insight/201204/12/t20120412_23235803.shtml

Xian, Z., Wang, D., O'Leary, J. T., & Fesenmaier, D. R. (2015). Adapting to the internet: Trends in traveller's use of the web for trip planning. *Journal of Travel Research*, *54*(4), 511–527. doi:10.1177/0047287514522883

ADDITIONAL READING

Achaerandio, R., Bigliani, R., Maldonado, G. G., & Curto, J. (2011). *Smart cities analysis in Spain*. IDC.

Anttiroiko, A. V., Valkama, P., & Bailey, S. J. (2014). Smart cities in the new service economy: Building platforms for smart services. *AI & Society*, *29*(3), 323–334. doi:10.100700146-013-0464-0

Arup. (2010). *Smart cities: Transforming the 21st century city via the creative use of technology*. Arup.

Falconer, G., & Mitchell, S. (2012). *Smart city framework. Cisco Internet Business Solutions Group*. IBSG.

Hedlund, J. (2012). *Smart city 2020: Technology and society in the modern city*. Microsoft Services.

Komninos, N. (2015). *The age of intelligent cities: smart environments and innovation-for-all strategies*. New York: Routledge, Taylor & Francis Group.

M. D. A.-S. A. BLamsfus. (2015). *C. and E. Torres-Manzanera, Smart Tourism Destinations: An extended conception of smart cities focusing on human mobility*. Cham: Springer International Publishing.

Reischl, C. (2013). *Data will drive smart cities' medical tourism success*. Denver: Mercury Advisory Group.

Wang, D., Park, S., & Fesenmaier, D. (2012). The role of smartphones in mediating the tourism experience. *Journal of Travel Research*, *51*(4), 371–387. doi:10.1177/0047287511426341

KEY TERMS AND DEFINITIONS

Area Based Development: The development which focuses on a particular area within a city/town for redecoration & reconstruction.

Smart City: A city or town aims to apply integrated technology and internet to administer and control infrastructural facilities & services.

Smart City Mission: The project pursued by any country/country for achieving development of urban areas by using digital technology.

Smart Destination: A tourist destination where tourist facilities & services such as accommodation, transportation, attraction booking and operated through technology.

Smart Waste Management: The management of waste/garbage through uses of technology enabled digital equipment in a sustainable manner.

Sustainability: Current use of economic, social, cultural and physical resources in such a manner to provide maximum utilization without compromising needs of future generations.

Tourist Traffic: The total number of tourist's foreign and domestic tourist's visits in a particular.

Chapter 21
Intelligent Tourist Destinations and Their Application to Public Policies: The Spanish Case

Luis Galindo Pérez-de-Azpillaga
https://orcid.org/0000-0001-5331-6496
Universidad de Sevilla, Spain

Alfonso Fernández-Tabales
Universidad de Sevilla, Spain

Concepción Foronda-Robles
Universidad de Sevilla, Spain

ABSTRACT

The chapter is divided into two differentiated parts. The first includes a brief technical review of the concepts of Territorial Intelligence and Intelligent Tourist Destination, situating them in the context of the new tourist destination planning paradigms. This part ends with a first approach to the Spanish case and the progress of its public tourism policies towards these concepts. The second part goes deeper into the Intelligent Tourist Destination model applied in Spain, driven by the public administration, paying special attention to its official standardisation process, and the indicators adopted to that end. Finally, the chapter ends with some brief conclusions.

DOI: 10.4018/978-1-7998-1989-9.ch021

INTRODUCTION

Spain is one of the world's major tourist destinations, featuring the hallmarks of a mature destination within the current tourism industry. It therefore presents high prices compared to newly emerging destinations. Given this situation and the inability to compete by means of low prices, Spanish tourism policies have been seeking new formulas to enhance competitiveness by boosting quality.

With this in mind, Spanish authorities have adopted a strategy of promoting smart tourism destinations. This strategy aims to try new destination management formulas ensuring an emphasis on innovation, technological development and competitiveness without disregarding the respective sustainability and good governance

The aim of the chapter is to specify and undertake a critical review of the concepts of territorial intelligence and their application to tourism areas such as smart destinations. It also supplies a more applied view, analysing public policies based on the smart destination concept in Spain, one of the countries which has endowed it with more administrative content, with numerous real cases whose results can be assessed and compared. In this regard, the Spanish case stands out owing to the establishment of an official methodology to achieve the smart destination qualification, with systems of measurable indicators that will likewise be appraised in the chapter. Finally, the internationalisation process of this Spanish methodology, which has already been exported to several destinations in Mexico, will be examined.

BACKGROUND

The concept of Territorial Intelligence (TI) is being increasingly used by scientists as the new 21st century progresses. Although TI has been subject to several definitions such as those coined by Dumas (2004) or Bertacchini (2912), possibly the most widely- disseminated and accepted is Jean-Jacques Girardot's definition. A scientific coordinator of the ENTI (European Network for Territorial Intelligence), his definition states that TI is a means for researchers, actors and territorial communities to acquire a better knowledge (of the territory) to better control its development. The appropriation of technologies of information and communication and of information itself is an essential step for the actors to accede to a training process, which will allow them to act in a pertinent and efficient way. TI is particularly useful in helping *local actors* to plan, define, animate and evaluate the policies and the actions of sustainable territorial development (Girardot, 2000).

One of the basic principles of the new approach is not to consider the territory as an enterprise or a market, but essentially as a cooperation space (Masselot, 2008), where the generation and transmission of information and knowledge take on a key role (Bozzano, 2013). The territory is considered an organisational reality with learning capacity (Devillet & Breuer, 2008). This interrelation between the concepts of territory, society and knowledge can be summed up in the following expression: TI aspires to be the multidisciplinary science the object of which is the sustainable development of territories in the knowledge society and the subject the territorial community (Girardot, 2008).

The TI approach is perceived suitable to deal with the problems of tourist destinations, as its integrating nature (territorial, institutional, social, economic, technological, etc.) corresponds to the holistic approach that tourist destination planning, and management must adopt. This management of the destination cannot be limited to an economic, infrastructural, urbanistic or environmental type sectoral practice, but rather it must include all the aforementioned aspects of the real situation, insofar as the product enjoyed by the visitor will be the territory-destination as a whole. The long-term competitiveness of the destination will depend on the comprehensive quality of this whole, and not on that reached by any of its components.

Approximations of the TI approach to the tourist activity have emerged in the scientific field over the last few years (Angelaccio, Basili & Buttarazzi, 2013; Hjalager, 2012; Ivars et al., 2016; Jovicic, 2016, Perea-Medina et al. 2018). Attempts by the public administration to carry out a practical implementation can also be observed (Gretzel et al., 2015). Thus, the concept of Smart Destination was born, understood as a destination whose management involves a strong innovation component, based on the use of information and communication technologies (ICTs), as tools to obtain a competitive destination, the sustainable development of the territory, and the participation of the citizens involved.

A review of scientific literature on Smart Destinations (Mora et al., 2017) indicates the existence of two viewpoints, one focussed on European academic fields, with a more social perspective, and another, more prevalent in the US, that has a more technological and technocratic approach. The proposed text combines both perspectives. The first refers to innovation in the tourist sector, which is deeply transforming the processes, products and organisational methods of the production fabric engaged in providing the tourist service (Santos-Júnior, et al, 2017). It is especially significant due to its role in improving user experiences, in increasing the amount of data obtained through online devices, and in terms of its influence on the citizens' relationship and their connection with this issue. The second perspective is based on the fact that all these technological aspects have a strong instrumental nature. However, it is also essential for them to respond to a social approach, helping local actors to improve their procedures and actions. Thus, they simultaneously appear as knowledge and information management tools, but they also make it possible to correctly implement governance in the territories (Fernández Tabales et al, 2017; Perea-Medina et al. 2018).

NEW PARADIGMS IN THE PLANNING OF TOURIST DESTINATIONS

Framework of Action

The use of new technologies has opened up new horizons in urban and tourist planning, related to the boom of smart cities (Del Chiappa & Baggio, 2015; La Rocca, 2014). Smart cities link into the theories of innovative territories (Moulaert & Sekia, 2003), and their adaptation to the digital economy, forming a renewed planning and management approach. ICTs have become a competitiveness factor (Buhalis & Matloka, 2013), as they provide tourist organisations with tools: competitive advantages in costs, differentiation or specialisation (Buhalis, 2003).

The recent history of smart cities is considered from two viewpoints (Angelidou, 2015). On the one hand, the viewpoint associated with their urban future, as technology has always played an important role (Schaffers, 2012). On the other hand, as the economy of knowledge and technological innovation have introduced a new level of management capacities into the urban context (Boulton et al., 2012; Komninos, 2008, 2015), to tackle the challenges of the 21st century, and gear the urban growth and renewal processes towards competitiveness and sustainability. This approach can be transferred to the planning and management of tourist destinations that are coping with a considerable structural change derived from the digitalisation of the tourist activity.

In the Spanish case, the Spanish "Horizon 2020" Tourism Plan (2008-2012) reformulated the development plans of the destinations according to the new market trends (Vera & Baños, 2010). This plan was a programmatic document of the state administration, whose objective was to commit to a "new tourist economy", based on innovation, and the sustainability of the model in an increasingly competitive environment that focusses on a leadership culture (Ministry of Industry, Tourism and Trade, 2007). However, there are also valid criticisms that stress that the contributions of this approach are already contained in other proposals (strategic planning, sustainable tourist management, etc.). Some authors doubt the transformation capacity of the destinations based on the new technological solutions (Ivars-Baidal et al, 2016; La Rocca, 2014), and declare that the technological bias can generate new dependences for the territories (Townsend, 2013) and even favour the privatisation of public services (Greenfield, 2013). The objective of the 2020 Plan was ambitious and had a good strategic and methodological structure, although not all the proposals could be developed due to the economic crisis, and to the lack of budget to undertake the proposed actions (López & Pulido, 2013).

In this changing environment, dominated by this new digital economy, the Ministry of Industry, Energy and Tourism decided, in 2012, to commit to the transformation of the Spanish tourist model, which would permit modernising the tourist destinations under the concept of Intelligent Tourist Destinations (ITDs), as initially proposed in the National Integrated Tourist Plan (Spanish acronym: PNIT) (2012-2015) (Secretary of State for Tourism, 2012), promoted by the Secretariat of State for the Management of Innovation and Tourism Technologies (Spanish acronym: SEGITTUR). It had two objectives. On the one hand, to improve the positioning of Spain, seeking mechanisms to boost innovation at the destinations, with the deployment of ICTs, to be able to create differentiated and highly competitive services. The white paper, "Intelligent Tourist Destinations: constructing the future" (SEGITTUR, 2015) is the result of this. This paper sums up the work carried out in that legislature, explaining what an intelligent tourist destination is, what are the four axes that comprise it: innovation, technology, sustainability and accessibility, and what they consist of. On the other hand, to create a homogeneous framework that will establish the minimum requirements to classify the ITDs.

The tourism field-related measures, promoted by SEGITTUR, coexisted with other programmes related to smart cities, until, in the framework of the Digital Agenda for Spain, the National Plan for Smart Cities was developed (Ministry of Industry, Energy and Tourism, 2015) to foster technological industry, and to help local entities in their transformation processes towards smart cities and destinations (Femenia-Serra & Ivars-Baidal, 2018). Red.es is a public corporate entity, responsible for executing the Digital Agenda for Spain. It addresses topics related to Smart Cities, starting up a programme of aid in three calls, geared towards favouring the technological transformation process of smart cities and smart islands. Some of their results were:

- First Call of Smart Cities (2014). €15 million budget. 11 projects were selected (Alcalá la Real, Lepe, Sevilla, Martos, Huelva, Granada, Costa del Sol, Valdepeñas, Toledo, Villanueva de la Serena, Almendralejo-Badajoz).
- Second Call for smart Cities (2015). €63 million budget. 14 projects from the Spanish geography were selected. (Alicante, Gijón, Santander, Las Palmas de Gran Canarias, Lugo, Valencia, Palencia, Cáceres, Valladolid, Murcia, Ponferrada, Segovia and Córdoba).
- First Call for Smart Islands (2015). €19 million budget. The initiatives had to be technology-intensive and be geared towards mitigating the possible negative impact of the insular factor on

the territory, through the use of ICTs. Three projects were selected (El Hierro, Mallorca and Fuerteventura).

The National Smart Cities Plan had established a strategy focussed on three areas: standardisation, support to industry and governance. In contrast, the new National Smart Territory Plan (2017-2020) (Ministry of Industry, Energy and Tourism, 2017), which aims to provide continuity to the previous plan, addresses three fields of action:

- "Territorial actions" integrated into six areas: Internal City Objects, 5G, Virtual Interoperability Laboratory, Smart Rural Territories, Intelligent Tourism and Public Services 4.0 in rural world and city platforms.
- "Support Actions" that facilitate territorial actions (Standardisation, Internationalisation, Dissemination, Capacity-building and Training); and
- "Complementary actions", that focus on two actions that are currently being consolidated: IoT to provide public services (privacy and security), and Mobility.

The model proposed by the Plan defines Intelligent Territory as a series of interrelated elements that must communicate with each other. The objectives of this model are: to offer the citizens better services, provide a response to the needs of the more underprivileged territories, and generate new business models that boost the economy (Ministry of Industry, Energy and Tourism, 2017). Making a territory intelligent entails heading innovative solutions in the tourist sector (D'Angelo et al., 2017; Garcia-Ayllon & Miralles, 2015; Herrera et al., 2019).

In short, we are in a new paradigm of urban and tourist planning and management, which responds to the needs of 21st century tourist destinations. The Spanish State promotes the formation of destinations as consolidated innovative tourist spaces on an avant-garde technological infrastructure, which will guarantee the sustainable development of the territory.

Intelligent Tourist Destination (ITD) Model 2012-2019

The definition of the Intelligent Tourist Destination (ITD) model, as interpreted today in Spain, is the result of an evolutionary process that has lasted for practically the entire second decade of the 21st century. The main elements of this process are indicated in Figure 1.

The process started off in 2012, when SEGITTUR undertook the first definition phases of the model, establishing, among other issues, the actual definition of the ITD concept. This was officially established for the Spanish Administration as an innovative tourist destination, consolidated on an avant-garde technological infrastructure, that guarantees the sustainable development of the tourist territory, is accessible for all, that facilitates the interaction and integration of the visitor in the environment, and increases the quality of their experience at the destination.

Based on this conceptual definition and on a first approximation to the specific characteristics that an ITD must have, two experimentation phases were carried out at destinations, in 2014 and 2015. These were considered as pilot projects (four in the 1st phase and five in the 2nd phase). In agreement with the results of these experimental phases, the methodology was redefined between 2015 and 2016. This was finally expressed in the first official standard on ITDs in the world. This standard, drafted in collaboration with AENOR (Spanish Standardisation Association), is the co-called Spanish Standard

Figure 1. Evolutionary process of ITDs in Spain
Source: Prepared by authors

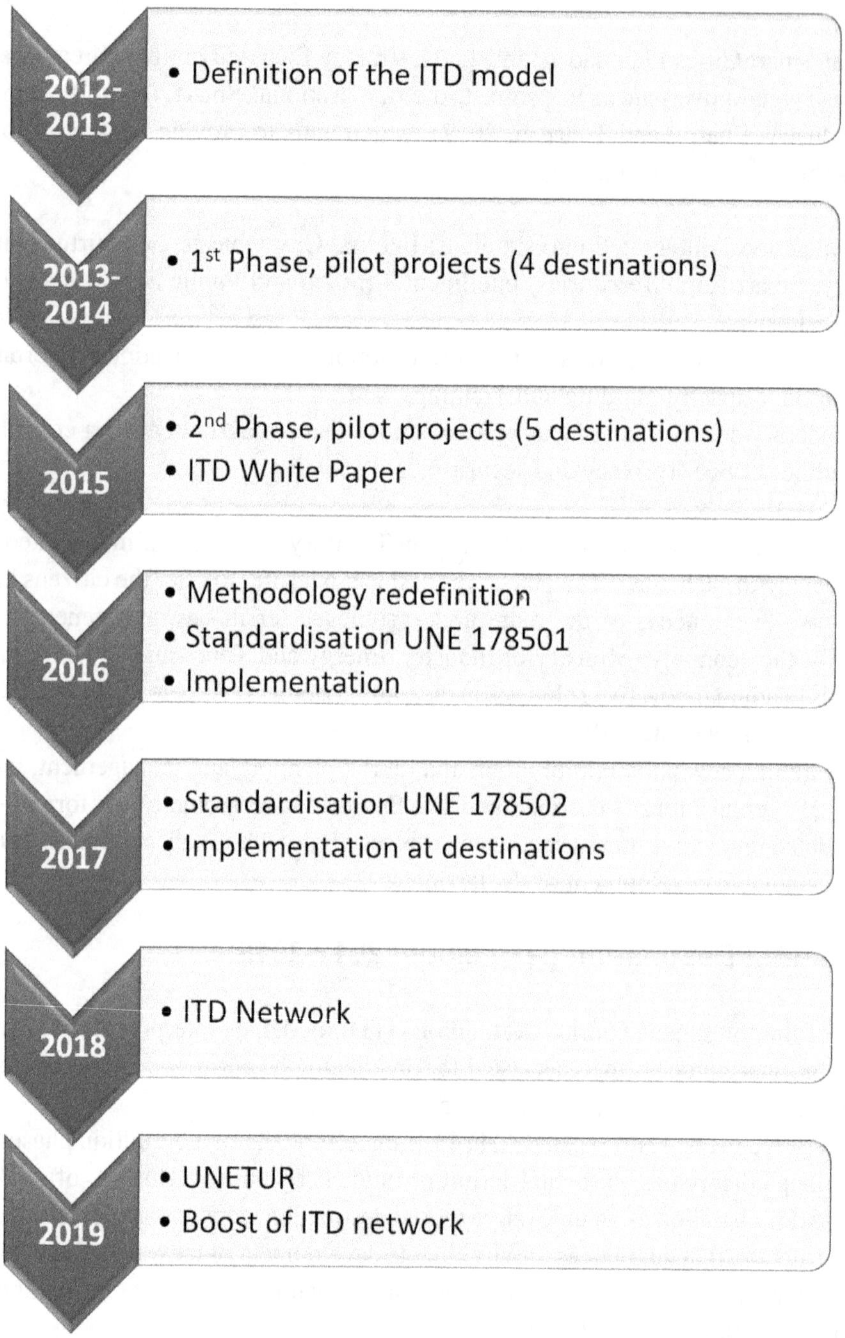

UNE 178501:2016, Management System of Intelligent Tourist Destinations. Requirements (April 2016), whose characteristics will be commented in the next section, but which, in synthesis, establishes the conditions that destinations must meet to be officially acknowledged as ITDs.

After establishing these methodological bases, the first ITDs, which have a Management System (MS) as required in the Standard, were tested. Thus, a basic element is established, i.e., the creation of an ITD network, to understand the dissemination of the model, which, by 2018, would already include 47 destinations and 10 affiliated institutions.

This Network is essential as its groups together destinations that, until then, had implemented measures specifically considered as typical of ITDs, receiving financing from public funds, mainly from EU structural funds.

Standardisation of the Intelligent Tourist Destination (IDT)

The evolution of the model continued, until finally, in 2018, a second version was approved: the Spanish standard UNE 178501:2018, Management system of Intelligent Tourist Destinations (June 2018), and a second Spanish Standard UNE 178502:2018, on Indicators and Tools of Intelligent Tourist Destinations (June 2018). This second standard comprises a system of indicators to measure the approximation of the destinations to the objectives established for the ITDs, representing the pinnacle of the Spanish ITD model, to date, and currently being developed and applied.

The UNE (*Una Norma Española;* in English, One Spanish Standard) regulatory documents are taken as reference for the standardisation of ITDs in Spain. These documents are a body of standards and reports created in Technical Standardisation Committees (Spanish acronym: CTN) of AENOR.

AENOR is a private non-profit association, legally acknowledged in Spain as a national standardisation organisation. This association has a considerable number of UNE standards in different subsectors, which include accommodation (hotels, rural, camp sites, etc.), catering, tourist information offices, golf, among others (Solsona, 2018).

The Ministry of Industry, Trade and Tourism has firmly fostered the definition of a homogenous framework, which has permitted the modernisation of the tourist destinations under this ITD concept, mentioned above, in line with the Smart Cities creation trends, as appeared in the aforementioned National Integrated Tourism Plan 2012-2015 (PNIT). This task, entrusted to SEGITTUR, has promoted, within the CTN 178 "Smart Cities", the creation of Subcommittee 5 as a forum for debate and consensus on ITDs (AENOR 2015). UNE 178501:2016 appeared as a result of the work of this Subcommittee and was tested at several destinations: Benidorm (Valencian Community), Val d'Aran (Catalonia), Sanxenxo (Galicia) and Jaca (Aragon).

Following these first actions, changes in the standard were considered advisable, to provide the governance of the ITDs with greater relevance, as well as to review some minor aspects, to adapt them to the real situation and diversity of the territories. The result was the current Standard UNE 178501:2018. This new standard, complemented by UNE 178502:2018, is an essential tool to assess compliance with the requirements, compare destinations, and monitor the evolution of intelligent tourist destinations in time.

Of the originally tested destinations, Benidorm is the location where this MS has already been implemented. This will be described in the next section on solutions and recommendations (AENOR, 2019).

The Intelligent Tourist Destination Management Entity (DME)

Based on this standardisation established by AENOR, the functioning of the MS highlights, in Standard UNE 178501:2018, the existence of a Destination Management Entity (DME). This DME is the local ter-

Table 1. Main competencies necessary for the DME

Grouping	Description
Tourist destination	Capacity to know the real situation of the destination and its resources
	Maximisation of synergies and use of resources
	Prioritisation of local and proximity resources
	Respect for the principles of sustainable tourist development and for the application of the principles of circular economy.
Intelligent management	Use of high value knowledge systems that facilitate decision-making processes
	Development of an activity on open and transparent destinations, establishing and maintaining fluent communication channels with visitors, the sector and the resident population
	Perfecting new models at the destination, directly interacting with the local business fabric
	Application of new technologies, and maintenance of an innovative mindset that favours flexibility to adapt to a constantly evolving industry and to the new challenges presented by each destination

Prepared by authors. Source: AENOR, 2018a

ritorial entity, local entity or uniprovincial autonomous community that governs and manages an ITD with sufficient responsibility and authority to establish, implement, maintain and improve the MS of an ITD.

A DME is an essential element in the application of the MS, and for a management element to be able to be considered as such, it must have sufficient competencies in its government capacity in the areas affected by tourism. The main ones are included in Table 1 below.

With the provision of these competencies, the DME must implement the MS, and therefore activate the necessary processes for a destination, and its services, to be able to be considered as an ITD. The stages associated with these processes are indicated in table 2.

Consistent with the perspective in the design of the different phases of the tourist experience (Jernsand, et al., 2015), the DME also undertakes to ensure the requirements that are considered in the delivery, use and treatment processes of the products and services of the ITD.

Planning and Operativity of the Intelligent Tourist Destination (IDT)

The drafting and start-up of the processes designed for the MS of the ITD, must be contained and classified in a schedule, through the existence of a Master Plan (MP) and an Operation Plan (OP). The former is the document that identifies the objectives and challenges to be reached by the ITD, it develops strategies to achieve these objectives, and assigns resources to carry them out. The latter refers to the

Table 2. Substantial stages in processes associated with the DME

Stages	Description
Definition	Establishment of necessary action criteria for the processes
Monitoring	Implementation of process controls in agreement with the criteria.
Verification	Storage of documented information to ensure that the processes have been carried out as defined

Prepared by authors. Source: AENOR, 2018a

Table 3. Essential purposes concerning the ITDP

Grouping	Description
Tourist destination	Be suitable for the purpose of the ITD
	Correspond to the economic, environmental and sociocultural context of the ITD
Intelligent management	Provide a reference framework to establish the ITD objectives
	Insert the agreement of providing resources to undertake the strategic and operational objectives
	Create a participation framework that will integrate all the stakeholders, to help to make the decisions
	Introduce collaboration for the continuous improvement of the MS of the ITD

Prepared by authors. Source: AENOR, 2018a

specific actions to be developed by the DME in a certain period, to achieve the objectives defined in the MP, assigning tasks, people responsible and deadlines, among others.

The DME must collaborate closely in the functioning of both the MD and the OP with other public institutions and private entities, through agreements or coordination actions, establishing consultation mechanisms with the stakeholders. Both elements form part of the so-called Intelligent Tourist Destination Policy (ITDP) within the defined scope of its MS. The purposes of the ITDP are indicated in table 3.

Intelligent Tourist Destination (ITD) Axes

The analysis of the ITDs comprise five axes (Figure 2), governance, innovation, technology, universal accessibility and sustainability. These axes include indicators that are all interrelated in the ITD and where all have an influence on each other. They are not understood as watertight compartments, but as supports that are connected to each other, and that have to be applied transversally in the management of the destination.

The axes that the DME must take into account in the management of the ITD and its schedule are explained in this section. To this end, a total of 84 indicators have been designed, organised under the following themes (Table 4).

The indicators can be quantitative or qualitative. The former has taken precedence in this standard used for standardisation, in order to able to obtain data that are comparable in time. As a general rule, all the indicators are of compulsory application. However, the DME can exclude the use of some of them due to conditions such as size, characteristics or others.

The final objective of the application is to obtain a global perspective, based on the compilation of evidence, to identify a series of improvement opportunities in specific fields and in a hierarchised manner. Moreover, the aim is to promote collaboration between different areas of the destination, promoting a favourable environment for the creation of positive synergies among all stakeholders.

Governance Axis indicators

The governance axis must take into account the principles that refer to the governability of the ITD (legitimacy for its management by its institutions), and the co-responsibility, collaboration and cohesion of all the components of the tourist sector (direct and indirect), studied from the list of indicators included in table 5.

Figure 2. Thematic axes of the ITD
Source: Prepared by authors

Table 4. Description of the blocks of indicators by thematic axes

Name	Description	Indicators
Governance	Tourist destination steering process through synergic and coordinated efforts by the governments at their different levels, and attributions of the civil society that delegates on the receiving communities and the business fabric related to the operation of the tourist system (World Tourism Organisation - UNWTO-).	11
Innovation	Innovative internal management proposals and external proposals that are translated into significant improvements aimed at the activities involved before, during and after the stay at the ITD, through the implementation of the innovation management tools, competitive intelligence being one of them	13
Technology	By incorporating technologies (information, communication, energy improvement, etc.) and technology surveillance that will permit the use and application of data and contents on markets, clients and products, aiming to increase efficiency and effectiveness of the processes and services of the ITD.	16
Universal accessibility	Way to afford value to all the initiatives developed by the stakeholders in the ITD, taking the awareness-raising, training and participation strategies as starting point, with mainstreaming criteria and grounded in human diversity and equal opportunities.	17
Sustainability	Rational and efficient management of the resources (environmental), quality of life of tourists and residents (sociocultural) and business competitiveness linked to the economic vector.	27

Source: AENOR, 2018a

Table 5. Governance Axis Indicators by sub-axes

Sub-axis	Code	Indicator
Government and Management	G1	ITD context (document that contains the description of its territorial scope)
	G2	ITD competence (document that sets out and justifies the competencies of the DME)
	G3	Coordination table
	G4	DME organisation chart
Planning	G5	MP and OP
	G6	Citizen participation
Transversal and coherent management	G7	Information promotion
Transparent management	G8	Transparency in recruitment
	G9	Transparency in management
Assessment of the undertaking	G10	Tourist satisfaction
	G11	Resident satisfaction

Source: AENOR, 2018b

As seen in the table, the governance requirements of an ITD are integrated into the definition of the actual MS, its policies, processes and documentary records, the majority of which have already been described in the previous sections on the DME, relating to the planning and operability of the ITD.

Innovation Axis Indicators

For the functioning of this axis, for the DME, the first two indicators indicate that it must acquire a permanent commitment to innovation and incorporate it into its strategy, by adding systematic innovation processes to the ITD. Further, it must have the (human, technical and financial) resources to develop this innovation strategy. Together with these two aspects, all the indicators of the block are included in table 6 below.

The abovementioned resources can be organised into an Innovation Management Unit (IMU), which will be able to permanently identify and analyse the problems and opportunities of the ITD, in discussion with all the public and private agents. Simultaneously, the planning, monitoring and control, and execution of the innovation action portfolio with the MP and its OP, can also be entrusted.

As described in the standard, to carry out the referenced functions, the DME must use the following technology surveillance, technology foresight and creativity innovation tools.

The results of the innovation actions will be monitored based on the indicators established, both inside each process and outside it. These results must provide sufficient information about the success or failure of each action, permitting learning and improvement of future actions.

Table 6. Innovation Axis Indicators by sub-axes

Sub-axis	Code	Indicator
Strategy	I1	Strategy and commitment
	I9	Public engagement
Resources	I2	Financial resources
Identification and analysis of risks and opportunities	I3	Communication with stakeholders
	I4	Tourist intelligence
Planning and assessment	I5	Monitoring and control of actions
	I6	Measurement, analysis and improvement (objectives)
Data protection and exploitation	I7	Measurement, analysis and improvement (ideas)
	I8	Measurement of profitability
	I10	Business innovation (companies created)
	I11	Business innovation (company incubators)
Identification of a knowledge map	I12	Generation of knowledge database
Identification of new tourist offers, tourist promotion and development plans/actions for the destination	I13	Tourist plan management

Source: AENOR, 2018b

Table 7. Technology Axis Indicators by sub-axes

Sub-axis	Code	Indicator
Telecommunications infrastructures	T1	Quality mobile connection
	T2	Free WiFi coverage at municipal facilities
	T7	Social networks (comments)
Business Intelligence Strategy and Tools	T3	Use of Business Intelligence tools and strategies
	T4	Open Data
	T5	Percentage of accesses to the open data portal
Two-directional communication applications or platforms between destination and tourist	T6	Visits to website and publication of the info
	T8	Virtual tourist offices
	T9	Social networks (followers)
	T10	Tourist applications
Technological tools and solutions for tourist interaction with the ITD	T11	Online tourist management
	T12	Geopositioning
Development of infrastructures, devices and instruments that will allow tourists to get to know the destination	T13	Online promotion
Implementation and promotion of technological solutions that will permit the security of tourists' personal data	T14	Security in communications
	T15	Promotion of data protection in tourist companies
Implementation and promotion of technological solutions that afford and improve the physical safety of the destination	T16	Promotion of physical safety

Source: AENOR, 2018b

Technology Axis Indicators

In terms of technologies applicable to an ITD, those that can be provided or managed by the DME (table 7) must, in principle, be taken into account. They must also act as a cross-cutting tool to provide the destination with intelligence in all its facets, in order to improve the tourist's experience at the destination.

As a first line of work, the DME is responsible for promoting the existence of telecommunication infrastructures to access the telecommunication services. These infrastructures will be as necessary and sufficient for tourists to be able to be connected by means of a great variety of devices, with quality connection and sufficient security guarantees. The possibilities offered by wireless communications are especially important in this regard.

The development of these infrastructures, devices or instruments must permit tourists to get to know the destination and all its components (transport, beaches, museums, areas of tourist interest, etc.), regardless of whether they have exceptional needs (universal accessibility).

The second line of work is to seek Business Intelligence and Tourist Intelligence tools that will help to compile, classify and process data about the tourists, their tastes, customs, movements, activities, etc., together with the development of applications or platforms that will bilaterally permit permanent communication between destination and tourist, in all phases of the tourist experience.

Both lines, the existence of telecommunication infrastructures and the search for Business Intelligence tools, are covered by these technological solutions, so their competitiveness must be improved, increasing security and trust in the tourists' personal data, integrating greater environmental sustainability, and affording and improving the physical safety of the destination.

Universal Accessibility Axis Indicators

The mission of the DME is to acquire a documented commitment to improve the universal accessibility of the destination, and watch over compliance with this type of requirements, both in the public and private management part. This will include not only new projects, and the partial and/or total remodelling of infrastructures, services, products, technologies, etc., but also the current state of existing ones. To this end, sufficient (human, technical and economic) means are necessary. These first two indicators, together with the rest of this block, are included in table 8 below for this axis.

The MP and the different OPs of the DME must include commitments to introduce actions to improve universal accessibility in the value chain of the tourist destination. These actions must be specific, measurable, coordinated and established in the short, medium and long term, so that improving accessibility is a permanent and cross-cutting effort in the ITD.

The scope of these actions must be agreed between both public and private stakeholders, in their areas of competence. It is recommended for this scope to cover at least one element of each link of the accessibility chain, in order to offer a complete and significant tourist experience of the ITD.

Establishing the participation mechanisms with the stakeholders will be essential, including entities that represent the users or other relevant entities, that might provide information about the accessibility needs and their priorities.

Likewise, the DME is responsible for ensuring that as many people as possible who participate in the entire execution process of these improvement actions in the ITD are trained in universal accessibility, design for all, and attention to the public in this type of needs.

Table 8. Universal Accessibility Axis Indicators by sub-axes

Sub-axis	Code	Indicator
Public Management	A1	Commitment to accessibility
	A2	Human and economic means
	A3	Diagnosis on accessibility
	A4	Action Plan
	A5	Number of training actions
	A6	Number of trained students
Planning, Information and Communication	A7	Information about accessibility of the destination
	A8	Available accessible information (supports)
	A9	Available accessible information (visual/auditory/touch/…)
	A10	Promotion of the destination's accessibility (budget earmarked to means and initiatives on accessible offer)
	A11	Promotion of the destination's accessibility (document)
Compliance with the destination's accessibility	A12	Compliance with the destination's accessibility
Information and Communication with the surrounding area	A13	Accessibility in terms of signage, orientation and guidance
Transport (rolling stock, vehicles)	A14	Accessible vehicles with driver
Tourist routes and itineraries	A15	Accessible tourist itineraries or routes
Public management	A16	Commitment to accessibility
Development of infrastructures, devices or instruments that will help the tourist to get to know the destination	A17	Website accessibility

Source: AENOR, 2018b

Finally, providing information about universal accessibility services (sign language interpreter, availability and repair of technologies and support products, etc.) will also be included in this DME work, in all the fields managed by it (transport, accommodation, catering, etc.), offering information about the accessibility conditions of the destination's tourist offer. To this end, promotion will play another important role, especially about the accessible offer of the ITD and, depending on the competencies of the DME, facilitating the commercialisation of accessible tourist facilities and service.

Sustainability Axis Indicators

The tourist interest of a region comes from its culture and traditions, and the identity that its citizens provide to the territory. People and their customs play an essential role in the preservation and sustainability of an ITD, which is why the series of indicators associated with its analysis are listed in table 9 below.

The DME has a commitment to improve the sustainability of the destination and integrate this commitment into its strategy in a cross-cutting manner, promoting compliance with dual protection for the ITD; on the one hand, protection of the present and future economic activity, and on the other, environmental protection.

Table 9. Sustainability Axis Indicators by sub-axes

Sub-axis	Code	Indicator
Environmental	S1	Energy efficiency
	S2	Electrical vehicle recharging points
	S3	Waste management
	S4	Water management
	S5	Climate change adaptation plan
	S6	Pollution map
	S7	Noise map
	S8	Water consumption in hotels
	S9	Bike lane
	S10	Bike points of use
	S11	Urban pedestrianisation
	S12	Public carpark sensors
	S13	User satisfaction with public transport
	S14	Water treatment and recycling
	S15	Green zones
	S16	Landscape protection
	S17	Protected natural spaces
	S18	Bathing water quality
	S19	Renewable energy-based energy consumption
Sociocultural	S20	Heritage protection
	S21	Impact on the local community
Economic	S22	Average stay
	S23	Monthly accommodation occupation
	S24	Total and average expenditure of tourists
	S25	Tourist employment
	S26	Tourism contribution to GNP
	S27	Unemployment rate

Source: AENOR, 2018b

With regards to economic aspects, worthy of note are the promotion of innovative job opportunities, generation of safety and security, motivation of resilience, and protection of traditional trade.

In terms of environmental management, the following aspects will be fostered: measuring environmental parameters, efficient energy management, promoting recycling and reuse, and improving passenger transport systems and pedestrian mobility.

Ethnic and social integration, gender equality and health-related topics will be taken into consideration within the area of sociocultural sustainability. It will also take into account the design of the provision of tourist services that apply technology and interact with the citizens and tourists, and the preservation of heritage. Finally, it will also be advisable to favour the culture of social responsibility and the engagement of all stakeholders.

SOLUTIONS AND RECOMMENDATIONS

Experiences of Intelligent Tourist Destinations in Spain

Once the degree of compliance with the requirements of the five axes of the ITD methodology has been analysed, and after drafting the action plan to reach excellence in these fields, the destinations receive the Intelligent Tourist Destination distinction, forming part of the project in the category of affiliates. They also join the ITD Network.

A total of 14 destinations joined the project between 2012 and 2017 (Almería, Badajoz, Castelldefels, El Hierro, Jaca, Valle de Aran, Villajoyosa, Las Palmas, Lloret de Mar, Marbella, Murcia, Noja, Palma and Valle del Jerte), with the number varying each year. This shows that the project was still at an embryonic stage with the destinations gradually joining up. There were even some that acted as pilot projects to implement the different regulations.

However, in 2018, the ITD project started to show signs of development, after the tourists and residents had verified its positive effect on competitiveness and coexistence, with the integration of 10 destinations (Arona, Avilés, Benidorm, Canal de Castilla, San Sebastián, El Ejido, León, Monfragüe, Puerto de la Cruz and Torrox). Furthermore, in 2019, many other destinations have expressed their interest in joining (Figure 3).

Figure 3. Map of the ITD in Spain
Source: Prepared by authors

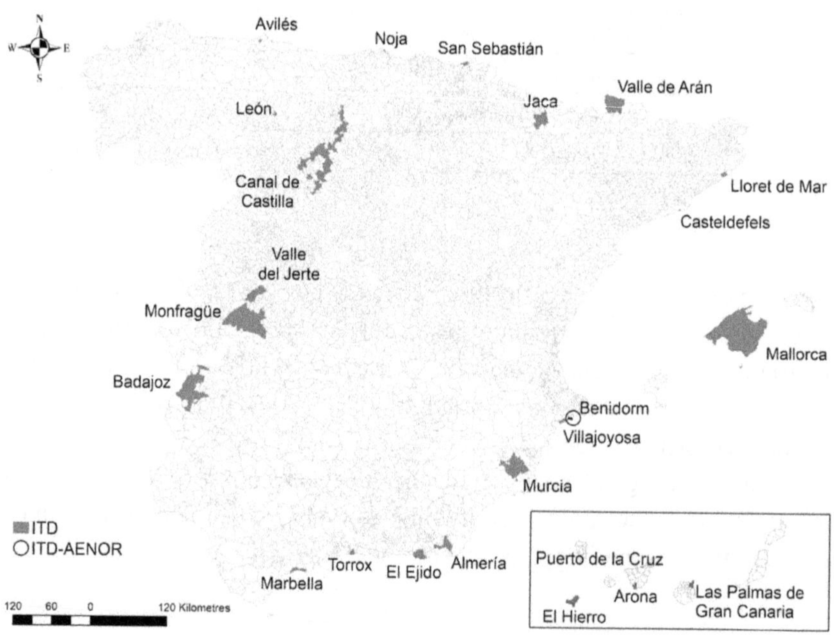

The scale of work of the ITDs varies, and this has generated advantages and limitations associated with the adaptation capacity in terms of destination management. The majority of the ITDs, 79%, have a municipal scale, and the remaining 21% are local entities (Communities, Provincial Councils or Island Councils). Of the examples already tested for Standard UNE 178501:2016, Benidorm, Sanxenxo and Jaca are destinations whose territorial scope coincides with them being one single municipality, so this simplifies the management to some extent. The Town Council is the DME considered for these studies. In the case of Val d'Aran, the destination tested was supra-municipal, so the DME considered was the Conselh Generau d'Aran (De Pablo, 2017).

Some of the ITDs (21%) have other tourist recognitions, such as the Q for Quality granted by the Institute for Spanish Tourist Quality. The same percentage of ITDs possess SICTED destination (Integrated Spanish Quality System for Tourist Destinations), a tool that permits controlling the quality of the services provided at the accommodations or tourist destinations, under the methodology of the integrated continuous improvement approach and system www.calidadendestino.org. Other destinations (8%) are Reserves of the Biosphere; 4% of the ITDs are certified according to ISO 9001, whilst 13% of the ITDs do not have any other certification.

The Spanish government's great commitment to transform the ITDs became clear when it granted aid to local entities though the call of the National Intelligent Territory Plan of the Digital Agenda for Spain, "Intelligent Tourist Destinations", tendered by Red.es. This initiative encourages the generation of suitable mechanisms to facilitate the fast incorporation of innovations into the destinations.

Addressing the projects published on the website, https://www.destinosinteligentes.es, a study has been conducted to verify the main axes that have been developed in the 24 Spanish ITDs. 105 actions, which foster specific actions in the five action axes: sustainability, governance, technology, innovation and accessibility, are analysed (Figure 4)

Figure 4. Chart of the actions per axis of the ITDs
Source: Prepared by authors

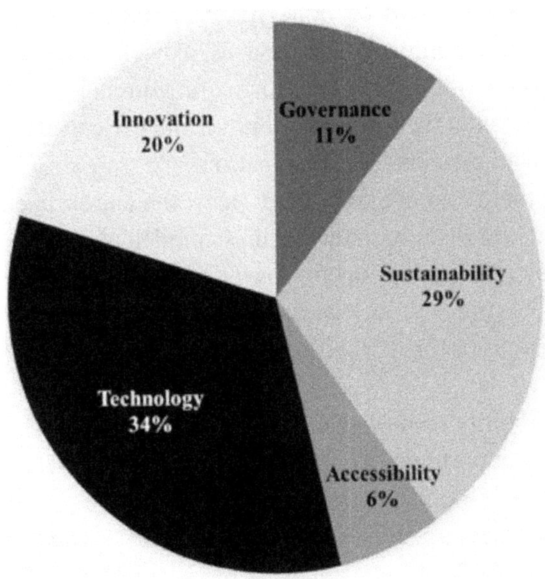

Most of the actions developed at the ITDs have a common characteristic, namely specialisation in technological advances, regardless of the axis where they operate. Although the tourist destinations undoubtedly face a context where technology (34%), sustainability (29%) and innovation (20%) are among the main axes on which their competitiveness is based. However, tourist destinations are far from being considered an example in terms of governance (11%) and accessibility (6%) within the framework of the ITD model. In other words, the ITDs have still not tapped into the facilities provided by the current technological development context, to obtain a greater degree of interaction and cooperation among the public and private actors, that would permit an accessible tourist environment.

Technology has placed itself at the service of tourist development at the ITDs to improve decision-making, the experiences offered to visitors, and the residents' quality of life. The two action lines carried out are:

- Providing telecommunication infrastructures, both mobile connection with optical fibre (Castelldefell and Las Palmas) and free WiFi in town centres, in city squares and on beaches (Almería, El Hierro, Jaca, Palma de Mallorca or Lloret de Mar).
- And the development of mobile applications or communication platforms between destinations and tourists, named "Spain in Apps". This platform was created to offer tourists updated and geo-localised information about destinations, tourist products and services, and also to generate applications that permit obtaining data in an anonymous manner. All of this will help to understand tourists' behaviour patterns. Some of these applications are:
    - Experience Spain contains tourist destination resources, where users create their own points of interest or routes. It also provides interaction with social networks. This App has been developed in Las Palmas, Badajoz or Villajoyosa.
    - Spain for Kids is geared towards families and is carried out by Castelldefels or Murcia.
    - Spain Feedback permits visitors to express their degree of satisfaction in categories such as security/safety, transport, cleanliness, etc. This App is available in El Hierro, Palma de Mallorca or Villajoyosa.

The 21st century tourist information office model has been redefined with the support of technology, leading to an in-depth change in the information services, as well as the tools and supports used (App, touch screen, totem, augmented reality, etc.). The Torrox or Murcia tourist offices have the capacity to inform and market the destination in real time: 24-hour virtual offices, use of social networks to attend to the information demands, and personalised information.

Regarding sustainability, there are several priority lines: the entrepreneurial business model or sustainable mobility. Those that have to do with the business model, and their capacity to create new companies, are developing capacity-building plans for tourist SMEs (Almería, Badajoz, Canal de Castilla), or launching new innovative businesses, such as the Innovation and Economic Development Centre of Noja (Spanish acronym: CIDEN) or the Technology-Based Company Centre, to support entrepreneurship in Leon.

In the case of sustainable mobility, the utility and efficiency of an intelligent transport system converge. The ITDs provide an excellent opportunity for the mobility sector to offer its best services and products to tourists. The ICTs applied to this area have the following objectives: reduce traffic congestion, smart management of urban parking, decrease contamination, use LED technology in public roads (Puerto de la Cruz), locate electrical vehicle recharging points (Jaca), or support the Biosphere certificate (Valley of Aran and Leon). Some of these actions are highly associated with the accessibility axis.

Accessibility must not be viewed by public administrations as a barrier, but rather as an advantage to improve tourist services, thanks to the development of accessible infrastructures and equipment, differentiating them from other less competitive destinations. The actions to foster accessibility in tourist destinations focus on specialised services, technical aids or infrastructures depending on their needs. Villamuseu (Villajoyosa), a European benchmark in accessibility, is designed according to universal principles, bearing in mind physical, organic, sensorial and mental disabilities. The administration is also fostering public works (pedestrianisation), evaluation and management protocols (Accessibility manual and inclusion in museums and National Heritage sites in Villajoyosa, or obtaining the Blue Flag), as well as incentives to companies that foster accessibility.

In terms of innovation, the Tourist Intelligence Systems (TIS) have been developed. Their objective is to use technological solutions that are able to compile, process and analyse large information flows, transforming data into knowledge, to improve the competitiveness and efficiency of both private enterprises and public organisations. The destination's success depends increasingly on information management, as the quality of the decision-making process is strongly conditional upon the quality of the information of the System. Controlling and monitoring visitors' behaviour, their geolocation, a knowledge of the visitors' flow in a territory in real time, the impact on economic activities, on the management of social, energy, and logistics resources, etc., is a key distinguishing factor of their technological and innovative component, such as that implemented in Badajoz, Benidorm or Las Palmas de Gran Canarias.

In the governance axis, the elements considered most important are those related to the role of the government, and its management, transparency and participation processes. In the first case, the Palma ITD has fostered the Smart Office, a team of professionals who design the sustainable city model, responsible for technology surveillance, searching for financing, public-private agreements, and communication with citizens. Other development offices aimed at promoting tourist strategies can be found in Jaca or Valle del Jerte. In the second case, generating transparency processes in the management of the destination information consolidates transparent destinations, and fosters the digitalisation of the administration (Aviles). The third issue is how the administration has partially modified the by-laws for the use and enjoyment of the beaches in Torrox, how it has prepared the candidacy of El Hierro to form part of the Global Geoparks Network, and of course, favoured citizen participation.

Technology alone does not convert a tourist destination into a smart or intelligent destination. It must be accompanied by a transformation process at all levels, starting with the destination strategy, and the tourist intelligence that must lead towards a new, innovative, accessible and sustainable destination model.

Internationalisation Process

One of the most relevant aspects of the Spanish experience in ITDs, is the international outreach of the model established, especially referring to its standardisation and definition of a management system. Thus, as tangible realities, there are now two foreign destinations included in the ITD network as recognised members. These are Mexican destinations (Tequila and Cozumel), which have assumed and fully developed the same model applied in the Spanish cases, so they are full members of the Network.

In the case of the island of Cozumel, it joined the Network in 2015, when they drafted the Evaluation Report and Action Plan to transform Cozumel into an Intelligent Tourist Destination.

On its part, Tequila joined the Network in 2016, with its respective Evaluation Report and Action Plan, which has been developed through an Intelligent Tourist Destination Master Plan managed in a public-private agreement by the Integrated Development Council of Tequila (CODIT).

This dissemination of the model seems to have continued, as other cooperation mechanisms have now been established to diagnose and collaborate in the presentation of the model in other Latin American destinations, such as Temuco (Chile) or Buenos Aires (Argentina).

The relevance attained by the Spanish policies in ITDs has received international backing, with different recognitions. The most noteworthy are:

- The World Tourism Organisation (UNWTO) Award to Excellence and Innovation in Tourism in 2018. More specifically, the first Prize in the Research and Technology category was granted to SEGITTUR, for its Smart Tourism System (SIT) project.
- Express mention of "Spain: Investing in 'Smart Destinations' for digitalisation and sustainable tourism" as Tourism investment case studies in the official publication of the OECD Tourism papers no. 2018/03, entitled "Effective Policy Approaches for Quality Investment in Tourism" (OECD, 2018).
- Global Champions Award 2019 of the World Travel & Tourism Council (WTTC), in the innovation and technology category. More specifically, this was granted to the Spanish Government for the development of the ITD network, as a global pioneer project to improve the governance of the tourist policy, favouring the incorporation of technology, and the digital transformation of the destinations, and promoting innovation in all areas of the tourist activity.

FUTURE RESEARCH DIRECTIONS

UNETUR (Committee of Interinstitutional Coordination of Standards and Tourist Distinctions) was created in 2019 for it application in Spain in the future. Its main aim will be to establish a coordination framework among the different organisations and associations involved in the standardisation or establishment of tourist quality distinctions. The ITD Network will be among these associations, so this concept and its practical application will be reinforced, as it will form part of an initiative whose main objectives include providing the Spanish tourist quality model with consistency, throughout the entire value chain.

In this sense, one of the future lines established by SEGITTUR also focuses on the international outreach of this model. In fact, the objectives expressed by UNETOR include exporting the Spanish model to other international destinations. This will be done through promotion, training and consultancy, fostering the Ibero-American coordination table on tourist quality, supporting European tourism policies, and establishing a dialogue mechanism with the European Commission (DG Grow). The ITDs will contribute to the Sustainable Development Goals, the UNE standards will be raised to international standards (ISO/IEC/ITU/CEN/CENELEC/ETSI), and agreements will be signed with the UNWTO, incorporating actions to disseminate the Spanish model.

Thus, the express will of the Spanish Administration to continue taking steps towards the consolidation of the ITD concept and policies can be observed. It can be said that, in terms of its innovative nature, it is an international benchmark, both in the definition of management systems and success measurement indicators.

CONCLUSION

The Spanish ITD model can be considered as a global pioneer public initiative, given the official transcendence granted to the concept, its continuity for almost a decade now, and its consideration as a strategic future line within the State tourist policies.

This transcendence is the result of the ITD model not being limited to a theoretical definition, or a statement of conceptual intentions, but rather, that the model is based on an official standardised standard. A standard that establishes what an ITD is, what technical requirements exist to receive this qualification, what its management must be like, which indicators are used to measure an ITD, and which are the specific measures that are going to receive public financing.

Due to this standardised nature, the Spanish model can be exportable to other areas in the international context, which represents one of its main distinguishing features. In fact, the model has already been adopted and has started to be applied in other countries, and its internationalisation is foreseen within the action lines of the Spanish administration.

ACKNOWLEDGMENT

This chapter is part of the research project entitled *"Inteligencia territorial vs. crecimiento turístico. La planificación y gestión de destinos ante el nuevo ciclo expansivo inmobiliario"* — "Territorial intelligence versus tourism growth. Planning and managing destinations for the new cycle of real estate expansion." — (PGC2018-095992-) by the Spanish Government Ministry of Science, Innovation and Universities, National R&D Programme (*Programa Nacional de I+D del Ministerio de Ciencia, Innovación y Universidades del Gobierno de España*).

REFERENCES

AENOR (Spanish Association for Standardization and Certification). (2015). *Standardization in smart cities – Spain*. Retrieved from https://portal.aenormas.aenor.com /descargasweb/normas/aenor-Spanish-standardization-on-Smart-Cities-CTN-178.pdf

AENOR (Spanish Association for Standardization and Certification). (2016). *Management system for smart tourist destinations. UNE 178501:2016*. Retrieved from https://www.une.org/encuentra-tu-norma/busca-tu-norma/norma?c=N0056506

AENOR (Spanish Association for Standardization and Certification). (2018a). *Management system for smart tourist destinations. UNE 178501:2018*. Retrieved from https://www.une.org/encuentra-tu-norma/busca-tu-norma/norma?c=N0060239

AENOR (Spanish Association for Standardization and Certification). (2018b). *Indicators and tools for smart tourism destinations. UNE 178502:2018*. Retrieved from https://www.une.org/encuentra-tu-norma/busca-tu-norma/norma/?c=N0060240

AENOR (Spanish Association for Standardization and Certification). (2019). *DTI Benidorm certificate.* Retrieved from https://www.aenor.com/salainformaciondocumentos/NP%20Certificado%20DTI%20Benidorm.pdf

Angelaccio, M., Basili, A., & Buttarazzi, B. (2013). Using geo-business intelligence and social integration for smart tourism cultural heritage platforms. *Proceedings of the Workshop on Enabling Technologies: Infrastructure for Collaborative Enterprises, WETICE*, 196-199. 10.1109/WETICE.2013.87

Angelidou, M. (2015). Smart cities: A conjuncture of four forces. *Cities (London, England)*, *47*, 95–106. doi:10.1016/j.cities.2015.05.004

Bertacchini, Y. (2012). Between Information and Communication Process, the Territorial Intelligence, as a Network Concept & a Framework to Shape Local Development. *International Journal of Humanities and Social Science*, *2*(18), 242–247.

Boulton, A., Brunn, S., & Devriendt, L. (2012). Cyberinfrastructures and "smart" world cities: Physical, human, and soft infrastructures. In P. Taylor, B. Derudder, M. Hoyler, & F. Witlox (Eds.), *International handbook of globalization and world cities*. UK: Edward Elgar.

Bozzano, H. (2013). Geography and Territorial Intelligence. Geo-grafein, Geo-explanans, Geo-transformare. *Revista Geográfica Digital. IGUNNE*, *10*(19), 1–24.

Buhalis, D. (2003). *E-tourism: Information technology for strategic tourism management*. London, UK: Pearson (Financial Times / Prentice-Hall).

Buhalis, D., & Matloka, J. (2013). Technology-enabled Tourism Destination Management and Marketing. In C. Costa, E. Panyik, & D. Buhalis (Eds.), *Trends in European tourism planning and organization* (pp. 339–350). Buffalo, NY: Channel View Publications.

D'Angelo, G., Ferretti, S., & Ghini, V. (2017). Multi-level simulation of internet of things on smart territories. *Simulation Modelling Practice and Theory*, *73*, 3–21. doi:10.1016/j.simpat.2016.10.008

De Pablo, F. (2017). Smart tourist destinations. *Revista de la Normalización y Evaluación de la conformidad*, 238, 20-26.

Del Chiappa, G., & Baggio, R. (2015). Knowledge transfer in smart tourism destinations: Analyzing the effects of a network structure. *Journal of Destination Marketing & Management*, *4*(3), 145–150. doi:10.1016/j.jdmm.2015.02.001

Devillet, G., & Breuer, C. (2008). Contribution to the applied territorial intelligence: reasoned catalog of territorial information available on internet and sources in Europe. *International Conference of Territorial Information. Papers on Tools and Methods of Territorial Intelligence*. Besançon, France: MSHE.

Dumas, P. (2004). Intelligence, Territoire, Décentralisation, ou la Région à la française. *Actes des 3ème rencontres TIC & Territoire: quels développements ? de Lille,* ISDM, 16(163).

Femenia-Serra, F., & Ivars-Baidal, J. A. (2018). Smart tourism: Implications for the management of cities and tourist destinations. In M. T. Cantó López, J. A. Ivars Baidal, & R. Martínez Gutiérrez (Eds.), *Gestión inteligente y sostenible de las ciudades: Gobernanza, smart cities y turismo* (pp. 129–151). Valencia, Spain: Tirant Lo Blanch.

Fernández Tabales, A., Foronda, C., Galindo, L., & García, A. (2017). Developing a system of territorial governance indicators for tourism destinations. *Journal of Sustainable Tourism*, *25*(9), 1275–1305. doi :10.1080/09669582.2016.1260136

Garcia-Ayllon, S., & Miralles, J. L. (2015). New strategies to improve governance in territorial management: Evolving from "smart cities" to "smart territories". *Procedia Engineering*, *118*, 3–11. doi:10.1016/j.proeng.2015.08.396

Girardot, J. (2000). Principes, Méthodes et Outils d'Intelligence Territoriale. Évaluation participative et Observation Coopérative. Conhecer melhor para agir melhor, Actes du séminaire européen de la Direction Générale de l'Action Sociale du Portugal, Évora, (7-17). Lisboa, Portugal: DGAS.

Girardot, J. (2008). Evolution of the concept of territorial intelligence within the coordination action of the European network of territorial intelligence. *Res-Ricerca e Sviluppo per le politiche sociali*, 1(2), 11-30.

Greenfield, A. (2013). *Against the smart city*. London: Kindle Edition.

Gretzel, U., Sigala, M., Xiang, Z., & Koo, C. (2015). Smart tourism: Foundations and developments. *Electronic Markets*, *25*(3), 179–188. doi:10.100712525-015-0196-8

Herrera, F., López, R., & Fajardo, C. (2019). Developing Smart Regions: Proposal and Application of a Model for Island Territories. In Smart Cities and Smart Spaces: Concepts, Methodologies, Tools, and Applications (pp. 360-385). Hershey, PA: IGI Global

Hjalager, A. (2012). Innovation policies for tourism. *International Journal of Tourism Policy*, *4*(4), 336–356. doi:10.1504/IJTP.2012.052565

Ivars-Baidal, J., Solsona Monzonís, F. J., & Giner Sánchez, D. (2016). Tourism management and information and communication technologies (ICTs): The new smart destinations approach. *Documents d'Analisi Geografica*, *62*(2), 327–346. doi:10.5565/rev/dag.285

Jernsand, E. M., Kraff, H., & Mossberg, L. (2015). Tourism experience innovation through design. *Scandinavian Journal of Hospitality and Tourism*, *15*(1), 98–119. doi:10.1080/15022250.2015.1062269

Jovicic, D. Z. (2016). Key issues in the conceptualization of tourism destinations. *Tourism Geographies*, *18*(4), 445–457. doi:10.1080/14616688.2016.1183144

Komninos, N. (2008). *Intelligent cities and globalisation of innovation networks*. London, UK: Routledge, Taylor & Francis Group. doi:10.4324/9780203894491

Komninos, N. (2015). *The age of intelligent cities: smart environments and innovation-for-all strategies*. New York: Routledge, Taylor & Francis Group.

La Rocca, R. A. (2014). The role of tourism in planning the smart city. *TeMA. Journal of Land Use. Mobility and Environment*, *7*(3), 269–284.

López, Y., & Pulido, J. I. (2013). Sustainability in Spanish tourism policy. Advances or setbacks? *Papers de Turisme*, *53*, 44–68.

Masselot, C. (2008). Territorial Intelligence Communicational and Community System (TICCS). *Res-Ricerca e Sviluppo per le politiche social*, 1(2), 90-104.

Ministry of Industry. Tourism and Commerce. (2007). *Spanish Tourism Plan Horizonte 2020. Executive Document.* Retrieved from https://www.tourspain.es/es-es/Conozcanos/Documents/HistoricoPolitica-Turistica/PlanTurismoEspanolHorizonte2020.pdf

Ministry of Industry. Energy and Tourism. (2015). *National Plan of Intelligent Territories- Spain.* Retrieved from http://www.upv.es/contenidos/CAMUNISO/info/U0684517.pdf

Ministry of Industry. Energy and Tourism. (2017). *National Plan of Intelligent Territories- Spain.* Retrieved from https://www.esmartcity.es/2018/01/08/ministerio-energia-turismo-agenda-digital-aprueba-plan-nacional-territorios-inteligentes

Mora, L., Bolici, R., & Deakin, M. (2017). The First Two Decades of Smart-City Research: A Bibliometric Analysis. *Journal of Urban Technology*, *24*(1), 3–27. doi:10.1080/10630732.2017.1285123

Moulaert, F., & Sekia, F. (2003). Territorial innovation models: A critical survey. *Regional Studies*, *37*(3), 289–302. doi:10.1080/0034340032000065442

OECD. (Organization for Economic Cooperation and Development). (2018). Spain: Investing in 'Smart Destinations' for digitalisation and sustainable tourism. In Effective Policy Approaches for Quality Investment in Tourism, OECD Tourism Papers, (pp. 87-92). Paris, No.2018/03, OECD Publishing. doi:10.1787/88ea780c-en

Perea-Medina, M. J., Navarro-Jurado, E., & Luque-Gil, A. M. (2018). Territorial Intelligence: Conceptualization and progress in the state of the question. Possible links with tourist destinations. *Cuadernos de Turismo*, (41): 535–554.

Santos-Júnior, A., Mendes-Filho, L., Almeida-García, F., & Manuel-Simões, J. (2017). Smart Tourism Destinations: A study based on the vision of the stakeholders. *Revista Turismo em Análise*, *28*(3), 358–379. doi:10.11606/issn.1984-4867.v28i3p358-379

Schaffers, H. (2012). *Empowering citizens to realizing smart cities: Results from FIREBALL Smart city case studies (presentation).* 2012 future internet assembly (FIA), "smart cities and internet of things". Retrieved from http://www.fi-aalborg.eu/downloads/Session_1.1_Hans_Schaffers.pdf

Secretary of State for Tourism. (2012). *National and Integral Tourism Plan. Madrid, Ministry of Industry, Energy and Tourism.* Retrieved from https://turismo.gob.es/es-es/servicios/Documents/Plan-Nacional-Integral-Turismo-2012-2015.pdf

SEGITTUR. (Spanish state company dedicated to the management of innovation and tourism technologies). (2015). *Libro blanco de Destinos turísticos inteligentes: construyendo el futuro.* Retrieved from http://www.segittur.es/opencms/export/sites/segitur/.content/galerias/descargas/proyectos/Libro-Blanco-Destinos-Tursticos-Inteligentes-construyendo-el-futuro.pdf

Solsona, M. J. (2018). Quality in the management of rural tourist accommodation. *Gestión Turística*, *5*, 65–89. doi:10.4206/gest.tur.2000.n5-0

Townsend, A. (2013). *Smart Cities: Big Data, Civic Hackers, and the Quest for a New Utopia*. New York: Norton W. & Co.

Vera, J. F., & Baños, C. J. (2010). Renewal and restructuring of the consolidated tourist destinations of the coast: Recreational practices in the evolution of the tourist area. *Boletín de la Asociación de Geógrafos Españoles*, *53*, 329–353.

ADDITIONAL READING

Fernández Tabales, A., Foronda, C., Galindo, L., & García, A. (2017). Developing a system of territorial governance indicators for tourism destinations. *Journal of Sustainable Tourism*, *25*(9), 1275–1305. doi:10.1080/09669582.2016.1260136

Fernández Tabales, A., Mercado Alonso, I., Villar Lama, A., & Bascarán Estévez, M. V. (2015). Territorial governance and management of tourist destinations in the context of strong pressure from the real estate sector. Analysis of best local practices along the Cadiz Coast (Andalusia, Spain). *Revista de Geografía Norte Grande*, (60): 173–194.

Galindo-Pérez-de-Azpillaga, L., & Foronda-Robles, C. (2018). Digital governance and information technologies in local action groups (LAGs). *Cogent Social Sciences*, *4*(1), 1528730. doi:10.1080/23311886.2018.1528730

Gretzel, U., Sigala, M., Xiang, Z., & Koo, C. (2015). Smart tourism: Foundations and developments. *Electronic Markets*, *25*(3), 179–188. doi:10.100712525-015-0196-8

Jovicic, D. Z. (2019). From the traditional understanding of tourism destination to the smart tourism destination. *Current Issues in Tourism*, *22*(3), 276–282. doi:10.1080/13683500.2017.1313203

Li, Y., Hu, C., Huang, C., & Duan, L. (2017). The concept of smart tourism in the context of tourism information services. *Tourism Management*, *58*, 293–300. doi:10.1016/j.tourman.2016.03.014

Romão, J., & Neuts, B. (2017). Territorial capital, smart tourism specialization and sustainable regional development: Experiences from Europe. *Habitat International*, *68*, 64–74. doi:10.1016/j.habitatint.2017.04.006

Vinodan, A., & Manalel, J. (2019). Examining the Linkages of Community Intervention Strategies and Destination Quality of Ecotourism Destinations. *Journal of Quality Assurance in Hospitality & Tourism*, *20*(3), 362–386. doi:10.1080/1528008X.2018.1537820

KEY TERMS AND DEFINITIONS

Destination Management Entity (DME): Is the local territorial entity, local entity or uniprovincial autonomous community that governs and manages an Intelligent Tourist Destination (ITD) with sufficient responsibility and authority to establish, implement, maintain and improve the Management System (MS) of an ITD.

Destination Management System: Set of elements of a DME that interacts to establish policies, objectives and processes to achieve those objectives.

Indicator: Data or set of data that help to objectively measure the evolution of a process or activity.

Public Destination Management: Set of actions for public entities related to tourism that have the purpose of developing at the local scale. It includes both tourism policy measures and other policies that affect tourism (environment, culture, mobility, etc.); and responds to a broad approach, more typical of a general destination management than of the strict application of tourism planning instruments.

Smart Tourist Destination: A destination whose management involves a strong innovation component, based on the use of information and communication technologies (ICTs), as tools to obtain a competitive destination, the sustainable development of the territory, and the participation of the citizens involved.

Territorial Intelligence: The multidisciplinary science the object of which is the sustainable development of territories in the knowledge society and the subject the territorial community (Girardot, 2008).

Tourist Destination: Singular territory receiver of tourist flows. It locates the tourist resources that generate the activity, the offer of lodging and complementary, as well as the infrastructures and equipment that make it possible, also having an image and marketing channels that make the destination recognizable and accessible by the demand.

Tourist Governance: Way of governing tourism characterized by the interrelation between public administrations, private agents and articulated civil society, in a context of interaction between different spatial and administrative scales. The concept of governance is based on the key idea of increasing the legitimacy and effectiveness of public management, by broadening and deepening citizen participation in decision-making, in order to achieve lasting economic, social and institutional development.

Chapter 22
Smart Tourism Planning:
Geographical Evidence From Poland

Katarzyna Leśniewska-Napierała
https://orcid.org/0000-0003-2998-6179
University of Lodz, Poland

Tomasz Napierała
https://orcid.org/0000-0002-6407-5197
University of Lodz, Poland

Sevda Sahilli Birdir
Mersin University, Turkey

Kemal Birdir
https://orcid.org/0000-0003-1353-3618
Mersin University, Turkey

ABSTRACT

This chapter discusses geographical context of smart tourism planning through the analysis of relations between spatial planning, tourism development, and technology. The main question addressed in this enquiry is whether development of technology and tourism respects sustainable spatial planning. Thus, geographical influence of tourism and ICT on spatial planning was investigated. Research is carried out in Poland as tourist destinations in the country are substantially different regarding their tourist attractions, infrastructure, and development. Geographically weighted regression (GWR) was applied to identify regions where spatial planning is significantly influenced by tourism, and technology development as well. Both mentioned phenomena were recognized as competing in terms of spatial planning. Tourism and technology development did not support spatial planning parallely. Hence, establishing and developing smart tourism planning was confirmed as challenging from the geographical perspective.

DOI: 10.4018/978-1-7998-1989-9.ch022

INTRODUCTION

The scientific discussion on smart tourism destinationsand smart cities has been growing in recent years (Del Chiappa&Baggio 2015). The rise of the tourist numbers all over the world seems to be one of the major contemporary concerns of the tourist attracting destinations, especially the most popular ones. Some of destinations are under threat of extreme visitor numbers which severely affect social, economic, cultural and especially environmental aspects of the destinations. Of course, the extreme rise of the visitor numbers is a threat not only for popular destinations but tourism destinations all over the World. Mentioned trends of tourism development strongly affect spatial planning of tourism destinations, and are related to development of technology supporting all contemporary development processes.

The main focus of the chapter is to discuss geographical context of smart tourism planning through the analysis of relations between spatial planning, tourism development, and technology. The main question addressed in this enquiry is whether development of technology and tourism respects sustainable spatial planning. Thus, geographical influence of tourism and ICT on spatial planning was investigated. Research is carried out in Poland as tourist destinations in the country are substantially different regarding their tourist attractions, infrastructure, and development. Geographically weighted regression (GWR) was applied to identify regions where spatial planning is significantly influenced by tourism, and technology development as well.

BACKGROUND

The word "smart" has become a popular term for describing the development assisted by information and communication technologies (ICTs). The relation between people and multi devices has occurred also in tourism as support of travel experience (Gretzel et al. 2016). The Internet has changed the channels of dissemination of travel information and enabled tourists to freely share knowledge (Munar& Jacobsen 2014). ICTs have changed the consumer behaviour as well. Smart tourism is a progression from traditional tourism to e-tourism, where the consumers are very familiar with the new technologies, and use them permanently. It contains multiple components, like reservations systems, social media, various communication and connectionapplications etc. (Gretzel et al. 2015). In tourism industry, technological developments and applications are important not only for the competitiveness of tourism destinations and tourism enterprises but also for the experience of tourists (Huang et al. 2017). The idea of smart tourism destinations derived from the concept of smart cities, as technology has become an enabler of development of tourism destinations (Boes et al. 2015).

The main focus of this chapter is to discuss geographical context of smart tourism planning. Tourism is a spatial complex of processes that shapes the tourism landscape and affect social, economic, cultural, and political relations (Shaw & Williams 2004). Moreover, smart tourism simultaneously uses and influences technology which make tourism destinations available. Thus, the relations between spatial planning, tourism development, and technology need to be investigated.The focus of this chapter is on spatial planning as it is expected that sustainable spatial planning of tourism destinations might enable spatial justice (Jamal & Camargo 2014). However, when considering smart tourism planning, tourism development and development of ICT industry need to improve the quality and effectiveness of spatial planning. Thus, following notions need to be discussed and related to reach other: spatial planning, tourism development, sustainability, spatial justice, and smart tourism.

Bagio& Del Chiappa (2014) emphasized that smart tourism destination is a networked system of stakeholders who deliver different services to tourists, suported by a technological infrastructure aimed at creating a digital environment for cooperation, knowledge and innovation.Conceptof smart tourism destination is closely related to smart city, which could be define as a city where new digital technologiesintegrate and coordinate a system of infrastructure and services. ICTs are mainly used for communication, dissemination, new forms of governance, and defining problems related to e.g. transportation or energy (Batty et.al. 2012, Massa &Campagna 2014, Sigalat-Signes et.al. 2019). Except ICTs, a holistic development with sustainability, governance, human-centricity, public value creation, or citizen participationare key elements of smart cities (Desdemoustier et.al. 2019; Sigalat-Signes et.al. 2019).

Technology is fundamental to develop modern and effective smart cities so that to reach spatially enabled urban population, and improve access to the city and its services (Roche et al. 2012). Applications and developments clearly show that technology is fundamental to reach or create smart cities so that the citizens or locals can access to what a city provides, the services the city offers and may improve their quality of life in a more efficient and effective way (Floris&Campagna 2014). A successful smart city program should help monitoring and improving a city's infrastructure, planning activities and enhnce the services provided to the locals (Hall 2000).Smart applications, smart cities and smart spatial planning are all interrelated concepts that need to be studied in a delicate way so that we can plan, organize and manage all the phenomenal growth of global tourism to make the cities, fragile destinations, monuments and anything related to tourism sustainable. Infact, various successful applications could be very insightful for policy makers to understand how smart applications may help overcome some of the pressing problems of urban cities, societies and anything related to human actions including planning tourism sites (March &Ribera-Fumaz 2016, Albino et.al. 2015, Caragliu et.al. 2011).

In addition, Higgins-Desbiolles (2008) defined the idea of justice tourism. She emphasized that justice tourism could be recognized not only as a contrary to the contemporary mass tourism, but also as a contribution to a just global order. Jamal & Camargo (2014: 12) defined just destination as "*a destination whose tourism planning, policy making and practices enable the fair treatment of its environmental and social-cultural resources (tangible and intangible), and facilitate the well-being of place, people and pasts*". During planning sustainable tourism development, tourism sustainable contribution in local community well-being should be recognized as priority (Barton & Leonard, 2010).Bramwell& Lane (2008), Gezon (2014), Higgins-Desbiolles (2008; 2018), and Ramon-Hidalgo & Harris (2018) noticed that sustainable tourism can exist only with the balanced of: 1) justice, 2) equity and democracy, including political empowerment of local people, organizations, and communities, 3) ecological and heritage preservation, and aesthetic landscape integrity, 4) and economic efficiency resulting in a decrease of negative visitors' impact, and in increase of local community benefits. Thus, spatial justice needs to be recognized as a goal of sustainable tourism development, including development of smart tourism destinations.

MAIN FOCUS OF THE CHAPTER

Methods, Data and Research Area

The main question discussed in this enquiry is whether the development of technology and tourism correspond to sustainable spatial planning, and might be recognized as a potential for development of smart tourism planning. To address mentioned question geographical influence of tourism and ICTs on spatial planning development was investigated. Research was carried out in Poland as tourist destinations in th ecountry are substantially different regarding their tourist attractions, infrastructure, and development. All mentioned phenomena: tourists mobilities, location of ICT enterprises, and development of spatial planning are geographically differentiated, and potentially spatially autocorrelated. Thus, traditional approach for regression analysis of investigated impact might fail. The solution for diagnosed issue recommended in the literature is geographically weighted regression (GWR).

GWRwas applied to identify regions where spatial planning is significantly influenced by tourism, and technology development as well. Moreover, this investigated impact is expected to be positive or negative, like the smart tourism might be considered as a positive or negative phenomenon regarding its geographical context. GWR enabled local models for every *j* Polish county considered as a polygon represented by geographical coordinates *Long* and *Lat* of polygon's centroid. GWR is based on Euclidean distance between the centroids (Gao& Li 2011). GWR allowed to estimate local spatial patterns of tourism and ICTs development impact on spatial planning, and solved reported issues of global statistics like ordinary least squares regression (Huang et al. 2015). The coverage of county area by valid local spatial development plans in 2017 was considered as a dependent variable *LocSpatDevPlansCov*, while the number of tourist nights sold by county in 2017 per 1 km^2, and number of ICT enterprises by county in 2017 per 1 km^2 - as predictors *TouristNights* and *IctEnterprises*. Investigated relation might be expressed as:

$$\begin{aligned} LocSpatDevPlansCov_j = \beta_0\left(Long_j, Lat_j\right) + +\beta_1\left(Long_j, Lat_j\right) \cdot TouristNights \\ +\beta_2\left(Long_j, Lat_j\right) \cdot IctEnterprises + Error \end{aligned} \quad (1)$$

GWR applies weighted distance decay function to find the bandwidth for every county spatially distributed around every considered location to be included in the local model estimation. Moreover, corrected Akaike information criterion was applied to delimit optimal, adaptive bandwidth (Brown et al. 2012).

The study area of this chapter is Poland which is a country located in Central-Eastern Europe. The area of the country is 322 575 km^2. With the population 38.4 mln people. Poland is the 6th most populated member of European Union. The neighboring countries of Poland are Germany in the west, the Czech Republic and Slovakia in the south, Ukraine and Belarus in the east, and Lithuania and Russia (Kaliningrad Oblast) in the northeast. On the north, Poland is bordered by the Baltic Sea. The capital city and also the biggest metropolis is Warsaw with a population of about 1.8 mln in 2018. The gross domestic product (GDP) of Poland in 2017 was 524,5 billion USD.

Trying to characterize and interpret contemporary phenomena in Poland in their spatial dimension it is necessary to emphasize the role of political changes which took place after 1989. That period saw a radical transformation of many elements of the country's social and economic life (Bański 2007). Since 1989 we may talk about the political transformation and transition from centrally planned economy to a market economy (Borowska-Stefańska et al. 2018). The most important changes in Poland after 1989 directly affecting land development include: (1) decentralization of the political authority and economy, (2) adjustment of legal regulations to the EU standards, and (3) obtaining access to structural funds and agricultural subsidies of the European Union the moment Poland joins the EU (Węcławowicz et al. 2006). As a result of decentralisation process in Poland, a new division has been created. Since the beginning

Table 1. Descriptive statistics of investigated variables

Abbreviation of the variable	Minimum	Maximum	Average	Standard deviation
LocSpatDevPlansCov	0.56	101.08	36.28	29.03
TouristNights	0.00	45,467.77	701.00	2,771.78
IctEnterprises	0,02	72.99	1.64	5.25

Source: Own elaboration.

of 1999 the administrative division of Poland has been based on 3 levels of subdivision (Wendt 2001). The country is divided into 16 regions (*województwo* in Polish), 379 counties (*powiat*) including 65 cities, and 2479 municipalities (*gmina*). In this chapter, the analysis included all counties in Poland.

All data for quantitative studies were received from two sources. Cartographic data on administrative division of Poland in shape files were gained from the Head Office of Geodesy and Cartography from State geodetic and cartographic resource. Statistical data were obtained from the Local Data Bank resources provided by the Statistics Poland. The following data were estimated based on this resource: (1) number of tourist nights sold by county in 2017 per 1 km^2, (2) number of ICT entities by county in 2017 per 1 km^2, and (3) the coverage of county area by valid local spatial development plans in 2017.

RESULTS AND DISCUSSION

The main consideration of this chapter is the sustainable spatial planning of tourism destinations towards spatial justice, thorough smart tourism.If spatial planning quality is supported mainly through tourism development, it is expected that such kind of regions has a potential to become sustainable tourism destinations. On the other hand, if the quality of spatial planning results mainly from technological development, it might be assumed that identified regions are expected to be characterized as smart. When both types of impact are identified in same place, there is a room for smart tourism destinations.The goal of this research is to identify regions where spatial planning is significantly influenced by tourism, and technology development as well. The values of all investigated variables are differentiated across the observations in Poland (Table 1). Thus, they might be considered as a potential dependent and independent factors in estimated models.

It needs to be underlined that spatial planning development is differentiated across the regions in Poland. The highest values of coverage of counties' areas by valid local spatial development plans in 2017 were noticed in the southern part of Poland (Figure 1). However, there is no clear evidence explaining spatial volatility of investigated spatial planning development.

City tourism destinations and both summer and winter tourism resorts located on the Baltic seaside and in the mountain regions in the south are recognized as the most popular destinations of tourist mobilities in Poland (Figure 2). All mentioned destinations are the areas recognized by Napierała (2013) as the most profitable from the perspective of tourism sector, especially lodging industry.

It is confirmed that ICT enterprises operate mainly in the biggest cities of Poland and their functional areas (Figure 3). The impact of ICT development on tourism in metropolitan areas in Poland was already discussed (Napierała 2017). It should be emphasized that the ICT development, including internetization

Figure 1. Coverage of counties by valid local spatial development plans in 2017
Source: Own elaboration based on Statistics Poland

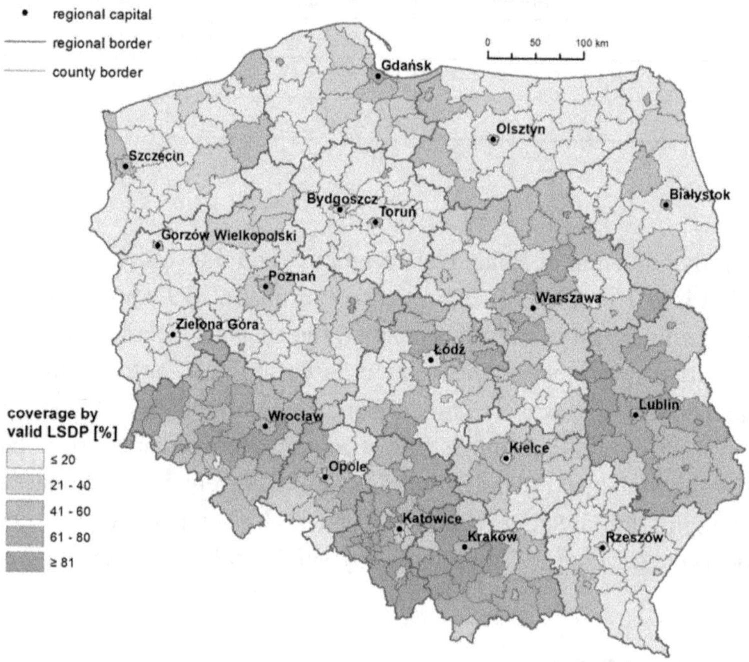

Figure 2. Number of tourists' nights sold by county in 2017 per 1 km²
Source: Own elaboration based on Statistics Poland

Table 2. Global model describing impact of tourism and ICT enterprises on spatial planning development

Abbreviation of the variable	Coefficient	Error	t-Student test value	p-value
Intercept	34.8400	1.5540	22.4190	Tends to 0
TouristNights	0.0007	0.0006	1.0310	0.3034
IctEnterprises	0.5958	0.3399	1.7530	0.0804

Source: Own elaboration.

of tourism industry, covered not only the core cities but also peripheries of metropolitan areas. Access to place-based technological resources seems to be crucial from the perspective of smart tourism planning.

Global impact of two investigated predictors (number of tourist nights sold by county in 2017 per 1 km^2, and number of ICT enterprises by county in 2017 per 1 km^2) of spatial planning development ws identified (Table 2). It need to be emphasized that the results of the analysis are completely unsatisfactory, as the global model based on Oridinary Least Square regression explained only 2.3% of variance of coverage of counties by valid local spatial development plans in 2017.

Map of local coefficients of determination, and maps of values and significance (results of t-Student test) of estimated local parameters of independent variables are the basic results of GWR implementation (Gao& Li 2011; Mennis 2006). Map of local coefficients of determination allow to identify how well local models fit to observations (Fig. 4). It needs to be emphasized that local models explained investigated phenomenon mainly in the northern part of Poland. Best results of geographically weighted estimation

Figure 3. Number of ICT enterprises by county in 2017 per 1 km^2
Source: Own elaboration based on Statistics Poland

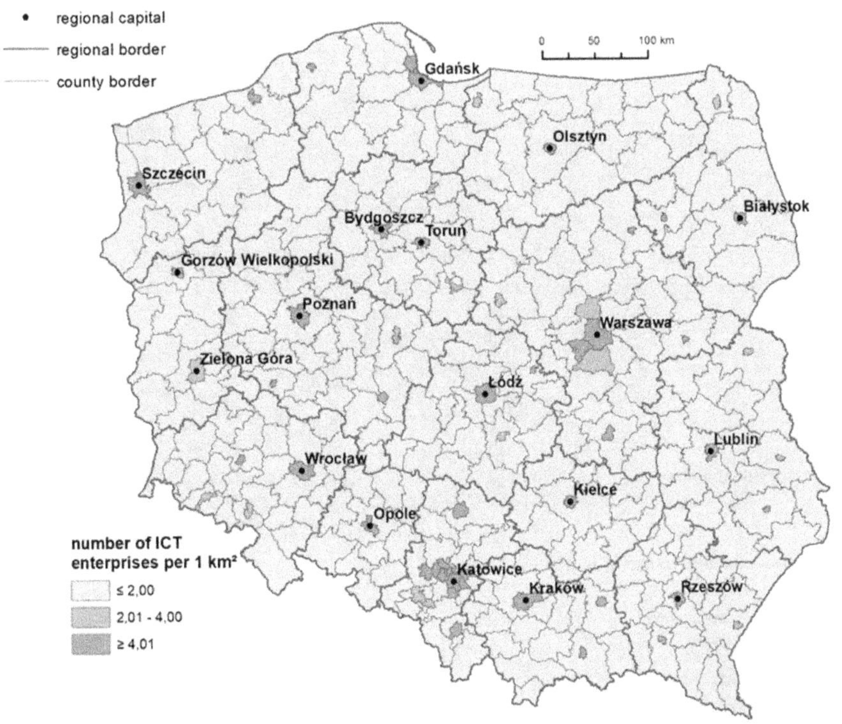

Figure 4. Coefficients of determination for local models describing impact of tourism and ICT enterprises on spatial planning development
Source: Own elaboration

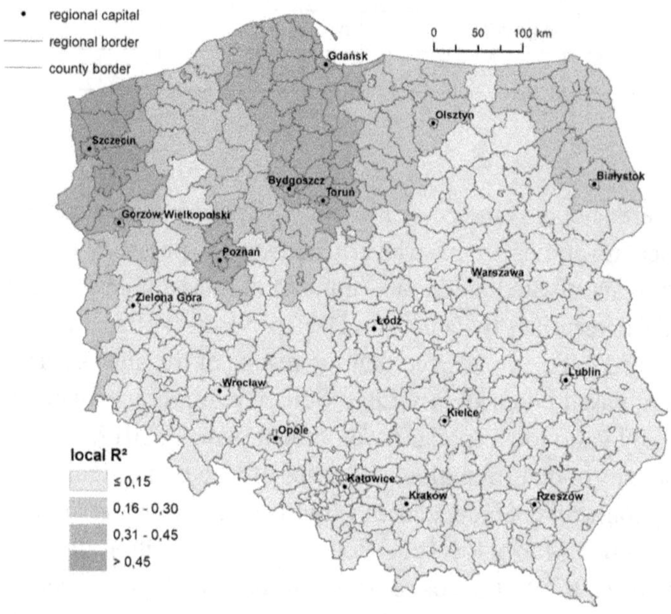

Figure 5. Impact of tourism on spatial planning development
Source: Own elaboration

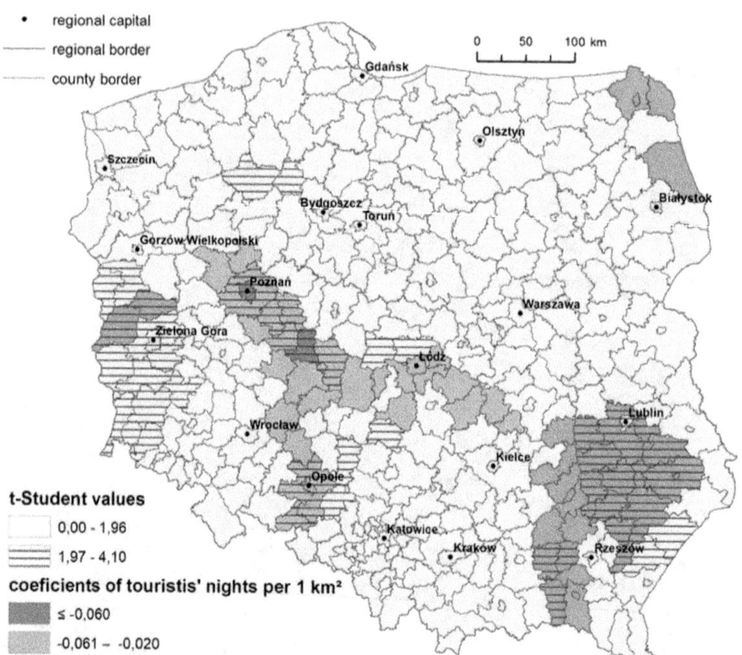

were obtained for Baltic coast, main Polish destination for leisure tourism. This confirms that the more tourism is developed and significant from the perspective of local economy, the easier understanding of tourism related phenomenon is. The limitation of applied method is that urban tourism and ICT influence on spatial planning development might not be detected as the phenomenon are spatially concentrated.

Maps of estimated values of local parameters present spatial varying impact of investigated tourism and ICT sector on spatial planning development (Fig. 5, and Fig. 6). Two underdeveloped rural regions need to be emphasized: lubuskie in the western part of Poland, and lubelskie in south-eastern one. In both regions significant and positive impact of tourism on spatial planning has been detected. In mentioned regions tourism is recognized as a positive phenomena from the perspective of spatial planning development. Both tourist mobilities and tourism economy development stimulate the development of spatial planning as the quality of space is recognized as one of the resources crucial for achieving competitive advantage by tourism industry (Maćkiewicz&Konecka-Szydłowska 2017; Marković&Klarić 2015; Petrović&Marković 2013). Interestingly counties in the lubelskie region are characterized by highest coverage from local development plans, while in the lubuskie region –bylowest one. Thus, we can observe the same process at different levels of spatial planning and rural tourism development.

On the other hand, Greater Poland is the region where two opposite but significant tendencies are observed. Spatial planning development is positively influenced by technology, and negatively by tourism. Higher number of ICT enterprises stimulate spatial planning development, mainly in the counties

Figure 6. Impact of ICT enterprises on spatial planning development
Source: Own elaboration

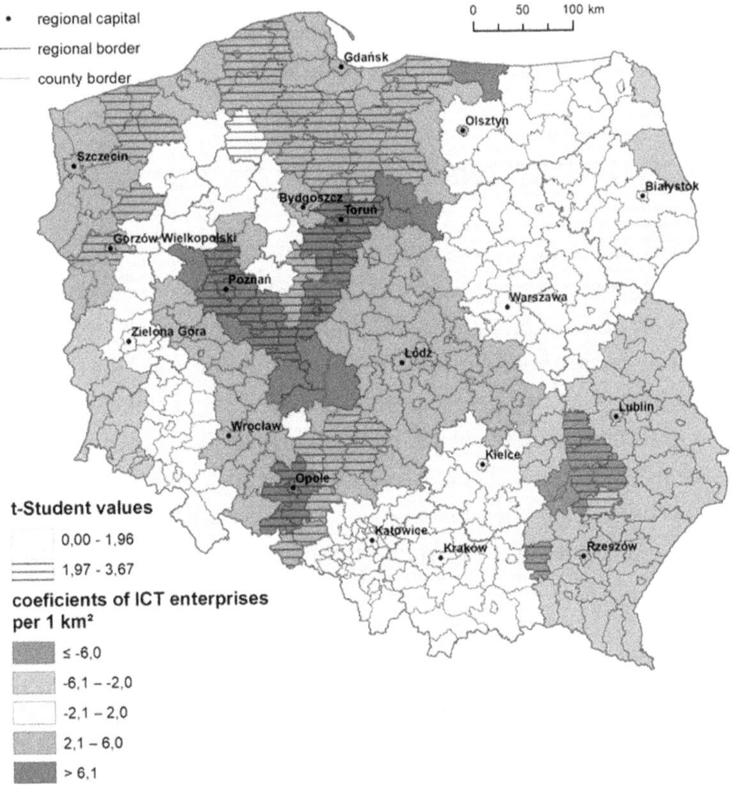

where the number of tourists is relatively lower. Thus, technology development is recognized as a driving force for spatial planning development in the mentioned region.

It needs to be emphasized that there is no area in Poland where spatial planning development is significantly influenced by both tourism and technology. Both phenomena are recognized as competing in terms of spatial planning. Thus, introducing the idea of smart tourism planning reveals as challenging from the geographical perspective as tourism and technology development does not support spatial planning in parallel.

SOLUTIONS AND RECOMMENDATIONS

Planning of tourism destinations on a spatial basis sems to be a very crucial and pressing issue all over the world. Uncontrolled, unplanned and not managed, tourism activities could be a huge threat to the sustainability of tourist hosting destinations. Breaching carrying capacity of a destination in terms of tourist numbers, man made constructions, roads, every forms of superstructures and accommodation capacities may result in a highly polluted, crowded, noisy, stressfull and economically degrading destination. The main question that was addressed in this enquiry is whether development of technology and tourism respects sustainable spatial planning. Thus, geographical influence of tourism and ICT on spatial planning was investigated. Even though establishing and developing smart tourism planning was confirmed as challenging from the geographical perspective, its important to point out that developments on ICT should carefully be monitored by destination managers to effectively manage all tourism movements at the destination. To decrease harmfull effects and to increase benefits of tourism, smart special planning methodologies could help destination managers in every aspect of the industry. Following best actions and applications in other destinations and sharing the successful experiences can also be very helpfull for sustainability and effective management of destinations.

FUTURE RESERACH DIRECTIONS

Developing tools and methods to plan the tourism movement seems to be a major challenge both for tourism destinations and managers/stakeholders of these places. Smart tools such as artificial intelegence, augmented reality, robotic applications and other developments all may be important to assess and manage the tourism activities both for benefit of communities and sustainability of destinations. Researchers should closely work with destination managers and other stakeholders to measure tourism's and tourists' impact on destinations and find and develop new tools and perspectives to manage tourism while keep the destinations sustainable. Therefore, studies that incorporates smart tools and methodologies to explore the effects of tourism phenomena on destinations and future of these mostly very fragile destinations could to be an important future direction for researchers.

CONCLUSION

Destination managers and locals should never forget that tourism presents both an opportunity and a threat to destination resources. Therefore, a smart spatial planning of tourism destinations must combine needs and wants, and expectations and interests of all the stakeholders. This focus on local stakeholders for tourism development must be a global priority. Hence, increase of the benefits of the local stakeholders may gain from the influx of tourists all over the world. Moreover, means, methods and technology to minimize the negative effects of tourism on destinations are needed. Smart tourism planning, which incorporates the latest technology to effectively manage the tourist flows and all the resources a destination presents may be an effective approach for the sustainability of the destinations and spatial justice.

For most of the popular tourist destinations, excess tourism numbers seems to be a major problem in the coming decades. Overtourism, as its is expressed, discussed and searched in the tourism academia has become a major concern for both the locals because of its negative effects on daily life of the people and sustainability of destinations and their fragile and must-preserved attractions such as natural beauties, historical remains, unique heritage and cultural richness. Tourism industry has been doubling the growth experienced in the world economy in the last decade and estimates show that the growth in the global tourism industry has no sign of slowing in the coming years. To be able to manage the negative effects of the global tourism boom and to increase the benefits from this boom for the locals and economy of the destinations, spatial planning benefiting from smart applications might be a powerful and smart action for the authorities and destination management organizations.

The main empirical question discussed in this chapter was whether the development of technology and tourism correspond to sustainable spatial planning in tourism destinations. To address mentioned question, geographical influence of tourism and ICTs on spatial planning in Poland was investigated. Thus, GWR was applied to identify regions where spatial planning is significantly influenced by tourism, and technology development as well. Interestingly, no region in Poland was identified where spatial planning development is significantly affected by both tourism and technology. Both phenomena were recognized as competing in terms of spatial planning. Both tourism and technology development did not support spatial planning parallely. Hence, establishing and developing smart tourism planning was confirmed as challenging from the geographical perspective.

REFERENCES

Albino, V., Berardi, U., & Dangelico, R. M. (2015). Smart Cities: Definitions, Dimensions, Performance, and Initiatives. *Journal of Urban Technology*, *2*(1), 3–21. doi:10.1080/10630732.2014.942092

Baggio, R., & Del Chiappa, G. (2014). Real and virtual relationships in tourism digital ecosystems. *Information Technology & Tourism*, *14*(1), 3–19. doi:10.100740558-013-0001-5

Bański, J. (2007). Koncepcje rozwoju struktury przestrzennej w Polsce – polaryzacja czy równoważenie? *PrzeglądGeograficzny*, *79*(1), 45–77.

Barton, A. W., & Leonard, S. J. (2010). Incorporating social justice in tourism planning: Racial reconciliation and sustainable community development in the Deep South. *Community Development (Columbus, Ohio)*, *41*(3), 298–322. doi:10.1080/15575330903444051

Batty, M., Axhausen, K. W., Giannotti, F., Pozdnoukhov, A., Bazzani, A., Wachowicz, M., & Portugali, Y. (2012). Smart cities of the future. *The European Physical Journal. Special Topics*, *214*(1), 481–518. doi:10.1140/epjst/e2012-01703-3

Boes, K., Buhalis, D., & Inversini, A. (2015), Conceptualising smart tourism destination dimensions. In I. Tussyadiah, & A. Inversini (Eds.), Information and Communication Technologies in Tourism 2015, Springer, 391–403.

Borowska-Stefanska, M., Lesniewska-Napierala, K., & Wisniewski, S. (2018). Land cover changes in Poland between 1990 and 2012. *Geografie (Utrecht)*, *123*(3), 63–83.

Bramwell, B., & Lane, B. (2008). Priorities in Sustainable Tourism Research. *Journal of Sustainable Tourism*, *16*(1), 1–4. doi:10.2167/09669580803489612

Brown, S., Versace, V. L., Laurenson, L., Ierodiaconou, D., Fawcett, J., & Salzman, S. (2012). Assessment of Spatiotemporal Varying Relationships Between Rainfall, Land Cover and Surface Water Area Using Geographically Weighted Regression. *Environmental Modeling and Assessment*, *17*(3), 241–254. doi:10.100710666-011-9289-8

Caragliu, A., Del Bo, C., & Nijkamp, P. (2011). Smart Cities in Europe. *Journal of Urban Technology*, *18*(2), 65–82. doi:10.1080/10630732.2011.601117

Del Chiappa, G., & Baggio, R. (2015). Knowledge transfer in smart tourism destinations: Analyzing the effects of a network structure. *Journal of Destination Marketing & Management*, *4*(3), 145–150. doi:10.1016/j.jdmm.2015.02.001

Desdemoustier, J., Crutzen, N., Cools, M., & Teller, J. (2019). Smart City appropriation by local actors: An instrument in the making. *Cities (London, England)*, *92*, 175–186. doi:10.1016/j.cities.2019.03.021

Floris, R., & Campagna, M. (2014). Social media geographic information in tourism planning. *TEMA-Journal of Land Use, Mobility and Environment,* Special Issue; Eight Int. *Conference INPUT Smart City-Planning for Energy, Transportation and Sustainability of the Urban System*. Naples, June 4-6, 2014.

Gao, J., & Li, S. (2011). Detecting spatially non-stationary and scale-dependent relationships between urban landscape fragmentation and related factors using Geographically Weighted Regression. *Applied Geography (Sevenoaks, England)*, *31*(1), 292–302. doi:10.1016/j.apgeog.2010.06.003

Gezon, L. L. (2014). Who wins and who loses? Unpacking the "local people" concept in ecotourism: A longitudinal study of community equity in Ankarana, Madagascar. *Journal of Sustainable Tourism*, *22*(5), 821–838. doi:10.1080/09669582.2013.847942

Gretzel, U., Sigala, M., Xiang, Z., & Koo, C. (2015). Smart tourism: Foundations and developments. *Electronic Markets*, *25*(3), 179–188. doi:10.100712525-015-0196-8

Gretzel, U., Zhong, L., & Koo, C. (2016). Application of smart tourism to cities. *International Journal of Tourism Cities*, *2*(2). doi:10.1108/IJTC-04-2016-0007

Hall, R. E. (2000). "The vision of a smart city", In *Proceedings of the 2nd International Life Extension Technology Workshop,* Paris, France. Available at http://ntl.bts.gov/lib/14000/14800/14834/DE2001773961.pdf

Harrison, C., Eckman, B., Hamilton, R., Hartswick, P., Kalagnanam, J., Paraszczak, J., & Williams, P. (2010). Foundations for Smarter Cities. *IBM Journal of Research and Development*, *54*(4), 1–16. doi:10.1147/JRD.2010.2048257

Higgins-Desbiolles, F. (2008). Justice tourism and alternative globalisation. *Journal of Sustainable Tourism*, *16*(3), 345–364. doi:10.1080/09669580802154132

Higgins-Desbiolles, F. (2018). Sustainable tourism: Sustaining tourism or something more? *Tourism Management Perspectives*, *25*, 157–160. doi:10.1016/j.tmp.2017.11.017

Huang, C. D., Goo, J., Nam, K., & Yoo, C. W. (2017). Smart tourism technologies in travel planning: The role of exploration and exploitation. *Information & Management*, *54*(6), 757–770. doi:10.1016/j.im.2016.11.010

Huang, J., Huang, Y., Pontius, R. G. Jr, & Zhang, Z. (2015). Geographically weighted regression to measure spatial variations in correlations between water pollution versus land use in a coastal watershed. *Ocean and Coastal Management*, *103*, 14–24. doi:10.1016/j.ocecoaman.2014.10.007

Jamal, T., & Camargo, B. A. (2014). Sustainable tourism, justice and an ethic of care: Toward the Just Destination. *Journal of Sustainable Tourism*, *22*(1), 11–30. doi:10.1080/09669582.2013.786084

Maćkiewicz, B., & Konecka-Szydłowska, B. (2017). Green Tourism: Attractions and Initiatives of Polish Cittaslow Cities. In N. Bellini, & C. Pasquinelli (Eds.), Tourism in the City: Towards an Integrative Agenda on Urban Tourism, Springer, 297–310.

March, H., & Ribera-Fumaz, R. (2016). Smart contradictions: The politics of making Barcelona a Self-sufficient city. *European Urban and Regional Studies*, *23*(4), 816–830. doi:10.1177/0969776414554488

Marković, I., & Klarić, Z. (2015). Attitudes of Local Population of Tourism Impacts on Destination Sustainability – Case of Croatia. *Turizam*, *19*(3), 98–110. doi:10.5937/Turizam1503098M

Massa, P., & Campagna, M. (2014). Social media geographic information: recent findings and opportunities for smart spatial planning. *TEMA- Journal of Land Use, Mobility and Environment.* Special Issue; *Eight Int. Conference INPUT Smart City-Planning for Energy, Transportation and Sustainability of the Urban System.*Naples, June 4-6, 2014.

Mennis, J. (2006). Mapping the Results of Geographically Weighted Regression. *The Cartographic Journal*, *43*(2), 171–179. doi:10.1179/000870406X114658

Munar, A. M., & Jacobsen, J. K. S. (2014). Motivations for sharing tourism experiences through social media. *Tourism Management*, *43*, 46–54. doi:10.1016/j.tourman.2014.01.012

Napierała, T. (2013). *Przestrzenne zróżnicowanie cen usług hotelowych w Polsce*. Wydawnictwo Uniwersytetu Łódzkiego. doi:10.18778/7525-929-2

Napierała, T. (2017). Internetization of selling hotel rooms in metropolitan area of Łódź (Poland). Journal of Geography. *Politics & Society*, *7*(3), 19–30.

Petrović, M. D., & Marković, J. J. (2013). Sustainable Tourism as a Part of Comprehensive Environmental Monitoring. A Study of Serbia. *Journal of Environmental and Tourism Analyses*, *I*, 1, 30–42.

Ramón-Hidalgo, A. E., & Harris, L. M. (2018). Social Capital, political empowerment and social difference: A mixed-methods study of an ecotourism project in the rural Volta region of Ghana. *Journal of Sustainable Tourism*, *26*(12), 2153–2172. doi:10.1080/09669582.2018.1546711

Roche, S., Nabian, N., Kloeckl, K., & Ratti, C. (2012). Are 'Smart Cities' Smart Enough? *Global Geospatial Conference 2012*, Global Spatial Data Infrastructure Association. Available at http://www.gsdi.org/gsdiconf/gsdi13/papers/182.pdf

Shaw, G., & Williams, A. M. (2004). Tourism and tourism spaces. Atlanta, GA: *Sage*.

Sigalat-Signes, E., Calvo-Palomares, R., Roig-Merino, B., & García-Adán, I. (2019). Transition towards a tourist innovation model: The smart tourism destination. Reality or territorial marketing? *Journal of Innovation & Knowledge.*

Węcławowicz, G., Bański, J., Degórski, M., Komornicki, T., Korcelli, P., & Śleszyński, P. (2006). *Przestrzenne Zagospodarowanie Polski na początku XXI* wieku. PAN IGiPZ.

Wendt, J. A. (2001). The administrative division of Poland and the prospects of the transborder cooperation. In M. Koter, & K. Heffner (Eds.), Changing Role of Border Areas and Regional Policies, Region and Regionalism No. 5, Wydawnictwo Uniwersytetu Łódzkiego,39–48.

ADDITIONAL READING

Alan, W. B., & Leonard, S. J. (2010). Incorporating social justice intourism planning: Racial reconciliation and sustainable community development in the Deep South. *Community Development (Columbus, Ohio)*, *41*(3), 298–322. doi:10.1080/15575330903444051

Brougham, D., & Haar, J. (2018). Smart technology, artificial intelligence, robotics, and algorithms (STARA): Employees' perceptions of our future workplace. *Journal of Management & Organization*, *24*(2), 239–257. doi:10.1017/jmo.2016.55

Di Bella, A., Petino, G., & Scrofani, L. (2019). The Etna macro region between peripheralization and innovation: Towards a smart territorial system based on tourism. *Regional Science Policy and Practice*, *11*(3), 493–507. doi:10.1111/rsp3.12176

Gezon, L. L. (2014). Who wins and who loses? Unpacking the "local people"concept in ecotourism: A longitudinal study of community equity in Ankarana, Madagascar. *Journal of Sustainable Tourism*, *22*(5), 821–838. doi:10.1080/09669582.2013.847942

Guo, Y., Liu, H., & Chai, Y. (2014). The embedding convergence of smart cities and tourism internet of things in China: An advance perspective. [AHTR]. *Advances in Hospitality and Tourism Research*, *2*(1), 54–69.

Ramón-Hidalgo, A. E., & Harris, L. M. (2018). Social capital, political empowerment and social difference: A mixed-methods study of an ecotourism project in the rural Volta region of Ghana. *Journal of Sustainable Tourism*, *26*(12), 2153–2172. doi:10.1080/09669582.2018.1546711

Schellhorn, M. (2010). Development for whom? Social justice and the business of ecotourism. *Journal of Sustainable Tourism*, *18*(1), 115–135. doi:10.1080/09669580903367229

Shamai, A., & Yousofi Babadi, S. (2018). An Analysis of Good Governance Role on Sustainable Tourism Revenues (Case: City of MasjedSeleyman). *Urban Economics and Management*, *6*(4), 477–496.

Yüncü, D. (2015). Relationships between perceptions of virtual destination environment, satisfaction and loyalty. *Journal of Hospitality and Tourism Technology*, *6*(2), 160–173. doi:10.1108/JHTT-07-2014-0021

KEY TERMS AND DEFINITIONS

Geographically Weighted Regression: Geographically Weighted Regression is a geographical method of analysis which allows to identify areas where the impact of predictors on dependent variable is significant, positive or negative. The method solves reported issues of global statistics like ordinary least squares regression.GWR enables local models for every considered spatial unit of analysis, and neighbouring unitis. In other words, GWR allows to estimate local spatial patterns of any investigated dependent variable (Gao& Li 2011; Huang et al. 2015).

Just Destination: Just destination is a tourism destination whose tourism planning, policy making and practices of all stakeholders of tourism development enable the fair treatment of its environmental, social, economic, cultural resources, both tangible and intangible. The focus of planning and development of just destination is on well-being of place, people and pasts(Barton & Leonard, 2010; Jamal & Camargo, 2014).

Smart Tourism Destination: The idea of smart tourism destination derived from the concept of smart city, as technology has become an enabler of development of tourism destinations (Boes et al. 2015).Smart tourism destination is focused on e-tourism, where the consumers are very familiar with new technologies, and use them permanently. It contains multiple components, like reservations systems, social media, various communication and connection applications etc. (Gretzel et al. 2015).

Sustainable Tourism: World Tourism Organisation's defination of sustainable tourism is; "Tourismwhichmeetstheneeds of present tourists and host regions while protecting and enhancing opportunity for the future" (Butler, 1999, p. 10). There are of course many other definitions which put great emphasis on environmental sustainability, local's welfare, ecological balance and being responsible to all the stakeholders of the tourism activities.

Compilation of References

Ab Hamid, N. R., & McGrath, G. M. (2005). The Diffusion of Internet's Interactivity on E-tail Web Sites: A Customer Relationship Model. *Communications of the International Information Management Association*, *2*, 45–70.

Abomhara, M., & Koien, G. M. (2015). Cyber Security and the Internet of Things: Vulnerabilities, Threats, Intruders, and Attacks. *Journal of Cyber Security and Mobility*. doi:10.13052/jcsm2245-1439.414

Addis, M. (2005). New technologies and cultural consumption–edutainment is born! *European Journal of Marketing*, *39*(7/8), 729–736. doi:10.1108/03090560510601734

Adonisotel.com. (2018). 5 Applications from smart hotels you will say "wow" when you see. Retrieved from https://www.adonisotel.com/Destek/Otel-Yonetimi-Makaleler/gorunce-vay-can%C4%B1na-diyecek-siniz

AENOR (Spanish Association for Standardization and Certification). (2015). *Standardization in smart cities – Spain*. Retrieved from https://portal.aenormas.aenor.com /descargasweb/normas/aenor-Spanish-standardization-on-Smart-Cities-CTN-178.pdf

AENOR (Spanish Association for Standardization and Certification). (2016). *Management system for smart tourist destinations. UNE 178501:2016*. Retrieved from https://www.une.org/encuentra-tu-norma/busca-tu-norma/norma?c=N0056506

AENOR (Spanish Association for Standardization and Certification). (2018a). *Management system for smart tourist destinations. UNE 178501:2018*. Retrieved from https://www.une.org/encuentra-tu-norma/busca-tu-norma/norma?c=N0060239

AENOR (Spanish Association for Standardization and Certification). (2018b). *Indicators and tools for smart tourism destinations. UNE 178502:2018*. Retrieved from https://www.une.org/encuentra-tu-norma/busca-tu-norma/norma/?c=N0060240

AENOR (Spanish Association for Standardization and Certification). (2019). *DTI Benidorm certificate*. Retrieved from https://www.aenor.com/salainformaciondocumentos/NP%20Certificado%20DTI%20Benidorm.pdf

Afzaal, M., & Usman, M. (2016). A novel framework for aspect-based opinion classification for tourist places. *The 10th International Conference on Digital Information Management, ICDIM 2015*, (Icdim), 1–9. 10.1109/ICDIM.2015.7381850

Afzaal, M., Usman, M., & Fong, A. (2019). Predictive aspect-based sentiment classification of online tourist reviews. *Journal of Information Science*, *45*(3), 341–363. doi:10.1177/0165551518789872

Ağaoğlu, O. K. (1992). Yerel yönetimler ve turizm. *Anatolia Turizm, Çevre ve Kültür Dergisi*, *3*(4), 18–23.

Agarwal, B., & Mittal, N. (2016). *Prominent feature extraction for sentiment analysis* (N. Mittal, Ed.)., doi:10.1007/978-3-319-25343-5

Ahmad, I., & Chowdhury, A. R. (2008). *Electronic Customer Relationship Management (eCRM) - customers' perception of value from eCRM features on airline e-ticketing Websites*. (Master thesis). Available from Diva-portal student theses database. (1020631)

AirbnbEng. (2010, Nov. 15). *MySQL in the cloud at Airbnb*. Retrieved from https://medium.com/airbnb-engineering/mysql-in-the-cloud-at-airbnb-336e5666bc94#.llrxogduu

Akbulut, D. (2018). *The use of "Gamification" approach in public relations*. International Symposium on Communication in the Digital Age Proceeding E-Book. Oct. 18-19, Mersin, Turkey. Academic Press.

Akgül, M. K. (2013). Kentlerin E-dönüşümü: akıllı kentler. *Kalkınmada Anahtar Verimlilik Dergisi, 25*(291). Retrieved from http://anahtar.sanayi.gov.tr, Date of Access: 12.10.2018.

Akın Robotics. (2019). *Waitress Robot Ada*. Retrieved from https://akinrobotics.com/tr/akinsoft_garson_robot_ada.php

Aksoy, M., & Akbulut, B. A. (2016). Restoranlardaki teknolojik yeniliklerin deneyim pazarlaması açısından değerlendirilmesi. In II Eurasia Tourism Congress, Konya. Retrieved from https://www. researchgate. net/publication/310994877

Aktan, E. (2018). Big Data: Application Areas, Analytics and Security Dimension. *Bilgi Yönetimi Dergisi, 1*(1), 1–22. doi:10.33721/by.403010

Aktaş, C. (2012). Medya Yakınsaması: Hızlı Yanıt Veren Kod Aracılığıyla Geleneksel Gazetenin, Çevrimiçi Gazete ile Artan Rekabet Potansiyeli Üzerine Bir Tartışma. *International New Media Conference*, 118–128.

Alabdulsalam, S., Schaefer, K., Kechadi, T., & Le-Khac, N. A. (2018). *Internet of things forensics – Challenges and a case study*. IFIP Advances in Information and Communication Technology; doi:10.1007/978-3-319-99277-8_3

Aladag, E. (2015). Makine Öğrenmesi Nedir? Retrieved from https://www.emrealadag.com/makine-ogrenmesi-nedir.html

Albino, V., Berardi, U., & Dangelico, R. M. (2015). Smart cities: Definitions, dimensions, performance, and initiatives. *Journal of Urban Technology, 22*(1), 3–21. doi:10.1080/10630732.2014.942092

Alexis, P. (2017). R-Tourism: Introducing the Potential Impact of Robotics and Service Automation in Tourism. *Ovidius University Annals. Economic Sciences Series, XVII*(1), 211–216.

Alexis, P. (2017). R-Tourism: Introducing the potential impact of robotics and service automation in tourism. *Ovidius University Annals. Series Economic Sciences, 17*(1), 211–216.

Ali, A., & Frew, A. J. (2013). Information and communication technologies for sustainable tourism. In Information and Communication Technologies for Sustainable Tourism. doi:10.4324/9780203072592

Alkalbani, A. M., Gadhvi, L., Patel, B., Hussain, F. K., Ghamry, A. M., & Hussain, O. K. (2017). Analysing cloud services reviews using opining mining. *Proceedings - International Conference on Advanced Information Networking and Applications, AINA*, 1124–1129. 10.1109/AINA.2017.173

Alkan, T. (2017). Akıllı kentler ya da 21. Yüzyıl şehirleri. Retrieved from http://www.bilisimdergisi.org, Date of Access: 29.11.2018.

Almotairi, M. (2009). A Framework for Successful CRM Implementation. In *Proceedings of the European and Mediterranean Conference on Information Systems 2009*. İzmir, Turkey.

Alsaadi, E., & Tubaishat, A. (2015). Internet of things: Features, challenges, and vulnerabilities. *Int. J. Adv. Comput. Sci. Inf. Technol, 4*(1), 1–13.

Altaş, A. (2018). Geographical information system applications utilized in museum in Turkey within the scope of the cultural heritage tourism: a case study of mobile application of museum asistance. In C. S., & R. N. (Eds.), GIS applications in the tourism and hospitality ındustry (pp. 42-60). Hershey, PA: IGI Global: Business Science References.

Aluri, A. (2017). Mobile Augmented Reality (MAR) Game As A Travel Guide: Insights From Pokemon Go. *Journal of Hospitality and Tourism Technology*, *8*(1), 55–72. doi:10.1108/JHTT-12-2016-0087

Amadeus. (2017). *Blockchain: harnessing its potential in travel.* Retrieved from https://amadeus.com/en/insights/research-report/blockchain-harnessing-its-potential-in-travel

Amalia, N., Putri, S., & Alamsyah, A. (2017). *Opinion Mining of Tripadvisor Review Towards Five-Star Hotels in Bandung City.*, *4*(1), 4.

Amazon. (2019). Amazon EC2. Retrieved from https://aws.amazon.com/tr/ec2/

Amstel, P., Eijk, P., Haasdijk, E., & Kuilman, D. (2000). An interchange format for cross-media personalized publishing. *Computer Networks*, *33*(1-6), 179–195. doi:10.1016/S1389-1286(00)00049-9

Amsterdam Smart City. (2019a). *Climate street (Klimaatstraat).* Retrieved from https://amsterdamsmartcity.com/projects/climate-street

Amsterdam Smart City. (2019b). *GridFriends.* Retrieved from https://amsterdamsmartcity.com/projects/grid-friends

Anand, P. B., & Navío-Marco, J. (2018). Governance and economics of smart cities: opportunities and challenges, Telecommunications Policy, Available online October 11, 2018 In Press, Corrected Proof, /. doi:10.1016/j.telpol.2018.10.001

Andersen, S., Chen, S., & Carter, C. (2000). Fundamental Human Needs: Making Social Cognition Relevant. *Psychological Inquiry*, *11*(4), 269–276.

Anderson, L. F., & Littrell, M. A. (1995). Souvenir-purchase behavior of women tourists. *Annals of Tourism Research*, *22*(2), 328–348. doi:10.1016/0160-7383(94)00080-8

Anderson, R. E., & Srinivasan, S. S. (2003). E-Satisfaction and E-Loyalty: A contingency framework. *Psychology and Marketing*, *20*(2), 123–138. doi:10.1002/mar.10063

Andreassen, T. W. (2000). Antecedents to satisfaction with service recovery. *European Journal of Marketing*, *34*(1/2), 156–175. doi:10.1108/03090560010306269

Andriotis, K. (2002). Local authorities in Crete and the development of tourism. *Journal of Tourism Studies*, *13*(2), 53–62.

Angelaccio, M., Basili, A., & Buttarazzi, B. (2013). Using geo-business intelligence and social integration for smart tourism cultural heritage platforms. *Proceedings of the Workshop on Enabling Technologies: Infrastructure for Collaborative Enterprises, WETICE*, 196-199. 10.1109/WETICE.2013.87

Angelakis, V., Tragos, E., Pöhls, H. C., & Kapovits, A. (2016). *Designing, Developing, and Facilitating Smart Cities.* Designing, Developing, and Facilitating Smart Cities; doi:10.1007/978-3-319-44924-1

Angelidou, M. (2015). Smart cities: A conjuncture of four forces. *Cities (London, England)*, *47*, 95–106. doi:10.1016/j.cities.2015.05.004

Angelidou, M. (2016). Four European smart city strategies. *International Journal of Social Science Studies*, *4*(4), 18–30. doi:10.11114/ijsss.v4i4.1364

Anokwa, Y., Borriello, G., Pering, T., & Want, R. (2007, March). A user interaction model for NFC enabled applications. In *Proceedings of the Fifth Annual IEEE International Conference on Pervasive Computing and Communications Workshops (PerComW'07),* White Plains, NY. IEEE.10.1109/PERCOMW.2007.18

Anthopoulos, L. G. (2015). Transforming City Governments for Successful Smart Cities. In Transforming City Governments for Successful Smart Cities. doi:10.1007/978-3-319-03167-5

Arcese, G., Campagna, G., Flammini, S., & Martucci, O. (2014). Near field communication: Technology and market trends. *Technologies*, *2*(3), 143–163. doi:10.3390/technologies2030143

Argandoña, A. (2010). Corporate social responsibility in the tourism industry: Some lessons from the Spanish experience. Retrieved from www.iese.edu/research/pdfs/DI-0844-E.pdf

Arıkan, M. (2019). *Yapay zeka nedir?* Retrieved from https://www.mediaclick. com.tr/blog/yapay-zeka-nedir

Armutlu, H., & Akçay, M. (2013).Bulut bilişimin bireysel kullanımı için örnek bir uygulama. *Akademik Bilişim Konferansı*, 23-25.

Asabere, N. Y., & Doku, V. (2013). Measuring Customer Relationship Management (CRM) in the Hospitality Industry of Some Selected Hotels in Accra, Ghana: The Role of Information and Communication Technologies (ICTs). [IJAIEM]. *International Journal of Application or Innovation in Engineering & Management*, *2*(3), 19–28.

Asgary, N., De Los Santos, G., Vincent, V., & Davila, V. (1997). The determinants of expenditures by Mexican visitors to the border cities of Texas. *Tourism Economics*, *3*(4), 319–328. doi:10.1177/135481669700300402

Ashton, K. (2009). That 'Internet of Things' thing. *RFID Journal.* Retrieved from http://www.itrco.jp/libraries/RFID-journal-That%20Internet%20of%20Things%20Thing.pdf

Ashworth, G. J., & Tunbridge, J. E. (1990). *The tourist-historic city Belhaven. AWS Case Study: Red Lion Hotels*. (2019, Aug. 13). Retrieved from https://aws.amazon.com/solutions/case-studies/red-lion/

Atalay, M., & Çelik, E. (2017). Artificial Intelligence And Machine Learning Applications In Big Data Analysis. *Mehmet Akif Ersoy Üniversitesi Sosyal Bilimler Enstitüsü Dergisi*, *9*(22), 155–172. doi:10.20875/makusobed.309727

Ateş, M. (2005). Avrupa'da sosyal diyalogun kuramsal yapısı: AB ekonomik ve sosyal komitesi. *Ankara Avrupa Çalışmaları Dergisi*, *5*(1), 45–66.

Avedeon, E. M., & Sutton Smith, B. (1971). *The study of games*. New York: John Wiley & Sons.

Ávila, A. L., Lancis, E., García, S., Alcantud, A., García, B., & Muñoz, N. (2015). Smart destinations report: Building the future. SEGITTUR. Retrieved from https://www.segittur.es/opencms/export/sites/segitur/.content/galerias/descargas/documentos/Report-on-smart-destinations-en.pdf

Awad, E. M. (2000). The structure of e-commerce in the banking industry. In *Proceedings of the 2000 ACM SIGCPR Conference on Computer Personnel Research.* Chicago, IL. New York: ACM. 10.1145/333334.333375

AWS Case Study: SETTour. (2019, Aug. 13). Retrieved from https://aws.amazon.com/solutions/case-studies/settour/

Axelsson, K., & Granath, M. (2018). Stakeholders' stake and relation to smartness in smart city development: Insights from a Swedish city planning project. *Government Information Quarterly*, *35*(4), 693–702. doi:10.1016/j.giq.2018.09.001

Ayaz, N., & Göktaş Kulualp, H. (2019). Strategic role of human resource management in tourism enterprises (pp. 293-308) in (R. Efe, I. Koleva, M. Öztürk, & R. Arabacı (Eds.), Recent Advances in Social Sciences, Chapter Eighteen, 1st Edition, Cambridge Scholars Publishing.

Ayaz, N., & Kulualp, H. G. (2019). Strategic role of human resource management in tourism enterprises. In R. Efe, I. Koleva, M. Öztürk, & R. Arabacı (Eds.), *Recent Advances in Social Sciences* (pp. 275–289). Newcastle, UK: Cambridge Scholars Publishing.

Aydoğan, D. (2017). Virtual Museums in the Context of Virtual Reality and Simulation. *E-Journal of New Media*, *1*(2), 137–148. doi:10.17932/IAU.EJNM.25480200.2017.1/2.137-148

Ay, L. (2009). Bilgi Teknolojisindeki Gelişmelerin Seyahat Acentalarına Yansımaları. *Online Seyahat Acentacılığı. Sosyal Ekonomik Araştırmalar Dergisi*, *9*(17), 117–136.

Azaklı, S., & Özgür, H. (2004). *Belediyelerin turizme yönelik ve turizmi etkileyen hizmetleri. A. Yüksel ve M. Hançer (Editörler). Turizm İlkeler ve Yönetim* (pp. 121–142). Ankara, Turkey: Turhan Kitabevi.

Azuma, R. T. (1997). A survey of augmented reality. *Presence (Cambridge, Mass.)*, *6*(4), 355–385. doi:10.1162/pres.1997.6.4.355

Azuma, R., Baillot, Y., Behringer, R., Feiner, S., Julier, S., & MacIntyre, B. (2001). Recent advances in augmented reality. *IEEE Computer Graphics and Applications*, *21*(6), 34–47. doi:10.1109/38.963459

Baggio, R., & Del Chiappa, G. (2014). Real and virtual relationships in tourism digital ecosystems. *Information Technology & Tourism*, *14*(1), 3–19. doi:10.100740558-013-0001-5

Bahga, A., & Madisetti, V. (2019). *Big Data Science and Analytics: A Hands-On Approach*. Published by VPT.

Bahrammirzaee, A. (2010). A comparative survey of artificial intelligence applications in finance: Artificial neural networks, expert system and hybrid intelligent systems. *Neural Computing & Applications*, *19*(8), 1165–1195. doi:10.100700521-010-0362-z

Baines, A. (1998). Technology and tourism. *Work Study*, *47*(5), 160–163. doi:10.1108/00438029810370492

Bakıcı, T., Almirall, E., & Wareham, J. (2013). A smart city initiative: The case of Barcelona. *Journal of the Knowledge Economy*, *4*(2), 135–148. doi:10.100713132-012-0084-9

Balakrishnan, A. (2016). Wynn Las Vegas to Add Amazon Alexa to all Hotel Rooms. Retrieved from http://www.cnbc.com/2016/12/14/wynn-las-vegas-to-add-amazon-alexa-to-all-hotel- rooms.html

Baliga, A. (2017). *Understanding blockchain consensus models*. Santa Clara, CA: Persistent Systems Ltd.

Balkan, E. (2019). *Yapay Zeka, Büyük Veri ve Pazarlama Devrimi*. Retrieved from https://kobitek.com/yapay-zeka-buyuk-veri-ve-pazarlama-devrimi

Bamodu, O., & Ye, X. (2013). Virtual Reality and Virtual Reality System Components. *Proceedings of the 2nd International Conference On Systems Engineering and Modeling* (ICSEM-13). 10.2991/icsem.2013.192

Bamoriya, H. (2014). QR Code Based Marketing in Indıa and Japan. *Singidunum Journal of Applied Sciences*, *11*(2), 20–29. doi:10.5937jas11-5504

Bani-hani, R. M., Wahsheh, Y. A., & Al-Sarhan, M. B. (2014). Secure QR Code System. In *Proceedings 2014 10th International Conference on Innovations in Information Technology (IIT)*. Doi:, 1–6.10.1109/Innovations.2014.6985772

Bański, J. (2007). Koncepcje rozwoju struktury przestrzennej w Polsce – polaryzacja czy równoważenie? *PrzeglądGeograficzny*, *79*(1), 45–77.

Barcelona City Website. (2019a). *Barcelona Digital City, Bicing*. Retrieved from https://ajuntament.barcelona.cat/digital/en/digital-transformation/urban-technology/bicing

Barcelona City Website. (2019b). *Info Barcelona*. Retrieved from https://www.barcelona.cat/infobarcelona/en/tema/mobility-and-transport/approval-to-move-bike-lanes-from-pavements-onto-the-road-surface-in-av-diagonal_784832.html

Barcelona City Website. (2019c). *Smart vehicles to reduce accidents*. Retrieved from https://ajuntament.barcelona.cat/en/noticia/smart-vehicles-to-reduce-accidents_735098 Accessed on July 4, 2019.

Bartle, R. (1996). Hearts, clubs, diamonds, spades: Players who suit MUDs. *Journal of MUD Research,1*(1).

Bartneck, C., & Forlizzi, J. (2004). A Design-Centred Framework for Social Human-Robot Interaction, *2004 IEEE International Workshop on Robot and Human Interactive Communication*, September 20-22, Kurashiki, Japan. 10.1109/ROMAN.2004.1374827

Barton, A. W., & Leonard, S. J. (2010). Incorporating social justice in tourism planning: Racial reconciliation and sustainable community development in the Deep South. *Community Development (Columbus, Ohio)*, *41*(3), 298–322. doi:10.1080/15575330903444051

Başer, G., & Doğan, O. (2018). Future Trends for Smart Tourism Destinations: Case of Antalya. In *Proceedings 1st International Congress of New Generation and New Trends in Tourism* pp: 694-697, (November 1-3, 2018), Sakarya, Turkey.

Baş, M., Kılıç, B., & Güçer, E. (2007). Türkiye'de yerel yönetimler ve turizm. *Mevzuat Dergisi*, *9*(119), 1–12.

Batty, M., Axhausen, K. W., Giannotti, F., Pozdnoukhov, A., Bazzani, A., Wachowicz, M., & Portugali, Y. (2012). Smart cities of the future. *The European Physical Journal. Special Topics*, *214*(1), 481–518. doi:10.1140/epjst/e2012-01703-3

Bauter Engel, J. (2018). *SoftBank Taps Affective to Boost Pepper Robot's Emotional IQ*. Retrieved from https://xconomy.com/boston/2018/08/28/softbank-taps-affectiva-to-boost-pepper-robots-emotional-iq/

Baydeniz, E. (2019). The Relation of tourism with technology. Retrieved from https://www.academia.edu/12421334/TUR%C4%B0ZM%C4%B0N_tEKNOLOJ%C4%B0_%C4%B0LE_%C4%B0L%C4%B0%C5%9EK%C4%B0S%C4%B0

Bayrak Meydanoglu, E. (2016). *Türk Üniversite Öğrencilerinin Kare Kod Reklamciliğa Karşi Tutumlari Attitudes Of Turkish University Students Towards QR Code Advertising. Journal of Yasar University* (Vol. 11). Retrieved from https://dergipark.org.tr/download/article-file/343663

Bayraktar, Ö. (2014). *Gamification as a communication model* (Master dissertation). University of Maltepe, Turkey.

Baytok, A., Emren, A., Gürel, N., Dalkıranoğlu, A., & Töre, H. ve Güney H. (2001). Afyonkarahisar Mutfağı. Ankara: Uyum Ajans.

Beck, R., Czepluch, J. S., Lollike, N., & Malone, S. (2016, June). *Blockchain-the gateway to trust-free cryptographic transactions.* Paper presented at the meeting of 24th European Conference on Information Systems, Istanbul, Turkey.

Beck, J., & Egger, R. (2018). Emotionalise me: Self-reporting and arousal measurements in virtual tourism environments. In *Information and Communication Technologies in Tourism 2018* (pp. 3–15). Cham, Switzerland: Springer. doi:10.1007/978-3-319-72923-7_1

Beck, J., Rainoldi, M., & Egger, R. (2019). Virtual reality in tourism: A state-of-the-art review. *Tourism Review*, *74*(3), 586–612. doi:10.1108/TR-03-2017-0049

Beck, R. (2019). Beyond bitcoin: The rise of blockchain World. *Computer*, *2018*(February), 54–58.

Bellenger, D. N., & Korgaonkar, P. K. (1980). Profiling the recreational shopper. *Journal of Retailing*, *56*(3), 77–92.

Bellini, P., Benigni, M., Billero, R., Nesi, P., & Rauch, N. (2014). Km4City ontology building vs data harvesting and cleaning for smart-city services. *Journal of Visual Languages and Computing*, *25*(6), 827–839. doi:10.1016/j.jvlc.2014.10.023

Benckendorff, P., Sheldon, P., & Fesenmaier, D. R. (2014). *Tourism information technology*. Oxford, UK: CAB International. doi:10.1079/9781780641850.0000

Benea, I. A. (2014). *Influences of social media on the tourism and hospitality industry*. Vienna, Austria: Modul Vienna University.

Benli, T., & Kızgın, Y. (2002, April). A Field study on the importance of use of information technology in hospitality enterprises: The case of Muğla province. *Journal of Legislation*, *5*(52).

Berezina, K., Bilgihan, A., Cobanoglu, C., & Okumus, F. (2016). Understanding Satisfied and Dissatisfied Hotel Customers: Text Mining of Online Hotel Reviews. *Journal of Hospitality Marketing & Management*, *25*(1), 1–24. doi:10.1080/19368623.2015.983631

Berget, Ø. (2008). *Investigation of security features in Near-field communication (NFC).* (Unpublished Master's thesis). University of Oslo, Norway.

Bernabeu, M. A. C., Mazón, J. N., Giner, D., & Baidal, J. I. (2016) Big Data and Smart Tourism Destinations: Challenges and opportunities from an industry perspective. In *School of Hospitality and Tourism Management Conference*, University of Surrey, UK.

Berryman, D. R. (2012). Augmented reality: A review. *Medical Reference Services Quarterly*, *31*(2), 212–218. doi:10.1080/02763869.2012.670604 PMID:22559183

Bertacchini, Y. (2012). Between Information and Communication Process, the Territorial Intelligence, as a Network Concept & a Framework to Shape Local Development. *International Journal of Humanities and Social Science*, *2*(18), 242–247.

Bertucci, G. (2002). Strengthening local governance in tourism-driven economies. In statement prepared for the International Colloquium on Regional Governance and Sustainable Development in Tourism-driven Economies Strengthening. Cancun, Mexico, Feb. 20-22.

Bhardwaj, S., Jain, L., & Jain, S. (2010). Cloud Computıng: A Study Of Infrastructure As A Servıce (IaaS). *International Journal of Engineering*.

Bhaumik, A. (2018). *From AI to Robotics: Mobile, Social, and Sentinent Robots.* Boca Raton, FL: CRC Press. doi:10.1201/9781315372549

Bi, L., Feng, Z., Liu, M., & Wang, W. (2008). Design and Implementation of the Airline Luggage Inspection System Base on Link Structure of QR Code. In *Proceedings 2008 International Symposium on Electronic Commerce and Security* (pp. 527–530). IEEE. 10.1109/ISECS.2008.200

Bilge, U. (2007). *Tıpta Yapay Zeka ve Uzman Sistemler*. Türkiye Bilsim Dernegi Kongresi.

Bilginer, B., & Ljunggren, P. L. (2011) Near Field Communication. (Unpublished Master's thesis). Lund University, Sweden.

Billard, A. (2003). Robota: Clever Toy and Educational Tool. *Robotics and Autonomous Systems*, *42*(3-4), 259–269. doi:10.1016/S0921-8890(02)00380-9

Binark, M., & Bayraktutan, G. (2011). ''E-status'' of participation: The virtual World of youth. Aslı Telli Aydemir (Ed.), Digital game culture map players: Digital player's habitus and career derivatives (pp. 303-330). Turkey: Alternative Information Publishing.

Birbil, I. (2018). Tahmin ve Çıkarım. Retrieved from http://www.veridefteri.com/2018/04/04/tahmin-ve-cikarim-makine-%0Aogrenmesi-yapay-ogrenme/

Bitcoin Transaction Fees. (2019, June 15). Retrieved from https://bitcoinfees.info/

Blair-Goldensohn, S., Neylon, T., Hannan, K., Reis, G. A., McDonald, R., & Reynar, J. (2008). Building a sentiment summarizer for local service reviews. *Workshop on NLP in the Information Explosion Era*.

Blaser, M. (2019). *Development of a general maturity model for Smart Tourism Destinations.*

Blog, U. (2017). *Top 10 Yapay Zeka (Artificial Intelligence) Uygulamaları.* Retrieved from http://blog.udentify.co/04/2017/en-cok-ilgi-ceken-yapay-zeka-uygulamalari/

Blowuphall5050. (2019). Innovations. Retrieved from http://www.blowuphall5050.com/hotel/

Boden, R. (2013a). Madhya Pradesh to offer NFC info. Retrieved from https://www.nfcworld.com/2013/04/29/323771/madhya-pradesh-to-offer-nfc-info/

Boden, R. (2013b). Spanish hotel delivers guest information via NFC. Retrieved from https://www.nfcworld.com/2013/07/24/325127/spanish-hotel-delivers-guest-information-via-nfc/

Boden, R. (2014). Barcelona equips 8,000 city locations with NFC and QR touchpoints. Retrieved from https://www.nfcworld.com/2014/12/05/333113/barcelona-equips-8000-city-locations-nfc-qr-touchpoints/

Boden, R. (2017). Carnival unveils NFC and BLE wearable for cruise ship passengers. Retrieved from https://www.nfcworld.com/2017/01/05/349260/carnival-unveils-nfc-ble-wearable-passengers-access-personalised-cruise-ship-services/

Bodhani, A. (2013). Getting a purchase on AR. *Engineering & Technology*, *8*(4), 46–49. doi:10.1049/et.2013.0408

Boes, K., Buhalis, D., & Inversini, A. (2015), Conceptualising smart tourism destination dimensions. In I. Tussyadiah, & A. Inversini (Eds.), Information and Communication Technologies in Tourism 2015, Springer, 391–403.

Boes, K., Buhalis, D., & Inversini, A. (2015). Conceptualising smart tourism destination dimensions. In L. Tussyadiah, & A. Inversini (Eds), *Information and communication technologies in tourism 2015: Proceedings of the international conference in Lugano, Switzerland*, February 3-6, 2015 (pp. 391-403). Cham, Switzerland: Springer.

Boes, K., Buhalis, D., & Inversini, A. (2016). Smart tourism destinations: ecosystems for tourism destination competitiveness. *International Journal of Tourism Cities.* doi:10.1108/IJTC-12-2015-0032

Boes, K., Buhalis, D., & Inversini, A. (2015). Conceptualizing Smart Tourism Destination Dimensions. In I. Tussyadiah, & A. Inversini (Eds.), *Information and Communication Technologies in Tourism* (pp. 391–403). Heidelberg, Germany: Springer.

Bogdanovych, A., Esteva, N., Gu, M., Simoff, S., Maher, M. L., & Smith, G. (2007). The role of online travel agents in the experience economy. In *Proceedings of the 14th international conference on information technology in tourism ENTER*, Ljubljana, Slovenia. Academic Press.

Bohanec, M. (2008). *DEXi: Program for Multi-Attribute Decision Making User's Manual.* Ljubljana, Slovenia: Institut Jozef Stefan.

Boiy, E., & Moens, M. F. (2009). A machine learning approach to sentiment analysis in multilingual web texts. *Information Retrieval*, *12*(5), 526–558. doi:10.100710791-008-9070-z

Bonetti, F., Pantano, E., Warnaby, G., Quinn, L., & Perry, P. (2018, June). Augmented Reality in Real Stores: Empirical Evidence from Consumers' Interaction with AR in a Retail Format. In 4th International AR &VR Conference 2018: The Power of AR & VR for Business.

Borgia, E. (2014). The internet of things vision: Key features, applications and open issues. *Computer Communications*, *54*, 1–31. doi:10.1016/j.comcom.2014.09.008

Borowska-Stefanska, M., Lesniewska-Napierala, K., & Wisniewski, S. (2018). Land cover changes in Poland between 1990 and 2012. *Geografie (Utrecht)*, *123*(3), 63–83.

Bostan, B. (2007). *Sanal Gerçeklikte Etkileşim*. Marmara Universitesi Sosyal Bilimler Enstitüsü İletişim Bilimleri Anabilim Dalı Yayınlanmamış Doktora Tezi.

Bostancı, S. H. (2011). Kent konseylerinin Eko-Belediyecilik vizyonuna katkıları. In E. B. Keskin (Ed.), Kent Konseyleri Sempozyum Bildiri Kitabı (pp. 250-264), Bursa, Turkey: Bursa Kent Konseyi Bilimsel Yayınlar Dizisi-1.

Boulton, A., Brunn, S., & Devriendt, L. (2012). Cyberinfrastructures and "smart" world cities: Physical, human, and soft infrastructures. In P. Taylor, B. Derudder, M. Hoyler, & F. Witlox (Eds.), *International handbook of globalization and world cities*. UK: Edward Elgar.

Bowen, J., & Morosan, C. (2018). Beware Hospitality Industry: The Robots are Coming. *Worldwide Hospitality and Tourism Themes, 10*(6), 726–733. doi:10.1108/WHATT-07-2018-0045

Bozic, J., Tazl, O. A., & Wotawa, F. (2019). Chatbot Testing Using AI Planning. In *Proceedings 2019 IEEE International Conference on Artificial Intelligence Testing (AITest)*, Newark, CA. 10.1109/AITest.2019.00-10

Bozzano, H. (2013). Geography and Territorial Intelligence. Geo-grafein, Geo-explanans, Geo-transformare. *Revista Geográfica Digital. IGUNNE, 10*(19), 1–24.

Brabazon, T., Winter, M., & Gandy, B. (2014). *Digital Wine: How QR Codes Facilitate New Markets for Small Wine Industries*. Springer. doi:10.1007/978-981-287-059-9

Bramwell, B., & Lane, B. (2008). Priorities in Sustainable Tourism Research. *Journal of Sustainable Tourism, 16*(1), 1–4. doi:10.2167/09669580803489612

Brar, K. (2019). *Hotels using big data to check out guests*. Retrieved from https://www.tnp.sg/news/views/hotels-using-big-data-check-out-guests

Broadway World. (2018). Order with your face at philly restaurant: facial recognition software debuts at fuel. Retrieved from https://www.broadwayworld.com/philadelphia/article/Order-With-Your-Face-At-Philly-Restaurant-Facial-Recognition-Software-Debuts-At-Fuel-20180404

Brody, S., & Elhadad, N. (2010). An Unsupervised Aspect-Sentiment Model for Online Reviews. *HLT '10 Human Language Technologies: The 2010 Annual Conference of the North American Chapter of the Association for Computational Linguistics*.

Brougham, D., & Haar, J. (2018). Smart Technology, Artificial Intelligence, Robotics, and Algorithms (STARA): Employees' Perceptions of our Future Workplace. *Journal of Management & Organization, 24*(2), 239–257. doi:10.1017/jmo.2016.55

Brown, C. (2011). Spanish city to lend NFC phones to tourists. Retrieved from https://www.nfcworld.com/2011/04/17/37061/spanish-city-to-lend-nfc-phones-to-tourists/

Brown, S., Versace, V. L., Laurenson, L., Ierodiaconou, D., Fawcett, J., & Salzman, S. (2012). Assessment of Spatiotemporal Varying Relationships Between Rainfall, Land Cover and Surface Water Area Using Geographically Weighted Regression. *Environmental Modeling and Assessment, 17*(3), 241–254. doi:10.100710666-011-9289-8

Buffa, F. (2015). Young Tourists and Sustainability: Profiles, Attitudes and Implications for Destination Strategies. *Sustainability, 7*(10), 14042–14062. doi:10.3390u71014042

Buhalis, D. & Amaranggana, A. (2014). Smart tourism destinations. In Z. Xiang, & L. Tussyadiah (Eds.), *Information and communication technologies in tourism 2014: Proceedings of the international conference in Dublin, Ireland*, January 21-24, 2014 (pp. 553-564). Cham, Switzerland: Springer.

Buhalis, D. (2003). *E-tourism: Information technology for strategic tourism management*. London, UK: Pearson (Financial Times / Prentice-Hall).

Buhalis, D. (2003). eTourism: Information technology for strategic tourism management. Pearson Education.

Buhalis, D., & Amaranggana, A. (2013). Smart Tourism Destinations. In Information and Communication Technologies in Tourism 2014. doi:10.1007/978-3-319-03973-2_40

Buhalis, D., & Amaranggana, A. (2013). Smart tourism destinations. Information and Communication Technologies in Tourism 2014 (pp. 553-564). Dublin, Ireland: Springer International Publishing.

Buhalis, D. (1998). Statejik Use of Information in the Tourism Industry. *Tourism Management*, *19*(5).

Buhalis, D. (1998). Strategic use of information technologies in the tourism industry. *Tourism Management*, *19*(5), 409–421. doi:10.1016/S0261-5177(98)00038-7

Buhalis, D. (2000). Marketing the competitive destination of the future. *Tourism Management*, *21*(1), 97–116. doi:10.1016/S0261-5177(99)00095-3

Buhalis, D., & Amaranggana, A. (2014). Smart Tourism Destinations. In Z. Xiang, & I. Tussyadiah (Eds.), *Information and Communication Technologies in Tourism* (pp. 553–564). Heidelberg, Germany: Springer.

Buhalis, D., & Amaranggana, A. (2015). Smart Tourism Destinations Enhancing Tourism Experience Through Personalisation of Services. In *Information and communication technologies in tourism 2015* (pp. 377–389). Cham, Switzerland: Springer. doi:10.1007/978-3-319-14343-9_28

Buhalis, D., & Law, E. (2001). Tourism Distribution. *Channels. Continuum Journal*, *12*(2), 207–220.

Buhalis, D., & Law, R. (2008). Progress in information technology and tourism management: 20 years on and 10 years after the Internet—The state of eTourism research. *Tourism Management*, *29*(4), 609–623. doi:10.1016/j.tourman.2008.01.005

Buhalis, D., & Leung, R. (2018). Smart hospitality—Interconnectivity and interoperability towards an ecosystem. *International Journal of Hospitality Management*, *71*, 41–50. doi:10.1016/j.ijhm.2017.11.011

Buhalis, D., & Matloka, J. (2013). Technology-enabled Tourism Destination Management and Marketing. In C. Costa, E. Panyik, & D. Buhalis (Eds.), *Trends in European tourism planning and organization* (pp. 339–350). Buffalo, NY: Channel View Publications.

Buterin, V. (2015). *On public and private blockchains*. Retrieved from https://blog.ethereum.org/ 2015/08/07/on-public-and-private-blockchains/

Buttle, F. (2009). *Customer Relationship Management Concepts and Technologies* (2nd ed.). Burlington, VT: Elsevier.

Buzova, D., Sanz-Blas, S., & Cervera-Taulet, A. (2018). Does culture affect sentiments expressed in cruise tours' eWOM? *Service Industries Journal*, 1–20. doi:10.1080/02642069.2018.1476497

Caber, M. (2010). *Tur Operatörü-Seyahat Acentesi Elektronik Müşteri İlişkileri Yönetiminde Web Sitesi Kalitesinin Memnuniyet Ve Güven Üzerine Etkisi*. (Doctoral dissertation). Available from Yükseköğretim Kurulu Ulusal Tez Merkezi. (250541)

Cackett, D. (2016). *Information Management and Big Data, A Reference Architecture*. White paper. Redwood Shores: Oracle Corporation.

Caillois, R. (2001). *Man, play, and games*. Chicago, IL: University of Illinois Press.

Çakırel, Y. (2016). İşletmelerde Büyük Veri, *Kırklareli Üniversitesi İktisadi ve İdari Bilimler Fakültesi Dergisi*, *5*(1).

Cambria, E., Schuller, B., Xia, Y., & Havasi, C. (2013). New avenues in opinion mining and sentiment analysis. *IEEE Intelligent Systems, 28*(2), 15–21. doi:10.1109/MIS.2013.30

Cameron, N. (2018). Domino's debuts augmented reality pizza ordering. Retrieved from www.cmo.com.au/article/649324/domino-debuts-augmented-reality-pizza-ordering/

Canadi, M., Höpken, W., & Fuchs, M. (2010). Application of QR Codes in Online Travel Distribution. In U. Gretzel, R. Law, & M. Fuchs (Eds.), *Information and Communication Technologies in Tourism 2010* (pp. 137–148). Vienna, Austria: Springer. doi:10.1007/978-3-211-99407-8_12

Canestrelli, E., & Costa, P. (1991). Tourist carrying capacity: A fuzzy approach. *Annals of Tourism Research, 18*(2), 295–311. doi:10.1016/0160-7383(91)90010-9

Can, U., & Alatas, B. (2017). Duygu Analizi ve Fikir Madenciliği Algoritmalarının İncelenmesi. *Int. J. Pure Appl. Sci, 3*(1), 75–111.

Caragliu, A., & Del Bo, C. F. (2018). Smart innovative cities: The impact of smart city policies on urban innovation. *Technological Forecasting and Social Change*. doi:10.1016/j.techfore.2018.07.022

Caragliu, A., Del Bo, C., & Nijkamp, P. (2011). Smart cities in Europe. *Journal of Urban Technology, 18*(2), 65–82. doi:10.1080/10630732.2011.601117

Çarkacıoğlu, A. (2016). *Crypto-money bitcoin*. Ankara, Turkey: Capital Market Board Research Department.

Carmigniani, J., Furht, B., Anisetti, M., Ceravolo, P., Damiani, E., & Ivkovic, M. (2011). Augmented reality technologies, systems and applications. *Multimedia Tools and Applications, 51*(1), 341–477. doi:10.100711042-010-0660-6

Castañón, J. (2019). 10 Machine Learning Methods that Every Data Scientist Should Know.

CastleW. (2019). Retrieved from https://www.wartburg.de/en/wartburg-castle.html

Castro, J. C., Quisimalin, M., Córdova, V. H., Quevedo, W. X., Gallardo, C., Santana, J., & Andaluz, V. H. (2017). Virtual reality on e-Tourism. In *International conference on information theoretic security*, Singapore: Springer, pp. 86-97.

Cata, T., Patel, P. S., & Sakaguchi, T. (2013). Article ID 813339, 7 pages. *Cancer, 2013*. doi:10.5171/2013

Cavada, D., Elahi, M., Massimo, D., Maule, S., Not, E., Ricci, F., & Venturini, A. (2018). Tangible tourism with the internet of things. In *Information and Communication Technologies in Tourism 2018* (pp. 349–361). Cham, Switzerland: Springer. doi:10.1007/978-3-319-72923-7_27

Cawley, M., Gaffey, S., & Gillmor, D. A. (2002). Localization and global reach in rural tourism: Irish evidence. *Tourist Studies, 2*(1), 63–86. doi:10.1177/1468797602002001097

Ceipidor, U., Medaglia, C. M., Volpi, V., Moroni, A., Sposato, S., Carboni, M., & Caridi, A. (2013). NFC technology applied to touristic-cultural field: a case study on an Italian museum. In *Proceedings 5th International Workshop on Near Field Communication*. 10.1109/NFC.2013.6482445

Çelik, P., & Topsakal, Y. (2017). *Akıllı Turizm Destinasyonları: Antalya Destinasyonunun Akıllı Turizm Uygulamalarının İncelenmesi Smart Tourism Destinations: Review of Smart Tourism Applications of Antalya Destination. Journal of Travel and Hospitality Management* (Vol. 14). Retrieved from https://dergipark.org.tr/download/article-file/389164

Çelik, P., & Topsakal, Y. (2017). Akıllı Turizm Destinasyonları: Antalya Destinasyonunun Akıllı Turizm Uygulamalarının İncelenmesi Smart Tourism Destinations: Review of Smart Tourism Applications of Antalya Destination. *Journal of Travel and Hospitality Management, 14*(3), 149–166. Retrieved from http://dergipark.gov.tr/download/article-file/389164

Çelik, K., Güleryüz, S., & Özköse, H. (2018). 4. Endüstri Devrimine Kuramsal Bakış. *Avrasya Sosyal ve Ekonomi Araştırmaları Dergisi, 5*(9), 86–95.

Çelik, P., & Topsakal, Y. (2017). Akıllı Turizm Destinasyonları: Antalya Destinasyonunun Akıllı Turizm Uygulamalarının İncelenmesi. *Seyahat ve Otel İşletmeciliği Dergisi, 14*(3), 149–166. doi:10.24010oid.369951

Çelik, P., & Topsakal, Y. (2017b). Smart Tourism Destinations: Review of Smart Tourism Applications of Antalya Destination. *Journal of Travel and Hospitality Management, 14*(3), 149–166.

Celtek, E. (2015). Augmented Reality Advertisements in Tourism Marketing. In N. Takıran, & R. Yılmaz (Eds.), Handbook of Research on Effective Advertising Strategies in the Social Media Age. pp. 125-146. Hershey, PA: IGI Global.

Çeltek, E. (2017). QR Code Advertisements in Tourism Marketing. In In Narrative Advertising Models and Conceptualization in the Digital Age (pp. 269–289). Hershey, PA: IGI Global. doi:10.4018/978-1-5225-2373-4.ch015

Çeltek, E. (2010). Mobile advergames in tourism marketing. *Journal of Vacation Marketing, 16*(4), 267–281. doi:10.1177/1356766710380882

Cerruela García, G., Luque Ruiz, I., & Gómez-Nieto, M. (2016). State of the art, trends and future of bluetooth low energy, near field communication and visible light communication in the development of smart cities. *Sensors (Basel), 16*(11), 1968. doi:10.339016111968 PMID:27886087

Çevik, S., & Saçılık, M. (2011). Destinasyonun Rekabet Avantajı Elde Etmesinde Gastronomi Turizminin Rolü: Erdek Örneği. 12. *Ulusal Turizm Kongresi Bildiriler Kitabı, 12*, 503–515.

Chalmeta, R. (2006). Methodology for customer relationship management. *Journal of Systems and Software, 79*(7), 1015–1024. doi:10.1016/j.jss.2005.10.018

Chang, C. K. (2010). Acceptability of an asynchronous learning forum on mobile devices. *Behaviour & Information Technology, 29*(1), 23–33. doi:10.1080/01449290701806337

Chang, H. H. (2013). *Everyday NFC: near field communication explained.* Bellevue, WA: Coach Seattle Incorporated.

Chaouchi, H. (2013). The Internet of Things: Connecting Objects to the Web. In *The Internet of Things*. Connecting Objects to the Web; doi:10.1002/9781118600146.ch1

Cheng, J., Chen, F., & Chang, Y. (2008). Airline relationship quality: An examination of Taiwanese passengers. *Tourism Management, 29*(3), 487–499. doi:10.1016/j.tourman.2007.05.015

Cheng, K. H., & Tsai, C. C. (2013). Affordances of augmented reality in science learning: Suggestions for future research. *Journal of Science Education and Technology, 22*(4), 449–462. doi:10.100710956-012-9405-9

Chen, I. J., & Popovich, K. (2003). Understanding customer relationship management (CRM): People, process and technology. *Business Process Management Journal, 9*(5), 672–688. doi:10.1108/14637150310496758

Chenlo, J. M., & Losada, D. E. (2014). An empirical study of sentence features for subjectivity and polarity classification. *Information Sciences, 280*, 275–288. doi:10.1016/j.ins.2014.05.009

Chen, Q., & Chen, H. (2004). Exploring the success factors of eCRM strategies in practice. *Journal of Database Marketing and Customer Strategy Management, 11*(4), 333–343. doi:10.1057/palgrave.dbm.3240232

Chen, R. H., Lin, R. J., & Yang, P. C. (2011). The relationships between eCRM, innovation, and customer value-An empirical study. In *Proceedings 2011 IEEE International Summer Conference of Asia Pacific Business Innovation and Technology Management (APBITM 2011)*. Dalian, China: IEEE Press. 10.1109/APBITM.2011.5996343

Chen, S., Pan, Z., & Zhang, M. (2012). A virtual informal learning system for cultural heritage. In Z. Pan, A. D. Cheok, W. Mueller, M. Chang, & M. Zhang (Eds.), *Transactions on edutainment VII* (pp. 180–187). Berlin, Germany: Springer International Publishing. doi:10.1007/978-3-642-29050-3_16

Chen, X., Sun, L., Zhu, H., Zhen, Y., & Chen, H. (2012). Application of internet of things in power-line monitoring. *Proceedings of the 2012 International Conference on Cyber-Enabled Distributed Computing and Knowledge Discovery, CyberC 2012*. 10.1109/CyberC.2012.77

Cheong, R. (1995). The virtual threat to travel and tourism. *Tourism Management*, *16*(6), 417–422. doi:10.1016/0261-5177(95)00049-T

Cherapanukorn, V. (2017). Development of eCRM Success: A Case Study of Hotel Industry. *International Journal of Trade, Economics, and Finance*, *8*(2), 90–95.

Chip Online. (2014). *Unutulmayacak Mekanik Türk Efsanesi*. Retrieved from https://www.chip.com.tr/haber/unutulmayacak-mekanik-turk-efsanesi_48558.html

Choi, B., Poon, S. K., & Davis, J. G. (2008). Effects of knowledge management strategy on organizational performance: A complementarity theory-based approach. *Omega*, *36*(2), 235–251. doi:10.1016/j.omega.2006.06.007

Chou, T.-S. (2013). Security Threats on Cloud Computing Vulnerabilities. *International Journal of Computer Science and Information Technology*. doi:10.5121/ijcsit.2013.5306

Chou, T.-L., & ChanLin, L.-J. (2012). Augmented Reality Smartphone Environment Orientation Application: A Case Study of the Fu-Jen University Mobile Campus Touring System. *Procedia: Social and Behavioral Sciences*, *46*, 410–416. doi:10.1016/j.sbspro.2012.05.132

Chou, Y.-K. (2016). *Actionable gamification beyond points, badges, and leaderboards*. Japan: Octalysis Media.

Cho, V. (2003). A comparison of three different approaches to tourist arrival forecasting. *Tourism Management*, *24*(3), 323–330. doi:10.1016/S0261-5177(02)00068-7

Christidis, K., & Devetsikiotis, M. (2016). Blockchains and smart contracts for the internet of things. *IEEE Access Journal*, *4*, 2292–2303. doi:10.1109/ACCESS.2016.2566339

Chuah, J. W. (2014). The internet of things: An overview and new perspectives in systems design. In *Proceedings 2014 International Symposium on Integrated Circuits (ISIC)*. Singapore: IEEE Computer Society. pp. 216–219. 10.1109/ISICIR.2014.7029576

Chung, H., Park, J., Lee, S., & Kang, C. (2012). Digital forensic investigation of cloud storage services. *Digital Investigation*, *9*(2), 81–95. doi:10.1016/j.diin.2012.05.015

Chung, N., Han, H., & Joun, Y. (2015). Tourists' intention to visit destination: Role of augmented reality applications for heritage site. *Computers in Human Behavior*, *50*, 588–599. doi:10.1016/j.chb.2015.02.068

Cianciarulo, D. (2015). From local traditions to "augmented reality". The MUVIG Museum of Viggiano (Italy). *Procedia: Social and Behavioral Sciences*, *188*, 138–143. doi:10.1016/j.sbspro.2015.03.349

Cinnioğlu, D., & Demirdelen, H. (2018). Kare Kodların (QR Code) Restoran İşletmelerinde Kullanımının Belirlenmesi: İstanbul Örneği (pp. 100–108).

Clark, S. (2010). NFC phones replace room keys and eliminate check-in at Swedish hotel. Retrieved from https://www.nfcworld.com/2010/11/03/34886/nfc-keys-hotel-sweden/

Clark, S. (2011). Museum of London adds NFC. Retrieved from https://www.nfcworld.com/2011/08/16/39129/museum-of-london-adds-nfc/

Clark, S. (2012a). Sydney picks NFC and QR codes to guide visitors around The Rocks. Retrieved from https://www.nfcworld.com/2012/07/02/316609/sydney-picks-nfc-and-qr-codes-to-guide-visitors-around-the-rocks/

Clark, S. (2012b). Grenoble builds NFC system. Retrieved from https://www.nfcworld.com/2012/04/04/314972/grenoble-builds-nfc-system/

Clark, S. (2014a). London taxi firm promotes app downloads with NFC and QR stickers. Retrieved from https://www.nfcworld.com/2014/05/20/329236/london-taxi-firm-promotes-app-downloads-nfc-qr-stickers/

Clark, S. (2014b). Rio gets 5,000 NFC tags. Retrieved from https://www.nfcworld.com/2014/06/20/329851/rio-gets-5000-nfc-tags/

Clark, S. (2019a). Chinese airlines to offer NFC baggage tags at Beijing airport. Retrieved from https://www.nfcworld.com/2019/08/19/363939/chinese-airlines-to-offer-nfc-baggage-tags-at-beijing-airport/

Clark, S. (2019b). California winemaker to use NFC to provide buyers with a 'farm-to-table' experience. Retrieved from https://www.nfcworld.com/2019/08/07/363802/california-winemaker-to-use-nfc-to-provide-buyers-with-a-farm-to-table-experience/

Clark, S. (2019c). Tourism Australia adds visitor card to Alipay. Retrieved from https://www.nfcworld.com/2019/02/04/360292/tourism-australia-adds-visitor-card-to-alipay/

Claveria, O., Monte, E., & Torra, S. (2015). A new forecasting approach for the hospitality industry. *International Journal of Contemporary Hospitality Management*, *27*(7), 1520–1538. doi:10.1108/IJCHM-06-2014-0286

Cloud-based environmental modelling. (2019). Retrieved from FORTISSIMO: https://www.fortissimo-project.eu/experiments/514

CNN. (2019). Retrieved from https://edition.cnn.com/travel/article/virgin-seat-to-seat-service/index.html

Cognizant. (2016). *Blockchain in banking: a measured approach.* Retrieved from https://www.cognizant.com/whitepapers/Blockchain-in-Banking-A-Measured-Approach-codex1809.pdf

Cohen, B. (2012). *Smart cities hub.* Retrieved July 22, 2013, from http://smartcitieshub.com/2012/11/11/smart-cities-ranking-methodology/

Cohen, B. (2012). *The smartest cities in the world 2015: Methodology.* Retrieved from https://www.fastcompany.com/3038818/the-smartest-cities-in-the-world-2015-methodology

Coleman, J. (2011). QR Codes: What Are They and Why Should You Care? *Kansas Library Association College and University Libraries Section Proceedings*, *1*(1), 16–23. doi:10.4148/culs.v1i0.1355

Collier, D. A., & Meyer, S. M. (1998). A service positioning matrix. *International Journal of Operations & Production Management*, *18*(12), 1223–1244. doi:10.1108/01443579810236647

Collins, B., Paquet, G., Roy, J., & Wilson, C. (2002). E-Governance and smart communities: A social learning challenge. In *Proceedings of SSHRC Knowledge Based Economy Conference* (pp. 1-23). Newfoundland, Canada. Academic Press.

Conroy, J. (2016). Hotel trends for 2016. Retrieved from http://www.telegraph.co.uk/travel/hotels/articles/Hotel-trends-for-2016/

Constantelou, A. (2002). Emerging Trends in Customer Relation Management Using ICT: the Travel Industry, Issue Report N. 22, STAR (SocioEconomic Trends Assessment for the DigitalRevolution) Retrieved May 10, 2019, from http://www.cerna.ensmp.fr/Enseignement/CoursModelesEcoNumerique/TD/4-Constantelou.pd

Cooper, C. (2006). Knowledge management and tourism. *Annals of Tourism Research*, *33*(1), 47–64. doi:10.1016/j.annals.2005.04.005

Cooper, M., & Macneil, N. J. (2015). Virtual Reality Mapping: IT Tools for the Divide Between Knowledge and Action in Tourism. *Tourism Recreation Research*, *30*(3), 61–68. doi:10.1080/02508281.2005.11081487

Corbin, J., & Strauss, A. (2008). *Basics of qualitative research: Techniques and procedures for developing grounded theory* (3rd ed.). Thousand Oaks, CA: Sage. doi:10.4135/9781452230153

Córdova, F. M., & Gutiérrez, F. A. (2018). Knowledge management system in service companies. *Procedia Computer Science*, *139*, 392–400. doi:10.1016/j.procs.2018.10.275

Correia, A., Kozak, M., & Ferradeira, J. (2013). From tourist motivations to tourist satisfaction. *International Journal of Culture, Tourism and Hospitality Research*, *7*(4), 411–424. doi:10.1108/IJCTHR-05-2012-0022

Coşkun, A. (2012). QR Kod Nedir? Retrieved from http://aylincsknn.blogspot.com/2012/08/qr-kod-cagn-barkodu-olarak-anlan-qr_11.html

Coşkun, C. (2017). Augmented reality as an exhibition method. *Art and Design Magazine* (20), 61-75.

Coskun, V., Ok, K., & Ozdenizci, B. (2011). *Near field communication (NFC): From theory to practice*. UK: John Wiley & Sons.

Craig, A. B. (2013). Understanding Augmented Reality, Concepts and Applications. (1st Ed.). Imprint: Morgan Kaufmann.

Craig, A. B., Sherman, W. R., & Will, J. D. (2009). *Developing Virtual Reality Applications*. Morgan Kaufmann Publishers-Elsevier.

Cranmer, E. E. (2017). *Developing an augmented reality business model for cultural heritage tourism: the case of Geevor Museum* (Unpublished Doctoral thesis). Manchester Metropolitan University.

Crompton, H., Lafrance, J., & Van 't Hooft, M. (2012). QR Codes 101. *Learning and Leading with Technology*, 22–25. Retrieved from www.sparqcode.com/static/maestro

Crook, J. (2014). *Starwood Introduces Robotic Butler at Aloft Hotel in Cupertino*. Retrieved from https://techcrunch.com/2014/08/13/starwood-introduces-robotic-butlers-at-aloft-hotel-in-palo-alto/

Crook, J. (2014). Starwood introduces robotic butlers at Aloft hotel in Cupertino. https://techcrunch.com/2014/08/13/starwood-introduces-robotic-butlers-at-aloft-hotel-inpalo-alto/

Croteau, A. M., & Li, P. (2003). Critical Success Factor at CRM Technological Initiative. *Canadian Journal of Administrative Sciences*, *20*(1), 21–30. doi:10.1111/j.1936-4490.2003.tb00303.x

Cruz-Neira, C., Sandin, D., DeFanti, T., Kenyon, R., & Hart, J. (1992). The CAVE: Audio visual experience automatic virtual environment. *Communications of the ACM*, *35*(6), 64–73. doi:10.1145/129888.129892

Csapodi, M., & Nagy, A. (2007, July). New Applications for NFC Devices. Paper presented at the meeting 16tj IST Mobile &Wireless Communications Summit, IEEE, and Budapest, Hungary. 10.1109/ISTMWC.2007.4299077

Csikszentmihalyi, M. (1990). *Flow: The psychology of optimal experience*. New York: Harper Perennial.

Csobanka, Z. E. (2016). The Z Generation. *Acta Technologica Dubnicae*, *6*(2), 63–76. doi:10.1515/atd-2016-0012

D'Angelo, G., Ferretti, S., & Ghini, V. (2017). Multi-level simulation of internet of things on smart territories. *Simulation Modelling Practice and Theory*, *73*, 3–21. doi:10.1016/j.simpat.2016.10.008

Dadwal, S. S., & Hassan, A. (2015). The Augmented Reality Marketing: A Merger of Marketing and technology in Tourism. In N. Ray (Ed.), *Emerging Innovative Marketing Strategies in the Tourism Industry* (pp. 63–80). Hershey, PA: IGI Global. doi:10.4018/978-1-4666-8699-1.ch005

Dalgıç, A., & Birdir, K. (2017). Otel işletmeciliğinde yeni trendler. In Y. Akgündüz (Ed.), *Otel İşletmeciliği* (pp. 311–328). Paradigma Akademi.

Damala, A., Cubaud, P., Bationo, A., Houlier, P., & Marchal, I. (2008). Bridging the gap between the digital and the physical: Design and evaluation of a mobile augmented reality guide for the museum visit. In *Proceedings of the 3rd International Conference on Digital Interactive Media in Entertainment and Arts (DIMEA '08)* (pp. 120-127). ACM, New York. 10.1145/1413634.1413660

Dameri, R. P. (2013). Searching for smart city definition: A comprehensive proposal. *International Journal of Computers and Technology*, *11*(5), 2544–2551. doi:10.24297/ijct.v11i5.1142

Dave, N. (2018). *digitaldoughnut*. Retrieved from https://www.digitaldoughnut.com/articles/2018/january/ways-in-which-iot-is-shaping-the-future-of-travel

Davenport, T. H. (2013). *At the Big Data Crossroads: turning towards a smarter travel experience*. Amadeus.

Davydova, O. (2017). Sentiment Analysis Tools Overview, Part 1. Positive and Negative Words Databases. Retrieved June 15, 2019, from https://medium.com/@datamonsters/sentiment-analysis-tools-overview-part-1-positive-and-negative-words-databases-ae35431a470c

De Avila, A. L. (2011). Tourism: What lies ahead?, World Tourism Organization (UNWTO) affiliate members AM-reports – Technology in Tourism (Vol. 1).

De Jong, M., Zhang, K., Roth, A. M., Rhodes, T., Schmucker, R., Zhou, C., . . . Veloso, M. (2018). Towards a Robust Interactive and Learning Social Robot, *AAMAS 2018,* July 10-15, Stockholm, Sweden.

De Mauro, A., Greco, M., & Grimaldi, M. (2016). A formal definition of Big Data based on its essential features. *Library Review*, *65*(3), 122–135. doi:10.1108/LR-06-2015-0061

De Pablo, F. (2017). Smart tourist destinations. *Revista de la Normalización y Evaluación de la conformidad*, 238, 20-26.

Dean, J., & Sanjay, G. (2016). *MapReduce: Simplified Data Processing on Large Clusters*. Google, Inc. Web

Deci, E. L., Connell, J. P., & Ryan, R. M. (1989). Self determination in a work organization. *The Journal of Applied Psychology*, *74*(4), 580–590. doi:10.1037/0021-9010.74.4.580

Decker, M., Fischer, M., & Ott, I. (2017). Service Robotics and Human Labor: A first technology assessment of substitution and cooperation. *Robotics and Autonomous Systems*, *87*, 348–354. doi:10.1016/j.robot.2016.09.017

Del Chiappa, G., & Baggio, R. (2015). Knowledge transfer in smart tourism destinations: Analyzing the effects of a network structure. *Journal of Destination Marketing & Management*, *4*(3), 145–150. doi:10.1016/j.jdmm.2015.02.001

Del Chiappa, G., & Fotiadis, A. (2018). Generation Y Perspective of Hotel Disintermediation and User-Generated Content. In D. Gursoy (Ed.), *The Routledge Handbook of Hospitality Marketing*. New York: Taylor & Francis Group.

Del Vecchio, P., Mele, G., Ndou, V., & Secundo, G. (2018). Creating value from social big data: Implications for smart tourism destinations. *Information Processing & Management*, *54*(5), 847–860. doi:10.1016/j.ipm.2017.10.006

Deloitte. (2015). *Blockchain distrupting the financial services industry*. Retrieved from https://www2.deloitte.com/content/dam/Deloitte/ie/Documents/FinancialServices/IE_Cons_Blockchain_1015.pdf

Demirezen, B. (2019). Artırılmış gerçeklik ve sanal gerçeklik teknolojisinin turizm sektöründe kullanılabilirliği üzerine bir literatür taraması. *Uluslararası Global Turizm Araştırmaları Dergisi*, *3*(1), 1–26.

Demirhan, A., Kılıç, Y. A., & İnan, G. (2010). Tıpta yapay zeka uygulamaları. *Yoğun Bakım Dergisi*, *9*(1), 31–41.

Denning, S. (2000). *The springboard: how storytelling ignites action in knowledge era organizations*. Oxford, UK: Butterworth-Heinemann.

Department of Tourism Rajasthan. (2018). *Annual Progress Report 2017-18*. Retrieved from Rajasthan Tourism Department website: http://www.tourism.rajasthan.gov.in/content/dam/rajasthan-tourism/english/others/tourism-department-annual-progress-report-2017-18.pdf

Desai, P. R., Desai, P. N., Ajmera, K. D., & Mehta, K. (2014). A review paper on oculus rift-a virtual. [IJETT]. *Int. J. Eng. Trends Technol*, *13*(4), 175–179. doi:10.14445/22315381/IJETT-V13P237

Desdemoustier, J., Crutzen, N., Cools, M., & Teller, J. (2019). Smart City appropriation by local actors: An instrument in the making. *Cities (London, England)*, *92*, 175–186. doi:10.1016/j.cities.2019.03.021

Deterding, S., Dixon, D., Khaled, R., & Nacle, L. (2011). From game design elements to gamefulness: Defining “Gamification”. In *Proceedings of the 15th International Academic MindTrek Conference on Envisioning Future Media Environments*, Tampere, Finland. 10.1145/2181037.2181040

Devillet, G., & Breuer, C. (2008). Contribution to the applied territorial intelligence: reasoned catalog of territorial information available on internet and sources in Europe. *International Conference of Territorial Information. Papers on Tools and Methods of Territorial Intelligence.* Besançon, France: MSHE.

Dhillon, J. S., Joshi, M., & Verma, R. (2012). Indian Hospitality Industry: Moving Towards Customer Oriented Information System (COIS). *Journal of Business Management & Social Sciences Research (JBM&SSR)*, *1*(1), 58-69.

Digital Tourism Think Tank. (2013). Retrieved from https://www.thinkdigital.travel/tag/2013/

Dilek, N. K., Kızılırmak, İ., & Dilek, S. E. (2018). Virtual Reality or Just Reality? A Swot Analysis of the Tourism Industry. *Journal of Tourismo*, *4*(1), 67–74.

Dilworth, J. (2010). Realistic Virtual Reality and Perception. *Philosophical Psychology*, *23*(1), 23–42. doi:10.1080/09515080903533942

Dolnicar, S., & Otter, T. (2003). Which Hotel attributes Matter? A review of previous and a framework for future research. *Proceedings of the 9th Annual Conference of the Asia Pacific Tourism Association (APTA)*. Academic Press.

Dönmez, M. (1996). *Mahalli İdarelerin Teşkilat Yapısı, Organları ve Görevleri*. Ankara, Turkey: Mahalli İdareler Derneği Yayınları.

Doolin, B., Burgess, L., & Cooper, J. (2002). Evaluating the use of the Web for tourism marketing: A case study from New Zealand. *Tourism Management*, *23*(5), 557–561. doi:10.1016/S0261-5177(02)00014-6

Dörner, R., Jung, B., Grimm, P., Broll, W., & Göbel, M. (2013). Einleitung. In R. Dörner, W. Broll, P. Grimm, & B. Jung (Eds.), *Virtual und Augmented Reality (VR/AR)* (pp. 1–32). Berlin, Germany: Springer. doi:10.1007/978-3-642-28903-3_1

Doswell, R. (1997). *Tourism: How Effective Management Make the Difference*. Butterworth-Heinemann.

DPT. (2006). *Dokuzcu Kalkınma Planı*. Ankara, Turkey: DPT Yayınları.

Dragović, N., Stankov, U., & Vasiljević, Đ. (2018). Contactless Technology as a Factor of Tourism Industry Development-A Review of Current Practices and Future Directions. *Economic Themes*, *56*(2), 179–202. doi:10.2478/ethemes-2018-0011

Drummond, M. (2019). 5 Great ways airlines are using the ınternet of things. Retrieved from https://w3.accelya.com/blog/5-great-ways-airlines-are-using-the-internet-of-things

Dubey, A. K. (2016). Future Technology and Service Industry: A Case study of Travel and Tourism Industry. *Global Journal of Enterprise Information System, 8*(3).

Dubey, A. K., Giri, M., Sahare, M., & Dubey, A. K. (2011, June). Step-up analysis and generalization approach for Trusted NFC application Development for enhancing real time use Location. In *Proceedings of the 2011 International Conference on Communication Systems and Network Technologies*, IEEE. Katra, Jammu. 10.1109/CSNT.2011.73

Dubihlela, J., & Khosa, P. M. (2014). Impact of e-CRM Implementation on Customer Loyalty, Customer Retention and Customer Profitability for Hoteliers along the Vaal Meander of South Africa. *Mediterranean Journal of Social Sciences*, *5*(16), 175–183.

Dülger, Ü. (2015). *Stratejik Büyük Veri Yönetiminin Yatırımlar Üzerindeki Etkileri* (Unpublished master dissertation). İstanbul University, Turkey.

Dumas, P. (2004). Intelligence, Territoire, Décentralisation, ou la Région à la française. *Actes des 3ème rencontres TIC & Territoire: quels développements ? de Lille,* ISDM, 16(163).

Duncan, T., & Moriarty, S. (2006). How integrated marketing communication's 'touchpoints' can operationalize the service-dominant logic. *The Service-Dominant Logic of Marketing: Dialog, Debate, and Directions*, 236–249.

Dyer, K. (2013a) French region provides multimedia content to tourists via NFC. Retrieved from https://www.nfcworld.com/2013/10/15/326406/french-region-provides-multimedia-content-tourists-via-nfc/

Dyer, K. (2013b). The Informed Tourist equips Sydney hotels with NFC touchpoints. Retrieved from https://www.nfcworld.com/2013/10/07/326228/informed-tourist-equips-sydney-hotels-nfc-touchpoints/

Dzhandzhugazova, E. A., Zaitseva, N. A., Larionova, A. A., & Pervunin, S. N. (2015). The Russian Hotel Market: condition and development under the crisis. *Mediterranean Journal of Social Sciences, 6*(3 S5), 289.

Dzhandzhugazova, E. A., Blinova, E. A., Orlova, L. N., & Romanova, M. M. (2016). Innovations in hospitality industry. *International Journal of Environmental and Science Education, 11*(17), 10387–10400.

EC. (1999). *Towards Quality Rural Tourism*. Brussels, Belgium: Enterprise Directorate General Tourism Unit.

Eğer, Ö. (2019). *Big Data'nın (Büyük Veri) Endüstriyel Kullanımı.* Retrieved from https://www.endustri40.com/big-datanin-buyuk-veri-endustriyel-kullanimi/

Egger, R. (2013). The impact of near field communication on tourism. *Journal of Hospitality and Tourism Technology*, *4*(2), 119–133. doi:10.1108/JHTT-04-2012-0014

Eken, S., & Sayar, A. (2014). A smart bus tracking system based on location-aware services and QR codes. *INISTA 2014 - IEEE International Symposium on Innovations in Intelligent Systems and Applications, Proceedings*, 299–303. IEEE. 10.1109/INISTA.2014.6873634

Ekonomihaber7.com. (2017). The waiters of this restaurant are robot! Retrieved from http://ekonomi.haber7.com/ekonomi/haber/2373365-bu-restoranin-garsonlari-robot

Elgendy, N., & Elragal, A. (2014). Big data analytics: a literature review paper. *Industrial Conference on Data Mining* (pp. 214–227). Springer. 10.1007/978-3-319-08976-8_16

Elisabeth, E., Nock, R., & Célimène, F. (2013). Demonstrator of a Tourist Recommendation System. In V. Bhatnagar, & S. Srinivasa (Eds.), Big Data Analytics. (pp. 171-175) BDA. Lecture Notes in Computer Science, vol. 8302. Cham, Switzerland: Springer. doi:10.1007/978-3-319-03689-2_11

Elliot, J. (1997). *Politics of Tourism*. London, UK: Routledge.

Elmastaş, L. (2019) Antalya Akıllı Şehir Mobil Uygulaması, 05.08.2019. Retrieved from https://rayhaber.com/2019/03/antalya-akilli-sehir-mobil-uygulamasi/

Emek, M. (2012). *Usage of QR Code in Tourism Industry. In 6th World Conference for Graduate Research in Tourism.* Fethiye, Turkey: Hospitality and Leisure; Retrieved from http://hdl.handle.net/11376/108

Emrealp, S. (2005). Yerel gündem 21 uygulamalarına yönelik kolaylaştırıcı bilgiler elkitabı. İstanbul, Turkey: IULA-EMME (UCLG-MEWA) Yayını.

Entertainment Software Association. (2018). *Essential facts about the computer and video game industry.* Retrieved from www.theesa.com/wp-content/uploads/2018/05/EF2018_FINAL.pdf

Erbay, M. (2017). Education in museums with new generation Technologies. *National Education Magazine, 46*(214), 255–268.

Eren, E. (2000). *İşletmelerde Stratejik Yönetim ve İşletme Politikası.* İstanbul, Turkey: Beta Basım Yayım.

Erickson, S., & Rothberg, H. (n.d.). Big Data and Knowledge Management: Establishing a Conceptual Foundation. Electronic Journal of Knowledge Management, 12(2), 108–116.

Ericsson. (2016). *Ericsson Mobility Report: On the pulse of the networked society.* Retrieved from https://www.ericsson.com/assets/local/mobility-report/documents/2016/Ericsson-mobility-report-june-2016.pdf

Eriksen, K. (2018). Benefits of Artificial Intelligence in the Restaurant Industry. Retrieved from https://www.deputy.com/blog/benefits-of-artificial-intelligence-in-the-restaurant-industry

Erkek, S. (2017). Akıllı Şehircilik anlayışı ve belediyelerin inovatif uygulamaları. *Medeniyet ve Toplum Dergisi, 1*(1), 55–72.

Erkiah, S., & Ladkoo, A. D. (2018). A Quantitative Study about Assessing the Effectiveness of Electronic Customer Relationship Management: A Case of Two Hotels in Mauritius. *International Journal of Social and Business Sciences, 12*(10), 1438–1446.

Erkiz, K. S. (2018) Yakın Alan İletişim Teknolojisinin Sağlık Alanında Kullanımı Üzerine Araştırma ve Bir Uygulama Örneği, *(NFC)* (Unpublished Master's thesis). Gazi University, Turkey.

Ersu, Ö. (2018). Profesyonel Turist Rehberliğinde Dijital Dönüşüm: Mesleğin Yakın Geleceği Arttırılmış Gerçeklik ve Sanal Gerçeklik Uygulamaları. *Uluslararası Turizm, İşletme. Ekonomi Dergisi, 2*(2), 578–586.

Ertuğral, S. M., Aslan, S., & Balık, M. (2015). Turizm İşletmelerinde İnsan Kaynakları Yönetimi. In O. Akova, İ. Kızılırmak, & H. Tanrıverdi (Eds.), *Turizm İşletmeciliği Temel Kavramlar ve Uygulamalar* (pp. 177–204). Ankara, Turkey: Detay Yayıncılık.

Etro, F. (2009). The economic impact of cloud computing on business creation, employment and output in Europe. *Review of Business and Economics, 54*(2), 179–208.

Etro, F. (2009). The Economic impact of cloud computing on business creation, employment, and output in Europe. *Review Business and Economics, 54*(2), 179–208.

European Parliament, Mapping Smart Cities in the EU. (2014). Retrieved from http://www.europarl.europa.eu/RegData/etudes/etudes/join/2014/507480/IPOL-ITRE_ET(2014)507480_EN.pdf

EverywhereN. F. C. (2017). Retrieved from https://www.nxp.com/docs/en/supporting-information/BL-STI-NFC-is-everywhere-Hannibal-Su.pdf

Eyal, N., & Hoover, R. (2014). *Hooked: How to build habit forming products*. Wood Dale, Ilinois: Penguin Group.

Falconer, G., & Mitchell, S. (2012). Smart city framework: A systematic process for enabling smart + connected communities, 1-11. Cisco Internet Business Solutions Group (IBSG). Retrieved from https://smartcitiescouncil.com/sites/default/files/public_resources/Smart%20city%20framework_a%20systematic%20process.pdf Accessed on July 10, 2019.

Falconer, G., & Mitchell, S. (2012). *Smart city framework. Cisco Internet Business SolutionsGroup*. IBSG.

FalkeC. (2012). Retrieved from http://www.uni-weimar.de/kunst-und-gestaltung/wiki/images/IFD_mobile-culture_2.01_ChristopherFalke_GhostsGames.pdf

Farshid, M., Paschen, J., Eriksson, T., & Kietzmann, J. (2018). Go boldly! Explore augmented reality (AR), virtual reality (VR), and mixed reality (MR) for business. *Business Horizons*. doi:10.1016/j.bushor.2018.05.009

Faulkner, B., & Valerio, P. (2000). An Integrative Approach to Tourism Demand Forecasting. In C. Ryan, & S. Page (Eds.), Tourism Management Towards the New Millennium (pp. 45-57). Pergamon.

Feinberg, R. A., Kadam, R., Hokama, L., & Kim, I. (2002). The state of electronic customer relationship management in retailing. *International Journal of Retail & Distribution Management*, *30*(10), 470–481. doi:10.1108/09590550210445344

Feinberg, R., & Kadam, R. (2002). E-CRM Web Service Attributes as Determinants of Customer Satisfaction with Retail Web Sites. *International Journal of Service Industry Management*, *13*(5), 432–451. doi:10.1108/09564230210447922

Feldman, R., & Sanger, J. (2006). The Text Mining Handbook. In The Text Mining Handbook. doi:10.1017/CBO9780511546914

Feliu, C. (2019). *Big data case study: 5 relevant examples from the airline industry*. Retrieved from https://blog.datumize.com/5-relevant-examples-of-a-big-data-case-study-from-the-airline-industry

Femenia-Serra, F., & Ivars-Baidal, J. A. (2018). Smart tourism: Implications for the management of cities and tourist destinations. In M. T. Cantó López, J. A. Ivars Baidal, & R. Martínez Gutiérrez (Eds.), *Gestión inteligente y sostenible de las ciudades: Gobernanza, smart cities y turismo* (pp. 129–151). Valencia, Spain: Tirant Lo Blanch.

Femenia-Serra, F., Neuhofer, B., & Ivars-Baidal, J. A. (2018). Towards a conceptualisation of smart tourists and their role within the smart destination scenario. *Service Industries Journal*, 1–25.

Fernandes, R. P. A., Almeida, J. E., & Rosseti, R. J. F. (2013). A collaborative tourist system using serious games. *Advances in Intelligent Systems and Computing*, *206*, 725–734. doi:10.1007/978-3-642-36981-0_67

Fernández Tabales, A., Foronda, C., Galindo, L., & García, A. (2017). Developing a system of territorial governance indicators for tourism destinations. *Journal of Sustainable Tourism*, *25*(9), 1275–1305. doi:10.1080/09669582.2016.1260136

Filieri, R. (2016). What makes an online consumer review trustworthy? *Annals of Tourism Research*, *58*, 46–64. doi:10.1016/j.annals.2015.12.019

Fino, E., Martín-Gutiérrez, J., Fernández, M., & Davara, E. (2013). Interactive Tourist Guide: Connecting Web 2.0, Augmented Reality and QR Codes. *Procedia Computer Science*, *25*, 338–344. doi:10.1016/j.procs.2013.11.040

Finžgar, L., & Trebar, M. (2011). Use of NFC and QR code identification in an electronic ticket system for public transport. In *Proceedings SoftCOM 2011, 19th International Conference on Software, Telecommunications, and Computer Networks*, (iii), 1–6.

Firoiu, D., & Croitoru, A.-G. (2013). Tourism and Tourism Infrastructure From the Perspective of Technological Changes. *Romanian Economic Business Review*, *8*(2), 93–103. Retrieved from http://search.proquest.com/docview/1448007631?accountid=35465%5Cnhttp://sfx.dbc.dk/dbc-45DBC_UCN?url_ver=Z39.88-2004&rft_val_fmt=info:ofi/fmt:kev:mtx:journal&genre=article&sid=ProQ:ProQ%3Apqrl&atitle=TOURISM+AND+TOURISM+INFRASTRUCTURE+FROM+THE+PERSPECTIVE

Fisher, A., & Krutilla, J. (1972). Determination of optimal capacity of resource-based recreation facilities. *Natural Resources Journal*, *12*, 417.

Floris, R., & Campagna, M. (2014). Social media geographic information in tourism planning. *TEMA- Journal of Land Use, Mobility and Environment,* Special Issue; Eight Int. *Conference INPUT Smart City-Planning for Energy, Transportation and Sustainability of the Urban System.* Naples, June 4-6, 2014.

Fodness, D. (1994). Measuring tourist motivation. *Annals of Tourism Research*, *21*(3), 555–581. doi:10.1016/0160-7383(94)90120-1

Fogg, B. (2009). A behavior model for persuasive design. In *Proceedings of the 4th International Conference of Persuasive Technology*, Claremont, CA. 10.1145/1541948.1541999

Forman, G. (2003). An Extensive Empirical Study of Feature Selection Metrics for Text Classification George. *Journal of Machine Learning Research.* doi:10.1162/153244303322753670

Foroudi, P., Gupta, S., Sivarajah, U., & Broderick, A. (2018). Investigating the effects of smart technology on customer dynamics and customer experience. *Computers in Human Behavior*, *80*, 271–282. doi:10.1016/j.chb.2017.11.014

Forster, J. (1964). The Sociological Consequences of Tourism. *International Journal of Comparative Sociology*, *5*(2), 217–227. doi:10.1177/002071526400500208

ForumN. F. C. (2019). Retrieved from https://nfc-forum.org/

Foss, N. J., Laursen, K., & Pedersen, T. (2011). Linking Customer Interaction and Innovation: The Mediating Role of New Organizational Practices. *Organization Science*, *22*(4), 980–999. doi:10.1287/orsc.1100.0584

Frechtling, D. C. (2000). Assessing the Impacts of Travel and Tourism-Measuring Economic Benefits. *International Library of Critical Writings in Economics*, *121*, 9–27.

Freeland, J. (2002). *The Ultimate CRM Handbook: Strategies and concepts for building enduring customer loyalty and profitability*. New York: McGraw-Hill.

Frew, A. J. (2000). A critical analysis of tourism ınformation technology research. In *Information and Communication Technologies in Tourism* (pp. 39–52). Vienna, Austria: Springer.

Frey, C. B., & Osborne, M. A. (2013). *The Future of Employment: How Susceptible are Jobs to Computerisaton?* Retrieved from https://www.oxfordmartin.ox.ac.uk/downloads/academic/The_Future_of_Employment.pdf

Fritz, F., Superregui, A., & Linaza, M. (2005). Enhancing Cultural tourism experiences with Augmented Reality technologies. In *Proceedings 6th International Symposium on Virtual Reality, Archaeology and Cultural Heritage VAST*, pp. 1-5, Pisa, Italy, November, San Sebastian, Spain, The Eurographics Association.

Furlonger, D., & Valdes, R. (2017). *Practical blockchain: a gartner trend insight report.* Retrieved from https://haas.campusgroups.com/htc/get_file?eid=139611897577441f06512fc062b0a63e

Fyall, A., & Garrod, B. (2005). From Competition to Collaboration in the Tourism Industry. In W. F. Theobald (Ed.), Global Tourism 3rd Edition (pp. 52-74). Elsevier Butterworth-Heinemann. doi:10.1016/B978-0-7506-7789-9.50009-1

Gagliordi, N. (2016). *This Watson-Powered Robot Concierge is Rethinking the Hotel Industry.* Retrieved from https://www.zdnet.com/article/this-watson-powered-robot-concierge-is-rethinking-the-hotel-industry/

Galeano, G. (2017). *Design and development of a Mixed Reality application in the automotive field.* (Unpublished Master Thesis). Department of Control and Computer Engineering, Politecnico di Torino.

Gao, J., & Li, S. (2011). Detecting spatially non-stationary and scale-dependent relationships between urban landscape fragmentation and related factors using Geographically Weighted Regression. *Applied Geography (Sevenoaks, England), 31*(1), 292–302. doi:10.1016/j.apgeog.2010.06.003

Garcia-Ayllon, S., & Miralles, J. L. (2015). New strategies to improve governance in territorial management: Evolving from "smart cities" to "smart territories". *Procedia Engineering, 118*, 3–11. doi:10.1016/j.proeng.2015.08.396

Garrido-Moreno, A., & Padilla-Meléndez, A. (2011). Analyzing the impact of knowledge management on CRM success: The mediating effects of organizational factors. *International Journal of Information Management, 31*(5), 437–444. doi:10.1016/j.ijinfomgt.2011.01.002

Garrido, P. C., Miraz, G. M., Ruiz, I. L., & Gómez-Nieto, M. Á. (2010). A near field communication tool for building intelligent environment using smart posters. *International Journal of Computers and Communications, 4*(1), 9–16.

Garrison, G., Kim, S., & Wakefield, R. L. (2012). Success factors for deploying cloud computing. *Communications of the ACM, 55*(9), 62–68. doi:10.1145/2330667.2330685

Gartner. (2017). Gartner Says 8.4 Billion Connected "Things" Will Be in Use in 2017, Up 31 Percent From 2016. doi:10.1017/CBO9781107415324.004

Gaudin, S. (2016). *Meet Connie, Hilton's Smart Robot Concierge.* Retrieved from https://www.computerworld.com/article/3042401/meet-connie-hilton-s-smart-robot-concierge.html

Gavalas, D., Konstantopoulos, C., Mastakas, K., & Pantziou, G. (2014). Mobile recommender systems in tourism. *Journal of Network and Computer Applications, 39*(1), 319–333. doi:10.1016/j.jnca.2013.04.006

Geissbauer, R., Schrauf, S., & Koch, V. (2014) Industry 4.0 – Opportunities and Challenges of the Industrial Internet assessment, PricewaterhouseCoopers. Retrieved from https://www.pwc.nl/en/assets/documents/pwcindustrie-4–0.pdf

Gentes, A., Guyot Mbodji, A., & Demeure, I. (2010). Gaming on the move: Urban experience as a new paradigm for mobile pervasive game design. *Multimedia Systems, 16*(1), 43–55. doi:10.100700530-009-0172-2

Gentile, C., Spiller, N., & Noci, G. (2007). How to sustain the customer experience: An overview of experience components that co-create value with the customer. *European Management Journal, 25*(5), 395–410. doi:10.1016/j.emj.2007.08.005

George, E. W., Mair, H., & Reid, D. G. (2009). *Rural Tourism Development Localism and Cultural Change.* Bristol, UK: Channel View Publications.

Gezon, L. L. (2014). Who wins and who loses? Unpacking the "local people" concept in ecotourism: A longitudinal study of community equity in Ankarana, Madagascar. *Journal of Sustainable Tourism, 22*(5), 821–838. doi:10.1080/09669582.2013.847942

Ghosh, S., Goswami, J., Kumar, A., & Majumder, A. (2015, May). Issues in NFC as a form of contactless communication: A comprehensive survey. In *Proceedings International Conference on Smart Technologies and Management for Computing, Communication, Controls, Energy and Materials (ICSTM)*, IEEE. Chennai, India. 10.1109/ICSTM.2015.7225422

Gibbs, T., & Yordchim, S. (2014). Thai perception on Bitcoin value. *International Journal of Social, Behavioral, Educational, Economic Business, and Industrial Engineering*, *8*(7), 2334–2336.

Gibson, A., & O'Rawe, M. (2018). Virtual reality as a travel promotional tool: Insights from a consumer travel fair. In T. Jung, & M. tom Dieck (Eds.), *Augmented reality and virtual reality, Progress in IS* (pp. 93–107). Cham, Switzerland: Springer. doi:10.1007/978-3-319-64027-3_7

Giffinger, R., Fertner, C., Kramar, H., Kalasek, R., Pichler-Milanovic, N., & Meijers, E. (2007). *Smart cities - Ranking of European medium-sized cities.* Final Report, Vienna University of Technology. Retrieved from http://curis.ku.dk/ws/files/37640170/smart_cities_final_report.pdf

GilaniNia, S., Balaei, S., & Niyari, F. S. (2013). Impact of ECRM on the Create Competitive Advantage in Hotels of Sarein. *Arabian Journal of Business and Management Review*, *3*(4), 77–85.

Girardot, J. (2000). Principes, Méthodes et Outils d'Intelligence Territoriale. Évaluation participative et Observation Coopérative. Conhecer melhor para agir melhor, Actes du séminaire européen de la Direction Générale de l'Action Sociale du Portugal, Évora, (7-17). Lisboa, Portugal: DGAS.

Girardot, J. (2008). Evolution of the concept of territorial intelligence within the coordination action of the European network of territorial intelligence. *Res-Ricerca e Sviluppo per le politiche sociali*, 1(2), 11-30.

Glebova, I. S. (2014). Assessment of Cities in Russia According to the Concept of "Smart City" in the Context of the Application of Information and Communication Technologies. *Mediterranean Journal of Social Sciences*. doi:10.5901/mjss.2014.v5n18p55

GlobalData.com. (2019). Cloud computing in tourism - thematic research. Retrieved from https://www.globaldata.com/https://hot-topics.globaldata.com/reports/cloud-computing-in-tourism-thematic-research/

Gockley, R., Bruce, A., Forlizzi, J., Michalowski, M., Mundell, A., Rosenthal, S., ... Wang, S. (2005). Designing Robots for Long-Term Social Interaction, *2005 IEEE/RSJ International Conference on Intelligent Robots and Systems*, August 2-6, Alberta, Canada. 10.1109/IROS.2005.1545303

Goeldner, C. R., Ritchie, J. R. B., & McIntosh, R. W. (2000). *Tourism: Principles, Practice, Philosophies*. New York: Wiley.

Gökalp, E., & Eren, E. (2016). Application of ıntelligent technologies in tourism and hospitality sector. In V. Tecim, C. Tarhan, & C. Aydın (Eds.), Smart Technology & Smart Management: Smart Technology & Intelligent Management, 278-287.

Gökalp, E., & Eren, P. E. (2016). Akıllı Teknolojilerin Turizm ve Otelcilik Sektöründe Uygulanması, Date of Access: 04.11.2018. Retrieved from www.researchgate.net/profile/Ebru_Goekalp/publication/309619452_

Gökkuş, B. (2018). First robot waiter restaurant of South Asia in service. Retrieved from https://tr.euronews.com/2018/12/09/guney-asya-nin-ilk-robot-garsonlu-restorani-hizmette

Göksu, C. (2014). *Datawarehouse Türkiye*. Retrieved from http://datawarehouse.gen.tr/big-datanedir-geleneksel-veri-yonetimine-etkisi-ne-olur/ https://proente.com

Göktaş Kulualp, H. (2016). Bilgi yönetimi olgunluk modellerine göre işletmelerin bilgi yönetimi düzeyleri: genel bir bakış. Akademik Bakış Uluslararası Hakemli Sosyal Bilimler Dergisi, (56), 69-91.

Goncharova, N. L., & Bezdenezhnykh, T. I. (2018) Employing the Elderly in the Service Sector in Conditions of Electronic and Fourth Innovation and Technology Revolution: Industry 4.0. *Proc. of the 31st Int. Business Information Management Association Conf., IBIMA 2018 - Innovation Management and Education Excellence through Vision 2020, IBIMA* 2018, pp. 2330 –36.

Goodhue, D. L., Wixom, B. H., & Watson, H. J. (2002). Realizing business benefits through CRM: Hitting the right target in the right way. *MIS Quarterly Executive*, *1*(2), 79–94.

Goodman, J. (1999). Basic facts on customer complaint behavior and the impact of service on the bottom line. *Competitive Advantage*, *8*(1), 1–5.

Gössling, S., Peeters, P., Hall, C., Ceron, J.-P., Dubois, G., Lehmann, L., & Scott, D. (2012). Tourism and water use: Supply, demand, and security. An international review. *Tourism Management*, *33*(1), 1–15. doi:10.1016/j.tourman.2011.03.015

Government of India, Ministry of Urban Development. (2015). *Mission Statement and Guidelines*. Retrieved from http://164.100.161.224/upload/uploadfiles/files/SmartCity Guidelines(1).pdf

Government of the Republic of Slovenia. (2015). *Slovenia's Smart Specialisation Strategy*. Retrieved from http://www.svrk.gov.si/fileadmin/svrk.gov.si/pageuploads/Dokumenti_za_objavo_na_vstopni_strani/S4_document_2015_ENG.pdf

Göymen, K. (2004). Yerel kalkınma önderi ve paydaşı olarak belediyeler. Yerel Kalkınmada Belediyelerin Rolü (pp. 21-30), İstanbul, Turkey: Pendik Belediyesi Kültür Yayınları 21.

Grassie, K. (2007). Easy handling and security make NFC a success. *Card Technology Today*, *19*(10), 12–13. doi:10.1016/S0965-2590(08)70134-8

Greenberg, P. (2001). Capturing and Keeping Customers in Internet Real Time. *The Mcgraw-Hill Companies*. Retrieved May 10, 2019, from https://www.amazon.com/CRM-Speed-Light-Capturing-Customers/dp/0072127821

Greenberg, P. (2002). *CRM at the Speed of Light*. New York: Mcgraw-Hill Companies.

Greenfield, A. (2013). *Against the smart city*. London: Kindle Edition.

Gretzel, U. (2011). Intelligent systems in tourism: A social science perspective. *Annals of Tourism Research*, *38*(3), 757–779. doi:10.1016/j.annals.2011.04.014

Gretzel, U., Sigala, M., Xiang, Z., & Koo, C. (2015a). Smart tourism: Foundations and developments. *Electronic Markets*, *25*(3), 179–188. doi:10.100712525-015-0196-8

Gretzel, U., Werthner, H., Koo, C., & Lamsfus, C. (2015). Conceptual foundations for understanding smart tourism ecosystems. *Computers in Human Behavior*, *50*, 558–563. doi:10.1016/j.chb.2015.03.043

Gretzel, U., Zhong, L., & Koo, C. (2016). Application of smart tourism to cities. *International Journal of Tourism Cities*, *2*(2). doi:10.1108/IJTC-04-2016-0007

Grewal, D. (2009). *Enhancing the Retail Customer Experience*. Elsevier.

Grover, D. (2011). Effective Customer Relationship Management through e-CRM. *Viewpoint*, *2*(1), 27–38.

GSMA. (2013). Guide to Smart Cities the Opportunity for Mobile Operators. Retrieved from https://www.gsma.com/iot/wp-content/uploads/2013/02/cl_sc_guide_wp_02_131.pdf, Date of Access: 04/11/2018

Güçlü, N., & Sotirofski, K. (2006). Bilgi yönetimi. *Türk Eğitim Bilimleri Dergisi*, *4*(4), 351–373.

Guirao, B., Eugenia López, M., & Comendador, J. (2015). *New QR Survey Methodologies to Analyze User Perception of Service Quality in Public Transport: The Experience of Madrid. Journal of Public Transportation* (Vol. 18). Retrieved from https://scholarcommons.usf.edu/cgi/viewcontent.cgi?article=1460&context=jpt

Gül, M. (2011). Yerel yönetimlerde sosyal sorumluluk faaliyetlerinin seçmen davranışlarına etkileri: Kayseri Büyükşehir Belediyesi örneği. (Unpublished Master's Thesis), Erciyes Üniversitesi Sosyal Bilimler Enstitüsü, Kayseri.

Güleri, T. (2000). CRM throughout the enterprise. *Call center solutions, 18*(12), 44-49.

Güneş, E., Bayram, Ş. B., Özkan, M., & Nizamlıoğlu, H. F. (2018). Gastronomy Four Zero (4.0). *International Journal of Environmental Pollution and Environmental Modelling, 1*(3), 77–84.

Güngör, Y. M., Güngör, O., & Doğan, S. (2017). Evaluation of food and beverage businesses in Kuşadası from the perspective of SoLoMo marketing. *International Management Economics and Business Journal, 13*(3), 683–704.

Gunn, J. (2004). *Encyclopedia of caves and karst science*. Taylor & Francis. doi:10.4324/9780203483855

Guo, Y., Liu, H., & Chai, Y. (2014). The embedding convergence of smart cities and tourism internet of things in China: An advance perspective. *Advances in Hospitality and Tourism Research, 2*(1), 54–69.

Gupta, A., & Govindarajan, V. (2000). Knowledge Management's Social Dimension: Lessons From Nucor Steel. *Sloan Management Review, 42*(1), 71–80.

Gupta, M. (2017). *Blockchain for dummies*. Hoboken, NJ: John Wiley & Sons.

Gurau, C. (2007). Virtual Reality Applications in Tourism, Information and Communication Technologies in Support of the Tourism Industry (pp. 180–197). Hershey, PA: IGI Global. doi:10.4018/978-1-59904-159-9.ch007

Gürbüz, S., & Şahin, F. (2014). Sosyal bilimlerde araştırma yöntemleri. In Seçkin Yayıncılık. Ankara, Turkey.

Gürlek, M., & Tuna, M. (2018). Reinforcing competitive advantage through green organizational culture and green innovation. *Service Industries Journal, 38*(7-8), 467–491. doi:10.1080/02642069.2017.1402889

Gürlek, M., & Tuna, M. (2019). Corporate social responsibility and work engagement: Evidence from the hotel industry. *Tourism Management Perspectives, 31*, 195–208. doi:10.1016/j.tmp.2019.05.004

Gutierrez, M., Vexo, F., & Thalmann, D. (2008). *Stepping into Virtual Reality*. Springer. doi:10.1007/978-1-84800-117-6

Guttentag, D. A. (2010). Virtual reality: Applications and implications for tourism. *Tourism Management, 31*(5), 637–651. doi:10.1016/j.tourman.2009.07.003

Hair, J. F. Jr, Black, W. C., Babin, B. J., & Anderson, R. E. (2010). *Multivariate Data Analysis* (7th ed.). Upper Saddle River, NJ: Pearson.

Haklay, M. E. (2003). Virtual Reality and GIS. In P. Fischer, & D. Unwin (Eds.), *Virtual Reality in Geography* (pp. 47–56). London, UK: Taylor and Francis.

Halarnkar, P., Shah, S., Shah, H., Shah, H., & Shah, A. (2012). A Review on Virtual Reality. *IJCSI International Journal of Computer Science Issues, 9*, 6–1.

Hall, R. E. (2000). "The vision of a smart city", In *Proceedings of the 2nd International Life Extension Technology Workshop,* Paris, France. Available at http://ntl.bts.gov/lib/14000/14800/14834/DE2001773961.pdf

Hall, R. E., Bowerman, B., Braverman, J., Taylor, J., Todosow, H., & Von Wimmersperg, U. (2000, September). *The vision of a smart city.* Paper presented at the 2nd International Life Extension Technology Workshop, Paris, France.

Hall, C. M. (1994). *Tourism and Politics, Policy, Power, and Place*. Chichester, UK: John Wiley & Sons.

Hall, C. M. (1994). *Tourism and politics: policy, power and place*. John Wiley & Sons.

Hall, M., & Sharples, L. (2003). *The consumption of experiences or the experience of consumption? An introduction to the tourism of taste in Food Tourism Around the World Development, management and markets*. Burlington, VT: Butterworth-Heinemann an Imprint of Elsevier Linacre.

Halvorsrud, R., Kvale, K., & Følstad, A. (2016). Improving service quality through customer journey analysis. *Journal of Service Theory and Practice*, *26*(6), 840–867. doi:10.1108/JSTP-05-2015-0111

Hambling, D. (2018). *We: Robot, The Robots that already Rule our World*. London, UK: Aurum Press.

Hamid, N. R., Cheng, A. Y., & Akhir, R. M. (2011). Dimensions of E-CRM: An Empirical Study on Hotels' Web Sites. Retrieved May 2, 2019, from IBIMA Publishing: http://www.ibimapublishing.com/journals/JSAR/2011/820820/820820.pdf

Hampton, N. (2016). *Understanding the blockchain hype: Why much of it is nothing more than snake oil and spin*. Retrieved from https://www.computerworld.com.au/article /606253/understanding-blockchain-hype-why-much-it-nothing-more-than-snake-oil-spin/

Hançer, M., & Ataman, C. (2006). Use of communication technology in travel agencies and evaluation of web sites: The Case of Aegean Region, Dokuz Eylül University. *Journal of Social Sciences*.

Han, D.-I., Jung, T., & Gibson, A. (2014). Dublin AR: implementing augmented reality in tourism. In Z. Xiang, & I. Tussyadiah (Eds.), *Information and Communication Technologies in Tourism* (pp. 511–523). New York: Springer International Publishing.

Hanson, C. III, & Marshall, B. (2001). Artificial intelligence applications in the intensive care unit. *Critical Care Medicine*, *29*(2), 427–435. doi:10.1097/00003246-200102000-00038 PMID:11269246

Hardy, R., Rukzio, E., Holleis, P., Broll, G., & Wagner, M. (2010, September). *MyState: Using NFC to share social and contextual information in a quick and personalized way*. Paper presented at the *12th ACM international conference adjunct papers on ubiquitous computing*, Copenhagen, Denmark iPads, Apple Pay and self-service at heart of LaGuardia Airport upgrades (2015). Retrieved from https://www.futuretravelexperience.com/2015/04/ipads-apple-pay-self-service-heart-laguardia-airport-upgrades/

Harrington, R. J., Chathoth, P., Ottenbacher, M., & Altinay, L. (2014). Strategic management research in hospitality and tourism: Past, present and future. *International Journal of Contemporary Hospitality Management*, *26*(5), 778–808. doi:10.1108/IJCHM-12-2013-0576

Harrison, C., Eckman, B., Hamilton, R., Hartswick, P., Kalagnanam, J., Paraszczak, J., & Williams, P. (2010). Foundations for smarter cities. *IBM Journal of Research and Development*, *54*(4), 1–16. doi:10.1147/JRD.2010.2048257

Hassan, A., & Rahimi, R. (2016). Consuming 'innovation' in tourism: Augmented Reality as an innovation tool in Digital Tourism Marketing. In N. Pappas, & I. Bregoli (Eds.), Global Dynamics in Travel, Tourism, and Hospitality (pp. 130–147). Hershey, PA: IGI Global. doi:10.4018/978-1-5225-0201-2.ch008

Haubensak, O. (2011). *Smart cities and Internet of things*. Zurich, Switzerland: ETH.

Hay, B. (2008). Fantasy Tourism and Second Life. In S. Richardson, L. Fredline, A. Patiar, & M. Ternel (Eds.), *CAUTHE 2008: Tourism and hospitality research, training, and practice; "Where the 'Bloody hell' are we?"* (pp. 345–348). Retrieved from https://search.informit.com.au/documentSumary;dn=970132315896676;res=IELBUS

Helkkula, A. (2011). Characterising the concept of service experience. *Journal of Service Management*, *22*(3), 367–389. doi:10.1108/09564231111136872

Hendawi, A. M., & El-Shishiny, H. (2008). A Data Warehouse Prototype for the Tourism Industry: A Case Study from Egypt. Retrieved May 9, 2019, from https://www.academia.edu/17324396/A_Data_Warehouse_Prototype_for_the_Tourism_Industry_A_Case_Study_from_Egypt

Herger, M. (2014). *Enterprise gamification engaging people by letting them have fun*. USA: CreateSpace Independent Publishing Platform.

Herrera, F., López, R., & Fajardo, C. (2019). Developing Smart Regions: Proposal and Application of a Model for Island Territories. In Smart Cities and Smart Spaces: Concepts, Methodologies, Tools, and Applications (pp. 360-385). Hershey, PA: IGI Global

He, W., Zha, S., & Li, L. (2013). Social media competitive analysis and text mining: A case study in the pizza industry. *International Journal of Information Management*, *33*(3), 464–472. doi:10.1016/j.ijinfomgt.2013.01.001

He, Z., Wu, L., & Li, X. (2018). (R) (2018). When art meets tech: The role of augmented reality in enhancing museum experiences and purchase intentions. *Tourism Management*, *68*, 127–139. doi:10.1016/j.tourman.2018.03.003

Higgins-Desbiolles, F. (2008). Justice tourism and alternative globalisation. *Journal of Sustainable Tourism*, *16*(3), 345–364. doi:10.1080/09669580802154132

Higgins-Desbiolles, F. (2018). Sustainable tourism: Sustaining tourism or something more? *Tourism Management Perspectives*, *25*, 157–160. doi:10.1016/j.tmp.2017.11.017

Hillenbrand, U., Ott, C., Brunner, B., Borst, C., & Hirzinger, G. (2005). *Towards Service Robots for the Human Environment: The Robutler*. Retrieved from https://www.dlr.de/rm/Portaldata/52/Resources/Roboter_und_Systeme/Robutler/abstract.pdf

Hjalager, A. (2012). Innovation policies for tourism. *International Journal of Tourism Policy*, *4*(4), 336–356. doi:10.1504/IJTP.2012.052565

Hobson, J. S., & Williams, A. P. (1995). Virtual Reality: A New Horizon for the Tourism Industry. *Journal of Vacation Marketing*, *2*(1), 125–135.

Hofmann, M., & Klinkenberg, R. (2013). RapidMiner: Data mining use cases and business analytics applications. Boca Raton, FL: CRC Press.

Höjer, M., & Wangel, J. (2015). Smart sustainable cities: definition and challenges. In *ICT innovations for sustainability* (pp. 333–349). Cham, Switzerland: Springer. doi:10.1007/978-3-319-09228-7_20

Höllerer, T. H., & Feiner, S. K. (2004). Mobile augmented reality. In H. Karimi, & A. Hammad (Eds.), *Telegeoinformatics: Location-Based Computing and Services*. Taylor & Francis Books Ltd.

Homburg, C., Jozić, D., & Kuehnl, C. (2017). Customer experience management: Toward implementing an evolving marketing concept. *Journal of the Academy of Marketing Science*, *45*(3), 377–401. doi:10.100711747-015-0460-7

Hong Kong Monetary Authority. (2016). *Whitepaper on distributed ledger technology*. Hong Kong: Hong Kong Monetary Authority.

Hospitality Technology. (2017). *Robots in Hospitality: Five Trends on the Horizon*. Retrieved from https://hospitalitytech.com/robots-hospitality-five-trends-horizon

Howard, E. (2007). New shopping centres: Is leisure the answer? *International Journal of Retail & Distribution Management, 35*(8), 661–672. doi:10.1108/09590550710758649

Hoy, M. B. (2017). An introduction to the blockchain and its implications for libraries and medicine. *Medical Reference Services Quarterly, 36*(3), 273–279. doi:10.1080/02763869.2017.1332261 PMID:28714815

HPC-Cloud-based prediction of air quality. (2019). Retrieved from https://www.fortissimo-project.eu/experiments/410

HPC-Cloud-based urban planning. (2019). Retrieved from https://www.fortissimo-project.eu/experiments/406

http://emenu.cc. (2018). No Title. Retrieved from http://emenu.cc/desktop/en/weixindiancan

http://led.co.za/sites/led.co.za/files/The_Role_of_Local_Authorities_in_Tourism.pdf

http://www.glc.org/announce/06/pdf/Schnook-small.pdf

http://www.smart-cities.eu/download/smart_cities_final_report.pdf

http://www.southhams.gov.uk/administration/committee_agendas/prosperity_pdg/23Jun05/Tourism%20and%20Marketing%20Strategy%20Review_Appendix%20A_23.06.05.pdf

http://www.stevenspoint.com/index.aspx?NID=620

https://unctad.org/meetings/en/Presentation/CSTD_2015_ppt11_Latvia_en.pdf

https://www.segittur.es/opencms/export/sites/segitur/.content/galerias/descargas/documentos-en/Smart-Destination.pdf

https://www2.deloitte.com/content/dam/Deloitte/tr/Documents/public-sector/deloitte-nl-ps-smart-cities-report.pdf

Huang, C. D., Goo, J., Nam, K., & Yoo, C. W. (2017). Smart tourism technologies in travel planning: The role of exploration and exploitation. *Information & Management, 54*(6), 757–770. doi:10.1016/j.im.2016.11.010

Huang, J., Huang, Y., Pontius, R. G. Jr, & Zhang, Z. (2015). Geographically weighted regression to measure spatial variations in correlations between water pollution versus land use in a coastal watershed. *Ocean and Coastal Management, 103*, 14–24. doi:10.1016/j.ocecoaman.2014.10.007

Huang, X. Kai, Yuan, J. Zheng, & Shi, M. Yu. (2012). Condition and Key Issues Analysis on the Smarter Tourism Construction in China. *Communications in Computer and Information Science.* doi:10.1007/978-3-642-35286-7_56

Huang, Y. C., Backman, K. F., Backman, S. J., & Chang, L. L. (2016). Exploring the implications of virtual reality technology in tourism marketing: An integrated research framework. *International Journal of Tourism Research, 18*(2), 116–128. doi:10.1002/jtr.2038

Huang, Y., Backman, S. J., Backman, K. F., & Moore, D. (2013). Exploring user acceptance of 3d virtual worlds in travel and tourism marketing. *Tourism Management, 36*, 490–501. doi:10.1016/j.tourman.2012.09.009

Hudson, J., Orviska, M., & Hunady, J. (2017). People's Attitudes to Robots in Caring for the Elderly. *International Journal of Social Robotics, 9*(2), 199–210. doi:10.100712369-016-0384-5

Hu, M., & Liu, B. (2004). Mining and summarizing customer reviews. *Proceedings of the 2004 ACM SIGKDD International Conference on Knowledge Discovery and Data Mining - KDD '04*, 168. 10.1145/1014052.1014073

Hung, P. (2016). *Big data applications and use cases.* Berlin, Germany: Springer. doi:10.1007/978-3-319-30146-4

Hunter, W. C., Chung, N., Gretzel, U., & Koo, C. (2015). Constructivist research in smart tourism. *Asia Pacific Journal of Information Systems, 25*(1), 105–120. doi:10.14329/apjis.2015.25.1.105

Hurcombe, J. (1998). *Developing strategic customers and key accounts*. Bedford, PA: Policy Publications.

Hwang, J., Park, H. Y., & Hunter, W. C. (2015). Constructivism in smart tourism research: Seoul destination image. *Asia Pacific Journal of Information Systems*, *25*(1), 163–178. doi:10.14329/apjis.2015.25.1.163

Hwang, J., Park, T., & Hwang, W. (2013). The Effects of Overall Robot Shape on the Emotions Invoked in Users and the Perceived Personalities of Robot. *Applied Economics*, *44*, 459–471. PMID:23157974

Hyperledger Whitepaper. (2016). *Hyperledger Whitepaper v2.0.0.* Retrieved from https://github.com/hyperledger/hyperledger/wiki/Whitepaper-WG

Hyperledger. (2018). *Hyperledger community*. Retrieved from https://www.hyperledger. org/community

Iansiti, M., & Lakhani, K. (2017). The truth about blockchain. *Harvard Business Review*, *95*(1), 118–127.

İbiş, S. (2019). Turizm Endüstrisinde Robotlaşma. *Türk Turizm Araştırmaları Dergisi*, *3*(3), 403–420. doi:10.26677/TR1010.2019.169

İçişleri Bakanlığı Mahalli İdareler Genel Müdürlüğü. (1998). *Yerel Yönetimlerde Halk Katılımı*. Ankara, Turkey.

Icten, T., & Bal, G. (2017). Artırılmış gerçeklik üzerine son gelişmelerin ve uygulamaların incelenmesi. *Gazi Üniversitesi Fen Bilimleri Dergisi Part C: Tasarım ve Teknoloji*, *5*(2), 111–136.

IDS. (2014). A hotel system for the future. Retrieved from https://sites.google.com/a/ids.software.ucv.ro/smart-hotel/about#section1

IFR. (2016). Service robots. Retrieved from www.ifr.org/service-robots/

İlhan, I., & Çeltek, E. (2016). Mobile Marketing : Usage of Augmented Reality in Tourism Mobile Marketing : Usage of Augmented Reality in Tourism Mobil Pazarlama: Turizmde Artırılmış Gerçeklik Kullanımı. *Gaziantep University Journal of Social Sciences*, (December). doi:10.21547/jss.256721

Ilyenkova, S. D., & Kuznetsov, V. I. (2009). *Innovation Management*. Moscow, Russia: Eurasian Open Institute.

Inkpen, G. (1998). Information technology for travel and tourism (No. Ed. 2). Addison Wesley Longman Ltd.

Innoactive. (2016). *Mobile VR in Marketing – Kawasaki presents the Ninja ZX-10R 360° App*. Retrieved from https://innoactive.de/mobile-vr-in-marketing-kawasaki-presents-the-ninja-zx-10r-360-app/

International Organization for Standardization. (2012). ISO 8373:2012(en) Robots and robotic devices—vocabulary. Retrieved from https://www.iso.org/obp/ui/#iso:std:iso:8373:ed-2:v1:en:term:2.2

Ip, C., Leung, R., & Law, R. (2010). Progress and Development of Information and Communication Tehnologies in Hospitality. *International Journal of Contemporary Hospitality Management*, *23*(4), 533–551. doi:10.1108/09596111111130029

I-scoop. (2018). The Internet of Things (IoT) - essential IoT business guide. Retrieved from https://www.i-scoop.eu/internet-of-things-guide/

Ishak, I., Sidi, F., Affendey, L. S., Sani, N. F. M., Hamzah, A. S., & Bawon, P. (2014). Mobile plant tagging system for urban forest eco-tourism using QR code. *Proceedings - 2013 International Conference on Advanced Computer Science Applications and Technologies, ACSAT 2013*, 37–41. 10.1109/ACSAT.2013.15

Işıkçı, Y. (2002). *Türk turizminde altyapı ve çevre, II. Turizm Şurası Bildirileri* (pp. 267–273). Ankara, Turkey: T.C. Turizm Bakanlığı Yayınları.

Ivanov, S., & Webster, C. (2017). Designing Robot-Friendly Hospitality Facilities. *Scientific Conference: Tourism. Innovations. Strategies*, October 13-14, Bourgas, Bulgaria. Academic Press.

Ivanov, S., & Webster, C. (2019a). Perceived Appropriateness and Intention to Use Service Robots in Tourism. 2019 Information and Communication Technologies in Tourism, January 30-February 1, Nicosia, Cyprus.

Ivanov, S., & Webster, C. (2019b). Perceived appropriateness and intention to use service robots in tourism. In Pesonen, J. & Neidhardt, J. (Eds.) *Information and Communication Technologies in Tourism 2019, Proceedings of the International Conference in Nicosia, Cyprus*, pp. 237-248. 10.1007/978-3-030-05940-8_19

Ivanov, S., & Webster, C. (2019b). What should Robots do? A Comparative Analysis of Industry Professionals, Educators and Tourists. 2019 Information and Communication Technologies in Tourism, January 30-February 1, Nicosia, Cyprus. Academic Press.

Ivanov, S., Gretzel, U., Berezina, K., Sigala, M., & Webster, C. (2019). Progress on Robotics in Hospitality and Tourism: A Review of the Literature. *Journal of Hospitality and Tourism Technology*. Retrieved from https://www.emeraldinsight.com/doi/pdfplus/10.1108/JHTT-08-2018-0087

Ivanov, S., & Webster, C. (2018). Adoption of robots, artificial intelligence and service automation by travel, tourism and hospitality companies. In V. Marinov, M. Vodenska, & M. Assenova (Eds.), *Traditions and Innovations in Contemporary Tourism* (pp. 190–203).

Ivanov, S., & Webster, C. (2019a). What should robots do? A comparative analysis of industry professionals, educators, and tourists. In *Information and Communication Technologies in Tourism 2019* (pp. 249–262). Cham, Switzerland: Springer. doi:10.1007/978-3-030-05940-8_20

Ivanov, S., Webster, C., & Berezina, K. (2017). Adoption of Robots and Service Automation by Tourism and Hospitality Companies. *Revista Tourismo & Desenvolvimento*, *27/28*, 1501–1517.

Ivanov, S., Webster, C., & Garekno, A. (2018). Young Russian adults' attitudes towards the Potential Use of Robots in Hotels. *Technology in Society*, *55*, 24–32. doi:10.1016/j.techsoc.2018.06.004

Ivanov, S., Webster, C., & Seyyedi, P. (2018). Consumers' Attitudes towards the Introduction of Robots in Accommodation Establishments. *Tourism: An International Interdisciplinary Journal*, *66*(3), 302–317.

Ivanovska, L. P. (2008). E-CRM in the Tourism Sector. *Revista de Turism*, *5*, 14–19.

Ivars Baidal, J. A., Solsona Monzonís, F. J., & Giner Sánchez, D. (2016). Gestión turística y tecnologías de la información y la comunicación (TIC): El nuevo enfoque de los destinos inteligentes. *Documents d'Analisi Geografica*, *62*(2), 327–346. doi:10.5565/rev/dag.285

Ixigo. (2019). Retrieved from https://www.ixigo.com/%20yoindia

Jackson, G., & Morpeth, N. (2000). Local Agenda 21 Reclaiming Community Ownership in Tourism or Stalled Process? In R. Greg, & D. Hall (Eds.), Tourism and Sustainable Community Development (pp. 119-135), London, UK: Routledge.

Jackson, K. T. (1996). All the world's a mall: Reflections on the social and economic consequences of the American shopping center. *The American Historical Review*, *101*(4), 1111–1121. doi:10.2307/2169636

Jain, R., Aagja, J., & Bagdare, S. (2017). Customer experience – a review and research agenda. *Journal of Service Theory and Practice*, *27*(3), 642–662. doi:10.1108/JSTP-03-2015-0064

Jamal, T., & Camargo, B. A. (2014). Sustainable tourism, justice and an ethic of care: Toward the Just Destination. *Journal of Sustainable Tourism*, *22*(1), 11–30. doi:10.1080/09669582.2013.786084

Jaremen, D. E., Jędrasiak, M., & Rapacz, A. (2016). The Concept of smart hotels as an ınnovation on the hospitality ındustry market - case study of PURO Hotel in Wrocław. *Economic Problems of Tourism, 36*(4), 65–75.

Jarvis, H. (2016*). Robot-Run Hotel Gets Guinness Approval.* Retrieved from https://standbynordic.com/robot-run-hotel-gets-guinness-approval/

Jeffries, D. (2001). *Governments and Tourism. (First published).* Oxford, UK: Butterworth- Heinemann.

Jernsand, E. M., Kraff, H., & Mossberg, L. (2015). Tourism experience innovation through design. *Scandinavian Journal of Hospitality and Tourism, 15*(1), 98–119. doi:10.1080/15022250.2015.1062269

Jevremovic, V., & Petrovski, S. (2012). MUZZEUM - Augmented Reality and QR codes enabled mobile platform with digital library, used to Guerrilla open the National Museum of Serbia. *Proceedings of the 2012 18th International Conference on Virtual Systems and Multimedia, VSMM 2012: Virtual Systems in the Information Society*, 561–564. 10.1109/VSMM.2012.6365977

Jia, Z., Shi, Y., Jia, Y., & Li, D. (2012). A framework of knowledge management systems for tourism crisis management. *Procedia Engineering, 29*, 138–143. doi:10.1016/j.proeng.2011.12.683

Jin, J., Gubbi, J., Marusic, S., & Palaniswami, M. (2014). An information framework for creating a smart city through internet of things. *IEEE Internet of Things Journal, 1*(2), 112-121.

Jindal, F., Jamar, R., & Churi, P. (2018). Future and Challenges of Internet of Things. *International Journal of Computer Science and Information Technology, 10*(2), 13–25. doi:10.5121/ijcsit.2018.10202

Jinendra, D. R., Bhagyashri, J. R., Pranav, G. Y., Seema, V. U., & Parag, A. N. (2012). Smart travel guide: Application for android mobile. International Journal of Electronics [IJECSCSE]. *Communication and Soft Computing Science & Engineering, 2*, 115.

Joan, B. (2012) Difference Between NFC and Bluetooth, Retrieved from http://www.differencebetween.net/technology/hardware-technology/difference-between-nfc-and-bluetooth/

Jones, M. A. (1999). Entertaining shopping experiences: An exploratory investigation. *Journal of Retailing and Consumer Services, 6*(3), 129–139. doi:10.1016/S0969-6989(98)00028-9

Jones, S. (2006). Reality© And Virtual Reality ©. *Cultural Studies, 20*(2-3), 211–226. doi:10.1080/09502380500495692 PMID:9548022

Jo, T. (2019). *Text Mining*. Cham, Switzerland: Springer; doi:10.1007/978-3-319-91815-0

Jovicic, D. Z. (2016). Key issues in the conceptualization of tourism destinations. *Tourism Geographies, 18*(4), 445–457. doi:10.1080/14616688.2016.1183144

Jovicic, D. Z. (2017). From the traditional understanding of tourism destination to the smart tourism destination. *Current Issues in Tourism*, 1–7. doi:10.1080/13683500.2017.1313203

Jo, Y., & Oh, A. H. (2011). Aspect and sentiment unification model for online review analysis. *Proceedings of the Fourth ACM International Conference on Web Search and Data Mining - WSDM '11*, 815. 10.1145/1935826.1935932

Jung, T., TomDieck, M. C., Lee, H., & Chung, N. (2016). Effects of virtual reality and augmented reality on visitor experiences in museum. In A. Inversini, & R. Schegg (Eds.), *Information and communication technologies in tourism 2016: Proceedings of the international conference in Bilbao, Spain*, pp. 621–635, February 2-5, 2016. Cham, Switzerland: Springer International Publishing. 10.1007/978-3-319-28231-2_45

Jung, T., TomDieck, M. C., Moorhouse, N., & TomDieck, D. (2017). Tourists' experience of virtual reality applications. Paper presented at 2017 *IEEE International Conference on Consumer Electronics (ICCE)*. doi:. 2017.788928710.1109/ICCE

Jung, T., Chung, N., & Leue, M. (2015). The determinants of recommendations to use augmented reality technologies: The case of a Korean theme park. *Tourism Management, 49*, 75–86. doi:10.1016/j.tourman.2015.02.013

Kahn, P. H., Kanda, T., Ishiguro, H., Freier, N. G., Severson, R. L., Gill, B. T., ... Shen, S. (2012). "Robovie, You'll Have to Go into the Closet Now": Children's Social and Moral Relationships With a Humanoid Robot. *Developmental Psychology, 48*(2), 303–314. doi:10.1037/a0027033 PMID:22369338

Kale, S. H. (2004). CRM failure and the seven deadly sins. *Marketing Management, 13*(5), 42–46.

Kaminska, I. (2015). *Blockchain promises back-office ledger revolution*. Retrieved from https://www.ft.com/content/7aad0826-638c-11e5-9846-de406ccb37f2

Kan, T.-W., Teng, C.-H., & Chou, W.-S. (2009). Applying QR code in augmented reality applications, *1*(212), 253. doi:10.1145/1670252.1670305

Kang, B. (2011). The Impact of channel knowledge on shopping orientations in consumer buying behaviour. *Anatolia- An International Journal of Tourism and Hospitality Research, 22*(1), 120-124.

Kansakar, P., Munir, A., & Shabani, N. (2017). Technology in hospitality industry: Prospects and challenges. *arXiv preprint arXiv:1709.00105*.

Kansakar, P., Munir, A., & Shabani, N. (2019). Technology in the Hospitality Industry: Prospects and Challenges. *IEEE Consumer Electronics Magazine, 8*(3), 60–65. doi:10.1109/MCE.2019.2892245

Kapiki, S. (2012). Current and Future Trends in Tourism and Hospitality: The Case of Greece. *International Journal of Economic Practices and Theories, 2*(1). Available at https://ssrn.com/abstract=2150562

Kapucugil, A., & Özdağoğlu, G. (2015). Text Mining as a Supporting Process for VoC Clarification. *Alphanumeric Journal, 3*(1). doi:10.17093/aj.2015.3.1.5000105108

Karaarslan, M. H., & Altuntaş, B. (2016). Investigation of the selected marketing cases in Turkey within the concept of gamification. *Mehmet Akif Ersoy University Journal of Social Sciences Institute, 8*(17), 433–447.

Karabegovic, I., & Dolecek, V. (2017). The Role of Service Robots and Robotic Systems in the Treatment of Patients in Medical Institutions. In M. Hadzikadic, & S. Avdakovic (Eds.), *Advanced Technologies, Systems, and Applications – Lecture Notes in Networks and Systems*. Switzerland: Springer International Publishing AG. doi:10.1007/978-3-319-47295-9_2

Karabegovic, I., Karabegovic, E., Mahmic, M., & Husak, E. (2015). The Application of Service Robots for Logistics in Manufacturing Processes. *Advances in Production Engineering & Management, 10*(4), 185–194. doi:10.14743/apem2015.4.201

Karabulut, Z. E. (2018). *Mobil sistemler üzerinde, biyometrik, NFC ve konum bilgilerini kullanarak kişi tanıma* (Unpublished Master's thesis). İstanbul Ticaret University, Turkey.

Karakoçak, K. (2007). *Bilgi yönetimi ve verimliliğe etkisi: Türkiye Büyük Millet Meclisi uygulaması*. Ankara, Turkey: Ankara Üniversitesi Sosyal Bilimler Enstitüsü Yayımlanmış Doktora Tezi.

Karampatsou, M. (2018) Big Data in Tourism, (Unpublished master's thesis). School Of Economics, Business Administration, & Legal Studies, Greece.

Kasa, M., & Hassan, Z. (2015). The role of flow between burnout and organizational citizenship behavior (OCB) among hotel employees in Malaysia. *Procedia: Social and Behavioral Sciences, 211*, 199–206. doi:10.1016/j.sbspro.2015.11.084

Katz, L. (2011). *NY hotel books Yobot the luggage-handling robot*. Retrieved from https://www.cnet.com/news/ny-hotel-books-yobot-the-luggage-handling-robot/

Kaur, A., & Chopra, D. (2016). Comparison of text mining tools. *2016 5th International Conference on Reliability, Infocom Technologies, and Optimization, ICRITO 2016: Trends and Future Directions*, 186–192. 10.1109/ICRITO.2016.7784950

Kaur, K., & Kaur, R. (2016). Internet of things to promote tourism: An insight into smart tourism. *International Journal of Recent Trends in Engineering & Research*, *2*(4), 357–361.

Kaushik, A. K., Agrawal, A. K., & Rahman, Z. (2015). Tourist behaviour towards self-service hotel technology adoption: Trust and subjective norm as key antecedents. *Tourism Management Perspectives*, *16*, 278–289. doi:10.1016/j.tmp.2015.09.002

Keleş, A., Keleş, A., & Akçetin, E. (2017). Pazarlama Alaninda Yapay Zekâ Kullanim Potansiyeli Ve Akilli Karar Destek Sistemleri. *Electronic Turkish Studies, 12*(11).

Kennedy-Eden, H., & Gretzel, U. (2012). A taxonomy of mobile applications in tourism. *Ereview of Tourism Research, 10*(20), 47–50.

Keown, C. F. (1989). A model of tourists' propensity to buy: The case of Japanese visitors to Hawaii. *Journal of Travel Research*, *27*(3), 31–34. doi:10.1177/004728758902700306

Keş, Y., & Akyürek, A. B. (2018). Interactive museums for the new generation growing with technology. *IMU Journal of Art, Design, and Architecture* *4*(2), 95-110.

Khan, M. S., Woo, M., Nam, K., & Chathoth, P. K. (2017). Smart city and smart tourism: A case of Dubai. *Sustainability (Switzerland)*, *9*(12). doi:10.3390u9122279

Khomsi, M. R., & Bédard, F. B. (2017). From smart city to smart destination: The case of three Canadian cities. *ARA: Revista de Investigaciónen Turismo*, *6*(2), 69–74.

Kieseberg, P., Leithner, M., Mulazzani, M., Munroe, L., Schrittwieser, S., Sinha, M., & Weippl, E. (2011). QR code security, 430. doi:10.1145/1971519.1971593

Kılıç, G. (2002). Belediyelerin bölge turizmine olan etkileri. (Unpublished Master's Thesis), Sakarya Üniversitesi Sosyal Bilimler Enstitüsü, Sakarya.

Kılıç, G. (2006). Belediyelerin turizme ilişkin faaliyetlerini değerlendirmeye yönelik bir uygulama. *Çağdaş Yerel Yönetimler Dergisi, 15*(3), 79-110.

Kim, D., Park, J., & Morrison, A. M. (2008). A Model of Traveller Acceptance of Mobile Technology 1, *407*, 393–407.

Kim, G. J. (2005). *Designing Virtual Reality Systems The Structered Approach*. London, UK: Springer.

Kim, H. M., & Han, S. S. (2012). City profile Seoul. *Cities (London, England)*, *29*(2), 142–154. doi:10.1016/j.cities.2011.02.003

Kim, H., Ritchie, J. R., & Mccormick, B. (2012). Development of a scale to measure memorable tourism experiences. *Journal of Travel Research*, *51*(1), 12–25. doi:10.1177/0047287510385467

Kımıloğlu, H., & Zaralı, H. (2008). What signifies success in e-CRM? *Marketing Intelligence & Planning*, *27*(2), 246–267. doi:10.1108/02634500910945011

Kim, J., Christodoulidou, N., & Brewer, P. (2012). Impact of individual differences and consumers' readiness on likelihood of using self-service technologies at hospitality settings. *Journal of Hospitality & Tourism Research (Washington, D.C.)*, *36*(1), 85–114. doi:10.1177/1096348011407311

Kim, K., & Justl, J. M. (2018). Potential antitrust risks in the developement and use of blockchain. *Journal of Taxation and Regulation of Financial Institutions*, *31*(3), 5–16.

Kim, K., Park, O., Yun, S., & Yunc, H. (2017). What makes tourists feel negatively about tourism destinations? Application of hybrid text mining methodology to smart destination management. *Technological Forecasting and Social Change*, *123*, 362–369. doi:10.1016/j.techfore.2017.01.001

Kim, S., & Littrell, M. A. (2001). Souvenir buying intentions for self versus others. *Annals of Tourism Research*, *28*(3), 638–657. doi:10.1016/S0160-7383(00)00064-5

Kim, W. G., & Cha, Y. (2002). Antecedents and consequences of relationship quality in hotel industry. *International Journal of Hospitality Management*, *21*(4), 321–338. doi:10.1016/S0278-4319(02)00011-7

Kincaid, J. W. (2003). *Customer Relationship Management: Getting it Right!* New Jersey: Prentice-Hall.

Kipper, G., & Rampolla, J. (2012). *Augmented Reality: An Emerging Technologies Guide to AR.*

Kıroğlu, P. (2012). Elektronik turizm analizi, uygulamaları ve beklentiler (Doctoral dissertation, İstanbul Kültür Üniversitesi/Sosyal Bilimler Enstitüsü/İktisat Anabilim Dalı/Yönetim Ekonomisi Bilim Dalı).

Kiryakova, G., Angelova, N., & Yordanova, L. (2013). QR Codes in The Business World. *Trakia Journal of Sciences*, *11*, 370–376.

Koçel, T. (2018). *İşletme Yöneticiliği*. İstanbul, Turkey: Beta.

Koçoğlu, C. M., & Acar, A. (2019). Safety perception in touristic destinations: the example of Istanbul province. In R. Efe, I. Koleva, M. Öztürk, & R. Arabacı (Eds.), *Recent Advances in Social Sciences* (pp. 305–319). Newcastle, UK: Cambridge Scholars Publishing.

Koçoğlu, C. M., & Kıycı, Ş. (2019). The evaluation level of foreign tourists for five-star hotels in Belek in terms of corporate reputation. *Journal of Tourism Theory and Research*, *5*(2), 112–122. doi:10.24288/jttr.523396

Köksal, G. (2004). *Napolyonu Mat Eden Türk*. Retrieved from https://bianet.org/bianet/print/31804-napolyonu-mat-eden-turk

Komninos, N. (2008). *Intelligent cities and globalisation of innovation networks*. London, UK: Routledge, Taylor & Francis Group. doi:10.4324/9780203894491

Komninos, N. (2015). *The age of intelligent cities: smart environments and innovation-for-all strategies*. New York: Routledge, Taylor & Francis Group.

Komninos, N., Pallot, M., & Schaffers, H. (2013). Special issue on smart cities and the future internet in Europe. *Journal of the Knowledge Economy*, *4*(2), 119–134. doi:10.100713132-012-0083-x

Koo, C., Shin, S., Gretzel, U., Hunter, W. C., & Chung, N. (2016). *Conceptualization of Smart Tourism Destination Competitiveness*. Asia Pacific Journal of Information Systems; doi:10.14329/apjis.2016.26.4.561

Köroglu, Y. (2017). *Yapay Zeka'nın Teorik ve Pratik Sınırları*.

Korstanje, M., & Busby, G. (2010). Understanding the bible as the roots of physical displacement: The origin of tourism. *Ereview of Tourism Research*, *8*(3).

Kostakos, V., & O'Neill, E. (2007, March). NFC on mobile phones: issues, lessons and future research. Paper presented at Fifth Annual IEEE International Conference on Pervasive Computing and Communications Workshops (PerComW'07). IEEE. New York. 10.1109/PERCOMW.2007.84

Kotler, P., & Armstrong, G. (2012). *Principles of Marketing* (14th ed.). Harlow, UK: Pearson Education Limited.

Kotler, P., & Keller, K. (2006). *Marketing Management* (12th ed.). New Jersey: Pearson Prentice Hall.

Kounavis, C., Kasimati, A., & Zamani, E. (2012). Enhancing the tourism experience through mobile augmented reality: Challenges and prospects. *International Journal of Engineering Business Management*, *4*, 10. doi:10.5772/51644

Kounavis, C., Kasimati, A., & Zamani, E. (2012). Enhancing the tourist experience through mobile augmented reality: Challenges and prospects. *International Journal of Engineering Business Management*, *4*(10), 1–6.

Kourouthanassis, P., Boletsis, C., Bardaki, C., & Chasanidou, D. (2015). Tourists responses to mobile augmented reality travel guides: The role of emotions on adoption behavior. *Pervasive and Mobile Computing*, *18*, 71–87. doi:10.1016/j.pmcj.2014.08.009

Kourtesopoulou, A., Papaioannou, A., & Kriemadis, T. (Eds.). 2011. Exploring E-crm application: A case study in outdoor recreation sector. In *Proceedings of the 19th International Congress on Physical Education and Sport, ICPES*. Komotini, Greece. Academic Press.

Kowal, J., & Fortier, M. S. (1999). Motivational determinants of flow: Contributions from self-determination theory. *The Journal of Social Psychology*, *139*(3), 355–368. doi:10.1080/00224549909598391

Kozak, N., Kozak, M. A., & Kozak, M. (2017). *Genel Turizm İlkeler Kavramlar*. Ankara, Turkey: Detay Yayıncılık.

Krombholz, K., Frühwirt, P., Kieseberg, P., Kapsalis, I., Huber, M., & Weippl, E. (2014). QR code security: A survey of attacks and challenges for usable security. Lecture Notes in Computer Science (Including Subseries Lecture Notes in Artificial Intelligence and Lecture Notes in Bioinformatics), 8533 LNCS, 79–90. doi:10.1007/978-3-319-07620-1_8

Kudyba, S. (2014). *Big Data, Mining and Analytics: Components of Strategic Decision Making. Publisher*. Taylor Francis. doi:10.1201/b16666

Kuflik, T., Wecker, A. J., Lamir, J., & Stock, O. (2015). An Integrative Framework for Extending the Boundaries of the Museum Visit Experience: Linking the Pre, During and Post Visit Phases. *Information Technology & Tourism*, *15*(1), 14–47. doi:10.100740558-014-0018-4

Kumar, N., & Malik, K. (2013). E-CRM golden opportunity for banks. *International Journal of Multidisciplinary and Academic Research*, *2*, 1–10.

Kumar, S. (2016). Developing India as Smart Tourism Destination-A Sap Lap Analysis. *South Asian Journal of Tourism and Heritage*, *9*(2), 124–136.

Kunduracıoğlu, İ. (2018). *A content analysis on gamification concept* (Master dissertation). University of Balıkesir, Turkey.

Kuo, C.-M., Chen, L.-C., & Tseng, C.-Y. (2017). Investigating an Innovative Service with Hospitality Robots. *International Journal of Contemporary Hospitality Management*, *29*(5), 1305–1321. doi:10.1108/IJCHM-08-2015-0414

Kurashima, T., Iwata, T., Irie, G., & Fujimura, K. (2013). Travel route recommendation using geotagged photos. *Knowledge and Information Systems*, *37*(1), 37–60. doi:10.100710115-012-0580-z

Kutlar, A., Yüksel, F., & Bakırcı, F. (2011). Türkiye'de Belediyelerin Ekonomik Etkinliği ve Etkinliğe Etki Eden faktörler Üzerine Bir Araştırma. Retrieved from http://www.tbb.gov.tr/onlineyayinlar

La Rocca, R. A. (2014). The role of tourism in planning the smart city. *TeMA. Journal of Land Use. Mobility and Environment, 7*(3), 269–284.

La Valle, S. M. (2016, Forthcoming). *Virtual Reality. Draft October 31*. Cambridge, UK: University Press.

Lampropoulos, G., Siakas, K., & Anastasiadis, T. (2018). Internet of Things (IoT) in Industry: Contemporary Application Domains, Innovative Technologies and Intelligent Manufacturing. *International Journal of Advances in Scientific Research and Engineering, 4*(10), 109–118. doi:10.31695/IJASRE.2018.32910

Lamsfus, C., & Alzua-Sorzabal, A. (2013). Theoretical framework for a tourism internet of things: Smart destinations. *Journal of Tourism and Human Mobility, 2*, 15–21.

Landman, P. (2011). How hotels can use QR Codes for marketing and customer service. Retrieved from https://www.phocuswire.com/How-hotels-can-use-QR-Codes-for-marketing-and-customer-service

Laney, D. (2001). 3D data management: Controlling data volume, velocity and variety. *META group research note, 6*(70), 1.

Lange, P., & Steck, T. (2014). Near Field Communication-Its adoption process and technology acceptance. (Unpublished Master's thesis). Lund University, Sweden.

Lang, J. C. (2001). Managerial concerns in knowledge management. *Journal of Knowledge Management, 5*(1), 43–59. doi:10.1108/13673270110384392

Lara, A. P., Da Costa, E. M., Furlani, T. Z., & Yigitcanlar, T. (2016). Smartness that matters: Towards a comprehensive and human-centred characterisation of smart cities. *Journal of Open Innovation: Technology, Market, and Complexity, 2*(1), 8. doi:10.118640852-016-0034-z

Larkin, C. M., Jancourt, M., & Hendrix, W. H. (2018). The Generation Z World: Shifts in Urban Design, Architecture and the Corporate Workplace. *Corporate Real Estate Journal., 7*(3), 230–242.

Lasek, M., & Jessa, S. (2013). Chatbots for customer service on hotels' websites. *Information Systems Management, 2*(2), 146–158.

Lashkari, A., Parhizkar, B., & Mohamedali, M. (2010). Augmented Reality Tourist Catalogue Using Mobile Technology. In *Proceedings IEEE, Second International conference on Computer Research and Development*, pp. 121-125, *Kuala Lumpur, Malaysia*. Washington, DC: IEEE.

Lasi, H., Fettke, P., Kemper, H. G., Feld, T., & Hoffmann, M. (2014). Industry 4.0. *Business & Information Systems Engineering, 6*(4), 239–242. doi:10.100712599-014-0334-4

Lastminute.com. Retrieved April 5, 2019 from https://www.lastminute.com

Lavrijssen, S., & Carrilo, A. P. (2017). Radical prosumer innovations in the electricity sector and the impact on prosumer regulation. *Sustainability, 9*(7), 1207. doi:10.3390u9071207

Law, R., & Au, N. (2000). Relationship modeling in tourism shopping: A decision rules induction approach. *Tourism Management, 21*(3), 241–249. doi:10.1016/S0261-5177(99)00056-4

Law, R., Buhalis, D., & Cobanoglu, C. (2014). Progress on information and communication technologies in hospitality and tourism. *International Journal of Contemporary Hospitality Management, 26*(5), 727–750. doi:10.1108/IJCHM-08-2013-0367

Law, R., Leung, R., & Buhalis, D. (2009). Information technology applications in hospitality and tourism: A Review of publications from 2005 to 2007. *Journal of Travel & Tourism Marketing, 26*(5-6), 599–623. doi:10.1080/10548400903163160

Lebedev, O. T., Mokeeva, T. V., & Rodionov, D. G. (2018) Matrix Structures of Science and Technology Innovations Development and Implementation Trajectory. In *Proc. of the 31st Int. Business Information Management Association Conf., IBIMA 2018 - Innovation Management and Education Excellence through Vision 2020, IBIMA 2018,* pp. 1759 –68.

Lee, I., Cai, G., & Lee, K. (2014). Exploration of geo-tagged photos through data mining approaches. *xpert Systems with Applications, 41(2)*, 397-405.

Lee, J., Kao, H. A., & Yang, S. (2014). Service innovation and smart analytics for industry 4.0 and big data environment. *Procedia Cirp*, *16*, 3–8. doi:10.1016/j.procir.2014.02.001

Lee, J., Lee, H., Chung, N., & Koo, C. (2017). An Integrative Model of the Pursuit of Happiness and the Role of Smart Tourism Technology: A Case of International Tourists in Seoul. In *Information and Communication Technologies in Tourism* (pp. 173–186). Cham: Springer.

Lee, K. (2012). Augmented reality in education and training. *TechTrends*, *56*(2), 13–21. doi:10.100711528-012-0559-3

Lee-Kelley, L., Gilbert, D., & Mannicom, R. (2003). How e-CRM can enhance customer loyalty. *Marketing Intelligence & Planning*, *21*(4), 239–248. doi:10.1108/02634500310480121

Lee, M. (2008). *A qualitative case study approach to define and ıdentify perceived challenges of knowledge management for casino hotel ındustry, Graduate College, University of Nevada Las Vegas William F.* Harrah College of Hotel Administration. Doktora Tezi.

Lee, Y.-C., & Lee, S.-K. (2006). Capabilities, processes, and performance of knowledge management: A structural approach. *Human Factors and Ergonomics in Manufacturing*, *17*(1), 21–41. doi:10.1002/hfm.20065

Lehr, T. (2018). Smart cities: Vision on-the-Ground. In S. McClellan, J. A. Jimenez, & G. Koutitas (Eds.), *Smart cities: Applications, technologies, standards, and driving factors* (pp. 3–17). Cham, Switzerland: Springer. doi:10.1007/978-3-319-59381-4_1

Lei, S., & Law, R. (2015). Content analysis of TripAdvisor reviews on restaurants: A case study of Macau. *Journal of Tourism*, *16*(1), 17–28. Retrieved from http://search.ebscohost.com/login.aspx?direct=true&db=hjh&AN=111435361&site=ehost-live

Lele, A. (2019). Cloud computing. In *Smart Innovation*. Systems and Technologies; doi:10.1007/978-981-13-3384-2_10

Lemieux, V. L. (2016). Trusting records: Is blockchain technology the answer? *Records Management Journal*, *26*(2), 110–139. doi:10.1108/RMJ-12-2015-0042

Lemon, K. N. (2016). The art of creating attractive consumer experiences at the right time: Skills marketers will need to survive and thrive. *GfK Marketing Intelligence Review*, *8*(2), 44–49. doi:10.1515/gfkmir-2016-0015

Lemon, K. N., & Verhoef, P. C. (2016). Understanding customer experience throughout the customer journey. *Journal of Marketing*, *80*(6), 69–96. doi:10.1509/jm.15.0420

Lenhart, A. (2015). *Teens, Social Media, & Technology. WWW document*, Retrieved from http://www.pewinternet.org/2015/04/09/teens-social-media-technology-2015/

Lenzen, M., Sun, Y.-Y., Faturay, F., Ting, Y.-P., Geschke, A., & Malik, A. (2018). The carbon footprint of global tourism. *Nature Climate Change*, *8*(6), 522–528. doi:10.103841558-018-0141-x

Leslie, D., & Hughes, G. (1997). Agenda 21, local authorities and tourism in the UK. *Managing Leisure*, *2*(3), 143–154. doi:10.1080/13606719708559316

Liao, S. H., & Wu, C. C. (2009). Knowledge management and ınnovation: the mediating effects of organizational learning. In *Proceedings IEEE International Conference on Industrial Engineering and Engineering Management*, pp. 1850-1854. IEEE.

Liberato, P., Alen, E., & Liberato, D. (2017). Smart Tourism Destination Triggers Consumer Experience: The Case of Porto. *European Journal of Management and Business Economics*, *27*(1), 6–25. doi:10.1108/EJMBE-11-2017-0051

Li, J. J., Bonn, M. A., & Ye, B. H. (2019). Hotel employee's artificial intelligence and robotics awareness and its impact on turnover intention: The moderating roles of perceived organizational support and competitive psychological climate. *Tourism Management*, *73*, 172–181. doi:10.1016/j.tourman.2019.02.006

Li, J., Xu, L., Tang, L., Wang, S., & Li, L. (2018). Big data in tourism research: A literature review. *Tourism Management*, *68*, 301–323. doi:10.1016/j.tourman.2018.03.009

Li, K., Sun, S., Zhao, X., Wu, J., & Tan, M. (2019). Inferring User Intent to Interact with a Public Service Robot Using Bimodal Information Analysis. *Advanced Robotics*, *33*(7-8), 369–387. doi:10.1080/01691864.2019.1599727

Lim, C., Kim, M., & Kim, S. B. (2018, January). *A Q & A application using the concept gamification.* ISER 98th International Conference, January 8-9, 2018, Osaka, Japan. Academic Press.

Lin, J. (2004). *University Unveils Chatty "Roboreceptionist"*. Retrieved from http://www.nbcnews.com/id/4306856/ns/technology_and_science-science/t/university-unveils-chatty-roboceptionist/#.XTstbegzbIU

Linaza, M. T., Gutierrez, A., & Garcia, A. (2014). Pervasive augmented reality games to experience tourism destinations. In Z. Xiang, & I. Tussyadiah (Eds.), *Information and Communication Technologies in Tourism* (pp. 497–509). Switzerland: Springer International Publishing.

Linaza, M., Marimon, D., Carrasco, P., Alvarez, R., Montesa, J., Aguilar, S. R., & Diez, G. (2012). Evaluation of mobile augmented reality applications for tourism destinations. In M. Fuchs, F. Ricci, & L. Cantoni (Eds.), *Information and Communication Technologies in Tourism*. Wien, Austria: Springer-Verlag.

Lindsay, A., Downs, D., & Lunn, K. (2003). Business processes—Attempts to find a definition. *Information and Software Technology*, *45*(15), 1015–1019. doi:10.1016/S0950-5849(03)00129-0

Lindsay, R., Buchanan, B., Feigenbaum, E., & Lederberg, J. (1980). *Applications of artificial intelligence for organic chemistry*. McGraw-Hill.

Lin, I. C., & Liao, T. (2017). A survey of blockchain security ıssues and challenges. *International Journal of Network Security*, *19*(5), 653–659.

Lin, N. (1976). *Foundations of social research*. USA: McGraw-Hill.

Li, S., Da Xu, L., & Zhao, S. (2015). The internet of things: A survey. *Information Systems Frontiers*, *17*(2), 243–259. doi:10.100710796-014-9492-7

Liu, B. (2012). Sentiment Analysis and Opinion Mining. *Synthesis Lectures on Human Language Technologies*, *5*(1), 1–167. doi:10.2200/S00416ED1V01Y201204HLT016

Liu, C., & Hung, K. (2019). Understanding Self-service Technology in Hotels in China: Technology Affordances and Constraints. In *Information and Communication Technologies in Tourism 2019* (pp. 225–236). Cham, Switzerland: Springer. doi:10.1007/978-3-030-05940-8_18

Li, Y., Hu, C., Huang, C., & Duan, L. (2017). The concept of smart tourism in the context of tourism information services. *Tourism Management*, *58*, 293–300. doi:10.1016/j.tourman.2016.03.014

Lloyd, V. (2014). *A brief history of gamification.* Retrieved from https://www.thehrdirector.com/features/gamification/a-brief-history-of-gamification/

London (2019a). *Green transport.* Retrieved from https://www.london.gov.uk/what-we-do/transport/green-transport

London (2019b). *The Mayor's ultra-low emission Zone for London.* Retrieved from https://www.london.gov.uk/what-we-do/environment/pollution-and-air-quality/mayors-ultra-low-emission-zone-london

London (2019c). *Our air and our vehicles.* Retrieved from https://www.london.gov.uk/sites/default/files/shorthand/cleaner_vehicles/

London (2019d). *Waste policy.* Retrieved from https://www.london.gov.uk/what-we-do/environment/waste-and-recycling/waste-policy

Long, J. (1994). Local authority tourism strategies-a British appraisal. *Journal of Tourism Studies*, *5*(2), 17–23.

López-de-Ipiña, D., Díaz-de-Sarralde, I., & Zubía, J. G. (2010). An ambient assisted living platform integrating RFID data-on-tag care annotations and twitter. *Journal of Universal Computer Science*, *16*(12), 1521–1538.

López, J., Pérez, D., Zalama, E., & Gómez-García-Bermejo, J. (2013). Bellbot-a hotel assistant system using mobile robots. *International Journal of Advanced Robotic Systems*, *10*(1), 40. doi:10.5772/54954

López, Y., & Pulido, J. I. (2013). Sustainability in Spanish tourism policy. Advances or setbacks? *Papers de Turisme*, *53*, 44–68.

Loureiro, A. (2018). There Is A Forth Industrial Revolution: The digital Revolution. *Worldwide Hospitality and Tourism Themes*, *10*(6), 740–744. doi:10.1108/WHATT-07-2018-0044

Lovett, E., & Longworth Dames, M. (1901). The ancient and modern game of astragals. *Folklore*, *12*(3), 280–293. doi:10.1080/0015587X.1901.9719634

Lowbrow, Y. (2014). *Livin' the dream with green stamps: A 1975 catalog.* Retrieved from https://flashbak.com/livin-the-dream-with-green-stamps-a-1975-catalog-26187/

Lukas, A. S. (2013). *The Immersive Worlds Handbook Designing Theme Parks and Consumer Spaces*. New York: Focal Press.

Lu, L., Cai, R., & Gürsoy, D. (2019). Developing and Validating a Service Robot Integration Willingness Scale. *International Journal of Hospitality Management*, *80*, 36–51. doi:10.1016/j.ijhm.2019.01.005

Luo, J., Wu, M., Gopukumar, D., & Zhao, Y. (2016). Big data application in biomedical research and health care: a literature review. *Biomedical informatics insights, 8, BII-S31559.*

Lu, X., Wang, C., Yang, J.-M., Pang, Y., & Zhang, L. (2010). Photo2trip: generating travel routes from geo-tagged photos for trip planning. *In Proceedings of the 18th ACM international conference on Multimedia* (pp. 143-152). Florence, Italy: ACM. 10.1145/1873951.1873972

Lu, Y. (2016). Industry 4.0: A survey on technologies, applications and open research issues. *Journal of Industrial Information Integration*, *6*, 1–10. doi:10.1016/j.jii.2017.04.005

Macdonald, E. K., Kleinaltenkamp, M., & Wilson, H. N. (2016). How business customers judge solutions: Solution quality and value in use. *Journal of Marketing*, *80*(3), 96–120. doi:10.1509/jm.15.0109

Maćkiewicz, B., & Konecka-Szydłowska, B. (2017). Green Tourism: Attractions and Initiatives of Polish Cittaslow Cities. In N. Bellini, & C. Pasquinelli (Eds.), Tourism in the City: Towards an Integrative Agenda on Urban Tourism, Springer, 297–310.

Madakam, S., Ramaswamy, R., & Tripathi, S. (2015). Internet of Things (IoT): A literature review. *Journal of Computer and Communications*, *3*(05), 164–173. doi:10.4236/jcc.2015.35021

Manyika, J., Chui, M., Brown, B., Bughin, J., Dobbs, R., Roxburgh, C., & Byers, A. H. (2011). *Big data: The next frontier for innovation, competition, and productivity. McKinsey Global Institute*. MGI.

Marakos, P. (2015). Implementing QR code in museums and archaeological sites. In *International Workshop on Virtual Archaeology* (pp. 23–26). Museums & Cultural Tourism; Retrieved from http://vamct.syros.aegean.gr/2015/abstracts2015.pdf#page=62

Marasco, A., Buonincontri, P., Niekerk, M., Orlowski, M., & Okumus, F. (2018). Exploring the role of next-generation virtual technologies in destination marketing. *Journal of Destination Marketing & Management*, *9*, 138–148. doi:10.1016/j.jdmm.2017.12.002

March, H., & Ribera-Fumaz, R. (2016). Smart contradictions: The politics of making Barcelona a Self-sufficient city. *European Urban and Regional Studies*, *23*(4), 816–830. doi:10.1177/0969776414554488

Marchiori, E., & Cantoni, L. (2015). Including Augmented Reality In Tourism Education Programs. In P. Sheldon, & C. H. Hsu (Eds.), Tourism Education: Global Issues and Trends (pp. 115-134). Emerald Insight (Published Online). doi:10.1108/S1571-504320150000021006

Marczewski, M. (2015). *Gamification user types*. Retrieved from http://www.gamified.uk/user-types/

Marczewski, A. (2013). *A simple introduction tips, advice and thoughts on gamification, United State of America*. Lulu Press.

Marimon, D., Sarasua, C., Carrasco, P., Alvarez, R., Montesa, J., Adamek, T., . . . Gasco, P. (2018). *Mobi AR: Tourist experiences through mobile augmented reality*. Retrieved from http://www.researchgate.net/publication/228979424_MobiAR_Tourist_Experiences_through_Mobile _Augmented_Reality

Marjani, M., Nasaruddin, F., Gani, A., Karim, A., Hashem, I. A. T., Siddiqa, A., & Yaqoob, I. (2017). Big IoT Data Analytics: Architecture, Opportunities, and Open Research Challenges. *IEEE Access: Practical Innovations, Open Solutions*, *5*, 5247–5261. doi:10.1109/ACCESS.2017.2689040

Markit, I. H. S. (2018). Telecommunications Industry Solutions - Market Data, Analysis; Forecast Reports; Services IHS Markit. Retrieved from https://ihsmarkit.com/industry/telecommunications.html

Marković, I., & Klarić, Z. (2015). Attitudes of Local Population of Tourism Impacts on Destination Sustainability – Case of Croatia. *Turizam*, *19*(3), 98–110. doi:10.5937/Turizam1503098M

Marrese-Taylor, E., Velásquez, J. D., & Bravo-Marquez, F. (2014). A novel deterministic approach for aspect-based opinion mining in tourism products reviews. *Expert Systems with Applications*, *41*(17), 7764–7775. doi:10.1016/j.eswa.2014.05.045

Martín Martín, J., Guaita Martínez, J., & Salinas Fernández, J. (2018). An Analysis of the Factors behind the Citizen's Attitude of Rejection towards Tourism in a Context of Overtourism and Economic Dependence on This Activity. *Sustainability*, *10*(8), 2851. doi:10.3390u10082851

Martínez-Graña, A., Goy, J., & Cimarra, C. (2013). A virtual tour of geological heritage: Valourising geodiversity using Google Earth and QR code. *Computers & Geosciences*, *61*(12), 83–93. doi:10.1016/j.cageo.2013.07.020

Martins, D., Ramos, C., Rodrigues, J. M., Cardoso, P., Lam, R., & Serra, F. (2015). Challenges in building a big data warehouse applied to the hotel business intelligence. In *6th Int. Conf. on Applied Informatics and Computing Theory (AICT'15), in Recent Research in Applied Informatics*. 110-117.

Masdar (2019a). *Sustainability*. Retrieved from https://masdar.ae/en/masdar-city/the-city/sustainability

Masdar (2019b). *Waste-to-Energy*. Retrieved from https://masdar.ae/en/masdar-clean-energy/technologies/waste-to-energy

Masdar (2019c). *Mobility*. Retrieved from https://masdar.ae/en/masdar-city/the-city/mobility

Masdar (2019d). *Research & Development*. Retrieved from https://masdar.ae/en/masdar-city/the-city/randd

Masdar (2019e). *Recreation*. Retrieved from https://masdar.ae/en/masdar-city/the-city/recreation

Massa, P., & Campagna, M. (2014). Social media geographic information: recent findings and opportunities for smart spatial planning. *TEMA- Journal of Land Use, Mobility and Environment*. Special Issue; *Eight Int. Conference INPUT Smart City-Planning for Energy, Transportation and Sustainability of the Urban System*.Naples, June 4-6, 2014.

Masselot, C. (2008). Territorial Intelligence Communicational and Community System (TICCS). *Res-Ricerca e Sviluppo per le politiche social*, 1(2), 90-104.

Matallaoui, A., Hanner, N., & Zarnekow, R. (2017). Introduction to gamification: Foundation and underlying theories. In S. Stieglitz, C. Lattemann, S. Robra-Bissantz, R. Zarnekow, & T. Brockmann (Eds.), Gamification using game elements in serious contexts (pp. 3-18) Switzerland: Springer International Publishing.

May, K. (2011). Hotel chain enhances dining experience with QR codes. Retrieved from https://www.phocuswire.com/Hotel-chain-enhances-dining-experience-with-QR-codes

Mayra, F. (2008). *Play in the mobile Internet: Towards contextual gaming*. Paper presented at Internet Research 9.0 Conference, October 15-18, Copenhagen, Denmark.

Mazuryk, T., & Gervautz, M. (1996). Virtual Reality-History, Applications, Technology, and Future. Retrieved from https://www.cg.tuwien.ac.at/research/publications/1996/mazuryk-1996-VRH

McColl-Kennedy, J. R., Zaki, M., Lemon, K. N., Urmetzer, F., & Neely, A. (2019). Gaining customer experience insights that matter. *Journal of Service Research*, *22*(1), 8–26. doi:10.1177/1094670518812182

McCool, S., & Moisey, R. (2001). *Tourism, recreation, and sustainability: Linking culture and the environment*. Wallingford, CT: Cabi. doi:10.1079/9780851995052.0001

Mccormick, T. (2013). *Gamification: A short history*. Retrieved from https://foreignpolicy.com/2013/06/24/gamification-a-short-history/

McKinsey. (2016). Internet of Things: The IoT opportunity - Are you ready to capture a once-in-a-lifetime value pool? *Hong Kong IoT Conference*, (June). 10.1007/978-1-4419-8237-7

Mclaughlin, D. (2012). *Game on! How video games and game design principles can help to make adult education more engaging for the learner* (Master dissertation). University of Ireland Maynooth, Ireland.

Medhat, W., Hassan, A., & Korashy, H. (2014). Sentiment analysis algorithms and applications: A survey. *Ain Shams Engineering Journal*, *5*(4), 1093–1113. doi:10.1016/j.asej.2014.04.011

Meena, A., & Prabhakar, T. V. (2007). Sentence Level Sentiment Analysis in the Presence of Conjuncts Using Linguistic Analysis. In Advances in Information Retrieval (pp. 573–580). doi:10.1007/978-3-540-71496-5_53

Meijer, A., & Bolívar, M. P. R. (2015). Governing the smart city: A review of the literature on smart urban governance. *International Review of Administrative Sciences*, *82*(2), 392–408. doi:10.1177/0020852314564308

Mellit, A., & Kalogirou, S. (2008). Artificial intelligence techniques for photovoltaic applications: A review. *Progress in Energy and Combustion Science*, *34*(5), 574–632. doi:10.1016/j.pecs.2008.01.001

Memiş, L. (2018). Smart technologies and reflections on local public issues in Turkey: A review of metropolitan municipalities. *AVRASYA Uluslararası Araştırmalar Dergisi*, *6*(14), 141–163. doi:10.33692/avrasyad.510640

Mencarelli, R., & Pulh, M. (2012). Museoparks and re-enchantment of the museum visits: An approach centred on visual ethnology. *Qualitative Market Research*, *15*(2), 148–164. doi:10.1108/13522751211215877

Mennis, J. (2006). Mapping the Results of Geographically Weighted Regression. *The Cartographic Journal*, *43*(2), 171–179. doi:10.1179/000870406X114658

METI. (2017). *Evaluation forms for blockchain-based systems*. Tokyo, Japan: Ministry of Economy, Trade, and Industry, Information Economy Division Commerce and Information Policy Bureau.

Mettler, M. (2016). Blockchain technology in healthcare: The revolution starts here. In *Proceedings of IEEE 18th International Conference on E-Health Networking, Applications and Services* (pp. 1-3). Munich, Germany. IEEE. 10.1109/HealthCom.2016.7749510

Meyer, C., & Schwager, A. (2007). Understanding customer experience. *Harvard Business Review*, *85*(2), 116. PMID:17345685

Miah, S. J., Vu, H. Q., Gammack, J., & McGrath, M. (2017). A big data analytics method for tourist behaviour analysis. *Information & Management*, *54*(6), 771–785. doi:10.1016/j.im.2016.11.011

Micera, R., Presenza, A., Splendiani, S., & Del Chiappa, G. (2013). *SMART Destinations: new strategies to manage tourism industry*. Shiuma et al.

Michalko, G., & Timothy, D. J. (2001). Cross-border shopping in Hungary: Causes and effects. *Visions in Leisure and Business*, *20*(1), 2.

Microsoft. (2019). *Microsoft Store*. Retrieved from https://www.microsoft.com/en-us/store/b/virtualreality

Middleton, V. T. C., & Hawkins, R. (1998). *Sustainable Tourism: A Marketing Perspective*. London, UK: Butterworth-Heinemann.

Mil, B., & Dirican, C. (2018). Endüstri 4.0 Teknolojileri ve Turizme Etkileri. *Disiplinlerarası Akademik Turizm Dergisi*, *3*(1), 1–9.

Miller, M. (2008). *Cloud computing: Web-based applications that change the way you work and collaborate online*. Que publishing.

Mills, D., Wang, K., Malone, B., Ravi, A., Marquardt, J., Chen, C., . . . Baird, M. (2016). Distributed ledger technology in payments, clearing, and settlement. *Finance and Economics Discussion Series, 2016*(95), 1-36.

Mingjun, W., Zhen, Y., Wei, Z., Xishang, D., Xiaofei, Y., Chenggang, S., ... & Jinghai, H. (2012, October). A research on experimental system for Internet of things major and application project. In *Proceedings 2012 3rd International Conference on System Science, Engineering Design and Manufacturing Informatization* (Vol. 1, pp. 261-263). IEEE.

Ministry of Industry. Energy and Tourism. (2015). *National Plan of Intelligent Territories- Spain.* Retrieved from http://www.upv.es/contenidos/CAMUNISO/info/U0684517.pdf

Ministry of Industry. Energy and Tourism. (2017). *National Plan of Intelligent Territories- Spain.* Retrieved from https://www.esmartcity.es/2018/01/08/ministerio-energia-turismo-agenda-digital-aprueba-plan-nacional-territorios-inteligentes

Ministry of Industry. Tourism and Commerce. (2007). *Spanish Tourism Plan Horizonte 2020. Executive Document.* Retrieved from https://www.tourspain.es/es-es/Conozcanos/Documents/HistoricoPoliticaTuristica/PlanTurismoEspanolHorizonte2020.pdf

Misner, I., & Devine, V. (1999). *The world's best-known marketing secret: building your business with word-of-mouth marketing*. Retrieved from https://scholar.google.com.tr/scholar?hl=tr&as_sdt=0%2C5&q=The+world's+best+known+marketing+secret%3A+Building+your+business+with+word-of-mouth+marketing.&btnG=

Mohammad, S., & Azam, A. (2015). A model for ordering in restaurant based on QR Code without presence of a waiter at the table Introduction, *2*(12), 1–9.

Monino, J.-L., & Sedkaoui, S. (2016). Big Data, Open Data, and Data Development: Vol. 3. *London, UK: ISTE, Ltd.* Hoboken, NJ: John Wiley.

Mooney, C. (2011). The truth about fracking. *Scientific American*, *305*(5), 80–85. doi:10.1038cientificamerican1111-80 PMID:22125868

Moorhouse, N., TomDieck, M. C., & Jung, T. (2018). Technological Innovations Transforming the Consumer Retail Experience: A Review of Literature. In T. Jung, & M. TomDieck (Eds.), Augmented Reality and Virtual Reality. Progress in IS. Cham, Switzerland: Springer.

Mora, L., Bolici, R., & Deakin, M. (2017). The first two decades of smart-city research: A bibliometric analysis. *Journal of Urban Technology*, *24*(1), 3–27. doi:10.1080/10630732.2017.1285123

Morch, A. (2018). How cloud computing will revolutionize modern hotel marketing and tap into new uncontested markets. Retrieved from https://aremorch.com/how-cloud-computing-will-revolutionize-modern-hotel-marketing-and-tap-into-new-uncontested-markets/

Morfikirler.com. (2018). World's first smart restaurant opened in Dubai. Retrieved from https://morfikirler.com/dubaide-dunyanin-ilk-akilli-restorani-acildi/

Morosan, C., Bowen, J. T., & Atwood, M. (2014). The evolution of marketing research. *International Journal of Contemporary Hospitality Management*, *26*(5), 706–726. doi:10.1108/IJCHM-11-2013-0515

Morozova, A. (2018). How Much Does It Cost to Build an AR App for Food Menu like KabaQNo Title. Retrieved from https://jasoren.com/how-much-does-it-cost-to-build-an-ar-app-for-food-menu-like-kabaq/

Mostafa, M. M. (2013). More than words: Social networks' text mining for consumer brand sentiments. *Expert Systems with Applications*, *40*(10), 4241–4251. doi:10.1016/j.eswa.2013.01.019

Moulaert, F., & Sekia, F. (2003). Territorial innovation models: A critical survey. *Regional Studies*, *37*(3), 289–302. doi:10.1080/0034340032000065442

Moumita. (2010). *Robovie II Helps You with your Grocery Shopping*. Retrieved from http://gadgether.com/robovie-ii-grocery-shopping/

Mugla, E. K., Akba, M. F., Katip, I., & Karaarslan, E. (2017, October). *Blockchain based cyber security systems.* Paper presented at the International Conference on Information Security & Cryptology, Ankara, Turkey.

Mukhopadhyay, U., Skjellum, A., Hambolu, O., Oakley, J., Yu, L., & Brooks, R. (2016). A brief survey of cryptocurrency systems. In *Proceedings of 14th Annual Conference on Privacy, Security and Trust* (pp. 745-752), Auckland, New Zealand

Müller, C., & Hasic, D. (2016). *Blockchain: Technology and Applications*. Salzburg, Austria: University of Salzburg Department of Computer Sciences.

Munar, A. M., & Jacobsen, J. K. S. (2014). Motivations for sharing tourism experiences through social media. *Tourism Management, 43*, 46–54. doi:10.1016/j.tourman.2014.01.012

Muro, M., Magutu, P., & Getembe, K. (2013). The strategic benefits and challenges in the use of customer relationship management system among commercial banks in Kenya. *European Scientific Journal, 9*, 327–349.

Murphy, J., Gretzel, U., & Pesonen, J. (2019). Marketing Robot Services in Hospitality and Tourism: The Role of Anthropomorphism. *Journal of Travel & Tourism Marketing*. Retrieved from https://www.tandfonline.com/doi/pdf/10.1080/10548408.2019.1571983?needAccess=true

Murphy, J., Hofacker, C., & Gretzel, U. (2017). Dawning of the Age of Robots in Hospitality and Tourism: Challenges for Teaching and Research. *European Journal of Tourism Research, 15*, 104–111.

Muthumanicam, S. K. (2018). The evolution of artificial intelligence (AI) in hotel industry. Retrieved from https://chatbotnewsdaily.com/the-evolution-of-artificial-intelligence-ai-in-hotel-industry-23298d1e8dd4

Nadaroğlu, H. (1978). *Mahalli İdareler Felsefesi Ekonomisi Uygulaması. (Birinci Baskı)*. İstanbul, Turkey: Sermet Matbaası.

Nakamoto, S. (2008). *Bitcoin: A peer-to-peer electronic cash system (technical report)*. Retrieved from www.bitcoin.org

Nakamoto, S. (2008). *Bitcoin: A Peer-to-Peer Electronic Cash System (White Paper)*. Retrieved from https://bitcoin.org/bitcoin.pdf

Nakamura, J., & Csikszentmihalyi, M. (2009). The concept of flow. In C. R. Snyder, & S. J. Lopez (Eds.), *Oxford handbook of positive psychology* (pp. 89–105). USA: Oxford University Press.

Nam, T. & Pardo, T. A. (2011). Conceptualizing smart city with dimensions of technology, people, and institutions. In *Proceedings of 12th Annual International Digital Government Research Conference: Digital Government Innovation in Challenging Times* (pp. 282–291). USA: ACM. 10.1145/2037556.2037602

Naphade, M., Banavar, G., Harrison, C., Paraszczak, J., & Morris, R. (2011). Smarter cities and their innovation challenges. *Computer, 44*(6), 32–39. doi:10.1109/MC.2011.187

Napierała, T. (2013). *Przestrzenne zróżnicowanie cen usług hotelowych w Polsce*. Wydawnictwo Uniwersytetu Łódzkiego. doi:10.18778/7525-929-2

Napierała, T. (2017). Internetization of selling hotel rooms in metropolitan area of Łódź (Poland). Journal of Geography. *Politics & Society, 7*(3), 19–30.

Narayanan, A. S. (2012). QR Codes and Security Solutions. *International Journal of Computer Science and Telecommunications, 3*(7), 1–4. Retrieved from http://www.ijcst.org/Volume3/Issue7/p13_3_7.pdf

Narayanan, A., Bonneau, J., Felten, E., Miller, A., & Goldfeder, S. (2015). *Bitcoin and Cryptocurrency Technologies*. Princeton, NJ: Princeton University.

Nasim, Z., & Haider, S. (2017). ABSA Toolkit: An Open Source Tool for Aspect Based Sentiment Analysis. *International Journal of Artificial Intelligence Tools, 26*(06). doi:10.1142/S0218213017500233

Natarajan, H., Krause, S., Karla, G., & Helen, L. (2017). *Distributed Ledger Technology (DLT) and blockchain*. Washington, D.C.: World Bank Group. doi:10.1596/29053

Neirotti, P., De Marco, A., Cagliano, A. C., Mangano, G., & Scorrano, F. (2014). Current trends in smart city initiatives: Some stylised facts. *Cities (London, England)*, *38*, 25–36. doi:10.1016/j.cities.2013.12.010

Neuhofer, B. (2014). The Technology Enhanced Tourist Experience. In Information and Communication Technologies in Tourism (pp. 90–96).

Neuhofer, B., Buhalis, D., & Ladkin, A. (2014). A typology of technology-enhanced tourism experiences. *International Journal of Tourism Research*, *16*(4), 340–350. doi:10.1002/jtr.1958

Newell, F., & Seth, G. (2003). *Why CRM Doesn't Work: How to Win By Letting Customers Manage the Relationship*. Princeton, NJ: Bloomberg Press.

NFC and Bluetooth: The Perfect Pair. (n.d.). Retrieved from https://nfc-forum.org/nfc-and-bluetooth/

NFC Reference Guide for Air Travel. (2013). Retrieved from http://nfc-forum.org/wp-content/uploads/2013/12/2013_10_28_NFC_Reference_Guide_for_Air_Travel_FINAL.pdf

NFC-Technology. (n.d.) Retrieved from https://www.smart-tec.com/en/auto-id-world/nfc-technology

Nick, G., Pongrácz, F., & Radács, E. (2018). Interpretation of Disruptive Innovation in the Era of Smart Cities of the Fourth Industrial Revolution. *Deturope*, *10*(1), 53–70.

Niklas, S., & Böhm, S. (2011). Applying Mobile Technologies for Personnel Recruiting –an Analysis of User-Sided Acceptance Factors. *International Journal of EBusiness and EGovernment Studies*, *3*(1), 169–178.

Nitti, M., Pilloni, V., Giusto, D., & Popescu, V. (2017). *IoT Architecture for a sustainable tourism application in a smart city environment*. Mobile Information Systems; doi:10.1155/2017/9201640

Noguera, J. M., Barranco, M. J., Segura, R. J., & Martínez, L. (2012). A mobile 3D-GIS hybrid recommender system for tourism. *Information Sciences*, *215*, 37–52. doi:10.1016/j.ins.2012.05.010

Novák-Marcinčin, J., & Kuzmiaková, M. (2009). Basic Components Of Virtual Reality. *Annals of the University of Petroşani Mechanical Engineering (New York, N.Y.)*, *11*, 175–182.

Nowacki, M. (2019). World Cities' Image in TripAdvisor Users' Reviews. *Ereview of Tourism Research*, *16*(2–3). Retrieved from https://journals.tdl.org/ertr/index.php/ertr/article/view/327

Ntv.com.tr. (2013). Robot waiters are at your service. Retrieved from https://www.ntv.com.tr/galeri/yasam/robot-garsonlarhizmetinizde,9SpVjbRyZkyEQgok65ZnCw

Nuckolls, J., & Long, P. (2009). *Organizing resources for tourism development in rural areas. Center for Sustainable Tourism East Carolina University*. Retrieved from www.sustainabletourism.org

Nysveen, H. (2003). The importance of risk-reducing value-added services in online environments: An exploratory study on various segments of leisure tourism. *Information Technology & Tourism*, *6*(2), 113–127. doi:10.3727/109830503773048228

Nysveen, H., & Lexhagen, M. (2002). Reduction of perceived risk through on-line value added services offered on tourism businesses websites. In K. W. Wöber, A. J. Frew, & M. Hitz (Eds.), *Proceedings of the ENTER 2002 Conference*. Wien, Austria: Springer-Verlag.

O'Connor, P., & Murphy, J. (2004). Research on information technology in the hospitality industry. *International Journal of Hospitality Management*, *23*(5), 473–484. doi:10.1016/j.ijhm.2004.10.002

OECD. (2003). Introduction and Draft Results of the Survey of Knowledge Management Practices in Ministries/Departments/Agencies of Central Government. Retrieved from www.oecd.org/officialdocuments/publicdisplaydocumentpdf/?cote=GOV/PUMA/HRM(2003)2&docLanguage=En

OECD. (Organization for Economic Cooperation and Development). (2018). Spain: Investing in 'Smart Destinations' for digitalisation and sustainable tourism. In Effective Policy Approaches for Quality Investment in Tourism, OECD Tourism Papers, (pp. 87-92). Paris, No.2018/03, OECD Publishing. doi:10.1787/88ea780c-en

Oğuzlar, A. (2011). *Temel Metin Madenciliği*. Bursa, Turkey: Dora.

Okazaki, S., Navarro, A., & Campo, S. (2013). Cross-Media Integration Of Qr Code: A Preliminary Exploration. Journal of Electronic Commerce Research, 14(2), 137–148.

Ok, K., Coskun, V., Aydin, M. N., & Ozdenizci, B. (2010, November). Current benefits and future directions of NFC services. In *Proceedings International Conference on Education and Management Technology*, IEEE. Cairo, Egypt. 10.1109/ICEMT.2010.5657642

Ökmen, M. (2009). *Sürekli değişme ve gelişme sürecinde kamu yönetimi ve yerel yönetimler. K. Görmez ve M. Ökmen (Editörler). Yerel Yönetimlerin Güncel Sorunları: Küresel, Bölgesel ve Yerel Perspektifler* (pp. 5–41). Ankara, Turkey: Beta Basım A. Ş.

Okuyama, K., & Yanai, K. (2013). A Travel Planning System Based on Travel Trajectories Extracted from a Large Number of Geotagged Photos on the Web. In *The era of interactive media* (pp. 657–670). New York: Springer. doi:10.1007/978-1-4614-3501-3_54

Oliveira, T., Thomas, M., & Espadanal, M. (2014). Assessing the determinants of cloud computing adoption: An analysis of the manufacturing and services sectors. *Information & Management, 51*(5), 497–510. doi:10.1016/j.im.2014.03.006

Olivieri, S. J. (2015). *An Investigation of Security In Near Field Communication Systems* (Unpublished doctoral dissertation). Northeastern University, Boston, MA.

Olsen, M. D., & Connolly, D. J. (2000). Experience-based travel. How technology is changing the hospitality industry. *The Cornell Hotel and Restaurant Administration Quarterly, 41*(1), 30–40. doi:10.1177/001088040004100121

Önaçan, M. B. K. (2015). *Organizasyonlar için bilgi yönetimi çerçevesi ve bilgi yönetim sistemi mimarisi önerisi: dOBLYN (Doküman ve bilgi yönetimi)*. Ankara Üniversitesi Sosyal Bilimler Enstitüsü İşletme A.B.D., Doktora Tezi.

Önder, I., & Treiblmaier, H. (2018). Blockchain and tourism: Three research propositions. *Annals of Tourism Research, 72*(C), 180–182. doi:10.1016/j.annals.2018.03.005

Osawa, H., Akiya, N., Koyama, T., Ema, A., Kanzaki, N., Ichise, R., . . . Kubo, A. (2017). What is Real Risk and Benefit on Work with Robots? From the Analysis of a Robot Hotel, *HRI2017 Companion of the 2017 ACM/IEEE International Conference on Human-Robot Interaction*, March 6-9, Vienna, Austria. 10.1145/3029798.3038312

Osawa, H., Ema, A., Hattori, H., Akiya, N., Kanzaki, N., Kubo, A., . . . Ichise, R. (2017, March). What is Real Risk and Benefit on Work with Robots?: From the Analysis of a Robot Hotel. In *Proceedings of the Companion of the 2017 ACM/IEEE International Conference on Human-Robot Interaction (pp. 241-242)*. ACM.

Özdamar, K. (2004). *Paket Programlar İle İstatistiksel Veri Analiz*. Eskişehir, Turkey: Kaan Kitabevi.

Özdenizci, B., Ok, K., Aydın, M. N., & Coşkun, V. (2011) Yakın Alan İletişimi Teknolojisi. *Türkiye Bilişim Vakfı Bilgisayar Bilimleri ve Mühendisliği Dergisi, 4*(1).

Özen, A. (2019). Kapadokya Bölgesinde Yaşanan Sıcak Hava Balon Deneyiminin Davranışsal Niyetler Üzerindeki Etkilerini Belirlemeye Yönelik Bir Araştırma. *Verimlilik Dergisi*, *1*, 165–198.

Özgüneş, R. E., & Bozok, D. (2017). Virtual competitor of tourism sector (is that so?): Augmented reality, 1st International Sustainable Tourism Congress, November 23-25, 2017, Kastamonu-Turkey.

Özgüneş, R. E., & Bozok, D. (2017). Turizm Sektörünün Sanal Rakibi (Mi?): Arttırılmış Gerçeklik. *Uluslararası Türk Dünyası Turizm Araştırmaları Dergisi*, *2*(2), 146–160.

Özgüven, N. (2011). Tüketicilerin Online Alışverişe Karşı Tutumları ile Demografik Özellikleri Arasındaki İlişki Analizi. *Karamanoğlu Mehmetbey Üniversitesi Sosyal ve Ekonomik Araştırmalar Dergisi*, *13*(21), 47–54.

Özkan, Ç. (2018). *Gamification use in the context of corporate communication strategies: A research on the banking industry* (Master dissertation). University of Istanbul, Turkey.

Öztürk, S. (2015). *Using gamification in effective team planning and team activity* (Master dissertation). University of Bahçeşehir, Turkey.

Öztürk, Y., & Ayaz, N. (2010). *Turizm olgusunda Yerel Gündem 21 uygulamaları üzerine bir araştırma. O. E. Çolakoğlu (Editör). 11. Ulusal Turizm Kongresi* (pp. 324–333). Ankara, Turkey: Detay Yayıncılık.

Pablo, P., Ester Gonzalez, S., Redouane, K., Jaime, R., Ignacio, B., Francisco, P., & Alvaro, V. (2019). Immersive Gastronomic Experience with Distributed Reality. *Workshop on Everyday Virtual Reality*. Osaka, Japan: WEVR 2019.

Pagallo, U., Corrales, M., Fenwick, M., & Forgo, N. (2018). The Rise of Robotics & AI: Technological Advances & Normative Dilemmas. In M. Corrales, M. Fenwick, & N. Forgo (Eds.), *Robotics, AI and the Future of Law*. Singapore: Springer Nature. doi:10.1007/978-981-13-2874-9_1

Palumbo, F., & Dominici, G. (2015) Unraveling the complexity of tourist experience with nfc technology and mobile wallets. In S. Erçetin, & S. Banerjee (Eds.), Chaos, complexity, and leadership 2013 (pp. 189-196). Springer Proceedings in Complexity. Cham, Switzerland: Springer. doi:10.1007/978-3-319-09710-7_16

Pan, S. L., & Lee, J. N. (2003). Using e-CRMfor a Unified View of the Customer. *Communications of the ACM*, *46*(4), 95–99. doi:10.1145/641205.641212

Pantano, E., & Timmermans, H. (2014). What is smart for retailing? *Procedia Environmental Sciences*, *22*, 101–107. doi:10.1016/j.proenv.2014.11.010

Pantona, E., & Servidio, R. (2001). An Exploratory Study of the Role of Pervasive Environments for Promotion of Tourism Destinations. *Journal of Hospitality and Tourism Technology*, *2*(1), 50–65. doi:10.1108/17579881111112412

Pan, X., & Hamilton, A. F. D. C. (2018). Why and how to use virtual reality to study human socialinteraction: The challenges of exploring a new research landscape. *British Journal of Psychology*, *109*(3), 1–23. doi:10.1111/bjop.12290 PMID:29313958

Parvatiyar, A., & Sheth, J. N. (2001). Customer relationship management: Emerging practice, process, and discipline. *Journal of Economic and Social Research, 3*(2), 01-34.

Patel, H., & Cardinali, R. (1994). Virtual Reality Technology in Business. *Management Decision*, *32*(7), 5–12. doi:10.1108/00251749410068111

Patkar, R. S., Singh, S. P., & Birje, S. V. (2013). Marker based augmented reality using Android os. [IJARCSSE]. *International Journal of Advanced Research in Computer Science and Software Engineering*, *3*(5).

Payne, A., & Frow, P. (2005). A Strategic Frameworkfor Customer Relationship Management. *Journal of Marketing*, *69*(4), 167–176. doi:10.1509/jmkg.2005.69.4.167

Peceny, U. S., Urbančič, J., Mokorel, S., Kuralt, V., & Ilijaš, T. (2019). Tourism 4.0: Challenges in Marketing a Paradigm Shift. In Consumer Behavior and Marketing. IntechOpen.

Peceny, U., Urbančič, J., Mokorel, S., Kuralt, V., & Ilijaš, T. (2019). Tourism 4.0: Challenges in Marketing a Paradigm Shift. In Consumer Behavior and Marketing. IntechOpen.

Pehlivan, B. (2019). *Dijital Pazarlamada yapay zeka.* Retrieved from http://www.yapayzekatr.com/2019/03/03/dijital_pazarlamada_yapay_zeka/

Pemberton, J. D., Stonehouse, G. H., & Barber, C. E. (2001). Competing with CRS-generated information in the airline industry. *The Journal of Strategic Information Systems*, *1*(1), 59–76. doi:10.1016/S0963-8687(01)00042-7

Pendergast, D. (2010). Getting to Know the Y Generation. In P. Benckendorff, G. Moscardo, & D. Pendergast (Eds.), *Tourism and Generation Y.* Oxford, UK: CABI.

Perea-Medina, M. J., Navarro-Jurado, E., & Luque-Gil, A. M. (2018). Territorial Intelligence: Conceptualization and progress in the state of the question. Possible links with tourist destinations. *Cuadernos de Turismo*, (41): 535–554.

Perkins, H. C., & Thorns, D. C. (2001). Gazing or performing? Introduction: The tourist performance. *International Sociology*, *16*(2), 185–204. doi:10.1177/0268580901016002004

Perry Hobson, J., & Williams, A. (1995). Virtual reality: A new horizon for the tourism industry. *Journal of Vacation Marketing*, *1*(2), 124–135. doi:10.1177/135676679500100202

Pesonen, J., & Horster, E. (2012). Near field communication technology in tourism. *Tourism Management Perspectives*, *4*, 11–18. doi:10.1016/j.tmp.2012.04.001

Peters, G., & Panayi, E. (2015). *Understanding modern banking ledgers through blockchain technologies: future of transaction processions and smart contracts on the internet of money.* Retrieved from https://arxiv.org/pdf/1511.05740.pdf

Petrevska, B., Cingoski, V., & Gelev, S. (2016). From smart rooms to smart hotels. *Zbornik radova sa XXI međunarodnog naučno-stručnog skupa Informacione tehnologije-sadašnjost i budućnost, Žabljak.* Feb. 29-March 5, 2016, 21, 201-204.

Petrović, M. D., & Marković, J. J. (2013). Sustainable Tourism as a Part of Comprehensive Environmental Monitoring. A Study of Serbia. *Journal of Environmental and Tourism Analyses*, *I*, 1, 30–42.

Piccione, P. A. (1980). In search of the meaning of Senet. *Archaeological Institute of America*, *33*(4), 55–58.

Pierce, A. (2015). *A Hotel Staffed by Robots.* Retrieved from http://www.technologytoday.us/columnPDF/A_Hotel_Staffed_by_Robots.pdf

Pieska, S., Luimula, M., Jauhiainen, J., & Spiz, V. (2012). Social Service Robots in Public and Private Environments, In D. Biolek, K. Volkov, & K. M. Ng (Eds.), Recent Researches in Circuits, Systems, Multimedia, and Automatic Control, Wisconsin: WSEAS Press.

Pilkington, M., & Crudu, R. (2016). *Tourism 2.0 and e-Governance in the Republic of Moldova - Can the Blockchain Help?* Retrieved from http://papers.ssrn.com/sol3/papers.cfm?abstract_id=2732350

Pilkington, M. (2016). Blockchain technology: principles and applications. In F. Xavier Olleros & M. Zhegu (Eds.), *Research handbook on digital transformations* (pp. 1–39). Northampton, MA: Edward Elgar.

Pine, B. J., Pine, J., & Gilmore, J. H. (1999). *The experience economy: work is theatre & every business a stage. Harvard Business Press.* doi:10.5860/choice.37-2254

Pinillos, R., Marcos, S., Feliz, R., Zalama, E., & Gomez-Garcia-Bermejo, J. (2016). Long-Term Assessment of a Service Robot in a Hotel Environment. *Robotics and Autonomous Systems*, *79*, 40–57. doi:10.1016/j.robot.2016.01.014

Pizam, A. (1978). Tourism's impacts: The social costs to the destination community as perceived by its residents. *Journal of Travel Research*, *16*(4), 8–12. doi:10.1177/004728757801600402

play.google.com. (2019). 05.08.2019 Retrieved from https://play.google.com/store/search?q=b%C3%BCy%C3%BCk%C5%9Fehir%20belediyesi&c=apps

Plymire, J. (1991). Complaints as opportunities. *Journal of Consumer Marketing*, *8*(2), 39–43. doi:10.1108/07363769110035009

Pochwatko, G., Giger, J.-C., Rozanska-Walzcuk, M., Swidrak, J., Kukielka, K., Mozaryn, J., & Piçarra, N. (2015). Polish Version of the Negative Attitude Toward Robots Scale (NARS-PL). *Journal of Automation, Mobile Robotics, & Intelligent Systems*, *9*(3), 65–72.

Poon, A. (1993). *Tourism, technology and competitive strategies*. Wallingford, CT: CABI.

Poon, A. (1993). *Tourism, Technology, and Competitive Strategies*. Oxford, UK: CAB International.

Porter, M. (2001). Strategy and the Internet. *Harvard Business Review*, *79*(3), 63–78. PMID:11246925

Porter, M. E., & Heppelmann, J. E. (2014). How smart, connected products are transforming competition. *Harvard Business Review*, *92*(11), 64–88.

Prameswari, P., Surjandari, I., & Laoh, E. (2017, November). Mining online reviews in Indonesia's priority tourist destinations using sentiment analysis and text summarization approach. In *Proceedings 2017 IEEE 8th International Conference on Awareness Science and Technology (ICAST)*, 121–126. 10.1109/ICAwST.2017.8256429

Prayag, G., Hosany, S., & Odeh, K. (2013). The role of tourists' emotional experiences and satisfaction in understanding behavioral intentions. *Journal of Destination Marketing & Management*, *2*(2), 118–127. doi:10.1016/j.jdmm.2013.05.001

Princes.com. (2019). Ocean medallion. Retrieved from https://www.princess.com/learn/faq_answer/pre_cruise/ocean_medallion.jsp

Probst, A. (2012). The Expectations of Quick Response (QR) Codes in Print Media: An Empirical Data Research Anthology Number of QR Codes Printed in the Top 100 Magazines (Jan-Dec 2011). *UW-L Journal of Undergraduate Research XV*, 1–13. Retrieved from https://www.uwlax.edu/urc/jur-online/pdf/2012/probst.ali.pdf

Proente. (2019). *Big Data (Büyük Veri) Nedir?* Retrieved from https://proente.com/big-data-buyuk-veri-nedir/

Put, V. W., & Smith, R. (2016). Smart Tourism Tools: Linking Technology to the Touristic Resources of a City, Retrieved from www.semanticscholar.org

Quick, D., Martini, B., & Choo, K. K. R. (2013). Cloud Storage Forensics. In Cloud Storage Forensics. doi:10.1016/C2013-0-09718-6

Rabotić, B. (2014). Special-purpose travel in ancient times: 'Tourism' before tourism? *Turističko poslovanje, (14)*, 5-17.

Raghupathi, W., & Raghupathi, V. (2014). Big data analytics in healthcare: promise and potential. *Health information science and systems, 2*(1), 3.

Rahimi, R., Nadda, V., & Wang, H. (2015). CRM in Tourism. In R. Nilanjan (Ed.), *Emerging Innovative Marketing Strategies in the Tourism Industry* (pp. 16–43). Hershey, PA: IGI Global. doi:10.4018/978-1-4666-8699-1.ch002

Rahyaputra, V., Muna, N. K., & Rizal, N. S. (2016). *Barcelona's smart city: The Frontrunner in digital transformation*. CfDS Case Study Series No. 4. Retrieved from http://cfds.fisipol.ugm.ac.id/uploads/files/posts/72/CFDS_CASESTUDIES_BARCELONA.pdf

Rajesh, M. (2015). *Inside Japan's First Robot-Staffed Hotel*. Retrieved from https://www.theguardian.com/travel/2015/aug/14/japan-henn-na-hotel-staffed-by-robots

Ramón-Hidalgo, A. E., & Harris, L. M. (2018). Social Capital, political empowerment and social difference: A mixed-methods study of an ecotourism project in the rural Volta region of Ghana. *Journal of Sustainable Tourism*, *26*(12), 2153–2172. doi:10.1080/09669582.2018.1546711

Rašula, J., Vukšić, V. B., & Štemberger, M. I. (2012). The impact of knowledge management on organisational performance. *EBR*, *14*(2), 147–168.

Ravì, D., Wong, C., Deligianni, F., Berthelot, M., Andreu-Perez, J., Lo, B., & Yang, G. (2017). Deep Learning for Health Informatics. *IEEE Journal of Biomedical and Health Informatics*, *21*(1), 4–21. doi:10.1109/JBHI.2016.2636665 PMID:28055930

RealityTechnologies.com. (2018). *The Ultimate Guide to Understanding Virtual Reality (VR) Technology*, Retrieved from https://www.realitytechnologies.com/virtual-reality/Reality

Reed, M. G. (1997). Power relations and community-based tourism planning. *Annals of Tourism Research*, *24*(3), 566–591. doi:10.1016/S0160-7383(97)00023-6

Reich-Stiebert, N., & Eyssel, F. (2015). Learning with Educational Companion Robots? Toward Attitudes on Education Robots, Predictors Of Attitudes, and Application Potentials for Education Robots. *International Journal of Social Robotics*, *7*(5), 875–888. doi:10.100712369-015-0308-9

Reid, C. (2018). *London launches 5-year plan to entice more people to fall in love with cycling*. Retrieved from https://www.forbes.com/sites/carltonreid/2018/12/17/london-launches-5-year-plan-to-make-more-people-fall-in-love-with-cycling/#6fb296e624fd

Reischl, C. (2013). *Data will drive smart cities' medical tourism success*. Denver, CO: Mercury Advisory Group.

Renkliyolculuk.com. (2019). Technological developments in tourism. Retrieved from https://www.renkliyolculuk.com/2019-turizm-alanindaki-teknolojik-gelismeler/

Rethink Music Initiative. (2015). *Rethink music report on transparency*. Retrieved from http://www.rethink-music.com/news/rethink-music-report-on-tranparency

Retz, M. (2019, May 1). *Facebook Wants to Bring Your Entire Body Into Virtual Reality*. Retrieved from https://edition.cnn.com/2019/05/01/tech/facebook-vr-avatar/index.html

Revfine. (2019a). How to use artificial intelligence in the hospitality industry. Retrieved from https://www.revfine.com/artificial-intelligence-hospitality-industry/

Revfine. (2019b). How augmented reality is transforming the hospitality industry. Retrieved from https://www.revfine.com/augmented-reality-hospitality-industry/

Revfine.com. (2019). *How to use Artificial Intelligence in the Hospitality Industry*. Retrieved from https://www.revfine.com/artificial-intelligence-hospitality-industry/

Rey-López, M., Barragáns-Martínez, A., Peleteiro, A., Mikic-Fonte, F., & Burguillo, J. (2011). MoreTourism: Mobile recommendations for tourism. In *IEEE International Conference on Consumer Electronics (ICCE)*. pp. 347-348. IEEE.

Richter, L. K. (1989). *The politics of tourism in Asia*. University of Hawaii Press.

Ritchie, J. R., & Crouch, G. I. (2005). *A model of destination competitiveness. Competitive destination: A sustainable tourism perspective*. Wallingford, CT: Cabi.

Riva Sanseverino, E., Riva Sanseverino, R., Vaccaro, V., Macaione, I., & Anello, E. (2017). Smart cities: Case studies. In E. Riva Sanseverino, R. Riva Sanseverino, & V. Vaccaro (Eds.), *Smart cities atlas: Western and eastern intelligent communities* (pp. 47–140). Cham, Switzerland: Springer. doi:10.1007/978-3-319-47361-1_3

Riva, G., Gagglioli, A., Villiani, D., Preziosa, A., Morganti, F., Strambi, L., ... Vezzadini, L. (2007). An Open Source Virtual Reality Platform for Clinical and Research Applications. *Virtual Reality Second International Conference ICVR* (pp. 699-707). Beijing, China. 10.1007/978-3-540-73335-5_76

Rizwan, Z. (2017). No Title. Retrieved from https://scanova.io/blog/blog/2017/01/18/qr-codes-hotels/

Robinson, P., Lück, M., & Smith, S. (2013). *Tourism* (1st ed.). Wallingford, UK: CABI. doi:10.1079/9781780642970.0000

Robots. (2019). *Kismet*. Retrieved from https://robots.ieee.org/robots/kismet/

RobotsR. O. S. (2019). *Maggie*. Retrieved from https://robots.ros.org/maggie/

Roche, S., Nabian, N., Kloeckl, K., & Ratti, C. (2012). Are 'Smart Cities' Smart Enough? *Global Geospatial Conference 2012*, Global Spatial Data Infrastructure Association. Available at http://www.gsdi.org/gsdiconf/gsdi13/papers/182.pdf

Rodriguez, J., Gutierrez, T., & Sanchez, J. E. (2012). Training of Procedural Tasks Through the Use of Virtual Reality and Direct Aids. In C. L. Lanyi (Ed.), Virtual Reality and Environments (pp. 43-68). Croatia: InTech. doi:10.5772/36650

Rodriguez-Lizundia, E., Marcos, S., Zalama, E., Gomez-Garcia-Bermejo, J., & Gordaliza, A. (2015). A bellboy Robot: Study of the Effects of Robot Behaviour on User Engagement and Comfort. *International Journal of Human-Computer Studies*, *82*, 83–95. doi:10.1016/j.ijhcs.2015.06.001

Romão, J., & Neuts, B. (2017). Territorial capital, smart tourism specialization and sustainable regional development: Experiences from Europe. *Habitat International*, *68*, 64–74. doi:10.1016/j.habitatint.2017.04.006

Rong, A. (2012). *China economic net*. Retrieved July 22, 2013 accessed from http://en.ce.cn/Insight/ 201204/12/t20120412_23235803.shtml

Rosenbaum, M. S., & Wong, I. A. (2015). If you install it, will they use it? Understanding why hospitality customers take "technological pauses" from self-service technology. *Journal of Business Research*, *68*(9), 1862–1868. doi:10.1016/j.jbusres.2015.01.014

Rosenbaum, M., Ostrom, A., & Kuntze, R. (2005). Loyalty programs and a sense of community. *Journal of Services Marketing*, *19*(4), 222–233. doi:10.1108/08876040510605253

Rouillard, J. (2008, July). *Contextual QR codes*. Paper presented at *2008 The Third International Multi-Conference on Computing in the Global Information Technology*, IEEE. Athens, Greece.

Royal-hospitality. (2016). Smart hotel. Smart return. Retrieved from royal hospitality.net/images/lighting/smart_hotel.pdf.

Ruiz, I. L., García, G. C., & Gómez-Nieto, M. Á. (2016). Red Thread. An NFC Solution for Attracting Students and Engaging Customers. In *Ubiquitous Computing and Ambient Intelligence* (pp. 263–274). Cham, Switzerland: Springer. doi:10.1007/978-3-319-48799-1_31

Rushdi Saleh, M., Martín-Valdivia, M. T., Montejo-Ráez, A., & Ureña-López, L. A. (2011). Experiments with SVM to classify opinions in different domains. *Expert Systems with Applications*, *38*(12), 14799–14804. doi:10.1016/j.eswa.2011.05.070

Ryan, R. M., & Deci, E. L. (2000). Self-determination theory and the facilitation of intrinsic motivation, social development, and well-being. *The American Psychologist*, *55*(1), 68–78. doi:10.1037/0003-066X.55.1.68 PMID:11392867

Sabella, R. R. (2016). *NFC for Dummies*. Hoboken, NJ: John Wiley & Sons.

Sadler, K. (2015). *SPENCER Robot to Assist Passengers at Schiphol Airport*. Retrieved from https://www.internationalairportreview.com/news/21360/spencer-robot-assists-passengers-schiphol-airport/

Saeidi, M., Bouchard, G., Liakata, M., & Riedel, S. (2016). *SentiHood: Targeted Aspect Based Sentiment Analysis Dataset for Urban Neighbourhoods*. 1546–1556. Retrieved from http://arxiv.org/abs/1610.03771

Şahin, E. (2016). *18. Yüzyılda Satranç Oynadığı Herkesi Madara Eden Türk Robot: Mekanik Türk*. Retrieved from https://www.webtekno.com/internet/mekanik-turk-h15195.html, Accessed on May 10, 2019.

Şahin, M., & Samur, Y. (2017). Instructional method of digital age: Gamification. *Journal of Ege Education Technologies*, *1*(1), 1–27.

Saleem, Y., Crespi, N., Member, S., Husain Rehmani, M., & Copeland, R. (2017). Internet of Things-aided Smart Grid: Technologies, Architectures, Applications, Prototypes, and Future Research Directions. 1–30. IEEE.

Salen, K., & Zimmerman, E. (2004). *Rules of play: Gaming Design Fundamentals*. Cambridge, MA: MIT Press.

Salichs, M. A., Barber, R., Khamis, A. M., Malfaz, M., Gorostiza, J. F., Pacheco, R., ... Garcia, D. (2006). Maggie: A Robotic Platform for Human-Robot Social Interaction. In *Proceedings 2006 IEEE Conference on Robotics, Automation, and Mechatronics*, June 1-3, Bangkok, Thailand. 10.1109/RAMECH.2006.252754

Sandry, E. (2015). Re-evaluating the Form and Communication of Social Robots. *International Journal of Social Robotics*, *7*(3), 335–346. doi:10.100712369-014-0278-3

Santosa, M., & Gook, L. B. (2012). ARkanoid: Development of 3D game and handheld augmented reality. *International Journal Of Computational Engineering Research*, *2*(4), 1053–1059.

Santos-Júnior, A., Mendes-Filho, L., Almeida-García, F., & Manuel-Simões, J. (2017). Smart Tourism Destinations: A study based on the vision of the stakeholders. *Revista Turismo em Análise*, *28*(3), 358–379. doi:10.11606/issn.1984-4867.v28i3p358-379

Sap. (2018). What is the Internet of Things? IoT Technology. SAP. Retrieved from https://www.sap.com/trends/internet-of-things.html#pdf-asset=e825c3a3-c27c-0010-82c7-eda71af511fa&page=1

Sarvary, M. (1999). Knowledge management and competition in the consulting industry. *California Management Review*, *41*(2), 95–107. doi:10.2307/41165988

Satranç Dünyası. (2009). *İlk Satranç Oynayan Makine (Türk)*. Retrieved from http://www.satranc.net/lk-satranc-oynayan-makinaturk/

Saulat, A. (2018) *Four Ways AI is Re-imagining the Future of Travel*. Retrieved from https://www.mindtree.com/blog/four-ways-ai-re-imagining-future-travel

Savaş, G. (2018). *Yapay Zeka İle Turizmde Dönüşümün 5 Örneği / Uygulaması*. Retrieved from http://www.turizmdosyasi.com/yapay-zeka-ile-turizmde-donusumun-5-ornegi-uygulamasi-h15034.html

Say2eat. (2017). Top 5 Reasons to Partner with Say2eat's Modernized Food Ordering Technology. Retrieved from https://medium.com/@say2eat/top-5-reasons-to-partner-with-say2eats-modernized-food-ordering-technology-4d5a3b70e6c1

Schaap, J. (2018). How Self-Service Is Replacing Traditional Guest Service In The Ideal Guest Experience. Retrieved from https://www.hotel-online.com/press_releases/release/how-self-service-is-replacing-traditional-guest-service-in-the-ideal-guest

Schaffers, H. (2012). *Empowering citizens to realizing smart cities: Results from FIREBALL Smart city case studies (presentation).* 2012 future internet assembly (FIA), "smart cities and internet of things". Retrieved from http://www.fi-aalborg.eu/downloads/Session_1.1_Hans_Schaffers.pdf

Schaffers, H., Komninos, N., Pallot, M., Trousse, B., Nilsson, M., & Oliveira, A. (2011). Smart cities and the future internet: Towards cooperation frameworks for open innovation. In *The Future internet assembly* (pp. 431–446). Heidelberg, Germany: Springer; doi:10.1007/978-3-642-20898-0_31

Schmalstieg, D., Langlotz, T., & Billinghurst, M. (2011). Augmented reality 2.0. In *Virtual Realities* (pp. 13–37). Vienna, Austria: Springer. doi:10.1007/978-3-211-99178-7_2

Schmidt-Belz, B., Nick, A., Poslad, S., & Zipf, A. (2002). Personalized and location-based mobile tourism services. *Workshop on "Mobile Tourism Support Systems" in Conjunction with Mobile HCI*, 14. 10.1016/j.canlet.2008.11.010

Schmunk, S., Höpken, W., Fuchs, M., & Lexhagen, M. (2013). Sentiment Analysis: Extracting Decision-Relevant Knowledge from UGC. In Information and Communication Technologies in Tourism 2014 (pp. 253–265). doi:10.1007/978-3-319-03973-2_19

Schönberger, V. M., & Cukier, K. (2013). *Büyük Veri - Yaşama, Çalışma ve Düşünme Şeklimizi Dönüştürecek Bir Devrim. Çev. Banu Erol*. İstanbul, Turkey: Paloma.

Schuckert, M., Liu, X., & Law, R. (2015). Hospitality and Tourism Online Reviews: Recent Trends and Future Directions. *Journal of Travel & Tourism Marketing, 32*(5), 608–621. doi:10.1080/10548408.2014.933154

Schwab, K. (2016) The Fourth Industrial Revolution: what it means, how to respond, World Economic Forum. Available at https://www.weforum.org/agenda/2 016/01/the-fourth-industrialrevolution- what-it-means-and-howto- respond/

Scullin, S., Allora, J., Lloyd, G. O., & Fjermestad, J. (2002). Electronic customer relationship management: benefits, considerations, pitfalls and trends. In *Proceedings of the IS One World Conference, 3-5*. Las Vegas, NV. Academic Press.

Seaborn, K., & Fels, D. I. (2015). Gamification in theory and action: A survey. *International Journal of Human-Computer Studies, 74*, 14–31. doi:10.1016/j.ijhcs.2014.09.006

Šebela, S. (2019). Postojna—Planina Cave System, Slovenia. In Encyclopedia of caves (pp. 812-821). Academic Press.

Secretary of State for Tourism. (2012). *National and Integral Tourism Plan. Madrid, Ministry of Industry, Energy and Tourism.* Retrieved from https://turismo.gob.es/es-es/servicios/Documents/Plan-Nacional-Integral-Turismo-2012-2015.pdf

Seebacher, S., & Schüritz, R. (2017). Blockchain technology as an enabler of service systems: A structured literature review. In *Proceedings 8th International Conference on Exploring Service Science, IESS 1.7* (pp. 12-23). Cham, Switzerland: Springer. 10.1007/978-3-319-56925-3_2

Segelström, F. (2013). *Stakeholder Engagement for Service Design: How service designers identify and communicate insights*. Linköping University Electronic Press. doi:10.3384/diss.diva-97320

SEGITTUR. (Spanish state company dedicated to the management of innovation and tourism technologies). (2015). *Libro blanco de Destinos turísticos inteligentes: construyendo el futuro*. Retrieved from http://www.segittur.es/opencms/export/sites/segitur/.content/galerias/descargas/proyectos/Libro-Blanco-Destinos-Tursticos-Inteligentes-construyendo-el-futuro.pdf

Sehra, A., Cohen, R., & Vic, A. (2018). On cyrptocurrencies, digital assets and private money. *Journal of Payments Strategy & Systems*, *12*(1), 13–32.

Şeker, Ş. (2013). Sınıflandırma. Retrieved June 2, 2019, from http://bilgisayarkavramlari.sadievrenseker.com/2013/03/31/siniflandirma-classification/

Şeker, A., Diri, B., & Balık, H. H. (2017). Derin Öğrenme Yöntemleri ve Uygulamaları Hakkında Bir İnceleme. [GMBD]. *Gazi Mühendislik Bilimleri Dergisi*, *3*(3), 47–64.

Self service smart solutions. (2018). Otellerin self service kiosk kullanması için 5 neden. Retrieved from https://selfservicesmartsolutions.com/makaleler/blogss/otellerin-self-service-kiosk kullanmasi-icin-5/

Selvam, A., Yap, T. T.-V., Ng, H., Tong, H.-L., & Ho, C.-C. (2016). Augmented Reality for Information Retrieval Aimed at Museum Exhibitions using Smartphones. *Journal of Engineering and Applied Sciences (Asian Research Publishing Network)*, *100*(3), 635–639.

Şener, S. (2019). *Yapay Zeka, Makine Öğrenimi ve Derin Öğrenme Arasındaki Farklar*. Retrieved from https://www.endustri40.com/yapay-zeka-makine-ogrenimi-ve-derin-ogrenme-arasindaki-farklar/

Şengöz, N. (2017). Yapay Sinir Ağları. Retrieved June 3, 2019, from http://www.derinogrenme.com/2017/03/04/yapay-sinir-%0Aaglari/

Serpanos, D., & Wolf, M. (2018). IoT System Architectures. In Internet-of-Things (IoT) Systems (pp. 7–15). doi:10.1007/978-3-319-69715-4_2

Serra, F. F., Ribes, J. F., & Baidal, J. A. (2019). Smart Destinations and Tech-Savvy Millennial Tourists: Hype versus Reality. *Tourism Review*, *74*(1), 63–81. doi:10.1108/TR-02-2018-0018

Serrano-Guerrero, J., Olivas, J. A., Romero, F. P., & Herrera-Viedma, E. (2015). Sentiment analysis: A review and comparative analysis of web services. *Information Sciences*, *311*, 18–38. doi:10.1016/j.ins.2015.03.040

Sertalp, E. (2016). The Usage of augmented reality (ag) in tourism areas. In *Proceedings 21st Internet Conference in Turkey* (pp. 1-7). TED University, Ankara, Turkey.

Sever, N. S., Sever, G. N., & Kuhzady, S. (2015). The evaluation of potentials of gamification in tourism marketing communication. *International Journal of Academic Research in Business and Social Sciences*, *5*(10), 188–202. doi:10.6007/IJARBSS/v5-i10/1867

Seyyar, A. (2008). Yerel siyasetin gelişiminde sosyal politikaların önemi. Retrieved from www.yerelsiyaset.com

Sha, C., Zhang, J., An, L., Zhang, Y., Wang, Z., Ilijaš, T., . . . Ji, Q. (2019). Facilitating HPC Operation and Administration via Cloud. *Supercomputing Frontiers and Innovations*, *6*(1).

Shamim, S., Cang, S., Yu, H., & Li, Y. (2016). Management approaches for Industry 4.0: A human resource management perspective. In *2016 IEEE Congress on Evolutionary Computation (CEC)* (pp. 5309-5316). IEEE. 10.1109/CEC.2016.7748365

Shamim, S., Cang, S., Yu, H., & Li, Y. (2017). Examining the feasibilities of Industry 4.0 for the hospitality sector with the lens of management practice. *Energies*, *10*(4), 499. doi:10.3390/en10040499

Shankar, V. (2018). How Artificial Intelligence (AI) is Reshaping Retailing. *Journal of Retailing*, *94*(4), vi–xi. doi:10.1016/S0022-4359(18)30076-9

Sharda, N. (2010). Tourism Informatics Visual Travel Recommender Systems, Social Communities, and User Interface Design. Hershey, PA: IGI Global. doi:10.4018/978-1-60566-818-5

Shaw, G., & Williams, A. M. (2004). Tourism and tourism spaces. Atlanta, GA: *Sage*.

Sheldon, P. (1997). *Tourism information technologies*. Oxford, UK: CAB.

Sheoran, S. K. (2017). Big data: A big boon for tourism sector. *International Journal of Research in Advanced Engineering and Technology*, *3*(1), 10–13.

SHGM. (2015). Nevşehir Kapadokya Balon Uçuş Sahası Slot Uygulama Talimatı. Retrieved from Nevşehir Kapadokya Balon Uçuş Sahası Slot Uygulama Talimatı website: http://web.shgm.gov.tr/documents/sivilhavacilik/files/mevzuat/sektorel/talimatlar/SHT-BALON-SLOT.pdf

SHGM. (2018). Sivil Havacılık Genel Müdürlüğü, Kapadokya Bölgesi Sıcak Hava Balon İstatistikleri, Resmi Yazısı (No. 29156034-401.04.99/E.136). Ankara, Turkey.

SHGM. (2019). Cappadocia Slot Service Center. Retrieved from http://shm.kapadokya.edu.tr/

Siciliano, B., & Khatib, O. (2019). Humanoid Robots: Historical Perspective, Overview, and Scope. In A. Goswami, & P. Vadakkepat (Eds.), Humanoid Robotics: A Reference. The Netherlands: Springer.

Sigala, M. (2006). Culture: The Software of Customer Relationship Management. *Journal of Marketing Communications*, *12*(3), 203–223. doi:10.1080/13527260600811787

Sigala, M. (2006). E-Customer Relationship Management in the hotel sector: Guests' perceptions of perceived e-service quality levels. *Turizam*, *54*(4), 333–344.

Sigala, M. (2015). Applying gamification and assessing its effectiveness in a tourism context: Behavioral and psychological outcomes of the TripAdvisor's gamification users. *Asia Pacific Journal of Information Systems*, *25*(1), 179–210. doi:10.14329/apjis.2015.25.1.179

Sigala, M., Christou, E., & Gretzel, U. (Eds.). (2012). *Social media in travel, tourism, and hospitality: Theory, practice, and cases*. Ashgate Publishing.

Sigala, M., & Connolly, D. (2004). In search of the next big thing: IT issues and trends facing the hospitality industry: A Review of the Sixth Annual Pan-EuropeanHospitality Technology Exhibition and Conference (EURHOTEC 2001). International Hotel and Restaurant Association, Feb. 19–21, 2001. Palais Des Congress, Paris, France. *Tourism Management*, *25*(6), 807–809. doi:10.1016/j.tourman.2003.08.009

Sigala, M., Lockwood, A., & Jones, P. (2001). Gaining advantage from yield management: Strategic implementation in the rapidly developing world of IT. *International Journal of Contemporary Hospitality Management*, *17*(3), 364–377. doi:10.1108/09596110110403956

Sigala, M., & Marinidis, D. (2012). E-Democracy and Web 2.0: A Framework Enabling DMOS to Engage Stakeholders in Collaborative Destination Management. *Tourism Analysis*, *17*(2), 105–120. doi:10.3727/108354212X13330406124052

Sigalat-Signes, E., Calvo-Palomares, R., Roig-Merino, B., & García-Adán, I. (2019). Transition towards a tourist innovation model: The smart tourism destination. Reality or territorial marketing? *Journal of Innovation & Knowledge.*

Siguaw, J. A., & Enz, C. A. (1999). Best practices in marketing. *The Cornell Hotel and Restaurant Administration Quarterly*, *40*(5), 31–43. doi:10.1177/001088049904000507

Sinha, S. R., & Park, Y. (2017). Building an effective IoT ecosystem for your business. In Building an Effective IoT Ecosystem for Your Business. doi:10.1007/978-3-319-57391-5

Sin, L. Y. M., Tse, A. C. B., & Yim, F. H. K. (2005). CRM: Conceptualization and scale development. *European Journal of Marketing*, *39*(11/12), 1264–1290. doi:10.1108/03090560510623253

Sirse, J., & Mihalic, T. (1999). Slovenian tourism and tourism policy: A case study, 54(3). *Tourism Review*, 34–47.

Smart Hotel Technology Guide. (2018). Using technology to navigate the guest experience journey. https://sha.org.sg/userfiles/ckeditor/Files/Smart%20Hotel%20Technology%20Guide%202018.pdf

Smith, R. (2015). *SMART tourism tools: Linking technology with the touristic resources of city destinations* (Unpublished doctoral dissertation). NHTV Breda University of Applied Sciences, The Netherlands.

Smith, A. D. (1995). Gastronomy or geology? The role of nationalism in the reconstruction of nations. *Nations and Nationalism*, *1*(1), 3–23. doi:10.1111/j.1354-5078.1995.00003.x

Sobocinski, M. (2017). I gamified my courses and I hate that. *World Journal of Science, Technology, and Sustainable Development*, *14*(2/3), 135–142.

Social Tables. (2019). *6 Hotel Brands Leading with the Way with Using Robot Technology*. Retrieved from https://www.socialtables.com/blog/hospitality-technology/hotel-brands-robot/

Söderström, O., Paasche, T., & Klauser, F. (2014). Smart cities as corporate storytelling. *City*, *18*(3), 307–320. doi:10.1080/13604813.2014.906716

Soegjobs. (2016). What are the top trends in hospitality industry in 2016? Retrieved from https://soegjobs.com/2016/05/21/top-trends-hospitality-industry 2016/#globalhportal

Solima, L. (2016). Smart Tourism Destination from IoT Perspective: Adaptive Orientation System. *International New Challenges and Boundaries In Tourism: Policies, Innovations, and Strategies*, (June), 654–666.

Solima, L., & Izzo, F. (2018). QR Codes in cultural heritage tourism: New communications technologies and future prospects in Naples and Warsaw. *Journal of Heritage Tourism*, *13*(2), 115–127. doi:10.1080/1743873X.2017.1337776

Solsona, M. J. (2018). Quality in the management of rural tourist accommodation. *Gestión Turística*, *5*, 65–89. doi:10.4206/gest.tur.2000.n5-0

Sorupia, E. (2005). Rethinking the Role of Transportation in Tourism. *Proceedings of the Eastern Asia Society for Transportation Studies*, 1767–1777. Retrieved from http://citeseerx.ist.psu.edu/viewdoc/download?doi=10.1.1.578.2905&rep=rep1&type=pdf

South Australian Tourism Commission and Local Government Association. (2006). Local government's engagement in tourism. Final Report, Retrieved from http://www.tourism.sa.gov.au/webfiles/tourismpolicy/LGEIT_Final_Report

Soylu, A. (2018). Industry 4.0 and new approaches in entrepreneurship. *Pamukkale University Journal of Institute of Social Sciences*, *32*, 43–57.

Spence, C., & Piqueras-Fiszman, B. (2013). Technology At The Dining Table. *Flavour (London)*, 1–13.

Sprenger, M., & Mettler, T. (2015). Service Robots. *Business & Information Systems Engineering*, *57*(4), 271–277. doi:10.100712599-015-0389-x

Steinicke, F., Bruder, G., Jerald, J., Frenz, H., & Lappe, M. (2010). Visiting Tourist Landmarks in Virtual Reality Systems by Real-Walking. In N. Sharda (Ed.), Tourism Informatics: Visual Travel Recommender Systems, Social Communities, and User Interface Design (pp. 180-193). Hershey, PA: IGI Global. doi:10.4018/978-1-60566-818-5.ch011

Sterling, G. (2008). Pew: Americans Increasingly Shop Online But Still Fear Identity Theft. Retrieved April 12, 2019, from https://searchengineland.com/pew-americans-increasingly-shop-online-but-still-fear-identity-theft-13366

Stipanuk, D. M. (1993). Tourism and technology: Interactions and applications. *Tourism Management*, *14*(4), 267–278. doi:10.1016/0261-5177(93)90061-O

Stockholms Stad. (2018). *Stockholm City Plan.* Retrieved from https://vaxer.stockholm/globalassets/tema/oversiktplanny_light/english_stockholm_city_plan.pdf

Stockholms Stad. (2019). *Stockholm, a sustainably growing city.* Retrieved from https://international.stockholm.se/globalassets/ovriga-bilder-och-filer/stockholm-a-sustainably-growing-city.pdf

Stone, R. (1995). The Reality of Virtual Reality. *World Class Design to Manufacture*, *2*(4), 11–17. doi:10.1108/09642369310091106

Strömgren, M., & Andersson, R. (2010). The usage of safety management tools in Swedish municipalities. *Safety Science*, *48*(3), 288–295. doi:10.1016/j.ssci.2009.11.009

Strommer, E., Jurvansuu, M., Tuikka, T., Ylisaukko-Oja, A., Rapakko, H., & Vesterinen, J. (2012, March). NFC-enabled wireless charging. In *Proceedings 4th International Workshop on Near Field Communication, IEEE.* Helsinki, Finland. 10.1109/NFC.2012.17

Styliani, S., Liarokapis, F., Kostas, L., & Petros, P. (2009). Virtual Museums, a Survey and Some Issues for Consideration. *Journal of Cultural Heritage*, *10*(4), 520–528. doi:10.1016/j.culher.2009.03.003

Suddrey, G., Jacobson, A., & Ward, B. (2018). *Enabling a Pepper Robot to provide Automated and Interactive Tours of a Robotics Laboratory.* Retrieved from https://arxiv.org/pdf/1804.03288.pdf

Su, K., Li, J., & Fu, H. (2011). Smart city and the applications. In *Proceedings of International Conference on Electronics, Communications and Control (ICECC)* (pp. 1028-1031). Ningbo, China: IEE.

Sultan, A. J. (2018). Orchestrating service brand touchpoints and the effects on relational outcomes. *Journal of Services Marketing*, *32*(6), 777–788. doi:10.1108/JSM-12-2016-0413

Sun Tung, V. W., & Au, N. (2018). Exploring Customer Experiences with Robotics in Hospitality. *International Journal of Contemporary Hospitality Management*, *30*(7), 2680–2697. doi:10.1108/IJCHM-06-2017-0322

Sun Tung, V. W., & Law, R. (2017). The Potential for Tourism and Hospitality Experience Research in Human-Robot Interactions. *International Journal of Contemporary Hospitality Management*, *29*(10), 2498–2513. doi:10.1108/IJCHM-09-2016-0520

Sunlu, U. (2003). Environmental impacts of tourism. In *Proceedings Conference on the Relationships between Global Trades and Local Resources in the Mediterranean Region*, (pp. 263-270). Academic Press.

Superdata Games & Interactive Media Intelligence. (2017). *2017 year in review, digital games, and interactive media.* Retrieved from https://www.superdataresearch.com/market-data/market-brief-year-in-review/

Sürücü, O., & Başar, M. E. (2016). Kültürel Mirası Korumada Bir Farkındalık Aracı Olarak Sanal Gerçeklik. *Artium*, *4*(1), 13–26.

Swan, M. (2016, October). *Decentralized finance: blockchains, prediction, and valuation.* Paper presented at the Economist and Finance Disrupted, New York, NY.

Swan, M. (2015). *Blockchain: blueprint for a new economy.* Sebastopol, CA: O'Reilly Media.

T. C. Ministry of Foreign Affairs. (2019) Retrieved from http://www.mfa.gov.tr/dunya-surdurulebilir-kalkinma-zirvesi_johannesburg_-26-agustos---4-eylul-2002_.tr.mfa. Date of access: 23/02/2019.

Tapscott, D., & Tapscott, A. (2017). *Realizing the potential of blockchain a multistakeholder approach to the stewardship of blockchain and cryptocurrencies.* Geneva, Switzerland: World Economic Forum.

Tasca, P., & Tessone, C. J. (2017). Taxonomy of blockchain technologies. principles of ıdentification and classification. *Ledger, 4*, 1–39.

Tavakoli, R., & Mura, P. (2015). Journeys in Second Life' – Iranian Muslim women's behaviour in virtual tourist destinations. *Tourism Management, 46*, 398–407. doi:10.1016/j.tourman.2014.07.015

Taylor, B. (2013). Augmented reality applications – from the consumer to the business. Retrieved from http://www.creativeguerrillamarketing. com/augmented-reality/augmented-reality-applications-from the -consumer-to-the-business/

TDK. (2018). Retrieved from www.tdk.gov.tr/index.php?option=com_gts&arama=gts&guid=TDK.GTS.5bdc1f11cbf6d6.64662412, Date of access 02.11.2018.

Tektaş, M., Akbaş, A., & Topuz, V. (2002). *Yapay zeka tekniklerinin trafik kontrolünde kullanilmasi üzerine bir inceleme.* Retrieved from http://www.trafik.gov.tr/icerik/bildiriler/ pdf/C4-7.pdf

Terry, L. (2016). 6 Mega-Trends in hotel technology. Retrieved from http://hospitalitytechnology.edgl.com/news/6-Mega-Trends-in-Hotel Technology105033

The Museum of Modern Art MOMA. Retrieved May 9, 2019 from https://www.moma.org/support/membership/

Thelwall, M. (2001). A web crawler design for data mining. *Journal of Information Science, 27*(5), 319–325. doi:10.1177/016555150102700503

Thet, T. T., Na, J. C., & Khoo, C. S. G. (2010). Aspect-based sentiment analysis of movie reviews on discussion boards. *Journal of Information Science*. doi:10.1177/0165551510388123

ThingSpeak. (2019). ThingSpeak Documentation. Retrieved from https://www.mathworks.com/help/thingspeak/index.html?s_tid=CRUX_lftnav

Thompson, S. (2019, March 12). *VR Applications: 21 Industries Already Using Virtual Reality.* Retrieved from https://virtualspeech.com/blog/vr-applications

Thrasher, J. (2013). RFID vs. NFC: What's the Difference? Retrieved from https://blog.atlasrfidstore.com/rfid-vs-nfc

Tian, F. (2016). An agri-food supply chain traceability system for China based on RFID blockchain technology. In *Proceedings of 13th International Conference on Service Systems and Service Management* (pp. 1-6), Kunming, China. Academic Press.

Tiffany, K. (2019). McDonald's new drive-thru menus will change based on the weather, traffic, and time of day. Retrieved from www.vox.com/the-goods/2019/3/27/18284045/mcdonalds-tech-startup-artificial-intelligence-menu-drive-thru

Timothy, D. J. (2005). *Shopping tourism, retailing and leisure.* Clevedon, UK: Channel View Publications. doi:10.21832/9781873150610

Timothy, D. J., & Butler, R. W. (1995). Cross-border shopping: A North American perspective. *Annals of Tourism Research, 22*(1), 16–34. doi:10.1016/0160-7383(94)00052-T

Ting, D. (2017). 10 hotel trends that will shape guest experience in 2017. Retrieved from https://skift.com/2017/01/03/10-hotel-trends-that-will-shape-guest-experience-in-2017/

Tirunillai, S., & Tellis, G. J. (2014). Mining marketing meaning from online chatter: Strategic brand analysis of big data using latent dirichlet allocation. *JMR, Journal of Marketing Research, 51*(4), 463–479. doi:10.1509/jmr.12.0106

Tkaczyk, C. (2017). Meet Rose, the sassy new bot concierge in Las Vegas. Retrieved from https://www.travelandleisure.com/hotels-resorts/cosmopolitan-las-vegas

Tokcan, H. (2015). *Bilginin üretimi ve kullanımı açısından bilgi yönetimi: üniversitede akademik yöneticilerin bilgi yönetimi algıları üzerine bir uygulama, Doktora Tezi.* T.C. Kırıkkale Üniversitesi Sosyal Bilimler Enstitüsü İşletme Anabilim Dalı, Kırıkkale.

Toksöz, D., & Aras, S. (2016). Turistlerin seyahat motivasyonlarında yöresel mutfağın rolü. *Journal of Tourism and Gastronomy Studies, 4*(1), 174–189.

tom Dieck, M. C., & Jung, T. H. (2017). Value of Augmented Reality at Cultural Heritage sites: A Stakeholder Approach. *Journal of Destination Marketing & Management, 6*(2), 110–117. doi:10.1016/j.jdmm.2017.03.002

Tongco, M. D. C. (2007). Purposive sampling as a tool for informant selection. *Ethnobotany Research and Applications, 5*, 147–158. doi:10.17348/era.5.0.147-158

Topsakal, Y. (2018). Akıllı Turizm Kapsamında Engelli Dostu Mobil Hizmetler: Türkiye 4.0 İçin Öneriler. *Journal of Tourism Intelligence and Smartness, 1*(1), 1–13.

Topsakal, Y., Yüzbaşıoğlu, N., Çelik, P., & Bahar, M. (2018). Turizm 4.0-Turist 5.0: İnsan Devriminin Neden Endüstri Devrimlerinden Bir Numara Önde Olduğuna İlişkin Bakış. *Journal of Tourism Intelligence and Smartness, 1*(2), 1–11.

Torabi, F., & El-Den, J. (2017, November). The impact of knowledge management on organizational productivity: a case study on koosar bank of ıran. Paper presented at the meeting of 4th Information Systems International Conference 2017 (ISICO 2017), Bali, Indonesia, *Procedia Computer Science, 124,* 300-310.

Torlak, Ö. (1999). Belediye hizmetlerinde pazarlama. *Çağdaş Yerel Yönetimler Dergisi, Ocak (1),* 96-114.

Tortop, N. (1988). Turizm ve belediyeler. *Amme İdaresi Dergisi, Haziran (2),* 3-10.

Tortop, N. (2008). Yerel yönetimlerin gelişmesinde turizmin önemi. *İller ve Belediyeler Dergisi, (721),* 17-18.

Tosun, C., & Fyall, A. (2005). Making tourism sustainable: prospects and pitfalls. *Environmentalism in Turkey: Between democracy and development*, 249-262.

Tosun, C. (2001). Challenges of sustainable tourism development in the developing world: The case of Turkey. *Tourism Management, 22*(3), 289–303. doi:10.1016/S0261-5177(00)00060-1

Tosun, C., Temizkan, S. P., Timothy, D. J., & Fyall, A. (2007). Tourist shopping experiences and satisfaction. *International Journal of Tourism Research, 9*(2), 87–102. doi:10.1002/jtr.595

Total Slovenia News. (2019, July 31). Retrieved from https://www.total-slovenia-news.com/travel/4215-strong-first-half-for-slovenian-tourism-in-2019

Tovar, C., & Lockwood, M. (2008). Social impacts of tourism: An Australian regional case study. *International Journal of Tourism Research, 10*(4), 365–378. doi:10.1002/jtr.667

Townsend, A. (2013). *Smart Cities: Big Data, Civic Hackers, and the Quest for a New Utopia*. New York: Norton W. & Co.

Travelzoo. (2016). Travellers Expect Robots on Their Holidays by 2020. Retrieved from https://press.travelzoo.com/robophiles--robophobes--britons-divided-over-use-of-robots-in-travel/

Trejos, N. (2016). *Introducing Connie, Hilton's New Robot Concierge*. Retrieved from https://www.usatoday.com/story/travel/roadwarriorvoices/2016/03/09/introducing-connie-hiltons-new-robot-concierge/81525924/

Tribe, J., & Liburd, J. J. (2016). The tourism knowledge system. *Annals of Tourism Research*, *57*, 44–61. doi:10.1016/j.annals.2015.11.011

Tromp, P. (2017). *How virtual reality will revolutionize the hospitality industry*. Retrieved from https://www.hospitalitynet.org/opinion/4080737.html

Tsytsarau, M., & Palpanas, T. (2012). Survey on mining subjective data on the web. *Data Mining and Knowledge Discovery*, *24*(3), 478–514. doi:10.100710618-011-0238-6

Tu, Q., & Liu, A. (2014). Framework of smart tourism research and related progress in China. In *Proceedings International Conference on Management and Engineering (CME 2014)*, pp. 140-146, Pennsylvania: DEStech Publications.

TÜBİTAK. (2003). *Vizyon 2023 Ulaştırma ve Turizm Paneli Raporu*. Retrieved from http://www.tubitak.gov.tr/tubitak_content_files/vizyon2023/ut/utp_son_surum.pdf

TÜBİTAK. (2016). *Yeni Sanayi Devrimi Akıllı Üretim Sistemleri Teknoloji Yol Haritası*. Retrieved from http://www.tubitak.gov.tr/sites/default/files/akilli_uretim_sistemleri_tyh_v27aralik2016.pdf

Tunga, Y., & İnceoğlu, M. M. (2016). Gamification design. In *Proceedings of the 3rd International Conference on New Trends in Education. April 26-29, 2016.*

Turizmgüncel.com. (2018). Akıllı ev konseptindeki ilk otelini İstanbulda açtı. Retrieved from https://www.turizmguncel.com/haber/akilli-ev-konseptindeki-ilk-otelini-istanbul'da-acti-15-otel-daha-geliyor-h27337.html

Turizmnews.com. (2018). Akıllı otel uygulaması Türkiye'de ilk kez Isparta'da. Retrieved from https://www.turizmnews.com/akilli-otel-uygulamasi-turkiye-de-ilk-kez-isparta-da/12773/

Turkish Airlines. Retrieved May 2, 2019 from https://www.turkishairlines.com/en-tr/any-questions/

Türkiye Bilişim Vakfı. (2017). 2016 Çalışma Raporu. Retrieved from http://www.tbv.org.tr/core/uploads/page/document/1123_24031710643.pdf, Date of Access 04/11/2018.

Türkiye, C. (2017). *Yapay Zeka ve Pazarlama – II*. Retrieved from https://www. campaigntr.com/yapay-zeka-ve-pazarlama-ii/

Turner, L. W., & Reisinger, Y. (2001). Shopping satisfaction for domestic tourists. *Journal of Retailing and Consumer Services*, *8*(1), 15–27. doi:10.1016/S0969-6989(00)00005-9

Tussyadiah, I. P., & Park, S. (2018). Consumer Evaluation of Hotel Service Robots. In B. Stangl, & J. Pesonen (Eds.), *Information and Communication Technologies in Tourism 2018*. Cham, Switzerland: Springer. doi:10.1007/978-3-319-72923-7_24

Tussyadiah, I. P., Wang, D., Jung, T. H., & tom Dieck, M. C. (2018). Virtual reality, presence, and attitude change: Empirical evidence from tourism. *Tourism Management*, *66*, 140–154. doi:10.1016/j.tourman.2017.12.003

Uçar, A., Şemşit, S., & Negiz, N. (2017). Avrupa Birliği Akıllı Kent Uygulamaları Ve Türkiye'deki Yansımaları. Suleyman Demirel University Journal of Faculty of Economics & Administrative Sciences, 22.

Ultramagic. (2008). Technical support - Ultramagic. Retrieved from https://ultramagic.com/technical-support/

Ünal, T. (1992). Yerel yönetimler ve turizm. *Anatolia Turizm, Çevre ve Kültür Dergisi*, (25-26), 5-7.

UNEP. (2003). Tourism and Local Agenda 21: The Role of Local Authorities in Sustainable Tourism. First edition, United Nations publication.

UNESCO. (2019). UNESCO World Heritage Centre - World Heritage List. Retrieved from http://whc.unesco.org/en/list/

United Nations Population Fund. (2007). *State of world population 2007: Unleashing the potential of urban growth.* Retrieved from www.unfpa.org/swp/

UNWTO. (2017). Innovation, technology and sustainability - pillars of smart destinations. Erişim Tarihi: 05.11.2018. Retrieved from http://media.unwto.org/press-release/2017-02-21/innovation-technology-and-sustainability-pillars-smart-destinations

Upadhay, A. K., & Khandelwal, K. (2016). Virtual Reality: Adding Immersive Dimension to Sales Training. *Human Resource Management International Digest*, *26*(4), 42–45. doi:10.1108/HRMID-01-2018-0014

Upton, L. (2014). The top 10 innovations that are shaping the future of rail and metro. Retrieved from https://www.smartrailworld.com/it-and-wifi/the-top-10-innovations-that-are-shaping-the-future-of-rail-and-metro

Ürey, A., & Akşit, K. (2018). Sanal Gerçeklik ve Artırılmış Gerçeklik Nedir? Retrieved May 20, 2019, from https://sarkac.org/2018/03/sanal-gerceklik-artirilmis-gerceklik-nedir/

Usta, A., & Doğantekin, S. (2017). *Blockchain 101*. İstanbul, Turkey: Kapital Medya Hizmetleri.

Usta, Ö. (2016). *Turizm Genel ve Yapısal Yaklaşım*. Ankara, Turkey: Detay Yayıncılık.

Uygunoğlu, T., & Yurtcu, Ş. (2006). Yapay Zeka Tekniklerinin İnşaat Mühendisliği Problemlerinde Kullanımı. *Yapı Teknolojileri Elektronik Dergisi*, *2*(1), 61–70.

Valentino, K., Christian, K., & Joelianto, E. (2017). Virtual Reality Flight Simulator. *Internetwoking Indonesia Journal*, *9*(1), 21–25.

Vanolo, A. (2014). Smartmentality: The Smart City as Disciplinary Strategy. *Urban Studies (Edinburgh, Scotland)*, *51*(5), 883–898. doi:10.1177/0042098013494427

Vardarlıer, P., & İnan, K. (2017). Gamification model proposal for the improvement of sales personnel performance. *Journal of Behavior at Work*, *2*(1), 8–19.

Velte, A. T., Velte, T. J., Elsenpeter, R. C., & Elsenpeter, R. C. (2010). *Cloud computing: A practical approach* (p. 44). New York: McGraw-Hill.

Ventrella, J. (2000). Avatar Physics and Genetics. In J. C. Heudin (Ed.), *VW virtual Worlds Second International Conference* (pp. 107-118). Paris, France: Springer. 10.1007/3-540-45016-5_11

Vera, J. F., & Baños, C. J. (2010). Renewal and restructuring of the consolidated tourist destinations of the coast: Recreational practices in the evolution of the tourist area. *Boletín de la Asociación de Geógrafos Españoles*, *53*, 329–353.

Verevka, T. (2018). Key performance indicators of high-tech enterprises. IV International Scientific Conference "The Convergence of Digital and Physical Worlds: Technological, Economic, and Social Challenges" (CC-TESC2018). In SHS Web of Conferences (Vol. 44, p. 00077). EDP Sciences. 10.1051hsconf/20184400077

Verevka, T. V. (2019). Development of Industry 4.0 in the Hotel and Restaurant Business, *IBIMA. Business Review (Federal Reserve Bank of Philadelphia)*. doi:10.5171/2019.324071

Verhoef, P. C., Kannan, P. K., & Inman, J. J. (2015). From multi-channel retailing to omni-channel retailing: Introduction to the special issue on multi-channel retailing. *Journal of Retailing*, *91*(2), 174–181. doi:10.1016/j.jretai.2015.02.005

Verhoef, P. C., Lemon, K. N., Parasuraman, A., Roggeveen, A., Tsiros, M., & Schlesinger, L. A. (2009). Customer experience creation: Determinants, dynamics, and management strategies. *Journal of Retailing*, *85*(1), 31–41. doi:10.1016/j.jretai.2008.11.001

Vitezić, V., Car, T., & Šimunić, M. (2015, April). *Managing Innovative Technology in the Hotel Industry--Response to Growing Consumer Preferences.* Paper presented at 3rd International Scientific Conference Tourism in Southern and Eastern Europe. Opatija, Croatia.

Vogel-Heuser, B., & Hess, D. (2016). Guest editorial industry 4.0–prerequisites and visions. *IEEE Transactions on Automation Science and Engineering*, *13*(2), 411–413. doi:10.1109/TASE.2016.2523639

Vogel, J., Bowers, C., Meehan, C., Hoeft, R., & Bradley, K. (2004). Virtual Reality for Life Skills Education: Program Evaluation. *Deafness & Education International*, *6*(1), 39–50. doi:10.1179/146431504790560636

Voorhees, C. M., Fombelle, P. W., Gregoire, Y., Bone, S., Gustafsson, A., Sousa, R., & Walkowiak, T. (2017). Service encounters, experiences, and the customer journey: Defining the field and a call to expand our lens. *Journal of Business Research*, *79*, 269–280. doi:10.1016/j.jbusres.2017.04.014

Vural, Z. B. A., & Coşkun, G. (2011). Kurumsal sosyal sorumluluk ve etik. *Gümüşhane Üniversitesi İletişim Fakültesi Elektronik Dergisi, (1),* 61-87.

Wagar, J. A. (1964). The carrying capacity of wild lands for recreation. *Forest Science, 10*(suppl_2), a0001-24.

Wakefield, K. L., & Baker, J. (1998). Excitement at the mall: Determinants and effects on shopping response. *Journal of Retailing*, *74*(4), 515–539. doi:10.1016/S0022-4359(99)80106-7

Waltz, S. P., & Ballagas, R. (2007). *Pervasive persuasive: A rhetorical design approach to a location-based spell-casting game for tourists in proceedings of situated play.* DiGRA 2007 - The 3rd International Digital Games Research Conference. Sept. 24-28, 2007, Tokyo.

Wang, C., Li, X., Liu, Y., & Wang, H. (2014). The research on development direction and points in IoT in China power grid. *Proceedings - 2014 International Conference on Information Science, Electronics, and Electrical Engineering, ISEEE 2014.* 10.1109/InfoSEEE.2014.6948106

Wang, C.-H. (2004). Predicting tourism demand using fuzzy time series and hybrid grey theory. *Tourism Management*, *25*(3), 367–374.

Wang, D., Li, X., & Li, Y. (2013). China's "smart tourism destination" initiative: A taste of the service-dominant logic. *Journal of Destination Marketing & Management*, *2*(2), 59–61. doi:10.1016/j.jdmm.2013.05.004

Wang, D., Park, S., & Fesenmaier, D. (2012). The role of smartphones in mediating the tourism experience. *Journal of Travel Research*, *51*(4), 371–387. doi:10.1177/0047287511426341

Wang, L., Von Laszewski, G., Younge, A., He, X., Kunze, M., Tao, J., & Fu, C. (2010). Cloud computing: A perspective study. *New Generation Computing*, *28*(2), 137–146. doi:10.100700354-008-0081-5

Wang, X., Kim, M., Love, P., & Kang, S.-C. (2013). Augmented Reality in built environment: Classification and implications for future research. *Automation in Construction*, *32*, 1–13. doi:10.1016/j.autcon.2012.11.021

Ward, H. (2001). CRM initiatives fail due to lack of integration. *ComputerWeekly.com.* Retrieved May 1, 2019 from https://www.computerweekly.com/news/2240042772/CRM-initiatives-fail-due-to-lack-of-integration

Washburn, D., Sindhu, U., Balaouras, S., Dines, R. A., Hayes, N., & Nelson, L. E. (2010). Helping CIOs understand "smart city" initiatives. Cambridge, MA: Forrester Research Inc. Report.

Wasko, C. (2013). What teachers need to know about augmented reality enhanced learning. *TechTrends*, *57*(1), 17–21. doi:10.100711528-013-0672-y

Watanabe, H., Fujimura, S., Nakadaira, A., Miyazaki, Y., Akutsu, A., & Kishigami, J. J. (2015). Blockchain contract: A complete consensus using blockchain. In *Proceedings of IEEE 4th Global Conference on Consumer Electronics* (pp. 577-578). Las Vegas, NV. 10.1109/GCCE.2015.7398721

Watson, C., McCarthy, J., & Rowley, J. (2013). Consumer attitudes towards mobile marketing in the smart phone era. *International Journal of Information Management*, *33*(5), 840–849. doi:10.1016/j.ijinfomgt.2013.06.004

Weber, J. (2014). Information and communication technologies in tourism 2014. In R. Baggio, M. Sigala, A. Inversini, & J. Pseonen (Eds.), *ENTER. Dublin, Ireland*. London, UK: Springer.

Węcławowicz, G., Bański, J., Degórski, M., Komornicki, T., Korcelli, P., & Śleszyński, P. (2006). *Przestrzenne Zagospodarowanie Polski na początku XXI* wieku. PAN IGiPZ.

Wei, W. (2019). Research progress on virtual reality (VR) and augmented reality (AR) in tourism and hospitality: A critical review of publications from 2000 to 2018. *Journal of Hospitality and Tourism Technology*.

Weinelt, B., & Moavenzadeh, J. (2017). Digital Transformation Initiative: Aviation, Travel and Tourism Industry. In Geneva, Switzerland: World Economic Forum.

Weismayer, C., Pezenka, I., & Gan, C. H.-K. (2018). Aspect-Based Sentiment Detection: Comparing Human Versus Automated Classifications of TripAdvisor Reviews. In Information and Communication Technologies in Tourism 2018 (pp. 365–380). doi:10.1007/978-3-319-72923-7_28

Wells, J. D., William, L. F., & Choobineh, J. (1999). Managing information technology for one-to-onecustomer interaction. *Information & Management*, *35*(1), 53–62. doi:10.1016/S0378-7206(98)00076-7

Wendt, J. A. (2001). The administrative division of Poland and the prospects of the transborder cooperation. In M. Koter, & K. Heffner (Eds.), Changing Role of Border Areas and Regional Policies, Region and Regionalism No. 5, Wydawnictwo Uniwersytetu Łódzkiego,39–48.

Werbach, K. (2015). *Gamification online via Wharton University of Pennsylvania*. Retrieved from https://www.coursera.org/learn/gamification

Werbach, K., & Hunter, D. (2012). *For the win how game thinking can revolutionize your business*. Pennsylvania: Wharton Digital Press.

Werbach, K., & Hunter, D. (2015). *The gamification toolkit dynamics, mechanics and components for the win*. Pennsylvania: Wharton Digital Press.

Werther, H., Koo, C., & Gretzel, U. (2015). Special issue on smart tourism systems: Convergence of information technologies, business models, and experiences. *Computers in Human Behavior*, *50*, 556–557. doi:10.1016/j.chb.2015.03.042

Werthner, H., & Klein, S. (1999). *Information technology and tourism: a challenging relationship*. Vienna, Austria: Springer. doi:10.1007/978-3-7091-6363-4

Werthner, H., & Ricci, F. (2004). E-Commerce and Tourism. *Communications of the ACM*, *47*(12), 101–105. doi:10.1145/1035134.1035141

What are the operating modes of NFC devices? (n.d.). Retrieved from https://nfc-forum.org/resources/what-are-the-operating-modes-of-nfc-devices/

What is a QR Code? QRcode.com DENSO WAVE. (n.d.). Retrieved from https://www.qrcode.com/en/about/

Whitley, D. (2018). *Chicago's Hotel EMC2 employs robot butlers to deliver room service to guests.* Retrieved from https://www.thesun.co.uk/uncategorized/7946799/chicagos-hotel-emc2-employs-robot-butlers-to-deliver-room-service-to-guests/

Whyte, J. (2002). *Virtual Reality and the Built Environment*. Architectural Press.

Wiebe, J., & Riloff, E. (2011). *Finding mutual benefit between subjectivity analysis and information extraction.* IEEE Transactions on Affective Computing; doi:10.1109/T-AFFC.2011.19

Will, D. (2009). Information Technology's Strategic Impact on the Air Travel Service Industry. *Journal of Information & Management for American Airlines*, *15*(2), 24–56.

Williams, A. P., & Hobson, J. S. P. (1995). Virtual reality and tourism: Fact or fantasy? *Tourism Management*, *16*(6), 423–427. doi:10.1016/0261-5177(95)00050-X

Wilson, S., Fesenmaier, D. R., Fesenmaier, J., & van Es, J. C. (2001). Factors for success in rural tourism development. *Journal of Travel Research*, *40*(2), 132–138. doi:10.1177/004728750104000203

Wilson, T., Wiebe, J., & Hoffmann, P. (2005). Recognizing contextual polarity in phrase-level sentiment analysis. *Proceedings of the Conference on Human Language Technology and Empirical Methods in Natural Language Processing - HLT '05*, 347–354. 10.3115/1220575.1220619

Windsor, M. (2018). *What is the future of service robots*? Retrieved from https://www.eenewspower.com/news/what-future-service-robots/page/0/1

Wind, Y. J., & Hays, C. F. (2016). *Beyond advertising: creating value through all customer touchpoints*. John Wiley & Sons.

Wirtz, J., Patterson, P. G., Kunz, W. H., Gruber, T., Lu, W. N., Paluch, S., & Martins, A. (2018). Brave New World: Service Robots in the Frontline. *Journal of Service Management*, *29*(5), 907–931. doi:10.1108/JOSM-04-2018-0119

World Tourism Organization (WTO). (1998). *Guidelines for the Transfer of New Technologies in the Field of Tourism.* Madrid, Spain.

World Tourism Organization and United Nations Development Programme. (2018). *Tourism and the Sustainable Development Goals – Journey to 2030*. Madrid, Spain: UNWTO.

WTTC. (2018). Retrieved from World Travel & Tourism Council: https://www.wttc.org/-/media/files/reports/economic-impact-research/regions-2019/world2019.pdf

Wu, M. (2017). *The magic potion of game dynamics*. Retrieved from https://community.khoros.com/t5/Science-of-Social-Blog/The-Magic-Potion-of-Game-Dynamics/ba-p/19260

Wu, B., & Wang, A. I. (2011). A pervasive game to know your city better. In *Proceedings of the 2011 IEEE International Games Innovation Conference*. Washington, DC: IEEE Computer Society. 10.1109/IGIC.2011.6115111

www.chinadaily.com.cn. (2017). No Title. Retrieved from http://www.chinadaily.com.cn/china/2017-07/11/content_30068314.htm

www.mercedes-benz.com. (n.d.). No Title. Retrieved from https://www.mercedes-benz.com/content/dam/brandhub/mercedes-benz/classic/generalimages/MBMUS_Information_2018_ENG_Update_ICv2_NETTO_x3.pdf

www.mots.go.th. (2019). No Title. Retrieved from https://www.mots.go.th/mots_en57/ewt_news.php?nid=3362&filename=index

www.radissonbluedwardian.com. (n.d.). No Title. Retrieved from https://www.radissonbluedwardian.com/feature.do?feature=press_qrcodes2011

www.scandit.com. (2019). No Title. Retrieved from https://www.scandit.com/the-qr-code-hotel-room/

Xianghua, F., Guo, L., Yanyan, G., & Zhiqiang, W. (2013). Multi-aspect sentiment analysis for Chinese online social reviews based on topic modeling and HowNet lexicon. *Knowledge-Based Systems*, *37*, 186–195. doi:10.1016/j.knosys.2012.08.003

Xiang, Z., Du, Q., Ma, Y., & Fan, W. (2017). A comparative analysis of major online review platforms: Implications for social media analytics in hospitality and tourism. *Tourism Management*, *58*, 51–65. doi:10.1016/j.tourman.2016.10.001

Xiang, Z., Magnini, V., & Fesenmaier, D. (2015). Information Technology and Consumer Behaviour in Travel and Tourism: Insights From Travel Planning Using the Internet. *Journal of Retailing and Consumer Services*, *22*, 244–249. doi:10.1016/j.jretconser.2014.08.005

Xiang, Z., Schwartz, Z., Gerdes, J. H. Jr, & Uysal, M. (2015). What can big data and text analytics tell us about hotel guest experience and satisfaction? *International Journal of Hospitality Management*, *44*, 120–130. doi:10.1016/j.ijhm.2014.10.013

Xian, Z., Wang, D., O'Leary, J. T., & Fesenmaier, D. R. (2015). Adapting to the internet: Trends in traveller's use of the web for trip planning. *Journal of Travel Research*, *54*(4), 511–527. doi:10.1177/0047287514522883

Xu, F., Buhalis, D., & Weber, J. (2017). Serious games and the gamification of tourism. *Tourism Management*, *60*, 244–256. doi:10.1016/j.tourman.2016.11.020

Xu, F., Nash, N., & Whitmarsh, L. Big data or small data? A methodological review of sustainable tourism. *Journal of Sustainable Tourism*. doi:10.1080/09669582.2019.1631318

Xu, F., Tian, F., Buhalis, D., Weber, J., & Zhang, H. (2016). Tourists as mobile gamers: Gamification for tourism marketing. *Journal of Travel & Tourism Marketing*, *33*(8), 1124–1142. doi:10.1080/10548408.2015.1093999

Xu, F., Weber, J., & Buhalis, D. (2014). Gamification in tourism. *Information and Communication Technologies in Tourism*, *2014*, 525–537.

Xu, J., Hou, Q., Niu, C., Wang, Y., & Xie, Y. (2018). Process optimization of the University-Industry-Research collaborative innovation from the perspective of knowledge management. *Cognitive Systems Research*, *52*, 995–1003. doi:10.1016/j.cogsys.2018.09.020

Xu, X., Pautasso, C., Zhu, L., Gramoli, V., Ponomarev, A., Tran, A. B., & Chen, S. (2016). The blockchain as a software connector. In *Proceedings of 13th Working IEEE/IFIP Conference on Software Architecture* (pp. 182-191). Venice, Italy. IEEE.

Xu, X., Weber, I., Staples, M., Zhu, L., Bosch, J., Bass, L., ... Rimba, P. (2017). A taxonomy of blockchain-based systems for architecture design. In *Proceedings of IEEE International Conference on Software Architecture* (pp. 243-252). Gothenburg, Sweden. 10.1109/ICSA.2017.33

Xu, Z., Frankwick, G., & Ramirez, E. (2016). Effects of big data analytics and traditional marketing analytics on new product success: A knowledge fusion perspective. *Journal of Business Research*, *69*(5), 1562–1566. doi:10.1016/j.jbusres.2015.10.017

Yalçın Kayıkçı, M., & Kutluk Bozkurt, A. (2018). Dijital Çağda Z & Alpha Kuşağı, Yapay Zeka Uygulamaları & Turizme Yansımaları. *Tekirdağ Namık Kemal Üniversitesi Sosyal Bilimler Enstitüsü Dergisi: Sosyal Bilimler Metinleri, 3*(1), 54–64.

Yalçınkaya, P., Atay, L., & Karakaş, E. (2018). Smart tourism applications. Gastroia: *Journal of Gastronomy and Travel Research,* Cilt 2. *Sayı, 2*, 34–52.

Yalçınkaya, P., Lütfi, A. T. A. Y., & Karakaş, E. Akıllı Turizm Uygulamaları. (2018). Gastroia. *Journal Of Gastronomy And Travel Research, 2*(2), 34–52.

Yang, E. C. L., Khoo-Lattimore, C., & Arcodia, C. (2017). A systematic literature review of risk and gender research in tourism. *Tourism Management, 58*, 89–100. doi:10.1016/j.tourman.2016.10.011

Yang, J., & Wan, C.-S. (2004). Advancing organizational effectiveness and knowledge management implementation. *Tourism Management, 25*(5), 593–601. doi:10.1016/j.tourman.2003.08.002

Yavuz, M. C. (2019). Smart destination: Tourism, ınnovation, entrepreneurship, challenges. *Journal of Tourism Theory and Research, 5*(2) Futourism, 203-211.

Yazici, M., Kamga, C., & Singhal, A. (2013). A big data driven model for taxi drivers' airport pick-up decisions in New York City. In *IEEE International Conference on Big Data*, pp. 37–44. IEEE. 10.1109/BigData.2013.6691775

Yell Robot. (2019). Robot Wedding Photographer can Recognize Guests. Retrieved from https://yellrobot.com/robot-wedding-photographer-eva/. Accessed on June 19, 2019.

Yen, B. T. H., Mulley, C., & Burke, M. (2019). Gamification in transport interventions: Another way to improve travel behavioral change. *Cities (London, England), 85*, 140–149. doi:10.1016/j.cities.2018.09.002

Yeter, E. (1993). Yerel yönetimler açısından turizmde denetim boyutu. *Anatolia Turizm, Çevre ve Kültür Dergisi, Mart-Nisan (2),* 25-27.

Yigitcanlar, T., Kamruzzaman, M., Buys, L., Ioppolo, G., Sabatini-Marques, J., da Costa, E. M., & Yun, J. J. (2018). Understanding 'smart cities': Intertwining development drivers with desired outcomes in a multidimensional framework. *Cities (London, England), 81*, 145–160. doi:10.1016/j.cities.2018.04.003

Yılmaz, D., & Yılmaz, S. (2005). Küreselleşme ve Bilgi Teknolojilerinin Turizm Endüstrisine, Etkileri. Retrieved from www.bilgiyönetimi.com

Yılmaz, E. (2016). Yeni bir hikâye konya akıllı şehir olabilir mi? Konya Ticaret Odası Araştırma Raporu. Retrieved from http://www.kto.org.tr/d/file/yeni-bir-hikaye-konya-akilli-sehir-olabilir-mi---emre-yilmaz.pdf

Yılmaz, E. (2016). *Yeni Bir Hikaye Konya Akıllı Şehir Olabilir mi?* Retrieved from http://www.kto.org.tr/yeni-bir-hikaye-konya-akilli-sehir-olabilir-mi

Yılmaz, Ö. (2015). Akıllı kentler ve bilgi toplumu stratejisi, T. C. Kalkınma Bakanlığı, Information Society Strategy and Action Plan (2015-2018) and Smart Cities. Retrieved from https://www.csb.gov.tr/db/cbsgunu/webmenu/webmenu15780.pdf, Erişim Tarihi: 17.11.2018.

Yılmaz, E. A. (2017). *Gamification*. Istanbul, Turkey: Abaküs Publishing.

Yılmaz, İ. A. (2009). *Ağırlama işletmeleri yöneticilerinin bilgiye ve bilgi yönetimine yönelik yaklaşımları: Antalya'daki beş yıldızlı otel işletmeleri üzerinde bir uygulama.* Mersin: Mersin Üniversitesi Sosyal Bilimler Enstitüsü Turizm İşletmeciliği ABD, Yayınlanmamış Yüksek Lisans Tezi.

Yli-Huumo, J., Ko, D., Choi, S., Park, S., & Smolander, K. (2016). Where is current research on blockchain technology? - a systematic review. *PLoS One, 11*(10), 1–27. doi:10.1371/journal.pone.0163477 PMID:27695049

Yolal, M. (2003). *Türkiyede'ki Küçük ve Orta Ölçekli Konaklama İşletmelerinde Bilgi Teknolojileri Kullanımı*. Eskişehir: Anadolu Üniversitesi Yayınları.

Young, G. (1973). *Tourism: Blessing or blight?*

Yovcheva, Z., Buhalis, D., & Gatzidis, C. (2013). Engineering Augmented Tourism Experiences. In L. Cantoni, & Z. Xiang (Eds.), Information and Communication Technologies in Tourism (pp. 24–35). Berlin, Germany: Springer.

Yovcheva, Z., Buhalis, D., & Gatzidis, C. (2012). Smartphone augmented reality applications for tourism. [ertr]. *Ereview of Tourism Research, 10*(2), 63–66.

Yovcheva, Z., Buhalis, D., Gatzidis, C., & Van Elzakker, C. (2014). Empirical evaluation of smartphone augmented reality browsers in an urban tourism destination context. *International Journal of Mobile Human Computer Interaction, 6*(2), 10–31. doi:10.4018/ijmhci.2014040102

Yu, C.-E., & Boyol Ngan, H. F. (2019). The Power of Head Tilts: Gender and Cultural Differences of Perceived Human vs Human-Like Robot Smile in Service. *Tourism Review, 74*(3), 428–442. doi:10.1108/TR-07-2018-0097

Yücel, D. (2012). *New Media Art and New Museum*. İstanbul, Turkey: İstanbul Kültür University Publications.

Yu, G., & Schwartz, Z. (2006). Forecasting short time-series tourism demand with artificial intelligence models. *Journal of Travel Research, 45*(2), 194–203. doi:10.1177/0047287506291594

Yüksek, G. (2013). Development of ınformation technologies, travel enterprises and global distribution system example. *Journal of Internet Applications & Management Journal, 4*(1). eturizm.org. (2019). Helsinki ve Lyon, 2019 Avrupa Akıllı Turizm Başkentleri ödüllerini aldı. Retrieved from https://eturizm.org/helsinki-ve-lyon-2019-avrupa-akilli-turizm-baskentleri-odullerini- aldi/

Yung, R., & Khoo-Lattimore, C. K. (2017). New realities: A systematic literature review on virtual reality and augmented reality in tourism research. *Current Issues in Tourism*. doi:10.1080/13683500.2017.1417359

Yusof, M. N., Bakar, A. H., & Tufail, M. A. (2012) Knowledge Management and Growth Performance: Construction industry perspective, DOI: , Conference: Sustaining the World with Better Structures & Construction Practice (APSEC-ICCER 2012), At Surabaya, Indonesia. Retrieved from www.researchgate.net/profile/Mohamad_Nizam_Yusof/publication/268331671_Knowledge_Management_and_Growth_Performance_Construction_industry_perspective/links/546894700cf20dedafd0d168.pdf doi:10.13140/2.1.3118.0482

Yüzbaşıoğlu, N., Çelik, P., Topsakal, Y., & Bahar, M. (2018). Endüstri 4.0 ve Akıllı Turizm: Antalya Destinasyonu Akıllı Turist Rehberi Uygulama Geliştirilmesi, Innovation and Global Issues in Social Sciences III (pp. 16-27.), April 26-29, Antalya-Turkey.

Zack, M., McKeen, J., & Singh, S. (2009). Knowledge management and organizational performance: An exploratory analysis. *Journal of Knowledge Management, 13*(6), 392–409. doi:10.1108/13673270910997088

Zagralı, E., & Akbaba, A. (2015). Turistlerin Destinasyon Seçiminde Yöresel Yemeklerin Rolü: İzmir Yarımadası'nı Ziyaret Eden Turistlerin Görüşleri Üzerine Bir Araştırma. *Journal of Yaşar University, 10*(40), 6633–6644.

Zaied, A. N. H., Hussein, G. S., & Hassan, M. M. (2012). The role of knowledge management in enhancing organizational performance, I. J. *Information Engineering and Electronic Business, 4*(5), 27–35. doi:10.5815/ijieeb.2012.05.04

Zaitseva, N. A. (2013). *Management in service industry: tourism and hospitality*. Moscow, Russia: Academia Publishing House.

Zanella, A., Bui, N., Castellani, A., Vangelista, L., & Zorzi, M. (2014). *Internet of things for smart cities*. IEEE Internet of Things Journal; doi:10.1109/JIOT.2014.2306328

Zasadna, K. O. (2014). *QR Coding and Alternative Technologies. Finansovyy prostir [Financial space]* (Vol. 3). Retrieved from http://privatbank.

Zhang, L., Li, N., & Liu, M. (2012). *On the basic concept of smarter tourism and its theoretical system*. Tribune Tourism.

Zhang, T., Wei, W., Fu, X., Hua, N., & Wang, Y. (2019). Exploring the roles of technology, people, and organization in building a tourism destination experience: Insights from the 2nd USA-China tourism research summit and Industry Dialogue. *Journal of Destination Marketing & Management*, *0–1*(February). doi:10.1016/j.jdmm.2019.03.001

Zhang, Z., Ye, Q., Zhang, Z., & Li, Y. (2011). Sentiment classification of Internet restaurant reviews written in Cantonese. *Expert Systems with Applications*. doi:10.1016/j.eswa.2010.12.147

Zhao, J., Zheng, X., Dong, R., & Shao, G. (2013). The planning, construction, and management toward sustainable cities in China needs the Environmental Internet of Things. *International Journal of Sustainable Development and World Ecology*, *20*(3), 195–198. doi:10.1080/13504509.2013.784882

Zheng, P., Wang, H., Sang, Z., Zhong, R. Y., Liu, Y., Liu, C., & Xu, X. (2018). Smart manufacturing systems for Industry 4.0: Conceptual framework, scenarios, and future perspectives. *Frontiers of Mechanical Engineering*, 1–14. doi:10.100711465-018-0499-5

Zheng, R., Zhang, T., Liu, Z., & Wang, H. (2016). An EIoT system designed for ecological and environmental management of the Xianghe Segment of China's Grand Canal. *International Journal of Sustainable Development and World Ecology*, *23*(4), 372–380. doi:10.1080/13504509.2015.1124470

Zheng, S., Xiong, X., Vause, J., & Liu, J. (2013). Real-time measurement of wind environment comfort in urban areas by Environmental Internet of Things. *International Journal of Sustainable Development and World Ecology*, *20*(3), 254–260. doi:10.1080/13504509.2013.779616

Zheng, Z., Xie, S., Dai, H., Chen, X., & Wang, H. (2018). Blockchain challenges and opportunities: A survey. *International Journal of Web and Grid Services*, *14*(4), 352–375. doi:10.1504/IJWGS.2018.095647

Zhou, Z. (2003). *E-Commerce and information technology in hospitality and tourism*. California: Thomson Delmar Learning.

Zhu, W., Zhang, L., & Li, N. (2014). Challenges, function changing of government and enterprises in Chinese smart tourism. *Information and Communication Technologies in Tourism, 10*.

Zichermann, G., & Cunningham, C. (2011). *Gamification by design: Implementing game mechanics in web and mobile apps*. Canada: O'Reilly Media.

Zichermann, G., & Linder, J. (2010). *Game-Based marketing inspire customer loyalty though rewards, challenges and contests*. Hoboken, NJ: John Wiley & Sons.

Zomerdijk, L. G., & Voss, C. A. (2010). Service design for experience-centric services. *Journal of Service Research*, *13*(1), 67–82. doi:10.1177/1094670509351960

Zuelow, E. (2015). *A history of modern tourism*. London, UK: Macmillan International Higher Education.

Zukin, S. (1998). Urban lifestyles: Diversity and standardisation in spaces of consumption. *Urban Studies (Edinburgh, Scotland)*, *35*(5-6), 825–839. doi:10.1080/0042098984574

Zunnurhain, K. (2016). Vulnerabilities with internet of things. *Proceedings of the International Conference on Security and Management (SAM)*, 83. The Steering Committee of The World Congress in Computer Science, Computer….

Zygiaris, S. (2013). Smart city reference model: Assisting planners to conceptualize the building of smart city innovation ecosystems. *Journal of the Knowledge Economy*, *4*(2), 217–231. doi:10.100713132-012-0089-4

About the Contributors

Evrim Çeltek, Ph.D., has been working as an Associate Professor at Gaziosmanpaşa University, Zile Dinçerler Tourism Business and Hotel Management College, Tourism Business and Hotel Management Department. Evrim Çeltek obtained her master of science in tourism business administration from Sakarya University (Turkey) and she obtained her PhD in tourism and hotel management from Anadolu University (Turkey). Her researches focused on tourism marketing. She has several articles, books and book chapters about, mobile marketing, e-commerce, mobile commerce, advergame, digital marketing, augmented reality, virtual reality, gamification and electronic customer relationship management.

* * *

Bayram Akay, Ph.D., has been working as an Assistant Professor at the Burdur Mehmet Akif Ersoy University, School of Tourism and Hotel Management. His research relates principally to tourism management, sustainable tourism.

Nurettin Ayaz, Ph.D., has been working as an Associate Professor at the Karabük University, Faculty of Tourism. His research relates principally to tourism policy, sustainable tourism, and rural tourism.

Şule Aydın completed her primary and secondary education in Çanakkale. She received her Master's degree from the Department of Tourism Administration at Çanakkale Onsekiz Mart University. In 2004, she received her doctoral degree from the Department of Tourism Administration at Dokuz Eylül University. She became an Associate Professor in the field of management and strategy in 2011. She has been working as a Full Professor at Nevşehir Hacı Bektaş Veli University and she is the dean of Tourism Faculty. She has many publications related to human resource management and organizational behaviour in tourism.

Orhan Batman was born in 1967 in Sakarya. Began his career in Akdeniz University as a research assistant. Completed the program of philosophy of doctorate in Istanbul University on "business politics and organization". Became assistant professor in 1996, associate professor in 2005 and professor in 2011 in Sakarya University. He is still Head of Tourism Management Department.

Kemal Birdir completed his B.S. degree in 1989 at Cukurova University Mersin School of Tourism and Hotel Management. The same year he was assigned as Research Assistant at the same school and awarded a scholarship to pursue his graduate studies in the USA. In 1984, he completed his MBA degree at Hawaii Pacific University at Travel Industry Management and in 1998; he received his Ph.D. from Purdue University's RHIT department at Hospitality Management. The same year, he returned to Mersin University, Turkey and since then he is working at Mersin University as a Faculty Member of Faculty of Tourism. Currently, he is serving as Dean of the Faculty. My research interest is mostly on Tourism Management Industry and Hospitality Management related topics.

Sevda Birdir was born in 1969 in Antakya-Turkey and completed her elemantary-middle school and high school education in Antakya. She earned a BS degree from School of Tourism and Hotel Management of Mersin University in 1998 and received her Master's degree in 2001 again from School of Tourism and Hotel Management at Mersin University. She worked as a lecturer between 2001 and 2006 in Nevsehir Vocational School of Erciyes University. She earned her PhD degree in 2009 in Erciyes University, Department of Production Management and Marketing. Currently, she is an associate professor in Department of Tourism Guding in Tourism Faculty of Mersin University.

Pınar Çelik, in 2003 completed her undergraduate degree in Tourism Management. She completed her master's and doctoral studies in the field of tourism management at the Akdeniz University. She started working at Muğla University in 2009 and has been working at Akdeniz University since 2010. She is currently working as an assistant professor in the Department of Tourism Management of the Serik Faculty of Business Administration at Akdeniz University. She works on tourism, destination management, intelligent tourism, intelligent tourism destination and intelligent destination management.

Gamze Coban is a research assistant at the Faculty of Tourism, Nevsehir Haci Bektas Veli University/Turkey. She has an MSc and a PhD in tourism management. Her research interests include destination management, gastronomy, smart tourism, and cycling.

Ali Dalgıç, Ph.D., was born in Antalya, Turkey in 1987. He is Research Assistant in the Department of Tourism Management at Mersin University. He completed his Master of Science Degree in Management at Adnan Menderes University, Aydin/Turkey in December 2013. He got his Ph.D. in Tourism Management at Mersin University, Faculty of Tourism. His research interests include event management, alternative tourism and strategic management.

Seden Doğan was born in 1978 in Karabuk-Turkey. Dr. Seden Dogan is an assistant professor at Ondokuz Mayis University since 2017. Before that, she worked as a lecturer at Adnan Menderes University between 2013-2017. She worked at different hotels and travel agencies between 1998-2013. She graduated from Akdeniz University. Her main study topics are consumer behavior, technology usage and digitalization in the hospitality industry.

Alfonso Fernández-Tabales is a senior lecturer at the Departamento de Geografía Física y Análisis Geográfico Regional at the Universidad de Sevilla. He has oriented his professional and academic career towards the relationships between tourism activity and development processes on the regional and local scales.

Concepción Foronda-Robles is an Associate Professor at the Departamento de Geografía Humana at the Universidad de Sevilla., where she got his Ph.D. Her scientific record focuses on rural development and territory, the planning and sustainable management of tourism destinations and the building of social capital in local development processes.

Luis Galindo-Pérez-de-Azpillaga is an Associate Professor at the Departamento de Geografía Humana at the Universidad de Sevilla. He has an extensive relationship with territorial development strategies and has been involved in a number of contracts and research projects linked to the subject. He also has wide experience in the application of technological tools to territory based on Geographical Information Systems and has undertaken stays at foreign research centres.

İbrahim İlhan is an Associate Professor at the Tourism Faculty, Nevşehir Hacı Bektaş Veli University, Nevşehir, Turkey. He obtained his MA in international hotel management from Schiller International University, London. He obtained his PhD in tourism management from Dokuz Eylül University, İzmir. His area of research includes human resources management, managerial ethics, regional tourism planning, destination marketing, mobile marketing and gastronomy.

Tomi Ilijaš, is founder and president of Arctur and he holds a MSC degree from Ljubljana University. Mr. Ilijaš is an entrepreneur with focus on Hi-Tech innovation and has shared his knowledge and experience to many start-ups and spin-offs in the region. Recently he is researching new business models in HPCaaS and successfully breaking the barriers in bringing HPC to manufacturing SMEs. He is a member of PRACE Industrial Advisory Committee, board member of Fortissimo Marketplace Limited and IEEE fellow.

Cemal İnce, Ph.D., has been working as an Assistant Professor at Gaziosmanpaşa University, Zile Dinçerler Tourism Business and Hotel Management College, Tourism Business and Hotel Management Department. His researches focus on tourism, management and organization in tourism.

Merve Yıldırım Kalem received her Bachelor's degree at the year 2014 at University of Yeditepe, Department of Business Administration. She has Master at the University of Karabük, Department of Tourism Management.

Eda Özgül Katlav is an Associate Professor at the Department of Tourism Management at Nevşehir Hacı Bektaş Veli University. Tourism businesses, tourism business management, organizational behavior, rural tourism is working on issues.

Vahit Oğuz Kiper was born in Canakkale in 1989. He was graduated from Tourism Management department in Anadolu University in 2015. He first started his master education on tourism management in Canakkale Onsekiz Mart University and completed his degree in Sakarya University of Applied Sciences in 2019. He began his career as a research assistant in Sakarya University of Applied sciences in 2018 and he still working there.

Cenk Murat Koçoğlu had undergraduate degree at the year 2006: University of Abant İzzet Baysal, Department of Tourism Management. Master degree at the year 2009: University of Düzce, Department of Tourism and Hospitality Management. Doctoral degree at the year 2016: University of Bülent Ecevit, Department of Business Administration. His research interests: Tourism Marketing and Research Methods

Halime Göktaş Kulualp completed her undergraduate degree in 2004 in İnönü University (Turkey), Faculty of Economics and Administrative Sciences, Department of Business Administration. She completed her master's and doctoral studies at Bulent Ecevit University in 2009 and 2015. Göktaş Kulualp started her academic career as a Research Assistant at Bulent Ecevit University, Turkey in 2007. She is presently a faculty member at the Tourism Management Department of the Safranbolu Tourism Faculty of Karabük University, Turkey. She has many publications in the field of Management and Organization.

Kamal Kumar is working as an Assistant Professor of Tourism in Pt. C.L.S. Government College, Karnal affiliated to Kurukshetra University, Kurukshetra, Haryana, India. He is Master of Philosophy with UGC-NET qualification in Tourism Administration & Management. He has extensive experience of teaching and industry in tourism and hospitality sector and has several research publications in his credit. He attended many research conferences and seminars of international & national repute. His areas of interest include tourist nature based tourism, culture tourism, impacts of tourism.

Vesna Kuralt is an Anthropologist with a Ph.D. in Market Communication. She has 10 years of experience working in communication, project management, and research. In the past few years, she has been active in the field of smart communities, consulting on smart city projects and researching topics of smart tourism with a focus on human-centered technology. She is most interested in emerging technologies and their social implications. More specifically, how qualitative insights can enrich data with a context of personal and nuanced narratives that data cannot reflect. By understanding the social context and connections between data points, one often finds unexpected solutions. For this reason, understanding the ecosystem of users is vital for any innovation process because the context is what gives meaning to data and turns it into information.

Katarzyna Leśniewska-Napierała, is a young researcher, geographer with the specialty socio-economic geography. Since 2014 she works at the Faculty of Geographical Sciences, University of Lodz in the Institute of the Built Environment and Spatial Policy and cooperating with CiTUR Centre for Tourism Research, Development and Innovation (Portugal). In her research she is mainly focused on social aspects of spatial planning and the use of qualitative methods in geographic sciences. She participated in educational and research projects, both Polish and international. She has been regularly participating in mobilities granted by Erasmus+ program and EEA Norway grants. She participated in bilateral, Polish-Norwegian project "SCHEDULE. Significance of Competitions in Higher Education" completed in the years 2014-2015 by the University of Lodz and Lillehammer University College. Since 2019 she is contracted as a project coordinator in the project "SPOT. Sustainable spatial planning of tourism destinations" granted under Erasmus+ Programme (2019-1-PL01-KA203-064946). She is a Sub-Editor in the European Spatial Research and Policy journal and a juror in the Regional Committee of the Geographical Olympic Games in Lodz.

Selen Levent was born in 1995 in Kastamonu, Turkey. She completed her high school education in Inebolu Anatolian Tourism and Vocational High School in Kastamonu. She received her Bachelor degree in Kocaeli University Hospitality Management Department in 2018. As of 2018, she has been continuing her graduate studies at Kocaeli University Tourism Management Department. She continues to work in the field of tourism marketing and tourism.

Simon Mokorel was born in Šempeter on 27th of March 1966. He has over twenty years of experience in the area of information technology and cybernetics. By working with research centres in Europe, he has gained knowledge about digital technologies, especially technologies of data acquisition, data exchange between heterogeneous systems and technology fusion. His combine knowledge of IoT (Internet of Things), analytics, cloud services and the development of user demand to develop new business models and to promote mental leap that environment adapt to social and technological changes. He was trained different target groups in the area of "Business Model Canvas", "Base Board" and "Design & Art Thinking" methodologies. He is a co-author of the Tourism 4.0 Enriched Tourist Experience.

Tomasz Napierala is working at the Faculty of Geographical Sciences, University of Lodz (Poland) and cooperating with CiTUR Centre for Tourism Research, Development and Innovation (Portugal). In his research he is focused on spatial aspects of competition in the hotel industry, mainly spatial volatility of hotel prices and its impact on hotel competitiveness. Cooperating with researchers representing other disciplines (spatial econometrics and ICT), he has already started investigation of competitiveness and attractiveness of tourism destinations. Moreover, he is studying social, political and economic effectiveness of municipal investment expenditures. He has been regularly participating in mobilities granted by Erasmus+ program and EEA Norway grants. He coordinated bilateral, Polish-Norwegian project "SCHEDULE. Significance of Competitions in Higher Education" completed in the years 2014-2015 by the University of Lodz and Lillehammer University College. Since 2017 he is contracted in the EU Horizon 2020 research project "RELOCAL. Resituating the local in cohesion and territorial development". Since 2019 he is contracted as a project manager in two projects: "Social media and the economic efficiency of hotels: Urban context" granted by The National Science Center of Poland, and "SPOT. Sustainable spatial planning of tourism destinations" granted under Erasmus+ Programme. He is Co-Editor-In-Chief in European Journal of Tourism, Hospitality and Recreation. He is holding a position of the Vice-Treasurer of the Polish Geographical Society.

İbrahim Özen obtained his master of science in Information Systems, from Sakarya University (Turkey-2013). He obtained his PhD in Tourism Management from Nevşehir University (2017). He is an Assistant Professor at Nevşehir University, Faculty of Tourism.

Emrah Özkul is an Associate Professor of Tourism at School of Tourism and Hotel Management, Kocaeli University, Turkey. He serves on the editorial boards of a number of journals. His doctorate and associate professorship field is tourism. His research primarily focuses on tourism sociology and tourism marketing.

Hande Mutlu Öztürk was born in Ordu on 14.06.1982. She completed her primary education in France and secondary and high school education in Istanbul, Turkey. In 2005, she graduated from Pamukkale University Food Engineering Department. In 2007, she completed her master's degree at Pamukkale

University, Institute of Science. In 2011, she completed her PhD at Ege University, Institute of Science and Technology. Between 2009-2013, she worked as a lecturer in Pamukkale University, School of Tourism and Hotel Management, Food and Beverage Management Department. In 2013, she started to work as Assistant Professor at Pamukkale University, Faculty of Tourism, Department of Gastronomy and Culinary Arts. She speaks fluent French and good English. She is married and has 2 children.

Urska Peceny, Ph.D., has gained an international education in Slovenia, Italy and Austria. Her Phd thesis in Business Communication with specialization in the New Media and Technologies entitled "Netlife" has paved the way in 2001 in research of collaboration design in the reality in which AI and big data play an increasing role. From Europe to Middle East, she has gained 20+ years of experience as entrepreneur by co-founding more companies and disruptive initiatives. #collaboration #design #SmartEcosystems. As Chief Innovation Officer at Arctur Ltd., she currently leads the Tourism 4.0 department. Living between Slovenia and Vienna, Austria, where she hacks the collaboration formats as lead of the NASA Space Apps Challenge in Austria, she is also a university lecturer in smart-related topics and proud mother of three children.

Hrvoje Ratkajec started his carrier by working as a researcher and a lecturer in the field of Economic and Business History at the University of Primorska (Slovenia) where he also obtained his PhD. As a researcher he was actively involved in several international projects. This inspired him to continued his career in project management, where he helped various stakeholders from industry, NGO, public institutons etc. to develop their ideas and obtain finances. Currently he is working for the company Arctur Ltd. in shaping the future of tourism with Tourism 4.0.

Gülmira Samatova is a faculty member of Krygyz-Turkish Manas, School of Tourism and Hotel Management. Her researches focus on Culinary Arts.

Ömer Sarı was born on 19-02-1990 in Osmaniye's Düziçi District. He completed his education at Gazi University Department of Tourism Management between 2012-2016. During his university education, he took Formation Education in Hospitality and Travel Services. Since 2017, he has been continuing his education as a graduate student in the field of Tourism Management in Karabuk University. Tourism also carries out scientific studies on current issues, business management and human resources.

Lakhvinder Singh is working as an Assistant Professor of Tourism in Government College, Kaithal affiliated to Kurukshetra University, Kurukshetra, Haryana, India. He holds Ph.D. Degree in with UGC-NET qualification in Tourism Administration & Management. He has more than three years of teaching experience and has more than 30 research publications in his credit. He is in the Editorial/Advisory Board of different Journals of International and National repute. His areas of interest include tourist behaviour, tourism marketing, tourism impact, culture & religious tourism, e-tourism and ICT in tourism.

Matevž Straus is a project leader at Arctur and holds a M.Sc. degree from Urban Studies (Vrije Universiteit Brussel, Université libre de Bruxelles, University of Copenhagen, Universidad Autónoma de Madrid, Universidad Complutense de Madrid), M.A. degree from Market Communication (University of Ljubljana) and a B.A. degree from Analytical Sociology (University of Ljubljana). Mr. Straus has been working at the crossroad of heritage and innovation for the past 5 years, has led several awardwinning projects on innovating heritaga and is the initiator of HeritageLab, a 2018 winner of EU Social Innovation Competition.

Yunus Topsakal is a Doctor in the Department of Tourism Management at the Adana Alparslan Türkes Science and Technology where he has been a faculty member since 2013. Topsakal completed his Master Degree at Akdeniz University and his bachelor degree at I.D. Bilkent University. He gets Ph.D. degree at Akdeniz University in 2017. His research interests lie in the area of Industry 4.0, smart tourism, digital tourism, innovation, destination competitiveness. He has collaborated actively with researchers in several other disciplines of statistic, economics, and agriculture.

İlhami Tuncer was born in 1987 in. He graduated from Marmara University, Faculty of Economics and Administrative Sciences, Department of Business Administration in 2010. He completed his master's degree at Selcuk University Department of Business Administration in 2012. In 2017, he received his PhD degree from Necmettin Erbakan University in Tourism Management. His scientific researches focus on consumer behavior, marketing communication and digital marketing.

Jurij Urbancic received his PhD in Physics from the University of Kaiserslautern, where he studied the interaction between matter and intense ultrafast laser pulses. He moved to Arctur in order to research the application of technological solutions to the tourism sector, with special emphasis on blockchain, internet of things and high performance data analytics. Among the areas if his interest are tourist flows optimization, applications for sustainable tourism and identity management.

Emre Uygun was born in 1995 in Istanbul, Turkey. He completed his high school education in Etiler Anatolian Tourism and Vocational High School in Istanbul. He received his Bachelor degree in Kocaeli University Hospitality Management Department in 2018. As of 2018, he has been continuing his graduate studies at Kocaeli University Tourism Management Department. He continues to work in the field of tourism marketing and tourism.

Alenka Vavros began to academic studies at the University of Ljubljana, Departments of East Asian Studies and Comparative Literature, where he majored in Chinese language and culture combined with Comparative Literature from 2005 to 2012. During this time he spent three semesters studying at two different universities in Shenyang, China as a foreign student. Major topics of his studies included: modern Chinese language and grammar, conversational Chinese language, cultural studies (Chinese history and evolution of Chinese characters, methodology of intercultural research etc.), literary theory and criticism. In the years following graduation he was working in customer service industry, until he decided in 2014 to improve his knowledge and obtain education degree in a completely different field of mathematical sciences. Major topics included: numerical analysis, probability and statistics, optimization algorithms, databases and data structures. Before graduating he completed his internship with a research about usage of graph databases in tourism sector. After obtaining Bachelor's Degree in Practical mathematics in 2019 he got employed by Arctur d.o.o., computer engineering company, where he is able to use both of my skills: using technical knowledge in Practical mathematics field in order to improve and enhance customer experience. Research areas he is interested in are big data, databases and Bayesian statistics.

Index

D

E

F

G

H

I

J

K

L

M

N

O

P

Q

R

S

T

U

V

W

Z

www.ingramcontent.com/pod-product-compliance
Lightning Source LLC
LaVergne TN
LVHW080618300926
849162LV00005B/1205